Surfactants Europa

*A Directory of Surface Active Agents
available in Europe*

THIRD EDITION

EDITED BY

GORDON L. HOLLIS MSc PhD

THE ROYAL SOCIETY OF CHEMISTRY

A catalogue record for this book is available from the British Library.

© **The Royal Society of Chemistry 1995**

ISBN 0–85404–804–9

Published by The Royal Society of Chemistry,
Thomas Graham House, The Science Park, Cambridge, CB4 4WF

Typeset by Datix International Ltd, Bungay, Suffolk
Printed by Staples Printers Ltd, Rochester

Production team

Nichole Gibson (Staff Editor)
Julie Hetherington
Sally Hawes
Angela Nicholls
Helen Thomas
Doug Hartley
Cheryl Beynon

James Butler (Design)
Alan Skull
Andrew Nightingale

Acknowledgements

The Editor is grateful to all the companies who have willingly supplied data for use in this directory and in particular for the many conversations with members of their staff which have helped in the presentation of their data in the various parts of the book. He is also appreciative of the work and enthusiasm of the many members of the production team, at Cambridge, who have combined to ensure publication in the minimum time.

Surfactants

at the interface of
ideas and success

Textiles

Detergents

Personal care

Polymerisation

Agricultural formulation additives

Paints and coatings

Polymer additives

Oilfield chemicals

Worldwide headquarters : ICI Surfactants • Wilton • Middlesbrough • UK

For further information • please contact : ICI Surfactants • Communications Department
Everslaan 45 • B-3078 Everberg • Belgium • Tel: 2 758 9361 • Fax: 2 758 9686

ICI Surfactants is part of ICI Chemicals & Polymers Limited

Contents

Also by Dr. Gordon Hollis:

A unique monthly newsletter

Focus on Surfactants

Focus on Surfactants provides data on surfactant intermediates and raw materials, associated products, detergent formulations, process and plant news, biodegradability and product safety and more every month. It is edited by Dr Gordon Hollis, an established consultant in detergents, surface active agents, their raw materials, markets and applications. Previously he was for many years employed by an important manufacturer in the development of detergent chemicals and intermediates.

Focus on Surfactants monitors information from both technical and commercial sources, including company literature, press releases, market research reports and journals. The newsletter is essential reading for anyone involved with this diverse and fascinating sector of the chemical industry.

Focus on Surfactants enables you to:

▶ gain information vital to your business, including
   * new product developments
   * market information
   * company news
▶ scan ALL the relevant news in one place
▶ be alerted to news you would otherwise miss
▶ keep up with legislation and environmental concerns
▶ read about forthcoming conferences and key events

Focus on Surfactants keeps you one step ahead!

For a FREE sample issue and further information, simply contact:

Kate Pearce, Senior Marketing Officer,
The Royal Society of Chemistry,
Thomas Graham House, Science Park,
Milton Road, Cambridge CB4 4WF, UK.

Tel: +44 (0) 1223 423821
Fax: +44 (0) 1223 423429
Toll free (US only): 1-800-473-9234

E-mail (Internet): marketing@rsc.org

THE ROYAL
SOCIETY OF
CHEMISTRY
Information
Services

K:\ADS\F\FOCSURF3.CDR5

Preface

The existence of soap, the original cleansing agent, has been known for many centuries and the active ingredient, the fatty acid carboxylate salt in various forms can rightly claim to represent the first category of surface active agents in regular use.

The name of turkey red oil has had a place in organic chemistry journals and text books for over a century in the context of mordant dyeing, while products such as alkyl sulfates, alkyl sulfonates and secondary olefine sulfonates achieved commercial significance in a number of industrial fields in the inter-war years. During World War II, ethoxylates, mostly derived from aromatic bases, contributed to alleviation of the shortage of naturally-derived soap.

Serious growth began in the 1940s in line with the introduction of synthetic detergents and continued rapidly both in quantities used and in chemical categories of products available as applications grew in other industries. By the 1960s, the industry as we know it today had become established. Subsequently the rate of increase of new surfactant types began to slow down but the number of manufacturers steadily increased on a me-too basis and also as producers increasingly established positions in countries outside their home-base. This inevitably led to re-organisation in and since the 1980s as companies began to recognise the need to concentrate on core activities and in appreciation of opportunities in the rapidly developing countries outside the traditionally so-called western world. The last few years in particular have witnessed changes in company ownership, many

on a large scale and most of international dimension. Subsequent rationalisation of product ranges has resulted in the disappearance of some long-familiar names both of companies and products.

The situation is reflected in the contents of this directory, the fifth in a series first published in 1976 with a second edition in 1979 listing surfactants available in the UK. In 1982 this was extended to include products of other European countries and further so extended in 1989.

This edition is the third under the title SURFACTANTS EUROPA. As in previous editions, listing is by over 40 surfactant classes together with an appreciable number of products of unspecified constitution. Properties of a product of a particular class and its trade name can be easily identified together with address details of companies marketing it across the continent of Europe.

Minor changes in presentation occur in particular with esters in which so-called food esters are no longer separately listed to avoid confusion where a given product can be used for food and other outlets, though probably with different levels of purity.

Readers are strongly advised to familiarise themselves with the introductory section and various cross-reference indexes as an essential background to studying the product lists.

G.L. Hollis

Introduction

and notes on the use of this directory

1. Nature of surfactants

The term SURFACTANTS is the contemporary name for surface active agents, the class of chemical products whose molecules are able to modify the properties of an interface, *e.g.* liquid/air or liquid/liquid by lowering the surface or interfacial tension, with associated changes occurring in other properties, *e.g.* wetting. Depending on the precise chemical nature of the product, the properties of, for example, emulsification, detergency and foaming may be exhibited in varying degrees.

Every surfactant possesses the fundamental characteristic of having two essential portions, one being water repellent, usually called hydrophobic (or lipophilic), the other being water attractive, usually called hydrophilic (lipophobic). The hydrophobic portion comprises a collection of hydrocarbon groups, some at least of which form a linear chain which may or may not be substituted to varying extents. The hydrophilic portion comprises a solubilising group such as sulfate, sulfonate or ethoxylate, for example.

The number and arrangement of the hydrocarbon groups together with the nature and position of the hydrophilic groups combine to determine the surface active properties of the molecule. C_{12} to C_{20} is generally regarded as the range covering optimum detergency properties while optimum wetting and foaming properties usually occur at somewhat shorter chain lengths.

Surfactants fall into four categories depending on the distribution of electrical charge on the molecule *viz.*

(i) Anionic in which the hydrophobic portion of the molecule carries a residual negative charge
e.g. sodium dodecyl sulfate:

$$CH_3(CH_2)_{10}CH_2OSO_3^-Na^+$$

(ii) Cationic in which the hydrophobic portion carries a residual positive charge
e.g. cetyltrimethyl ammonium bromide:

$$CH_3(CH_2)_{14}CH_2\overset{\displaystyle CH_3}{\underset{\displaystyle CH_3}{N^+-CH_3}} \quad Br^-$$

(iii) Nonionic in which there is no residual electrical charge
e.g. dodecylalcohol ethoxylate:

$$CH_3(CH_2)_{10}CH_2(OCH_2CH_2)_n OH$$

(iv) Amphoteric in which both positive and negative centres are to be found in the molecule
e.g. alkyldimethylbetaine:

$$R-\overset{\displaystyle CH_3}{\underset{\displaystyle CH_3}{N^+-CH_2CO_2^-}}$$

Further examples of products in the above categories are to be found in the product lists and in the glossary of chemical formulae.

Each of the above types finds application in industry though anionic and nonionic types are used in much greater quantities than cationic and amphoteric.

Depending on the conditions prevailing, cationic, anionic or nonionic type properties may be exhibited. The following, however, should be noted:

(i) Long chain amines are not strictly cationic in neutral solution but do exhibit that property in the form of salts, *e.g.* acetates. Amines are nevertheless listed under the heading 'cationics' for convenient reference.

(ii) Amine ethoxylates are listed in this directory under ethoxylates (*i.e.* nonionics) even though relatively short chain ethoxylates exhibit cationic properties.

(iii) Amine oxides in neutral or alkaline solution are essentially nonionics but weakly cationic below about pH 3. They are listed under nonionics.

Anionics and nonionics form the organic active constituent of detergents and also find widespread usage in wetting, spreading, emulsification, dispersing, foaming and other applications in a whole host of manufacturing industries including textiles, plastics, paints, paper, pharmaceuticals, disinfectants, agricultural preparations and many more.

Cationics are mainly used for bactericidal and softening applications but also find outlets in, for example, ore flotation and road making. They are not of importance in providing detergency as such.

Amphoterics have many actual and potential applications including detergency but hitherto have attracted relatively little commercial interest on account of their high cost of manufacture. Their usage, however, is now increasing, in part at least, because of their extra mildness to the skin providing real advantages in cosmetic and toiletry preparations.

2. Scope

The technical criterion for inclusion of a product in this directory is that it is considered to be a surfactant as described in section 1 of this introduction.

There are however many definitions of 'surfactant' depending on personal preference and field of usage, *e.g.* textiles. Some operators regard a dispersant, for example, as a surfactant while others would not. Some chemicals which fall into this wider definition of surfactants have been included for completeness.

Therefore, products listed in the directory should not necessarily be regarded as surfactants under any of those definitions used in section 1. This should be noted particularly in a litigation context.

The commercial criterion for inclusion is that the product listed should be on the product range of a company possessing a selling organisation or an approved agent in a European country, irrespective of where the product is manufactured. Data have, in all cases, been provided by the supplier.

In theory, any product is potentially available worldwide regardless of where it is manufactured. Imports are nevertheless an important part of the surfactant scene. However, the majority of products used in a particular country, especially in the larger countries of the developed areas of the world, will normally be made in that country.

It should be noted that often the product range publicised by a manufacturer represents only a fraction of those surfactants potentially available. By varying the quantity and types of hydrophobes and hydrophiles it is possible to create a vast number of surfactants. This is particularly so in the case of the versatile alkoxylates where the number of moles of ethylene oxide or propylene oxide per mole of hydrophobe is theoretically almost limitless. Values between 1 and over 100 occur in practice. In these cases, where specific ranges are offered, others can frequently be tailor-made by arrangement with the manufacturer.

It will be appreciated that considerable time is needed for compilation, editing and printing of this directory following receipt of product information. During this time, changes in products may be taking place. Such changes will be relatively small. However, there may be significant changes, especially in expanding companies or those experiencing product rationalisation.

While the directory is believed to provide an accurate overall picture of product and product range availability at the beginning of 1995, it is imperative that, when contemplating the use of a particular product, the supplier should be consulted regarding current availability or suitable alternatives.

3. Presentation of data

Surfactants are listed on the basis of the four types described in section 1, with subdivision where appropriate, together with a miscellaneous section covering:

(i) Certain blends of surfactants, often including products of more than one class.
(ii) Products of undisclosed composition.

For ease of cross-reference, an alphabetical list of trade names is included. An alphabetical list of the suppliers' name, address, telephone and fax numbers throughout Europe is provided and there is a chemical formula glossary covering the many surfactant types listed.

4. Company names

The name used to identify companies in the product listings is normally a suitably short name sufficient to identify the company. The full official name appears in the address index for those countries in Europe in which the company is represented for commercial purposes either by an associate company or by an appointed agent. Address details are confined to Europe. Many companies operate worldwide and details of addresses outside Europe can normally be obtained from the headquarters or local addresses.

This edition reflects the extensive changes that have taken place over recent years in company ownership. Many one-time familiar names both of companies and products no longer appear. In many cases the products are still offered, but by a different company. The products may be available under the same trade name as before, but are often available under a new trade name.

5. Terminology

As with organic compounds in general, there is more than one way of naming most surfactants. In addition, these names become distorted as a result of commercialisation. In many cases more than one of these names is in regular use for a given product or class of products.

The situation may be illustrated by the case of the product having the following chemical formula:

$$R(OCH_2CH_2)_3OSO_3^-Na^+$$

where R is a mixture of alkyl groups in the region of $C_{12}H_{25}/C_{14}H_{29}$.

This product will be recognised as a key ingredient of many liquid detergent formulations. Names by which it may be known include:

lauryl ether sulfate
lauryl 3-ethoxysulfate
lauryl 3EO sulfate
sodium salt of linear alcohol ethoxysulfate
alkyl polyglycol ether sulfate

There are thus many ways of naming the above product. In addition the term lauryl is capable of more than one interpretation. In the strict text book sense, this alcohol has the formula:

$$CH_3(CH_2)_{10}CH_2OH$$

Commercially, however, lauryl alcohol is normally offered in two grades. The purer form is usually called narrow cut and comprises a mixture of lauryl alcohol ($CH_3(CH_2)_{10}CH_2OH$) and myristyl ($CH_3(CH_2)_{12}CH_2OH$) alcohol in ratio of approximately 2:1.

The broad cut grade in common use comprises the analogous products having even numbers of carbon atoms between 8 and 18 inclusive, peaking in the $C_{12/14}$ range and sometimes known as coconut alcohol.

In applications where the true lauryl alcohol is used *e.g.* in dentrifice formulations, it is called *n*-dodecyl alcohol to distinguish it from the lauryl as commercially understood.

Over the years, a number of alcohols made by so-called synthetic processes using petrochemically derived feed-stocks have appeared on the market for uses similar to those of the traditional lauryl alcohol derived from natural sources, *viz* coconut oil and palm kernel oil. In actual product descriptions the word lauryl is sometimes used in a generic sense implying lauryl type alcohols.

The chemical names used in this directory are those used by the suppliers in their literature. In a number of cases products are described by the Cosmetic, Toiletry and Fragrance Association (CTFA) designation instead of or in addition to their traditional names. This has been done when the appropriate data have been provided by the supplier but no attempt has been made to introduce the CTFA designation when not supplied, in order to avoid possible misrepresentation.

It follows therefore that essentially the same product will be named in different ways in different places. To introduce chemical names other than those long established by a given supplier could lead to unnecessary confusion.

6. Surfactant application

For a directory such as this to be of maximum value, the properties of each product must be related to industries in which it finds or is likely to find application.

Whilst some products are restricted in usage to specific applications for which they may have been specially tailored, most are more likely to possess a variety of properties which may be usefully employed in many fields.

For reasons of space and sometimes availability of data, key uses only have been mentioned. Such information varies from supplier to supplier. Some companies state properties and applications for each product within a series while others prefer to summarise for a group of products.

Further, one company's information may differ from another's for what is apparently the same product. It must be emphasised that, while some of these variations may be as a result of company preference and space constraints, there may be good reasons why a product should not be used in place of an apparently similar one.

There might be important effects on, for example, physical properties and compatibility which are of great importance in application *e.g.* in formulations. These differences can arise, for example, through variations in the feedstock from which the hydrophobe is made, the method of conversion to the surfactant, the reagent and/or catalyst used, or the degree of purification.

It is essential therefore that before any product is used for a given application, its suitability be confirmed by the would-be user, consulting with the supplier if in any doubt whatever.

7. Patent liability

Mention in this directory of the use of a given product in a particular application or of its manufacture by a particular method does not imply freedom from patent restrictions of any kind. Manufacturers and users should, in their own interests, clarify their positions in this respect in advance of any operation or application involving the surfactants listed.

8. Environmental

In recent years, as we are all aware, environmental considerations have assumed an increasingly prominent place in the whole of life. In the post-war years, particularly from the 1960's onwards, increases in living standards throughout the world, particularly in the developed countries, have been associated with ever-improving quality and efficiency in product types, usage and production using new branches of science and technology that have been developed so comprehensively.

This situation has brought great benefits to society as a whole but, as so often in such situations, associated adverse (or allegedly adverse) factors have to be taken into account. The time for the latter has now been with us for some time and the chemical industry is under increasing scrutiny in respect of toxicity, waste disposal and environmental effects in general.

As those concerned with the industry at the time will recall, surfactants experienced their own problems starting over 40 years ago in the field of biodegradability. Extensive investigations carried out in the '50's and '60's brought acceptable solutions to this problem which has steadily receded from the spotlight following adoption of recommendations or compliance with regulations as appropriate throughout the world. Should this compliance ever, for any reason, diminish, then the problem could be expected to return to the headlines and probably with even greater urgency bearing in mind the level of environmental awareness now prevalent.

It is therefore the responsibility of all those concerned with surfactant production and usage to become and remain acquainted with the biodegradability requirements and any other regulations applicable in all areas throughout the world where their products are liable to be used. In cases where usage involves direct contact with the person, *e.g.* in cosmetic, toiletry and food applications, a precise knowledge of all necessary regulations and requirements is obviously of even greater importance.

Since this directory is expected to be used, like its predecessors, throughout the world, where the precise requirements for surfactant applications differ from country to country, reference is not made to the suitability or otherwise of any particular surfactant in any aspect of usage including biodegradability.

However, for the convenience of the reader and particularly those (relatively) new to the subject, the established principles of biodegradability are summarised below.

8.1. Biodegradability

Biodegradability is a word with which those connected with the manufacture and use of surface active agents have been familiar for many years.

Following the initial highlighting, in the early 1950's, of the problem of limited biodegradability of branched chain alkylbenzene sulfonates work was initiated in several countries to identify the cause of the problem and provide a suitable remedy.

This led eventually to the adoption in the UK from the beginning of 1965 of a voluntary agreement between the detergent manufacturers and chemical suppliers that branched chain alkylbenzene sulfonates should no longer be used in household detergents. Corresponding arrangements followed in other European countries, the USA, Japan and Australia (West Germany had introduced a legal standard of biodegradability in October 1964).

The EEC introduced two Directives covering the usage of detergents and surface agents on 22 November 1973; the first Directive (73/404/EEC) is a general Directive stipulating that the average biodegradability of each class of surfactant (anionic, nonionic, cationic and ampholytic) in a detergent formulation shall be not less than 90%; the second Directive (73/405/EEC) specified test procedures for determination of the biodegradability of anionic surfactants but, in order to take into account the unreliability of test methods, specifies a minimum biodegradability of 80% which has effectively become the standard used.

With regard to nonionic surfactants, a third Directive (82/242/EEC) was introduced on 31 March 1982. This Directive is broadly similar to the anionic surfactant Directive referred to above, but with the important difference that it includes some exemptions. These relate to low-foaming nonionic surfactants used in dish-washing products, and may be marketed until adequately degradable substitutes are found.

Directive (73/405/EEC) was later amended by Directive (82/243/EEC), and further specifies methods of testing the biodegradability of anionic surfactants.

So far there have been no official moves in respect of EEC Directives specifically concerned with the cationic or ampholytic surfactants.

A further EEC Directive (79/831/EEC) which became effective on 18 September 1981, concerns the notification of 'new' substances with regard to their classification, packaging and labelling. The very wide-ranging notification dossier required for a new substance includes information on biodegradability, though the requirements are somewhat different in principle to those of the specific surfactant directives. Those surfactants registered before 18 September 1981 are regarded as 'existing' chemicals and are listed on the EINECS list. Any 'new' surfactants not on this list have to be tested for both primary and ultimate biodegradability as described by Directive (92/69/EEC).

It is widely accepted that products having linear hydrocarbon chains are acceptably biodegradable whereas those with significantly branched chains are not. For example, linear alkylbenzene sulfonates as opposed to the branched chain variety give acceptable biodegradability under normal user conditions and, as is well known, have been used now for many years as leading ingredients of the household detergent formulations.

Linear alcohol sulfates and ether sulfates, linear olefine sulfonates and paraffin sulfonates are further examples of acceptable anionic types.

In the case of alcohol ethoxylates, products having more than about 15 moles of ethylene oxide per mole of hydrophobe are generally regarded as being less readily biodegraded. In those with less than this number, the position as with anionics, depends on the hydrophobe.

Thus linear or straight chain alcohols generally give rise to biodegradable products. In the various product lists in this directory, the alcohols used as hydrophobes are described in many different ways by the different suppliers. In many cases, it may not be readily apparent to the uninitiated which products are, in fact, linear or near-linear and therefore biodegradable and which are not. If in any doubt the prospective user should consult the supplier.

The position with alkylphenol ethoxylates is less clear-cut. There is undoubtedly evidence to show that, under appropriate conditions, these products can be caused to degrade. Their consistent degradability at an acceptable rate under normal user conditions, however, is not universally accepted (1).

These notes are intended as guidelines to the likely biodegradability of the main types of anionic and nonionic surfactants in this directory where sufficient information is given of the products' constitution. In other cases, the supplier may nevertheless refer to the products' biodegradability but in many cases this is not so.

Finally, it should be remembered that biodegradability requirements, as with others, differ from country to country. Thus, products may be used for certain export formulations which would not necessarily be acceptable if the same formulations were to be used in their country of origin.

9. Precautions in handling

Surfactants are not in general classed as dangerous chemicals but like all chemicals, they can give rise to adverse effects if not handled with proper precautions and common sense. It is not, however, regarded as within the scope of this directory to provide advice or suggestions in this field.

It is the responsibility of all concerned with production and usage of surfactants to ensure that all necessary precautions are observed in the handling of the surfactants and their intermediates in manufacture and usage at all stages including transportation, thereby playing their part in the maintenance of a fully safe chemical industry.

References

1 Swisher, R. D. *Surfactant Biodegradation, Surfactant Science Series* 2nd ed., 1987, **18**, Marcel Dekker AG, Basel, Switzerland.

Glossary

This section is included to provide a basic guide to the chemical structure of products listed in the various sections with particular reference to the nature of the hydrophilic groups used in the different surfactant classes. A knowledge of elementary organic chemistry is assumed.

In the majority of products listed, the hydrophobic group is easily identified by its chemical name. In a number of cases however, mainly in the miscellaneous section, chemical structures are highly complex and are often mixtures. In these cases no attempt is made to identify particular chemical types.

1 Amphoterics

Surfactants of this class carry both a positive and a negative charge. The cationic or anionic character of the molecule can be changed by varying the pH of an aqueous solution. Such products are sometimes referred to as ampholytes or amopholytic surfactants.

This class of surfactants includes a very large number of chemical types and the subject is one of considerable complexity. However, the principal structures on which most products are based are as follows:

(i) **Betaines:**

$$R^1 \!-\! \overset{\overset{\displaystyle R}{|}}{\underset{\underset{\displaystyle R^2}{|}}{N^+}} \!-\! CH_2CO_2^-$$

(ii) **Glycinates:**

$$RNHCH_2CO_2H$$

(iii) **Propionates:**

$$RNHCH_2CH_2CO_2H$$

(iv) **Imidazolines:**

$$\underset{R}{\,}\begin{array}{c} N\!-\!\!-\! \\ \| \quad \quad \\ N\!-\!CH_2CH_2OH \end{array}$$

These are essentially intermediates but become amphoterics on reaction with for example, chloroacetic acid. There is now evidence that the ring structure breaks down during formation of the amphoteric so the imidazoline ring as such is not part of the surfactant molecule. For more detailed information see reference 1, page 258.

2 Anionics

(i) **Alkylaryl sulfonates**
General formula:

$$C_nH_{2n+1}\!-\!\!\!\left\langle \bigcirc \right\rangle\!\!\!-\!SO_3^-M^+$$

The C_nH_{2n+1} (alkyl) portion may be branched, as when derived from propylene tetramer in biologically 'hard' surfactants or substantially linear as when derived from n-paraffins or alpha-olefines in biologically 'soft' surfactants. M is most commonly sodium but other metal ions, ammonium or amine derivatives are available. The most common average value for n is around 12 for detergent applications. Toluene, xylene and cumene sulfonates are also offered primarily for solubilisation purposes.

(ii) **Alkyl sulfates**
General formula:

$$CH_3(CH_2)_xCH_2OSO_3^-M^+$$

These products are also called alcohol sulfates.
The value of x is normally in the $C_{10/16}$ range. Again the most usual cation (M) is sodium but other metals and also ammonium and ethanolamines are offered commercially.

(iii) **Ether sulfates**
General formula:

$$CH_3(CH_2)_xCH_2(OCH_2CH_2)_n\,OSO_3^-M^+$$

The most usual values of n are 2 and 3 but others do occur in practice. The alkyl group is normally substantially or entirely linear ($x = 10-12$ being the most common value). Alkylaryl groups are also used. This class of product is known commercially by a wide variety of names.

(iv) **Ether carboxylates**
General formula:

$$R(OCH_2CH_2)_nOCH_2CO_2^-M^+$$

where R may be an alkyl or alkylaryl group.

(v) **Phosphate esters**
General formula:

$$\text{Mixture of} \quad \underset{\overset{|}{OH}}{RO\!-\!\overset{\overset{\displaystyle O}{\|}}{P}\!-\!OH} \quad \text{and} \quad \underset{\overset{|}{OH}}{RO\!-\!\overset{\overset{\displaystyle O}{\|}}{P}\!-\!OR}$$

depending on the conditions of reaction between alcohols (ROH) and either phosphoric acid or phosphorus pentoxide. In addition to alcohols, alkylphenols are used and also the ethoxylate derivatives (varying proportions of ethylene oxide) in both cases. These products are generally of a specialised nature with often relatively little information about their detailed molecular structure being available.

(vi) **Sulfosuccinates**
General formula:

$$Na^+SO_3^-\!-\!\underset{\overset{|}{CHCO_2^-Na^+}}{\overset{\overset{\displaystyle CH_2CO_2R}{|}}{\,}} \quad \text{or} \quad Na^+SO_3^-\!-\!\underset{\overset{|}{CHCO_2R}}{\overset{\overset{\displaystyle CH_2CO_2R}{|}}{\,}}$$

Derived from ROH where R may be an alkyl group or an ethoxylate group derived from alcohol, alkylphenol or alkylolamide.

(vii) **Sulfosuccinamates**
Derivatives of succinamic acid:

$$CH_2CONH_2$$
$$|$$
$$CH_2CO_2H$$

e.g. Disodium *N*-octadecylsulfosuccinamate:

$$CH_2CONHC_{18}H_{37}$$
$$|$$
$$CH_2CO_2{}^-Na^+$$
$$|$$
$$SO_3{}^-Na^+$$

(viii) **Paraffin sulfonates**

General formula: $RSO_3{}^-Na^+$

in which R is an alkyl group, normally a mixture within the $C_{13/18}$ range and the sulfonate linkage is predominantly of the secondary type. Also known as alkane sulfonates.

(ix) **Olefine sulfonates**
These products are normally mixtures of hydroxyalkane sulfonates and alkene sulfonates, the hydrocarbon chain being in the $C_{14/18}$ range.

(x) **Sarcosinates**
General formula:

$$RCO-N-CH_2CO_2H$$
$$|$$
$$CH_3$$

where R is an alkyl group.
Products are offered commercially in either the free acid form (as above) or as a metallic salt, usually sodium.

(xi) **Isethionates**
Derived from isethionic acid:

$$HOCH_2CH_2SO_3H$$

Formula of typical derived surfactant:

$$RCO_2CH_2CH_2SO_3{}^-Na^+$$

where R is an alkyl group.

(xii) **Taurates**
Derived from taurine:

$$NH_2CH_2CH_2SO_3H$$

Formula of typical derived surfactant:

$$RCO-N-CH_2CH_2SO_3{}^-Na^+$$
$$|$$
$$CH_3$$

(xiii) **Lignin products**
Lignin products are prepared from the waste liquor of the sulfite pulping process.

3 Cationics

(i) **Quaternaries**
General formula:

$$CH_3$$
$$|$$
$$R-N^+-CH_3 \quad Br^-$$
$$|$$
$$CH_3$$

Other types include pyridine derivatives in which the nitrogen atom assumes a quaternary form in, for example, an alkyl pyridinium bromide structure.

(ii) **Imidazolines**
Based on the structure:

Imidazolines can also form quaternaries.

(iii) **Amines, diamines**

General formula: RNH_2 and $RNHCH_2CH_2CH_2NHR^1$

Not strictly cationic in neutral solution.

(iv) **Amine Salts**
For example,

$$RN^+H_3CH_3CO_2{}^-$$

4 Nonionics

(i) **Alkoxylates**
General formula:

$$R(OCH_2CH_2)_nOH$$

where R is an alkyl derivative containing a reactive hydrogen atom, for example, alkylphenol, alcohol, amine, fatty acid, ester, glyceride (including castor oil), or amide.
The ethoxy portion (OCH_2CH_2) may be replaced by propoxy groups in which a hydrogen atom is replaced by a methyl group, to give propoxylates. Sometimes mixed ethoxy and propoxy groups are found in the same molecule.

(ii) **Block Polymers**
General formula:

$$CH_3$$
$$|$$
$$HO(CH_2CH_2O)_x(CHCH_2O)_y(CH_2CH_2O)_zH$$

(iii) **Alkylolamides**
Often also referred to as alkanolamides.
General formula:
(a) Monoethanolamides:

$$RCONHCH_2CH_2OH$$

(b) Diethanolamides:

$$RCON \begin{cases} CH_2CH_2OH \\ CH_2CH_2OH \end{cases}$$

When made using two moles of ethanolamine per mole of fatty acid, a mixture is formed comprising about 60% of diethanolamide together with ester amines and ester amides in which either or both of the OH groups of the ethanolamine are reacted with the fatty acid.

The acids used are normally in the fatty series ranging from lauric to stearic but other types *e.g.* undecylenic and ricinoleic are also to be found in practice.

(iv) **Esters**

General formula: RCO_2R^1

R represents the acid portion, *e.g.* lauric, oleic, stearic or more complex types. R^1 represents the hydroxyl portion, *e.g.* monohydric alcohols of a wide variety of chain lengths, and polyhydric alcohols, *e.g.* polyglycols, glycerol. In the case of the polyhydric alcohols, one or more of the hydroxyl groups may be esterified.

(v) **Amine oxides**

General formula:

$$R-\overset{\overset{\displaystyle R^1}{|}}{\underset{\underset{\displaystyle R^2}{|}}{N}} \rightarrow O$$

where R is an alkyl group, normally within the range $C_{10/16}$ R^1 and R^2 are also alkyl derivatives. Frequently R^1 and R^2 are both methyl.

Further Reading:

1   Porter, M.R. *Handbook of Surfactants*, Blackie Academic & Professional, Glasgow and London.
2   Davidsohn, A. and Milwidsky, B.M. *Synthetic Detergents*, 7th ed., Longman Group UK Ltd., London.
3   Chalmers, L. and Bathe, P. *Chemical Specialities*, 2nd ed., 1978 and 1979, **I** and **II**, Micelle Press, Weymouth.
4   Tadros, T.F., *Surfactants*, Academic Press Ltd., London.
5   *Surfactant Science Series*, Marcel Dekker AG, Basel, Switzerland.

Definitions

Acid Value

the number of mg of potassium hydroxide needed to neutralise the free acids in 1g of the substance.
Units: mg KOH/g

Cloud Point

the temperature at which a solid starts to precipitate from solution when cooled under standard conditions.
The solubility of ethoxylates decreases with temperature and hence the cloud point of ethoxylates is the temperature at which the product emerges from solution on heating. This is a characteristic of water soluble members of this class. Cloud points reported in this directory refer to measurements at room temperature unless otherwise specified.
Units: Degree Centigrade (°C)

HLB Value

represents the hydrophilic-lipophilic balance of the molecule. The lower the HLB value the more lipophilic (oil loving) the material is, and vice versa. For fatty esters of polyalcohols and their alkoxylates, the HLB value is given by:
$HLB = 20 \times (1 - S/A)$, using the Griffin method.
S = saponification number of the ester
A = acid value of the esterified fatty acid
There is an optimum HLB value for performing any particular surfactant function such as emulsifying, wetting, detergency *etc*. In general, emulsifiers with an HLB value between 1 and 6 will favour water-in-oil (w/o) emulsions whereas surfactants with an HLB between 6 and 17 will favour oil-in-water (o/w) emulsions. The HLB value of emulsifier blends can be obtained by adding together the contribution each makes to the total HLB value.

Molecular Weight

the mass of one molecule referred to the standard of ^{12}C.

Melting Point

the temperature at which a solid changes to a liquid, and the liquid and solid phases are in equilibrium under a pressure of 760 mm Hg.
Units: Degree Centigrade (°C)

'n' Number

the average number of moles of ethylene oxide per mole of hydrophobe.
Using nonylphenol ethoxylates as an example, products up to approximately 'n' = 5 are oil soluble but water insoluble. For higher values of 'n' water solubility increases.

OH Value

the number of mg of potassium hydroxide equivalent to the acetic acid used for the acetylation of 1 g of the substances.
Units: mg KOH/g

pH Value

a measure of the acidity or alkalinity of a solution and is defined as the negative logarithm (base 10) of the concentration of oxonium ions H_3O^+ (mol/dm³)

Pour Point

this is the lowest temperature of flow under standard conditions.
Units: Degree Centigrade (°C)

Saponification Number

the number of mg of potassium hydroxide needed to neutralise the free and bonded acids in 1 g of the substance.
Units: mg KOH/g

Surface Tension

the contractile surface force of a liquid which makes it tend to assume a spherical form *e.g.* to form a meniscus.
Units: dynes/cm

Viscosity

describes the property of a liquid to resist laminar shear between two layers (sliding friction). Viscosities reported in this directory refer to measurements at room temperature unless otherwise specified.
Units: Centipoise (cPs) or Centistokes (cSt)

Abbreviations

aq.	aqueous		min	minute
BDG	butyldiglycol		min.	minimum
ca.	circa		ml	millilitre
cm	centimetre		mm	millimetre
cPs	centipoise		mol	moles
cSt	centistokes		mmHg	millimetres of mercury
dm	decimetre		m.p.	melting point
EO	ethylene oxide		M.W.	molecular weight
g	gram		no.	number
HLB	hydrophilic-lipophilic balance		°C	degree centigrade
IMS	industrial methylated spirits		OH value	hydroxyl value
in.	inch		PO	propylene oxide
IPA	isopropyl alcohol		POE	polyoxyethylene
kg	kilogram		POP	polyoxypropylene
m	metre		sapon.	saponification
max.	maximum		sec	second
mg	milligram		soln.	solution

1 Amphoterics

Acetates

Supplier	Trade name	Chemical description	Composition	General properties	Functionality / *Applications*
Albright & Wilson	EMPIGEN CDL60/P	lauroamphoacetate	45% active	liquid	
	EMPIGEN CDR30	alkyl amphoacetate	50% active		
	EMPIGEN CDR40	cocoampho(di)acetate	45% active	liquid	wetting agent *textile industry*
	EMPIGEN CDR60	cocoamphoacetate	40% active	liquid	
Henkel	DEHYTON G	cocoamphodiacetate	29-31% active	liquid	*shampoos; skin cleansing preparations; cosmetics*
	VELVETEX CDC	disodium cocoamphodiacetate	50% active	light amber viscous liquid	*shampoos; bath preparations; shower preparations*
McIntyre Group	MACKAM 1C	sodium cocoamphoacetate	45% active	liquid; pH 11	
	MACKAM 1L	sodium lauroamphoacetate	44% active	liquid; pH 10	
	MACKAM 2C	disodium cocoamphodiacetate	50% active	liquid; pH 8.5	wetting agent *baby shampoos; caustic cleaners*
	MACKAM 2CY	disodium capryloamphodiacetate	50% active	liquid; pH 11	
	MACKAM 2L	disodium lauroamphoacetate	50% active	liquid; pH 9	
	MACKAM 2W	disodium wheatgermamphodiacetate	35% active	liquid; pH 9.5	
Rhone-Poulenc	MIRAPON EXCEL	cocoamphoacetate		liquid	foaming agent (high); emulsifier; wetting agent
Surfachem	SURFAC CAA30	cocoamphoacetate			
	SURFAC CAA60	cocoamphoacetate			
	SURFAC CAD40	cocoamphoacetate and cocoamphodiacetate			
Zschimmer & Schwarz	AMPHOTENSID GB 2009	disodium cocoamphodiacetate	38% active	fluid	*shampoos; bath preparations; skin care products; baby care products; liquid soaps*

Betaines

Supplier	Trade name	Chemical description	Composition	General properties	Functionality / *Applications*
Akcros Chemicals	AMPHOLAN D197	alkylamido betaine; toiletry grade	30% active; NaCl 5%	clear pale yellow liquid; viscosity 20 cP's; pH 5.0 (1% aq.)	*toiletries; bath preparations; shampoos; shower gels*
	AMPHOLAN E210	alkyl betaine; toiletry grade	30% active; NaCl 7.5%	clear water-white liquid; viscosity 40 cP's; pH 7.0 (1% aq.)	foam booster *toiletries; bath preparations; shampoos; shower gels; dishwashing agents*
Akzo Nobel	AMPHOTEEN 24	$C_{12/14}$ alkyldimethyl betaine	dry content 36-38%	surface tension 33 dynes/cm (0.1% soln.)	*shampoos; washing-up liquids; hard-surface cleaners; vehicle cleaners*
	AMPHOTEEN BCA-30	cocoamidopropylbetaine	dry content 34-36%	surface tension 31 dynes/cm (0.1% soln.)	foaming agent (high); foam booster; thickener *liquid soaps; washing-up liquids; toiletries*

Supplier	Trade name	Chemical description	Composition	General properties	Functionality / Application
Akzo Nobel	AMPHOTEEN BTH-35	tallow bis(hydroxyethyl)betaine	dry content 39-41%	surface tension 30 dynes/cm (0.1% soln.)	thickener; *household cleaners*
Albright & Wilson	ARMOTERIC 16	hexadecyldimethylbetaine	28-30% active	liquid	
	ARMOTERIC LB	cocodimethylbetaine (fractionated coco-alkyl)	29-31% active	liquid	
	EMPIGEN 5509	alkyl amido propyl sulfo betaine	45% active		foam booster; *polymer industry*
	EMPIGEN BB	alkyl dimethyl amine betaine	30% active	liquid	
	EMPIGEN BS/C	coco amido propyl dimethyl betaine	30% active		
	EMPIGEN BS/F	alkyl amido propyl dimethyl amine betaine	30% active	liquid	
	EMPIGEN BS/FA	alkyl amido propyl dimethyl amine betaine	30% active	liquid	
	EMPIGEN BS/J	alkyl amido propyl dimethyl amine betaine	30% active	liquid	
	EMPIGEN BS/P	alkyl amido propyl dimethyl amine betaine	30% active	liquid	
Auschem Cesalpinia	CHIMIN AX	cocamidopropyl betaine	30% active	liquid	detergent; foaming agent; conditioner; *liquid detergents; cosmetics*
	CHIMIN CB	lauryl betaine	30% active	liquid	
	CHIMIN LX	laurylamidopropyl betaine	30% active	liquid	
Croda	INCRONAM 30	cocoamidopropyl betaine supplied as a 30% aq. soln.	29.5-31.5% active; NaCl 4.5-5.2%	pale clear liquid; viscosity 15 cPs; colour 2 max. (Gardner); pH 5.5-7.5 (10% aq. soln.)	substantivity agent; foaming agent (high); viscosity modifier; *cosmetics; toiletries; skin care products; household products*
Dac International Surfactants	ANFODAC CB	coco amido betaine			foam booster
	ANFODAC LB	lauryl amido betaine			foam booster
Ellis & Everard	CAFLON ADB30	alkyl amino betaine	30% active		
	CAFLON PCA30	alkyl amido betaine	30% active		
Fina Chemicals	RADIATERIC 6860	coco betaine	30% active; carbon chain composition $C_{8/10}$ 3%, C_{12} 48-58%, C_{14} 18-24%, C_{16} 8-12%, C_{18} 10-13%	liquid; colour 2 max. (Gardner); pH 5-8 (10% in H_2O)	antistatic agent; bacteriostatic agent; detergent; foaming agent (high); coupling agent; wetting agent; *cosmetics; toiletries; hair care products; personal care products; shampoos; liquid soaps; shaving products; all-purpose cleaners; carpet cleaners; dishwashing agents; laundry products*
	RADIATERIC 6864	lauryl betaine	30% active; carbon chain composition $C_{8/10}$ 2%, C_{12} 68-73%, C_{14} 25-30%, C_{16} 2%	liquid; colour 2 max. (Gardner); pH 5-8 (10% in H_2O)	
Th. Goldschmidt	ABIL B 9950	dimethicone propyl PG-betaine		yellow liquid	conditioner; cleansing agent; film former; *hair care products; skin cleansing preparations*
	TEGO BETAIN F	cocamidopropyl betaine	ca. 30% active; NaCl content ca. 5%	yellow liquid; pH 5-7	
	TEGO BETAIN HS	cocamidopropyl betaine and glyceryl laurate	ca. 30% active; NaCl content ca. 5%	yellow liquid; pH 6-7	*bath preparations; shower preparations; baby cleansing preparations; intimate hygiene products; shampoos; baby shampoos; hair conditioners; liquid soaps; shower gels; bubble baths; skin cleansing preparations*
	TEGO BETAIN L 7, SPRAY DRIED	cocamidopropyl betaine	78-85% active; NaCl content ca. 15%	ivory powder; pH ca. 5 (10% soln. in H_2O)	
	TEGO BETAIN L 7	cocamidopropyl betaine	ca. 30% active; NaCl content ca. 5%	yellow liquid; pH ca. 5	
	TEGO BETAIN L 5351	cocamidopropyl betaine	ca. 30% active; NaCl content 1.2% max.	yellow liquid; pH ca. 6	
Henkel	DEHYTON AB 30	coco-betaine	29-31% active	liquid	*cosmetics; shampoos; skin cleansing preparations; dishwashing agents; cleaners*
	DEHYTON CB	coco-betaine	31% active	liquid	*cosmetics*

Supplier	Trade name	Chemical description	Composition	General properties	Functionality / Application
Henkel	DEHYTON G	fatty acid amide derivative with betaine structure	29-31% active	liquid	*dishwashing agents; cleaners*
	DEHYTON K	cocamidopropyl betaine	29-32% active	liquid	*shampoos; skin cleansing preparations; shower and bath preparations; dishwashing agents; cleaners*
	DEHYTON KE 3016 B	coconut amido propyl betaine	29-32% active	liquid	*dishwashing agents*
	EMERY 6744	cocamidopropyl betaine	35% active	clear yellow liquid	*foaming agent* / *shampoos; hair care products; skin cleansing preparations; baby care products*
	EMERY 6748	cocamidopropyl betaine	35% active	clear amber liquid	
	VELVETEX AB 45	coco betaine	43% active	clear liquid	*viscosity modifier; gelling agent; conditioner* / *hair care products; skin care products*
	VELVETEX BA 35	cocamidopropyl betaine	35% active	clear yellow liquid	*antistatic agent; foaming agent; conditioner* / *hair care products; skin care products; bath preparations*
	VELVETEX BK 35	cocamidopropyl betaine	35% active	clear amber liquid	
	VELVETEX OLB 50	oleyl betaine	50% active	amber translucent gel	*shampoos; bath preparations; cleansing preparations*
Hickson Manro	MANROTERIC CAB	cocoamidopropyl betaine	30% active	liquid	*hair care products; bath preparations; household detergents; industrial cleaners*
	MANROTERIC NAB	N-alkyldimethyl betaine	30% active	liquid	
	MANROTERIC SAB	N-alkyldimethyl betaine	30% active	liquid	
Hoechst	DODICOR CAB-50	coco amido betaine	50% active		*detergents; oils*
	DODIGEN 3558	coco biguamide derivative	78% active		*disinfectants*
	DP121	C$_8$ amido betaine	40% active		*specialty products*
	GENAGEN CAB PDR	coco amido betaine	80% active		*toiletries; construction industry*
	GENAGEN CAB	coco amido betaine	30% active		*toiletries; detergents; cleaners*
	GENAGEN DAB	C$_{12}$ amido betaine	30% active		*toiletries; detergents*
	GENAGEN KBT	coco amido betaine	30% active		*toiletries; detergents; cleaners*
	GENAGEN LAB	lauryl dimethyl betaine	30% active		*toiletries; detergents*
Hüls	AMPHOLYT JB 130	coconut fatty acid amidopropylbetaine	active detergent 30%	liquid	*shampoos; liquid soaps; washing-up liquids*
Inolex	LEXAINE C	cocoamidopropyl betaine			
	LEXAINE CG-30	cocoamidopropyl betaine			
Kao Corporation	BETADET HR	alkylamide propyl betaine	27.5-31.5% active	liquid; pH 4.3-5.8 (5%)	
Lakeland Laboratories	CAB	amido betaine	30% active; salt content 5%	pale yellow liquid	*foaming agent (high); solubiliser; hydrotrope; wetting agent* / *detergents; toiletries*
	CTA/N	alkyl betaine	30% active; salt content 6.5%	pale yellow to water white liquid	*foaming agent; solubiliser; hydrotrope; wetting agent* / *detergents; toiletries*
Lonza	LONZAINE 12C	coco betaine	solids 35%; NaCl 2%	liquid; pour point 3°C; viscosity 14 cPs; surface tension 34.4 dynes/cm (0.1% active); Draves wetting 11 sec; pH 7.5 (3%)	*foaming agent; viscosity modifier; irritation mitigant* / *personal care products; industrial applications*
	LONZAINE 16SP	cetyl betaine	solids 35%; NaCl 7%	solid; viscosity 720 cPs; surface tension 32 dynes/cm (0.1% active); Draves wetting 11 sec; pH 7.5	

Supplier	Trade name	Chemical description	Composition	General properties	Functionality Application
Lonza	LONZAINE C	cocamidopropyl betaine	solids 35%; NaCl 5%	liquid; pour point 3°C; viscosity 29 cPs; surface tension 33.6 dynes/cm (0.1% active); Draves wetting 28 sec; pH 5 (10%)	foaming agent; viscosity modifier; irritation mitigant *personal care products; industrial applications*
	LONZAINE CO	cocamidopropyl betaine	solids ca. 36%; NaCl 5%	liquid; pour point 3°C; viscosity 28 cPs; surface tension 33.9 dynes/cm (0.1% active); Draves wetting 16 sec; pH 7 (10%)	foaming agent; viscosity modifier; irritation mitigant *personal care products; industrial applications*
McIntyre Group	MACKAM 35	cocamidopropyl betaine (via glyceride)	35% active	liquid; pH 6	
	MACKAM 35HP	cocamidopropyl betaine	35% active	liquid; pH 6	
	MACKAM CB-35	coco betaine	35% active	liquid; pH 8	
	MACKAM CET	cetyl betaine	33% active	gel; pH 7	
	MACKAM HV	oleamidopropyl betaine	35% active	liquid; pH 7	*industrial cleaners; personal care products*
	MACKAM ISA	isostearamidopropyl betaine	33% active	liquid; pH 7.5	
	MACKAM J	cocamidopropyl betaine	35% active	liquid; pH 6	
	MACKAM L	cocamidopropyl betaine	35% active	liquid; pH 5	
	MACKAM LMB	lauramidopropyl betaine	35% active	liquid; pH 8	
	MACKAM OB-30	oleyl betaine	30% active	liquid; pH 7	
	MACKAM WGB	wheatgermamidopropyl betaine	34% active	liquid; pH 6.5	
Millchem	MAPROLYTE C	cocamidopropyl betaine	solids 33-37%	liquid	foaming agent; stabiliser; wetting agent *shampoos; bubble baths*
	SURCO COCO BETAINE	cocamidopropyl betaine	solids 40-44%	liquid	foaming agent; stabiliser; wetting agent *industrial and household cleaners; dishwashing agents; liquid soaps*
Mona Industries	MONATERIC ADA	cocamidopropyl betaine	total solids 38%; NaCl 5%; 33% active	clear amber liquid; Draves wetting 13 sec (1% active, 3 g hook); pH 7.7 (at 10%)	foaming agent (high); detergent; solubiliser; antistatic agent; anti-corrosion *metal working; cosmetics; personal care products; paper industry; textile industry; detergents; cleaners; softening agents; lubricants; agriculture*
	MONATERIC CAB	cocamidopropyl betaine	total solids 35%; NaCl 5%; 30% active	clear light yellow liquid; Draves wetting 18 sec (1% active, 3 g hook); pH 7.1 (at 10%)	
	MONATERIC COAB	cocamidopropyl betaine	total solids 37%; NaCl 5%; 32% active	clear yellow liquid; Draves wetting 10 sec (1% active, 3 g hook); pH 7.9 (at 10%)	
	MONATERIC LMAB	lauramidopropyl betaine	total solids 35%; NaCl 5%; 30% active	clear light yellow liquid; Draves wetting 9 sec (1% active, 3 g hook); pH 8.3 (at 10%)	
	MONATERIC MCB	cocamidopropyl betaine	total solids 33%; NaCl 3%; 30% active	clear light yellow liquid; Draves wetting 18 sec (1% active, 3 g hook); pH 4.8 (at 10%)	
Nikko Chemicals	NIKKOL AM-101	2-alkyl-N-carboxymethyl-N-hydroxyethyl imidazolinium betaine; aq. soln.		brown liquid	
	NIKKOL AM-102EX	2-alkyl-N-carboxyethyl-N-hydroxyethyl imidazolinium betaine; aq. soln.		brown liquid	
	NIKKOL AM-103EX	2-alkyl-N-sodium carboxymethyl-N-carboxymethyl oxyethyl imidazolinium betaine; aq. soln.		brown liquid	foaming agent; cleansing agent *cosmetics*
	NIKKOL AM-301	lauryl betaine; aq. soln.		colourless liquid	
	NIKKOL AM-3130N	cocamidopropyl betaine; aq. soln.		pale yellow liquid	
	NIKKOL AM-3130T	cocamidopropyl betaine; aq. soln.		pale yellow liquid	

Supplier	Trade name	Chemical description	Composition	General properties	Functionality Application
Pentagon	PENTATERIC 24B	alkyl (C₁₂/₁₄) dimethyl betaine	total solids 38-40%; NaCl 7-8%	clear pale yellow liquid; pH 6.0-8.0 (10% aq. soln.)	
	PENTATERIC B	cocoamidopropyl betaine	total solids 35-36%; NaCl 4-5%	clear pale yellow liquid; colour 3 max. (Gardner); pH 6.0-8.0 (10% aq. soln.)	shampoos; foam baths; industrial and household detergents
	PENTATERIC BLG	cocoamidopropyl betaine, low glycerol	total solids 33-35%; NaCl 6% max.; free amine 2% max.	clear pale yellow liquid; pH 4.0-6.0	
	PENTATERIC BSG	cocoamidopropyl betaine	total solids 33-35%; NaCl 6% max.; free amine 2% max.	clear pale yellow liquid; colour 16Y;3R max. (6 in. lovibond cell); pH 4.0-6.0	
PPG	MAFO CAB SP	cocamidopropyl betaine			chelating agent; solubiliser; foam booster; detergent; lubricant; viscosity modifier; metal working
	MAFO CAB	cocamidopropyl betaine	solid matter 35%		
	MAFO CB-40	coco-betaine	solid matter 40%		viscosity modifier; foam booster personal care products
	MAFO LMAB	lauramidopropyl betaine	solid matter 35%		
Raschig	RALUFON CA	N-alkyl acid amidopropyl-N,N-dimethyl-N-(3-sulfopropyl)-ammonium-betaine			
	RALUFON CAS-OH	amidosulfobetaine			
	RALUFON DCH	N-alkyl-N,N-dimethyl-N-(3-sulfopropyl)-ammonium-betaine			foaming agent; antistatic agent; bactericide detergents; textile industry; personal care products; photography; fire-fighting; electroplating; oil recovery
	RALUFON DL	N-alkyl-N,N-dimethyl-N-(3-sulfopropyl)-ammonium-betaine			
	RALUFON DM	N-alkyl-N,N-dimethyl-N-(3-sulfopropyl)-ammonium-betaine			
	RALUFON DP	N-alkyl-N,N-dimethyl-N-(3-sulfopropyl)-ammonium-betaine			
	RALUFON DS	N-alkyl-N,N-dimethyl-N-(3-sulfopropyl)-ammonium-betaine			
	RALUFON DT	N-alkyl-N,N-dimethyl-N-(3-sulfopropyl)-ammonium-betaine			
	RALUFON MDS	sulfobetaine surfactant			
	RALUFON TA	N-alkyl acid amidopropyl-N,N-dimethyl-N-(3-sulfopropyl)-ammonium-betaine			
Rhone-Poulenc	MIRATAINE D40	cocodimethyl betaine	solids 40%; NaCl 5.0%	liquid	foaming agent (high); viscosity modifier; foam booster; emulsifier; wetting agent; dispersant hard-surface cleaners
Scher Chemicals	SCHERCOTAINE APAB	apricotamidopropyl betaine	dry solids 35% min.; salt content 4.0%	clear amber liquid; pH 5-7	detergent; conditioner; emollient; viscosity modifier
	SCHERCOTAINE CAB-A	cocamidopropyl betaine and ammonium chloride	dry solids 45% min.; salt content 5% max.	clear light yellow liquid; pH 5-7	foaming agent
	SCHERCOTAINE CAB-K	cocamidopropyl betaine and potassium chloride	dry solids 45% min.; salt content 3.5% max.	clear light yellow liquid; pH 5-7	viscosity modifier soaps

Supplier	Trade name	Chemical description	Composition	General properties	Functionality / Application
Scher Chemicals	SCHERCOTAINE CAB	cocamidopropyl betaine	dry solids 45% min.; salt content 6.5% max.	clear light yellow liquid; pH 5-7	detergent; wetting agent; foaming agent; *shampoos; bubble baths*
	SCHERCOTAINE IAB	isostearamidopropyl betaine	dry solids 35% min.; salt content 4.5% max.	soft amber gel; pH 5-7	detergent; conditioner; *skin care products; shampoos*
	SCHERCOTAINE MAB	myristamidopropyl betaine	dry solids 30% min.; salt content 5.0% max.	clear light yellow liquid; pH 5-7	detergent; wetting agent; thickener; antistatic agent; *cosmetics; toiletries*
	SCHERCOTAINE PAB	palmitamidopropyl betaine	dry solids 35% min.; salt content 5.5% max.	soft light yellow gel; pH 5-7	thickener; conditioner; *skin care products; hair care products*
	SCHERCOTAINE SCAB-A	cocamidopropyl hydroxy sultaine and ammonium chloride	dry solids 50% min.; salt content 5.0% max.	clear light yellow liquid; pH 5-7	foaming agent
	SCHERCOTAINE SCAB-K	cocamidopropyl hydroxy sultaine and potassium chloride	dry solids 50% min.; salt content 6.0% max.	clear light yellow liquid; pH 5-7	viscosity modifier; *soaps*
	SCHERCOTAINE SCAB	cocamidopropyl hydroxy sultaine	dry solids 50% min.; salt content 6.0% max.	clear light amber liquid; pH 5-7	detergent; wetting agent; foaming agent
	SCHERCOTAINE UAB	undecylenamidopropyl betaine	dry solids 35% min.; salt content 5.5% max.	clear amber liquid; pH 5-7	germicide; bactericide; *shampoos*
	SCHERCOTAINE WOAB	wheatgermamidopropyl betaine	dry solids 35% min.; salt content 4.0% max.	clear amber liquid; pH 5-7	conditioner; *hair care products*
Seppic	AMONYL 265 BA	coco-betaine	30% active	liquid	foaming agent; detergent; *hair care products*
	AMONYL 380 BA	cocoamidopropyl betaine	30% active	liquid	
	AMONYL 440 NI	cocoamidopropyl betaine	34% active	liquid	
	AMONYL 673 BA	cocosulfobetaine	30% active	liquid	
	AMONYL 675 BA	cocoamidosulfobetaine	30% active	liquid	
Stepan Europe	AMPHOSOL CA	alkyl amido betaine C$_{12/18}$	30% active	pale yellow liquid	foaming agent; conditioner; thickener; wetting agent; detergent; foam booster; *shampoos; bubble baths; shower gels; liquid soaps; baby shampoos; household, institutional and industrial cleaners*
	AMPHOSOL CB3	alkyl amido betaine C$_{8/18}$	30% active	water white to pale yellow liquid	
	AMPHOSOL DM	alkyl betaine	30% active	water white to pale yellow liquid	
Surfachem	SURFAC B4	alkyl amido propyl dimethyl amine betaine			
	SURFAC LB3	alkyl betaine			
	SURFAC SB09	sulfobetaine			
Witco	EMCOL 6748	cocamidopropyl betaine	35% solids	liquid; colour 3 (Gardner)	detergent; foam stabiliser; foaming agent; viscosity modifier; wetting agent; *personal care products; household and industrial applications*
	EMCOL COCO BETAINE	cocamidopropyl betaine	42% solids	liquid; colour 3 (Gardner)	
	EMCOL NA-30	cocamidopropyl betaine	36% solids	liquid; colour 3 (Gardner)	
	REWOTERIC AM B 13	coconut amidobetaine	35% active	liquid	foam booster; *foam baths; shower gels; shampoos; liquid soaps; all-purpose cleaners*
	REWOTERIC AM B 14	special coconut amidobetaine	35% active	liquid	

Supplier	Trade name	Chemical description	Composition	General properties	Functionality Application
Witco	REWOTERIC AM CAS	coconut sulfobetaine	50% active	liquid	*shampoos; baby care products; foam baths; shower gels; cleaners*
	REWOTERIC AM DML	lauryldimethyl betaine	40% active	liquid	*shampoos; foam baths; hard-surface cleaners*
	REWOTERIC AM R 40	ricinol amidobetaine	40% active	liquid	*skin cleansing agents; baby shampoos*
	REWOTERIC AM TEG	tallow aminobetaine	40% active	liquid	*shampoos; cleaners*
Zschimmer & Schwarz	AMPHOTENSID B 4	fatty acid amido alkyl betaine	30% active	liquid	*shampoos; bath preparations; cleansing preparations; baby care products; liquid soaps*

Glycinates

Supplier	Trade name	Chemical description	Composition	General properties	Functionality Application
Akzo Nobel	AMPHOLAK 7CX/C	cosmetic grade of Ampholak 7CX	dry content 39-41%	surface tension 38 dynes/cm (0.1% soln.)	*cosmetics*
	AMPHOLAK 7CX	cocoamphocarboxyglycinate	dry content 39-41%	surface tension 38 dynes/cm (0.1% soln.)	*foaming agent cleaners; toiletries; washing-up liquids*
	AMPHOLAK 7TX/C	cosmetic grade of Ampholak 7TX	dry content 39-41%	surface tension 40 dynes/cm (0.1% soln.)	*cosmetics*
	AMPHOLAK 7TX	tallowamphocarboxyglycinate	dry content 39-41%	surface tension 40 dynes/cm (0.1% soln.)	*foaming agent (medium) detergents; hard-surface cleaners; toiletries*
	AMPHOLAK 7TY	tallowamphocarboxyglycinate salt-free	dry content 30-32%	surface tension 45 dynes/cm (0.1% soln.)	*low foam cleaners; shampoos; bath and shower preparations*
	AMPHOLAK X00-30P	oleoamphocarboxyglycinate	dry content 32-34%	surface tension 32 dynes/cm (0.1% soln.)	*foaming agent (medium) household detergents; industrial cleaners; hard-surface cleaners*
	AMPHOLAK XCE	complex cocoiminodiglycinate	dry content 38-40%	surface tension 31 dynes/cm (0.1% soln.)	*foaming agent (medium): hydrotrope industrial applications*
	AMPHOLAK XCO-30	capryloamphocarboxyglycinate	dry content 38-41%	surface tension 34 dynes/cm (0.1% soln.)	*foaming agent (medium) toiletries; shampoos; hard-surface cleaners*
	AMPHOLAK XJO	capryloamphocarboxyglycinate	dry content 35-38%	surface tension 38 dynes/cm (0.1% soln.)	*low foam; wetting agent; hydrotrope industrial hard-surface cleaners*
	AMPHOLAK XO7/C	cosmetic grade of Ampholak XO7	dry content 39-41%	surface tension 44 dynes/cm (0.1% soln.)	*cosmetics*
	AMPHOLAK XO7-SD55	oleoamphocarboxyglycinate	dry content ca. 94%	surface tension 45 dynes/cm (0.1% soln.)	*household detergents; hard-surface cleaners*
	AMPHOLAK XO7	oleoamphocarboxyglycinate	dry content 39-41%	surface tension 44 dynes/cm (0.1% soln.)	*foaming agent (medium) cleaners; toiletries; liquid soaps*
Auschem Cesalpinia	CHIMIN IMB	cocoamphocarboxy glycinate	30% active	liquid	*detergent; foaming agent; conditioner liquid detergents; cosmetics*

Supplier	Trade name	Chemical description	Composition	General properties	Functionality / Application
Hickson Manro	MANROTERIC 1202	bis-2-hydroxyethyl tallow glycinate	40% active	viscous liquid	*hair care products*
Hoechst	DP1325	lauryl amphoglycinate	27% active		
	DP1358	tallow polyamphoglycinate	30% active		*toiletries; detergents*
	HOE S 4995	coco amphoglycinate	40% active		
	HUK 030	coco amphoglycinate	40% active		
	HUK 036	lauryl amphoglycinate	40% active		
	HUK 048	oleic polyamphoglycinate			*detergents*
Hüls	AMPHOLYT JA 120	N-C$_{10/12}$-fatty acid amidoethyl-N-(2-hydroxyethyl)-glycinate	active detergent 33%	liquid	*industrial cleaners*
	AMPHOLYT JA 140	N-C$_{12/18}$-fatty acid amidoethyl-N-(2-hydroxyethyl)-glycinate	active detergent 33%	liquid	*shampoos; baby shampoos*
McIntyre Group	MACKAM TM	dihydroxyethyl tallow glycinate	40% active	liquid; pH 5	*thickener; conditioner household cleaners; shampoos; hair conditioners*
Rhone-Poulenc	MIRATAINE TM	dihydroxyethyl tallow glycinate	40% solids; NaCl 5.0%	liquid	*foaming agent (moderate); viscosity modifier; wetting agent; thickener hard-surface cleaners*

Imidazolines

Supplier	Trade name	Chemical description	Composition	General properties	Functionality / Application
Croda	CRODATERIC CYNA 50-AM 3578	2-alkyl-1-(ethyl-beta-oxipropanoic acid) imidazoline sodium salt based on caprylic acid	50% active (aq. soln.)	clear amber liquid; surface tension 30 dynes/cm (0.1% in deionised H$_2$O); pH 10-11.5	*antistatic agent; wetting agent; detergent; foaming agent (moderate)* *detergents; hard-surface cleaners; textile industry; glass cleaners*
	CRODAZOLINE C	1-hydroxyethyl-2-alkylimidazoline	alkali no. 201-211; imidazoline 85%	M.W. 288; m.p. 42°C; pH 11.5 (10% H$_2$O)	*dispersant; water repellent; emulsifier; biocide; fungicide; plasticiser; anti-corrosion; lubricant; detergent; viscosity modifier; wetting agent; softener; antistatic agent; solubiliser; flotation aid; thickener; levelling agent; foaming agent*
	CRODAZOLINE O	1-hydroxyethyl-2-alkylimidazoline	alkali no. 163-176; imidazoline 85%	M.W. 345; pour point − 12°C; pH 11.1 (10% H$_2$O)	*adhesives; agriculture; construction industry; car washes; cleaners; fibreglass; food industry; food packaging; leather industry; lubricants; metal working; mining; paint industry; inks; paper industry; plastics industry; oil industry; textile industry*
	CRODAZOLINE S	1-hydroxyethyl-2-alkylimidazoline	alkali no. 161-171; imidazoline 85%	M.W. 338; m.p. 52°C; pH 11.0 (10% H$_2$O)	
Hoechst	DP1408	coco imidazoline			*chemical intermediate*
	DP1409	tall oil imidazoline			
	HUK 025	lauryl imidazoline			
	HUK 049	coco imidazoline			
Kao Corporation	BETADET SHC-2	coco imidazoline dicarboxymethylated	38-40% active	liquid; pH 8-9 (5%)	

Supplier	Trade name	Chemical description	Composition	General properties	Functionality / *Application*
Kao Corporation	BETADET THC-2	coco imidazoline dicarboxymethylated	38-40% active	liquid; pH 8-9 (5%)	
Lonza	ALKAWET CF	modified imidazoline amphoteric	solids 83%	liquid	foaming agent; detergency enhancer; *industrial applications*
	ALKAWET LF	modified imidazoline amphoteric	solids 68%	liquid	low foam; detergency enhancer; *industrial applications*
	AMPHOTERGE J-2	substituted imidazoline amphoteric	solids 50%; NaCl 10%	liquid; viscosity 236 cPs; surface tension 26.4 dynes/cm (0.1% active); Draves wetting 30 sec; pH 8.5	foaming agent; detergency enhancer; *personal care products; industrial applications*
	AMPHOTERGE K-2	substituted imidazoline amphoteric	solids 40%; NaCl 0.02%	liquid; pour point 0°C; viscosity 76 cPs; surface tension 38.6 dynes/cm (0.1% active); Draves wetting 180 sec; pH 9.6	
	AMPHOTERGE K	substituted imidazoline amphoteric	solids ca. 38%; NaCl 0.02%	liquid; pour point 2°C; viscosity 186 cPs; surface tension 31.9 dynes/cm (0.1% active); Draves wetting 48 sec; pH 9.8	
	AMPHOTERGE KJ-2	substituted imidazoline amphoteric	solids 40%; NaCl 0.05%	liquid; pour point -2°C; viscosity 50 cPs; surface tension 27.5 dynes/cm (0.1% active); Draves wetting 83 sec; pH 9.6	
	AMPHOTERGE L SPECIAL	substituted imidazoline amphoteric	solids 37%	liquid	
	AMPHOTERGE NX	substituted imidazoline amphoteric	solids 40%	liquid	
	AMPHOTERGE SB	substituted imidazoline amphoteric	solids 45%; NaCl 8%	pour point -5°C; viscosity 26 cPs; surface tension 33.2 dynes/cm (0.1% active); Draves wetting 56 sec; pH 7.5 (1%)	
	AMPHOTERGE W-2	substituted imidazoline amphoteric	solids 50%; NaCl 11%	liquid; viscosity 96,000 cPs; surface tension 28.5 dynes/cm (0.1% active); Draves wetting 22 sec; pH 8.2 (20%)	
	AMPHOTERGE W	substituted imidazoline amphoteric	solids 46%; NaCl 8%	liquid; pour point 8°C; viscosity 564 cPs; surface tension 28.5 dynes/cm (0.1% active); Draves wetting 17 sec; pH 9.8	
Mona Industries	MONATERIC 1000	2-alkylimidazoline-derived product	total solids 50%; 50% active	clear to hazy amber liquid; Draves wetting instant (1% active, 3 g hook); pH 11.8 (at 10%)	foaming agent (high); detergent; solubiliser; antistatic agent; anti-corrosion; *metal working; cosmetics; personal care products; paper industry; textile industry; detergents; cleaners; softening agents; lubricants; agriculture*
	MONATERIC 810-A-50	2-alkylimidazoline-derived product	total solids 50%; 50% active	clear brown liquid; Draves wetting 5 sec (1% active, 3 g hook); pH 4.4 (at 10%)	
	MONATERIC 811	2-alkylimidazoline-derived product	total solids 50%; 50% active	clear to hazy amber liquid; Draves wetting instant (1% active, 3 g hook); pH 11.4 (at 10%)	
	MONATERIC CA-35	2-alkylimidazoline-derived product	total solids 35%; 35% active	clear amber liquid; Draves wetting 11 sec (1% active, 3 g hook); pH 5.7 (at 10%)	
	MONATERIC CAM-40	2-alkylimidazoline-derived product	total solids 40%; 40% active	clear amber liquid; Draves wetting 40 sec (1% active, 3 g hook); pH 9.3 (at 10%)	
	MONATERIC CDX-38 MOD	2-alkylimidazoline-derived product	total solids 50%; NaCl 11%; 39% active	viscous yellow liquid; Draves wetting 20 sec (1% active, 3 g hook); pH 8.8 (at 10%)	
	MONATERIC CDX-38	2-alkylimidazoline-derived product	total solids 50%; NaCl 11%; 39% active	viscous yellow liquid; Draves wetting 18 sec (1% active, 3 g hook); pH 8.5 (at 10%)	

Supplier	Trade name	Chemical description	Composition	General properties	Functionality / Application
Mona Industries	MONATERIC CEM-38	2-alkylimidazoline-derived product	total solids 39%; 39% active	clear to hazy amber liquid; Draves wetting 4.5 min (1% active, 3 g hook); pH 8.6 (at 10%)	foaming agent (high); detergent; solubiliser; antistatic agent; anti-corrosion
	MONATERIC CEM-38CG	2-alkylimidazoline-derived product	total solids 38%; 38% active	clear to hazy amber liquid; Draves wetting 10 min (1% active, 3 g hook); pH 9.8 (at 10%)	*metal working; cosmetics; personal care products; paper industry; textile industry; detergents; cleaners; softening agents; lubricants; agriculture*
	MONATERIC CM-36S	2-alkylimidazoline-derived product	total solids 42%; NaCl 6%; 36% active	clear amber liquid; Draves wetting 5 sec (1% active, 3 g hook); pH 11.9 (at 10%)	
	MONATERIC CNA-40	2-alkylimidazoline-derived product	total solids 40%; 40% active	clear amber liquid; Draves wetting 6 sec (1% active, 3 g hook); pH 10.9 (at 10%)	
	MONATERIC CSH-32	2-alkylimidazoline-derived product	total solids 40%; NaCl 8%; 32% active	clear yellow liquid; Draves wetting 20 sec (1% active, 3 g hook); pH 8.4 (at 10%)	
	MONATERIC CY	2-alkylimidazoline-derived product	total solids 50%; 50% active	dark amber liquid; Draves wetting 29 sec (1% active, 3 g hook); pH 10.6 (at 10%)	
	MONATERIC CYA-50	2-alkylimidazoline-derived product	total solids 50%; 50% active	clear dark brown liquid; Draves wetting 9 sec (1% active, 3 g hook); pH 5.6 (at 10%)	
	MONATERIC CYMM-40	2-alkylimidazoline-derived product	total solids 40%; 40% active	clear amber liquid; Draves wetting 4.5 min (1% active, 3 g hook); pH 9.8 (at 10%)	
	MONATERIC ISA-35	2-alkylimidazoline-derived product	total solids 35%; 35% active	clear amber liquid; Draves wetting 10 min (1% active, 3 g hook); pH 5.4 (at 10%)	
	MONATERIC LF NA-50	2-alkylimidazoline-derived product	total solids 50%; 50% active	clear brown liquid; Draves wetting 10 min (1% active, 3 g hook); pH 11.5 (at 10%)	
	MONATERIC LF-100	2-alkylimidazoline-derived product	total solids 100%; 100% active	clear brown liquid; Draves wetting 25 sec (1% active, 3 g hook); pH 11.7 (at 10%)	
	MONATERIC LMM-30	2-alkylimidazoline-derived product	total solids 36%; NaCl 6%; 30% active	viscous amber liquid; Draves wetting 5 sec (1% active, 3 g hook); pH 9.2 (at 10%)	
	MONATERIC TA-35	2-alkylimidazoline-derived product	total solids 35%; 35% active	dark brown gel; Draves wetting 7 min (1% active, 3 g hook); pH 5.2 (at 10%)	
Protex	SURFARON A4112 DN38	sodium copra dicarboxylic imidazoline			foaming agent; detergent *shampoos*
Rhône-Poulenc	MIRANOL C₂M CONC. NP	imidazoline derivative	50% solids; NaCl 11.5%	liquid	foaming agent (high); emulsifier; wetting agent *hard-surface cleaners; household cleaners*
	MIRANOL C₂M-SF CONC.	imidazoline derivative	39% solids	liquid	foaming agent (high); emulsifier; wetting agent *hard-surface cleaners*
	MIRANOL CM CONC. NP	imidazoline derivative	44% solids; NaCl 7.0%	liquid	
	MIRANOL CM-SF CONC.	imidazoline derivative	37% solids	liquid	foaming agent (high) *hard-surface cleaners*
	MIRANOL CS CONC.	imidazoline derivative	45% solids; NaCl 7.2%	liquid	
	MIRANOL J₂M CONC.	imidazoline derivative	49% solids; NaCl 11.3%	liquid	low foam; emulsifier; wetting agent *hard-surface cleaners*
	MIRANOL JEM CONC.	imidazoline derivative	34% solids; NaCl 5.5%	liquid	low foam; emulsifier *hard-surface cleaners*
	MIRANOL JS CONC.	imidazoline derivative	45% solids; NaCl 8.7%	liquid	low foam *hard-surface cleaners*

Supplier	Trade name	Chemical description	Composition	General properties	Functionality / Application
Scher Chemicals	SCHERCOTERIC CY-2	imidazolinium amphoteric	dry solids 70% min.; salt content 11% max.	clear amber liquid	low foam / household and industrial cleaners
	SCHERCOTERIC I-AA	imidazolinium amphoteric	dry solids 34% min.; salt content 0%	amber viscous liquid	cosmetics; industrial cleaners
	SCHERCOTERIC MS-2	imidazolinium amphoteric	dry solids 45% min.; salt content 10% max.	clear amber viscous liquid	detergent / industrial cleaners; shampoos
	SCHERCOTERIC MS-EP	imidazolinium amphoteric	dry solids 45% min.; salt content 6% max.	clear amber liquid	foaming agent
	SCHERCOTERIC MS	imidazolinium amphoteric	dry solids 45% min.; salt content 9% max.	clear amber viscous liquid	detergent / industrial cleaners; shampoos
	SCHERCOTERIC O-AA	imidazolinium amphoteric	dry solids 80% min.; salt content 0%	clear amber liquid	industrial cleaners; dry cleaning
Thor Chemicals	SOVATEX MP/1	oleyl imidazoline	38% active	liquid	antistatic agent; lubricant; anti-corrosion
Zschimmer & Schwarz	AMPHOTENSID CT	alkyl imidazoline	38% active	fluid	cleansing preparations; all-purpose cleaners; car washes

Propionates

Supplier	Trade name	Chemical description	Composition	General properties	Functionality / Application
Akzo Nobel	AMPHOLAK YCA/P	cocoiminodipropionate, half sodium salt	dry content ca. 30%	surface tension 37 dynes/cm (0.1% soln.)	foaming agent / cleaners
	AMPHOLAK YCE	complex cocoimidipropionate salt-free	dry content 29-31%	surface tension 40 dynes/cm (0.1% soln.)	foaming agent (medium); anti-corrosion; hydrotrope / industrial applications; cosmetics
	AMPHOLAK YJH-40	octyliminodipropionate	dry content 38-40%	surface tension 38 dynes/cm (0.1% soln.)	low foam; hydrotrope / cleaners
	AMPHOLYTE KKDP-60	coco alkyl amino propionic acid	dry content 60%	surface tension 29 dynes/cm (0.1% soln.)	emulsifier; dispersant
	AMPHOLYTE KKE-70	coco alkyl amino propionic acid	dry content 70%	surface tension 29 dynes/cm (0.1% soln.)	emulsifier; dispersant
Henkel	DEHYTON G-SF	cocoamphodipropionate	40% active	liquid	shampoos; hair care products
	DERIPHAT 151C	lauraminopropionic acid	40% active	clear liquid	wetting agent; emulsifier / hair care products; skin care products
	DERIPHAT 154	disodium N-tallow-β-iminodipropionate	98% active	white powder	
	DERIPHAT 160 C-KPC	monosodium-N-lauryl β-iminodipropionic acid	ca. 30% active	liquid	cleaners
	DERIPHAT 160	disodium lauriminodipropionate	98% active	white powder	wetting agent; emulsifier / hair care products; skin care products
	DERIPHAT 160C	sodium lauriminopropionic acid	30% active	clear amber liquid	
Hoechst	DP1217	2-ethylhexylamino dipropionate	30% active		heavy-duty cleaners
	DP122	coco amino dipropionate	30% active		detergents; cleaners

Supplier	Trade name	Chemical description	Composition	General properties	Functionality Application
Lakeland Laboratories	ACP-70	salt free monopropionate	70% active	amber coloured liquid	anti-corrosion; lubricant; emulsifier / *water treatment*
	AMA 100	salt free dipropionate		white powder	foaming agent (high); solubiliser; hydrotrope; anti-corrosion; biostatic agent; lubricant; wetting agent; detergent / *detergents; metal finishing; textile industry; water treatment; agriculture; laundry products*
	AMA LF40	salt free dipropionate	40% active	pale straw coloured liquid	low foam; solubiliser; hydrotrope; detergent; anti-corrosion / *detergents; metal finishing; textile industry; water treatment; dishwashing agents*
	AMA LF70	salt free dipropionate	70% active	pale straw coloured liquid	low foam; solubiliser; hydrotrope; detergent; anti-corrosion / *detergents; metal finishing; textile industry; water treatment*
	AMA	salt free dipropionate	30% active	pale straw coloured liquid	foaming agent (high); solubiliser; hydrotrope; anti-corrosion; biostatic agent; lubricant; wetting agent; detergent / *detergents; metal finishing; textile industry; water treatment; agriculture; laundry products*
	ODA	salt free dipropionate	30% active	amber coloured liquid	solubiliser; hydrotrope; lubricant / *detergents; metal finishing; textile industry; water treatment*
McIntyre Group	MACKAM 151C	cocaminopropionic acid	40% active	liquid; pH 5	conditioner
	MACKAM 151L	lauraminopropionic acid	40% active	liquid; pH 5	*shampoos*
	MACKAM 160C-30	sodium lauriminodipropionate	30% active	liquid; pH 7	
	MACKAM 2CSF-70	disodium cocoamphodipropionate and propylene glycol	70% active	liquid; pH 6	
	MACKAM 2CSF	disodium cocoamphodipropionate	39% active	liquid; pH 10	emulsifier / *aerosol preparations*
	MACKAM 2CYSF	disodium capryloamphodipropionate	50% active	liquid; pH 10	
	MACKAM 2LSF	disodium lauroamphodipropionate	39% active	liquid; pH 10	
	MACKAM CSF	sodium cocoamphopropionate	39% active	liquid; pH 10	
Rhone-Poulenc	MIRATAINE H2C-HA	sodium lauriminodipropionate	30% solids	liquid	foaming agent (high); foam booster; wetting agent; dispersant / *hard-surface cleaners*
	MIRATAINE JC-HA	sodium alkyliminopropionate	47% solids	liquid	low foam; dispersant / *hard-surface cleaners*
Witco	REWOTERIC AM KSF 40	salt-free propionate	40% active	liquid	*baby care products; intimate hygiene products; industrial cleaners*
	REWOTERIC AM VSF	salt-free propionate	50% active	liquid	wetting agent / *cleaners*

Miscellaneous amphoterics

Supplier	Trade name	Chemical description	Composition	General properties	Functionality Application
Akcros Chemicals	AGRILAN TKA114	amphoteric	35% active; H₂O content 65%	straw liquid; pour point − 5°C; viscosity 45 cSt; pH 4.0–6.0 (1% aq.)	foaming agent agrochemicals
	AMPHOLAN U203	partial sodium salt of a complex amphoteric surfactant containing both carboxyl and amino groups	30% active; H₂O content 70%	pale yellow liquid; pour point < 0°C; viscosity 132 cSt; Ross-Miles foam (0.1% soln., 50°C, pH 7 and H₂O hardness 50 ppm): initial 160 mm, 5 min 155 mm; pH 6.0–8.0 (1% aq.)	foaming agent (high); wetting agent; solubiliser; dispersant; anti-corrosion industrial cleaners; industrial detergents; shampoos; germicides; textile industry
	VERSILAN MX244	high-foaming amphoteric surfactant	solids 35% H₂O content 65%	clear yellow liquid; pour point − 5°C; viscosity 25 cSt; pH 5.0 (1% aq.)	detergent; foaming agent (high) industrial cleaners; carpet cleaners; fire-fighting; plaster board; concrete; carpet industry
Akzo Nobel	AMPHOLAK MDX-1	blend of Ampholak 7TX, Ampholak XCO-30 and Amphoteen BCA-30 (1:1:1)	dry content 38–40%	surface tension 38 dynes/cm (0.1% soln.)	
	AMPHOLAK MDX-2	blend of Ampholak 7TX, Ampholak XCO-30 and Amphoteen 24 (45:20:35)	dry content 38–40%	surface tension 35 dynes/cm (0.1% soln.)	washing-up liquids
	AMPHOLAK MDX-3	blend of Ampholak 7TX and Amphoteen 24 (55:45)	dry content 38–40%	surface tension 30 dynes/cm (0.1% soln.)	
	AMPHOLAK MSX-1	cosmetic grade of Ampholak MDX-1	dry content 38–40%	surface tension 38 dynes/cm (0.1% soln.)	
	AMPHOLAK MSX-2	cosmetic grade of Ampholak MDX-2	dry content 38–40%	surface tension 35 dynes/cm (0.1% soln.)	cosmetics
	AMPHOLAK MSX-3	cosmetic grade of Ampholak MDX-3	dry content 38–40%	surface tension 30 dynes/cm (0.1% soln.)	
	ARMEEN SZ	N-coco-3-aminobutyric acid, sodium salt	39–41% active	liquid	
	ARMEEN Z	N-coco-3-aminobutyric acid	51–55% active	paste	
	ARMEEN Z9 SPECIAL	N-coco-3-aminobutyric acid	46–50% active	liquid	
	ARMEEN Z9	N-coco-3-aminobutyric acid	46–50% active	liquid	
Albright & Wilson	EMPIGEN XDR302	cocoamphoacetate/anionic blend	50% active	liquid	
Ellis & Everard	CAFLON AA30	sodium salt of an alkylamine dicarboxylate	30% active		
	CAFLON BA40	alkylamine dicarboxylate	40% active		
	CAFLON DP806	amphoteric acid thickening agent	34% active		thickener
Th. Goldschmidt	TEGO-PEARL B 48	mixture of pearlescent substances with surfactants	ca. 44% active	white liquid; pH 5–7 (10% soln. in H₂O)	pearlescent agent; re-fatting agent baby shampoos; bath preparations; shower preparations; shampoos; hair conditioners; shower gels; bubble baths; liquid soaps
	TEGO-PEARL S 33	mixture of pearlescent substances with surfactants	ca. 33% active	white liquid; pH 5.5–7.5 (10% soln. in H₂O)	
Henkel	AQUASOFT 22	combination of betaine and quaternary components			
	AQUASOFT HEC	quaternary fatty acid ester			softener textile industry
	AQUASOFT WH	combination of betaine and quaternary components			
Hickson Manro	MANRO AT1200	blended amphoteric acid thickener	40% active	viscous liquid	household detergents; industrial cleaners; metal working
Kao Corporation	BETADET 1211	ethoxylated fatty alcohol carboxymethylated	> 19.5% active	liquid; pH 6.5–7.5 (5%)	
	SK POL 10 CA	N-coco β-amino butyric acid	49–51% active	liquid; pH ca. 7 (5%)	

Supplier	Trade name	Chemical description	Composition	General properties	Functionality / *Application*
Libra Chemicals	LIBRATERIC AA-30	N,N dicarboxyethyl alkylamine, sodium salt	30% active	clear liquid	hydrotrope; *traffic film removers; alkaline cleaners*
	LIBRATERIC BA-40	N,N dicarboxyethyl alkylamine, sodium salt	40% active	clear liquid	hydrotrope; low foam; *acid and alkaline cleaners*
Lonza	LONZAINE CS	cocamidopropyl hydroxy sultaine	solids 50%; NaCl 6%	liquid; pour point 9°C; viscosity 189 cPs; surface tension 35.2 dynes/cm (0.1% active); Draves wetting > 2 min; pH 8.0 (10%)	foaming agent; viscosity modifier; irritation mitigant; *personal care products; industrial applications*
	LONZAINE JS	cocamidopropyl hydroxy sultaine	solids 50%	liquid	
McIntyre Group	MACKAM 2CT	disodium cocoamphodiacetate, sodium trideceth sulfate and hexylene glycol	50% active	liquid; pH 6.5	
	MACKAM 2MCA	disodium cocoamphodiacetate, sodium lauryl sulfate and hexylene glycol	47% active	liquid; pH 8	
	MACKAM 2MCAS	disodium cocoamphodiacetate, sodium lauryl sulfate, sodium laureth sulfate and propylene glycol	47% active	liquid; pH 7.5	
	MACKAM 2MHT	disodium lauroamphodiacetate, sodium trideceth sulfate and hexylene glycol	50% active	liquid; pH 8	
	MACKAM CBS-50	cocamidopropyl hydroxysultaine	50% active	liquid; pH 8	thickener; foam booster
	MACKAM CBS-50G	cocamidopropyl hydroxysultaine	50% active	liquid; pH 8	thickener; foam booster
	MACKAM CS	sodium cocoamphohydroxypropyl sulfonate	45% active	liquid; pH 8	
	MACKAM JS	sodium capryloamphohydroxypropyl sulfonate	49% active	liquid; pH 8	
	MACKAM MEJ	mixed alkylamphocarboxylate	34% active	liquid; pH 10	
	MACKAM MLT	sodium lauroamphoacetate and sodium trideceth sulfate	35% active	liquid; pH 10	
	MACKAM OS	sodium oleoamphohydroxypropyl sulfonate	78% active	viscous liquid; pH 8	
Mona Industries	MONATERIC 805	blend of cocoamphocarboxyglycinate and cocamido MIPA-sulfosuccinate	total solids 42%; NaCl 2%; 40% active	clear amber liquid; Draves wetting 5 sec (1% active, 3 g hook); pH 7.7 (at 10%)	foaming agent (high); detergent; solubiliser; antistatic agent; anti-corrosion; *metal working; cosmetics; personal care products; paper industry; textile industry; detergents; cleaners; softening agents; lubricants; agriculture*
	MONATERIC 985A	blend of lauroamphoglycinate and sodium trideceth sulfate	total solids 39%; NaCl 3%; 36% active	clear amber liquid; Draves wetting instant (1% active, 3 g hook); pH 9.3 (at 10%)	
	MONATERIC ADFA	proprietary compound	total solids 34%; NaCl 3%; 31% active	clear amber liquid; Draves wetting 10 sec (1% active, 3 g hook); pH 7.8 (at 10%)	
	MONATERIC CDL	blend of cocoamphocarboxyglicinate, sodium laureth sulfate and sodium lauryl sulfate	total solids 37%; NaCl 6%; 31% active	clear yellow liquid; Draves wetting 4 sec (1% active, 3 g hook); pH 8.5 (at 10%)	
	MONATERIC CDS	blend of cocoamphocarboxyglycinate and sodium lauryl sulfate	total solids 37%; NaCl 6%; 31% active	clear yellow liquid; Draves wetting 3 sec (1% active, 3 g hook); pH 8.5 (at 10%)	
	MONATERIC CDTD	blend of cocoamphocarboxyglycinate and sodium trideceth sulfate	total solids 50%; NaCl 6%; 44% active	clear yellow liquid; Draves wetting instant (1% active, 3 g hook); pH 8.3 (at 10%)	
PPG	MAFO 13 MOD 1	potassium salt of a complex amine carboxylate			lubricant; chelating agent; solubiliser; detergent; *metal working*
	MAFO 13	potassium salt of a complex amine carboxylate			
	MAFO CSB-50	cocamidopropyl hydroxysultaine	solid matter 50%		viscosity modifier; foam booster; *personal care products*
Rhone-Poulenc	AMPHIONIC 25 B	biocidal ampholyte		liquid	foaming agent (high); wetting agent; *biocides*

Supplier	Trade name	Chemical description	Composition	General properties	Functionality / *Application*
Rhone-Poulenc	AMPHIONIC SFB	alkylaminocarboxylate	35% solids; NaCl 5.0%	liquid	foaming agent (high); wetting agent / *biocides; hard-surface cleaners*
	AMPHIONIC XL	polycarboxylated surfactant	40% solids; NaCl 10%	liquid	foaming agent (high); emulsifier; wetting agent; detergent; chelating agent / *hard-surface cleaners*
	MIRATAINE CBS	cocamidopropyl hydroxysultaine	50% solids; NaCl 6.5%	liquid	foaming agent (high); viscosity modifier; foam booster; emulsifier; wetting agent / *hard-surface cleaners; toilet bowl cleaners*
Sandoz	CERANINE AS	fatty amide derivative in aq. dispersion		liquid	softener; re-wetting agent
	DERMAGEN PN	modified fatty amino polyglycol ether in aq. soln.		liquid	penetrant; levelling agent; substantivity agent / *dyes; leather industry*
	ELFUGIN V	fatty alcohol polyglycol ether carboxylate/fatty amine polyglycol ether		liquid	antistatic agent
	LANASAN WAS	fatty amine derivative and inorganic halogenated oxidiser in H_2O		liquid	lubricant; softener / *textile industry; wools*
	LYOCOL V	nitrogenous condensation product		liquid	levelling agent; stripping agent
	LYOGEN TP	aq. soln. of polyglycol ethers		liquid	levelling agent / *dyeing; textile industry*
	SANDOBET SC	cocoamidohydroxy sultaine in aq. soln.		liquid	foaming agent / *toiletries; industrial products*
	SANDOGEN WAF	fatty acid aminoalkyl derivative/polyglycol ether		liquid	levelling agent / *textile industry; dyeing of polyamides; dyes*
	SANDOTAN DSN	aromatic sulfo acid derivative		powder	biocide; detergent / *food industry; surface cleaners*
	SANDOTERIC ABD LIQUID	biocidal amphoteric		liquid	
	SANDOTERIC CDP	aq. soln. of an alkylamine dicarboxylate		liquid	detergent / *industrial cleaners; hard-surface cleaners*
	SANDOTERIC SC	cocoamidohydroxy sultaine in aq. soln.		liquid	
	SANDOTERIC TFL PASTE	fatty acid derivative containing sulfo groups		yellow/brown paste	viscosity modifier; foaming agent (high) / *shampoos; hard-surface cleaners; industrial detergents*
	SANDOTERIC TIT LIQUID	dicarboxyethyl alkylamine sodium salt		clear liquid	*shampoos; industrial cleaners*
	TERGOLIX AL	fatty acid derivative in aq. soln.		liquid	dispersant / *natural fats*
	TERGOLIX E	alkyl phenol polyglycol ether		liquid	emulsifier; dispersant / *natural fats*
Surfachem	SURFAC BH30	alkylamine dicarboxylate sodium salt			
	SURFAC BH40	alkylamine dicarboxylate sodium salt - low foam			
Tomah Products	ALKALI SURFACTANT	amphoteric surfactant	35% active min.	light amber liquid	hydrotrope; coupling agent; detergent; foaming agent; wetting agent / *hard-surface and all-purpose cleaners; transportation, catering, household and institutional cleaning*
	AMPHOTERIC 400	amphoteric surfactant	50% active min.	clear liquid	
	AMPHOTERIC SC	amphoteric surfactant	35-39% active	slightly amber liquid	
Troy Chemicals	TROYSOL 98C			liquid	wetting agent; dispersant / *paint industry*

Supplier	Trade name	Chemical description	Composition	General properties	Functionality Application
Union Carbide	TRITON QS-15	sulfate	100% active	liquid; pour point − 4°C	*metal cleaning*
Warwick International	MYKON NRW-3				*wetting agent; foaming agent textile industry*
Witco	NATURAL EXTRACT AP	trimethylglycine	99% active	crystalline	*moisturiser; conditioner; anti-irritant*
	REWOTERIC AM 2 L	amphoteric glycine derivative	50% active	liquid	*shampoos; foam baths; intimate hygiene products; baby care products*
	REWOTERIC AM 2 C NM	amphoteric glycine derivative	50% active	liquid	*foam baths; shampoos; intimate hygiene products; baby care products*
	REWOTERIC AM CA	modified amphoteric glycine derivative	30% active	liquid	*foam baths; shampoos; baby care products*
	REWOTERIC AM G 30	modified amphoteric glycine derivative	38% active	liquid	*baby care products; shower baths; shampoos*
	REWOTERIC AM V	amphoteric glycine derivative	35% active	liquid	*wetting agent pickling baths; cleaners*
	REWOTERIC QAM 50		50% active	liquid	*hard-surface disinfectants; skin disinfectants*
	SOCHAMINE A 7525	coco-dicarboxylate	40% active	liquid	*shampoos*
	SOCHAMINE A 7527	coco-dicarboxylate, salt free	40% active	liquid	*shampoos*
	SOCHAMINE A 756	coco-hydroxy sulfonate	43% active	liquid	*softener household detergents*
	SOCHAMINE AC 721	coco-dicarboxylate lauryl sulfate complex	35% active	liquid	*shampoos*
Zschimmer & Schwarz	AMPHOTENSID 9 M	disodium cocoamphodiacetate blended with fatty alcohol ether sulfate	30% active	liquid	*shampoos; bath preparations; skin care products*
	AMPHOTENSID D 1	n-alkyl amino acid triethanolamine salt	40% active	liquid	*cleansing preparations; all-purpose cleaners; high-pressure cleaners; industrial cleaners*

2 Anionics

Alkyl sulfates

Supplier	Trade name	Chemical description	Composition	General properties	Functionality Applications
Akcros Chemicals	LANKROPOL WA	sulfated monoester of fatty acid, ammonium salt	fat content 50%; H$_2$O content ca. 35%	clear amber liquid; pour point <0°C; viscosity 55 cSt; pH 6.5 (1% aq.)	wetting agent; low foam; dispersant; emulsifier *textile industry; paint industry; degreasing formulations; herbicides*
	LANKROPOL WN	sulfated monoester of fatty acid, sodium salt	fat content 50%; H$_2$O content ca. 40%	clear amber liquid; pour point <0°C; viscosity 63 cSt; pH 6.5 (1% aq.)	
	PERLANKROL ATL40	lauryl sulfate, triethanolamine salt; toiletry grade	40% active	pale straw liquid; pour point <0°C; viscosity 60 cSt; pH 6.5-7.5 (1% aq.)	
	PERLANKROL DAF25	lauryl sulfate, ammonium salt; toiletry grade	25% active	pale yellow mobile gel; pour point 5°C; viscosity 1600 cSt; pH 6.2-6.7 (1% aq.)	*cosmetics; toiletries; bubble baths; shampoos*
	PERLANKROL DSA	lauryl sulfate, sodium salt; toiletry grade	28% active	white slurry; pour point 12°C; viscosity 220 cSt (40°C); pH 7.5-8.5 (1% aq.)	
Akzo Nobel	ELFAN 240 M	monoethanolamine lauryl myristyl sulfate	29% active	liquid	
	ELFAN 240 MG	magnesium lauryl myristyl sulfate	30% active	liquid	
	ELFAN 240 T	triethanolamine lauryl myristyl sulfate	40% active	liquid	
	ELFAN 240	sodium lauryl myristyl sulfate	30% active	liquid	
	ELFAN 280	sodium coco alkyl sulfate	36% active	paste	
	ELFAN 680	sodium oleyl cetyl sulfate	50% active	paste	
	ELFAN KT 550	sodium coco tallow alkyl sulfate	39% active	paste	
Albright & Wilson	EMPICOL 0045	sodium lauryl sulfate; toiletry grade	95% active	powder	*toiletries*
	EMPICOL 0045V	sodium lauryl sulfate; toiletry grade	88% active	needles	
	EMPICOL 0185	sodium lauryl sulfate	94% active		
	EMPICOL 0585/A	sodium ethyl hexyl sulfate; toiletry grade	40% active	liquid	
	EMPICOL 0758	sodium decyl sulfate; toiletry grade	40% active	liquid	*toiletries*
	EMPICOL 0775/55	sodium lauryl/tallow sulfate; toiletry grade	55% active	liquid	
	EMPICOL 0775	sodium lauryl/tallow sulfate; toiletry grade	42% active	liquid	
	EMPICOL 6017/23H	mixed amine lauryl sulfate	23% active		
	EMPICOL 6017/T/23	mixed amine lauryl sulfate	23% active		
	EMPICOL AL30/T	ammonium lauryl sulfate; toiletry grade	27% active	liquid/paste	*toiletries*
	EMPICOL AL30/TP	ammonium lauryl sulfate	28% active		
	EMPICOL AL30/TPA	ammonium lauryl sulfate	27% active		
	EMPICOL AL30/TX	ammonium lauryl sulfate	27% active		
	EMPICOL AL70	ammonium lauryl sulfate; toiletry grade	68% active	liquid	*toiletries*

Supplier	Trade name	Chemical description	Composition	General properties	Functionality/Application
Albright & Wilson	EMPICOL LB40	sodium octyl/decyl sulfate	40% active	liquid	
	EMPICOL LM45/Y	sodium lauryl sulfate	45% active		
	EMPICOL LM45	sodium lauryl sulfate	45% active		
	EMPICOL LQ33/F	monoethanolamine lauryl sulfate; toiletry grade	33% active	liquid	*toiletries*
	EMPICOL LQ33/TX	monoethanolamine lauryl sulfate	33% active		
	EMPICOL LQ33T	monoethanolamine lauryl sulfate	33% active		
	EMPICOL LX	sodium lauryl sulfate; toiletry grade	89% active	powder	emulsifier; dispersant; wetting agent; air entraining agent; flotation aid
	EMPICOL LX100	sodium lauryl sulfate; toiletry grade	96% active	powder	*toiletries; construction; emulsion and suspension polymerisation;*
	EMPICOL LX28	sodium lauryl sulfate; toiletry grade	29% active	liquid/paste	*ore flotation*
	EMPICOL LXSV/S	sodium lauryl sulfate	93% active		
	EMPICOL LXV	sodium lauryl sulfate; toiletry grade	85% active	needles	emulsifier; dispersant; wetting agent; air entraining agent; flotation aid *toiletries; construction; emulsion and suspension polymerisation; ore flotation*
	EMPICOL LXV100/C	sodium lauryl sulfate	95% active		
	EMPICOL LXV100	sodium lauryl sulfate; toiletry grade	95% active	needles	emulsifier; dispersant; wetting agent; air entraining agent; flotation aid
	EMPICOL LZ/D	sodium lauryl sulfate; B.P. grade	90% active	powder	*toiletries; construction; emulsion and suspension polymerisation; ore flotation*
	EMPICOL LZ/PVC	sodium lauryl sulfate	87% active	powder	
	EMPICOL LZ	sodium lauryl sulfate; toiletry grade	89% active		emulsifier; dispersant; wetting agent; air entraining agent; flotation aid *toiletries; construction; emulsion and suspension polymerisation; ore flotation*
	EMPICOL LZG30/B	sodium lauryl sulfate	30% active		
	EMPICOL LZV/D	sodium lauryl sulfate; B.P. grade	90% active	needles	emulsifier; dispersant; wetting agent; air entraining agent; flotation aid
	EMPICOL LZV	sodium lauryl sulfate; toiletry grade	85% active	needles	*toiletries; construction; emulsion and suspension polymerisation; ore flotation*
	EMPICOL O298	sodium lauryl sulfate	28% active		
	EMPICOL O585/A	sodium ethyl hexyl sulfate	40% active		
	EMPICOL OO3IT	diethanolamine lauryl sulfate	36% active		
	EMPICOL TA30	sodium tallow sulfate	30% active		
	EMPICOL TL40/T	triethanolamide lauryl sulfate; toiletry grade	40% active	liquid	*toiletries*
	EMPIMIN LR28	sodium alkyl sulfate	28% active	liquid	
	EMPIMIN LX SERIES	lauryl sulfate; B.P. grade		viscous liquid	compounding aid *pharmaceuticals*
	EMPIMIN LZ SERIES	lauryl sulfate; B.P. grade			
Auschem	COSMOPON AM	ammonium lauryl sulfate	28% active	viscous liquid	foaming agent *liquid detergents; laundry products; household detergents;*
Cesalpinia	COSMOPON MO	MEA lauryl sulfate	33% active	liquid	*dishwashing agents; polymerisation; bubble baths; liquid soaps; shower gels; shampoos*
	COSMOPON TR	TEA lauryl sulfate	40% active	liquid	*shampoos; bubble baths; liquid soaps; shower gels*

Supplier	Trade name	Chemical description	Composition	General properties	Functionality / Application
Auschem Cesalpinia	MADEOL LSP	sodium lauryl sulfate	90% active	powder	wetting agent / *pesticides*
	POLIROL LSA	sodium lauryl sulfate	28% active	liquid	emulsifier / *emulsion polymerisation*
	ROLPON 42 P	sodium alkyl sulfate	42% active	paste	foaming agent / *liquid detergents; laundry products; household detergents; dishwashing agents; polymerisation; bubble baths; liquid soaps; shower gels; shampoos*
	ROLPON LSA	sodium lauryl sulfate	28% active	liquid	
Bayer	LEVAPON OL	alkyl sulfate	33% active	highly viscous, brownish liquid	scouring agent: softener / *textile industry; wools; cellulosic fibres; feathers*
Croda	CRODEX A	sodium lauryl sulfate and cetearyl alcohol		white to pale yellow wax	o/w emulsifier / *skin care products; pharmaceuticals*
Elf Atochem	LAURAL D	triethanolamine lauryl sulfate		liquid	
	LAURAL P	sodium lauryl sulfate		paste	
	MELIORAN 118	sodium cetyl-stearyl sulfate		paste	
Ellis & Everard	CAFLON HB7	alcohol sulfate	30% active		
	CAFLON MS30	monoethanolamine lauryl sulfate	28% active		
	CAFLON SS28	sodium lauryl sulfate	90% active		
Henkel	LANETTE E	sodium cetearyl sulfate	90% active	white powder	o/w emulsifier / *ointments; creams; lotions*
	LAVIRON WA 1 SPEZ	fatty alcohol sulfate			foaming agent / *textile industry; finishing*
	STANDAPOL A-HV	ammonium lauryl sulfate	28% active	water-white to very pale yellow hazy liquid	foaming agent / *bubble baths; cleansing preparations*
	STANDAPOL A	ammonium lauryl sulfate	28% active	water-white to very pale yellow hazy liquid	foaming agent / *shampoos; bubble baths; cleansing preparations*
	STANDAPOL DEA	DEA-lauryl sulfate	37% active	clear, light yellow liquid	foaming agent / *shampoos; bubble baths*
	STANDAPOL MG	magnesium lauryl sulfate	30% active	liquid	foaming agent / *personal care products*
	STANDAPOL T	TEA-lauryl sulfate	40% active	water white liquid	foaming agent / *shampoos; bubble baths*
	STANDAPOL WA-AC	sodium lauryl sulfate	29% active	white to off-white pearlescent paste	pearlescent agent / *shampoos; bubble baths; cleansing preparations*
	STANDAPOL WAQ SPECIAL	sodium lauryl sulfate	29% active	water white liquid	*shampoos; bubble baths; cleansing preparations*
	STANDAPOL WAQ-115	sodium $C_{12/15}$ alcohol sulfate	30% active	water white viscous liquid	*shampoos; bubble baths; cleansing preparations*
	STANDAPOL WAQ-LC	sodium lauryl sulfate	29% active	water white liquid	*shampoos; cleansing preparations; bath preparations*
	SULFOPON 101 SPECIAL	sodium lauryl sulfate	29-31% active	liquid/paste	wetting agent; foaming agent; emulsifier; air-entraining agent / *cosmetics; emulsion polymerisation; latexes; carpet cleaners; mortars; detergents*

Supplier	Trade name	Chemical description	Composition	General properties	Functionality / Application
Henkel	SULFOPON 101	sodium lauryl sulfate	28-30% active	liquid/paste	wetting agent; foaming agent; cosmetics
	SULFOPON 1218 G	sodium $C_{12/18}$ sulfate	85-90% active	granules	laundry products; LAS substitute
	SULFOPON K 115-50	sodium $C_{12/18}$ sulfate	49-52% active	paste	laundry products; hand cleansing preparations
	SULFOPON K 115-55	sodium $C_{12/18}$ sulfate	54-58% active	paste	
	SULFOPON K 35	sodium $C_{12/18}$ sulfate	24.5-36.5% active	paste	
	SULFOPON O 680	sodium cetyl oleyl sulfate $C_{16/18}$	ca. 50% active	paste	co-emulsifier; plastics industry; coatings
	SULFOPON T 55	sodium $C_{16/18}$ sulfate	≥55% active	paste	laundry products; LAS substitute
	SULFOPON T POWDER	sodium $C_{16/18}$ sulfate	≥91% active	powder	laundry products; LAS substitute
	TEXAPHOR	highly concentrated solvent soln. of fatty alcohol sulfates		liquid; cloud point <5°C; viscosity 100-150 mPas; pH 6.5-7.5 (1% aq. soln.)	anti-settling agent; paint industry
	TEXAPON 1030	sodium decyl sulfate	29-31% active	liquid	wetting agent; solubiliser; hydrotrope; cleaners; dishwashing agents
	TEXAPON 842	sodium n-octyl sulfate	ca. 42% active	liquid	wetting agent; co-emulsifier; solubiliser; hydrotrope; plastics industry; coatings; cleaners; dishwashing agents
	TEXAPON 890	sodium octyl sulfate	≥90% active	powder	wetting agent; solubiliser; hydrotrope; cleaners; dishwashing agents
	TEXAPON A	ammonium lauryl sulfate	33-36% active	liquid	emulsifier; foaming agent; air-entraining agent; shampoos; emulsion polymerisation; latexes; carpet cleaners; mortars; dishwashing agent; cleaners
	TEXAPON ALS	ammonium lauryl sulfate	29-31% active	liquid	foaming agent; shampoos; foam baths; skin cleansing preparations; dishwashing agents; cleaners
	TEXAPON CS PASTE	mixture of fatty alcohol sulfates	57-63% active	paste	pearlescent agent; shampoos; bath preparations; skin cleansing preparations
	TEXAPON EHS	sodium 2-ethylhexyl sulfate	39.7-47.3% active	liquid	wetting agent; solubiliser; hydrotrope; co-emulsifier; cleaners; plastics industry; coatings
	TEXAPON K 12 POWDER	sodium lauryl sulfate	90.5% active	white powder	foaming agent; wetting agent; emulsifier; air-entraining agent
	TEXAPON K 12 GRANULES	sodium lauryl sulfate	86% active	white granules	toothpastes; bath preparations; mouthwashes; mortars; emulsion polymerisation; latex; carpet cleaners
	TEXAPON K 12 NEEDLES	sodium lauryl sulfate	87% active	white needles	
	TEXAPON K 1296	sodium lauryl sulfate	96% active	white powder	
	TEXAPON LLS	lithium lauryl sulfate	29-30% active	liquid	carpet cleaners
	TEXAPON LS 100 F	sodium lauryl sulfate	98% active	white powder	dentifrices; oral dosage forms
	TEXAPON LS 35	sodium $C_{12/14}$ sulfate	34-35.5% active		dishwashing agents; cleaners
	TEXAPON LS HIGHLY CONCN. NEEDLES	sodium lauryl sulfate	87% active	needles	foaming agent; emulsifier; air-entraining agent; shampoos; baby shampoos; bath preparations; mortars; emulsion polymerisation; latex; carpet cleaners

Supplier	Trade name	Chemical description	Composition	General properties	Functionality / Application
Henkel	TEXAPON MGS	magnesium lauryl sulfate	29-31% active	liquid	foaming agent / *shampoos; skin cleansing preparations*
	TEXAPON MLS	MEA-lauryl sulfate	31-33% active	liquid	dispersant / *shampoos; bubble baths; essential oils; detergents*
	TEXAPON NAG	sodium ammonium lauryl sulfate	28-30% active	liquid	*shampoos; shower and bath preparations; skin cleansing preparations; cosmetics*
	TEXAPON OT HIGHLY CONC. NEEDLES	sodium lauryl sulfate	91-92% active	needles	foaming agent / *cosmetics*
	TEXAPON T 42	TEA-lauryl sulfate	41-43% active	liquid	foaming agent; solubiliser; emulsifier; air-entraining agent / *shampoos; bubble baths; essential oils; cosmetics; emulsion polymerisation; latex; carpet cleaners; mortars; glass cleaners*
	TEXAPON TH	TEA-lauryl sulfate	47-49% active	liquid	foaming agent; solubiliser / *shampoos; bubble baths; essential oils; cosmetics; glass cleaners*
	TEXAPON V HIGHLY CONC. NEEDLES	sodium lauryl sulfate	86-90% active	white needles	foaming agent; emulsifier; air-entraining agent / *cosmetics; shampoos; bath preparations; skin cleansing preparations; emulsion polymerisation; latex; carpet cleaners; mortars; detergents*
	TEXAPON Z HIGHLY CONC. POWDER	sodium lauryl sulfate	90% active	white powder	foaming agent / *toothpastes; shampoos; bath preparations; bubble baths; skin cleansing preparations; detergents*
	TEXAPON Z HIGHLY CONC. NEEDLES	sodium lauryl sulfate	88-90% active	nondusting white needles	foaming agent / *toothpastes; shampoos; bath preparations; skin cleansing preparations; detergents*
	TEXAPON Z POWDER	sodium lauryl sulfate	60% active	powder	foaming agent / *toothpastes; shampoos; bath preparations; detergents*
Hickson Manro	MANRO ADS35	ammonium alkyl sulfate	35% active	liquid	*household detergents; industrial cleaners*
	MANRO ALS27	ammonium alkyl sulfate	27% active	liquid	*hair care products; bath preparations; household detergents*
	MANRO ALS30	ammonium alkyl sulfate	30% active	liquid	*hair care products; bath preparations; household detergents*
	MANRO DL28	sodium alkyl sulfate	28% active	liquid	
	MANRO HB7	sodium alkyl sulfate	40% active	liquid	*household detergents; industrial cleaners*
	MANRO ML33	monoethanolamine alkyl sulfate	33% active	viscous liquid	*hair care products; bath preparations*
	MANRO SLS28	sodium alkyl sulfate	28% active	liquid	*hair care products; bath preparations; household detergents; industrial cleaners*
	MANRO TL40	triethanolamine alkyl sulfate	40% active	liquid	*hair care products; bath preparations; household detergents*
	MANRO TMA23	mixed amine alkyl sulfate	23% active	liquid	*hair care products; bath preparations*
	TENSOPOL 528LS	sodium alkyl sulfate	28% active	liquid	
	TENSOPOL A79	sodium alkyl sulfate	90% active	needles	
	TENSOPOL A795	sodium alkyl sulfate	95% active	needles	*hair care products; bath preparations; household detergents; industrial cleaners*
	TENSOPOL ACL79	broadcut sodium alkyl sulfate	90% active	needles	
	TENSOPOL ACL795	broadcut sodium alkyl sulfate	95% active	needles	
	TENSOPOL AG	magnesium alkyl sulfate	80% active	needles	
	TENSOPOL LTS	triethanolamine alkyl sulfate	40% active	liquid	*hair care products; bath preparations; household detergents*

Supplier	Trade name	Chemical description	Composition	General properties	Functionality / Application
Hickson Manro	TENSOPOL SPK	sodium potassium alkyl sulfate	96% active	powder	hair care products; bath preparations; household detergents;
	TENSOPOL USP 94	sodium alkyl sulfate	94% active	powder	industrial cleaners
	TENSOPOL USP 97	sodium alkyl sulfate	97% active	powder	
Hoechst	GENAPOL AMG	magnesium-PEG 3 cocoamide sulfate	30% active		toiletries
	GENAPOL AMS	TEA-PEG 3 cocoamide sulfate	40% active		toiletries
	GENAPOL ARO LIQUID	2 mole sodium lauryl sulfate	28% active	liquid	shampoos; detergents
	GENAPOL ARO PASTE	2 mole synthetic sodium lauryl sulfate	70% active	paste	shampoos; detergents
	GENAPOL CRT-40	TEA-lauryl sulfate	40% active		shampoos
Hüls	MARLINAT DFK 30	lauryl sulfate, sodium salt	active detergent 30%	liquid	shampoos; liquid cleaners
	MARLINAT DFL 40	lauryl alcohol sulfate, TEA salt	active detergent 40%	liquid	
	MARLINAT DFN 30	lauryl sulfate, ammonium salt	active detergent 30%	liquid	shampoos; liquid cleaners
	MARLINAT KT 50	coconut-tallow-alkyl sulfate, sodium salt	active detergent 50%	paste	hand-washing pastes
	NA-BUTYLMONO GLYCOLSULPHATE 50	sodium butylmonoglycolsulfate	active detergent 50%	liquid	hydrotrope liquid cleaners
Kao Corporation	EMAL 10 N	sodium lauryl sulfate	> 94% active	needles; M.W. 304; pH 6.8-8 (1%)	
	EMAL 10 P	sodium lauryl sulfate	> 94% active	powder; M.W. 304; pH 8.5-10 (1%)	
	SAPANOL ACB	sodium cetyl-stearyl sulfate	29-31% active	paste; M.W. 357; pH 6-8 (5%)	
	SAPANOL ACE	sodium cetyl-stearyl sulfate	32-34% active	paste; M.W. 357; pH 6-8 (5%)	
	SAPANOL AS-30	sodium lauryl sulfate	29-31% active	liquid; M.W. 299; pH 7-8 (5%)	
	SAPANOL AT	triethanolamine lauryl sulfate	39-41% active	liquid; M.W. 427; pH 7-7.5 (5%)	
	SAPANOL NS	sodium decyl sulfate	34.5-35.5% active	liquid; M.W. 262; pH 7-10 (5%)	
Lonza	CARSONOL ALS-R	ammonium lauryl sulfate	29% active	liquid	foaming agent; wetting agent; emulsifier personal care products; household and industrial applications
	CARSONOL DLS	DEA lauryl sulfate	29% active	liquid	foaming agent; wetting agent; emulsifier personal care products
	CARSONOL MLS	magnesium lauryl sulfate	30% active	liquid	foaming agent; wetting agent; emulsifier carpet cleaners
	CARSONOL SHS	sodium octyl sulfate	40% active	liquid	foaming agent; wetting agent; emulsifier food industry
	CARSONOL SLS PASTE B	sodium lauryl sulfate	29% active	paste	foaming agent; wetting agent; emulsifier household and industrial applications
	CARSONOL SLS-R	sodium lauryl sulfate	29% active	liquid	foaming agent; wetting agent; emulsifier
	CARSONOL SLS-S	sodium lauryl sulfate	29% active	liquid	personal care products; household and industrial applications
	CARSONOL TLS	TEA lauryl sulfate	40% active	liquid	
Millchem	MAPROFIX 563	sodium lauryl sulfate; recrystallised dentifrice grade	95% active	powder	
	MAPROFIX DLS-35	DEA lauryl sulfate	33-35% active	liquid	shampoos; bubble baths
	MAPROFIX MG	magnesium lauryl sulfate	27-29% active	liquid	carpet cleaners

Supplier	Trade name	Chemical description	Composition	General properties	Functionality / Application
Millchem	MAPROFIX NH-54	ammonium lauryl sulfate	49-51% active	liquid	foaming agent / *industrial applications*
	MAPROFIX NH	ammonium lauryl sulfate	27-30% active	viscous liquid	*shampoos; bubble baths; detergents*
	MAPROFIX TLS-106	TEA lauryl sulfate	79% active	paste	detergent / *shampoos*
	MAPROFIX TLS-500	TEA lauryl sulfate	40% active	liquid	*cosmetics; industrial formulations*
	MAPROFIX TLS-513	TEA lauryl sulfate	39% active	liquid	*cosmetics; industrial formulations*
	MAPROFIX WA	sodium lauryl sulfate	28-30% active	paste	*shampoos; bubble baths; detergents*
	MAPROFIX WAC-LA	sodium lauryl sulfate	28-30% active	liquid; viscosity low	*personal care products; industrial formulations*
	MAPROFIX WAC	sodium lauryl sulfate	28-30% active	liquid	*shampoos; bubble baths; detergents*
	MILLIFOAM Q-127-1	blend of alkyl sulfates	solids 26%	liquid	foaming agent / *plasterboard*
Niacet Corporation	NIAPROOF ANIONIC SURFACTANT (NAS) 08	sodium 2-ethyl hexyl sulfate	39% active	essentially colourless liquid	wetting agent / *textile industry; household and industrial cleaners; agriculture; metal working; pharmaceuticals; white wall tyre cleaners*
	NIAPROOF ANIONIC SURFACTANT (NAS) 4	sodium tetradecyl sulfate	27% active	essentially colourless liquid	penetrant; wetting agent / *textile industry; cleaners; metal working; pharmaceuticals*
Nikko Chemicals	NIKKOL ALS-25	ammonium lauryl sulfate; aq. soln.		yellow viscous liquid	foaming agent; cleansing agent / *cosmetics*
	NIKKOL KLS	potassium lauryl sulfate		white crystalline powder	
	NIKKOL SCS	sodium cetyl sulfate		white powder	emulsifier / *cosmetics*
	NIKKOL SLS-30	sodium lauryl sulfate; aq. soln.		yellow liquid	foaming agent; cleansing agent / *cosmetics*
	NIKKOL SLS	sodium lauryl sulfate		white crystalline powder	
	NIKKOL SMS	sodium myristyl sulfate		white crystalline powder	
	NIKKOL SSS	sodium stearyl sulfate		white powder	emulsifier / *cosmetics*
	NIKKOL TEALS-42	triethanolamine lauryl sulfate; aq. soln.		yellow liquid	foaming agent; cleansing agent / *cosmetics*
	NIKKOL TEALS	triethanolamine lauryl sulfate; aq. soln.		pale yellow liquid	
Protex	SURFARON A6015 N30	triethanolamine alkyl sulfate			foaming agent; detergent
Rhone-Poulenc	RHODAPON BOS	sodium 2-ethylhexyl sulfate	40% solids	liquid; surface tension 33 dynes/cm	emulsifier; stabiliser; wetting agent; low foam / *emulsion polymerisation*
Sandoz	LYOCOL SDL	fatty alkyl sulfates in aq. emulsion		paste	dispersant; low foam
	SANDOPAN MW MOD	sulfated fatty alcohol		paste	detergent; scouring agent / *wool; hair*
	THIOTAN RMFN	sulfated fatty alcohol		liquid	reserving agent / *textile industry; dyeing of wools*

Supplier	Trade name	Chemical description	Composition	General properties	Functionality Application
Seppic	MELANOL API 70 SUPRA	isopropanolamine lauryl sulfate	60% active	paste	wetting agent; foaming agent
	MELANOL LP 20	ammonium lauryl sulfate	28% active	liquid	
	MELANOL LPI 45	isopropanolamine lauryl sulfate	60% active	gel	
	MELANOL V 90	sodium lauryl sulfate	95% active	powder	
Servo Delden	SERDET DFK 30	sodium lauryl sulfate	30% active	liquid	
	SERDET DSK 40	sodium 2-ethylhexyl sulfate	39% active	liquid	
Shell Chemicals	DOBANOL 25S/70	alcohol sulfate	70% active	M.W. 309; colour 15 (5%; Klett); pH 11 (5%)	washing-up liquids; laundry products; hard-surface cleaners; textile industry; industrial applications; cosmetics; toiletries
	TEEPOL HB7	alcohol sulfate	39.0% active	M.W. 266; colour 175 max. (5%; Klett); pH 7.5-9.0 (5%)	
	TEEPOL PB	alcohol sulfate	32% active	M.W. 372; colour 150 max. (5%; Klett); pH 5-8.0 (5%)	
Stepan Europe	POLYSTEP B 24	sodium lauryl alcohol sulfate; made-to-order	29% active	pale yellow liquid	emulsifier emulsion polymerisation
	POLYSTEP B 25	sodium decyl sulfate; made-to-order	38% active	pale yellow liquid	
	POLYSTEP B 29	sodium octyl sulfate; made-to-order	33% active	water white to pale yellow liquid	low foam; emulsifier emulsion polymerisation
	POLYSTEP B 3	sodium lauryl sulfate; made-to-order	96% active	white powder	emulsifier emulsion polymerisation
	POLYSTEP B 5	sodium lauryl sulfate; made-to-order	29% active	pale yellow liquid	
	STEPANOL SPT	TEA lauryl sulfate	40% active	yellow viscous liquid	foaming agent; detergent shampoos; liquid soaps; bubble baths; household cleaners
	STEPANOL WA 100	sodium lauryl sulfate; made-to-order	97% active	white powder	foaming agent toothpastes; shampoos; bubble baths
	STEPANOL WAC	sodium lauryl sulfate	29% active	water white to yellow liquid	foaming agent; detergent shampoos; liquid soaps; bubble baths; household cleaners
Surfachem	SURFAC ALS30	ammonium lauryl sulfate			
	SURFAC EH40	sodium 2-ethyl hexyl sulfate			
	SURFAC MLS30	monoethanolamine lauryl sulfate			
	SURFAC SLS 38 PASTE	sodium lauryl sulfate (preservative free)		paste	
	SURFAC SLS/BP POWDER	sodium lauryl (medium cut) sulfate BP/USP		powder	
	SURFAC SLS/BP NEEDLES	sodium lauryl (medium cut) sulfate BP		needles	
	SURFAC SLS/N POWDER	sodium lauryl (narrow cut) sulfate		powder	
	SURFAC SLS/N NEEDLES	sodium lauryl (narrow cut) sulfate		needles	
	SURFAC SLS100/N NEEDLES	sodium lauryl (narrow cut) sulfate USP/BP		needles	
	SURFAC SLS28/S	sodium alkyl sulfate			
	SURFAC SLS28	sodium lauryl (narrow cut) sulfate soln.			

Supplier	Trade name	Chemical description	Composition	General properties	Functionality / Application
Surfachem	SURFAC TLS40	triethanolamine lauryl sulfate		liquid	detergent; wetting agent
Thor Chemicals	SOVATEX DPW (B)	ether sulfate		paste	detergent *textile industry; leather industry; industrial applications*
	SUFATOL A PASTE	sulfated fatty alcohol blend		paste	dispersant *textile industry; leather industry; industrial applications*
	SUFATOL CL PASTE	sulfated fatty alcohol, sodium salt		paste	detergent; wetting agent; levelling agent *textile industry; leather industry; industrial applications*
	SUFATOL K POWDER	sulfated fatty alcohol, sodium salt		powder	detergent *textile industry; leather industry; industrial applications*
	SUFATOL K2 PASTE	sulfated fatty alcohol, sodium salt		paste	detergent *textile industry*
	SUFATOL LS3	combined sulfated fatty alcohol and nonionic		liquid	detergent; wetting agent; levelling agent; softener *textile industry; leather industry; industrial applications*
	SUFATOL LX/B PASTE	sulfated fatty alcohol, sodium salt		paste	detergent *textile industry; leather industry; industrial applications*
	SUFATOL PD/BA PASTE	sulfated fatty alcohol, potassium salt		paste	
	SUFATOL TL/B	sulfated fatty alcohol blend		liquid	detergent *textile industry; leather industry; industrial applications*
Unger Fabrikker	UFAROL AM 30	ammonium lauryl sulfate	30% active; carbon chain distribution $C_{12/14}$	pale yellow liquid; M.W. 291; pH 6-7 (1% soln.)	detergent; emulsifier
	UFAROL AM 70	ammonium lauryl sulfate	70% active; carbon chain distribution $C_{12/14}$	pale yellow liquid; M.W. 291; pH 6-7 (1% soln.)	*laundry products; both preparations; textile industry*
	UFAROL NA-30	sodium lauryl sulfate	30% active; carbon chain distribution $C_{12/14}$	pale yellow liquid; M.W. 296; pH 7-8 (1% soln.)	
	UFAROL TA-40	triethanolamine lauryl sulfate	40% active; carbon chain distribution $C_{12/14}$	pale yellow liquid; M.W. 423; pH 6-7 (1% soln.)	
Witco	NEOPON LAM	lauryl sulfate, ammonium salt	26% active	liquid	*shampoos; laundry products*
	NEOPON LS	lauryl sulfate, sodium salt	29% active	liquid/paste	*shampoos; carpet cleaners*
	NEOPON LT	lauryl sulfate, TEA-salt	41% active	liquid	*shampoos; fire fighting*
	REWOPOL ALS 30	ammonium lauryl sulfate	28% active	liquid	*shampoos; foam baths*
	REWOPOL MLS 30	monoethanolammonium lauryl sulfate	30% active	liquid	*shampoos; foams baths*
	REWOPOL NEHS 40	sodium 2-ethylhexyl sulfate	40% active	liquid	wetting agent; low foam; hydrotrope *cleaners; electroplating*
	REWOPOL NLS 15 L	sodium lauryl sulfate	15% active	liquid	*emulsion polymerisation*
	REWOPOL NLS 28	sodium lauryl sulfate	28% active	liquid	*light-duty detergents; skin cleansing preparations; emulsion polymerisation*
	REWOPOL NLS 30 L	sodium lauryl sulfate	30% active	liquid	*emulsion polymerisation*
	REWOPOL NLS 90	sodium lauryl sulfate	95% active	powder	*detergents; soaps*
	REWOPOL TLS 40	triethanolammonium lauryl sulfate	40% active	liquid	*shampoos; foam baths*
	REWOPOL TLS 90 L	trialkylolammonium lauryl sulfate	75% active	liquid	*personal care products; foam baths*
	SUPRALATE C	sodium lauryl sulfate	90-96% active	white to cream-coloured powder; pH 9-10 (3% in H_2O)	foaming agent; emulsifier *food industry; cosmetics; pharmaceuticals; waxes*
	SUPRALATE D	sodium lauryl/oleyl sulfate	38% active	yellow to pale brown viscous paste; viscosity 17,000 cPs; pH 8-9 (3% in H_2O)	foaming agent; detergent; emulsifier *textile industry; detergents; leather industry; paper industry; metal cleaning; waxes*
	SUPRALATE EP	diethanolamine salt of lauryl sulfate	33-36% active	pale yellow slightly viscous paste; cloud point 5°C; viscosity 50-150 cPs; pH 7-8 (3% in H_2O)	foaming agent *cosmetics; textile industry; detergents*

Supplier	Trade name	Chemical description	Composition	General properties	Functionality / Application
Witco	SUPRALATE G	amine salt of lauryl sulfate	92% active	yellow to light amber viscous paste; pH 9-10 (3% in H_2O)	foaming agent; antistatic agent; emulsifier *waxes*
	SUPRALATE LS	sodium oleyl/lauryl sulfate	26% active	yellow to tan fluid paste; cloud point 20°C; viscosity 6000 cP's; pH 8-9 (3% in H_2O)	foaming agent; detergent; emulsifier *textile industry; detergents; leather industry; paper industry; metal cleaning; waxes*
	SUPRALATE ME DRY	sodium lauryl sulfate	90-96% active	white to cream-coloured powder; pH 9-11 (3% in H_2O)	foaming agent; emulsifier *food industry; cosmetics; textile industry; detergents; waxes*
	SUPRALATE QC	sodium lauryl sulfate	29-31% active	slightly viscous, almost colourless liquid; cloud point 18°C; viscosity 500 cP's max.; pH 7.5-8.5 (3% in H_2O)	foaming agent *food industry; cosmetics*
	SUPRALATE SP	sodium octyl sulfate	33-35% active	pale yellow liquid; cloud point <0°C; viscosity 30 cP's; pH 9-11 (3% in H_2O)	wetting agent; low foam *emulsion polymerisation; textile industry; paper industry; leather industry; metal cleaning*
	SUPRALATE WA PASTE	sodium lauryl sulfate	29-31% active	pale yellow viscous liquid; cloud point 25°C; viscosity 79,000 cP's; pH 7.5-8.5 (3% in H_2O)	foaming agent; emulsifier *cosmetics; waxes*
	SUPRALATE WAQ	sodium lauryl sulfate	28-30% active	pale yellow, noticeably viscous liquid; cloud point 20°C; viscosity 3000 cP's; pH 7.5-8.5 (3% in H_2O)	foaming agent; wetting agent *emulsion polymerisation; food industry; cosmetics; textile industry*
	SUPRALATE WAQE	sodium lauryl sulfate	29-31% active	pale yellow, slightly viscous liquid; cloud point 18°C; viscosity 500 cP's max.; pH 7.5-8.5 (3% in H_2O)	foaming agent; emulsifier *emulsion polymerisation; food industry; waxes*
	SUPRALATE WN	sodium octyl/decyl sulfate	33-35% active	pale yellow liquid; cloud point <0°C; viscosity 30 cP's; pH 9-11 (3% in H_2O)	wetting agent; low foam *emulsion polymerisation; textile industry; leather industry*
	SUPRALATE XL	amine salt of lauryl sulfate plus amphoteric salt	36% active	yellow, moderately viscous paste; cloud point 10°C; viscosity 2500-4500 cP's; pH 7-8 (3% in H_2O)	foaming agent; wetting agent; detergent *cosmetics*
	WITCOLATE 6462	sodium alkyl sulfate	38% active	liquid	detergent; dispersant; wetting agent *personal care products; household and industrial applications*
	WITCOLATE 7031	sodium alkyl sulfate	39% active	liquid	
	WITCOLATE A PWD	sodium lauryl sulfate	93% active	powder	detergent; foaming agent; wetting agent *personal care products; household and industrial applications*
	WITCOLATE C	sodium lauryl sulfate	29% active	paste	
	WITCOLATE D-510	sodium 2-ethylhexyl sulfate	39% active	liquid	detergent; dispersant; wetting agent *personal care products; household and industrial applications*
	WITCOLATE LCP	sodium lauryl sulfate	29% active	liquid	
	WITCOLATE NH	ammonium lauryl sulfate	28.5% active	liquid	detergent; foaming agent; wetting agent *personal care products; household and industrial applications*
	WITCOLATE TLS-500	TEA-lauryl sulfate	40% active	liquid	
	WITCOLATE WAC-GL	sodium lauryl sulfate	29% active	liquid	
	WITCOLATE WAC-LA	sodium lauryl sulfate	29% active	liquid	
Zschimmer & Schwarz	NEWALOL ME	combination of alcohol sulfates		yellow to brown liquid	wetting agent *textile industry*
	SULFETAL 4069	sodium alkyl sulfate ($C_{8/10}$)	40% active	liquid	wetting agent *high-pressure cleaners; industrial cleaners; metal cleaning*

Supplier	Trade name	Chemical description	Composition	General properties	Functionality / *Application*
Zschimmer & Schwarz	SULFETAL 4105	sodium isooctyl sulfate	40% active	liquid	low foam; wetting agent / *high-pressure cleaners; industrial cleaners; metal cleaning*
	SULFETAL 4187	modified sodium isooctyl sulfate	40% active	liquid	wetting agent
	SULFETAL AF	sodium fatty alcohol sulfate (C$_{16/14}$)	42% active	paste	*cleansing preparations; flotation agents; detergents*
	SULFETAL C 38	sodium fatty alcohol sulfate (C$_{12/18}$)	38% active	paste	*light-duty detergents; dishwashing agents*
	SULFETAL CJOT 38	monoisopropanolamine fatty alcohol sulfate (C$_{12/14}$)	38% active	liquid	*shampoos; cleansing preparations; bath preparations*
	SULFETAL CJOT 60	monoisopropanolamine fatty alcohol sulfate (C$_{12/14}$)	60% active	liquid	*shampoos; cleansing preparations; bath preparations*
	SULFETAL FA 40	modified sodium alkyl sulfate (C$_8$)	40% active	liquid	wetting agent; cleansing agent / *high-pressure cleaners*
	SULFETAL KT 400	triethanolamine fatty alcohol sulfate (C$_{12/14}$)	40% active	liquid	*shampoos*
	SULFETAL MF	monoethanolamine fatty alcohol sulfate (C$_{12/14}$)	30% active	liquid	*shampoos; cleansing preparations; bath preparations*
	SULFETAL TC 50 W	sodium tallow/coco fatty alcohol sulfate (C$_{12/18}$)	55% active	paste	*detergents; cleansing preparations*
	SULFETAL TC 50	sodium tallow/coco fatty alcohol sulfate (C$_{12/18}$)	50% active	paste	*detergents; cleansing preparations*

Alkylaryl sulfonates

Supplier	Trade name	Chemical description	Composition	General properties	Functionality / *Application*
Akcros Chemicals	AGRILAN X98	branched chain dodecyl benzene sulfonate, calcium salt	58% active; H$_2$O content <1%	viscous liquid; pour point −13°C; viscosity 1080 cSt; pH 5-8 (1% aq.)	emulsifier / *agrochemicals*
	ARYLAN CA	straighter chain dodecyl benzene sulfonate, calcium salt	70% active; H$_2$O content <1%	viscous liquid; pour point −4°C; viscosity 3025 cSt; pH 5-8 (1% aq.)	emulsifier / *degreasers; agrochemicals; waxes; hydrocarbon solvents*
	ARYLAN DA95	specially blended liquid detergent concentrate based on alkylaryl sulfonates blended with foam stabilisers	98% active; H$_2$O content ca. 2%	amber liquid; pour point 5°C; viscosity 1650 cSt; pH 7.0 (1% aq.)	foaming agent (high); detergent; foam stabiliser / *liquid detergents*
	ARYLAN HAL	specially blended liquid detergent concentrate based on alkylaryl sulfonates blended with foam stabilisers	87% active; H$_2$O content ca. 5%	clear amber liquid; pour point <0°C; viscosity 300 cSt; pH 7.0 (1% aq.)	
	ARYLAN LQ	specially blended liquid detergent concentrate based on alkylaryl sulfonates blended with foam stabilisers	40% active; H$_2$O content ca. 60%	clear light amber liquid; pour point <0°C; viscosity 350 cSt; pH 8.0 (1% aq.)	
	ARYLAN PWS	straighter chain dodecyl benzene sulfonate, amine salt	90% active; H$_2$O content ca. 4%	viscous liquid; pour point <0°C; viscosity 4200 cSt; pH 3.5-4.5 (1% aq.)	emulsifier / *hand cleaning gels; degreasers; waxes; organic solvents; mineral oils*
	ARYLAN SBC ACID	straighter chain dodecyl benzene sulfonic acid based on a broad cut alkylate	96% active	brown viscous liquid; pour point 0°C; viscosity 4300 cSt; pH 2 (1% aq.)	intermediate; detergent; foaming agent / *liquid detergents; dishwashing agents; emulsifying agents*
	ARYLAN SBC25	straighter chain dodecyl benzene sulfonate, sodium salt based on a broader cut alkylate	25% active; H$_2$O content ca. 75%	clear liquid; pour point <0°C; viscosity 150 cSt; pH 6.5-9.0 (1% aq.)	detergent; wetting agent; emulsifier; foaming agent / *liquid detergents; cleaners; emulsion polymerisation*
	ARYLAN SC ACID	straighter chain dodecyl benzene sulfonic acid based on a narrow cut alkylate	96% active	brown viscous liquid; pour point 5°C; viscosity 2000 cSt; pH 2 (1% aq.)	intermediate; detergent; foaming agent / *liquid detergents; dishwashing agents; emulsifying agents*
	ARYLAN SC15	straighter chain dodecyl benzene sulfonate, sodium salt	15% active; H$_2$O content ca. 85%	clear liquid; pour point 5°C; viscosity 300 cSt; pH 6.5-8.0 (1% aq.)	detergent; wetting agent; emulsifier; foaming agent / *emulsion polymerisation*

Supplier	Trade name	Chemical description	Composition	General properties	Functionality / *Application*
Akcros Chemicals	ARYLAN SC30	straighter chain dodecyl benzene sulfonate, sodium salt	30% active; H_2O content ca. 70%	liquid/soft paste; pour point 12°C; viscosity 5100 cSt; pH 6.5-8.0 (1% aq.)	detergent; wetting agent; emulsifier; foaming agent / *liquid detergents; cleaners; emulsion polymerisation*
	ARYLAN SE ACID	straighter chain dodecyl benzene sulfonic acid based on a narrow cut alkylate	97% active	brown viscous liquid; pour point 0°C; viscosity 1800 cSt; pH 2 (1% aq.)	intermediate; detergent; foaming agent / *liquid detergents; dishwashing agents; emulsifying agents*
	ARYLAN SX85	straighter chain dodecyl benzene sulfonate, sodium salt	85% active; H_2O content ca. 2%	off-white powder; pH 9.0-11.0 (1% aq.)	detergent; wetting agent; emulsifier; foaming agent / *industrial cleaners; hard-surface cleaners; industrial detergents; textile industry; powder detergents*
	ARYLAN SY ACID	straighter chain dodecyl benzene sulfonic acid based on a narrow cut alkylate	96% active	brown viscous liquid; pour point 0°C; viscosity 1970 cSt; pH 2 (1% aq.)	intermediate; detergent; foaming agent / *liquid detergents; dishwashing agents; emulsifying agents*
	ARYLAN SY30	straighter chain dodecyl benzene sulfonate, sodium salt	30% active; H_2O content ca. 70%	clear yellow liquid; pour point 0°C; viscosity 6900 cPs; pH 6.5-8.0 (1% aq.)	detergent; emulsifier; wetting agent; foaming agent / *liquid detergents; cleaners; emulsion polymerisation*
Akzo Nobel	BEROL 822	linear alkylbenzene sulfonate, calcium salt	60-62% active	paste	wetting agent; emulsifier / *industrial solvent cleaners*
Albright & Wilson	DEHSCOFIX 911	naphthalene sulfonic acid, formaldehyde condensate	concentration 40%	paste	
	DEHSCOFIX 914/AS	sodium naphthalene sulfonate, formaldehyde condensate	concentration 40%	liquid	
	DEHSCOFIX 914/ASL	sodium naphthalene sulfonate, formaldehyde condensate	concentration 40%	liquid	dispersant / *coatings*
	DEHSCOFIX 914	sodium naphthalene sulfonate, formaldehyde condensate	concentration 45%	liquid	dispersant; superplasticiser / *leather industry; concrete; construction*
	DEHSCOFIX 915/AS	sodium naphthalene sulfonate, formaldehyde condensate	concentration 92%	powder	
	DEHSCOFIX 915	sodium naphthalene sulfonate, formaldehyde condensate	concentration 92%	powder	superplasticiser / *construction; concrete*
	DEHSCOFIX 920	sodium naphthalene sulfonate, formaldehyde condensate	concentration 92%	powder	wetting agent; dispersant / *agrochemicals; leather industry*
	DEHSCOFIX 923	sodium dimethylnaphthalene sulfonate, formaldehyde condensate	concentration 92%	powder	wetting agent; dispersant / *agrochemicals*
	DEHSCOFIX 926	sodium dimethylnaphthalene sulfonate, formaldehyde condensate	concentration 92%	powder	
	DEHSCOFIX 930	ammonium naphthalene sulfonate, formaldehyde condensate	concentration 92%	powder	dispersant; emulsifier / *leather industry; emulsion polymerisation; rubber industry*
	ELTESOL 4200 SERIES	blends of toluene/xylene/benzene sulfonic acids		liquids to pastes	catalysts / *foundry industry*
	ELTESOL 4443M	benzene sulfonic acid blend	concentration 72%	liquid	
	ELTESOL 5400 SERIES	phenol sulfonic acid condensates		liquids to powders	*leather industry*
	ELTESOL 7200 SERIES	dihydroxy diphenyl sulfonates		liquids to powders	
	ELTESOL AC60	ammonium cumene sulfonate	concentration 60%	liquid	hydrotrope / *agrochemicals*
	ELTESOL AX40	ammonium xylene sulfonate	concentration 40%	liquid	

Supplier	Trade name	Chemical description	Composition	General properties	Functionality / Application
Albright & Wilson	ELTESOL CA65	cumene sulfonic acid	concentration 65%	liquid	catalyst; *foundry industry*
	ELTESOL PA65	phenol sulfonic acid	concentration 65%	liquid	*tin plating*
	ELTESOL PX40	potassium xylene sulfonate	concentration 40%	liquid	
	ELTESOL PX93	potassium xylene sulfonate	concentration 93%	powder	
	ELTESOL SC PELLETS	sodium cumene sulfonate	concentration 88%	pellet	
	ELTESOL SC40	sodium cumene sulfonate	concentration 40%	liquid	
	ELTESOL SC93	sodium cumene sulfonate	concentration 93%	powder	
	ELTESOL ST PELLETS	sodium toluene sulfonate	concentration 85%	pellet	
	ELTESOL ST40	sodium toluene sulfonate	concentration 40%	liquid	
	ELTESOL ST40/F	sodium toluene sulfonate	concentration 40%		
	ELTESOL ST90	sodium toluene sulfonate	concentration 90%	powder	
	ELTESOL SX PELLETS	sodium xylene sulfonate	concentration 88%	pellet	*polymer industry*
	ELTESOL SX30	sodium xylene sulfonate	concentration 30%	liquid	
	ELTESOL SX40	sodium xylene sulfonate	concentration 40%	liquid	
	ELTESOL SX93	sodium xylene sulfonate	concentration 93%	powder	
	ELTESOL TA SERIES	toluene sulfonic acids		liquids	hydrotrope; catalyst; *agrochemicals; foundry industry*
	ELTESOL TSX SERIES	toluene sulfonic acids monohydrates	concentration 99%	crystals	*polymer industry*
	ELTESOL XA SERIES	xylene sulfonic acids		liquids	hydrotrope; catalyst; *agrochemicals; foundry industry*
	NANSA 1169/A	sodium dodecyl benzene sulfonate	concentration 30%		
	NANSA 1395	sodium dodecyl benzene sulfonate	concentration 25%		
	NANSA AS40	ammonium dodecyl benzene sulfonate	concentration 40.5%		
	NANSA EVM SERIES	calcium alkyl benzene sulfonates		liquids	dispersant; emulsifier; antistatic agent; *emulsion and suspension polymerisation; coatings; agrochemicals*
	NANSA HS80/S SERIES	sodium alkyl benzene sulfonates	concentration 80%	powders	dispersant; emulsifier; antistatic agent; wetting agent; grinding agent; detergent; *emulsion and suspension polymerisation; textile industry; construction; coatings*
	NANSA HS85/S SERIES	sodium alkyl benzene sulfonates	concentration 85%	flakes	dispersant; emulsifier; antistatic agent; wetting agent; detergent; *emulsion and suspension polymerisation; textile industry; coatings*
	NANSA HS90/S SERIES	sodium alkyl benzene sulfonates	concentration 90%	powders	dispersant; emulsifier; antistatic agent; wetting agent; detergent; *emulsion and suspension polymerisation; textile industry; coatings*

Supplier	Trade name	Chemical description	Composition	General properties	Functionality / Application
Albright & Wilson	NANSA SB30	sodium alkyl benzene sulfonate (branched)	concentration 30%	paste	dispersant; emulsifier; antistatic agent / emulsion and suspension polymerisation
	NANSA SBA	alkyl benzene sulfonic acid (branched)	concentration 96%	liquid	
	NANSA SS50	sodium alkyl benzene sulfonate	concentration 50%	paste	dispersant; emulsifier; antistatic agent; detergent; air entraining agent / emulsion and suspension polymerisation; textile industry; agrochemicals
	NANSA SSA SERIES	alkyl benzene sulfonic acids	concentration 97%	liquids	dispersant; emulsifier; antistatic agent; air entraining agent / emulsion and suspension polymerisation; agrochemicals; construction
	NANSA TS 50/F	triethanolamine alkyl benzene sulfonate	concentration 50%	liquid	dispersant; emulsifier; antistatic agent / emulsion and suspension polymerisation
	NANSA YS94	isopropylamine dodecyl benzene sulfonate	concentration 92%	liquid	
Auschem Cesalpinia	ALARSOL AL	linear dodecylbenzene sulfonic acid	96% active	liquid	foaming agent; detergent / liquid and solid detergents; laundry products; dishwashing agents
	BETA MSHF POLV.	di-naphthalenemethane sulfonate sodium salt	93% active	powder	dispersant; anti-redepositing agent; fluidising agent
	BETA MSHF	di-naphthalenemethane sulfonate sodium salt	45% active	liquid	dispersant; anti-redepositing agent; fluidising agent
	BETA SHF DESCA	betanaphthalenemethane sulfonate salt	40% active; low salt content	liquid	slump agent; super-fluidising agent / concrete
	BETA SHF EI	betanaphthalenemethane sulfonate salt	40% active	liquid	
	BETA SHF POLV.	polynaphthalenemethane sulfonate sodium salt	93% active	powder	dispersant; anti-redepositing agent; fluidising agent; slump agent / concrete
	BETA SHF	betanaphthalenemethane sulfonate salt	40% active	liquid	
	CHIMIPON BAC	linear dodecylbenzene sulfonate, sodium salt	60% active	paste	foaming agent; detergent / liquid and solid detergents; laundry products; dishwashing agents
	CHIMIPON GT	alkyl aryl polyether sulfonate, sodium salt	40% active	liquid	
	CHIMIPON TSB	linear dodecylbenzene sulfonate, TEA salt	70% active	paste	
	EMULSON 255	triethanolamine dodecylbenzene sulfonate	70% active	liquid	
	EMULSON 270	triethylamine dodecylbenzene sulfonate	80% active	liquid	
	EMULSON 660 HB	calcium dodecylbenzene sulfonate	60% active	liquid	emulsifier / pesticides
	EMULSON CAL HB	calcium dodecylbenzene sulfonate	60% active	liquid	
	EMULSON CAL	calcium dodecylbenzene sulfonate	60% active	liquid	
	EMULSON LDT	triethanolamine dodecylbenzene sulfonate	70% active	liquid	
	MADEOL BX	naphthalenemethanesulfonate dialkyl derivative	70% active	powder	wetting agent / pesticides
	MADEOL OR 95 L	naphthalenemethanesulfonate condensed with formaldehyde	40% active	liquid	dispersant / pesticides
	MADEOL OR 95	naphthalenemethanesulfonate condensed with formaldehyde	88% active	powder	dispersant / pesticides
	MADEOL PWA	naphthalenemethanesulfonate dialkyl derivative	70% active	powder	wetting agent / pesticides
	MADEOL W 90	naphthalenemethanesulfonate condensed with formaldehyde	94% active	powder	wetting agent; dispersant / pesticides
	POLIROL DS	linear dodecylbenzene sulfonic acid	100% active	liquid	emulsifier / latex production; polymerisation

Supplier	Trade name	Chemical description	Composition	General properties	Functionality / Application
BASF	LUTENSIT A-LBA	dodecylbenzenesulfonate, amine salt	55% active		wetting agent; dispersant; adjuvant; additive agrochemicals; paint industry; plastics industry; emulsion polymerisation; latexes; metal cleaning; hard-surface cleaners
	NEKAL BX CONC. PASTE	sodium alkylnaphthalene sulfonate	60% active		
	NEKAL BX DRY	sodium alkylnaphthalene sulfonate	68% active		
	TAMOL NH 7519	napthalenesulfonic acid condensation product, sodium salt (high degree of condensation)	75% active		
	TAMOL NHC 3001	napthalenesulfonic acid condensation product, calcium salt (high degree of condensation)	30% active		
	TAMOL NN 2406	napthalenesulfonic acid condensation product, sodium salt (high degree of condensation)	24% active		
	TAMOL NN 2901	napthalenesulfonic acid condensation product, sodium salt (low degree of condensation)	29% active		
	TAMOL NN 4501	napthalenesulfonic acid condensation product, sodium salt (low degree of condensation)	45% active		
	TAMOL NN 7718	napthalenesulfonic acid condensation product, sodium salt (low degree of condensation)	77% active		
	TAMOL NN 8906	napthalenesulfonic acid condensation product, sodium salt (low degree of condensation)	89% active		
	TAMOL NN 9104	napthalenesulfonic acid condensation product, sodium salt (low degree of condensation)	91% active		
	TAMOL NN 9401	napthalenesulfonic acid condensation product, sodium salt (low degree of condensation)	94% active		
	TAMOL NNA 4109	napthalenesulfonic acid condensation product, ammonium salt (low degree of condensation)	41% active		
Cytec	AEROSOL OS	sodium diisopropyl naphthalene sulfonate	75% active	powder; surface tension 35 dynes/cm (in H$_2$O)	
Dac International Surfactants	SOLFODAC ACL	linear alkylbenzene sulfonic acid			detergents
Elf Atochem	BEYCOPON 345	amine-based alkylaryl sulfonate		liquid	
	BEYCOPON EC	sodium/triethanolamine alkylaryl sulfonate		liquid	
	BEYCOPON S 50	calcium alkylaryl sulfonate		liquid	
	ORAPRET WTNB 25	alkylaryl sulfonate and mineral oil		paste	
Ellis & Everard	CAFLON MIS	monoisopropylamine salt of dodecyl benzene sulfonic acid	92% active		
	CAFLON NAS25	sodium dodecyl benzene sulfonate	25% active		
	CAFLON NAS30	sodium dodecyl benzene sulfonate	30% active		
	CAFLON SBA	linear alkyl benzene sulfonic acid (broad cut)	96% active		
	CAFLON SNA	linear alkyl benzene sulfonic acid (narrow cut)	96% active		
	CAFLON SXS30	sodium xylene sulfonate	30% active		
EniChem Augusta		linear alkylbenzene sulfonic acid based on Isorchem 83		M.W. 305	laundry products; dry cleaning; industrial applications

Supplier	Trade name	Chemical description	Composition	General properties	Functionality Application
EniChem Augusta		linear alkylbenzene sulfonic acid based on Isorchem 93		M.W. 310	
		linear alkylbenzene sulfonic acid based on Isorchem 112		M.W. 315	
		linear alkylbenzene sulfonic acid based on Isorchem 113		M.W. 320	
		linear alkylbenzene sulfonic acid based on Isorchem 134		M.W. 342	laundry products; dry cleaning; industrial applications
		linear alkylbenzene sulfonic acid based on Sirene X 11		M.W. 315	
		linear alkylbenzene sulfonic acid based on Sirene X 12-L		M.W. 321	
		linear alkylbenzene sulfonic acid based on Sirene 113		M.W. 321	
Hays Colours	DYSPERSE NS LIQ	aromatic sulfonate			levelling agent; dispersant textile industry; dyeing
	DYSPERSE NS PDR	aromatic sulfonate			
	METAPEX A	alkylaryl sulfonate			detergent; wetting agent textile industry
Henkel	BREVIOL DS	naphthalene sulfonic acid-condensation product			dispersant; low foam textile industry; dyeing
	LAMEPON WPA	naphthalene sulfonic acid derivate			dispersant textile industry; dyeing
	LOMAR D	high molecular weight sulfonated naphthalene condensate	55% active	powder	dispersant paper industry; pigments
	LOMAR LS LIQUID	sodium salt of formaldehyde-naphthalene sulfonic acid condensate	ca. 46% active	liquid	
	LOMAR LS	sodium salt of formaldehyde-naphthalene sulfonic acid condensate	ca. 95% active	powder	dispersant plastics industry; coatings; pigments; polymers
	LOMAR PWA LIQUID	ammonium salt of formaldehyde-naphthalene sulfonic acid condensate	ca. 44% active	liquid	
	LOMAR PWA	ammonium salt of formaldehyde-naphthalene sulfonic acid condensate	ca. 92% active	powder	
	MARANIL A 25	sodium dodecyl benzene sulfonate	ca. 25% active	liquid	emulsifier plastics industry
	MARANIL DBS	n-dodecyl benzene sulfonic acid	ca. 97% active	liquid	intermediate detergents; emulsifiers; textile industry
	MARANIL PASTE A 55	sodium dodecyl benzene sulfonate	ca. 55% active	paste	
	MARANIL PASTE A 75	sodium dodecyl benzene sulfonate	ca. 75% active	paste	emulsifier plastics industry
	MARANIL POWDER A	sodium dodecyl benzene sulfonate	ca. 90% active	powder	
	OSIMOL 109	alkylarylsulfonate			levelling agent textile industry; dyeing

Supplier	Trade name	Chemical description	Composition	General properties	Functionality / Application
Henkel	SYNKANOL KE 2780	dipotassium oleic acid sulfonate	ca. 50% active	liquid	dispersant; emulsifier; low foam cleaners
Hickson Manro	MANRO CUSA65	cumene sulfonic acid	65% active	liquid	catalyst
	MANRO FCM100 SERIES	xylene sulfonic acid	90% active	liquid	catalyst
	MANRO FCM90LV	xylene sulfonic acid	90% active	liquid	
	MANRO HA ACID	branched dodecylbenzene sulfonic acid	96% active	viscous liquid	household detergents; industrial cleaning; catalysts; metal working
	MANRO HCS	isopropylamine dodecylbenzenesulfonate, narrow cut	90% active	liquid	industrial cleaning
	MANRO LA ACID	linear narrow cut dodecylbenzene sulfonic acid, low 2-phenyl	96% active	viscous liquid	household detergents; industrial cleaning; catalysts; metal working
	MANRO NA ACID	linear narrow cut dodecylbenzene sulfonic acid, high 2-phenyl	96% active	viscous liquid	household detergents; industrial cleaning; catalysts; metal working
	MANRO PTS/C	toluene sulfonic acid	98% active	white crystals	
	MANRO PTSA/C99	para-toluene sulfonic acid	98% active	white crystals	
	MANRO PTSA/E	toluene sulfonic acid	65% active	liquid	catalyst
	MANRO PTSA/LS	toluene sulfonic acid	65% active	liquid	
	MANRO SDBS 25/30	sodium dodecylbenzene sulfonate, narrow cut	25/30% active	liquid	
	MANRO SDBS 60	sodium dodecylbenzene sulfonate, narrow cut	60% active	paste	household detergents; industrial cleaning
	MANRO TDBS 60	triethanolamine dodecylbenzene sulfonate, narrow cut	60% active	liquid	
	MANROSOL ACS60	ammonium cumene sulfonate	60% active	liquid	hydrotrope household detergents; industrial cleaning
	MANROSOL AXS40	ammonium xylene sulfonate	40% active	liquid	hydrotrope hair care products; bath preparations; household detergents; industrial cleaning
	MANROSOL SCS40	sodium cumene sulfonate	40% active	liquid	
	MANROSOL SCS93	sodium cumene sulfonate	93% active	powder	hydrotrope household detergents; industrial cleaning
	MANROSOL STS40	sodium toluene sulfonate	40% active	liquid	
	MANROSOL STS90	sodium toluene sulfonate	90% active	liquid	
	MANROSOL SXS30	sodium xylene sulfonate	30% active	liquid	hydrotrope hair care products; bath preparations; household detergents; industrial cleaning
	MANROSOL SXS40	sodium xylene sulfonate	40% active	liquid	
	MANROSOL SXS93	sodium xylene sulfonate	93% active	powder	hydrotrope household detergents; industrial cleaning
	TENSARYL SB-LA	linear narrow cut dodecylbenzene sulfonic acid, low 2-phenyl	96% active	viscous liquid	household detergents; industrial cleaning; catalysts; metal working
	TENSARYL SB	linear narrow cut dodecylbenzene sulfonic acid, high 2-phenyl	96% active	viscous liquid	household detergents; industrial cleaning; catalysts; metal working
Hoechst	PHENYL SULPHONATE 1387	calcium salt of alkylbenzene sulfonate	35% active		dispersant oil slick dispersants

Supplier	Trade name	Chemical description	Composition	General properties	Functionality / Application
Hoechst	PHENYL SULPHONATE CA	calcium salt of alkylbenzene sulfonate; branched chain	60% active		emulsifier / agrochemicals
	PHENYL SULPHONATE CAL	calcium salt of alkylbenzene sulfonate; straight chain	60% active		emulsifier / agrochemicals
Hüls	KNA-CUMENE-SULPHONATE 40	potassium/sodium cumene sulfonate	active detergent 40%	liquid	hydrotrope / liquid detergents; cleaners
	MARLON A 350	n-C$_{10/13}$-alkylbenzene sulfonate, sodium salt	active detergent 50%	liquid/paste	
	MARLON A 360	n-C$_{10/13}$-alkylbenzene sulfonate, sodium salt	active detergent 60%	liquid/paste	liquid and powder detergents; cleaners; textile industry
	MARLON A 365	n-C$_{10/13}$-alkylbenzene sulfonate, sodium salt	active detergent 65%	liquid/paste	
	MARLON A 375	n-C$_{10/13}$-alkylbenzene sulfonate, sodium salt	active detergent 75%	paste	
	MARLON A 390	n-alkylbenzene sulfonate, sodium salt	active detergent 90%	powder	powder cleaners
	MARLON AFO 40	blend based on n-alkylbenzene sulfonate, ether sulfate and fatty alcohol ethoxylate	active detergent 40%	liquid	washing-up liquids; cleaners
	MARLON AFO 50	blend based on n-alkylbenzene sulfonate, ether sulfate and fatty alcohol ethoxylate	active detergent 50%	liquid/paste	
	MARLON AFR	blend of n-alkylbenzene sulfonate, sodium salt with urea	active detergent 30%	liquid	
	MARLON AM 80	blend of n-alkylbenzene sulfonate, sodium salt with Marlophen and dipropylene glycol	active detergent 80%	liquid	degreaser / washing-up liquids
	MARLON AMX	n-alkylbenzene sulfonate-methoxypropylamine salt and alkylbenzene	active detergent 90%	paste	dry cleaning
	MARLON ARL	n-alkylbenzene sulfonate, sodium salt	active detergent 80%	powder	
	MARLON AS$_3$-R	n-C$_{10/13}$-alkylbenzene sulfonic acid	active detergent 98%	liquid	liquid and powder detergents; cleaners; textile industry
	MARLON AS$_3$	n-C$_{10/13}$-alkylbenzene sulfonic acid	active detergent 98%	liquid	liquid and powder detergents; cleaners; textile industry
	MARLOPON ADS 65	n-alkylbenzene sulfonate, DEA salt	active detergent 65%	liquid	anti-corrosion / liquid detergents
	MARLOPON AMS 60	blend of n-alkylbenzene sulfonate, MEA salt with Marlophen and carboxamide	active detergent 60%	liquid	liquid cleaners
	MARLOPON AT 50	n-alkylbenzene sulfonate, TEA salt	active detergent 50%	liquid	fine detergents; washing-up liquids; shampoos
	MARLOPON CA	blend of n-alkylbenzene sulfonate, TEA salt with carboxamide and alkanediol	active detergent 60%	liquid	fine detergents
	NA-CUMENE-SULPHONATE 40	sodium cumene sulfonate	active detergent 40%	liquid	hydrotrope / liquid detergents; cleaners
	NA-CUMENE-SULPHONATE POWDER	sodium cumene sulfonate	active detergent 96%	spray-dried powder	
ICI	ATLAS G-3298	isopropylamine dodecylbenzene sulfonate		HLB 11.7; amber liquid	household and industrial applications; agrochemicals; textile industry
	ATLAS G-3300B	alkylaryl sulfonate		HLB 11.4; red-brown liquid	
	ATLOX 3300B	alkylaryl sulfonate		HLB 11.4; red-brown liquid	
	ATLOX 4842	sodium naphthalene sulfonic acid condensate		light brown powder	
	ATLOX 4861B	alkylaryl sulfonate		HLB 8.6; red-amber liquid	emulsifier; dispersant / agrochemicals
	ATLOX 4862	disodium salt of methylene dinaphthalene sulfonic acid		reddish-yellow powder	

Supplier	Trade name	Chemical description	Composition	General properties	Functionality / Application
Kao Corporation	DANOX 35 ST	dodecylbenzene sulfonic acid, mix salt	33-35% active	liquid; M.W. 396; pH 6.5-7.5 (5%)	emulsifier; solubiliser / *mineral oils; kerosene; waxes; chlorinated solvents; hand gels*
	MELIOSOL 50 X	dodecylbenzenesulfonate sodium salt	48-50% active	paste; M.W. 343; pH 6.5-7.5 (5%)	
	SULFONAX	dodecylbenzene sulfonic acid	> 95% active	liquid; M.W. 323	
Libra Chemicals	LIBRAMUL IPA	dodecylbenzene sulfonate, monoisopropylamine salt	95% active	clear amber liquid	
	LIBRATEX AS-40	dodecylbenzene sulfonate, ammonium salt	40% active	viscous paste	detergent; wetting agent / *cleaners*
	LIBRATEX AT-60	dodecylbenzene sulfonate, triethanolamine salt	60% active	clear yellow liquid	
Lonza	CARSONOL T-60-L	TEA dodecyl benzene sulfonate	60% active	liquid; pH 7.3	foaming agent; wetting agent; emulsifier / *detergents; industrial cleaners*
Protex	SURFARON A1500 SERIES	range of naphthalene formaldehyde sulfonates in acid and neutralized form			dispersant
	SURFARON A1632 N40	sodium di-isopropyl naphthalene sulfonate			hydrotrope
Rhone-Poulenc	BEVALOID 35	condensed naphthalene sulfonate	solids 93%	powder: M.W. 750	dispersant / *paint industry; oil industry; agrochemicals*
	BEVALOID 35L	condensed naphthalene sulfonate	solids 42%	M.W. 750; viscosity 25 cP	dispersant / *oil industry*
	BEVALOID 36	condensed naphthalene sulfonate	solids 93%	powder: M.W. 1500	
	RHODACAL RM/77 LIQ.	dinaphthalenemethane sulfonate sodium salt	solids 45%	liquid	emulsifier; dispersant / *emulsion polymerisation*
	RHODACAL RM/77-D	dinaphthalenemethane sulfonate sodium salt	solids 95%	powder	emulsifier; dispersant / *emulsion polymerisation*
	RHODACAL WP	sodium diisopropyl naphthalene sulfonate		powder	low foam; emulsifier; wetting agent; dispersant
	RHODACOL DS-10	sodium dodecyl benzene sulfonate	solids 98%	flake; surface tension 32 dynes/cm	emulsifier / *emulsion polymerisation*
	RHODACOL DS-4	sodium dodecyl benzene sulfonate	solids 23%	liquid; surface tension 32 dynes/cm	
Sandoz	LYOCOL O	aryl sulfonate in aq. emulsion		liquid	dispersant; low foam
	NYLOFIXAN P	aryl sulfonate in aq. soln.		liquid	fixative / *textile industry; dyeing*
	NYLOFIXAN PSA	aryl sulfonate in aq. soln.		liquid	
	NYLOFIXAN PSTM	aryl sulfonate in aq. soln.		liquid	
	SANDOPAN AS	alkyl aryl sulfonic acid		liquid	*industrial applications*
	SANDOTAN NE 22	mixture of naphthalene sulfonate formaldehyde condensation product		powder	
	SANDOTAN WZ	phenol sulfonic acid condensate		powder	
	SANDOZIN CE	alkylaryl sulfonate in aq. soln.		liquid	wetting agent; detergent; foaming agent (high)
	THIOTAN CAS	aryl sulfonate in aq. soln.		liquid	
	THIOTAN HW	aromatic sulfonate in aq. soln.		liquid	reserving agent / *textile industry; dyeing of wools*
	THIOTAN WPN	aryl sulfonate in aq. soln.		liquid	reserving agent
Seppic	PEROLENE NT	ammonium/triethanolamine dodecylbenzene sulfonate	36% active	liquid	detergent
Servo Delden	SERMUL EA 27	calcium tetrapropylenebenzene sulfonate	65% active	liquid	

Supplier	Trade name	Chemical description	Composition	General properties	Functionality / Application
Servo Delden	SERMUL EA 88	calcium dodecyl benzene sulfonate	65% active	liquid	detergent intermediate; *household, institutional and industrial cleaners*
Stepan Europe	BIO-SOFT S 100	alkylbenzene sulfonic acid; made-to-order	96% active	brown viscous liquid	detergent; foaming agent; emulsifier; wetting agent; *laundry products; emulsion polymerisation*
	NACCONOL 90G	sodium alkylbenzene sulfonate; made-to-order	93% active	white to beige granules	
	NINATE 401 SERIES	calcium alkylbenzene sulfonates	60-65% active	HLB 7.5; dark brown viscous liquid	emulsifier; dispersant; *agrochemicals*
	POLYSTEP A 11	isopropylamine alkylbenzene sulfonate; made-to-order	90% active	yellow viscous liquid	emulsifier; *emulsion polymerisation*
Surfachem	SURFAC BW940	monoisopropylamine dodecyl benzene sulfonate			
	SURFAC CABS70	calcium dodecyl benzene sulfonate			
	SURFAC PS	dodecyl benzene sulfonic acid			
	SURFAC PTSA CRYSTALS	para toluene sulfonic acid		crystals	
	SURFAC SCS PELLETS	sodium cumene sulfonate		pellets	
	SURFAC SCS40	sodium cumene sulfonate		liquid	
	SURFAC SDBS25	sodium dodecyl benzene sulfonate		powder	
	SURFAC SDBS80	sodium dodecyl benzene sulfonate		flake	
	SURFAC SDBS85	sodium dodecyl benzene sulfonate		pellets	
	SURFAC STS PELLETS	sodium toluene sulfonate		powder	
	SURFAC STS90	sodium toluene sulfonate		pellets	
	SURFAC SXS PELLETS	sodium xylene sulfonate		liquid	
	SURFAC SXS30	sodium xylene sulfonate		powder	
	SURFAC SXS93	sodium xylene sulfonate			
	SURFAC TS60	triethanolamine dodecyl benzene sulfonate			
	SURFAC TSA65	toluene sulfonic acid			
	SURFAC XSA65	xylene sulfonic acid			
Thor Chemicals	ATOLEX DA/25	naphthalene sulfonate		liquid	dispersant; levelling agent; *textile industry; dyeing*
	SOVATEX C1	alkylaryl sulfonate		liquid	scouring agent; milling aid
Unger Fabrikker	UFACID K	linear alkylbenzene sulfonic acid	97% active	brown liquid; M.W. 322	emulsifier; detergent ⎡ *liquid and powder detergents; plastics industry; metal working;*
	UFACID TPB	branched dodecyl benzene sulfonic acid	97% active	brown liquid; M.W. 324	⎣ *agrochemicals; polishes; textile industry; mining industry; oil industry; cement industry*
	UFARYL DB 80	branched sodium dodecylbenzene sulfonate	80% active	white/pale yellow free-flowing powder; M.W. 346	⎡ *powder detergents; agrochemicals; textile industry; cement industry*
	UFARYL DL 80 CW	linear sodium alkylbenzene sulfonate	80% active	white/pale yellow free-flowing powder; M.W. 344	⎣

Supplier	Trade name	Chemical description	Composition	General properties	Functionality Application
Unger Fabrikker	UFARYL DL 80	linear sodium alkylbenzene sulfonate	80% active	white/pale yellow free-flowing powder; M.W. 344	*powder detergents; agrochemicals; textile industry; cement industry*
	UFARYL DL 85	linear sodium alkylbenzene sulfonate	85% active	white/pale yellow free-flowing powder; M.W. 344	
	UFARYL DL 90	linear sodium alkylbenzene sulfonate	90% active	white/pale yellow free-flowing powder; M.W. 344	
	UFASAN 35	linear sodium alkylbenzene sulfonate	35% active	golden viscous liquid; M.W. 344	*emulsifier; detergent* *liquid and powder detergents; plastics industry; metal working; agrochemicals; polishes; textile industry; mining industry; oil industry; cement industry*
	UFASAN 50	linear sodium alkylbenzene sulfonate	50% active	pumpable liquid/paste; M.W. 344	
	UFASAN 60 A	linear sodium alkylbenzene sulfonate	60% active	pumpable liquid/paste; M.W. 338	
	UFASAN 62 B	branched sodium dodecylbenzene sulfonate	62% active	white paste; M.W. 346	
	UFASAN 65	linear sodium alkylbenzene sulfonate	65% active	white paste; M.W. 344	
	UFASAN IPA	linear isopropylamine alkylbenzene sulfonate	95% active	golden viscous liquid; M.W. 381	
	UFASAN TEA	linear triethanolamine alkylbenzene sulfonate	50% active	golden viscous liquid; M.W. 471	*textile industry*
Warwick International	WARCODET K54	based on sodium alkyl aryl sulfonate			
Witco	ACS 60	ammonium cumene sulfonate	60% active	liquid	*hydrotrope* *powder detergents*
	HEXARYL D 60 L	dodecyl benzene sulfonate, TEA-salt	60% active	liquid	*liquid detergents*
	MORWET 3008 POWDER	sodium alkyl aryl sulfonate	95.0% active min.; moisture 2.5% max.; inorganic (sodium sulfate) 2.5% max.	water-dispersible granules; surface tension 29.9 dynes/cm (1.0% in H_2O); Draves wetting 18 sec (1.0% in H_2O) (5% solids)	
	MORWET 3028 POWDER	sodium alkyl aryl sulfonate	93.5% active min.; moisture 2.0% max.; inorganic (sodium sulfate) 2.5% max.; unsulfonated organics 1.5% max.	water-dispersible granules; surface tension 37.6 dynes/cm (0.1% in H_2O); Draves wetting 13-23 sec (0.1% in H_2O); pH 7.5-10.0 (5% solids)	*wetting agent* *pesticides*
	MORWET B POWDER	sodium n-butyl naphthalene sulfonate	75.0% active min.; moisture 2.5% max.; unsulfonated oil 8.0% max.; inorganic (sodium sulfate) 20.0% max.	water-dispersible granules; surface tension 36.2 dynes/cm (1.0% in H_2O); Draves wetting 4.0 sec (1.0% in H_2O); pH 7.5-10.0 (5% solids)	
	MORWET D-425 POWDER	sodium naphthalene formaldehyde condensate	88.0% active min.; moisture 2.0% max.; ether extractables 2.5% max.; inorganic (sodium sulfate) 7.0-9.0% max.	water-dispersible granules; surface tension 48.3 dynes/cm (1.0% in H_2O); Draves wetting 3 min (1.0% in H_2O); pH 7.5-10.0 (5% solids)	*dispersant; viscosity modifier* *pesticides*
	MORWET DB POWDER	sodium di-n-butyl naphthalene sulfonate	75.0% active min.; moisture 2.5% max.; ether extractables 9.0% max.; inorganic (sodium sulfate) 20.0% max.	water-dispersible granules; surface tension 35.9 dynes/cm (1.0% in H_2O); Draves wetting 1 sec (1.0% in H_2O); pH 7.5-10.0 (5% solids)	
	MORWET IP POWDER	sodium di-isopropyl naphthalene sulfonate	75.0% active min.; moisture 2.5% max.; ether extractables 8.0% max.; inorganic (sodium sulfate) 20.0% max.	water-dispersible granules; surface tension 39.1 dynes/cm (1.0% in H_2O); Draves wetting 8 sec (1.0% in H_2O); pH 7.5-10.0 (5% solids)	*wetting agent* *pesticides*
	MORWET M POWDER	sodium mono and di-methyl naphthalene sulfonate	95.0% active min.; moisture 2.0% max.; ether extractables 1.5% max.; inorganic (sodium sulfate) 2.5% max.	water-dispersible granules; surface tension 32.2 dynes/cm (1.0% in H_2O); Draves wetting 3 sec (1.0% in H_2O); pH 7.5-10.0 (5% solids)	
	REWORYL K	dodecyl benzene sulfonic acid	97% active	liquid	*anionic detergents*
	REWORYL NKS 10	sodium dodecyl benzene sulfonate	98% active	powder	*powder detergents; powder cleaners; textile industry*
	REWORYL NKS 50	sodium dodecyl benzene sulfonate	50% active	paste	*liquid detergents; liquid cleaners; textile industry*
	REWORYL NXS 40	sodium xylene sulfonate	40% active	liquid	*hydrotrope* *detergents; cleaners*

Supplier	Trade name	Chemical description	Composition	General properties	Functionality / Application
Witco	REWORYL TKS 90 F	trialkanolammonium dodecyl benzene sulfonate	90% active	liquid	dishwashing agents; light-duty cleaners; general-purpose cleaners; lubricants; polishes
	SCS 40	sodium cumene sulfonate	40% active	liquid	hydrotrope; powder detergents
	STS 40	sodium toluene sulfonate	40% active	liquid	hydrotrope; powder detergents
	SULFRAMIN 1250	dodecyl benzene sulfonate, sodium salt	50% active	paste	emulsifier; liquid detergents; emulsion polymerisation
	SULFRAMIN ACIDE B	linear dodecyl benzene sulfonic acid	97% active	liquid	production of biodegradeable sulfonates
	SULFRAMIN ACIDE TPB	tetrapropylene benzene sulfonic acid	97% active	liquid	production of branched sulfonates; emulsion polymerisation; agrochemicals
	SXS 40	sodium xylene sulfonate	40% active	liquid	hydrotrope; coupling agent; liquid detergents
	WITCO 1298 H	branched dodecyl benzene sulfonic acid	97% active	liquid	detergent; dispersant; o/w emulsifier; wetting agent; personal care products; household and industrial applications
	WITCO 1298	linear dodecyl benzene sulfonic acid	97% active	liquid	detergent; dispersant; o/w emulsifier; wetting agent; personal care products; household and industrial applications
	WITCONATE 1240	sodium dodecylbenzenesulfonate	40% active; sulfate 0.7%	slurry; pH 7.5	detergent; foaming agent; wetting agent; personal care products; household and industrial applications
	WITCONATE 1250	sodium dodecylbenzenesulfonate	53% active; sulfate 0.9%	slurry; pH 7.5	
	WITCONATE 1260	sodium dodecylbenzenesulfonate and sodium xylene sulfonate	60% active; sulfate 1.1%	slurry; pH 7.5	
	WITCONATE 30DS	sodium dodecylbenzenesulfonate	30% active; sulfate 0.8%	liquid; pH 7.5	
	WITCONATE 45 LIQ	sodium dodecylbenzenesulfonate and sodium xylene sulfonate	45% active; sulfate 1%	liquid; pH 7	
	WITCONATE 45 LX	sodium dodecylbenzenesulfonate and sodium xylene sulfonate	43% active; sulfate 1.2%	liquid; pH 7.5	
	WITCONATE 605A	calcium dodecylbenzene sulfonate	60% active; sulfate 1.3%	liquid; pour point -12.2°C; pH 5-6	detergent; dispersant; o/w emulsifier; w/o emulsifier; lubricant; wetting agent; demulsifier; paraffin inhibition; personal care products; household and industrial applications; petroleum industry
	WITCONATE 60T	TEA-dodecylbenzenesulfonate	58% active; sulfate 2.5%	liquid; pH 6.5	detergent; foaming agent; wetting agent; personal care products; household and industrial applications
	WITCONATE 703	alkylaryl sulfonate		liquid; pour point -15°C; pH 7	demulsifier; slugging compound; petroleum industry
	WITCONATE 705	alkylaryl sulfonate		liquid; pour point -17.8°C; pH 6.5	demulsifier; slugging compound; petroleum industry
	WITCONATE 79S	TEA-dodecylbenzenesulfonate	52% active; sulfate 1.5%	liquid; pH 7	detergent; dispersant; foaming agent; wetting agent; personal care products; household and industrial applications
	WITCONATE 90 DENSE	sodium dodecylbenzenesulfonate	91% active; sulfate 6%	powder; pH 7.5	
	WITCONATE 90F	sodium dodecylbenzenesulfonate	91% active; sulfate 3%	flake; pH 8	
	WITCONATE 90F, H	branched sodium dodecylbenzenesulfonate	91% active; sulfate 3%	flake; pH 8	
	WITCONATE LX, F	sodium dodecylbenzenesulfonate and sodium xylenesulfonate	91% active; sulfate 6%	flake; pH 8	
	WITCONATE LXH	TEA-branched dodecylbenzenesulfonate	53% active; sulfate 1.5%	liquid; pH 7.2	detergent; dispersant; foaming agent; wetting agent; personal care products; household and industrial applications
	WITCONATE P10-59	amine dodecylbenzenesulfonate	90% active; sulfate 1%	liquid; pour point -15°C; pH 5	coupling agent; detergent; dispersant; o/w emulsifier; wetting agent; demulsifier; personal care products; household and industrial applications; petroleum industry

Supplier	Trade name	Chemical description	Composition	General properties	Functionality / Application
Witco	WITCONATE SK	sodium dodecylbenzenesulfonate	40% active; sulfate 55%	flake; pH 8	detergent; foaming agent; wetting agent *personal care products; household and industrial applications*
	WITCONATE SXS, 40%	sodium xylene sulfonate	41% active; sulfate 2%	liquid; pH 8	cloud point depressant; coupling agent *personal care products; household and industrial applications*
	WITCONATE SXS, 90%	sodium xylene sulfonate	92% active; sulfate 4%	powder; pH 8	cloud point depressant; coupling agent *personal care products; household and industrial applications*
	WITCONATE TX ACID	modified toluene sulfonic acid	96% active	liquid	coupling agent; wetting agent *personal care products; household and industrial applications*
	WITCOR PC200	amine alkylaryl sulfonate		liquid; pour point − 3.9°C; pH 4.5	parafin inhibition *petroleum industry*
	WITCOR PC205	amine alkylaryl sulfonate		liquid; pour point − 12.2°C; pH 7.5	
	WITCOR PC210	amine alkylaryl sulfonate		liquid; pour point − 15°C; pH 5.0	
Zschimmer & Schwarz	SUPRALAN TA	alkylaryl sulfonate		yellowish viscous liquid	washing agent; wetting agent; detergent *textile industry*
	ZETESAN NP LIQUID	naphthalene sulfonic acid condensate		brown liquid	dispersant *textile industry*

Alkylarylether carboxylates

Supplier	Trade name	Chemical description	Composition	General properties	Functionality / Application
CHEM-Y	AKYPO NP 70	nonoxynol-8 carboxylic acid	active matter 89% min.; H_2O 6.0-10.0%; chloride (as NaCl) 1.0% max.; acid value 84-102	clear, light yellow liquid; M.W. ca. 607; colour 200 max. (Hazen)	solubiliser *perfumes; cleaners*
Hüls	MARLOWET 4530 LF	dinonylphenol polyglycol ether carboxylic acid	active detergent 90%	liquid	emulsifier; low foam *lubricants; drilling oils*
	MARLOWET 4530	dinonylphenol polyglycol ether carboxylic acid	active detergent 90%	liquid	emulsifier *lubricants; drilling oils*
	MARLOWET 4536	nonylphenol polyglycol ether carboxylic acid	active detergent 90%	liquid	emulsifier *textile industry; lubricants; drilling oils*

Alkylarylether sulfates

Supplier	Trade name	Chemical description	Composition	General properties	Functionality / Application
Akros Chemicals	PERLANKROL FD63	alkyl phenol ether sulfate, ammonium salt	63% active; H_2O content ca. 21%; alcohol content ca. 9%	clear straw liquid; viscosity 132 cSt; pH 7.5 (1% aq.)	emulsifier; wetting agent *emulsion polymerisation*
	PERLANKROL FF	alkyl phenol ether sulfate, ammonium salt	90% active; H_2O content ca. 1%; alcohol content ca. 9%	hazy viscous amber liquid; pour point 5°C; viscosity 4200 cSt; pH 7.5 (1% aq.)	wetting agent; foam booster; foam stabiliser
	PERLANKROL FN65	alkyl phenol ether sulfate, sodium salt	65% active; H_2O content ca. 24%; alcohol content ca. 11%	clear straw liquid; viscosity 398 cSt; pH 7.0-8.5 (1% aq.)	emulsifier; wetting agent; foam stabiliser *emulsion polymerisation*
	PERLANKROL FX35	alkyl phenol ether sulfate, sodium salt	35% active; H_2O content ca. 65%	clear pale yellow liquid; pour point < 0°C; viscosity 152 cSt; pH 7.0-8.5 (1% aq.)	emulsifier *emulsion polymerisation*
	PERLANKROL PA CONC.	alkyl phenol ether sulfate, ammonium salt	90% active; H_2O content ca. 1%; alcohol content ca. 9%	hazy viscous amber liquid; pour point 7°C; viscosity 6000 cSt; pH 7.5 (1% aq.)	emulsifier; wetting agent; foam stabiliser; foaming agent *emulsion polymerisation; detergents*

Supplier	Trade name	Chemical description	Composition	General properties	Functionality Application
Akcros Chemicals	PERLANKROL RN75	alkyl phenol ether sulfate, sodium salt	75% active; H_2O content ca. 15%; alcohol content ca. 10%	clear straw liquid; viscosity 347 cSt; pH 7.0-8.5 (1% aq.)	emulsifier; wetting agent; foaming agent *emulsion polymerisation*
Auschem Cesalpinia	POLIROL 23	alkyl aryl ethoxy sulfate, ammonium salt	23% active	liquid	
	POLIROL 4	alkyl aryl ammonium ethoxy sulfate	30% active	liquid	emulsifier
	POLIROL 425	ammonium alkyl aryl ethoxy sulfate in aq. alcohol soln.	25% active	liquid	*emulsion polymerisation*
	POLIROL 6	alkyl aryl ethoxy sulfate, ammonium salt	22% active	liquid	emulsifier
BASF	EMULPHOR OPS 25	sodium alkylphenol ether sulfate	34% active		
	LUTENSIT A-ES	sodium alkylphenol ether sulfate	40% active		
Croda	SURFACTANT SP0063	alkylarylpolyglycol ether sulfate ammonium salt		pale yellow liquid	detergent; wetting agent; foaming agent *window cleaners; windscreen washes; detergents*
Cytec	AEROSOL NPES-2030	ammonium salt of sulfated nonylphenoxy poly(ethyleneoxy) ethanol	30% in H_2O	liquid; surface tension 43 dynes/cm min. (in H_2O)	emulsifier; detergent *latexes*
	AEROSOL NPES-3030	ammonium salt of sulfated nonylphenoxy poly(ethyleneoxy) ethanol	30% in H_2O	liquid; surface tension 43 dynes/cm min. (in H_2O)	emulsifier; detergent *emulsion polymerisation; latexes*
	AEROSOL NPES-428	sodium salt of sulfated nonylphenoxy poly(ethyleneoxy) ethanol	28% in H_2O	liquid; surface tension 31 dynes/cm min. (in H_2O)	detergent; scouring agent *emulsion polymerisation; textile industry*
	AEROSOL NPES-430	ammonium salt of sulfated nonylphenoxy poly(ethyleneoxy) ethanol	30% in H_2O	liquid; surface tension 31 dynes/cm min. (in H_2O)	detergent; scouring agent *emulsion polymerisation; textile industry; germicides*
	AEROSOL NPES-458	ammonium salt of sulfated nonylphenoxy poly(ethyleneoxy) ethanol	58% in alcohol and H_2O	liquid; surface tension 31 dynes/cm min. (in H_2O)	foaming agent (high); detergent; scouring agent *emulsion polymerisation; detergents; germicides; textile industry*
	AEROSOL NPES-930	ammonium salt of sulfated nonylphenoxy poly(ethyleneoxy) ethanol	30% in H_2O	liquid; surface tension 33 dynes/cm min. (in H_2O)	emulsifier; detergent *latexes*
Henkel	DISPONIL AES 13	sodium alkyl aryl ether sulfate	31-33% active	liquid	
	DISPONIL AES 21	sodium alkyl aryl ether sulfate	30-32% active	liquid	
	DISPONIL AES 48	ammonium alkyl aryl ether sulfate	ca. 70% active	liquid	emulsifier *plastics industry; coatings*
	DISPONIL AES 60	sodium alkyl aryl ether sulfate	ca. 34% active	liquid	
	DISPONIL AES 63	sodium alkyl aryl ether sulfate	ca. 31% active	liquid	
	DISPONIL AES 72	sodium alkyl aryl ether sulfate	32.5-34.5% active	liquid	
Hüls	MARLOWET 4540	nonylphenol polyglycol ether sulfate, sodium salt	active detergent 35%	liquid	emulsifier *polymer dispersions*
Kao Corporation	SAPANOL 0640	sodium alkylarylether sulfate	34-36% active	liquid; M.W. 586; pH 6-7 (5%)	
Millchem	NEUTRONYX S-60	ammonium salt of a sulfated alkyl phenol polyglycol ether		liquid	detergent *surgical scrubs*
Nikko Chemicals	NIKKOL SNP-4N	sodium nonoxynol-4 sulfate; aq. soln.		pale yellow liquid	foaming agent; cleansing agent *cosmetics*
	NIKKOL SNP-4T	triethanolamine POE (4) nonylphenyl ether sulfate; aq. soln.		pale yellow liquid	
Rhone-Poulenc	RHODAPEX CO-433	sodium salt of sulfated alkylphenol ethoxylates	28% solids; 'n' no. 4	liquid; surface tension 32 dynes/cm	*emulsion polymerisation*
	RHODAPEX CO-436	ammonium salt of sulfated alkylphenol ethoxylates	58% solids; 'n' no. 4	liquid; surface tension 33 dynes/cm	emulsifier; wetting agent; dispersant; foaming agent (high); foam booster *emulsion polymerisation*

Supplier	Trade name	Chemical description	Composition	General properties	Functionality *Application*
Rhone-Poulenc	RHODAPEX EP/120	ammonium salt of sulfated alkylphenol ethoxylates	30% solids; 'n' no. 30	liquid; surface tension 42 dynes/cm	emulsifier *emulsion polymerisation*
	RHODAPEX EP/110	ammonium salt of sulfated alkylphenol ethoxylates	30% solids; 'n' no. 9	liquid; surface tension 37 dynes/cm	emulsifier; stabiliser *emulsion polymerisation*
	RHODAPEX LIV/2330	ammonium salt of sulfated alkylphenol ethoxylates	30% solids; 'n' no. 23	liquid; surface tension 41 dynes/cm	emulsifier *emulsion polymerisation*
	RHODAPEX LIV/30	ammonium salt of sulfated alkylphenol ethoxylates	30% solids; 'n' no. 5	liquid; surface tension 33 dynes/cm	emulsifier *emulsion polymerisation*
Seppic	EMULSIFIANT 33 AD	ammonium nonylphenyl ether sulfate	40% active; 'n' no. 25	liquid	emulsifier
	OCTARON PS 20	ammonium nonylphenyl ether sulfate	20% active; 'n' no. 4.5	liquid	detergent
	OCTARON PS 80	ammonium nonylphenyl ether sulfate	80% active; 'n' no. 4.5	gel	detergent
Servo Delden	SERMUL EA 146	sodium nonylphenol ether (15EO) sulfate	32-35% active	liquid	
	SERMUL EA 151	sodium nonylphenol ether (10EO) sulfate	33-36% active	liquid	
Stepan Europe	POLYSTEP B 27	sodium alkylphenol ether sulfate; made-to-order	30% active	pale yellow liquid	emulsifier *emulsion polymerisation*
	POLYSTEP C-OP 3S	sodium octyl phenol ethoxylate sulfate; made-to-order	20% active; 'n' no. 3	white viscous dispersion	emulsifier *emulsion polymerisation*
Union Carbide	TRITON 770 CONC	alkylarylether sulfate	30% active	liquid; pour point −29°C	*liquid detergents; textile industry*
	TRITON W-30 CONC	alkylarylether sulfate	27% active	liquid; pour point −26°C	*liquid detergents; textile industry*
	TRITON X-301	alkylarylether sulfate	20% active	paste; pour point −1°C	*emulsion polymerisation; liquid detergents*
Witco	REWOPOL NOS 10	nonyl phenol polyglycol ether sulfate	35% active	liquid; surface tension 36 dynes/cm; pH 7-8 (1% in H$_2$O)	
	REWOPOL NOS 25	nonyl phenol polyglycol ether sulfate	35% active	liquid; surface tension 40 dynes/cm; pH 7-8 (1% in H$_2$O)	emulsifier; wetting agent *emulsion polymerisation*
	REWOPOL NOS 5	nonyl phenol polyglycol ether sulfate	30% active	liquid; surface tension 29 dynes/cm; pH 6-8 (1% in H$_2$O)	
	REWOPOL NOS 8	nonyl phenol polyglycol ether sulfate	33% active	liquid; surface tension 35 dynes/cm; pH 7-8 (1% in H$_2$O)	emulsifier; wetting agent; co-emulsifier *emulsion polymerisation*
Zschimmer & Schwarz	SETAVIN PT	aryl ether sulfates		viscous clear liquid	levelling agent *textile industry*

Alkylether carboxylates

Supplier	Trade name	Chemical description	Composition	General properties	Functionality *Application*
Auschem Cesalpinia	POLIROL 200	alkyl carboxylated ether sodium salt	28% active	liquid	emulsifier; wetting agent; dispersant; solubiliser; penetrant *polymerisation*
CHEM-Y	AKYPO LF 2	capryleth-9 carboxylic acid	active matter 87.5% min.; H$_2$O 8.0-11.0%; chloride (as NaCl) 1.5% max.; acid value 90-102	clear, almost colourless to slightly yellow liquid; M.W. ca. 547; colour 200 max. (APHA); pH 1.5-2.5 (10% in water)	hydrotrope; co-emulsifier; low foam *household and industrial cleaners; metal working*

Supplier	Trade name	Chemical description	Composition	General properties	Functionality Application
CHEM-Y	AKYPO LF 4 N	mixture of sodium capryleth-9 carboxylate and sodium hexeth-4 carboxylate	active matter 74.0% min.; H_2O 23-25%; chloride (as NaCl) 1.0% max.	clear, colourless to light yellow liquid; M.W. ca. 490; viscosity 500 mPas max.; colour 350 max. (Hazen); pH 7.0-8.5 (10% in water)	solubiliser; hydrotrope; co-emulsifier perfumes
	AKYPO LF 4	mixture of capryleth-9 carboxylic acid and hexeth-4 carboxylic acid	active matter 86.5% min.; H_2O 8.0-12.0%; chloride (as NaCl) 1.5% max.	clear, light yellow liquid; colour 150 max. (Hazen); pH 1.6-2.8 (10% in water)	co-emulsifier; solubiliser lubricants
	AKYPO LF 6	mixture of capryleth-9 carboxylic acid and buteth-2 carboxylic acid	active matter 85-89%; H_2O 10-13%; chloride (as NaCl) 2.0% max.	clear, almost colourless to pale yellow liquid; colour 2 max. (Gardner)	anti-corrosion; solubiliser; co-emulsifier metal working; perfumes; cleaners
	AKYPO RLM 25	laureth-4 carboxylic acid	active matter 92.0% min.; H_2O 7.5% max.; chloride (as NaCl) 0.5% max.	clear, colourless to light yellow liquid; M.W. ca. 356; colour 200 max. (Hazen); pH 2.0-3.5 (1% in water)	emulsifier metal working; lubricants
	AKYPO RLM 45 CA	laureth-6 carboxylic acid	active matter 90-94%; H_2O 6.0-9.0%; chloride (as NaCl) 1.0% max.; sapon. no. 120	clear, almost colourless to slightly yellowish liquid; M.W. ca. 457; colour 125 max.(Hazen); pH 2.7-3.2 (1% in water)	shampoos; shower preparations; liquid soaps
	AKYPO RLM 45 N	sodium laureth-6 carboxylate	active matter 80.0-84.0%; H_2O 16.0-19.0%; chloride (as NaCl) 0.8% max.	hazy, almost colourless to pale yellow paste; M.W. ca. 479; pH 4.0-5.0 (10% in water)	emulsifier; cleansing agent shampoos; shower preparations; liquid soaps; hair care products
	AKYPO RO 20	oleth-3 carboxylic acid	active matter 94% min.; H_2O 5% max.; chloride (as NaCl) 1.0% max.; acid value 70-90	clear, brownish, oily liquid; M.W. ca. 411; pH 2.0-3.5 (10% in water)	
	AKYPO RO 50	oleth-6 carboxylic acid	active matter 91% min.; H_2O 8% max.; chloride (as NaCl) 1% max.; acid value 80-95	clear yellow-brown liquid; M.W. ca. 544; colour 7 max. (Gardner); pH 2.5-3.5 (1% in water)	emulsifier metal working
	AKYPO RO 90	oleth-10 carboxylic acid	active matter 89% min.; H_2O 6.0-10.0%; chloride (as NaCl) 1.5% max.	yellow gelatinous liquid; M.W. ca. 698; colour 6 max. (Gardner); pH 1.6-2.8 (10% in water)	
	AKYPO SOFT 100 NV	sodium laureth-11 carboxylate	active matter 21-23%; H_2O 76.5-78.5%; chloride (as NaCl) 0.5% max.	clear, colourless to pale yellow liquid; M.W. ca. 707; viscosity 100 mPas max.; colour 150 max. (Hazen); pH 6.5-7.5	synergist baby care products; shampoos; shower preparations; bubble baths; liquid soaps
	AKYPO SOFT 45 NV	sodium laureth-6 carboxylate	active matter 21.0% min.; H_2O 77.5-78.5%; chloride (as NaCl) 0.5% max.	clear, light yellow liquid; M.W. ca. 457; colour 125 max. (Hazen); pH 6.5-7.5	synergist shampoos; shower preparations; liquid soaps
	AKYPO TEC AM	mixture of oleth-10 carboxylic acid, capryleth-9 carboxylic acid and hexeth-4 carboxylic acid	active matter 88% min.; H_2O 8.0-11.5%; chloride (as NaCl) 1.2% max.; acid value 85-95	clear yellow liquid; colour 5 max. (Gardner)	emulsifier; cleaning agent metal working
Hoechst	DP1219	laureth-7 carboxylic acid	90% active		detergents; toiletries
	DP1220	laureth-3 carboxylic acid	90% active		
	HOSTAPON LEC	laureth-3 carboxylic acid	90% active		
Hüls	MARLINAT CM 100	$C_{12/14}$-fatty alcohol polyglycolether carboxylic acid 10EO	active detergent 90%	liquid	
	MARLINAT CM 105/80	$C_{12/14}$-fatty alcohol polyglycolether carboxylate 10EO, sodium salt	active detergent 80%	liquid	
	MARLINAT CM 105	$C_{12/14}$-fatty alcohol polyglycolether carboxylate 10EO, sodium salt	active detergent 23%	liquid	detergents
	MARLINAT CM 20	$C_{12/14}$-fatty alcohol polyglycolether carboxylic acid 2EO	active detergent 90%	liquid	
	MARLINAT CM 40	$C_{12/14}$-fatty alcohol polyglycolether carboxylic acid 4EO	active detergent 90%	liquid	
	MARLINAT CM 45	$C_{12/14}$-fatty alcohol polyglycolether carboxylate 4EO, sodium salt	active detergent 23%	liquid	
	MARLOWET 1072	$C_{12/14}$-alcohol polyglycol ether carboxylic acid	active detergent 90%	liquid	emulsifier textile industry; lubricants; drilling oils
	MARLOWET 4538	C_{13}-oxo alcohol polyglycol ether carboxylic acid	active detergent 90%	liquid	

Supplier	Trade name	Chemical description	Composition	General properties	Functionality / Application
Hüls	MARLOWET 4539	C_9-oxo alcohol polyglycol ether carboxylic acid	active detergent 90%	liquid	emulsifier; *lubricants; drilling oils*
	MARLOWET 4541	$C_{11/14}$-alcohol polyglycol ether carboxylic acid	active detergent 92%	liquid	emulsifier; *textile industry; lubricants; drilling oils*
Kao Corporation	DANOX AM	sodium carboxylate alkyl ether	ca. 71% active	paste; pH 6.5-7.5 (5%)	
Nikko Chemicals	NIKKOL AKYPO RLM 45 NV	sodium POE (4.5) lauryl ether carboxylate; aq. soln.		pale yellow liquid	foaming agent; cleansing agent; *cosmetics*
	NIKKOL ECT-3	POE (3) tridecyl ether carboxylic acid; free acid form		pale yellow liquid	*cosmetics*
	NIKKOL ECT-3NEX	sodium POE (3) tridecyl ether carboxylate		pale yellow viscous liquid	foaming agent; cleansing agent; *cosmetics*
	NIKKOL ECT-7	trideceth-7 carboxylic acid; free acid form		pale yellow liquid	*cosmetics*
	NIKKOL ECTD-3NEX	sodium POE (3) tridecyl ether carboxylate		pale yellow viscous liquid	foaming agent; cleansing agent; *cosmetics*
	NIKKOL ECTD-6NEX	sodium POE (6) tridecyl ether carboxylate		pale yellow viscous liquid	foaming agent; cleansing agent; *cosmetics*
Witco	REWOPOL CL 30	fatty alcohol ether carboxylic acid	90% active	liquid	*household cleaners; toiletries*
	REWOPOL CLN 100	sodium fatty alcohol ether carboxylate	23% active	liquid	*toiletries; shampoos; shower gels; liquid soaps*
	REWOPOL CT 65	fatty alcohol ether carboxylic acid	90% active	liquid	*household and industrial cleaners; textile industry; oil recovery; mineral oils*

Alkylether sulfates

Supplier	Trade name	Chemical description	Composition	General properties	Functionality / Application
Akcros Chemicals	PERLANKROL ADP3	lauryl ether sulfate, sodium salt; toiletry grade	27% active	clear water-white liquid; pour point <0°C; viscosity 28 cSt; pH 6.5-7.5 (1% aq.)	*cosmetics; toiletries; bubble baths; shampoos*
	PERLANKROL ASC2	lauryl ether sulfate, sodium salt; toiletry grade	27.5% active	clear water-white liquid; pour point <0°C; viscosity 600 cSt; pH 6.5-7.5 (1% aq.)	
	PERLANKROL ASC82	lauryl ether sulfate, sodium salt; toiletry grade	27.5% active; 'n' no. 2	clear water-white liquid; pour point <0°C; viscosity 2100 cSt; pH 6.5-7.5 (1% aq.)	
	PERLANKROL EAD60	primary alcohol ether sulfate, ammonium salt, in aq. alcohol; detergent grade: made-to-order	60% active; H_2O content ca. 30%; alcohol content ca. 10%	clear to hazy pale yellow liquid; pour point <0°C; viscosity 240 cSt; pH 6.0-8.0 (1% aq.)	foam booster; foaming agent (high); wetting agent; foam stabiliser; detergent; dispersant; emulsifier
	PERLANKROL EP12	fatty alcohol ether sulfate, ammonium salt	20% active; H_2O content ca. 80%	clear amber viscous liquid; pour point 0°C; viscosity 800 cPs (40°C); pH 7.0-8.5 (1% aq.)	*liquid detergents; industrial and household cleaners; fire-fighting; chlorinated solvents*
	PERLANKROL EP24	fatty alcohol ether sulfate, sodium salt	28% active; H_2O content ca. 70%	clear straw liquid; pour point <0°C; viscosity 85 cPs; pH 7.0-8.5 (1% aq.)	emulsifier; wetting agent; emulsion polymerisation
	PERLANKROL EP36	fatty alcohol ether sulfate, sodium salt	27% active; H_2O content ca. 70%	clear straw liquid; pour point <0°C; viscosity 70 cPs; pH 7.0-8.5 (1% aq.)	
	PERLANKROL EP48	fatty alcohol ether sulfate, sodium salt	25% active; H_2O content ca. 72%	clear straw liquid; pour point <0°C; viscosity 45 cPs; pH 7.0-8.5 (1% aq.)	

Supplier	Trade name	Chemical description	Composition	General properties	Functionality/Application
Akcros Chemicals	PERLANKROL EP60	fatty alcohol ether sulfate, sodium salt	27% active; H$_2$O content ca. 70%	clear straw liquid; pour point <0°C; viscosity 26 cPs; pH 7.0-8.5 (1% aq.)	emulsifier; wetting agent *emulsion polymerisation*
	PERLANKROL ESD	primary alcohol ether sulfate, sodium salt, in H$_2$O; detergent grade	27% active; H$_2$O content ca. 70%	clear pale yellow liquid; pour point <0°C; viscosity 26 cSt; pH 7.5-8.5 (1% aq.)	foam booster; foaming agent (high); wetting agent; foam stabiliser; detergent; dispersant; emulsifier *liquid detergents; industrial and household cleaners; fire-fighting; chlorinated solvents*
	PERLANKROL ESD60	primary alcohol ether sulfate, sodium salt, in aq. alcohol; detergent grade	60% active; H$_2$O content ca. 23.5%; alcohol content ca. 16.5%	clear to hazy pale yellow liquid; pour point <0°C; viscosity 210 cSt; pH 7.5-8.5 (1% aq.)	
	PERLANKROL ESK29	synthetic primary alcohol ether sulfate, sodium salt, in H$_2$O	30% active; H$_2$O content ca. 67%	clear yellow liquid; pour point −2°C; viscosity 19 cSt; pH 6.5-7.5 (1% aq.)	foam booster; foaming agent; wetting agent; foam stabiliser; detergent; dispersant; emulsifier *liquid detergents; industrial and household cleaners; fire-fighting; chlorinated solvents*
	PERLANKROL ESK32	synthetic primary alcohol ether sulfate, sodium salt	32% active; H$_2$O content 67%	clear viscous amber liquid; viscosity 225 cPs; pH 6.5-8.0 (1% aq.)	foam booster; foam stabiliser; foaming agent *detergents; fire-fighting; plaster board production; industrial and household applications*
Akzo Nobel	ELFAN NS 232 S CONC.	sodium dodecyl tridecyl ether (2.5) sulfate	70% active	paste	
	ELFAN NS 233 SL CONC.	sodium dodecyl tridecyl ether (3) sulfate	70% active	paste	
	ELFAN NS 242 CONC.	sodium lauryl myristyl ether (2) sulfate	70% active	paste	
	ELFAN NS 242 A	sodium lauryl myristyl ether (2) sulfate (alkaline)	27% active	liquid	
	ELFAN NS 242	sodium lauryl myristyl ether (2) sulfate	27% active	liquid	
	ELFAN NS 243 S CONC.	sodium lauryl myristyl ether (3) sulfate	70% active	paste	
	ELFAN NS 243 S MG CONC.	magnesium lauryl myristyl ether (3) sulfate	70% active	paste	
	ELFAN NS 243 S	sodium lauryl myristyl ether (3) sulfate	28% active	liquid	
	ELFAN NS 243 S MG	magnesium lauryl myristyl ether (3) sulfate	27% active	liquid	
	ELFAN NS 248 S MG	magnesium lauryl myristyl ether (8) sulfate	27% active	liquid	
	ELFAN NS 252 S CONC.	sodium dodecyl pentadecyl ether (2.5) sulfate	70% active	paste	
	ELFAN NS 252 SL CONC.	sodium dodecyl pentadecyl ether (2.5) sulfate	70% active	paste	
	ELFAN NS 252 S CONC.	sodium dodecyl pentadecyl ether (2.5) sulfate	25.5% active	liquid	
Albright & Wilson	EMPICOL 0251/70	sodium lauryl ethoxy (3EO) sulfate	70% active		
	EMPICOL 0399	sodium lauryl ethoxy sulfate	27% active		
	EMPICOL 0411	modified sodium lauryl ether sulfate	50% active		
	EMPICOL BSD	sodium/magnesium alkyl ethoxy sulfate; toiletry grade	26% active	liquid	*toiletries*
	EMPICOL BSD52	sodium/magnesium alkyl ethoxy sulfate; toiletry grade	52% active	liquid	
	EMPICOL EAA 25	ammonium lauryl ethoxy sulfate; toiletry grade	27% active	liquid	

Supplier	Trade name	Chemical description	Composition	General properties	Functionality / *Application*
Albright & Wilson	EMPICOL EAB/CG	ammonium lauryl ethoxy (2EO) sulfate	24% active		
	EMPICOL EAB/XH	ammonium lauryl ethoxy sulfate; toiletry grade	24% active	liquid	*toiletries*
	EMPICOL EAC/T	ammonium lauryl ethoxy (3EO) sulfate	25% active		
	EMPICOL EAC/TP	ammonium lauryl ethoxy sulfate; toiletry grade	25% active	liquid	*toiletries*
	EMPICOL EAC70	ammonium lauryl ethoxy (3EO) sulfate	70% active		
	EMPICOL EAX25	ammonium lauryl ethoxy (3EO) sulfate	25% active		
	EMPICOL EGC 70	magnesium lauryl ethoxy sulfate; toiletry grade	70% active	liquid	*toiletries*
	EMPICOL EGC	magnesium lauryl ethoxy sulfate; toiletry grade	27% active	liquid	*toiletries*
	EMPICOL EMB/FL	monoethanolamine lauryl ethoxy (2.2EO) sulfate	28% active		
	EMPICOL ESA	sodium lauryl ethoxy sulfate; toiletry grade	25% active	liquid	*toiletries*
	EMPICOL ESA70	sodium lauryl ethoxy (1EO) sulfate	70% active		
	EMPICOL ESB3/M	sodium lauryl ethoxy sulfate; toiletry grade	27% active	liquid	dispersant; emulsifier; detergent; air entraining agent; *toiletries; textile industry; leather industry; construction; emulsion polymerisation; suspension polymerisation*
	EMPICOL ESB70	sodium lauryl ethoxy sulfate; toiletry grade	70% active	liquid	
	EMPICOL ESC3/G2	sodium lauryl ethoxy (3EO) sulfate	28% active		
	EMPICOL ESC3	sodium lauryl ethoxy sulfate; toiletry grade	28% active	liquid	
	EMPICOL ESC70	sodium lauryl ethoxy sulfate; toiletry grade	70% active	liquid	
	EMPICOL ETB/T	triethanolamine lauryl ethoxy sulfate; toiletry grade	29% active	liquid	*toiletries*
	EMPICOL ETIB90	tri-isopropylamine lauryl ethoxy sulfate	89% active	liquid	
	EMPIMIN KSN27/LA	sodium alkyl ethoxy sulfate; detergent/industrial grade	27% active	liquid	detergent; dispersant; emulsifier; *textile industry; leather industry; emulsion and suspension polymerisation*
	EMPIMIN KSN27/XW	sodium alkyl ethoxy sulfate; detergent/industrial grade	27% active	liquid	detergent; dispersant; emulsifier; *leather industry; emulsion and suspension polymerisation*
	EMPIMIN KSN70/XW	sodium alkyl ethoxy (3EO) sulfate	70% active		
	EMPIMIN KSN70/LA	sodium alkyl ethoxy sulfate; detergent/industrial grade	70% active	liquid	detergent; dispersant; emulsifier; *textile industry; leather industry; emulsion and suspension polymerisation*
	EMPIMIN LSM30	sodium alkyl ethoxy sulfate; detergent/industrial grade	30% active	liquid	
Auschem Cesalpinia	COSMOPON AEC	ammonium lauryl ether sulfate	80% active	paste	foaming agent; *liquid detergents; laundry products; household detergents; dishwashing agents; polymerisation*
	COSMOPON ME	MEA laureth sulfate	30% active	liquid	foaming agent; *liquid detergents; laundry products; household detergents; dishwashing agents; polymerisation; bubble baths; liquid soaps: shower gels; shampoos*
	EMULSON 230	sodium laureth sulfate	28% active	liquid	wetting agent; *pesticides*
	POLIROL 24 A	ammonium lauryl ethoxy sulfate	25% active	liquid	emulsifier; *emulsion polymerisation*

Supplier	Trade name	Chemical description	Composition	General properties	Functionality / Application
Auschem Cesalpinia	ROLPON 230	sodium laureth-2 sulfate	28% active	liquid	foaming agent; *liquid detergents; laundry products; household detergents; dishwashing agents; polymerisation; bubble baths; liquid soaps; shower gels; shampoos*
	ROLPON 330 N	sodium laureth-3 sulfate	28% active	liquid	
	ROLPON 330	sodium laureth-3 sulfate	28% active	liquid	
	ROLPON 370	sodium laureth-3 sulfate	70% active	viscous liquid	
	ROLPON LE 50 S	sodium laureth-3 sulfate	50% active	viscous liquid	
	ROLPON LE 50	sodium laureth-2 sulfate	50% active	viscous liquid	
CHEM-Y	CHEMSALAN RLM 28	sodium laureth sulfate	active matter 27.5-29.5%; anionic active matter 27.0% min.	clear, colourless to pale, yellowish liquid; M.W. ca. 385; viscosity 200 mPas max.; colour 100 max. (Hazen); pH 6.0-7.0 (10% in water)	*shampoos; bubble baths; liquid detergents*
	CHEMSALAN RLM 30-70	sodium laureth sulfate	active matter 70-74%; anionic active matter 69.0% min.	almost colourless to pale yellow paste; M.W. ca. 423; colour 75 max. (Hazen; 40% in IPA 25%); pH 7.0-8.5 (10% in water)	
	CHEMSALAN RLM 56	sodium laureth sulfate	active matter 55% min.; anionic active matter 54.0% min.	slightly turbid, colourless to pale yellowish, viscous liquid; M.W. ca. 385; viscosity 5000-15000 mPas; colour 100 max. (Hazen; 50% in water); pH 6.5-8.0 (10% in water)	
	CHEMSALAN RLM 70	sodium laureth sulfate	active matter 70-73%; anionic active matter 69% min.	slightly turbid, colourless to pale yellowish gel; M.W. ca. 385; colour 100 max. (Hazen; 40% in water); pH 7.0-8.5 (10% in water)	
Dac International Surfactants	DACLOR L70	lauryl ether sulfate based on synthetic alcohol			
	DACLOR N70	lauryl ether sulfate based on natural alcohol			
	DACPON L	lauryl ether sulfate based on synthetic alcohol			
	DACPON N	lauryl ether sulfate based on natural alcohol			
	DACPON T	alcohol sulfate based on tallow alcohol			
Elf Atochem	LAURAL EC	ammonium lauryl ether sulfate		liquid	
	LAURAL ED	sodium/triethanolamine lauryl ether sulfate		liquid	
	LAURAL LS	sodium lauryl ether sulfate		gel	
Ellis & Everard	CAFLON 2L28M	sodium lauryl ethoxy (2EO) sulfate	28% active		
	CAFLON 2L70	sodium lauryl ethoxy (2EO) sulfate	70% active		
	CAFLON 3S27	sodium alkyl ethoxy (3EO) sulfate	27% active		
	CAFLON 3S60A	ammonium alkyl ethoxy (3EO) sulfate	60% active		
	CAFLON 3S70	sodium alkyl ethoxy (3EO) sulfate	70% active		
	CAFLON NLV2	sodium lauryl ethoxy (2EO) sulfate	28% active		
Henkel	DISPONIL FES 32	sodium laureth sulfate	ca. 30% active	liquid	emulsifier; *plastics industry; coatings*
	DISPONIL FES 61	sodium laureth sulfate	ca. 30% active	liquid	
	DISPONIL FES 77	sodium laureth sulfate	ca. 30% active	liquid	
	DISPONIL FES 993	sodium laureth sulfate	ca. 30% active	liquid	
	EUMULGIN MPK 850	dispersion of fatty alcohol ether sulfates with pearl-shine imparting substances		liquid	pearlescent agent; *shampoos; foam baths; shower preparations*
	EUPERLAN PK 771	dispersion of fatty alcohol ether sulfates with pearl-shine imparting substances	45-47% active	pasty	pearlescent agent

Supplier	Trade name	Chemical description	Composition	General properties	Functionality Application
Henkel	EUPERLAN PK 776	dispersion of fatty alcohol ether sulfates with pearl-shine imparting substances	41-43% active	pasty	*pearlescent agent*
	EUPERLAN PK 789	dispersion of fatty alcohol ether sulfates with pearl-shine imparting substances	29-31% active	viscous	
	EUPERLAN PK 810	dispersion of fatty alcohol ether sulfates with pearl-shine imparting substances	35-39% active	liquid	
	EUPERLAN PK 900	dispersion of fatty alcohol ether sulfates with pearl-shine imparting substances	ca. 40% active	liquid	
	STANDAPOL EA-1	ammonium laureth sulfate	26% active	water white to pale yellow viscous liquid	foaming agent *shampoos; bath preparations; cleansing preparations*
	STANDAPOL EA-2	ammonium laureth sulfate	25% active	water white to pale yellow viscous liquid	
	STANDAPOL EA-3	ammonium laureth sulfate	27% active	water white to pale yellow viscous liquid	
	STANDAPOL EA-40	ammonium myreth sulfate	59% active	pale yellow viscous liquid	*shampoos; bath preparations; cleansing preparations*
	STANDAPOL ES-1	sodium laureth sulfate	25% active	water white viscous liquid	foaming agent *shampoos; cleansing preparations; bath preparations*
	STANDAPOL ES-2	sodium laureth sulfate	25% active	water white viscous liquid	
	STANDAPOL ES-250	sodium laureth sulfate	53% active	pale yellow clear liquid	*shampoos; bath preparations; cleansing preparations*
	STANDAPOL ES-3	sodium laureth sulfate	28% active	water white viscous liquid	foaming agent *shampoos; cleansing preparations; bath preparations*
	STANDAPOL ES-350	sodium laureth sulfate	53% active	clear pale yellow liquid	*shampoos; bath preparations; cleansing preparations*
	STANDAPOL ES-40	sodium myreth sulfate	59% active	pale yellow clear liquid	*shampoos; bath preparations*
	TEXAPON ASV 70 SPECIAL	mixture of sodium laureth sulfate, sodium laureth-8 sulfate and sodium oleth sulfate	67% active	paste	
	TEXAPON ASV	mixture of fatty alcohol ether sulfates	28-32% active	liquid	*shampoos; baby shampoos; bubble baths; intimate hygiene products*
	TEXAPON EVR	sodium laureth sulfate with special additives	35-37% active	liquid	pearlescent agent *shampoos; bubble baths*
	TEXAPON K 14 S SPECIAL	sodium myreth sulfate	27-29% active	liquid	foaming agent *shampoos; baby shampoos; bath preparations; cosmetics*
	TEXAPON K 14 S 70 SPECIAL	sodium myreth sulfate	68% active	paste	foaming agent *shampoos; baby shampoos; bath preparations; cosmetics*
	TEXAPON M	MEA-laureth sulfate	27.5-28.5% active	liquid	*shampoos; shower preparations; skin cleansing preparations; cosmetics*
	TEXAPON MG	magnesium laureth sulfate	29-31% active	liquid	*shampoos; baby shampoos; bath preparations*
	TEXAPON N 103	sodium laureth sulfate	27-29% active	liquid	*shampoo; shower preparations; skin cleansing preparations*
	TEXAPON N 25	sodium laureth sulfate	27-29% active	liquid	*shampoos; bubble baths; cosmetics*
	TEXAPON N 40	sodium laureth sulfate	27-29% active	liquid	*shampoos; bubble baths; cosmetics*
	TEXAPON N 70 LS	sodium laureth sulfate	72-75% active	paste	*shampoos; bubble baths; cosmetics; dishwashing agents; cleaners; detergents*
	TEXAPON N 70	sodium laureth sulfate	68-73% active	paste	
	TEXAPON NA	ammonium laureth sulfate	23-25% active	liquid	*shampoos; shower and bath preparations; skin cleansing preparations; cosmetics*

Supplier	Trade name	Chemical description	Composition	General properties	Functionality Application
Henkel	TEXAPON NSO	sodium laureth sulfate	27–29% active	liquid	emulsifier *shampoos; shower and bath preparations; skin cleansing preparations; cosmetics; polymer industry; coatings; dishwashing agents; cleaners; detergents*
	TEXAPON NT	TEA-laureth sulfate	34–36% active	liquid	*shampoos; shower and bath preparations; skin cleansing preparations*
	TEXAPON SG	sodium laureth sulfate with special additives		thick liquid	pearlescent agent *shampoos; bubble baths*
Hickson Manro	MANRO ALEC25	ammonium lauryl (3 mole) ether sulfate; natural based	25% active	liquid	*hair care products; bath preparations; household detergents*
	MANRO ALES60	ammonium lauryl (3 mole) ether sulfate; synthetic based	60% active	liquid	*hair care products; household detergents; industrial cleaning*
	MANRO BEC28	sodium lauryl (3 mole) ether sulfate; natural based	28% active	liquid	*hair care products; bath preparations*
	MANRO BEC70	sodium lauryl (3 mole) ether sulfate; natural based	68% active	liquid	*hair care products; bath preparations*
	MANRO BES27	sodium lauryl (3 mole) ether sulfate; synthetic based	27% active	liquid	*hair care products; bath preparations; household detergents; industrial cleaning*
	MANRO BES60	sodium lauryl (3 mole) ether sulfate; synthetic based	60% active	liquid	*hair care products; household detergents; industrial cleaning*
	MANRO BES70	sodium lauryl (3 mole) ether sulfate; synthetic based	68% active	mobile gel	*hair care products; bath preparations; household detergents; industrial cleaning*
	MANRO DES32	sodium alcohol (2.5 mole) ether sulfate; synthetic based	32% active	liquid	*household detergents; industrial cleaning*
	MANRO NEC28	sodium lauryl (2 mole) ether sulfate; natural based	28% active	liquid	*hair care products; bath preparations*
	MANRO NEC70	sodium lauryl (2 mole) ether sulfate; natural based	70% active	mobile gel	*hair care products; bath preparations*
	TENSAGEX DLM927	sodium lauryl (3 mole) ether sulfate; synthetic based	27% active	liquid	*hair care products; bath preparations; household detergents; industrial cleaning*
	TENSAGEX DLM970	sodium lauryl (3 mole) ether sulfate; synthetic based	68% active	mobile gel	*hair care products; bath preparations; household detergents; industrial cleaning*
	TENSAGEX EOC628	sodium lauryl (2 mole) ether sulfate; natural based	28% active	liquid	*hair care products; bath preparations*
	TENSAGEX EOC670	sodium lauryl (2 mole) ether sulfate; natural based	70% active	mobile gel	*hair care products; bath preparations*
Hoechst	GENAPOL LRO LIQUID	2 mole natural sodium lauryl ether sulfate	28% active	liquid	*toiletries; detergents*
	GENAPOL LRO PASTE	2 mole natural sodium lauryl ether sulfate	70% active	paste	*toiletries; detergents*
	GENAPOL ZRO LIQUID	3 mole synthetic sodium lauryl ether sulfate	28% active	liquid	*toiletries; detergents; construction industry*
	GENAPOL ZRO PASTE	3 mole synthetic sodium lauryl ether sulfate	70% active	paste	*toiletries; detergents; construction industry*
Hüls	MARLINAT 242/28	$C_{12/14}$-fatty alcohol ether sulfate 2EO, sodium salt	active detergent 28%; dioxane content < 10 ppm	liquid	foaming agent (high) *detergents; cleaners; shampoos; liquid soaps*

Supplier	Trade name	Chemical description	Composition	General properties	Functionality / Application
Hüls	MARLINAT 242/70 S	C$_{12/14}$-fatty alcohol ether sulfate 2EO, sodium salt	active detergent 70%; dioxane content <10 ppm	liquid/paste	foaming agent (high); *detergents, cleaners; shampoos; liquid soaps*
	MARLINAT 242/70	C$_{12/14}$-fatty alcohol ether sulfate 2EO, sodium salt	active detergent 70%; dioxane content <10 ppm	liquid/paste	
	MARLINAT 243/28	C$_{12/14}$-fatty alcohol ether sulfate 3EO, sodium salt	active detergent 28%	liquid	foaming agent (high); *detergents, cleaners*
	MARLINAT 243/70	C$_{12/14}$-fatty alcohol ether sulfate 3EO, sodium salt	active detergent 70%	liquid/paste	foaming agent (high); *detergents, cleaners*
Kao Corporation	EMAL 3 CR 70	sodium alkyl ether sulfate	69–71% active	paste; M.W. 470; pH 7–8.5 (5%)	
	EMAL E 70 CM	sodium lauryl ether sulfate	69–71% active	paste; M.W. 390; pH 7–8 (1%)	
	SAPANOL DSS	sodium lauryl ether sulfate	26–27% active	liquid; M.W. 390; pH 6.5–7.5 (5%)	
Lonza	CARSONOL SES-A	ammonium laureth sulfate	59% active	liquid	foaming agent; wetting agent; emulsifier *household and industrial applications*
	CARSONOL SES-S	sodium laureth sulfate	59% active	liquid	
	CARSONOL SLES-2	sodium laureth sulfate	27% active	liquid	
	CARSONOL SLES-3	sodium laureth sulfate	29% active	liquid	
Millchem	MAPROFIX ES-1	sodium laureth sulfate	24.5–25.5% active	liquid	*shampoos; bubble baths; cosmetics*
	MAPROFIX ES-2	sodium laureth sulfate	25.0–26.5% active	liquid	
	MAPROFIX ES-3	sodium laureth sulfate	27% active min.	liquid	*shampoos; dishwashing agents*
	MAPROFIX LES-60A	ammonium laureth sulfate	57–60% active	liquid	*shampoos; bubble baths; dishwashing agents*
	MAPROFIX LES-60C	sodium laureth sulfate	56–59% active	liquid	
Nikko Chemicals	NIKKOL NES-203-27	sodium POE (3) alkyl ether sulfate; aq. soln.		colourless liquid	
	NIKKOL NES-303-36	triethanolamine POE (3) alkyl ether sulfate; aq. soln.		colourless liquid	
	NIKKOL SBL-2A-27	ammonium POE (2) lauryl ether sulfate; aq. soln.		pale yellow viscous liquid	foaming agent; cleansing agent *cosmetics*
	NIKKOL SBL-2N-27	sodium POE (2) lauryl ether sulfate; aq. soln.		pale yellow liquid	
	NIKKOL SBL-2T-36	triethanolamine POE (2) lauryl ether sulfate; aq. soln.		pale yellow liquid	
	NIKKOL SBL-3N-27	sodium POE (3) lauryl ether sulfate; aq. soln.		pale yellow liquid	
	NIKKOL SBL-4N	sodium laureth sulfate; aq. soln.		pale yellow liquid	
	NIKKOL SBL-4T	TEA-laureth sulfate; aq. soln.		pale yellow liquid	
Rhône-Poulenc	RHODAPEX AB/20	ammonium lauryl ether sulfate	30% solids	liquid; surface tension 33 dynes/cm	*emulsion polymerisation*
	RHODAPEX EST-30	sodium tridecyl ether sulfate	30% solids	liquid; surface tension 33 dynes/cm	wetting agent; dispersant; foaming agent (moderate) *emulsion polymerisation*
Sandoz	LYOGEN DFT SULPHATE	alkyl polyglycol ether sulfate		liquid	levelling agent; stripping agent *textile industry; dyeing*
Seppic	MONTELANE KRO	sodium lauryl ether sulfate	28% active	liquid	
	MONTELANE LT 4088	triethanolamine lauryl ether sulfate	30% active	liquid	
	MONTELANE MG	magnesium lauryl ether sulfate	26% active	liquid	wetting agent; foaming agent
	MONTELANE PSA 80	ammonium lauryl ether sulfate	80% active	gel	

Supplier	Trade name	Chemical description	Composition	General properties	Functionality Application
Seppic	ORONAL BLD	sodium laureth sulfate	33% active	liquid	hair care products
	ORONAL LCG	sodium coceth sulfate and PEG-40 glyceryl cocoate	50% active	liquid	hair care products
Servo Delden	SERMUL EA 266	tridecylether sulfate (15EO)	26% active min.	liquid	
Shell Chemicals	DOBANOL 23-25/70	alcohol ethoxysulfate	68-72% active	M.W. 384; colour 25 max. (5%; Klett); pH 7.0-8.0 (5%)	washing-up liquids; laundry products; hard-surface cleaners; textile industry; industrial applications; cosmetics; toiletries
	DOBANOL 25-25/70	alcohol ethoxysulfate	68-72% active	M.W. 419; colour 25 max. (5%; Klett); pH 7.0-8.0 (5%)	
	DOBANOL 25-3S/27	alcohol ethoxysulfate	26.5% active	M.W. 441; colour 50 max. (5%; Klett); pH 6.5-8.0 (5%)	
Stepan Europe	POLYSTEP B 11	ammonium lauryl ether sulfate; made-to-order	59% active	pale yellow liquid	emulsifier
	POLYSTEP B 12	sodium lauryl ether sulfate; made-to-order	60% active	pale yellow liquid	emulsion polymerisation
	POLYSTEP B 23	sodium lauryl ether sulfate; made-to-order	58% active	amber turbid viscous liquid	
	STEOL OS 28	sodium lauryl ether sulfate	28% active	water white to yellow liquid to paste	foaming agent; detergent shampoos; liquid soaps; bubble baths; household, institutional and industrial cleaners
	STEOL OS 70	sodium lauryl ether sulfate	70% active	water white to yellow liquid to paste	foaming agent shampoos; liquid soaps; bubble baths
Surfachem	SURFAC LC	sodium lauryl ethoxy (2 mole) sulfate			
	SURFAC LC3	sodium lauryl ethoxy (3 mole) sulfate (natural)			
	SURFAC LC70	sodium lauryl ethoxy (2 mole) sulfate - high active			
	SURFAC LES27	sodium lauryl ethoxy (3 mole) sulfate (synthetic)			
	SURFAC LES70	sodium lauryl ethoxy (3 mole) sulfate - high active			
Unger Fabrikker	UNGEROL AM 3-75	ammonium lauryl ether sulfate	75% active; carbon chain distribution $C_{13/15}$	water white liquid; M.W. 445; pH 7-8 (1% soln.)	
	UNGEROL LES 2-28	sodium lauryl ether sulfate	28% active; carbon chain distribution $C_{12/13}$	water white liquid; M.W. 384; pH 7-8 (1% soln.)	detergent; emulsifier
	UNGEROL LES 2-70	sodium lauryl ether sulfate	70% active; carbon chain distribution $C_{12/13}$	water white viscous liquid; M.W. 384; pH 7-9 (1% soln.)	shampoos; liquid detergents; bath preparations; textile industry
	UNGEROL LES 3-28	sodium lauryl ether sulfate	28% active; carbon chain distribution $C_{13/15}$	water white liquid; M.W. 450; pH 7-8 (1% soln.)	
	UNGEROL N 3-70	sodium lauryl ether sulfate	70% active; carbon chain distribution $C_{12/14}$	water white viscous liquid; M.W. 425; pH 7-9 (1% soln.)	
	UNGEROL N2-28	sodium lauryl ether sulfate	28% active; carbon chain distribution $C_{12/14}$	water white liquid; M.W. 382; pH 7-8 (1% soln.)	detergent; emulsifier
	UNGEROL N2-70	sodium lauryl ether sulfate	70% active; carbon chain distribution $C_{12/14}$	water white viscous liquid; M.W. 382; pH 7-9 (1% soln.)	shampoos; liquid detergents; bath preparations; textile industry
Witco	NEOPON LOS 70	lauryl ether sulfate, sodium salt	70% active	pumpable gel	shampoos; bubble baths
	NEOPON LOS	lauryl ether sulfate, sodium salt	28% active	liquid	shampoos; bubble baths; liquid detergents
	NEOPON LOT	lauryl ether sulfate, TEA-salt	28% active	liquid	shampoos; bubble baths
	REWOPOL NL 3-28	sodium lauryl ether sulfate	28% active	liquid	shampoos; shower gels; foam baths; liquid soaps; dishwashing agents; emulsion polymerisation; textile industry
	REWOPOL NL 3-70	sodium lauryl ether sulfate	70% active	paste	
	SUPRALATE FAS	alkyl ether sulfate plus nonionic surfactant	50% active	pale yellow liquid; cloud point <0°C; viscosity 30 cPs; pH 7-8 (3% in H_2O)	foaming agent textile industry; paper industry

Supplier	Trade name	Chemical description	Composition	General properties	Functionality Application
Witco	SUPRALATE RA	sodium alkyl ether sulfate	34% active	light amber liquid; cloud point 20°C; viscosity 250 cPs; pH 7.5-8.5 (3% in H_2O)	foaming agent; detergent; wetting agent; *textile industry; detergents*
	WITCOLATE 1050	sodium pareth-25 sulfate	39% active	liquid	detergent; foaming agent; wetting agent; *personal care products; household and industrial applications*
	WITCOLATE 1276	ammonium alcohol ether sulfate	solids 60%	liquid	detergent; foaming agent; wetting agent; *personal care products; household and industrial applications; petroleum industry*
	WITCOLATE 7093	sodium alcohol ether sulfate	38.5% active	liquid	
	WITCOLATE 7103	ammonium alcohol ether sulfate	58% active	liquid	
	WITCOLATE AE-3	ammonium pareth-25 sulfate	59% active	liquid	
	WITCOLATE ES-1	sodium laureth sulfate	25% active	liquid	detergent; foaming agent; wetting agent; *personal care products; household and industrial applications*
	WITCOLATE ES-2	sodium laureth sulfate	26% active	liquid	
	WITCOLATE ES-270	sodium laureth sulfate	70% active	paste	
	WITCOLATE ES-3	sodium laureth sulfate	28% active	liquid	
	WITCOLATE ES-370	sodium laureth sulfate	70% active	paste	
	WITCOLATE L 33 70	alkylether sulfate, sodium salt	70% active	liquid	*liquid detergents*
	WITCOLATE LES-60A	ammonium laureth sulfate	59% active	liquid	detergent; foaming agent; wetting agent; *personal care products; household and industrial applications*
	WITCOLATE LES-60C	sodium laureth sulfate	58% active	liquid	
	WITCOLATE SE-5	sodium pareth-25 sulfate	59% active	liquid	
Zschimmer & Schwarz	CEFATEX SPECIAL	combination of fatty alcohol ether sulfates		light brown, viscous liquid	wetting agent; softener; anticrease agent *textile industry*
	ZETESOL 2056	monoisopropanolamine fatty alcohol ether sulfate ($C_{12/14}$)	59% active	liquid	*bath preparations; shampoos; cleansing preparations; liquid soaps*
	ZETESOL 856 T	monoisopropanolamine fatty alcohol ether sulfate ($C_{12/14}$) with fatty acid amido alkyl betaine	58% active	liquid	*bath preparations; shampoos; liquid soaps*
	ZETESOL 856	monoisopropanolamine fatty alcohol ether sulfate ($C_{12/14}$)	58% active	liquid	*bath preparations; shampoos; liquid soaps*
	ZETESOL AP	ammonium alkyl ether sulfate blended with 1,2-propylene glycol	60% active	liquid	*bath preparations; shampoos; dishwashing agents; liquid soaps; floor cleaners*
	ZETESOL MS	sodium/magnesium fatty alcohol ether sulfate	28% active	liquid	*baby shampoos; baby care products; liquid soaps*
	ZETESOL NL	sodium lauryl ether sulfate ($C_{12/14}$)	28% active	liquid	*bath preparations; shampoos; cleansing preparations; all-purpose cleaners; liquid soaps*

Lignin products

Supplier	Trade name	Chemical description	Composition	General properties	Functionality / Application
Borregaard	ADDITIVE-A TYPE 1	calcium/barium lignosulfonate-based clay conditioner	cation 5.2%; calcium 4.0%; bulk density 1260 kg/m³; reducing sugars 20.0%	light brown liquid; pH 4.0 (10% soln.)	plasticiser; lubricant; binder; anti-scumming agent / *brick and tile manufacture*
	ADDITIVE-A TYPE 2	calcium lignosulfonate-based clay conditioner	cation 4.0%; calcium 4.0%; bulk density 1250 kg/m³; reducing sugars 21.0%	dark brown liquid; pH 4.0 (10% soln.)	
	ADDITIVE-A TYPE 3	calcium/barium lignosulfonate-based clay conditioner	cation 6.0%; calcium 5.0%; bulk density 1300 kg/m³; reducing sugars 16.0%	brown liquid; pH 5.5 (10% soln.)	
	ADDITIVE-A TYPE 373	sodium lignosulfonate	cation 6.0%; calcium 0.5%; bulk density 600 kg/m³; reducing sugars 4.0%	light brown liquid; pH 7.5 (10% soln.)	deflocculant / *tile manufacture*
	AMERIBOND 2000	calcium lignosulfonate	cation 8.7%; bulk density 400 kg/m³	free-flowing brown powder; pH 8.0 (10% soln.)	pellet binder / *animal feeds*
	BORREBOND	calcium lignosulfonate	cation 5.8%; bulk density 500 kg/m³	brown powder; pH 4.3 (10% soln.)	binder; dispersant / *animal feeds*
	BORRECHEL	lignosulfonate containing chelated trace elements		brown powder	micronutrient
	BORRESPERSE CA	desugared calcium lignosulfonate	cation 5.2%; calcium 5.2%; bulk density 500 kg/m³; reducing sugars 3.0%	light brown powder; pH 4.5 (10% soln.)	*concrete*
	BORRESPERSE CAF	desugared calcium lignosulfonate	cation 5.2%; calcium 5.2%; bulk density 1250 kg/m³; reducing sugars 4.5%	brown liquid; pH 4.5 (10% soln.)	binder / *pesticides; ceramics*
	BORRESPERSE NA	sugar-free sodium lignosulfonate	cation 8.0%; calcium 0.4%; bulk density 500 kg/m³; reducing sugars 0.5%	brown powder; pH 8.0 (10% soln.)	dispersant; filler / *building industry*
	BORREWELL C	chrome lignosulfonate			conditioner / *oilwell drilling*
	BORREWELL FC	ferrochrome lignosulfonate	cation 10.8%; calcium 0.4%; bulk density 500 kg/m³	brown powder; pH 3.2 (10% soln.)	
	BORREWELL FE	iron lignosulfonate	cation 7.5%; calcium 0.4%; bulk density 500 kg/m³	brown powder; pH 3.2 (10% soln.)	
	COLLEX G	fermented calcium lignosulfonate	cation 4.8%; calcium 4.8%; bulk density 500 kg/m³; reducing sugars 9.0%	brown powder; pH 4.5 (10% soln.)	additive / *concrete; ceramics*
	CURBETON 0550	hardwood calcium/magnesium lignosulfonate (desugared)	cation 3.0%; calcium 3.0%; bulk density 1260 kg/m³; reducing sugars 4.1%	brown powder; pH 7.0 (10% soln.)	water reducer; strength increaser / *concrete*
	CURTEXIL 100S	hardwood calcium/magnesium lignosulfonate	cation 4.3%; calcium 3.0%; bulk density 400 kg/m³; reducing sugars 23.0%	brown powder; pH 3.5 (10% soln.)	binder; dispersant / *animal feeds*
	DIWATEX 30 FK	fractionated and sulfonated kraft lignin derivative	cation 9.5%; calcium 0.02%; bulk density 1150 kg/m³	brown liquid; pH 10.0 (10% soln.)	dispersant
	DIWATEX 30	sulfonated kraft lignin derivative	cation 8.0%; bulk density 500 kg/m³	brown powder; pH 10.0 (10% soln.)	dispersant / *dyes; pesticides*
	DIWATEX 40	sulfonated kraft lignin derivative			
	DURABOND	calcium/magnesium lignosulfonate	cation 5.0%; bulk density 500 kg/m³	brown powder; pH 4.0 (10% soln.)	pellet binder; contributes nutritive value / *animal feeds*
	DUSTEX	calcium lignosulfonate-based product	cation 4.0%; calcium 4.0%; bulk density 1250 kg/m³; reducing sugars 20.0%	brown liquid; pH 6.0 (10% soln.)	dust binder; road stabiliser / *road and soil improvement*
	GOULAC	calcium lignosulfonate	cation 7.0%; calcium 7.0%; bulk density 600 kg/m³; reducing sugars 14.0%	brown powder; pH 7.0 (10% soln.)	industrial binder / *refractory brick manufacture*

Supplier	Trade name	Chemical description	Composition	General properties	Functionality / Application
Borregaard	KELIG 32	sodium lignosulfonate with high content of polyhydroxycarboxylic acid salts	cation 8.3%; calcium 0.3%; bulk density 650 kg/m³; reducing sugars 0.6%	brown powder; pH 8.5 (10% soln.)	high temperature oil well cement retarder / oilwell drilling
	KELIG FS	lignosulfonate	cation 16.1%; calcium 0.2%; bulk density 700 kg/m³; reducing sugars 0.1%	brown powder; pH 8.8 (10% soln.)	carrier for micronutrient metals; chelating agent / road and soil improvement
	LIGNORIT	unfermented calcium lignosulfonate	cation 3.1%; calcium 3.1%; bulk density 1300 kg/m³; reducing sugars 22.0%	brown liquid; pH 3.5 (10% soln.)	extender for U/F resin; binder / particle board manufacture
	LIGNOSOL SD-60	kraft lignin	cation 10.0%; calcium 0.10%; bulk density 500 kg/m³	brown powder; pH 10.3 (10% soln.)	dispersant / dyes
	MARABOND 21	calcium lignosulfonate	cation 6.4%; calcium 6.4%; bulk density 650 kg/m³; reducing sugars 0.3%	brown powder; pH 8.6 (10% soln.)	low temperature oil well cement retarder / oilwell drilling
	MARACARB N-1	low molecular weight, highly sulfonated lignin	cation 11.5%; calcium 0.1%; bulk density 600 kg/m³; reducing sugars 5.0%	dark brown powder; pH 8.2 (10% soln.)	slime control agent; dispersant; humectant / paper industry; dyes
	MARACELL XC-2	high purity desulfonated oxylignin	cation 7.0%; calcium 0.3%; bulk density 550 kg/m³	brown powder; pH 9.3 (10% soln.)	expander; dispersant / lead acid batteries
	MARACELL XE	partially de-sulfonated sodium lignosulfonate	cation 2.0%; calcium 0.1%; bulk density 550 kg/m³	brown powder; pH 12.0 (10% soln.)	sludge conditioner; dispersant / boilers; industrial cleaners
	MARASPERSE 52CP	polymerised sodium lignosulfonate	cation 11.6%; calcium 0.3%; bulk density 650 kg/m³	brown powder; pH 9.9 (10% soln.)	dispersant / dyes; paper industry
	MARASPERSE CBA-1	sodium lignosulfonate	cation 7.0%; calcium 0.2%; bulk density 550 kg/m³	dark brown powder; pH 9.6 (10% soln.)	dispersant / ceramics
	MARASPERSE CBOS-6	highly purified oxylignin with a high degree of sulfonation	cation 10.8%; calcium 0.1%; bulk density 550 kg/m³; reducing sugars 0.2%	brown powder; pH 9.3 (10% soln.)	dispersant / oilwell drilling
	MARASPERSE GFC	sugar-free calcium lignosulfonate	cation 4.0%; calcium 4.0%; bulk density 1260 kg/m³; reducing sugars 1.3%	brown liquid; pH 7.5 (10% soln.)	dispersant / concrete
	MARASPERSE GNS	calcium-based sugar-free polymerised lignosulfonate	cation 4.4%; calcium 4.4%; bulk density 1200 kg/m³; reducing sugars 1.3%	brown liquid; pH 3.8 (10% soln.)	dispersant / building industry
	MARASPERSE N-22	purified, oxidised sodium lignosulfonate	cation 6.7%; calcium 0.6%; bulk density 700 kg/m³; reducing sugars 0.8%	brown powder; pH 8.0 (10% soln.)	dispersant / inorganics; pesticides; dyes
	MARASPERSE N-3	purified carboxylated sodium lignosulfonate	cation 13.1%; calcium 0.1%; bulk density 650 kg/m³	light brown powder; pH 9.0 (10% soln.)	dispersant / herbicides; pesticides
	NORLIG 11 DA	calcium lignosulfonate	cation 4.3%; calcium 4.3%; bulk density 550 kg/m³; reducing sugars 3.5%	light brown powder; pH 4.0 (10% soln.)	binder; dispersant / pesticides; herbicides
	NORLIG 24CL	low sugar calcium lignosulfonate	cation 8.5%; calcium 8.5%; bulk density 1260 kg/m³; reducing sugars 4.0%	brown liquid; pH 8.0 (10% soln.)	dispersant / concrete
	NORLIG 415	sodium-based sulfonated carboxylated lignosulfonate	cation 7.2%; calcium 0.8%; bulk density 650 kg/m³; reducing sugars 0.2%	brown powder; pH 7.2 (10% soln.)	dispersant / building industry
	NORLIG A	calcium lignosulfonate	cation 4.0%; calcium 4.0%; bulk density 600 kg/m³; reducing sugars 16.0%	light brown powder; pH 3.0 (10% soln.)	binder / agrochemicals; pesticides; herbicides; dust binder for unpaved roads and parking lots
	NORLIG BD	calcium lignosulfonate	cation 5.8%; calcium 5.8%; bulk density 600 kg/m³; reducing sugars 16.0%	light brown powder; pH 5.0 (10% soln.)	binder; tanning agent / leather industry
	NORLIG HP2	calcium lignosulfonate	cation 6.4%; calcium 3.9%; bulk density 1310 kg/m³; reducing sugars 9.5%	brown liquid; pH 7.0 (10% soln.)	hardener / urea; road and soil improvement
	TRAFFAID 30B	barium/calcium lignosulfonate	cation 14.0%; calcium 5.0%; bulk density 1230 kg/m³; reducing sugars 3.0%	light brown liquid; pH 4.3 (10% soln.)	anti-scumming agent / brick and tile manufacture

Supplier	Trade name	Chemical description	Composition	General properties	Functionality / Application
Borregaard	UFOXANE 2	purified, de-sulfonated and fractioned high molecular weight sodium lignosulfonate	cation 8.5%; calcium 0.02%; bulk density 650 kg/m³	brown powder; pH 10.0 (10% soln.)	dispersant / textile industry; dyes
	UFOXANE 3A	similar to Ufoxane 2 but with lower pH and higher degree of sulfonation	cation 8.0%; calcium 0.02%; bulk density 650 kg/m³	brown powder; pH 8.5 (10% soln.)	dispersant / pesticides
	ULTRAZINE CA	calcium lignosulfonate	cation 4.2%; calcium 4.2%; bulk density 500 kg/m³; reducing sugars 2.0%	light brown powder; pH 7.0 (10% soln.)	gypsum board manufacture
	ULTRAZINE NA	purified, high molecular weight sodium lignosulfonate	cation 5.5%; calcium 0.02%; bulk density 500 kg/m³; reducing sugars 1.0%	light brown powder; pH 8.5 (10% soln.)	dispersant / textile industry; dyes; pesticides; gypsum board manufacture; concrete
	ULTRAZINE NAS	low price variant of Ultrazine NA	cation 5.8%; calcium 0.4%; bulk density 500 kg/m³; reducing sugars 1.0%	light brown powder; pH 8.0 (10% soln.)	dispersant / concrete; wettable powder formulations
	VANISPERSE CB	fractionated sodium salt of oxylignin	cation 8.2%; calcium 0.004%; bulk density 650 kg/m³	dark brown powder; pH 8.5 (10% soln.)	dispersant / oilwell drilling
	WAFEX	unfermented calcium lignosulfonate	cation 4.0%; calcium 4.0%; bulk density 500 kg/m³; reducing sugars 22.0%	light yellow powder; pH 4.0 (10% soln.)	binder; dispersant; emulsifier / ceramics; building industry
	WAFOLIN K	unfermented calcium lignosulfonate	cation 4.0%; bulk density 500 kg/m³	brown powder; pH 5.0 (10% soln.)	binder / animal feeds
	WAFOLIN	calcium/magnesium lignosulfonate	cation 5.0%; bulk density 500 kg/m³	brown powder; pH 4.0 (10% soln.)	binder; contributes nutritive value / animal feeds
	WANIN AM	unfermented ammonium lignosulfonate	cation 3.5%; calcium 0.3%; bulk density 500 kg/m³; reducing sugars 22.0%	light brown powder; pH 4.0 (10% soln.)	binder; dispersant; tanning agent / leather industry
	WANIN S	unfermented, purified sodium lignosulfonate	cation 4.0%; calcium 0.3%; bulk density 500 kg/m³; reducing sugars 21.0%	blond powder; pH 3.5 (10% soln.)	tanning agent; dispersant / leather industry
	WARGONIN COMPACT	desugared sodium/calcium lignosulfonate	cation 12.0%; calcium 4.0%; bulk density 550 kg/m³; reducing sugars 0.3%	brown powder; pH 11.0 (10% soln.)	water reducer; strength increaser; air excluder / concrete
	WARGOTAN	unfermented calcium lignosulfonate	cation 4.0%; calcium 4.0%; bulk density 500 kg/m³; reducing sugars 22.0%	blond powder; pH 4.0 (10% soln.)	tanning agent; dispersant / leather industry
	WELLTEX 300 F	fermented calcium lignosulfonate	cation 5.0%; calcium 5.0%; bulk density 1230 kg/m³; reducing sugars 4.0%	brown liquid; pH 6.5 (10% soln.)	sizing agent; binder / paper industry
	ZEWA SL 2	virtually desugared sodium lignosulfonate	cation 12.0%; bulk density 600 kg/m³; reducing sugars 0.3%	brown powder or liquid; pH 11.0 (10% soln.)	water reducer; dispersant / concrete; pesticides
	ZEWAKOL CA 253N	calcium lignosulfonate	cation 4.0%; calcium 4.0%; bulk density 1260 kg/m³; reducing sugars 22.0%	brown liquid; pH 6.0 (10% soln.)	extender for U/F resins; binder / particle board manufacture
	ZEWAKOL SK70	lignosulfonate	cation 3.5%; calcium 0.3%; bulk density 1220 kg/m³; reducing sugars 3.0%	brown liquid; pH 3.5 (10% soln.)	extender for U/F resins / particle board manufacture
	ZEWALON FN	sodium lignosulfonate	cation 6.9%; calcium 0.1%; bulk density 500 kg/m³; reducing sugars 5.0%	brown powder; pH 6.5 (10% soln.)	tanning agent; dispersant / leather industry

Olefine sulfonates

Supplier	Trade name	Chemical description	Composition	General properties	Functionality Application
Akzo Nobel	ELFAN OS 46 A	sodium olefine ($C_{14/16}$) sulfonate (alkaline)	37% active	liquid	
	ELFAN OS 46	sodium olefine ($C_{14/16}$) sulfonate	37% active	liquid	
Albright & Wilson	NANSA LSS38/AS	sodium $C_{14/16}$ olefine sulfonate	38% active	liquid	dispersant; emulsifier; antistatic agent; wetting agent
	NANSA LSS480	sodium $C_{14/16}$ alpha olefine sulfonate	80% active	powder	
Hoechst	HOSTAPUR OS LIQUID	olefine sulfonate	40% active	liquid	detergents
	HOSTAPUR OSB	sodium olefine sulfonate			construction industry
Kao Corporation	ALFANOX 46	sodium alpha olefine sulfonate	36-38% active	liquid; M.W. 323; pH 6-8 (5%)	
Lonza	CARSONOL AOS	sodium $C_{14/16}$ olefine sulfonate	40% active	liquid; pH 8.0 (5% soln.)	foaming agent; wetting agent; emulsifier shampoos; liquid soaps; industrial cleaners
Nikko Chemicals	NIKKOL OS-14	sodium alpha-olefine sulfonate		white powder	foaming agent; cleansing agent cosmetics
Stepan Europe	POLYSTEP A 18	sodium alpha olefine sulfonate; made-to-order	39% active	pale yellow liquid	emulsifier emulsion polymerisation
Witco	SULFRAMIN AOS	$C_{14/16}$ alpha olefine sulfonate, sodium salt	37% active	liquid	liquid detergents; plastics industry; bubble baths; shampoo
	WITCONATE AOK	sodium $C_{14/16}$ olefine sulfonate	90% active; sulfate 6.5%	flake; pH 8.5	detergent; foaming agent; wetting agent personal care products; household and industrial applications
	WITCONATE AOS-PC	sodium $C_{14/16}$ olefine sulfonate	39% active; sulfate 0.7%	slurry; pH 7.7	detergent; foaming agent; viscosity modifier; wetting agent personal care products; household and industrial applications
	WITCONATE AOS	sodium $C_{14/16}$ olefine sulfonate	ca. 38% active; sulfate 0.7%	liquid/slurry; pour point 0°C; pH 7.5-7.7	detergent; foaming agent; wetting agent personal care products; household and industrial applications; petroleum industry

Paraffin sulfonates

Supplier	Trade name	Chemical description	Composition	General properties	Functionality Application
Hoechst	HOSTAPUR SAS-30	secondary alkane sulfonate	30% active	clear liquid	wetting agent; foaming agent
	HOSTAPUR SAS-60	secondary alkane sulfonate	60% active	soft, yellowish paste	detergents; toiletries; heavy-duty cleaners; textile industry; flotation
	HOSTAPUR SAS-93	secondary alkane sulfonate	93% active	flake	
Hüls	MARLON PF 40	blend of paraffin sulfonate with fatty alcohol ether sulfate	active detergent 40%	liquid	washing-up liquids; cleaners
	MARLON PS 30	$C_{13/17}$-alkane sulfonate, sodium salt	active detergent 30%	liquid	
	MARLON PS 60 W	$C_{13/17}$-alkane sulfonate, sodium salt	active detergent 60%	liquid/paste	liquid cleaners; washing-up liquids; liquid detergents; textile industry
	MARLON PS 60	$C_{13/17}$-alkane sulfonate, sodium salt	active detergent 60%	liquid/paste	
	MARLON PS 65	$C_{13/17}$-alkane sulfonate, sodium salt	active detergent 65%	liquid/paste	
Witco	WITCONATE NAS 8	sodium alcane sulfonate	39% active	liquid	detergent; wetting agent

Phosphate esters

Supplier	Trade name	Chemical description	Composition	General properties	Functionality *Application*
Akcros Chemicals	AGRILAN F513	phosphate ester-based additive	100% active	viscous pale amber liquid; pour point 9°C; viscosity 4300 cSt; pH 6.5 (1% aq.)	wetting agent; additive *agrochemicals*
	AGRILAN F535	phosphate ester-based additive	100% active	viscous pale amber liquid; pour point 14°C; viscosity 1600 cSt (60°C); pH 6.5 (1% aq.)	wetting agent; additive *agrochemicals*
	AGRILAN F546	free acid form of a complex phosphate ester	100% active	viscous pale amber liquid; pour point 10°C; viscosity 3200 cSt; pH 2.6 (1% aq.)	emulsifier; viscosity modifier; additive *agrochemicals*
	AGRILAN F557	phosphate ester-based additive	60% active	viscous pale amber liquid; pour point 0°C; viscosity 15,000 cSt; pH 6.5 (1% aq.)	wetting agent; additive *agrochemicals*
	AGRILAN TKA125	complex phosphate ester, free acid	70% active; H_2O content 15%	amber liquid; pour point 15°C; viscosity 475 cSt; pH 2.5 (1% aq.)	compatibility agent *agrochemicals*
	AGRILAN TKA147	complex phosphate ester, potassium salt	65% active; H_2O content 35%	pale straw liquid; pour point −18°C; viscosity 180 cSt; pH 7.2 (1% aq.)	compatibility agent *agrochemicals*
	LANKROSOL HS101	potassium salt of a complex phosphate ester	50% active	clear liquid; pour point < 0°C; viscosity 39 cSt; pH 7.0 (1% aq.)	hydrotrope; low foam; solubiliser *dishwashing agents; industrial and household detergent*
	PHOSPHOLAN ALF16	phospholan ester on inorganic carrier	50% active	fine white powder; pH 10.3 (1% aq.)	foam limiter
	PHOSPHOLAN ALF5	complex phosphate ester	100% active	off-white solid; pour point 65°C; viscosity 120 cSt (80°C); pH 2.5 (1% aq.)	foam limiter
	PHOSPHOLAN KPE4	complex phosphate ester, potassium salt	65% active	pale straw liquid; pour point −18°C; viscosity 180 cSt; pH 7.0 (1% aq.)	hydrotrope *dishwashing agents; industrial and household detergents; industrial and household cleaners*
	PHOSPHOLAN PA52	complex phosphate ester	100% active	pale straw liquid; pour point −18°C; viscosity cSt; pH 7.0 (1% aq.)	wetting agent; detergent; dispersant; emulsifier; lubricant *heavy-duty detergents; hard-surface cleaners; metal cleaning; metal working; hydraulic fluids; dry cleaning; textile industry; pigments; emulsion polymerisation; agriculture*
	PHOSPHOLAN PDB3	complex phosphate ester	100% active	pale straw liquid; pour point 10°C; viscosity 2400 cSt; pH 2.5 (1% aq.)	wetting agent; detergent; emulsifier; lubricant *industrial detergents; textile industry; dyes; dry cleaning; metal cleaning; emulsion polymerisation; agrochemicals; hand cleaners; metal working; surface coating stripping; iodophors*
	PHOSPHOLAN PE26K	complex phosphate ester	80% active; H_2O content 20%	pale straw liquid; pour point 12°C; pH 6.5 (1% aq.)	
	PHOSPHOLAN PE39	complex phosphate ester	100% active	pale straw liquid; pour point −15°C; pH 2.5 (1% aq.)	wetting agent; detergent; dispersant; emulsifier; lubricant *heavy-duty detergents; hard-surface cleaners; metal cleaning; metal working; hydraulic fluids; dry cleaning; textile industry; pigments; emulsion polymerisation; agriculture*
	PHOSPHOLAN PE65	complex phosphate ester	100% active	dark amber viscous liquid; pour point 5°C; pH 2.5 (1% aq.)	
	PHOSPHOLAN PE78	complex phosphate ester	100% active	pale straw liquid; pour point −10°C; pH 2.5 (1% aq.)	
	PHOSPHOLAN PHB14	complex phosphate ester, free acid	100% active	clear viscous yellow liquid; pour point 2°C; viscosity 4200 cSt (40°C); pH 2.0 (1% aq.)	hydrotrope
	PHOSPHOLAN PNP9	complex phosphate ester	100% active	dark amber viscous liquid; pour point 15°C; viscosity 11,500 cSt (40°C); pH 2.5 (1% aq.)	wetting agent; detergent; emulsifier; lubricant *industrial detergents; textile industry; dyes; dry cleaning; metal cleaning; emulsion polymerisation; agrochemicals; hand cleaners; metal working; surface coating stripping; iodophors*

Supplier	Trade name	Chemical description	Composition	General properties	Functionality / Application
Akcros Chemicals	PHOSPHOLAN PR13T	complex phosphate ester	100% active	pale straw liquid; pour point 9°C; pH 6.5 (1% aq.)	wetting agent; detergent; dispersant; emulsifier; lubricant *heavy-duty detergents; hard-surface cleaners; metal cleaning; metal working; hydraulic fluids; dry cleaning; textile industry; pigments; emulsion polymerisation; agriculture*
	PHOSPHOLAN PR91	complex phosphate ester	100% active	amber viscous liquid; pour point 10°C; pH 2.5 (1% aq.)	
	PHOSPHOLAN PRP5	neutral salt of a complex organic phosphate ester	100% active	yellow liquid; pour point 9°C; viscosity 4300 cSt; pH 5.5-7.5 (1% aq.)	wetting agent; dispersant; grinding aid; viscosity modifier *aq. pigment systems; optical brightening agents*
Akzo Nobel	BEROL 521	phosphate ester	ca. 40% active	surface tension 33.5 dynes/cm (0.1% soln.)	hydrotrope; anti-corrosion *soft metals; hard-surface cleaners*
	BEROL 522	phosphate ester	ca. 40% active	surface tension 31 dynes/cm (0.1% soln.)	hydrotrope; solubiliser *hard-surface cleaners; laundry products*
	BEROL 525	phosphate ester/nonionic blend	100% active	surface tension 33 dynes/cm (0.1% soln.)	hydrotrope *detergent slurries*
	BEROL 733	phosphate ester based on nonylphenol ethoxylate	ca. 35% active	surface tension 43 dynes/cm (0.1% soln.)	hydrotrope; solubiliser *hard-surface cleaners; detergent slurries*
	DAPRAL N 423	potassium lauryl phosphate	75% active	paste	
	ELFAN A 720	lauryl mystyl phosphate	100% active	solid	
	ELFAN A 913	lauryl mystyl ether (3) phosphate	100% active	liquid	
Albright & Wilson	DEHSCOFIX 900	alkylphenol ethoxy phosphate ester	concentration 80%	liquid	emulsifier; dispersant *agrochemicals*
	DEHSCOFIX 904	substituted phenol ethoxylate phosphate, triethanolamine salt	concentration 96%	liquid	
	DEHSCOFIX 905	substituted phenol ethoxylate phosphate, triethanolamine salt	concentration 96%	liquid	
	EMPIMIN TM	fatty alcohol phosphate ester	concentration 100%	solid	dispersant *coatings; polymer industry*
	EMPIPHOS 03D	fatty alcohol ethoxylate phosphate ester	concentration 100%	liquid	emulsifier; dispersant *agrochemicals; coatings; polymer industry*
	EMPIPHOS DF SERIES	alkyl phenol ethoxylate and fatty alcohol ethoxylate phosphate esters	concentration 100%	liquids to flakes	detergent; wetting agent; emulsifier; dispersant *textile industry*
Allied Colloids	ALCOPOL PPE	phosphate ester		liquid	
Auschem Cesalpinia	CHIMIN P1 A	alkylpolyglycolether phosphoric acid	97% active	viscous liquid	suspending agent; dispersant *hard-surface cleaners*
	CHIMIN P1	sodium alkylpolyglycolether phosphate	30% active	liquid	
	CHIMIN P13	alkylpolyglycolether phosphoric acid	100% active	viscous liquid	emulsifier *polar oils and solvents*
	CHIMIN P2	alkylpolyglycolether phosphoric acid	97% active	waxy	emulsifier *waxes; oils*
	CHIMIN P20	alkylpolyglycolether phosphoric acid	100% active	flakes	antifoaming agent *industrial liquid detergents*
	CHIMIN P40	sodium alkylpolyglycolether phosphate	75% active	liquid	low foam; wetting agent; hydrotrope *metal degreasing*
	CHIMIN P7	sodium alkylpolyglycolether phosphate	60% active	liquid	detergent; wetting agent; hydrotrope *hard-surface cleaners*
	EMULSON 3000	fatty alcohol polyethoxy phosphate	100% active	liquid	emulsifier *pesticides*
	EMULSON 7000	nonylphenol polyethoxy phosphate	100% active	liquid	
	EMULSON 7744	nonylphenol polyethoxy phosphate	100% active	liquid	

Supplier	Trade name	Chemical description	Composition	General properties	Functionality Application
Auschem Cesalpinia	EMULSON 7760 A	polyarylphenol polyethoxy phosphate	60% active	liquid	dispersant; *pesticides; fertilisers*
	EMULSON 7760	polyarylphenol polyethoxy phosphate	100% active	liquid	
	EMULSON FLS	polyarylphenol polyethoxy phosphate	100% active	paste	dispersant; *pesticides*
	EMULSON TNM	polyarylphenol polyethoxy phosphate	100% active	liquid	
	POLIROL 10 M	modified alkyl aryl ethoxy phosphate (acid)	98% active	liquid	emulsifier; *paint industry; polymerisation*
	POLIROL 10	alkyl aryl ethoxy phosphate (acid)	100% active	liquid	
BASF	LUTENSIT A-EP	phosphoric acid ester	100% active		
Dr. Th. Boehme	SYNTHESIN 4182	phosphoric acid ester			spin finish; *textile industry; fibre production*
	SYNTHESIN 4467	phosphoric acid ester			
Chemax	CHEMFAC NB-041	POE butyl ether phosphate	acid value 610 (to pH 9.5)	liquid	wetting agent; detergent; hydrotrope; emulsifier; anti-corrosion; *metal cleaning; industrial cleaners; textile industry; lubricants; dry cleaning; emulsion polymerisation; agrochemicals*
	CHEMFAC NB-042	POE butyl ether phosphate	acid value 515 (to pH 9.5)	liquid	
	CHEMFAC NC-004K	neutralised phosphate ester 50%		liquid	
	CHEMFAC NC-0910	POE alkyl phenol phosphate	acid value 145 (to pH 9.5)	liquid	
	CHEMFAC NF-200	glycol phosphate	acid value 500 (to pH 9.5)	liquid	
	CHEMFAC PA-080	alcohol phosphate	acid value 345 (to pH 9.5)	liquid	
	CHEMFAC PB-082	POE alkyl ether phosphate	acid value 195 (to pH 9.5)	liquid	
	CHEMFAC PB-106	POE alkyl ether phosphate	acid value 120 (to pH 9.5)	liquid	
	CHEMFAC PB-135	POE alkyl ether phosphate	acid value 115 (to pH 9.5)	liquid	
	CHEMFAC PB-184	POE alkyl ether phosphate	acid value 140 (to pH 9.5)	liquid	wetting agent; detergent; hydrotrope; emulsifier; anti-corrosion; *metal cleaning; industrial cleaners; textile industry; lubricants; dry cleaning; emulsion polymerisation; agrochemicals*
	CHEMFAC PB-264	POE alkyl ether phosphate	acid value 165 (to pH 9.5)	liquid	
	CHEMFAC PC-006	POE phenol phosphate	acid value 180 (to pH 9.5)	liquid	
	CHEMFAC PC-099E	POE alkyl phenol phosphate	acid value 115 (to pH 9.5)	liquid	
	CHEMFAC PC-188	POE alkyl phenol phosphate	acid value (to pH 9.5) 85	liquid	
	CHEMFAC PD-600	POE alkyl ether phosphate	acid value 210 (to pH 9.5)	liquid	
	CHEMFAC PF-623	POE alkyl ether phosphate	acid value 170 (to pH 9.5)	liquid	
	CHEMFAC PF-636	POE alkyl phosphate 90%	acid value 400 (to pH 9.5)	liquid	
	CHEMFAC PN-322	neutralised phosphate ester 50%		liquid	
Ciba	PRETOLON L	modified phosphoric acid ester		clear, pale brown low viscosity liquid	lubricant; *textile industry; synthetic fibres*
Croda	CRODAFOS 810A	phosphated $C_{8/10}$ alcohols		colourless to pale yellow liquid	emulsifier; antistatic agent; anti-corrosion; *detergents; cleaners; polishes*
	CRODAFOS 810D	phosphated $C_{8/10}$ alcohols neutralised		pale yellow to white soft gelatinous mass to hard solid	
	CRODAFOS CS10A	ceteareth-10 phosphate		white waxy solid	emulsifier; antistatic agent; anti-corrosion; *skin care products; antiperspirants; detergents; cleaners; polishes*
	CRODAFOS CS2A	ceteareth-2 phosphate		white to off-white waxy solid	
	CRODAFOS CS5A	ceteareth-5 phosphate		off-white waxy solid	
	CRODAFOS MCA	cetyl phosphate		white to pale yellow solid	emulsifier; stabiliser; *cosmetics; toiletries*

Supplier	Trade name	Chemical description	Composition	General properties	Functionality / Application
Croda	CRODAFOS N10A	oleth-10 phosphate		colourless to pale yellow viscous fluid to paste	emulsifier; antistatic agent; anti-corrosion / *skin care products; antiperspirants; detergents; cleaners; polishes*
	CRODAFOS N10N	phosphated distilled oleyl ether (POE 10) neutralised		buff soft solid	emulsifier; antistatic agent; anti-corrosion / *detergents; cleaners; polishes*
	CRODAFOS N3A	oleth-3 phosphate		clear yellow viscous liquid	emulsifier; conditioner; antistatic agent; anti-corrosion / *shampoos; detergents; skin care products; antiperspirants; cleaners; polishes*
	CRODAFOS N3N	phosphated distilled oleyl ether (POE 3) neutralised		amber gelatinous semi-solid	emulsifier; antistatic agent; anti-corrosion / *detergents; cleaners; polishes*
	CRODAFOS N5A	oleth-5 phosphate		colourless to pale yellow viscous fluid to paste	emulsifier; antistatic agent; anti-corrosion / *skin care products; antiperspirants; detergents; cleaners; polishes*
	CRODAFOS SG	PPG-5 ceteth-10 phosphate		clear yellow liquid	
Elf Atochem	BEYCOSTAT 148 K	phosphate ester, potassium salt		cream	
	BEYCOSTAT 231	phosphate ester, potassium salt		liquid	
	BEYCOSTAT 256 A	phosphate ester based on fatty alcohol C_8		liquid	
	BEYCOSTAT 273 A	phosphate ester based on fatty alcohol C_{13}		liquid	
	BEYCOSTAT 273 P	phosphate ester, potassium salt		fluid paste	
	BEYCOSTAT 319 A	phosphate ester based on fatty alcohol C_{13}		liquid	
	BEYCOSTAT 319 P	phosphate ester, potassium salt		liquid	
	BEYCOSTAT 656 A	phosphate ester based on alkylphenol		liquid	
	BEYCOSTAT 714 A	phosphate ester based on fatty alcohol C_{13}		liquid	
	BEYCOSTAT 714 P	phosphate ester, potassium salt		liquid	
	BEYCOSTAT B 706	phosphate ester based on fatty alcohol C_{13}		liquid	
	BEYCOSTAT DA	phosphate ester based on alkylphenol		liquid	
	BEYCOSTAT DP	phosphate ester, potassium salt		liquid	
	BEYCOSTAT LA	phosphate ester based on fatty alcohol C_{12}		solid	
	BEYCOSTAT LP 4 A	phosphate ester based on fatty alcohol C_{12}		liquid	
	BEYCOSTAT LP 9 A	phosphate ester based on fatty alcohol C_{12}		paste	
	BEYCOSTAT NA	phosphate ester based on alkylphenol		liquid	
	BEYCOSTAT NE	phosphate ester, triethanolamine salt		liquid	
	BEYCOSTAT QA	phosphate ester based on alkylphenol		liquid	
Ellis & Everard	CAFLON PE100	aromatic phosphate ester, free acid form	100% active		
	CAFLON PE288	aromatic phosphate ester, free acid form	100% active		
	CAFLON PE299	aromatic phosphate ester, free acid form	70% active		
	CAFLON PE65K	potassium salt of aromatic phosphate ester	65% active		
Henkel	ARBYL SFR	modified phosphoric acid ester			deaerating agent / *textile industry; dyeing*
	CRAFOL AP 261	sodium $C_{12/18}$ 10EO ether phosphate	28-32% active	liquid	wetting agent / *metal degreasers; cleaners*
	DEFINDOL EN	phosphoric acid ester with additives			deaerating agent / *textile industry; pretreatment*

Supplier	Trade name	Chemical description	Composition	General properties	Functionality Application
Henkel	DISPONIL AEP 5300	alkyl aryl ether phosphate, acid ester	ca. 100% active	liquid	emulsifier; dispersant plastics industry; coatings
	DISPONIL AEP 5302	alkyl aryl ether phosphate, acid ester	ca. 20% active	liquid	
	DISPONIL AEP 8100	alkyl aryl ether phosphate, acid ester	ca. 100% active	liquid	
	DISPONIL AEP 9525	alkyl aryl ether phosphate, sodium salt	ca. 24% active	liquid	o/w emulsifier
	FORLANIT E	hydroxycetyl phosphate	99-100% active	wax-like flakes	emulsifier; dispersant plastics industry; coatings
	FORLANIT P	fatty alcohol ether phosphate, sodium salt	ca. 30% active	liquid	
	KATAX AL	phosphoric acid ester			antistatic agent textile industry; spinning
	OSIMOL 728	modified phosphoric acid ester			crease inhibitor textile industry; dyeing
	OSIMOL LVP	modified phosphoric acid ester			
Hoechst	HOSTAPHAT KL340	trilaureth-4 phosphate	100% active		
	HOSTAPHAT KL340N	trilaureth-4 phosphate	100% active		
	HOSTAPHAT KW340	trilaureth-4 phosphate	100% active		emulsifier cosmetics
	HOSTAPHAT KW340N	trilaureth-4 phosphate	100% active		
Hüls	MARLOPHOR CS-ACID	isopropylphosphoric acid partial ester	active detergent 100%	liquid	industrial cleaners
	MARLOPHOR DS-ACID	n-butylphosphoric acid partial ester	active detergent 100%	liquid	industrial cleaners
	MARLOPHOR F 1-ACID	n-hexyl polyethyleneglycol ether phosphoric acid partial ester	active detergent 100%	liquid	low foam; wetting agent textile industry
	MARLOPHOR FC NA-SALT	2-ethylhexypolyethylene glycolether phosphoric acid partial ester, sodium salt	active detergent 85%	liquid	low foam; wetting agent cleaners; degreasers
	MARLOPHOR FC-ACID	2-ethylhexylpolyethylene glycolether phosphoric acid partial ester	active detergent 100%	liquid	low foam; wetting agent cleaners; degreasers
	MARLOPHOR HS-ACID	n-octyl-phosphoric acid partial ester	active detergent 100%	liquid	production of silicone oil emulsions
	MARLOPHOR IH-ACID	2-ethylhexylphosphoric acid partial ester	active detergent 100%	liquid	low foam; wetting agent textile industry
	MARLOPHOR MO 3-ACID	lauryl polyethyleneglycol ether phosphoric acid partial ester	active detergent 100%	liquid	textile industry; dry cleaning
	MARLOPHOR N5-ACID	isononyl polyethyleneglycol ether phosphoric acid partial ester	active detergent 100%	liquid	low foam; wetting agent textile industry
	MARLOPHOR ND-ACID	blend of iso-alkylphosphoric acid partial ester and alkylphenol polyethyleneglycol ether	active detergent 100%	liquid	wetting agent; antistatic agent textile industry; paper industry; dry cleaning
	MARLOPHOR ND-DEA-SALT	blend of iso-alkylphosphoric acid partial ester and alkylphenol polyethyleneglycol ether	active detergent 100%	liquid	wetting agent; antistatic agent textile industry; paper industry; dry cleaning
	MARLOPHOR NP 5-ACID	nonylphenol polyethyleneglycol ether phosphoric acid partial ester	active detergent 100%	liquid	low foam; wetting agent; emulsifier; antistatic agent textile industry

Supplier	Trade name	Chemical description	Composition	General properties	Functionality / Application
Hüls	MARLOPHOR NP 7-ACID	nonylphenol polyethyleneglycol ether phosphoric acid partial ester	active detergent 100%	liquid	low foam; wetting agent; emulsifier; antistatic agent / *textile industry*
	MARLOPHOR OC 5-ACID	olein polyethyleneglycol ether phosphoric acid partial ester	active detergent 100%	liquid	*textile industry; dry cleaning*
	MARLOPHOR ON 3-ACID	isotridecyl polyethyleneglycol ether phosphoric acid partial ester	active detergent 100%	liquid	
	MARLOPHOR ON 5-ACID	isotridecyl polyethyleneglycol ether phosphoric acid partial ester	active detergent 100%	liquid	low foam; wetting agent; emulsifier; antistatic agent / *textile industry*
	MARLOPHOR ON 7-ACID	isotridecyl polyethyleneglycol ether phosphoric acid partial ester	active detergent 100%	liquid	
	MARLOPHOR T6-ACID	tallow polyethyleneglycol ether phosphoric acid partial ester	active detergent 100%	liquid	*textile industry; dry cleaning*
	MARLOPHOR T10-ACID	tallow polyethyleneglycol ether phosphoric acid partial ester	active detergent 100%	paste	
	MARLOPHOR UW 12-ACID	alkyl polyethyleneglycol ether phosphoric acid partial ester	active detergent 100%	paste	
	MARLOWET 5301	coconut polyethylene glycol ether phosphoric acid partial ester	active detergent 100%	liquid	emulsifier / *mineral oils; metal working*
	MARLOWET 5311	isononanol phosphoric acid partial ester	active detergent 100%	liquid	wetting agent / *chlorohydrocarbons; textile industry; paint strippers*
	MARLOWET 5320	nonylphenol polyethylene glycol ether phosphoric acid partial ester	active detergent 100%	liquid	emulsifier / *mineral oils; chloroparaffins; metal working*
	MARLOWET 5324	phenol polyethylene glycol ether phosphoric acid partial ester	active detergent 100%	liquid	*lubricants; metal working*
	MARLOWET 5361	alkyl polyethylene glycol ether phosphoric acid partial ester	active detergent 100%	liquid	*lubricants; metal working*
ICI	ATLAS G-2203	potassium salt of phosphoric acid ester		pale yellow liquid	*household and industrial applications; agrochemicals; textile industry*
	ATLAS G-2207	potassium salt of phosphoric acid ester		pale yellow liquid	
Kao Corporation	FOSFODET 1214 N/16	ethoxylated fatty alcohol, phosphate ester	100% active; P_2O_5 content 6.2-7.2%	liquid	
	FOSFODET 20 D	fatty alcohol, phosphate ester	59-61% active; acid value 31-34	paste	
	FOSFODET 9Q/22	ethoxylated alkyl-aryl, phosphate ester	100% active; acid value 58-65, 95-107; P_2O_5 content 3-3.5%	liquid	
	FOSFODET T-17	ethoxylated fatty alcohol, phosphate ester	100% active; acid value 205-215; P_2O_5 content 13.7-14.7%	paste	
Lakeland Laboratories	PA100	phosphated alcohol	100% active; acid value 950	clear amber liquid; pH 2-3	low foam; solubiliser; hydrotrope / *detergents; metal cleaning; water treatment*
	PA800	phosphated alcohol ethoxylate	100% active; acid value 360	clear amber liquid; pH 2-3	low foam; solubiliser; hydrotrope; emulsifier / *textile industry; agriculture; oil industry; water treatment*
	PAE 106	phosphated alcohol ethoxylate	100% active; acid value 175; 'n' no. 6	clear amber liquid; pH 2-3	wetting agent; solubiliser; hydrotrope; emulsifier / *detergents; textile industry; agriculture*
	PAE 126	phosphated alcohol ethoxylate	100% active; acid value 190; 'n' no. 6	clear amber liquid; pH 2-3	
	PAE 136	phosphated alcohol ethoxylate	100% active; acid value 215; 'n' no. 6	clear amber liquid; pH 2-3	wetting agent; solubiliser; hydrotrope; anti-corrosion; emulsifier / *detergents; textile industry; agriculture; oil industry; water treatment; lubricants*

Supplier	Trade name	Chemical description	Composition	General properties	Functionality Application
Lakeland Laboratories	PAE 147	phosphated alcohol ethoxylate	100% active; acid value 220; 'n' no. 7	clear amber liquid; pH 2-3	wetting agent; solubiliser; hydrotrope; emulsifier; *metal working; metal finishing; textile industry; agriculture*
	PAE 1780	phosphated alcohol ethoxylate	100% active; acid value 44; 'n' no. 80	off-white waxy solid; pH 2-3 (soln.)	wetting agent; solubiliser; hydrotrope; emulsifier; *emulsion polymerisation*
	PPE 156	phosphated phenol ethoxylate	100% active; acid value 107; 'n' no. 6	clear amber viscous liquid; pH 2-3	wetting agent; solubiliser; hydrotrope; emulsifier; *dry cleaning; detergents; textile industry*
	PPE 159	phosphated phenol ethoxylate	100% active; acid value 100; 'n' no. 9	clear amber viscous liquid; pH 2-3	wetting agent; solubiliser; hydrotrope; emulsifier; *detergents; textile industry; lubricants; emulsion polymerisation*
	PPE 604	phosphated phenol ethoxylate	100% active; acid value 320; 'n' no. 4	clear amber viscous liquid; pH 2-3	wetting agent; solubiliser; hydrotrope; low foam; emulsifier; *metal finishing; detergents; textile industry; agriculture*
Libra Chemicals	LIBRAPHOS AD	sodium salt of Libraphos EP 120	80% active	clear yellow liquid	emulsifier; *lubricants*
	LIBRAPHOS EP 120	ethoxy alkyl aryl phosphate ester, acid form	100% active	clear yellow liquid	emulsifier; *lubricants*
	LIBRAPHOS EP	alkyl phosphate ester, acid form	100% active	clear yellow liquid	wetting agent; *acid cleaners*
	LIBRAPHOS ER 150	ethoxy alkyl aryl phosphate ester, acid form	100% active	clear yellow liquid	emulsifier; *lubricants*
	LIBRAPHOS HC 2A	ethoxy alkyl aryl phosphate ester, acid form	100% active	clear yellow liquid	emulsifier; wetting agent; *textile industry*
	LIBRAPHOS HC 2K	potassium salt of Libraphos HC 2A	80% active	clear yellow liquid	
	LIBRAPHOS HC 3A	alkyl phosphate ester, acid form	100% active	clear yellow liquid	
	LIBRAPHOS L66	potassium salt of Libraphos P4 Ester	65% active	clear yellow liquid	hydrotrope; solubiliser; *alkaline cleaners*
	LIBRAPHOS MR 100	phosphate ester, acid form	100% active	clear yellow liquid	emulsifier; *lubricants*
	LIBRAPHOS P4 ESTER	phosphate ester, acid form	100% active	clear yellow viscous liquid	hydrotrope; solubiliser; *acid and alkaline cleaners*
	LIBRAPHOS TL 100 P4	ethoxy alkyl phosphate ester, acid form	100% active	clear yellow liquid	emulsifier; *lubricants*
Lonza	ALKAWET N	mixed phosphate ester	100% active	pH 9.2 (0.1% soln.)	*industrial cleaners*
Nikko Chemicals	NIKKOL DDP-2	di POE (2) alkyl ether phosphate; free acid form		yellow liquid	
	NIKKOL DDP-4	di POE (4) alkyl ether phosphate; free acid form		yellow liquid	
	NIKKOL DDP-6	di POE (6) alkyl ether phosphate; free acid form		yellow liquid	*cosmetics*
	NIKKOL DDP-8	di POE (8) alkyl ether phosphate; free acid form		yellow liquid	
	NIKKOL DDP-10	di POE (10) alkyl ether phosphate; free acid form		yellow paste	
	NIKKOL DLP-10	dilaureth-10 phosphate		pale yellow paste	solubiliser; *cosmetics*
	NIKKOL DNPP-4	dinonoxynol-4 phosphate		yellow liquid	*cosmetics*
	NIKKOL DOP-8N	dioleth-8 phosphate		pale yellow liquid	solubiliser; *cosmetics*
	NIKKOL PHOSTEN HLP-N	sodium lauryl phosphate		white powder	cleansing agent; *cosmetics*

Supplier	Trade name	Chemical description	Composition	General properties	Functionality / *Application*
Nikko Chemicals	NIKKOL PHOSTEN HLP-1	POE (1) lauryl ether phosphate		pale yellow paste	cleansing agent / *cosmetics*
	NIKKOL PHOSTEN HLP	lauryl phosphate; free acid form		white solid	*cosmetics*
	NIKKOL SLP-N	sodium lauryl phosphate		white crystalline powder	foaming agent; cleansing agent / *cosmetics*
	NIKKOL TCP-5	triceteth-5 phosphate		white paste	emulsifier; dispersant / *cosmetics*
	NIKKOL TDP-2	tri POE (2) alkyl ether phosphate; free acid form		yellow liquid	
	NIKKOL TDP-6	tri POE (6) alkyl ether phosphate; free acid form		yellow liquid	*cosmetics*
	NIKKOL TDP-8	tri POE (8) alkyl ether phosphate; free acid form		yellow liquid	
	NIKKOL TDP-10	tri POE (10) alkyl ether phosphate; free acid form		pale yellow paste	
	NIKKOL TLP-4	trilaureth-4 phosphate		pale yellow liquid	emulsifier; dispersant / *cosmetics*
	NIKKOL TOP-0	sodium oleyl phosphate		pale yellow liquid	*cosmetics*
PPG	LAROSTAT 265-199	free acid phosphate ester	100% active	liquid	antistatic agent / *textile industry; fibre production*
	LAROSTAT 300 I-325	phosphate ester			antistatic agent; anti-corrosion
	LAROSTAT 300-1	neutralised phosphate ester	50% active	liquid	antistatic agent / *textile industry; fibre production*
	LAROSTAT 60A	phosphate ester			antistatic agent / *rubber industry; textile industry; carpet industry*
	MAPHOS 1130	complex phosphorylated nonionic surfactant	100% active		low foam / *metal cleaning; textile industry; detergents; dry cleaning*
	MAPHOS 14	complex phosphorylated nonionic surfactant	acid value 166 (to pH 5.2), 333 (to pH 9.5); 9.1% phosphorus	clear yellow liquid	coupling agent; antistatic agent; low foam / *liquid detergents; metal cleaning; metal working*
	MAPHOS 14A	complex phosphorylated nonionic surfactant	7.5% phosphorus	clear yellow liquid	coupling agent; anti-corrosion; antistatic agent; low foam / *liquid detergents; metal cleaning; metal working*
	MAPHOS 15	complex phosphorylated nonionic surfactant	acid value 55 (to pH 5.2), 98 (to pH 9.5); 3.0% phosphorus	clear pale yellow liquid	wetting agent; emulsifier; foaming agent (moderate) / *emulsion polymerisation; printing*
	MAPHOS 155	complex phosphorylated nonionic surfactant	80% active		metal cleaning; liquid detergents
	MAPHOS 17	complex phosphorylated nonionic surfactant	acid value 70 (to pH 5.2), 138 (to pH 9.5); 3.9% phosphorus	clear yellow liquid	emulsifier; foaming agent (high) / *emulsion polymerisation; printing*
	MAPHOS 2099FA	complex phosphorylated nonionic surfactant	100% active		detergent; emulsifier; anti-corrosion; antistatic agent; low foam / *cosmetics*
	MAPHOS 241-56	complex phosphorylated nonionic surfactant	acid value 55 (to pH 5.2), 90 (to pH 9.5); 2.1% phosphorus	clear yellow liquid	
	MAPHOS 30	complex phosphorylated nonionic surfactant	acid value 190 (to pH 5.2), 370 (to pH 9.5); 11.1% phosphorus		hydrotrope / *metal working; industrial cleaners; liquid detergents*
	MAPHOS 41A	complex phosphorylated nonionic surfactant	acid value 73 (to pH 5.2), 155 (to pH 9.5); 4.2% phosphorus	clear, pale yellow solid	foaming agent (high) / *emulsion polymerisation; printing; textile dyeing; metal working; industrial cleaners; liquid detergents*

Supplier	Trade name	Chemical description	Composition	General properties	Functionality / Application
PPG	MAPHOS 54	complex phosphorylated nonionic surfactant	acid value 70 (to pH 5.2), 115 (to pH 9.5); 4.1% phosphorus	clear yellow liquid	detergent; wetting agent; emulsifier; antistatic agent; foaming agent (high) / *metal cleaning; liquid detergents*
	MAPHOS 55	complex phosphorylated nonionic surfactant	acid value 110 (to pH 5.2), 225 (to pH 9.5); 8.2% phosphorus	clear to hazy amber liquid	wetting agent; coupling agent; anti-corrosion; antistatic agent; detergent; foaming agent (moderate); hydrotrope; emulsifier / *metal working; metal cleaning; liquid detergents; industrial cleaners*
	MAPHOS 56	complex phosphorylated nonionic surfactant	acid value 130 (to pH 5.2), 255 (to pH 9.5); 7.5% phosphorus	clear amber liquid	wetting agent; coupling agent; antistatic agent; foaming agent (moderate); dispersant; hydrotrope / *pigment dispersion; metal working; metal cleaning; chemical processing; industrial cleaners; liquid detergents*
	MAPHOS 58	complex phosphorylated nonionic surfactant	acid value 130 (to pH 5.2), 255 (to pH 9.5); 7.4% phosphorus		hydrotrope; dispersant; lubricant / *metal working; industrial cleaners; liquid detergents*
	MAPHOS 60A	complex phosphorylated nonionic surfactant	acid value 175 (to pH 5.2), 325 (to pH 9.5); 10.1% phosphorus		hydrotrope; wetting agent / *metal working; industrial cleaners; liquid detergents*
	MAPHOS 6591	complex phosphorylated nonionic surfactant	acid value 75 (to pH 5.2), 150 (to pH 9.5); 4.5% phosphorus	clear yellow liquid	detergent; wetting agent; emulsifier; coupling agent; antistatic agent; foaming agent (high) / *dry cleaning; textile industry; metal working; liquid detergents*
	MAPHOS 66	complex phosphorylated nonionic surfactant	4.7% phosphorus	clear yellow liquid	wetting agent; coupling agent; foaming agent (moderate) / *liquid detergents; metal cleaning; metal working*
	MAPHOS 66H	complex phosphorylated nonionic surfactant	4.5% phosphorus		hydrotrope / *metal working; industrial cleaners; liquid detergents*
	MAPHOS 76 NA	complex phosphorylated nonionic surfactant	2.6% phosphorus	clear, amber liquid	detergent; emulsifier; foaming agent (high) / *dry cleaning; metal working; liquid detergents; emulsion polymerisation*
	MAPHOS 8135	complex phosphorylated nonionic surfactant	acid value 95 (to pH 5.2), 145 (to pH 9.5); 5.5% phosphorus		hydrotrope; lubricant / *metal working; industrial cleaner; liquid detergents*
	MAPHOS 91	complex phosphorylated nonionic surfactant	acid value 65 (to pH 5.2), 120 (to pH 9.5); 3.1% phosphorus		emulsifier; detergent; wetting agent; foaming agent (high) / *metal working; industrial cleaners; liquid detergents*
	MAPHOS 96	complex phosphorylated nonionic surfactant	acid value 125 (to pH 5.2), 255 (to pH 9.5); 6.8% phosphorus		emulsifier; detergent; foaming agent (moderate) / *metal working; industrial cleaners; liquid detergents*
	MAPHOS E2000	complex phosphorylated nonionic surfactant	100% active		*textile industry*
	MAPHOS E2044	complex phosphorylated nonionic surfactant	100% active		*textile industry*
	MAPHOS E2099K	complex phosphorylated nonionic surfactant	60% active		
	MAPHOS FDEO	complex phosphorylated nonionic surfactant	acid value 105 (to pH 5.2), 187 (to pH 9.5); 6.0% phosphorus	clear pale amber liquid	coupling agent; anti-corrosion; low foam; detergent; wetting agent / *metal working; metal cleaning*
	MAPHOS IL4	complex phosphorylated nonionic surfactant	100% active		lubricant; foaming agent (high) / *textile industry; emulsion polymerisation*
	MAPHOS JA-60	complex phosphorylated nonionic surfactant	acid value 110 (to pH 5.2), 220 (to pH 9.5); 6.3% phosphorus		hydrotrope; detergent; wetting agent / *metal working; industrial cleaners; liquid detergents*
	MAPHOS JH1500	complex phosphorylated nonionic surfactant	100% active		wetting agent; emulsifier / *textile industry*
	MAPHOS JH1800	complex phosphorylated nonionic surfactant	100% active		wetting agent; emulsifier / *textile industry*

Supplier	Trade name	Chemical description	Composition	General properties	Functionality / Application
PPG	MAPHOS JM-51	complex phosphorylated nonionic surfactant	acid value 50 (to pH 5.2), 85 (to pH 9.5); 2.6% phosphorus	clear amber liquid	detergent; emulsifier; anti-corrosion; foaming agent (moderate) *metal working; metal cleaning; dry cleaning*
	MAPHOS JM-71	complex phosphorylated nonionic surfactant	acid value 37 (to pH 5.2), 70 (to pH 9.5); 2.0% phosphorus	clear amber liquid	detergent; emulsifier; anti-corrosion; foaming agent (high) *metal working; metal cleaning; dry cleaning*
	MAPHOS JP-70	complex phosphorylated nonionic surfactant	acid value 97 (to pH 5.2), 165 (to pH 9.5); 5.1% phosphorus		hydrotrope; lubricant *metal working; industrial cleaners; liquid detergents*
	MAPHOS L-4	complex phosphorylated nonionic surfactant	acid value 77 (to pH 5.2), 130 (to pH 9.5); 3.3% phosphorus		emulsifier; detergent *metal working; industrial cleaners; liquid detergents*
	MAPHOS L-6	complex phosphorylated nonionic surfactant	acid value 70 (to pH 5.2), 105 (to pH 9.5); 3.7% phosphorus	clear amber liquid	detergent; emulsifier; anti-corrosion; foaming agent (moderate) *emulsion polymerisation; metal working*
	PRM-98	complex phosphorylated nonionic surfactant	3.7% phosphorus	clear yellow liquid	detergent; wetting agent; emulsifier; foaming agent (high) *metal cleaning; liquid detergents; textile wetting*
Protex	PHOSFAC SERIES	phosphate esters in acid and neutralized forms			wetting agent; detergent; emulsifier; hydrotrope
	PROTE-PON P 2 EHA-02-Z	alkyl ether phosphoric acid		liquid	
	PROTE-PON P-0101-02-Z	alkyl ether phosphoric acid		liquid	
	PROTE-PON P-2 EHA-02-K30	alkyl ether potassium phosphate		liquid; pH 7.2 ± 0.3	
	PROTE-PON P-2 EHA-Z	alkyl ether phosphoric acid		liquid	
	PROTE-PON P-L 201-02-K30	alkyl ether potassium phosphate		emulsion; pH 7.0 ± 0.5	
	PROTE-PON P-L 201-02-Z	alkyl ether phosphoric acid		wax	
	PROTE-PON P-NP-06-K30	alkyl ether potassium phosphate		liquid; pH 7.2 ± 0.3	
	PROTE-PON P-NP-06-Z	alkyl ether phosphoric acid		liquid	
	PROTE-PON P-NP-10-K30	alkyl ether potassium phosphate		liquid; pH 7.2 ± 0.3	
	PROTE-PON P-NP-10-MZ	alkyl ether phosphoric acid		liquid	
	PROTE-PON P-NP-10-Z	alkyl ether phosphoric acid		liquid	
	PROTE-PON P-OX 101-02-K75	alkyl ether potassium phosphate		liquid; pH 6.5 ± 0.5	wetting agent; detergent; hydrotrope; emulsifier; anti-corrosion *detergents metal working; hard-surface cleaners; dry cleaning; textile industry; emulsion polymerisation; pesticides; herbicides*
	PROTE-PON P-TD-06-K60	alkyl ether potassium phosphate		gel; pH 7.5 ± 0.5	
	PROTE-PON P-TD-06-K13	alkyl ether potassium phosphate		liquid; pH 7.5 ± 0.5	
	PROTE-PON P-TD-06-K30	alkyl ether potassium phosphate		liquid; pH 7.2 ± 0.3	

Supplier	Trade name	Chemical description	Composition	General properties	Functionality / Application
Protex	PROTE-PON P-TD-06-Z	alkyl ether phosphoric acid		liquid	
	PROTE-PON P-TD-09-Z	alkyl ether phosphoric acid		paste	wetting agent; detergent; hydrotrope; emulsifier; anti-corrosion
	PROTE-PON P-TD-12-Z	alkyl ether phosphoric acid		paste	*detergents metal working; hard-surface cleaners; drycleaning; textile industry; emulsion polymerisation; pesticides; herbicides*
	PROTE-PON TD-09-K30	alkyl ether potassium phosphate		liquid; pH 7.0 ± 0.5	
Rhone-Poulenc	LUBRHOPHOS LP 700	phosphate ester with aromatic base		liquid	low foam
	RHODAFAC BG 510	phosphate ester with aliphatic base		liquid	foaming agent (high); emulsifier; wetting agent; dispersant
	RHODAFAC PA 15	phosphate ester with aliphatic base		liquid	low foam; emulsifier; wetting agent
	RHODAFAC PA/17	alkylphenoxypolyethyl ester of phosphoric acid	100% solids; 'n' no. 6	liquid; surface tension 32 dynes/cm	emulsifier; rust inhibitor; *emulsion polymerisation*
	RHODAFAC PA/19	alkylphenoxypolyethyl ester of phosphoric acid	100% solids; 'n' no. 10	liquid; surface tension 34 dynes/cm	
	RHODAFAC PE/610	alkylphenoxypolyethyl ester of phosphoric acid	100% solids; 'n' no. 9	liquid; surface tension 34 dynes/cm	emulsifier; rust inhibitor; wetting agent; dispersant; foaming agent (moderate); *emulsion polymerisation*
	RHODAFAC PE/960	alkylphenoxypolyethyl ester of phosphoric acid	90% solids; 'n' no. 50	paste; surface tension 43 dynes/cm	emulsifier; rust inhibitor; *emulsion polymerisation*
	RHODAFAC RA 600	phosphate ester with aliphatic base		liquid	foaming agent (high); emulsifier; wetting agent
Sandoz	DERMALIX C	alkyl phosphate in aq./organic medium		liquid	fat liquoring agent
	ELFUGIN AKT 300%	phosphate ester in aq. soln.		liquid	antistatic agent; anti-corrosion
	SANDOCLEAN SKBB	alkyl phosphate esters		liquid	dispersant; emulsifier; wetting agent; detergent
	SANDOCORIN 8160	phosphate esters		liquid	
	SANDOCORIN LF 10	phosphate esters in aq. soln.		liquid	
	SANDOCORIN LF 20	phosphate esters and triazoles in aq. soln.		liquid	anti-corrosion; low foam
	SANDOCORIN LF 30	phosphate esters, triazoles and sequestering agents		liquid	
	SANDOPAN BFN	mixture of phosphate esters		liquid	wetting agent; degreaser; low foam
	SANDOPAN SF	phosphate ester		liquid	wetting agent; dispersant; detergent
	SANDOZIN EH	phosphate ester in aq.-organic soln.		liquid	wetting agent; low foam; *textile industry; dyeing*
Servo Delden	SERVOXYL VPAZ 100	lauryl phosphate	acid no. 225-245	solid	
	SERVOXYL VPAZ 3/100	lauryl polyglycol ether (3EO) phosphate	acid no. 150-170	liquid	
	SERVOXYL VPBZ 5/100	C_{12}-C_{18} fatty alcohol (5EO) phosphate	acid no. 110-120	liquid	
	SERVOXYL VPDZ 100	tridecyl phosphate	acid no. 215-240	liquid	

Supplier	Trade name	Chemical description	Composition	General properties	Functionality Application
Servo Delden	SERVOXYL VPDZ 3/100	tridecyl polyglycol ether (3EO) phosphate	acid no. 140-155	liquid	
	SERVOXYL VPDZ 6/100	tridecyl polyglycol ether (6EO) phosphate	acid no. 110-125	liquid	
	SERVOXYL VPDZ 20/100	tridecyl polyglycol ether (20EO) phosphate	acid no. 48-58	solid	
	SERVOXYL VPEZ 100	iso-butyl phosphate	acid no. 440-510	liquid	
	SERVOXYL VPFZ 5/100	oleyl polyglycol ether (5EO) phosphate	acid no. 105-115	liquid	
	SERVOXYL VPFZ 7/100	oleyl polyglycol ether (7EO) phosphate	acid no. 83-100	liquid	
	SERVOXYL VPGZ 6/100	phenol polyglycol ether (6EO) phosphate	acid no. 140-180	liquid	
	SERVOXYL VPHZ 100	n-octyl phosphate	acid no. 300-320	liquid	
	SERVOXYL VPI 55	n-butyl phosphate, Na-salt	active matter 55%	liquid	
	SERVOXYL VPIZ 100	n-butyl phosphate	acid no. 450-480	liquid	
	SERVOXYL VPIZ 4/100	n-butyl polyglycol ether (4EO) phosphate	acid no. 195-260	liquid	
	SERVOXYL VPKZ 100	oleyl phosphate	acid no. 173-193	paste	
	SERVOXYL VPNZ 5/100	nonylphenol polyglycol ether (5EO) phosphate	acid no. 125-135	liquid	
	SERVOXYL VPNZ 6/100	nonylphenol polyglycol ether (6EO) phosphate	acid no. 83-91	liquid	
	SERVOXYL VPNZ 7/100	nonylphenol polyglycol ether (7EO) phosphate	acid no. 105-115	liquid	
	SERVOXYL VPNZ 9/100	nonylphenol polyglycol ether (9EO) phosphate	acid no. 86-92	liquid	
	SERVOXYL VPNZ 14/100	nonylphenol polyglycol ether (14EO) phosphate	acid no. 67-72	paste	
	SERVOXYL VPNZ 20/100	nonylphenol polyglycol ether (20EO) phosphate	acid no. 50-60	solid	
	SERVOXYL VPPZ 100	isopropyl phosphate	acid no. 470-520	liquid	
	SERVOXYL VPRZ 100	cetyl/stearyl phosphate	acid no. 175-195	solid	
	SERVOXYL VPRZ 6/100	cetyl/stearyl polyglycol ether (6EO) phosphate	acid no. 93-105	solid	
	SERVOXYL VPRZ 11/100	cetyl/stearyl polyglycol ether (11EO) phosphate	acid no. 66-76	solid	

Supplier	Trade name	Chemical description	Composition	General properties	Functionality / Application
Servo Delden	SERVOXYL VPT 3/85	2-ethylhexyl polyglycol ether (3EO) phosphate, Na-salt	active matter 85%	liquid	
	SERVOXYL VPTZ 100	2-ethylhexyl phosphate	acid no. 310-330	liquid	
	SERVOXYL VPTZ 3/100	2-ethylhexyl polyglycol ether (3EO) phosphate	acid no. 170-200	liquid	
	SERVOXYL VPUZ 100	methyl phosphate	acid no. 670-700	liquid	
	SERVOXYL VPXZ 100	isononyl phosphate	acid no. 280-310	liquid	
	SERVOXYL VPXZ 5/100	isononyl polyglycol ether (5EO) phosphate	acid no. 135-150	liquid	
Stepan Europe	POLYSTEP PN 209	alkylphenol ethoxylate phosphoric ester acid; made-to-order	99% active	water white to yellow liquid	emulsifier; dispersant; anti-corrosion *emulsion polymerisation*
	STEPFAC PN SERIES	phosphoric ester; made-to-order		water white to yellow viscous liquid	emulsifier; wetting agent; dispersant *agrochemicals*
Union Carbide	TRITON H-55	phosphate ester	50% active	liquid; pour point −23°C	hydrotrope
	TRITON H-66	phosphate ester	50% active	liquid; pour point −20°C	hydrotrope *rinsing aids*
	TRITON QS-44	phosphate ester	80% active	liquid; pour point 1°C	hydrotrope *metal cleaning; emulsion polymerisation*
Warwick International	MYKON 8402	phosphated nonyl phenol ethoxylate			wetting agent; degreaser *textile industry*
Witco	EMPHOS CS-1361	aromatic hydrophobic base	acid no. 28 (to pH 9.5)	liquid; colour 1 (Gardner); pH 5 (10% aq.)	antistatic agent; coupling agent; detergent; dispersant; o/w emulsifier; foaming agent; viscosity modifier; wetting agent *personal care products; household and industrial applications*
	EMPHOS CS-141	aromatic hydrophobic base	30% active; acid no. 65 (to pH 5.5), 110 (to pH 9.5)	liquid; colour 1 (Gardner); pH 2 (3% aq.)	coupling agent; detergent; o/w emulsifier; foaming agent; wetting agent; antistatic agent *personal care products; household and industrial applications; textile industry*
	EMPHOS CS-147	aromatic hydrophobic base	acid no. 87 (to pH 5.5), 150 (to pH 9.5)	liquid; colour 2 (Gardner); pH 2 (3% aq.)	coupling agent; detergent; o/w emulsifier; foaming agent; wetting agent *personal care products; household and industrial applications*
	EMPHOS PS-21A	aliphatic hydrophobic base	100% active; acid no. 130 (to pH 9.5)	liquid; colour 1 (Gardner); pH 1.8 (3% aq.)	detergent; dispersant; o/w emulsifier; wetting agent *personal care products; household and industrial applications; metal working*
	EMPHOS PS-220	aliphatic hydrophobic base	100% active; acid no. 110 (to pH 6.5), 185 (to pH 9.5)	liquid; colour 5 (Gardner); pH 2 (10% aq.)	anti-corrosion; detergent; dispersant; o/w emulsifier; lubricant *personal care products; household and industrial applications; floor cleaners*
	EMPHOS PS-222	aliphatic hydrophobic base	100% active; acid no. 105 (to pH 5.5), 55 (to pH 9.5)	liquid; colour 1 (Gardner); pH 2.4 (3% aq.)	anti-corrosion; detergent; dispersant; o/w emulsifier; lubricant *personal care products; household and industrial applications*

Supplier	Trade name	Chemical description	Composition	General properties	Functionality / Application
Witco	EMPHOS PS-236	aliphatic hydrophobic base	100% active; acid no. 90 (to pH 5.5), 140 (to pH 9.5)	liquid; colour 2 (Gardner); pH 2 (3% aq.)	anti-corrosion; detergent; o/w emulsifier; foaming agent; lubricant; wetting agent; additive / *personal care products; household and industrial applications; fuel oils*
	EMPHOS PS-331	aliphatic hydrophobic base	acid no. 140 (to pH 5.5), 280 (to pH 9.5)	liquid; colour 2 (Gardner); pH 1.8 (3% aq.)	detergent; dispersant; foaming agent; wetting agent / *personal care products; household and industrial applications*
	EMPHOS PS-400	aliphatic hydrophobic base	acid no. 220 (to pH 5.5), 320 (to pH 9.5)	liquid; colour 1 (Gardner); pH 2 (3% aq.)	dispersant; o/w emulsifier; lubricant; wetting agent / *personal care products; household and industrial applications*
	EMPHOS TS-230	aromatic hydrophobic base	acid no. 90 (to pH 5.5), 160 (to pH 9.5)	liquid; colour 3 (Gardner); pH 2 (3% aq.)	coupling agent; lubricant / *personal care products; household and industrial applications*
	REWOPHAT E 1027	alkyl phenol polyglycol ether phosphate	100% active	medium viscosity liquid; surface tension 32 dynes/cm; pH 2-4 (1%)	emulsifier; antistatic agent; anti-corrosion; low foam / *emulsion polymerisation*
	REWOPHAT EAK 8190	lauryl alcohol polyglycol ether phosphate	100% active	liquid	emulsifier / *metal working; textile industry; cleaners*
	REWOPHAT NP 90	alkyl phenol polyglycol ether phosphate	100% active	viscous liquid; surface tension 34 dynes/cm; pH 2-4 (1%)	emulsifier; antistatic agent; anti-corrosion; low foam / *emulsion polymerisation*
	SYNTOPHOS 06 N	based on ethoxylated alcohol	30% active	liquid	wetting agent; antistatic agent / *textile industry*
	ULTRAPHOS 59	based on ethoxylated phenol	100% active	liquid	additive / *metal working*
	WITCOR SI-3065	complex phosphate ester		liquid; pour point 0°C; pH 4.5	scale inhibitor / *petroleum industry*
Zschimmer & Schwarz	COLLASTAT AP	phosphoric acid ester		clear colourless liquid	sizing agent; antistatic agent / *textile industry*
	ELACTIV K	phosphoric acid ester		clear, straw-coloured liquid	antistatic agent / *textile industry*
	ELACTIV PS	phosphoric acid ester		clear viscous liquid	
	ELACTIV PV	phosphoric acid ester		straw-coloured, low viscous liquid	
	FLEROL TP 3000	modified phosphoric acid ester		clear colourless liquid	antistatic agent / *lubricants; textile industry*
	PHOSFETAL 201 K	potassium alkyl polyglycol ether phosphate	90% active	liquid	o/w emulsifier / *cosmetics; all-purpose cleaners*
	PHOSFETAL 201	alkyl polyglycol ether phosphate, acid	100% active	liquid	*cleansing preparations; all-purpose cleaners*
	PHOSFETAL 205	alkyl polyglycol ether phosphate, acid	100% active	wax	
	PHOSFETAL 600	alkylaryl polyglycol ether phosphate, acid	100% active	liquid	
	PHOSFETAL 601	alkylaryl polyglycol ether phosphate, acid	100% active	liquid	
	PHOSFETAL 602	alkylaryl polyglycol ether phosphate, acid	100% active	liquid	
	PHOSFETAL 603	alkylaryl polyglycol ether phosphate, acid	100% active	liquid	

Sarcosinates

Supplier	Trade name	Chemical description	Composition	General properties	Functionality *Application*
Auschem Cesalpinia	CHIMIN L	sodium lauryl sarcosinate	30% active	liquid	foaming agent; detergent *liquid detergents; textiles industry; laundry products; carpet cleaners; toiletries*
Dr. Th. Boehme	SYNTHESIN 7090	sarcosine acid			spin finish *textile industry; fibre production*
	SYNTHESIN CSS	sarcosine acid			
Croda	CRODASINIC L	lauroyl sarcosine		HLB 13.1; white waxy solid	detergent; wetting agent; conditioner; bacteriostatic agent; enzyme inhibitor; anti-corrosion; foaming agent *shampoos; skin cleansing preparations; oral hygiene products*
	CRODASINIC LS30/35	sodium lauroyl sarcosine and H_2O		clear colourless foaming liquid	detergent; wetting agent; conditioner; bacteriostatic agent; enzyme inhibitor; anti-corrosion; foaming agent *shampoos; skin cleansing preparations; oral hygiene products; hard-surface and carpet cleaners*
	CRODASINIC LS95	sodium lauroyl sarcosine		white powder	detergent; wetting agent; conditioner; bacteriostatic agent; enzyme inhibitor; anti-corrosion; foaming agent *shampoos; skin cleansing preparations; oral hygiene products*
	CRODASINIC LT40	TEA-lauroyl sarcosine		clear colourless to very pale yellow liquid	detergent; wetting agent; conditioner; bacteriostatic agent; enzyme inhibitor; anti-corrosion; foaming agent *shampoos; skin cleansing preparations; oral hygiene products; household products*
Hampshire Chemical	HAMPOSYL C-30	sodium cocoyl sarcosinate	active ingredient 30 ± 1%; sodium soap 2% max.	HLB 27 ± 0.5; colourless to very pale yellow liquid; M.W. 301; colour 100 max. (30% soln.; APHA); freezing point − 1°C; pH 7.5-8.5 (10% soln.)	conditioner; foaming agent; wetting agent; detergent; foam stabiliser; anti-corrosion; viscosity modifier *carpet cleaners; laundry products; dishwashing agents; personal care products; cosmetics; pesticides; metal working; textile industry; polymer industry; leather industry; petroleum industry*
	HAMPOSYL C	cocoyl sarcosine	active ingredient 94% min.; free fatty acid 6% max.	HLB 10 ± 0.5; pale yellow liquid; M.W. 280; colour 3 max. (Gardner); softening point 18-22°C	
	HAMPOSYL L-30	sodium lauroyl sarcosinate	active ingredient 30 ± 1% sodium soap 2% max.	HLB 29.8; colourless liquid; M.W. 292; colour 60 max. (APHA); freezing point − 1°C; pH 7.5-8.5 (10% soln.)	
	HAMPOSYL L-95	sodium lauroyl sarcosinate	active ingredient 94% min.; sodium soap 4% max.	HLB 28.9; dry white powder; M.W. 292; colour 60 max. (APHA); freezing point − 1°C; pH 7.5-8.5 (10% soln.)	
	HAMPOSYL L	lauroyl sarcosine	active ingredient 94% min.; free fatty acid 6% max.	HLB 13.1; white waxy solid; M.W. 270; colour 2 max. (Gardner); softening point 34-37°C	
	HAMPOSYL M-30	sodium myristoyl sarcosinate	active ingredient 30 ± 1% sodium soap 2% max.	HLB 29.8; colourless liquid; M.W. 320; colour 100 max. (APHA); freezing point − 1°C; pH 7.5-8.5 (10% soln.)	
	HAMPOSYL M	myristoyl sarcosine	active ingredient 94% min.; free fatty acid 6% max.	HLB 12.1; white waxy solid; M.W. 298; colour 2 max. (Gardner); softening point 48-53°C	
	HAMPOSYL O	oleyl sarcosine	active ingredient 94% min.; free fatty acid 6% max.	HLB 9.6; yellow liquid; M.W. 349; colour 4 max. (Gardner)	
	HAMPOSYL S	stearyl sarcosine	active ingredient 94% min.; free fatty acid 6% max.	HLB 9.6; white waxy solid; M.W. 338; colour 4 max. (Gardner); softening point 53-58°C	

Supplier	Trade name	Chemical description	Composition	General properties	Functionality Application
Hoechst	ARKOMON A CONC	sodium oleoyl sarcoside	60% active	paste	*textile industry*
	ARKOMON SO	oleoyl sarcoside	100% active	liquid	anti-corrosion *detergents; metal working*
Millchem	MEDIALAN KF	TEA-palm kernel sarcosinate	40% active	clear pale yellow liquid	*cosmetics; toiletries*
	MEDIALAN LD	sodium lauroyl sarcosinate	30% active	clear, slightly yellowish liquid	*cosmetics; toothpaste*
	MAPROSYL 30	sodium lauroyl sarcosinate	29.5–30.5% active; solids 30%	liquid	detergent; wetting agent; foaming agent *personal care products; household products*
	MAPROSYL 95	sodium lauroyl sarcosine	94% active	powder	detergent; anti-corrosion
	MAPROSYL C	cocoyl sarcosine	94% active	liquid	
	MAPROSYL L	lauroyl sarcosine	94% active	waxy solid	
	MAPROSYL O	oleoyl sarcosine	94% active	liquid	
Nikko Chemicals	NIKKOL SARCOSINATE CN-30	sodium cocoyl sarcosinate; aq. soln.		pale yellow liquid	foaming agent; cleansing agent *shampoos*
	NIKKOL SARCOSINATE LH	lauroyl sarcosine; free acid form		pale yellow crystalline solid	*cosmetics*
	NIKKOL SARCOSINATE LK-30	potassium N-lauroyl sarcosinate; aq. soln.		pale yellow liquid	
	NIKKOL SARCOSINATE LN-30	sodium lauroyl sarcosinate; aq. soln.		pale yellow liquid	foaming agent; cleansing agent *shampoos*
	NIKKOL SARCOSINATE LN	sodium lauroyl sarcosinate		white crystalline powder	
	NIKKOL SARCOSINATE MN	sodium myristoyl sarcosinate		white crystalline powder	cleansing agent *cosmetics; soaps*
	NIKKOL SARCOSINATE OH	N-oleoyl sarcosinate; free acid form		yellow liquid	*cosmetics*
	NIKKOL SARCOSINATE PN	sodium N-palmitoyl sarcosinate		white crystalline powder	cleansing agent *cosmetics; soaps*
Protex	SURFARON A1712 N30	sodium lauryl sarcosinate			anti-corrosion; detergent
	SURFARON A7217 Z	oleoyl sarcosine			anti-corrosion
Stepan Europe	SECOSYL	sodium lauroyl sarcosinate	30% active	water white to pale yellow liquid	foaming agent; detergent; anti-corrosion *shampoos; bubble baths; dishwashing agents; carpet cleaners; household cleaners*
Witco	REWOPOL SK 275	oleyl sarcosinic acid	100% active	liquid	emulsifier; anti-corrosion *mineral oils*

Sulfosuccinates/sulfosuccinamates

Supplier	Trade name	Chemical description	Composition	General properties	Functionality Application
Akcros Chemicals	AGRILAN AEC266	sodium dioctyl sulfosuccinate (contains ethanol)	58% active; H$_2$O content 23%	clear pale yellow liquid; pour point <0°C; viscosity 43 cSt; pH 5-8 (1% aq.)	emulsifier *agrochemicals*
	AGRILAN AEC299	sodium dioctyl sulfosuccinate (contains Solvesso 150)	68% active; H$_2$O content <0.3%	clear pale yellow liquid; pour point <0°C; viscosity 360 cSt; pH 6-8 (1% aq.)	emulsifier *agrochemicals*
	LANKROPOL KN51	sodium diester sulfosuccinate	solids 50%; H$_2$O content 50%	amber liquid; pour point −8°C; viscosity 200 cSt; pH 5.0-7.0 (1% aq.)	emulsifier; low foam *emulsion polymerisation*
	LANKROPOL KO	sodium dioctyl sulfosuccinate (contains mineral oil)	solids 60%; H$_2$O content 10%	clear/slightly hazy pale straw liquid; pour point <0°C; viscosity 1250 cSt; pH 5.0-7.0 (1% aq.)	wetting agent; emulsifier; dispersant; dewatering agent *inorganic pigments; dry cleaning; mineral oils*
	LANKROPOL KO2	sodium dioctyl sulfosuccinate (contains ethanol)	solids 60%; H$_2$O content 23%	clear pale straw liquid; pour point <0°C; viscosity 43 cSt; pH 5.0-8.0 (1% aq.)	wetting agent; emulsifier; co-dispersant *inorganic pigments; dry cleaning*
	LANKROPOL KPH70	sodium dioctyl sulfosuccinate in propylene glycol/H$_2$O	solids 70%; H$_2$O content 11%	clear/slightly hazy pale straw liquid; pour point −10°C; pH 6.5 (5% aq.)	wetting agent; emulsifier; dispersant *oil slick dispersants*
	LANKROPOL KRJ60	sodium dioctyl sulfosuccinate in glycol ether/ H$_2$O	solids 60%; H$_2$O content 23%	clear/slightly hazy pale straw liquid; pour point −10°C; pH 6.5 (5% aq.)	
	LANKROPOL KRJ70	sodium dioctyl sulfosuccinate in glycol ether/ H$_2$O	solids 70%; H$_2$O content 11%	clear/slightly hazy pale straw liquid; pour point −5°C; pH 6.5 (5% aq.)	
	LANKROPOL KSG72	sodium monoester sulfosuccinate of ethoxylated alkanolamide	solids 45%; H$_2$O content 55%	pale yellow liquid; pour point <0°C; viscosity 125 cSt; pH 5.5-7.5 (1% aq.)	*toiletries; cosmetics; shampoos; bubble baths*
	LANKROPOL KTK70	sodium dioctyl sulfosuccinate in odourless kerosene	solids 70%; H$_2$O content <1%	clear/slightly hazy pale straw liquid; pour point −10°C; pH 7.5 (1% in 50/50 IMS/H$_2$O)	wetting agent; emulsifier; dispersant *oil slick dispersants*
Akzo Nobel	ELFANOL 616	disodium lauryl myristyl ether (3) sulfosuccinate	40% active	liquid	
	ELFANOL 850	disodium sulfosuccinate coco amide ethoxylate (3) ester	45% active	liquid	
Albright & Wilson	EMPICOL SDD/Y	di-sodium lauryl ethoxy (3EO) sulfosuccinate	33% active	liquid	
	EMPICOL SDD	di-sodium lauryl ethoxy sulfosuccinate	22% active		
	EMPICOL SFF	di-sodium lauryl ethoxy (5EO) sulfosuccinate	63% active		
	EMPIMIN MA	sodium di-hexyl sulfosuccinate	40% active	liquid	*polymer industry*
	EMPIMIN MH	di-sodium N-cocoyl sulfosuccinamate	33% active	paste	*polymer industry*
	EMPIMIN MK/B	di-sodium N-cetyl stearyl sulfosuccinamate	70% active	liquid	wetting agent; dispersant *agrochemicals; coatings; emulsion polymerisation*
	EMPIMIN OP70	sodium di-octyl sulfosuccinate	61% active	liquid	wetting agent; detergent; dispersant *agrochemicals; leather industry; coatings; emulsion polymerisation*
	EMPIMIN OT	sodium di-octyl sulfosuccinate	75% active	liquid	
	EMPIMIN OT75	sodium di-octyl sulfosuccinate		liquid	
Allied Colloids	ALCOPOL O 20	sodium dioctyl sulfosuccinate	20% active	liquid	
	ALCOPOL O 60 PG	sodium dioctyl sulfosuccinate; as Alcopol O Conc 60 but with a much higher flash point	60% active	liquid	wetting agent *textile industry*
	ALCOPOL O 70 PG	sodium dioctyl sulfosuccinate; as Alcopol O 20 but with a much higher flash point	70% active	liquid	

Supplier	Trade name	Chemical description	Composition	General properties	Functionality / Application
Allied Colloids	ALCOPOL O CONC 60	sodium dioctyl sulfosuccinate	60% active	liquid	wetting agent; *textile industry*
Auschem Cesalpinia	IMBIROL OT 70	dioctylsulfosuccinate sodium salt	70% active	liquid	detergent; wetting agent
	MADEOL VA 40	dioctylsulfosuccinate sodium salt	40% active	powder	wetting agent; *pesticides*
	POLIROL 147	sodium N-alkyl sulfosuccinate	35% active	paste	emulsifier; wetting agent; dispersant; solubiliser; penetrant *polymerisation*
	POLIROL LNA	sodium di-alkyl sulfosuccinate	60% active	liquid	
	POLIROL OT 70	sodium di-alkyl sulfosuccinate	70% active	liquid	
	POLIROL SE 301	di-sodium sulfosuccinate semi-ester	30% active	liquid	
	ROLPON BL	alkanolamide sulfosuccinate sodium salt	40% active	liquid	foaming agent; detergent; degreaser; conditioner *liquid detergents; textile industry; laundry products; carpet cleaners; cosmetics*
	ROLPON BT	alkylether-sulfosuccinate semiester sodium salt	35% active	liquid	
BASF	LUTENSIT A-BO	sodium dioctylsulfosuccinate	60% active	liquid	
Chemax	CHEMAX DOSS/70E	sodium dioctylsulfosuccinate	70% active	liquid	wetting agent; dispersant; solubiliser *textile industry; agriculture; detergents; emulsion polymerisation; paint industry; inks; dry cleaning*
	CHEMAX DOSS/70HFP	sodium dioctylsulfosuccinate	70% active	liquid	
	CHEMAX DOSS/75E	sodium dioctylsulfosuccinate	75% active	liquid	
Croda	CROPOL 70	sodium dioctyl sulfosuccinate supplied as a soln. in either propylene glycol or odourless kerosene	anionic activity 68-72%	white to pale yellow, clear to slightly hazy, viscous liquid; M.W. 444; colour 225-250 max. (Hazen); pH 4-7 (10%)	wetting agent; spreading agent; dispersant; emulsifier *adhesives; agriculture; cleaners; electroplating; emulsion and suspension polymerisation; fire-fighting; paint industry; inks; photography; rubber industry; textile industry; polishes*
Cytec	AEROSOL 18	disodium N-octadecyl sulfosuccinamate	35% in H_2O	paste; surface tension 39 dynes/cm min. (in H_2O)	foaming agent; emulsifier; dispersant; detergent *emulsion and suspension polymerisation; latexes; plastics industry*
	AEROSOL 22	tetrasodium N-(1,2-dicarboxyethyl)-N-octadecyl sulfosuccinamate	35% in H_2O and alcohol	liquid; surface tension 41 dynes/cm min. (in H_2O)	emulsifier; dispersant; hydrotrope; solubiliser *emulsion polymerisation; latexes; agrochemicals; industrial and household cleaners; metal cleaning; soaps*
	AEROSOL 501	proprietary composite (US patent 3,947,400)	50% in H_2O	liquid; surface tension 28 dynes/cm min. (in H_2O)	emulsifier *emulsion and suspension polymerisation; paint industry; textile industry; paper industry*
	AEROSOL A-102	disodium ethoxylated alcohol half ester of sulfosuccinic acid	31% in H_2O	liquid; surface tension 33 dynes/cm min. (in H_2O)	solubiliser; foaming agent; dispersant; emulsifier *polymers; latexes; germicides; cosmetics; shampoos; adhesives*
	AEROSOL A-103	disodium ethoxylated nonyl phenol half ester of sulfosuccinic acid	34% in H_2O	liquid; surface tension 34 dynes/cm min. (in H_2O)	emulsifier; dispersant; foaming agent; solubiliser; stabiliser *latexes; polymers; germicides; cosmetics; shampoos; adhesives*
	AEROSOL A-196-40	sodium dicyclohexyl sulfosuccinate	40% in H_2O	liquid; surface tension 39 dynes/cm min. (in H_2O)	low foam; adhesion agent *emulsion polymerisation; latexes*
	AEROSOL A-196-97	sodium dicyclohexyl sulfosuccinate	97% active	flaky solid; surface tension 39 dynes/cm min. (in H_2O)	
	AEROSOL A-268	disodium isodecyl sulfosuccinate	50% in H_2O	liquid; surface tension 28 dynes/cm min. (in H_2O)	emulsifier *emulsion and suspension polymerisation*
	AEROSOL AY-100	sodium diamyl sulfosuccinate	100% active	waxy solid; surface tension 30 dynes/cm min. (in H_2O)	wetting agent; dispersant; adjuvant *emulsion polymerisation; leaching; electroplating; agrochemicals*
	AEROSOL AY-65	sodium diamyl sulfosuccinate	65% in H_2O and alcohol	liquid; surface tension 30 dynes/cm min. (in H_2O)	wetting agent; dispersant; adjuvant *emulsion polymerisation; leaching; electroplating; agrochemicals*

Supplier	Trade name	Chemical description	Composition	General properties	Functionality / Application
Cytec	AEROSOL GPG	sodium dioctyl sulfosuccinate	70% in H$_2$O and alcohol	liquid; surface tension 26 dynes/cm min. (in H$_2$O)	dispersant; emulsifier; wetting agent / *degreasing agents; dry cleaning; lubricants; coolants; fire fighting; dust control*
	AEROSOL IB-45	sodium diisobutyl sulfosuccinate	45% in H$_2$O	liquid; surface tension 42 dynes/cm min. (in H$_2$O)	stabiliser; dispersant / *emulsion polymerisation; leaching; electroplating*
	AEROSOL MA-80I	sodium dioctyl sulfosuccinate	80% in H$_2$O and alcohol	liquid; surface tension 28 dynes/cm min. (in H$_2$O)	penetrant; emulsifier; dispersant; solubiliser; wetting agent / *emulsion polymerisation; electroplating; leaching*
	AEROSOL OT-100	sodium dioctyl sulfosuccinate	100% active	waxy solid; surface tension 26 dynes/cm min. (in H$_2$O)	emulsifier; dispersant; lubricant; wetting agent; mould release aid; antistatic agent; adjuvant / *plastics industry; paint industry; rust inhibitors; lubricants; polymers; agrochemicals*
	AEROSOL OT-70-PG	sodium dioctyl sulfosuccinate	70% in propylene glycol and H$_2$O	liquid; surface tension 26 dynes/cm min. (in H$_2$O)	wetting agent; emulsifier; re-wetting agent / *emulsion and suspension polymerisation; latexes; paint industry; textile industry; adhesives; paper industry; rubber industry; cosmetics; metal working; mining industry; agrochemicals*
	AEROSOL OT-75-PG	sodium dioctyl sulfosuccinate	75% in propylene glycol and H$_2$O	liquid; surface tension 26 dynes/cm min. (in H$_2$O)	
	AEROSOL OT-75	sodium dioctyl sulfosuccinate	75% in H$_2$O and alcohol	liquid; surface tension 26 dynes/cm min. (in H$_2$O)	
	AEROSOL OT-B	sodium dioctyl sulfosuccinate	85% active; sodium benzoate 15%	powder; surface tension 26 dynes/cm min. (in H$_2$O)	dispersant; wetting agent; solubiliser; penetrant; adjuvant / *agrochemicals; plastics industry*
	AEROSOL OT-MSO	sodium dioctyl sulfosuccinate	62% in mineral seal oil	liquid; surface tension 26 dynes/cm min. (in H$_2$O)	detergent; lubricant; wetting agent; emulsifier / *plastics industry; dry cleaning; rust inhibitors; degreasing agents; lubricants*
	AEROSOL OT-S	sodium dioctyl sulfosuccinate	70% in light petroleum distillate	liquid; surface tension 26 dynes/cm min. (in H$_2$O)	
	AEROSOL TR-70	sodium bistridecyl sulfosuccinate	70% in H$_2$O and alcohol	liquid; surface tension 27 dynes/cm min. (in H$_2$O)	dispersant; wetting agent / *emulsion and suspension polymerisation; latexes; pigments; printing inks; rust inhibitors; plastics industry*
Grillo-Werke	GRILLOSOL SB 0/12	disodium lauryl sulfosuccinate	solids 39-45%; surface-active content 35% min.	white/light yellow paste; m.p. 40-50°C; pH 6.0-7.5 (1% in H$_2$O)	*soaps; detergents; baby shampoos; bubble baths*
	GRILLOSOL SB 3/12	disodium laureth sulfosuccinate	solids 37-44%; surface-active content 31-35%	pale yellow to colourless liquid; pH 6.0-8.0 (1% in H$_2$O)	*hair care products; skin care products; baby shampoos; bubble baths; skin cleansing preparations*
	GRILLOSOL SB 8	sodium di-isooctylsulfosuccinate	solids 09-74%; surface-active content 67-71%; ethanol ca. 5%	clear, almost colourless; pH 4.0-7.0 (1% in H$_2$O)	wetting agent; solubiliser / *textile industry; household products; metal cleaners*
Henkel	ARBYL ASN	sulfosuccinate			wetting agent / *textile industry; dyeing*
	DISPONIL SUS 65	disodium fatty alcohol ether sulfosuccinate	ca. 40% active	liquid	emulsifier / *plastics industry; coatings*
	DISPONIL SUS 87 SPEC.	disodium fatty alcohol ether sulfosuccinate	ca. 30% active	liquid	
	DISPONIL SUS 90	disodium alkyl aryl ether sulfosuccinate	ca. 30% active	liquid	wetting agent; co-emulsifier / *plastics industry; coatings*
	DISPONIL SUS IC 680	dihexyl sodium sulfosuccinate, sodium salt	ca. 80% active	liquid	wetting agent; co-emulsifier / *plastics industry; coatings*
	DISPONIL SUS IC 875	dioctyl sodium sulfosuccinate, sodium salt	ca. 75% active	liquid	
	STANDAPOL SH-100	disodium oleamido PEG-2 sulfosuccinate	29% active	light amber liquid	anti-irritant / *shampoos; baby shampoos*
	STANDAPOL SH-135	disodium oleamido PEG-2 sulfosuccinate	34% active	light amber liquid	

Supplier	Trade name	Chemical description	Composition	General properties	Functionality Application
Henkel	STANDAPOL SHC-101	disodium oleamido PEG-2 sulfosuccinate and sodium lauryl sulfate	30% active	light amber liquid	anti-irritant *shampoos; baby shampoos*
	TEXAPON SB 3	disodium laureth sulfosuccinate	38–42% active	liquid	detergent base *cosmetics; pharmaceuticals*
	TEXIN 128 POWDER	disodium sulfosuccinic acid mono C$_{12/18}$ ester	ca. 90% active	powder	foaming agent *syndet bars*
	TEXIN DOS 75	sodium sulfosuccinic acid diisooctyl ester with ethanol	≥75% active	liquid	low foam; wetting agent *cleaners*
Hickson Manro	TENSUCCIN HM935	sodium sulfosuccinate, half ester	40% active	liquid	*hair care products; bath preparations*
Hoechst	DP1360	disodium 2-ethylhexyl sulfosuccinate	70% active	liquid	detergents; *textile industry*
	GENAPOL SBE	disodium sulfosuccinate	40% active	liquid	*cosmetics; detergents*
	HUK 037	disodium alkylether sulfosuccinate	22% active	liquid	*detergents*
Hüls	MARLINAT DF 8	di-isooctyl sulfosuccinate, sodium salt	active detergent 65%	liquid	wetting agent
	MARLINAT SL 3/40	C$_{12/14}$ fatty alcohol polyglycol ether sulfosuccinate 3EO, sodium salt	active detergent 30%	liquid	*baby shampoos*
	MARLINAT SRN 30	blend containing sulfosuccinate	active detergent 30%	liquid	*carpet cleaners*
Kao Corporation	SUCCIDET NES	sodium lauryl ether sulfosuccinate	> 30% active	liquid; M.W. 542; pH 6-7 (5%)	
McIntyre Group	MACKANATE A-102	disodium deceth-6 sulfosuccinate	30% active	liquid; pH 5	anti-irritant *baby shampoos; liquid soaps; bath preparations; skin cleansing preparations; personal care products*
	MACKANATE A-103	disodium nonoxynol-10 sulfosuccinate	35% active	liquid; pH 6	
	MACKANATE AY-65TD	diamyl sodium sulfosuccinate and tridecyl alcohol	65% active	liquid; pH 6	wetting agent; antifogging agent; antigelling agent; antistatic agent; dispersant; mould release aid *emulsion polymerisation; household detergents; dry cleaning; car washes; dust control*
	MACKANATE CM-100	disodium cocamido MEA sulfosuccinate	100% active	powder; pH 6	
	MACKANATE CM	disodium cocamido MEA sulfosuccinate	40% active	liquid; pH 6	
	MACKANATE CP	disodium cocamido MIPA sulfosuccinate	40% active	liquid; pH 6	
	MACKANATE DC-30	disodium dimethicone copolyol sulfosuccinate	30% active	liquid; pH 5	anti-irritant *baby shampoos; liquid soaps; bath preparations; skin cleansing preparations; personal care products*
	MACKANATE DC-30A	diammonium dimethicone copolyol sulfosuccinate	30% active	liquid; pH 5	
	MACKANATE DC-50	disodium dimethicone copolyol sulfosuccinate	50% active	liquid; pH 5	
	MACKANATE DOS-70	dioctyl sodium sulfosuccinate	70% active	liquid; pH 6	wetting agent; antifogging agent; antigelling agent; antistatic agent; dispersant; mould release aid *emulsion polymerisation; household detergents; dry cleaning; car washes; dust control*
	MACKANATE DOS-70BC	dioctyl sodium sulfosuccinate and butyl carbitol	70% active	liquid; pH 6	
	MACKANATE DOS-70DEG	dioctyl sodium sulfosuccinate and diethylene glyol	70% active	liquid; pH 6	
	MACKANATE DOS-70MS	dioctyl sodium sulfosuccinate and mineral spirits	70% active	liquid; pH 6	

Supplier	Trade name	Chemical description	Composition	General properties	Functionality Application
McIntyre Group	MACKANATE DOS-70N	dioctyl sodium sulfosuccinate and nonoxynol-9	70% active	liquid; pH 6	wetting agent; antifogging agent; antigelling agent; antistatic agent; dispersant; mould release aid
	MACKANATE DOS-70PG	dioctyl sodium sulfosuccinate and propylene glycol	70% active	liquid; pH 6	emulsion polymerisation; household detergents; dry cleaning; car washes; dust control
	MACKANATE DOS-75	dioctyl sodium sulfosuccinate	75% active	liquid; pH 6	
	MACKANATE EL	disodium laureth sulfosuccinate	40% active	liquid; pH 6	
	MACKANATE LA	diammonium lauryl sulfosuccinate	40% active	liquid; pH 6	
	MACKANATE LM-40	disodium lauramido MEA sulfosuccinate	40% active	liquid; pH 6	
	MACKANATE LO-100	disodium lauryl sulfosuccinate	100% active	powder; pH 6	
	MACKANATE LO-SPECIAL	disodium lauryl sulfosuccinate	40% active	paste; pH 6	anti-irritant baby shampoos; liquid soaps; bath preparations; skin cleansing
	MACKANATE LO	disodium lauryl sulfosuccinate	40% active	paste; pH 6	preparations; personal care products
	MACKANATE OD-35	disodium oleamido PEG-2 sulfosuccinate	35% active	liquid; pH 6	
	MACKANATE OM	disodium oleamido MEA sulfosuccinate	35% active	liquid; pH 6	
	MACKANATE OP	disodium oleamido MIPA sulfosuccinate	38% active	liquid; pH 6	
	MACKANATE RM	disodium ricinoleamido MEA sulfosuccinate	40% active	liquid; pH 6	
	MACKANATE UM-50	disodium undecylenamido MEA sulfosuccinate	50% active	liquid; pH 6	
	MACKANATE WGD	disodium wheatgermamido PEG-2 sulfosuccinate	35% active	liquid; pH 6	
Millchem	ANIONYX 12S	disodium mono-oleamide sulfosuccinate	19-21% active; solids 28-30%	liquid	shampoos; bubble baths; dishwashing agents
Mona Industries	MONAMATE CPA-100	disodium cocamido MIPA sulfosuccinate	100% active	powder	
	MONAMATE CPA-40	disodium cocamido MIPA sulfosuccinate	40% active	liquid	
	MONAMATE LA-100	disodium lauryl sulfosuccinate	100% active	powder	foaming agent personal care products; household products; laundry products;
	MONAMATE LNT-40	ammonium lauryl sulfosuccinate	40% active	liquid	dishwashing agents; carpet cleaners; shaving products
	MONAMATE OPA-100	disodium oleamido PEG-2 sulfosuccinate	100% active	powder	
	MONAMATE OPA-30	disodium oleamido PEG-2 sulfosuccinate	30% active	liquid	
	MONATERIC C-1142	disodium cocamido MIPA sulfosuccinate	40% active	liquid	
	MONAWET MB-45	sodium disobutylsulfosuccinate	45% active; H$_2$O 55%; acid no. 2.5 max.	clear colourless liquid; cloud point 13°C; M.W. 332; colour 75 max. (APHA); surface tension 54 dynes/cm (0.1%); pH 6 ± 1	wetting agent; dispersant; emulsifier; penetrant; solubiliser emulsion and suspension polymerisation; textile industry; agriculture; cosmetics; detergents; paint industry; inks; dry cleaning; food industry

Supplier	Trade name	Chemical description	Composition	General properties	Functionality Application
Mona Industries	MONAWET MM-80	sodium dihexylsulfosuccinate	80% active; isopropanol 5%; H$_2$O 15%; acid no. 2.5 max.	clear colourless liquid; cloud point <0°C; M.W. 388; colour 75 max. (APHA); surface tension 46 dynes/cm (0.1%); pH 6 ± 1	
	MONAWET MO-70	sodium dioctylsulfosuccinate	70% active; butylcarbitol 10%; H$_2$O 20%; acid no. 2.5 max.	clear colourless liquid; M.W. 444; colour 100 max. (APHA); surface tension 29 dynes/cm (0.1%); pH 6 ± 1	
	MONAWET MO-70E	sodium dioctylsulfosuccinate	70% active; ethanol 11%; H$_2$O 19%; acid no. 2.5 max.	clear colourless liquid; cloud point ≦ −5°C; M.W. 444; colour 100 max. (APHA); surface tension 29 dynes/cm (0.1%); pH 6 ± 1	
	MONAWET MO-70R	sodium dioctylsulfosuccinate	70% active; propylene glycol 15%; H$_2$O 15%; acid no. 2.5 max.	clear colourless liquid; M.W. 444; colour 100 max. (APHA); surface tension 29 dynes/cm (0.1%); pH 6 ± 1	
	MONAWET MO-75E	sodium dioctylsulfosuccinate	75% active; ethanol 10%; H$_2$O 15%; acid no. 2.5 max.	clear colourless liquid; M.W. 444; colour 100 max. (APHA); surface tension 29 dynes/cm (0.1%); pH 6 ± 1	wetting agent; dispersant; emulsifier; penetrant; solubiliser emulsion and suspension polymerisation; *textile industry; agriculture; cosmetics; detergents; paint industry; inks; dry cleaning; food industry*
	MONAWET MO-84R2W	sodium dioctylsulfosuccinate	83% active; propylene glycol 16%; acid no. 2.5 max.	light yellow viscous liquid; M.W. 444; colour 450 max. (APHA); surface tension 29 dynes/cm (0.1%); pH 5.5 ± 1.0 (10%)	
	MONAWET MT-70	sodium ditridecylsulfosuccinate	70% active; hexylene glycol 12%; H$_2$O 12%; acid no. 2.5 max.	clear light straw-coloured liquid; cloud point −2°C; M.W. 584; colour 150 max. (APHA); surface tension 29 dynes/cm (0.1%); pH 6 ± 1	
	MONAWET MT-70E	sodium ditridecylsulfosuccinate	70% active; ethanol 12%; butyl carbitol 6%; H$_2$O 12%; acid no. 2.5 max.	light straw clear liquid; cloud point −15°C; M.W. 584; colour 150 max. (APHA); surface tension 29 dynes/cm (0.1%); pH 6 ± 1	
	MONAWET MT-80H2W	sodium ditridecylsulfosuccinate	80% active; 20% hexylene glycol; acid no. 2.5 max.	light yellow viscous liquid; cloud point <0°C; M.W. 584; colour 250 max. (APHA); surface tension 29 dynes/cm (0.1%); pH 5.5 ± 1.0 (10%)	
Nikko Chemicals	NIKKOL OTP-100	dioctyl sodium sulfosuccinate		white sponge	wetting agent; penetrant; dispersant
	NIKKOL OTP-75	dioctyl sodium sulfosuccinate		colourless liquid	*cosmetics*
Pentagon	PENTASOL DO-70PG	sodium dioctyl sulfosuccinate	total solids 68-72%; solvent propylene glycol/H$_2$O	clear to pale yellow liquid; pH 5.0-7.0 (1% aq. soln.)	wetting agent
Protex	SURFARON A 5108 N 65	dialkyl sulfosuccinate		viscosity 80 mPas; surface tension 26 dynes/cm (1%), 33 dynes/cm (0.1%)	wetting agent; o/w emulsifier; solubiliser; foaming agent; antistatic agent; demulsifier; bactericide emulsion and suspension polymerisation; *industrial and household detergents; cosmetics; shampoos; hair care products; skin care products; dry cleaning; photography; pharmaceuticals; fire-fighting; inks; paper, paint, plastics, rubber, oil and mining industries; metal cleaning*
	SURFARON A 5113 N 65	sodium di-tridecyl sulfosuccinate		viscosity 355 mPas	w/o emulsifier; solubiliser; dispersant emulsion and suspension polymerisation; *industrial and household detergents; cosmetics; shampoos; inks; mining industry; fire-fighting*
	SURFARON A 5216 N 35	sodium monoalkyl sulfosuccinamate		viscosity 150 mPas; surface tension 50 dynes/cm (1%), 65 dynes/cm (0.1%)	o/w emulsifier; solubiliser; foaming agent emulsion and suspension polymerisation; latex foams; *industrial and household detergents; cosmetics; skin cleansing products; shampoos; shaving products; oil industry; fire-fighting*

Supplier	Trade name	Chemical description	Composition	General properties	Functionality / *Application*
Protex	SURFARON A 5217 N 35	sodium monoalkyl sulfosuccinamate		viscosity 65 mPas; surface tension 50 dynes/cm (1%), 65 dynes/cm (0.1%)	wetting agent; o/w emulsifier; solubiliser; foaming agent / *emulsion and suspension polymerisation; latex foams; industrial and household detergents; cosmetics; skin cleansing products; shampoos; shaving products; fire-fighting*
	SURFARON A 5417 N 35	alkanolamide sulfosuccinate		viscosity 50 mPas; surface tension 43 dynes/cm (1%), 64 dynes/cm (0.1%)	o/w emulsifier; solubiliser; foaming agent / *emulsion and suspension polymerisation; industrial and household detergents; cosmetics; carpet cleaners; latex foams; shampoos; shaving products; bubble baths; liquid soaps; inks; paper industry; oil industry; fire-fighting*
	SURFARON A5113N	sodium bis tridecyl sulfosuccinate			emulsifier / *emulsion polymerisation*
Rhone-Poulenc	GEROPON ACR/4	disodic laurylethoxy sulfosuccinate	30% solids; 'n' no. 4	liquid; surface tension 29 dynes/cm	stabiliser / *emulsion polymerisation*
	GEROPON ACR/9	disodic alkylarylethoxy sulfosuccinate	30% solids; 'n' no. 9	liquid; surface tension 34 dynes/cm	emulsifier / *emulsion polymerisation*
	GEROPON BIS SODICO 2	sodium bistridecyl sulfosuccinate	60% solids	liquid; surface tension 26 dynes/cm	emulsifier; viscosity modifier / *emulsion polymerisation*
	GEROPON CYA/DEP	sodium dioctyl sulfosuccinate	75% solids	liquid; surface tension 26 dynes/cm	emulsifier; stabiliser / *emulsion polymerisation*
	GEROPON CYA-60	sodium dioctyl sulfosuccinate	60% solids	liquid; surface tension 26 dynes/cm	emulsifier; viscosity modifier / *emulsion polymerisation*
	GEROPON DOS	sodium dioctyl sulfosuccinate		liquid	foaming agent (high); emulsifier; wetting agent; dispersant
	GEROPON TW/99	amino sodium alkylsulfosuccinamate	38% solids	liquid	emulsifier; viscosity modifier; stabiliser / *emulsion polymerisation*
	GEROPON TX/99	amino sodium alkylsulfosuccinamate	38% solids	liquid	
Scher Chemicals	SCHERCOPOL CMS-NA	disodium cocamido MEA sulfosuccinate	dry solids 29% min.; sodium bisulfite 0.3% max.	clear yellow liquid; pH 5-7	detergent / *shampoos; bubble baths; dishwashing agents; carpet cleaners*
	SCHERCOPOL DOS-70	dioctyl sodium sulfosuccinate	dry solids 70% min.; sodium bisulfite 0.3% max.	viscous liquid; pH 6-8	wetting agent
	SCHERCOPOL DOS-PG-85	dioctyl sodium sulfosuccinate	dry solids 85% min.; sodium bisulfite 0.3% max.	viscous liquid; pH 6-8	wetting agent
	SCHERCOPOL LPS	disodium laureth sulfosuccinate	dry solids 39% min.; sodium bisulfite 0.3% max.	clear yellow liquid; pH 5-7	foaming agent (high); viscosity modifier
	SCHERCOPOL OMS-NA	disodium oleamido MEA sulfosuccinate	dry solids 34% min.; sodium bisulfite 0.3% max.	light amber liquid; pH 5-7	foaming agent (high) / *shampoos; bubble baths; skin cleansing preparations*
	SCHERCOPON 2WD	ethoxylated sulfosuccinate	90% active	liquid	detergent; wetting agent / *industrial cleaners; dry cleaning*
	SCHERCOWET DOS-70	sulfosuccinate	70% active	liquid	wetting agent / *textile industry*
Seppic	ORAMIX	isopropanolamine coco hemisulfosuccinate	40% active	liquid	detergent
	SUPERMONTALINE SLT 70	dioctyl sodium sulfosuccinate	70% active	liquid	emulsifier; wetting agent
Servo Delden	SERMUL EA 162	diammonium nonylphenol (10EO) monosulfosuccinate	21-23% active	liquid	
	SERMUL EA 176	di-sodium nonylphenol (10EO) monosulfosuccinate	21-24% active	liquid	
	SERWET WH 170	sodium di-2-ethylhexyl sulfosuccinate	65% active	liquid	

Supplier	Trade name	Chemical description	Composition	General properties	Functionality / Application
Servo Delden	SERWET WH 172	sodium di-2-ethylhexyl sulfosuccinate	60% active	liquid	dispersant; wetting agent; emulsifier / *agrochemicals*
Stepan Europe	NINATE DS 70	sodium dioctylsulfosuccinate		water white to pale yellow liquid	emulsifiers / *agrochemicals*
	PESTILIZER B SERIES	alkylsulfosuccinates; made-to-order		viscous liquid	emulsifiers / *agrochemicals*
	SECOSOL AL 959	sodium monolauryl sulfosuccinate; made-to-order	24% active	white to beige paste	foaming agent / *shampoos: bubble baths; liquid soaps; soaps; shower gels; bath preparations*
	SECOSOL ALL 40	sodium monolaurylether sulfosuccinate	31% active	pale yellow liquid	foaming detoxifying agent; emulsifier / *shampoos; bubble baths; liquid soaps; shower gels; bath preparations; emulsion polymerisation*
	SECOSOL DOS 70	sodium diethylhexylsulfosuccinate	70% active	water white to pale yellow liquid	wetting agent; dispersant; emulsifier; low foam / *household, institutional and industrial cleaners; emulsion polymerisation; textile industry; oil industry*
Surfachem	SURFAC OT/IPA	sodium di-octyl sulfosuccinate in isopropanol			
	SURFAC OT	sodium di-octyl sulfosuccinate 60%	60% active		
	SURFAC OT75	sodium di-octyl sulfosuccinate 75%	60% active		
	SURFAC SSD	disodium sulfosuccinate			
	SURFAC SSM40	di-sodium-N-cocoyl sulfosuccinate			
Thor Chemicals	SOVATEX WA/20	sulfosuccinate	20% active	liquid	wetting agent
	SOVATEX WA/60	sulfosuccinate	60% active	liquid	wetting agent
Union Carbide	TRITON GR-5M	sulfosuccinate	60% active	liquid; pour point −51°C	adhesion improver / *emulsion polymerisation; textile industry; coatings*
	TRITON GR-7M	sulfosuccinate	64% active	liquid; pour point −57°C	adhesion improver / *coatings; dry cleaning*
Warwick International	WARCOWET 020	sodium dioctyl sulfosuccinate			wetting agent / *textile industry*
	WARCOWET 060	sodium dioctyl sulfosuccinate; concentrated version of Warcowet 020			
Witco	EMCOL 4161L	disodium oleamide MIPA-sulfosuccinate	solids 38%	liquid; pH 6.5 (3% aq.)	conditioner; dispersant; foam stabiliser; foaming agent; wetting agent / *personal care products; household and industrial applications*
	EMCOL 4500	dioctyl sodium sulfosuccinate	solids 70%	liquid; pH 6.5 (3% aq.)	detergent; dispersant; wetting agent / *personal care products; household and industrial applications*
	EMCOL K8300	disodium oleamide sulfosuccinate	solids 38%	liquid; pH 6.5 (3% aq.)	dispersant; foam stabiliser; foaming agent; wetting agent / *personal care products; household and industrial applications*
	REWOCID SB U 185	undecylenic acid alkylolamide sulfosuccinate	50% active	liquid	anti-dandruff shampoos
	REWOCID SB U 185 P	undecylenic acid alkylolamide sulfosuccinate	95% active	powder	anti-dandruff shampoos
	REWODERM S 1333 P	ricinoleic acid alkylolamide sulfosuccinate	95% active	powder	soaps: powder detergents
	REWODERM S 1333	ricinoleic acid alkylolamide sulfosuccinate	40% active	liquid	shampoos: intimate hygiene products: skin care products
	REWODERM SPS	β-sitosterol polyglycol ether sulfosuccinate	35% active	liquid	shampoos: shower preparations; foam baths; skin cleansing preparations; baby care products

Supplier	Trade name	Chemical description	Composition	General properties	Functionality / *Application*
Witco	REWOPOL B 1003	alkyl sulfosuccinamate	35% active	paste; surface tension 42 dynes/cm; pH 8-10 (1%)	emulsifier; adhesion agent / *emulsion polymerisation*
	REWOPOL B 2003	special sulfosuccinamate	35% active	low viscosity liquid; surface tension 40 dynes/cm; pH 7-8 (1%)	emulsifier; adhesion agent / *emulsion polymerisation*
	REWOPOL SB C 212	coconut acid alkylolamide sulfosuccinate	40% active	liquid	cleaners; light-duty detergents; shampoos; foam baths
	REWOPOL SB C 212 P	coconut acid alkylolamide sulfosuccinate	95% active	powder	cleaners; light-duty detergents; shampoos
	REWOPOL SB CS 50	disodium citric acid polyethoxy lauryl ether sulfosuccinate	40% active	liquid	shampoos; shower preparations; foam baths; baby shampoos; skin cleansing preparations
	REWOPOL SB DB 45	di-isobutyl sulfosuccinate	45% active	liquid; 49 dynes/cm; pH 5-7 (1% in H_2O)	emulsifier; stabiliser; dispersant / *emulsion polymerisation; pigments*
	REWOPOL SB DC 40	dicyclohexyl sulfosuccinate	40% active	paste; surface tension 40 dynes/cm; pH 5-7 (1%)	
	REWOPOL SB DD 65	di-isodecyl sulfosuccinate	65% active	liquid; surface tension 31.5 dynes/cm; pH 5-6 (1%)	
	REWOPOL SB DO 75	di-isooctyl sulfosuccinate	75% active	liquid; surface tension 27 dynes/cm; pH 5-7 (1%)	
	REWOPOL SB F 12	lauryl alcohol sulfosuccinate	40% active	paste	spray drying additive / *detergents; soaps*
	REWOPOL SB F 12 P	lauryl alcohol sulfosuccinate	95% active	powder	
	REWOPOL SB FA 30	lauryl alcohol polyglycol ether sulfosuccinate	40% active	liquid	shampoos; foam baths; liquid soaps
	REWOPOL SB FA 50	fatty alcohol polyglycol ether sulfosuccinate	30% active	liquid	emulsion polymerisation
	REWOPOL SB L 203	lauric acid alkylolamide sulfosuccinate	40% active	paste	detergents; soaps; shampoos; carpet cleaners
	REWOPOL SB L 203 P	lauric acid alkylolamide sulfosuccinate	95% active	powder	carpet cleaners; soaps; household cleaners
	REWOPOL SB MB 80	di-isohexyl sulfosuccinate	80% active	viscous liquid; surface tension 30 dynes/cm; pH 5-7 (1%)	emulsifier; stabiliser; dispersant / *emulsion polymerisation; pigments*
	REWOPOL SB V	modified sulfosuccinate	28% active	liquid	foam baths; shampoos; dishwashing agents; light-duty detergents
	REWOPOL SB Z	coconut acid isopropanolamide polyglycol ether sulfosuccinate	50% active	liquid	foam baths; shampoos; light-duty detergents
Zschimmer & Schwarz	FLEROGUM 2317	sulfosuccinamate		straw-coloured to white dispersion	stabiliser / *textile industry; carpets*
	FLEROGUM S 35	derivative of sulfosuccinic acid		straw-coloured dispersion	foam stabiliser / *textile industry*
	SETACIN 103 SPEZIAL	disodium semi-sulfosuccinate of ethoxylated fatty alcohol ($C_{12/14}$)	40% active	liquid	bath preparations; shampoos; cleansing preparations; liquid soaps
	SETACIN F SPEZIAL PASTE	disodium semi-sulfosuccinate of fatty alcohol ($C_{12/14}$)	40% active	paste	skin care products; cleansing preparations; bath preparations
	SETACIN M	semi-sulfosuccinate of a blend of ethoxylated fatty alcohols and fatty acid derivatives, neutralised with organic base	42% active	liquid	bath preparations; shampoos; cleansing preparations; liquid soaps
	TRIUMPHNETZER SPECIAL	sulfosuccinate		viscous liquid	wetting agent / *textile industry*

Supplier	Trade name	Chemical description	Composition	General properties	Functionality / *Application*
Zschimmer & Schwarz	TRIUMPHNETZER ZSG	sulfosuccinic acid diisooctyl ester	72% active	liquid	low foam; wetting agent

Taurates/isethionates

Supplier	Trade name	Chemical description	Composition	General properties	Functionality / *Application*
Akzo Nobel	ELFAN AT 84 G	sodium cocoyl isethionate	80% active	granulate	
	ELFAN AT 84	sodium cocoyl isethionate	80% active	powder	
	ELFAN AT 8430	sodium cocoyl isethionate	30% active	paste	
Croda	ADINOL CT95	sodium methyl cocoyl taurate		white powder	detergent; wetting agent / *oral hygiene products; facial washes*
	ADINOL OT16	sodium methyl oleoyl taurate and H$_2$O		clear to hazy colourless liquid	detergent; wetting agent / *shampoos; foam baths; skin cleansing preparations*
	ADINOL OT32	sodium methyl oleoyl taurate and H$_2$O		white slurry	
Hoechst	ARKOPON T	sodium methyl oleoyl taurate	64% active	yellowish white powder	wetting agent; dispersant / *construction industry; agriculture*
	HOSTAPON CT PASTE	sodium methyl cocoyl taurate	30% active	soft white paste	detergent; foaming agent / *toiletries*
	HOSTAPON KTW	sodium lauroyl taurate	50% active	white powder	detergent; foaming agent / *toothpaste; toiletries*
	HOSTAPON SCI GRANULES	sodium cocoyl isethionate fatty acid	85% active	granules	*toiletries*
	HOSTAPON SCLD	sodium cocoyl isethionate fatty acid	66% active		
	HOSTAPON SCI	sodium cocoyl isethionate	83.5% active		
	HOSTAPON T	sodium methyl oleoyl taurate	64% active	white powder	*toiletries; textile industry; agrochemicals*
ICI	ARLATONE SCI-70	sodium cocoyl isethionate and stearic acid		white flakes	o/w emulsifier; stabiliser / *personal care products*
	ARLATONE SCI	sodium cocoyl isethionate		white powder	o/w emulsifier; solubiliser / *personal care products*
Nikko Chemicals	NIKKOL CMT-30	sodium methyl cocoyl taurate; aq. soln.		colourless liquid	
	NIKKOL LMT	sodium N-lauroyl methyltaurate		white crystalline solid	foaming agent; cleansing agent / *cosmetics*
	NIKKOL MMT	sodium N-myristoyl methyltaurate		white crystalline solid	
	NIKKOL PMT	sodium N-palmitoyl methyltaurate		white crystalline solid	thickener; cleansing agent / *cosmetics*
	NIKKOL SMT	sodium N-stearoyl methyltaurate		white crystalline solid	
PPG	JORDAPON ACI-30	ammonium cocoyl isethionate	27% active; moisture 70%	clear liquid; viscosity 50 cPs; colour 3 (Gardner); surface tension 32 dynes/cm (0.1%); pH 6.8	foaming agent (high); wetting agent / *liquid soaps; skin cleansing preparations; shampoos; bath preparations; shower gels*
	JORDAPON CI-60	sodium cocoyl isethionate and stearic acid	50% active; total fatty acid 45%: stearic acid 40%: moisture 0.3%	white flakes; colour 25 (10% soln.; APHA); surface tension 27 dynes/cm (0.1%); pH 5.0 (10% soln.)	foaming agent (high); wetting agent / *liquid soaps; skin cleansing preparations; shaving products*
	JORDAPON CI-75	sodium cocoyl isethionate and stearic acid	65% active; total fatty acid 30%; stearic acid 25%; moisture 0.3%	white flakes; colour 25 (10% soln.; APHA); surface tension 27 dynes/cm (0.1%); pH 5.5 (10% soln.)	foaming agent (high); wetting agent / *liquid soaps; skin cleansing preparations; shaving products*

Supplier	Trade name	Chemical description	Composition	General properties	Functionality / *Application*
PPG	JORDAPON CI POWDER	sodium cocoyl isethionate	85% active; total fatty acid 5%; moisture 0.3%	white powder; colour 15 (10% soln.; APHA); surface tension 27 dynes/cm (0.1%); pH 5.8 (10% soln.)	foaming agent (high); wetting agent; *liquid soaps; skin cleansing preparations; shampoos; bath preparations; shower gels*
	JORDAPON CI-UP	sodium cocoyl isethionate	85% active; total fatty acid 5%; moisture 0.3%	white granular; colour 15 (10% soln.; APHA); surface tension 27 dynes/cm (0.1%); pH 5.8 (10% soln.)	foaming agent (high); wetting agent; *liquid soaps; skin cleansing preparations; shampoos; bath preparations; shower gels*
Rhone-Poulenc	GEROPON AS 200	sodium cocoyl isethionate		powder	foaming agent (moderate); wetting agent
	GEROPON T 77	sodium N-methyloleyl taurate		flake	foaming agent (high); foam booster; emulsifier; wetting agent; dispersant
Witco	WITCONATE NIS	sodium isethionate	56% active	liquid; pH 8.5	detergent; foaming agent; wetting agent; *personal care products; household and industrial applications*

Miscellaneous anionics

Supplier	Trade name	Chemical description	Composition	General properties	Functionality / *Application*
Air Products	SURFYNOL CT-141	low molecular weight dispersant			dispersant; viscosity modifier; *pigments*
Akcros Chemicals	AGRILAN DG102		90% active	hazy viscous amber liquid; pour point 7°C; viscosity 6000 cSt; pH 7.8 (1% aq.)	wetting agent; *agrochemicals*
	AGRILAN DG113		96% active; H_2O content 4%	slightly hazy colourless liquid; pour point −7°C; viscosity 66 cSt; pH 7.0 (1% aq.)	wetting agent; *agrochemicals*
	AGRILAN F524	neutral salt of a complex sulfonated carboxylic acid	55% active	clear dark amber liquid; pour point <0°C; viscosity 270 cSt; pH 6.0 (1% aq.)	anti-caking agent; viscosity modifier; additive; *agrochemicals*
	LANKROPEARL T101	liquid pearlising concentrate	40% active	opaque creamy viscous liquid; pour point <0°C; viscosity 8000 cPs; pH 7.5 (1% aq.)	pearlescent agent; viscosity modifier; *shampoos*
	LANKROPEARL T112	liquid pearlising concentrate	40% active	pearly white viscous liquid; pour point 0°C; viscosity 21,000 cPs; pH 7.5 (1% aq.)	
	LANKROPEARL T123	liquid pearlising concentrate	40% active	pearly white viscous liquid; pour point <0°C; viscosity 7500 cPs; pH 7.5 (1% aq.)	
	LANKROSOL HS112	potassium salt of carboxylic acid containing compound	36% active	clear liquid; pour point <0°C; viscosity 70 cSt; pH 7.5 (1 aq.)	hydrotrope; solubiliser; *dishwashing agents; industrial and household detergents*
Albright & Wilson	EMPIWAX CL	self-emulsifying wax	concentration 100%	flake	
	EMPIWAX SK/BP	self-emulsifying wax; B.P. grade	concentration 100%	flake	*pharmaceuticals*
	EMPIWAX SK	self-emulsifying wax	concentration 100%	flake	
Allied Colloids	ALCOFIX SX	high molecular weight polycondensation product		liquid	afterfixing agent; *textile industry; polyamide fibres; dyes*
	ALCOPOL CWA			liquid	wetting agent; penetrant; detergent; low foam; *textile industry*
	ALCOPOL KBW			liquid	wetting agent; dispersant; detergent; emulsifier; scouring agent; *textile industry*
	ALCOPOL NFW			liquid	wetting agent; de-aerating agent; anti-foaming agent; *textile industry*

Supplier	Trade name	Chemical description	Composition	General properties	Functionality / *Application*
Allied Colloids	ALCOPRINT PDN			liquid	dispersant / *textile industry; pigments*
	ALCOSIST AWA	surfactant blend		liquid	wetting agent; anti-frosting agent / *textile industry; dyeing; carpet industry*
	ALCOSIST NBA	high molecular weight polycondensation product		liquid	reserving agent / *textile industry; dyeing of blends*
	ALCOSPERSE AD	polymeric dispersant and protective colloid		liquid	dispersant; protective colloid; sequestration agent / *textile industry; fabric preparation; bleaching*
	ALCOSPERSE LAB	surfactant blend		liquid	levelling agent; dispersant; buffer / *textile industry; dyeing of polyesters*
	SAPOLIB S	blend of solvents and detergents		liquid	scouring agent; detergent; solvent / *textile industry; stain removal*
Auschem Cesalpinia	BETA MEL	melamine sulfate, sodium salt	40% active	liquid	super-fluidising agent / *concrete*
	BIODET B/D	dilaureth-7 citric acid	100% active	viscous liquid	
	BIODET B/TA	monolaureth-7 tartaric acid	100% active	viscous liquid	
	BIODET D	sodium dilaureth-7 citrate	25% active	liquid	detergent / *dishwashing agents; laundry products*
	BIODET T	sodium trilaureth-7 citrate	80% active	viscous liquid	
	BIODET TA	sodium monolaureth-7 tartrate	25% active	liquid	
	CEROL AG 2017	blend of anionic surfactants in oil	80% active	liquid; solidification point ca. <0°C	anti-caking agent; anti-dusting agent / *fertilisers*
	CEROL AG 4043	blend of anionic surfactants	80% active	liquid; solidification point ca. <0°C	
	CEROL AG 5717	blend of anionic surfactants	70% active	liquid; solidification point ca.	
	EMULSON B/D	POE (7) dialkyl citric acid	100% active	liquid	emulsifier; stabiliser / *pesticides*
	EMULSON B/M	POE (7) monoalkyl citric acid	100% active	liquid	
	EMULSON B/TA	POE (7) monoalkyl tartaric acid	100% active	liquid	
	EMULSON D	POE (7) sodium dialkyl citrate	25% active	liquid	emulsifier; viscosity modifier; pH modifier / *pesticides*
	EMULSON M	POE (7) sodium monoalkyl citrate	25% active	liquid	
	EMULSON TA	POE (7) sodium trialkyl citrate	25% active	liquid	
	EUCAROL B/D	dilaureth-7 citrate (acid form)	100% active	liquid	detergent; wetting agent / *bubble baths; shampoos; liquid soaps*
	EUCAROL B/M	laureth-7 citrate (acid form)	100% active	liquid	
	EUCAROL B/TA	laureth-7 tartrate (acid form)	100% active	liquid	
	EUCAROL D	sodium dilaureth-7 citrate	25% active	liquid	
	EUCAROL LS	sodium laureth-7 citrate and SLS	27% active	liquid	
	EUCAROL M	sodium laureth-7 citrate	25% active	liquid	
	EUCAROL TA	sodium laureth-7 tartrate	25% active	liquid	
	MADEOL AG 1319	blend of anionic surfactants	100% active	powder	anti-caking agent; anti-dusting agent / *fertilisers*
	ROLPON C 200	sodium trideceth-7 carboxylate	28% active	liquid	foaming agent; detergent / *liquid detergents; textile industry; laundry products; carpet cleaners; toiletries*
BASF	LUTENSIT A-FK	fatty acid condensation product, sodium salt	55% active		

Supplier	Trade name	Chemical description	Composition	General properties	Functionality Application
BASF	LUTENSIT A-PS	sodium alkylsulfonate	65% active	clear, yellow low-viscosity liquid	emulsifier; wetting agent; scouring agent *textile industry; cellulosic fibres*
Bayer	LEVAPON TH LIQUID	mixture of alkyl sulfonate with polyglycol ether	70% active	yellowish wax	o/w emulsifier
CHEM-Y	AKYPO MULS 400	PEG-9 stearamide carboxylic acid	active matter 83% min.; H_2O 11-15%; chloride (as NaCl) 2.0% max.; acid value 50-70	clear, almost colourless liquid; M.W. ca. 707; viscosity 800-2000 mPas; colour 125 max. (Hazen); pH 7.0-8.0 (10% in H_2O)	*household and industrial cleaners*
	AKYPO SOFT 100 BVC	sodium laureth-11 carboxylate and laureth-10	active matter 69-72%; H_2O 23-28%; chloride (as NaCl) 3.0-5.0%	clear, slightly yellowish liquid; viscosity ca. 500 mPas; colour 250 max. (APHA); pH 7.0-8.0	foaming agent *shampoos; bubble baths; shower preparations; liquid soaps*
	AKYPO SOFT KA 250 BV	mixture of ethoxylated and carboxylated coconut fatty acid monoethanolamide and glycerin derivatives	active matter 29% min.; H_2O 66.5% max.; chloride (as NaCl) 4.5-5.5%	clear, yellowish, slightly viscous liquid; viscosity ca. 600 mPas; colour 300 max. (APHA); pH 7.0-8.0 (10% in H_2O)	foaming agent *shampoos; bubble baths; shower preparations; liquid soaps*
	AKYPO SOFT KA 250 BVC	mixture of ethoxylated and carboxymethylated coconut fatty acid monoethanolamide and glycerin derivatives	active matter 50% min.; H_2O 38.0-41.0%; chloride (as NaCl) 7.5-9.0%		
	AKYPO TFC-S	mixture of laureth-5 carboxylic acid and sodium octyl sulfate	H_2O 40.0-43.0%	light yellow liquid; viscosity 200-400 mPas; colour 3 max. (Gardner); pH 2-3	disinfectant; cleaning agent; solubiliser *industrial cleaners; perfumes*
	AKYPO TFC-SN	mixture of laureth-5 carboxylic acid, sodium octyl sulfate and isostearic acid	H_2O 37.0-40.0%	clear to slightly hazy, pale yellow liquid; viscosity 3000-9000 mPas; colour 4 max. (Gardner); pH 3.0-4.0 (1% in H_2O)	thickener; disinfectant; cleaning agent *industrial cleaners*
	AKYPOSAL 2010 SD	a mixture of ether carboxylate and pearling agents	active matter 39% min.; H_2O 54-56%; chloride (as NaCl) 3.5-4.5%	opaque, colourless to pale yellowish, thixotropic liquid; viscosity 1500-7000 mPas; pH 6.0-7.0	pearlescent agent
Chemax	SURMAX CS-504	alkaline stable surfactant designed for use in formulating alkaline detergent concentrates	80% active; H_2O 20%	clear amber to brown liquid; cloud point > 100°C (1% aq.); viscosity 1390 cP's; surface tension 38 dynes/cm (0.05% in distilled H_2O); pH 7 (5% aq.)	wetting agent; detergent; synergist; foaming agent (moderate) *metal working; paint industry; industrial cleaners; paper industry; sanitisers; liquid detergents; powder detergents*
	SURMAX CS-515	alkaline stable surfactant designed for use in formulating alkaline detergent concentrates	80% active; H_2O 20%	clear amber to brown liquid; cloud point > 100°C (1% aq.); viscosity 1490 cP's; surface tension 40 dynes/cm (0.05% in distilled H_2O); pH 7 (5% aq.)	wetting agent; detergent; synergist; foaming agent (moderate) *metal working; paint industry; industrial cleaners; paper industry; sanitisers; liquid detergents; powder detergents*
	SURMAX CS-521	alkaline stable surfactant designed for use in formulating alkaline detergent concentrates	80% active; H_2O 20%	clear amber to brown liquid; cloud point > 100°C (1% aq.); viscosity 1280 cP's; surface tension 57 dynes/cm (0.05% in distilled H_2O); pH 7 (5% aq.)	wetting agent; detergent; synergist; low foam *metal working; paint industry; industrial cleaners; paper industry; sanitisers; liquid detergents; powder detergents*
	SURMAX CS-522	alkaline stable surfactant designed for use in formulating alkaline detergent concentrates	80% active; H_2O 20%	clear amber to brown liquid; cloud point > 100°C (1% aq.); viscosity 1550 cP's; surface tension 51 dynes/cm (0.05% in distilled H_2O); pH 7 (5% aq.)	wetting agent; detergent; synergist; low foam *metal working; paint industry; industrial cleaners; paper industry; sanitisers; liquid detergents; powder detergents*
Croda	CROLACTIL SISL	sodium isostearoyl lactylate	acid value 60-90; sodium content 3.5-5.0%; sapon. no. 210-280	HLB ca. 6.5; yellow viscous liquid	emulsifier; emollient *food industry; skin care products; deodorants; colognes*
	CROLACTIL CSL	calcium stearoyl lactylate	acid value 50-90; calcium content 4.0-5.0%; sapon. no. 175-250	HLB ca. 5.0; off-white powder	
	CROLACTIL SSL	sodium stearoyl lactylate	acid value 60-90; sodium content 3.5-5.0%; sapon. no. 210-280	HLB ca. 6.5; off-white powder	
Cytec	AEROSOL DPOS-45	disodium mono- and didodecyl diphenyl oxide disulfonate	45% in H_2O	liquid; surface tension 34 dynes/cm min. (in H_2O)	emulsifier; dispersant; solubiliser; coupling agent *emulsion polymerisation; latexes; soaps; cleaners; agrochemicals; dyes*

Supplier	Trade name	Chemical description	Composition	General properties	Functionality Application
Dow Chemical	DOWFAX 2A0	disulfonated surfactant with tetrapropylene hydrophobe source; acid form	40% active	amber liquid; M.W. 524; viscosity 235 cPs; surface tension 30 dynes/cm (0.1% in H_2O); Ross-Miles foam height: initial 170 mm, 5 min 143 mm	
	DOWFAX 2A1	disulfonated surfactant with tetrapropylene hydrophobe source	45% active	clear amber liquid; M.W. 576; viscosity 145 cPs; surface tension 31 dynes/cm (0.1% in 0.1M NaCl); Ross-Miles foam height (1.0% in H_2O): initial 150 mm, 5 min 140 mm	detergent; solubiliser *bleach formulations; hard-surface cleaners*
	DOWFAX 3B0	disulfonated surfactant with C_{10} alpha-olefin hydrophobe source; acid form	40% active	amber liquid; M.W. 498; viscosity 110 cPs; surface tension 36 dynes/cm (0.1% in H_2O); Ross-Miles foam height (1.0%): initial 173 mm, 5 min 140 mm	
	DOWFAX 3B2	disulfonated surfactant with C_{10} alpha-olefin hydrophobe source	45% active	clear amber liquid; M.W. 542; viscosity 120 cPs; surface tension 35 dynes/cm (0.1% in 0.1M NaCl); Ross-Miles foam height (1.0% in H_2O): initial 155 mm, 5 min 145 mm	dispersant; wetting agent; solubiliser; detergent *bleach formulations; hard-surface cleaners*
	DOWFAX 8390	disulfonated surfactant with C_{16} alpha-olefin hydrophobe source	35% active	clear amber liquid; M.W. 643; viscosity 10 cPs; surface tension 47 dynes/cm (0.1% in 0.1M NaCl); Ross-Miles foam height (1.0% in H_2O): initial 120 mm, 5 min 40 mm	detergent; solubiliser *bleach formulations; hard-surface cleaners*
	DOWFAX C10L	disulfonated surfactant with C_{10} alpha-olefin hydrophobe source	45% active	clear amber liquid; M.W. 555; viscosity 120 cPs; surface tension 35 dynes/cm (0.1% in 0.1M NaCl); Ross-Miles foam height (1.0% in H_2O): initial 150 mm, 5 min 140 mm	solubiliser; detergent *bleach formulations; hard-surface cleaners*
	DOWFAX C6L	disulfonated surfactant with C_6 alpha-olefin hydrophobe source	45% active	clear amber liquid; M.W. 474; viscosity 130 cPs; surface tension 32 dynes/cm (0.1% in 0.1M NaCl); Ross-Miles foam height (1.0% in H_2O): initial 195 mm, 5 min 170 mm	hydrotrope; detergent; solubiliser; wetting agent *bleach formulations; hard-surface cleaners*
Elf Atochem	BEYCOPON S 3A	alkylsulfate alkylaryl sulfonate complex, sodium salt		liquid	
	BEYCOPON TB	alkylsulfate alkylaryl sulfonate complex, sodium salt		liquid	
	BEYCOPON TL	alkylsulfate alkylaryl sulfonate complex, sodium salt		liquid	
	DETERPAL 832	alkylether sulfonate and solvent		liquid	
	DETERPAL LC	alkyl sulfonate and solvent		liquid	
	OLEINE 549	antioxidised self-emulsifiable oleine			
	OLEINE D	oleic acid			
Hays Colours	KENALEV CSC				levelling agent *textile industry; dyeing*
	ZYMOLENE EM				detergent *textile industry*
Henkel	BREVIOL DO	partially sulfated fatty alcohol polyglycolether			dispersant; levelling agent *textile industry; dyeing*
	COTTOCLARIN AS	blend of surfactants			wetting agent; boiling-off agent; bleaching and dyeing
	COTTOCLARIN CL	blend of surfactants			auxiliary; dispersant *textile industry; pretreatment*

Supplier	Trade name	Chemical description	Composition	General properties	Functionality / Application
Henkel	COTTOCLARIN F	blend of surfactants			wetting agent; boiling-off agent; bleaching and dyeing auxiliary; dispersant / textile industry; pretreatment
	COTTOCLARIN KD	blend of surfactants			
	COTTOCLARIN OK	blend of surfactants			
	COTTOCLARIN SV	blend of surfactants			
	COTTOCLARIN TC	blend of surfactants			
	COTTOCLARIN VK	blend of surfactants			
	CUTINA LE	glyceryl stearate and sodium cetearyl sulfate		granules	cream base / creams; ointments
	DOITTOL 891		20% active	liquid	anti-blocking agent / rubber; latex
	DOITTOL K-21		25% active	liquid	
	DYMSOL 31-P		55% active	liquid	stabiliser / rubber; latex
	DYMSOL 36-A		24.5% active	liquid	emulsifier / rubber; latex
	EMULGADE CL SPECIAL	mixture of isopropylmyristate, stearic acid, cetyl palmitate, coconut acid, cetearyl alcohol, glyceryl stearate, potassium stearate, potassium cocoate, sodium stearate and sodium cocoate		lard-like mass	o/w base / o/w creams; o/w emulsions
	EMULGADE CL	mixture of octylstearate, stearic acid, cetyl palmitate, coconut acid, cetearyl alcohol, glyceryl stearate, potassium hydroxide and NaOH		lard-like mass	o/w base / o/w creams; o/w emulsions; skin care products
	EMULGADE F	cetearyl alcohol, PEG-40 castor oil and sodium cetearyl sulfate		flaked solid	personal care products
	EW-POL 9110	ammonium oleate	ca. 50% active	liquid	emulsifier / polymers; synthetic rubber
	FLORANIT 4028	derivative of alcohols			wetting agent / textile industry; pretreatment
	FLORANIT LT	derivative of alcohols			wetting agent; low foam / textile industry; pretreatment
	FUMAN L	combination of solvent and detergents without chlorinated hydrocarbons			cleaning agent / textile industry; pretreatment
	GERBASOL 4133	combination of alkylbenzene sulfonates and anionic detergents			scouring agent / textile industry; pretreatment
	KATAX 570	alkylamine polyglycolether			antistatic agent / textile industry; spinning
	LAMECREME KSM	glycerin mono/distearate with potassium stearate		pastilles	emulsion base / o/w creams; o/w ointments
	LAMECREME LPM	combination of glycerin mono/distearates with lipopeptides and fatty alcohols		pastilles	cream base / o/w creams; o/w ointments
	LANETTE N	cetearyl alcohol and sodium cetearyl sulfate		white to pale yellow granules	cream base / o/w creams; o/w ointments; liquid liniments
	LANETTE SX	cetearyl alcohol and sodium lauryl sulfate		white granules	cream base / o/w creams; o/w emulsions

Supplier	Trade name	Chemical description	Composition	General properties	Functionality / Application
Henkel	LANETTE W	cetearyl alcohol and sodium lauryl sulfate		granules	cream base *creams; ointments; liniments*
	MILTOPAN D 503	combination of anionic detergents with solvents			detergent *textile industry; pretreatment*
	MILTOPAN SE	detergents with emulsifiers and fibre protecting agent			degumming agent *textile industry; pretreatment*
	NOPCOFLOC 131-L			emulsion	machine drying and retention aid; sludge conditioner
	NOPCOFLOC 144-L			emulsion	*paper industry*
	NOPCOSANT K		34% active	liquid	low foam; dispersant *paint industry; carpet backing; paper industry*
	NOPCOTE 1338		70% active	liquid	flow levelling aid *paper industry; pigments; coatings*
	NOPCOWET ENA 307			liquid	wetting agent *adhesives*
	SELBANA 4243	combination of antistatics and adhesive agent			lubricant *textile industry; spinning*
	TEXAPHOR 873	aliphatic ester in di-2-ethylhexyl phthalate	47–50% active	clear, brown liquid; viscosity 1200–2400 mPas	additive *pigment pastes; coatings; printing inks; plastics industry*
	TEXAPHOR SPECIAL	highly concentrated soln. of surface-active anionic compounds		yellowish to reddish liquid; pH 5.5–7.0 (10% aq. soln.)	anti-settling agent *coatings; lacquer; decorative paints*
	TEXAPON MG 3	mixture of magnesium lauryl sulfate and disodium laureth sulfosuccinate	30–32% active	liquid	*shower and bath preparations*
	TEXAPON SBN	sodium laureth sulfate and disodium laureth sulfosuccinate	28–30% active	liquid	*shampoos; baby shampoos; bath preparations; cosmetics*
	TEXIN OSS 50 K	dipotassium oleic acid sulfonate	ca. 50%	liquid	emulsifier *emulsion polymerisation*
Hoechst	HOSTAPON KCG	sodium cocyl glutamate	25% active		*cosmetics; toiletries*
Hüls	MARLOWET 5609	carboxylic acid, amine salt	active detergent 100%	liquid	*cleaners*
	MARLOWET 5622	TEA salt of oleic acid	active detergent 100%	liquid	emulsifier
	MARLOWET 5635	TEA salt of oleic acid	active detergent 90%	liquid	*mineral oils; bitumen; white spirit; metal working; floor*
	MARLOWET T	TEA salt of oleic acid	active detergent 100%	paste	*cleaners*
ICI	ARLAMOL 801	mixture of nonionic surfactants		white solid	emollient *personal care products*
	ATLOX 5320	mixture of anionic surfactants		yellow liquid	emulsifier; dispersant *agrochemicals*
	TENSIANOL 399 ISL	syndet based on alkylsulfate and cocoylisethionate		white to ivory flakes	
	TENSIANOL 399 KS1	syndet based on alkylsulfate		white to ivory flakes	*syndet bars*
	TENSIANOL 399 KSS	syndet based on alkylsulfate and lauryl sulfosuccinate		white to ivory flakes	
	TENSIANOL 399 N1	syndet based on alkylsulfate		white to ivory flakes	

Supplier	Trade name	Chemical description	Composition	General properties	Functionality / Application
ICI	TENSIANOL 399 SCI/LSA	syndet based on cocoylisethionate and lauryl sulfosuccinate		white to ivory flakes	*syndet bars*
	TENSIANOL 399 SCIL	syndet based on cocoylisethionate		white to ivory flakes	
Kao Corporation	ICTEOL K-50	potassium soap	47-49% active	paste; pH 8-10.5 (5%)	
Millchem	MILLIFOAM	formulated anionic	solids 55%	liquid	foaming agent; *plasterboard*
Mona Industries	MONA NF-10		solids 50%	clear; viscosity 1000 cP; pH 9.2	low foam; solubiliser; detergent; wetting agent; *spray, soak tank, in-place pipeline cleaners; floor cleaners*
	MONA NF-15		solids 50%	clear; viscosity 900 cP; pH 7.5	
	MONA NF-25		solids 50%	clear; viscosity 900 cP; pH 7.9	
Nikko Chemicals	NIKKOL LSA	sodium lauryl sulfoacetate		white crystalline solid	foaming agent; cleansing agent; *cosmetics*
	NIKKOL SGC-80N	sodium cocomonoglyceride sulfate		white powder	
Olin Chemicals	POLY-TERGENT 2A1	branched C_{12} alkyl disulfonate, sodium salt	45% active	viscosity 120 cP; surface tension 45.1 dynes/cm (0.1%); Draves wetting 152 sec (0.25%); Ross-Miles foam (0.25%): initial 120 mm, 5 min 120 mm	
	POLY-TERGENT 2EP	branched C_{12} alkyl disulfonate, sodium salt	45% active	viscosity 120 cP; surface tension 45.1 dynes/cm (0.1%); Draves wetting 152 sec (0.25%); Ross-Miles foam (0.25%): initial 120 mm, 5 min 120 mm	
	POLY-TERGENT 3B2	linear C_{10} alkyl disulfonate, sodium salt	45% active	cloud point 75°C (1% aq. soln.); viscosity 150 cP; surface tension 43 dynes/cm (0.1%); Draves wetting 278 sec (0.25%); Ross-Miles foam (1.0%): initial 135 mm, 5 min 130 mm	foaming agent
	POLY-TERGENT 4C3	linear C_{16} alkyl disulfonate, sodium salt	35% active	viscosity 150 cP; surface tension 47.6 dynes/cm (0.25%); Draves wetting 7300 sec (0.25%); Ross-Miles foam (0.25%): initial 140 mm, 5 min 120 mm	
PPG	AVANEL S-30	sodium $C_{12/15}$ pareth-3 sulfonate	solids 35%	white paste; M.W. 420; pour point 25°C; viscosity 360 cPs (35°C)	*personal care products; household, industrial and institutional applications*
	AVANEL S-35	sodium octoxynol-2 ethane sulfonate	solids 35%	white liquid; M.W. 424	
	AVANEL S-70	sodium $C_{12/15}$ pareth-7 sulfonate	solids 35%	clear liquid; M.W. 600; pour point −1°C; viscosity 270 cPs	
	AVANEL S-74	sodium C_8 pareth-3 sulfonate	solids 35%	clear liquid; M.W. 260; pour point −8°C; viscosity 30 cPs	
	AVANEL S-150	sodium $C_{12/15}$ pareth-15 sulfonate	solids 35%	clear liquid; M.W. 950; pour point −1°C; viscosity 70 cPs	
	AVANEL S-150 CG	sodium $C_{12/15}$ pareth-15 sulfonate; cosmetic grade	solids 35%	white liquid; M.W. 424; pour point −1°C; viscosity 70 cPs	
	LAROSOL DBL-3 CONC.		65% active	opaque liquid	lubricant; *dyeing; textile industry*
	LAROSOL DBL		24% active	opaque liquid	lubricant; *dyeing; textile industry*
	LAROSOL NRL CONC.	complex anionic		liquid	levelling agent; migrating agent; *dyeing; carpet industry; textile industry*

Supplier	Trade name	Chemical description	Composition	General properties	Functionality / Application
PPG	LAROSOL NRL-40			liquid	levelling agent; migrating agent / *dyeing; carpet industry; textile industry*
	LAROSOL PRM-98		88% active	liquid	scouring agent; levelling agent / *dyeing; textile industry*
	MAZON 60T				detergent; foaming agent (high) / *metal working; metal cleaning; textile industry; household detergents; dishwashing agents*
	MAZON 70				detergent; low foam / *metal working; industrial and institutional cleaning*
	MAZON 85				degreaser; detergent; emulsifier / *metal working*
	MAZOX KCAO	phosphated amine oxide		liquid	cleaning agent / *textile industry and dyeing equipment*
	SAM 211A-80	surface active monomer	80% active	HLB 43.8; beige viscous liquid; viscosity 1120 cPs	emulsifier / *latex products*
Protex	SURFARON 1712 N45	sodium tetrapropyl diphenyl oxide sulfonate			dispersant
	SURFARON A6008 N40	sodium ethyl hexyl sulfonate			low foam; wetting agent
Raschig	RALUFON EA 15-90	sulfopropylated alkylalkoxylate			low foam / *electroplating; metal cleaning; oil recovery; emulsion polymerisation; paper industry; textile industry*
	RALUFON F 11-13	polyethylene glycol-alkyl-(3-sulfopropyl)-diether, potassium salt			
	RALUFON F 3-13	polyethylene glycol-alkyl-(3-sulfopropyl)-diether, potassium salt			
	RALUFON F 4-I	polyethylene glycol-alkyl-(3-sulfopropyl)-diether, potassium salt			
	RALUFON F 5-13	polyethylene glycol-alkyl-(3-sulfopropyl)-diether, potassium salt			
	RALUFON F 7-13	polyethylene glycol-alkyl-(3-sulfopropyl)-diether, potassium salt			foaming agent (high) / *electroplating; metal cleaning; oil recovery; emulsion polymerisation; paper industry; textile industry*
	RALUFON N 3.5	polyethylene glycol-(4-nonyl-phenyl)-(3-sulfopropyl)-diether, potassium salt			
	RALUFON N 6	polyethylene glycol-(4-nonyl-phenyl)-(3-sulfopropyl)-diether, potassium salt			
	RALUFON N 9	polyethylene glycol-(4-nonyl-phenyl)-(3-sulfopropyl)-diether, potassium salt			
	RALUFON N 10	polyethylene glycol-(4-nonyl-phenyl)-(3-sulfopropyl)-diether, potassium salt			
	RALUFON N 20	polyethylene glycol-(4-nonyl-phenyl)-(3-sulfopropyl)-diether, potassium salt			
	RALUFON NAPE 14-90	sulfoalkylated polyalkoxylated β-naphtol, potassium salt			low foam / *electroplating; metal cleaning; oil recovery; emulsion polymerisation; paper industry; textile industry*

Supplier	Trade name	Chemical description	Composition	General properties	Functionality / *Application*
Rhone-Poulenc	COPS 1	sodium salt of allyl ether sulfonate	40% solids	liquid	low foam / *emulsion polymerisation*
	MIRANATE B	sodium butoxyethoxyacetate	52% solids; NaCl 3.0%	liquid	hard-surface cleaners
	MIRANATE T/36-DF	sodium salt of polycarboxylic acid	25% solids	liquid	dispersant
	MIRANATE TA/72-S	sodium salt of polycarboxylic acid	90% solids	powder	
	MIRANATE TA/K 30	potassium salt of polycarboxylic acid	30% solids	liquid	*emulsion polymerisation*
	RHODACAL DSB	sodium dodecyl diphenyloxide disulfonate	45% solids	liquid; surface tension 32 dynes/cm	emulsifier; wetting agent; dispersant; foaming agent (high); foam booster / *emulsion polymerisation*
Rudolf Chemicals	RUCOGEN DWA-P	combination of fatty alcohol ethoxylate, hydrocarbons and fatty acid salts		clear, colourless liquid	detergent; low foam; dispersant; emulsifier; fat-dissolving agent / *textile industry*
Sandoz	CERANINE VRE			paste	softener; lubricant
	DEGREASER ESM	blend of solvents and anionic emulsifier		liquid	emulsifier; dispersant; solvent / *grease removal*
	DERMAGEN A-PRN	chlorotriazine derivative in aq. medium		paste	levelling agent; dispersant / *dyes*
	DILATIN 2AP	aliphatic phosphoric acid ester and aryl sulfonate in aq. soln.		liquid	wetting agent; detergent; low foam
	DILATIN LFR	blend of solvents, alkyl ester and emulsifier		liquid	solvent; emulsifier / *leather industry*
	DILATIN OFBM	aromatic ester, alkyl polyglycol ether and alkyl polyglycol ether sulfate		liquid	levelling agent / *textile industry; dyes; polyester fibres*
	DILATIN TPC	phenol derivative in an anionic aq. organic soln.		liquid	carrier / *textile industry; dyes; polyester fibres*
	DRIMAGEN ER	aromatic polyether sulfonate in aq. soln.		liquid	levelling agent / *textile industry; dyes*
	IMEROL AN	alkyl acid salt in aq. solvent system; contains pine oil		liquid	dispersant / *fats*
	IMEROL RT	blend of solvents and anionic emulsifier		liquid	emulsifier; detergent; solvent; spotting agent / *textile industry; industrial applications*
	IMEROL XN	polyglycol ether in aq. organic soln.		liquid	scouring agent; stain removal agent
	LYOCOL RDN	polyaryl ether sulfate		liquid	dispersant / *textile industry; dyeing*
	LYOGEN P	sulfonated oil in aq. soln.		liquid	levelling agent / *textile industry; dyeing*
	LYOGEN PLM	aromatic ester and fatty acid ester		liquid	levelling agent; migrating agent
	LYOGEN SU	modified polyglycol ether in aq. soln.		liquid	levelling agent / *textile industry; dyeing*
	LYOGEN TT	blend of emulsified esters		liquid	levelling agent
	MERCEROL QWLF	alkyl sulfonate		liquid	wetting agent; low foam
	MERCEROL SAW	sodium salt of a partial phosphate ester and alkyl sulfate in aq. soln.		liquid	wetting agent / *textile industry; mercerising liquors; dyeing*

Supplier	Trade name	Chemical description	Composition	General properties	Functionality Application
Sandoz	SANDOFLEX A	sulfuric ester/fatty acid derivative alkyl sulfonate		liquid	emulsifier *white spirit*
	SANDOGEN L2P	modified polyglycol ether in aq. soln.		liquid	
	SANDOLIX ON	sulfated fatty ester in aq. soln.		liquid	
	SANDOLIX SNE	sulfated animal oil		liquid	
	SANDOLIX SPE	sulfated natural oil		liquid	liquoring agent *fats*
	SANDOLIX WWL	fatty acid derivative in aq. soln.		paste	
	SANDOPAN 2N MOD	modified polyglycol ether		liquid	emulsifier *white spirit*
	SANDOPAN 9412 E	modified polyglycol ether		liquid	wetting agent; low foam
	SANDOPAN B MOD	modified glycol ether		liquid	wetting agent; dispersant; degreaser; low foam
	SANDOPAN BLF	polyglycol ether, fatty acid salt and aliphatic alcohol		liquid	
	SANDOPAN CBN	phosphate esters and aryl sulfonate		liquid	wetting agent; detergent; low foam
	SANDOPAN D-B LIQUID	alkoxylate containing an anionic group, sodium salt		clear, light amber liquid	*cleaners*
	SANDOPAN DKS	alkyl polyglycol ether		solid	emulsifier *cosmetics; toiletries; oils*
	SANDOPAN DLS 115%	alkyl polyglycol ether		liquid	*shampoos*
	SANDOPAN DPA	carboxylated derivative of alkylphenol ethoxylate		liquid	*toiletries; industrial applications*
	SANDOPAN DPB	carboxylated derivative of alkylphenol ethoxylate		liquid	*bleach-based products*
	SANDOPAN DPC	carboxylated derivative of alkylphenol ethoxylate		liquid	
	SANDOPAN DTA	modified polyglycol ether		liquid	*toiletries; industrial applications*
	SANDOPAN DTCL	modified polyglycol ether		liquid	detergent; wetting agent *textile industry; industrial applications*
	SANDOPAN DTD	modified polyglycol ether		liquid	*toiletries; industrial applications*
	SANDOPAN KDC	modified polyglycol ether		liquid	detergent; wetting agent; dispersant *toiletries; textile industry; industrial applications*
	SANDOPAN KST SOLID	carboxymethylated alkoxylate, sodium salt		white wax	emulsifier *skin care products*
	SANDOPAN PSC	aliphatic phosphate ester/aryl sulfonate and polyglycol ether		liquid	detergent; low foam
	SANDOPEROL PE	polyacrylic acid in aq. soln.		liquid	sequestering agent; non-foaming *polymers*
	SANDOPHOR SHA	alkyl polyglycol ether phosphate		liquid	wetting agent *detergents; cutting fluids*
	SANDOPUR RSK	polycarbonic acid in aq. soln.		liquid	sequestering agent; non-foaming *polymers*

Supplier	Trade name	Chemical description	Composition	General properties	Functionality / *Application*
Sandoz	SANDOPUR SR	blend of surfactant and phosphate builder		liquid	detergent; *textile industry; dyes*
	SANDOZIN AMP	alkyl sulfonate and organic phosphate		paste	wetting agent; de-aerating agent; low foam
	SANDOZIN N	carboxylic acid ester, solvent containing		liquid	wetting agent
	SANDOZIN NE	fatty acid derivative		liquid	wetting agent; *textile industry*
	SANDOZIN PI	carboxylate of aryl phenol ethoxylate		liquid	levelling agent; *textile industry*
	SANDOZOL KB	sulfonated oil in aq. soln.		liquid	
	SIRRIX AK	phosphoric acid derivative in aq. soln.		liquid	sequestering agent; *textile industry*
	SIRRIX CRC	sodium salt of alkyl sulfate and aliphatic carboxylic acid in aq. organic soln.		liquid	low foam; complexing agent; *textile industry*
	STABILIZER AWN	aq. soln. of inorganic salt and surfactant		liquid	stabiliser; low foam; *peroxide bleaching*
	TERGOTAN EFL	polypetide and aromatic sulfonate in aq. soln.		liquid	re-tanning agent; *leather industry*
Servo Delden	SERVO BRILLANT OLIE BAZ 75	castor oil sulfonate	77% active	liquid	
	SERVON XJS 112	cocoamine polyglycol ether (45EO) sulfate	52.5% active	liquid	
Stepan Europe	ALPHA-STEP ML40	sodium alkyl alpha sulfomethylester; made-to-order	37% active	pale yellow liquid	foaming agent; detergent; *household cleaners; liquid soaps*
	BIO-TERGE PAS 8 S	sodium octyl sulfonate	33% active	water white to pale yellow liquid	solubiliser; wetting agent; hydrotrope; low foam; *household, institutional and industrial cleaners; textile industry*
	LATHANOL LAL	sodium lauryl sulfoacetate	65% active	white powder or flakes	foaming agent; thickener; *toothpastes; shampoos; bath preparations; shower preparations*
	NINATE PA	sodium polycarboxylate; made-to-order		orange viscous liquid	dispersant; fluidifying agent; stabiliser; *agrochemicals*
Troy Chemicals	TROYSOL LAC			liquid	spreading agent; wetting agent; anti-cratering agent; *coatings; paint industry*
	TROYTHIX ANTI-SAG 4			liquid	rheology control agent
Unger Fabrikker	UFABLEND DC	anionic concentrated blended products	67% active	yellow viscous liquid; pH 7–8 (1% soln.)	dishwashing agents; *hard-surface cleaners*
Union Carbide	TRITON DF-20	low-foam surfactant	100% active	liquid; pour point −15°C	low foam; *metal cleaning; industrial cleaners*
	TRITON X-200	sulfonate	28% active	aq. dispersion; pour point −4°C	*cosmetics; emulsion polymerisation; metal cleaning*
Warwick International	WARCOSOFT 8190		100% active	liquid	softener; *textile industry*
Witco	REWOCOROS B 3010	alkenyl succinate, sodium salt	40% active	liquid	anti-corrosion
	REWOCOROS RA B 90	modified boric acid ester	100% active	liquid	anti-corrosion
	REWOCOROS RA BE	boric acid amine ester	100% active	liquid	anti-corrosion

Supplier	Trade name	Chemical description	Composition	General properties	Functionality / *Application*
Witco	REWOCOROS TPAC 100	tallow propylene diammonium caprylate	100% active	liquid	anti-corrosion / *oils; boiler feed water systems*
	REWOCOROS TS 25	surfactant blend	25% active	liquid	*carpet and upholstery cleaners*
	REWOCOROS TS 35	surfactant blend	35% active	low viscosity liquid	*carpet and upholstery cleaners*
	REWOPOL CHT 12	tricarboxymethyl diamino alkylamide	40% active	liquid	sequestration agent; complexing agent / *detergents*
	REWOPOL HD 50 L	surfactant blend	50% active	liquid	pre-concentrate / *liquid heavy-duty detergent*
	REWOPOL HM 14	surfactant blend	28% active	liquid	*shampoos; bath preparations*
	REWOPOL HM 28	surfactant blend	28% active	liquid	*dishwashing agents; liquid soaps; all-purpose cleaners*
	REWOPOL HM 30	surfactant blend	28% active	liquid	*shampoos; bath preparations*
	REWOPOL PGK 2000	surfactant blend with pearlising agent	35% active	liquid	pearlescent agent / *shampoos; foam baths; household cleaners*
	REWOPOL S 1954	surfactant blend	45% active	liquid	*dishwashing agents; all-purpose cleaners*
	REWOPOL S 2311	surfactant blend	40% active	liquid	*dishwashing agents; all-purpose cleaners*
	REWOPOL TS 100	surfactant blend	95% active	powder	*carpet and upholstery cleaners*
	REWOPOL TS 40 P	surfactant blend and polymers	30% active	emulsion	*carpet cleaners*
	REWOPOL TS K 30	surfactant blend and polymers	30% active	liquid	
	REWOPOL TS SP 25	surfactant blend and polymers	25% active	emulsion	
	REWOPOL WS 11	surfactant blend	30% active	liquid	*wool washing*
	REWOPOL WS 12	surfactant blend	30% active	liquid	*wool washing*
	SPONTO 168-D	anionic surfactant blend		liquid	compatibility agent / *agrochemicals*
	SPONTO 169-T	anionic surfactant blend		liquid	*agrochemicals; insecticides*
	WITBREAK 1390	alcohol ether sulfonate		liquid; pH 7.8	foaming agent / *petroleum industry*
	WITBREAK 770	surfactant blend		liquid; pH 6.5	demulsifier; slugging compound / *petroleum industry*
	WITCO 1298 HA	anionic surfactant blend		liquid	*agrochemicals; insecticides*
	WITCOLATE 1247H	alcohol ether sulfonate		liquid; pH 7.8	
	WITCOLATE 1259	alcohol ether sulfonate		liquid; pH 7.5	foaming agent / *petroleum industry*
	WITCOLATE 3220	alkyl ether sulfonate		liquid; pH 8.8	
	WITCOMUL 3126	surfactant blend		liquid; pour point -17.8°C; pH 8.5	lubricant / *petroleum industry*
	WITCONATE 3203	sulfonate		liquid; pH 7.5	foaming agent / *petroleum industry*
	WITCONATE 605-A	anionic surfactant blend		liquid	*agrochemicals*
	WITCONATE 79S	anionic surfactant blend		liquid	*agrochemicals*
	WITCONATE SE-5	sodium alcohol ether sulfonate		liquid; pH 7.5	foaming agent / *petroleum industry*

Supplier	Trade name	Chemical description	Composition	General properties	Functionality / *Application*
Witco	WITCOR PC100	surfactant blend		liquid; pH 7.5	paraffin inhibition / *petroleum industry*
Zschimmer & Schwarz	BUFFER SN	combination of organic acids and high molecular condensation products		brown liquid	buffering agent / *textile industry*
	CEFATEX AF	combination of sulfonates		clear yellowish liquid	anti-frosting agent / *textile industry*
	CEFATEX EL	combination of phosphoric acid esters and defoamers		clear viscous liquid	wetting agent; defoaming agent / *textile industry*
	CEFATEX ENN	combination of phosphoric acid esters and defoamers		white viscous liquid	wetting agent; defoaming agent / *textile industry*
	CEFATEX FN	anionic surfactants and phosphoric acid esters		clear, light yellow liquid	wetting agent / *textile industry*
	CEFATEX NR	combination of alkyl sulfonates		clear yellow liquid	low foam; wetting agent; dispersant; levelling agent; penetrant / *textile industry*
	COTTAVIN T	alkyl sulfonates		yellowish paste	softener / *textile industry*
	DEPICOL TLK	detergents combined with organic solvents		clear colourless liquid	fat dissolving agent; detergent / *textile industry*
	MARSEILLE SOAP	fatty acid salt		yellow needles	milling aid; detergent / *textile industry*
	NEWALOL DC	modified alkyl polyglycol ether		yellow paste	wetting agent; emulsifier / *textile industry*
	NEWALOL EBF	combination of ethoxylates and phosphoric acid esters		clear, slightly turbid liquid	low foam; wetting agent / *textile industry*
	NEWALOL SWF	combination of alkylethoxylates and alkyl sulfonates		clear colourless liquid	scouring agent; wetting agent; washing agent; dispersant / *textile industry*
	SANFOROL KN	vegetable oil sulfonate		clear brown liquid	wetting agent; softener / *textile industry*
	SETAVIN AP	cyclic compound		clear brown liquid	carrier / *textile industry*
	SETAVIN CA	phthalic acid derivative		clear yellowish liquid	
	SETAVIN CAW	carbonic acid ester		clear, paste yellow liquid	
	SETAVIN ER	fatty acid derivative		white viscous liquid	anti-crease agent; lubricant / *textile industry*
	SETAVIN ES	combination of modified phosphoric acid esters and alkylamine polyglycol ethers		yellow viscous liquid	low foam; anti-crease agent; levelling agent; dispersant / *textile industry*
	SETAVIN MO	hydrocarbons and anionic emulsifiers		clear yellowish liquid	anti-crease agent / *textile industry*
	SETAVIN PW	ethylene oxide adducts and sulfonates		yellowish viscous liquid	levelling agent / *textile industry*
	SETAVIN SF	combination of ethylene oxide adducts with alkylaryl sulfonates		clear, brown viscous liquid	levelling agent / *textile industry*
	SINCAL F	combination of alkyl sulfonates and ethylene oxide adducts		straw-coloured oily liquid	scouring agent; wetting agent; milling aid; detergent / *textile industry*

Supplier	Trade name	Chemical description	Composition	General properties	Functionality *Application*
Zschimmer & Schwarz	SINCAL KS	combination of ethylene oxide adducts and alkylaryl sulfonates		straw-coloured clear liquid	scouring agent; wetting agent; detergent *textile industry*
	SINCAL KSB	combination of alkylpolyglycol ethers and alkylaryl sulfonates		colourless liquid	scouring agent; wetting agent; detergent *textile industry*
	SINCAL OE	modified alkylene oxide condensation product		viscous yellowish liquid	low foam; wetting agent; scouring agent; detergent *textile industry*
	TORSINOL RF	hydrocarbons with emulsifier		clear yellow liquid	lubricant *textile industry*
	TURKISCHROTOL 100%	sodium sulforicinoleate	82% active	liquid	solubiliser; refatting agent; emulsifier *cosmetics; perfumes*
	ZETESAL NY CONC.	sulfonic acid derivatives		clear brown liquid	aftertreatment agent *textile industry*
	ZETESAL UR	combination of organic reducing agents and anionic surfactants		brown, low viscous liquid	reducing agent; aftertreatment agent *textile industry*

3 Cationics

Amides

Supplier	Trade name	Chemical description	Composition	General properties	Functionality / *Applications*
Akzo Nobel	ARMID E PASTILLES	erucamide	> 95% active	pastilles	softener; lubricant / *textile industry; all fibres*
	ARMID HT FLAKES	tallowamide (hydrogenated tallow-alkyl)	> 95% active	flakes	softener; processing aid; lubricant / *textile industry; synthetic and natural fibres*
	ARMID O PASTILLES	oleamide	> 95% active	pastilles	softener / *textile industry; synthetic fibres*
Allied Colloids	ALCAMINE 544 SPECIAL 100%	fatty acid amide		flakes	softener / *textile industry; all fibres*
	ALCAMINE CA NEW	substituted stearamide		emulsion	
	ALCAMINE DAS	substituted amide		emulsion	
	ALCAMINE FPS	substituted amide		emulsion	
Henkel	BELFASIN 44	fatty acid amide			softener; antistatic agent / *textile industry*
	BELFASIN 84	fatty acid amide			softener / *textile industry*
	BELFASIN 296	fatty acid amide			
	BELFASIN 615	fatty acid amide			
Sandoz	CATALIX L	fatty acid amide in aq. soln.		liquid	dispersant / *dyes*
	CERANINE PNP	modified fatty amide in aq. soln.		liquid	softener / *polyamide and acrylic fibres*
	CERANINE RSF (LF)	fatty amide derivative, weakly cationic surfactant and wax in aq. soln.		liquid	re-wetting agent; softener; low foam / *textile industry*
	CERANINE RSF	fatty amide derivative, weakly cationic surfactant and wax in aq. soln.		liquid	re-wetting agent; softener / *textile industry*
	SANDOFIX FFN	aliphatic polyamide derivative		liquid	fixative / *dyeing; textile industry*
	SANDOFIX WES	methylolamide		liquid	
	SANDOZOL CAT	aliphatic polyamide derivative		liquid	*dyes*
Scher Chemicals	COLSOL	fatty amide	25% active	paste	softener; lubricant; finishing agent / *textile industry*
Witco	WITCOMUL 3230	fatty acid amide		liquid; pour point 0°C; pH 7	foaming agent / *petroleum industry*

Amidoamines

Supplier	Trade name	Chemical description	Composition	General properties	Functionality / Application
Auschem Cesalpinia	INDAMIN 20	amido-amine	100% active	liquid	*waxes; bitumens*
	INDAMIN TO	amido-amine	100% active	liquid	*waxes; bitumens*
Th. Goldschmidt	TEGO-AMID S 18	stearamidopropyl dimethylamine	acid value 4 max.	ivory flakes	emulsifier; conditioner / *hair care products; shampoos; hair conditioners*
McIntyre Group	MACKALENE 116	cocamidopropyl dimethylamine lactate	25% active	liquid; pH 5	conditioner / *skin care products; hair care products; shampoos*
	MACKALENE 117	cocamidopropyl dimethylamine propionate	40% active	liquid; pH 6.5	
	MACKALENE 216	ricinoleamidopropyl dimethylamine lactate	95% active	liquid; pH 6	
	MACKALENE 316	stearamidopropyl dimethylamine lactate	25% active	liquid; pH 4.5	
	MACKALENE 326	stearamidopropyl morpholine lactate	25% active	liquid; pH 4.5	conditioner / *skin care products; hair care products; shampoos; baby shampoos; hair conditioners*
	MACKALENE 416	isostearamidopropyl dimethylamine lactate	25% active	liquid; pH 6	conditioner / *skin care products; hair care products; shampoos*
	MACKALENE 426	isostearamidopropyl morpholine lactate	25% active	liquid; pH 4	conditioner / *skin care products; hair care products; shampoos; baby shampoos; hair conditioners*
	MACKALENE 616	behenamidopropyl dimethylamine lactate	25% active	liquid; pH 4.5	conditioner / *skin care products; hair care products; shampoos*
	MACKALENE 716	wheatgermamidopropyl dimethylamine lactate	95% active	paste; pH 6	
	MACKINE 101	cocamidopropyl dimethylamine	100% active	liquid	conditioner; anti-corrosion; emulsifier / *hair conditioners*
	MACKINE 201	ricinoleamidopropyl dimethylamine	100% active	liquid	
	MACKINE 301	stearamidopropyl dimethylamine	100% active	flake	conditioner; anti-corrosion; emulsifier / *skin cleansing preparations; baby care products; hair conditioners*
	MACKINE 321	stearamidopropyl morpholine	100% active	flake	
	MACKINE 401	isostearamidopropyl dimethylamine	100% active	liquid	conditioner; anti-corrosion; emulsifier / *hair conditioners*
	MACKINE 421	isostearamidopropyl morpholine	100% active	liquid	conditioner; anti-corrosion; emulsifier / *skin cleansing preparations; baby care products; hair conditioners*
	MACKINE 501	oleamidopropyl dimethylamine	100% active	liquid	
	MACKINE 601	behenamidopropyl dimethylamine	100% active	flake	
	MACKINE 701	wheatgermamidopropyl dimethylamine	100% active	paste	conditioner; anti-corrosion; emulsifier / *hair conditioners*
	MACKINE 801	lauramidopropyl dimethylamine	100% active	solid	
	MACKINE 901	soyamidopropyl dimethylamine	100% active	paste	
Millchem	AMMONYX PG 90	ethoxylated ditallow amidoamine	89-91% active	liquid	softener / *laundry products*
Nikko Chemicals	NIKKOL AMIDOAMINE S	stearamidoethyl diethylamine		pale yellow powder	*cosmetics*

Supplier	Trade name	Chemical description	Composition	General properties	Functionality Application
Pentagon	PENTAMINE CDP	cocoamidopropyl dimethylamine	tertiary amine 88% min.	hazy amber liquid; colour 4 max. (Gardner)	chemical intermediate
Sandoz	CERANINE BASE 6532 FLAKES	stearamidoethyl ethanolamine		pale brown-coloured flakes	emulsifier *skin care products; hair care products*
	CERANINE HC BASE FLAKES	stearamidoethyl diethylamine		pale brown-coloured flakes	emulsifier *skin care products; hair care products*
Scher Chemicals	KATEMUL IG-70	isostearamidopropyl dimethylamino glycolate	dry solids 70% min.	amber viscous liquid; pH 7	emulsifier; conditioner *hair care products; skin care products*
	KATEMUL IGU-70	isostearamidopropyl dimethylamino gluconate	dry solids 70% min.	amber viscous liquid; pH 7	*skin care products; bath preparations*
	SCHERCODINE B	behenamidopropyl dimethylamine	amide 98% min.; free amine 1% max.; alkali value 135-145	tan hard flakes; M.W. 394; m.p. 63-68°C	conditioner *hair care preparations; skin care products*
	SCHERCODINE C	cocamidopropyl dimethylamine	amide 98% min.; free amine 1% max.; alkali value 177-187	tan soft solid; M.W. 304	foaming agent *bath preparations; hair care products*
	SCHERCODINE I	isostearamidopropyl dimethylamine	amide 98% min.; free amine 1% max.; alkali value 150-160	light amber liquid; M.W. 394	o/w emulsifier; lubricant *hair conditioners; hair care products*
	SCHERCODINE L	lauramidopropyl dimethylamine	amide 98% min.; free amine 1% max.; alkali value 96-206	light tan solid; M.W. 284; m.p. 35-40°C	intermediate for betaine amphoterics
	SCHERCODINE M	myristamidopropyl dimethylamine	amide 98% min.; free amine 1% max.; alkali value 180-190	light tan wax; M.W. 312; m.p. 45-50°C	o/w emulsifier; conditioner; viscosity modifier
	SCHERCODINE O	oleamidopropyl dimethylamine	amide 98% min.; free amine 1% max.; alkali value 150-160	amber liquid; M.W. 366	emollient; conditioner; lubricant; moisturiser *hair care products; skin care products*
	SCHERCODINE S	stearamidopropyl dimethylamine	amide 98% min.; free amine 1% max.; alkali value 145-155	tan hard flakes; M.W. 368; m.p. 65-70°C	softener; emulsifier; conditioner *hair care products; skin care products*
	SCHERCODINE T	tallamidopropyl dimethylamine	amide 98% min.; free amine 1% max.; alkali value 150-160	amber liquid; M.W. 366	conditioner; thickener
Tomah Products	TOMAH CATAMINE 101	amidoamine	100% active	dark viscous liquid	emulsifier *slurry seal asphalt emulsification*
Witco	EMCOL 1655	cocamidopropyl dimethylamine propionate	40% solids	pale yellow liquid; pH 6.5	antistatic agent; conditioner; emollient; substantivity agent *personal care products; householand industrial applications*
	EMCOL 3780	stearamidopropyl dimethylamine lactate	25% solids	pale yellow liquid; pH 4	
	EMCOL ISML	isostearamidopropyl morpholine lactate	25% solids	light yellow liquid; pH 4.5	
	WITCAMIDE 210	alkyl amidoamine		paste; pour point 32.8°C; pH 10	anti-corrosion *petroleum industry*

Amine salts

Supplier	Trade name	Chemical description	Composition	General properties
Akzo Nobel	ARMAC C	cocoamine acetate	neutralisation 99-104%	paste
	ARMAC HT	tallowamine acetate (hydrogenated tallow-alkyl)	neutralisation 99-104%	solid
	ARMAC T	tallowamine acetate	neutralisation 99-104%	paste
	DUOMAC C	N-coco-1,3-diaminopropane diacetate	neutralisation 99-104%	paste
	DUOMAC T	N-tallow-1,3-diaminopropane diacetate	neutralisation 99-104%	paste
	DUOMAC TDO-82	N-tallow-1,3-diaminopropane dioleate	neutralisation 92-98%	liquid

Supplier	Trade name	Chemical description	Composition	General properties	Functionality / *Application*
Akzo Nobel	DUOMAC TDO	N-tallow-1,3-diaminopropane dioleate	neutralisation 92-98%	paste	
	DUOMEEN CDA-50	N-coco-1,3-diaminopropane adipate		liquid	
Allied Colloids	ALCOFIX CU	amine polycondensation product with inorganic salts		powder	afterfixing agent *textile industry; dyes and prints*
Elf Atochem	NORAMAC C26	cocoamine acetate in aq. soln.	neutralisation 95-100%	liquid	bactericide; emulsifier; anti-caking agent; stabiliser; anti-corrosion; flocculant
	NORAMAC SH	hydrogenated tallow amine acetate	neutralisation 95-100%	solid/flakes	
Fina Chemicals	RADIAMAC 6149	acetate of $C_{16/18}$ primary amine	total amine value 170; acid value 160	solid; softening point 55-65°C	anti-caking agent; film former *mineral extraction*
	RADIAMAC 6159	proprietary primary amine acetate	total amine value 165; acid value 155	solid; softening point 55-65°C	
	RADIAMAC 6169	coconut oil amine acetate	total amine value 210; acid value 200	paste	
Hoechst	GENAMIN CC500A	coco amine acetate		solid	
	GENAMIN CC500E	coco amine oleate		liquid	
	GENAMIN OL500A	oleyl amine acetate		solid	
Kao Corporation	ACETAMIN C	coco amine acetate	amine value 192-210; carbon chain composition C_8 3%, C_{10} 5% C_{12} 50%, C_{14} 20%, C_{16} 10%, C_{18} 3%, $C_{18'}$ 9%; iodine value 10 max.	liquid/solid; M.W. 265; colour 5 max. (Gardner)	
	ACETAMIN O	oleyl amine acetate	amine value 162-172; carbon chain composition C_{14} 2% C_{16} 10%, C_{18} 8% $C_{18'}$ 80%; iodine value 58-67	solid; M.W. 330; colour 8 max. (Gardner)	
	ACETAMIN T	tallow amine acetate	amine value 165-175; carbon chain composition C_{12} 1% C_{14} 3%, C_{16} 35% C_{18} 18%, $C_{18'}$ 42%, $C_{20/22}$ 1%; iodine value 35-42	solid; M.W. 325; colour 5 max. (Gardner)	
	ACETAMIN TH	hydrogenated tallow amine acetate	amine value 160-165; carbon chain composition C_{14} 4%, C_{16} 32%, C_{18} 60% $C_{18'}$ 3%, $C_{20/22}$ 1%; iodine value 4 max.	solid; M.W. 325; colour 5 max. (Gardner)	
Pentagon	PENTONATE LAA	lauryl amine acetate	amine value 220-230; acid value 220-230	white flakes; colour 1 max. (Gardner)	processing aid *manufacture of explosives*
Seppic	AMONYL SP	dialkylestramine methosulfate	85% active	paste	softener; conditioner *hair conditioners*
	SOLAMINE HY	diaminester methosulfate	100% active	liquid	softener *textile industry*
Stepan Europe	CATISOL AO 100	oleylamine acetate; made-to-order	92% active	orange to yellow wax	emulsifier; wetting agent; antistatic agent; anti-corrosion; lubricant *textile industry*
Tomah Products	TOMAH PA-14 ACETATE	acetic acid salt of isodecyloxypropylamine	95% active	light amber liquid	emulsifier; clay additive

Amines

Supplier	Trade name	Chemical description	Composition	General properties	Functionality / Application
Akzo Nobel	ARMEEN 12D	n-dodecylamine, distilled	primary amine content > 98%	solid	
	ARMEEN 16D	n-hexadecylamine, distilled	primary amine content > 98%	solid	
	ARMEEN 18D	n-octadecylamine, distilled	primary amine content > 98%	solid or pastilles	
	ARMEEN 2C	dicocoamine	secondary amine content > 90%	solid	
	ARMEEN 2HT	ditallowamine (hydrogenated tallow-alkyl)	secondary amine content > 90%	solid or pastilles	
	ARMEEN C	cocoamine	primary amine content > 95%	liquid	
	ARMEEN CD	cocoamine, distilled	primary amine content > 98%	liquid	
	ARMEEN DM12D	dodecyldimethylamine, distilled	tertiary amine content > 98%	liquid	
	ARMEEN DM14D	tetradecyldimethylamine, distilled	tertiary amine content > 98%	liquid	
	ARMEEN DM16D	hexadecyldimethylamine, distilled	tertiary amine content > 98%	liquid	
	ARMEEN DM18D	octadecyldimethylamine, distilled	tertiary amine content > 98%	liquid	
	ARMEEN DMCD	cocodimethylamine, distilled	tertiary amine content > 98%	liquid	
	ARMEEN DMHTD	tallowdimethylamine, distilled (hydrogenated tallow-alkyl)	tertiary amine content > 98%	liquid	
	ARMEEN DMMCD	cocodimethylamine, distilled (fractionated coco-alkyl)	tertiary amine content > 98%	liquid	
	ARMEEN DMOD	oleyldimethylamine, distilled	tertiary amine content > 96%	liquid	
	ARMEEN DMTD	tallowdimethylamine, distilled	tertiary amine content > 98%	liquid	
	ARMEEN HT	tallowamine (hydrogenated tallow-alkyl)	primary amine content > 98%	solid or pastilles	
	ARMEEN HTD	tallowamine, distilled (hydrogenated tallow-alkyl)	primary amine content > 98%	solid or pastilles	
	ARMEEN M2HT	ditallowmethylamine (hydrogenated tallow-alkyl)	tertiary amine content > 96%	solid	
	ARMEEN MC2	dicocomethylamine	tertiary amine content > 96%	liquid	
	ARMEEN NCMD	N-cocomorpholine, distilled		liquid	
	ARMEEN O	oleylamine	primary amine content 95%	liquid	
	ARMEEN OD	oleylamine, distilled	primary amine content 98%	liquid	
	ARMEEN T	tallowamine	primary amine content > 98%	paste	
	ARMEEN TD	tallowamine, distilled	primary amine content > 98%	paste	
Albright & Wilson	EMPIGEN A SERIES	alkyl dimethylamine	concentration 100%	liquid	flotation aid / ore flotation
Allied Colloids	ALCOFIX FD 100% POWDER	concentrated version of Alcofix FD		powder	afterfixing agent / textile industry: cellulosic fibres; dyes
	ALCOFIX FD	amine polycondensation product		liquid	
Croda	INCROMINE SB	stearamidopropyl dimethylamine	acid value 4 max.	cream-coloured waxy solid; M.W. 369-379; colour 2 max. (Gardner); m.p. ca. 63°C	hair care products

Supplier	Trade name	Chemical description	Composition	General properties	Functionality / Application
Elf Atochem	NORAM 12D	distilled N-dodecylamine	≥98% active	solid	
	NORAM 2C	dicocoamine	≥95% active	solid	
	NORAM 2SH	dihydrogenated tallowamine	≥95% active	solid	
	NORAM 42	technical grade arachidyl behenyl amine	≥95% active	solid/flakes	
	NORAM C	technical grade cocoamine	≥95% active	liquid	
	NORAM C96	high grade cocoamine	≥97% active	liquid	
	NORAM CD	distilled cocoamine	≥98.5% active	liquid	
	NORAM DMCD	distilled dimethyl cocoamine	≥97.5% active	liquid	
	NORAM DMSD	distilled dimethyl tallow amine	≥97.5% active	liquid	
	NORAM DMSHD	distilled dimethyl hydrogenated tallow amine	≥97.5% active	liquid	
	NORAM M2C	methyl dicocoamine	≥95% active	solid	
	NORAM M2SH	methyl dihydrogenated tallow amine	≥96% active	solid	
	NORAM O	technical grade oleic amine	≥95% active	liquid	
	NORAM OD	distilled oleic amine	≥97% active	liquid	
	NORAM S	technical grade tallow amine	≥95% active	paste	
	NORAM S96	high grade tallow amine	≥97% active	paste	
	NORAM SD	distilled tallow amine	≥98.5% active	paste	
	NORAM SH	technical grade hydrogenated tallow amine	≥95% active	solid/flakes	
	NORAM SH96	high grade hydrogenated tallow amine	≥97% active	solid/flakes	
	NORAM SHD	distilled hydrogenated tallow amine	≥98.5% active	solid/flakes	
	POLYRAM S	tallow polypropylene polyamine	≥95% active	stiff paste	emulsifier; anti-stripping agent bitumen; road making
	TRINORAM C	coco dipropylene triamine		stiff paste	
	TRINORAM S	tallow dipropylene triamine		stiff paste	
Fina Chemicals	RADIAMINE 6138	technical octadecyl amine	primary amine 95% min.; total amine value 204; iodine value < 3; carbon chain composition C_{18} > 97%	solid; M.W. 275; colour 2 max. (Gardner); softening point 52°C	agrochemicals; mining industry; oil industry; polymers; rubber industry; textile industry
	RADIAMINE 6139	distilled octadecyl amine	primary amine 98% min.; total amine value 207; iodine value < 3; carbon chain composition C_{18} > 97%	solid; M.W. 271; colour 1 max. (Gardner); softening point 53°C	
	RADIAMINE 6140	technical hydrogenated tallow amine	primary amine 95% min.; total amine value 209; iodine value < 3; carbon chain composition C_{16} 30%, C_{18} 65%	solid; M.W. 268; colour 3 max. (Gardner); softening point 46°C	
	RADIAMINE 6141	distilled hydrogenated tallow amine	primary amine 98% min.; total amine value 212; iodine value < 3; carbon chain composition C_{14} 5%, C_{16} 30%, C_{18} 65%	solid; M.W. 265; colour 1 max. (Gardner); softening point 48°C	
	RADIAMINE 6144	technical hydrogenated tallow amine	primary amine 90% min.; total amine value 203; iodine value < 10; carbon chain composition C_{14} 5%, C_{16} 30%, C_{18} 65%	solid; M.W. 276; colour 3 max. (Gardner); softening point 45°C	
	RADIAMINE 6145	technical stearyl amine	primary amine 95% min.; total amine value 206; iodine value < 3; carbon chain composition C_{18} > 88%	solid; M.W. 272; colour 3 max. (Gardner); softening point 52°C	

Supplier	Trade name	Chemical description	Composition	General properties	Functionality Application
Fina Chemicals	RADIAMINE 6146	distilled stearyl amine	primary amine 98% min.; total amine value 209; iodine value <3; carbon chain composition C_{18} >88%	solid; M.W. 268; colour 1 max. (Gardner); softening point 50°C	
	RADIAMINE 6147	technical cetyl amine	primary amine 95% min.; total amine value 230; iodine value <2; carbon chain composition C_{16} >90%	solid; M.W. 244; colour 3 max. (Gardner); softening point 41°C	
	RADIAMINE 6148	distilled cetyl amine	primary amine 98% min.; total amine value 234; iodine value <2; carbon chain composition C_{16} >90%	solid; M.W. 240; colour 1 max. (Gardner); softening point 40°C	
	RADIAMINE 6160	technical coconut amine	primary amine 95% min.; total amine value 275; iodine value <13; carbon chain composition C_8 5%, C_{10} 7%, C_{12} 48%, C_{14} 18%, C_{16} 12%, C_{18} 6%, $C_{18'}$ 4%	liquid; M.W. 204; colour 3 max. (Gardner); softening point 16°C	
	RADIAMINE 6161	distilled coconut oil amine	primary amine 98% min.; total amine value 282; iodine value <12; carbon chain composition C_8 5%, C_{10} 7%, C_{12} 48%, C_{14} 18%, C_{16} 12%, C_{18} 6%, $C_{18'}$ 4%	liquid; M.W. 199; colour 1 max. (Gardner); softening point 15°C	*agrochemicals; mining industry; oil industry; polymers; rubber industry; textile industry*
	RADIAMINE 6164	distilled dodecyl amine	primary amine 98% min.; total amine value 300; iodine value <2; carbon chain composition C_{10} 1%, C_{12} 98%, C_{14} 1%	liquid; M.W. 187; colour 1 max. (Gardner); softening point 23°C	
	RADIAMINE 6170	technical tallow amine	primary amine 95% min.; total amine value 211; iodine value 40; carbon chain composition C_{14} 5%, C_{16} 30%, C_{18} 20%, $C_{18'}$ 45%	paste; M.W. 266; colour 3 max. (Gardner); softening point 32°C	
	RADIAMINE 6171	distilled tallow amine	primary amine 98% min.; total amine value 214; iodine value 40; carbon chain composition C_{14} 5%, C_{16} 30%, C_{18} 20%, $C_{18'}$ 45%	paste; M.W. 262; colour 1 max. (Gardner); softening point 30°C	
	RADIAMINE 6172	technical oleyl amine	primary amine 95% min.; total amine value 207; iodine value 82; carbon chain composition C_{14} 3%, C_{16} 9%, C_{18} 10%, $C_{18'}$ 70%, $C_{18''}$ 8%	liquid; cloud point <20°C; M.W. 271; colour 2 max. (Gardner)	
	RADIAMINE 6173	distilled oleyl amine	primary amine 97% min.; total amine value 209; iodine value 80; carbon chain composition C_{10} 3%, C_{16} 9%, C_{18} 10%, $C_{18'}$ 70%, $C_{18''}$ 8%	liquid; cloud point <20°C; M.W. 268; colour 2 max. (Gardner)	
	RADIAMINE 6240	secondary amine with hydrogenated tallow alkyl type	amine 98% min.; total amine value 112; iodine value <3; carbon chain composition C_{14} 5%, C_{16} 30%, C_{18} 65%	solid; M.W. 501; colour 2 max. (Gardner); softening point 68°C	
	RADIAMINE 6260	secondary amine with coconut oil alkyl type	amine 98% min.; total amine value 148; iodine value <10; carbon chain composition C_8 5%, C_{10} 7%, C_{12} 48%, C_{14} 18%, C_{16} 12%, C_{18} 6%, $C_{18'}$ 4%	liquid; M.W. 379; colour 2 max. (Gardner)	
	RADIAMINE 6270	secondary amine with tallow alkyl type	amine 98% min.; total amine value >38; carbon chain composition C_{14} 5%, C_{16} 30%, C_{18} 20%, $C_{18'}$ 45%	paste; M.W. 501; colour 2 max. (Gardner)	*paint industry; inks; metal working; oil industry; polymers; textile industry*
	RADIAMINE 6308	di-n-octylmethyl amine	amine 98% min.; total amine value 220; iodine value <2; carbon chain distribution C_8 >90%	liquid; M.W. 255; colour 1 max. (Gardner)	
	RADIAMINE 6310	di-n-decylmethyl amine	amine 98% min.; total amine value 180; iodine value <3; carbon chain composition C_{10} >90%	liquid; M.W. 312; colour 1 max. (Gardner)	
	RADIAMINE 6343	di-hydrogenated tallow methyl amine	amine 98% min.; total amine value 108; iodine value <3; carbon chain composition C_{14} 5%, C_{16} 30%, C_{18} 65%	solid; M.W. 520; colour 3 max. (Gardner); softening point 33°C	

Supplier	Trade name	Chemical description	Composition	General properties	Functionality Application
Fina Chemicals	RADIAMINE 6346	dioctadecylmethyl amine	amine 98% min.; total amine value 105; iodine value <3; carbon chain distribution C_{18} >96%	solid; M.W. 534; colour 3 max. (Gardner); softening point 35°C	
	RADIAMINE 6360	di-coconut oil methyl amine	amine 98% min.; total amine value 140; iodine value <12; carbon chain composition C_8 5%, C_{10} 7%, C_{12} 48%, C_{14} 18%, C_{16} 12%, C_{18} 6%, C_{18} 4%	liquid; M.W. 401; colour 3 max. (Gardner); softening point 15°C	paint industry; inks; metal working; oil industry; polymers; textile industry
	RADIAMINE 6670	proprietary blend of alkyl polyamines	amine 98% min.; total amine value 375; iodine value 35	paste; colour 6 max. (Gardner); softening point 30°C	emulsifier; adhesion agent; wetting agent road construction
	RADIAMINE 6675	proprietary blend of alkyl polyamines	amine 98% min.; total amine value 430; iodine value 35	paste; colour 6 max. (Gardner); softening point 30°C	emulsifier; adhesion agent; wetting agent road construction
	RADIAMINE 6739	octadecyldimethyl amine	amine 98% min.; total amine value 181; iodine value <3; carbon chain composition C_{16} 2%, C_{18} 98%, C_{20} 2%	solid; M.W. 310; colour 50 max. (APHA)	
	RADIAMINE 6743	stearyldimethyl amine	amine 98% min.; total amine value 190; iodine value <3; carbon chain composition C_{14} 1%, C_{16} 25-30%, C_{18} 60-75%, C_{20} 2%	liquid; M.W. 295; colour 50 max. (APHA)	
	RADIAMINE 6748	cetyldimethyl amine	amine 98% min.; total amine value 203; iodine value <3; carbon chain composition C_{14} 2%, C_{16} 94%, C_{18} 4%	liquid; M.W. 276; colour 50 max. (APHA)	
	RADIAMINE 6760	coconut oil dimethyl amine	amine 98% min.; total amine value 235; iodine value <3; carbon chain composition $C_{8/10}$ 3%, C_{12} 48-58%, C_{14} 18-24%, C_{16} 8-12%, C_{18} 10-13%	liquid; M.W. 238; colour 50 max. (APHA)	paint industry; inks; metal working; oil industry; polymers; textile industry
	RADIAMINE 6764	dodecyldimethyl amine	amine 98% min.; total amine value 255; iodine value <3; carbon chain composition C_{10} 1%, C_{12} 98%, C_{14} 1%	liquid; M.W. 220; colour 50 max. (APHA)	
	RADIAMINE 6765	lauryldimethyl amine	amine 98% min.; total amine value 247; iodine value <3; carbon chain composition $C_{8/10}$ 2%, C_{12} 68-73%, C_{14} 25-30%, C_{16} 2%	liquid; M.W. 227; colour 50 max. (APHA)	
Henkel	ARAPHEN G 2D	tertiary linear hydroxyalkyl amine	99.8-100% active	paste	solubiliser cleaners
	BREVIOL SCN	amine condensate			levelling agent textile industry; dyeing
Hoechst	GENAMIN 8R100D	distilled C_8 amine	99% active	liquid	
	GENAMIN 12R100D	distilled C_{12} amine	99% active	liquid	
	GENAMIN 12R302D	distilled C_{12} dimethylamine	99% active	liquid	
	GENAMIN 14R100D	distilled C_{14} amine	99% active	liquid	
	GENAMIN 14R302D	distilled C_{14} dimethylamine	99% active	liquid	
	GENAMIN 16R302D	distilled C_{16} dimethylamine	99% active	liquid	
	GENAMIN 18R100D	distilled C_{18} amine	99% active	solid	intermediate
	GENAMIN 18R302D	distilled C_{18} dimethylamine	99% active	liquid	
	GENAMIN CC100	coco amine	96% active	liquid	
	GENAMIN CC100D	distilled coco amine	99% active	liquid	
	GENAMIN CC302D	distilled cocodimethyl amine	99% active	liquid	
	GENAMIN CS302D	distilled cocodimethyl amine	99% active	liquid	

Supplier	Trade name	Chemical description	Composition	General properties	Functionality Application
Hoechst	GENAMIN MY302D	distilled myristyldimethylamine	99% active	liquid	
	GENAMIN OL100D	distilled oleyl amine	99% active	liquid	
	GENAMIN SH100	stearyl amine	96% active	solid	intermediate
	GENAMIN SH100D	distilled stearyl amine	99% active	solid	
	GENAMIN SH302D	distilled stearyldimethylamine	99% active	liquid	
	GENAMIN SO302D	soyadimethylamine	99% active	liquid	
	GENAMIN TA100	tallow amine	96% active	paste	
	GENAMIN TA100D	distilled tallow amine	99% active	paste	
	GENAMIN TA302D	distilled tallowdimethylamine	99% active	liquid	
Kao Corporation	FARMIN 20 D	distilled lauryl amine	amine 98% min.; amine value 297-303; carbon chain composition C_{10} 1%, C_{12} 98%, C_{14} 1%; iodine value 1 max.	liquid/solid; M.W. 200; colour 100 max. (APHA)	
	FARMIN 20	lauryl amine	amine 95% min.; amine value 278-303; carbon chain composition C_{10} 3%, C_{12} 95%, C_{14} 2%; iodine value 1 max.	liquid/solid; M.W. 200; colour 3 max. (Gardner)	
	FARMIN 80 D	distilled stearyl amine	amine 98% min.; amine value 204-209; carbon chain composition C_{16} 1%, C_{18} 98%, $C_{18'}$ 1%; iodine value 1 max.	solid; M.W. 270; colour 100 max. (APHA)	
	FARMIN 80	stearyl amine	amine 95% min.; amine value 192-209; carbon chain composition C_{16} 1%, C_{18} 98%, $C_{18'}$ 1%; iodine value 1 max.	solid; M.W. 270; colour 3 max. (Gardner)	*dyeing of fabrics; hair conditioners; wood preservation; germicides; bactericides; dishwashing agents; oil recovery agents*
	FARMIN AB	arachidyl behenyl amine	amine 95% min.; amine value 180-190; carbon chain composition C_{16} 10%, C_{18} 30%, $C_{20/22}$ 60%; iodine value 3 max.	solid; M.W. 290; colour 3 max. (Gardner)	
	FARMIN C	coco amine	amine 95% min.; amine value 254-267; carbon chain composition C_8 3%, C_{10} 5%, C_{12} 50%, C_{14} 20%, C_{16} 10%, C_{18} 3%, $C_{18'}$ 9%; iodine value 15 max.	liquid/solid; M.W. 210; colour 4 max. (Gardner)	
	FARMIN CD	distilled coco amine	amine 98% min.; amine value 275-285; carbon chain composition C_8 1%, C_{10} 3%, C_{12} 55%, C_{14} 22%, C_{16} 12%, C_{18} 2%, $C_{18'}$ 5%; iodine value 10 max.	liquid/solid; M.W. 200; colour 100 max. (APHA)	
	FARMIN DC	coco diamine	amine 90% min.; amine value 135-150; carbon chain composition C_8 3%, C_{10} 5%, C_{12} 50%, C_{14} 20%, C_{16} 10%, C_{18} 3%, $C_{18'}$ 9%; iodine value 5 max.	solid/paste; M.W. 395; colour 5 max. (Gardner)	
	FARMIN DM 08	dimethyl octylamine	total amine value 330-350; carbon chain composition C_8 95%, C_{10} 4%, C_{12} 1%; tertiary amine 95% min.	clear, colourless liquid; colour 30 max. (APHA)	
	FARMIN DM 10	dimethyl decylamine	total amine value 280-300; carbon chain composition C_8 2%, C_{10} 95%, C_{12} 3%; tertiary amine 95% min.	clear, colourless liquid; colour 30 max. (APHA)	
	FARMIN DM 1012	dimethyl decylamine	total amine value 287-307; carbon chain composition C_8 9%, C_{10} 82%, C_{12} 9%; tertiary amine 95% min.	clear, colourless liquid; colour 30 max. (APHA)	

Supplier	Trade name	Chemical description	Composition	General properties	Functionality Application
Kao Corporation	FARMIN DM 1214	dimethyl lauryl amine	total amine value 235-255; carbon chain composition C_{10} 1%, C_{12} 65%, C_{14} 33%, C_{16} 1%; tertiary amine 95% min.	clear, colourless liquid; colour 30 max. (APHA)	*dyeing of fabrics; hair conditioners; wood preservation; germicides; bactericides; dishwashing agents; oil recovery agents*
	FARMIN DM 1214M	dimethyl lauryl amine	total amine value 243-263; carbon chain composition C_{12} 68%, C_{14} 32%; tertiary amine 95% min.	clear, colourless liquid; colour 30 max. (APHA)	
	FARMIN DM 1218	dimethyl lauryl amine	total amine value 240-260; carbon chain composition C_{12} 67%, C_{14} 25%, C_{16} 7%, C_{18} 1%; tertiary amine 95% min.	clear, colourless liquid; colour 30 max. (APHA)	
	FARMIN DM 20	dimethyl lauryl amine	total amine value 246-266; carbon chain composition C_{10} 2%, C_{12} 96%, C_{14} 2%; tertiary amine 95% min.	clear, colourless liquid; M.W. 225; colour 40 max. (APHA)	
	FARMIN DM 24	dimethyl lauryl amine	total amine value 233-253; carbon chain composition C_{12} 58%, C_{14} 35%, C_{16} 7%; tertiary amine 95% min.	clear, colourless liquid; colour 30 max. (APHA)	
	FARMIN DM 4-26	dimethyl myristyl amine	total amine value 226-246; carbon chain composition C_{12} 40%, C_{14} 50%, C_{16} 10%; tertiary amine 95% min.	clear, colourless liquid; colour 30 max. (APHA)	
	FARMIN DM 4-60	dimethyl myristyl amine	total amine value 218-238; carbon chain composition C_{12} 25%, C_{14} 60%, C_{16} 15%; tertiary amine 95% min.	clear, colourless liquid; colour 30 max. (APHA)	
	FARMIN DM 40	dimethyl myristyl amine	total amine value 217-237; carbon chain composition C_{12} 3%, C_{14} 95%, C_{16} 2%; tertiary amine 95% min.	clear, colourless liquid; M.W. 255; colour 40 max. (APHA)	
	FARMIN DM 42	dimethyl myristyl amine	total amine value 220-240; carbon chain composition C_{12} 35%, C_{14} 55%, C_{16} 10%; tertiary amine 95% min.	clear, colourless liquid; colour 30 max. (APHA)	
	FARMIN DM 46	dimethyl myristyl amine	total amine value 209-229; carbon chain composition C_{12} 4%, C_{14} 62%, C_{16} 30%, C_{18} 4%; tertiary amine 95% min.	clear, colourless liquid; colour 30 max. (APHA)	
	FARMIN DM 5-24	dimethyl lauryl amine	total amine value 234-254; carbon chain composition C_{12} 63%, C_{14} 30%, C_{16} 7%; tertiary amine 95% min.	clear, colourless liquid; colour 30 max. (APHA)	
	FARMIN DM 60	dimethyl palmityl amine	total amine value 197-217; carbon chain composition C_{14} 3%, C_{16} 95%, C_{18} 2%; tertiary amine 95% min.	clear, colourless liquid; M.W. 280; colour 40 max. (APHA)	
	FARMIN DM 68	dimethyl palmityl amine	total amine value 193-213; carbon chain composition C_{14} 6%, C_{16} 74%, C_{18} 20%; tertiary amine 95% min.	clear, colourless liquid; colour 30 max. (APHA)	
	FARMIN DM 80	dimethyl stearyl amine	total amine value 177-197; carbon chain composition C_{14} 1%, C_{16} 4%, C_{18} 95%; tertiary amine 95% min.	clear, colourless liquid; M.W. 310; colour 40 max. (APHA)	
	FARMIN DM 86	dimethyl stearyl amine	total amine value 183-197; carbon chain composition C_{14} 3%, C_{16} 17%, C_{18} 80%; tertiary amine 95% min.	clear, colourless liquid; colour 30 max. (APHA)	
	FARMIN DMC	dimethyl coconut amine	total amine value 230-250; carbon chain composition C_8 8%, C_{10} 6%, C_{12} 50%, C_{14} 19%, C_{16} 9%, C_{18} 8%; tertiary amine 95% min.	clear, colourless liquid; M.W. 230; colour 40 max. (APHA)	

Supplier	Trade name	Chemical description	Composition	General properties	Functionality / Application
Kao Corporation	FARMIN DMO	dimethyl oleyl amine	amine 98.5% min.; amine value 180-192; carbon chain composition C_{14} 2%, C_{16} 10%, C_{18} 8%, $C_{18'}$ 80%	liquid/solid; M.W. 305	
	FARMIN DMT	dimethyl tallow amine	amine 98.5% min.; amine value 180-195; carbon chain composition C_{12} 1%, C_{14} 3%, C_{16} 35%, C_{18} 18%, $C_{18'}$ 42%, $C_{20/22}$ 1%	liquid/solid; M.W. 300; colour 300 max. (APHA)	
	FARMIN DMTH	hydrogenated dimethyl tallow amine	amine 98.5% min.; amine value 185-195; carbon chain composition C_{14} 4%, C_{16} 32%, C_{18} 60%, $C_{18'}$ 3%, $C_{20/22}$ 1%	liquid/solid; M.W. 308; colour 50 max. (APHA)	
	FARMIN DTH	hydrogenated tallow diamine	amine 88% min.; amine value 109-119; carbon chain composition C_{14} 4%, C_{16} 32%, C_{18} 60%, $C_{18'}$ 3%, $C_{20/22}$ 1%; iodine value 3 max.	solid; M.W. 499; colour 2 max. (Gardner)	
	FARMIN M2C	methyl dicoco amine	amine 97% min.; amine value 130-140; carbon chain composition C_8 3%, C_{10} 5%, C_{12} 50%, C_{14} 20%, C_{16} 10%, C_{18} 3%, $C_{18'}$ 9%	liquid; M.W. 415; colour 4 max. (Gardner)	
	FARMIN M2TH	methyl distearyl amine	amine 97% min.; amine value 105-115; carbon chain composition C_{14} 4%, C_{16} 32%, C_{18} 60%, $C_{18'}$ 3% $C_{20/22}$ 1%	solid; M.W. 510; colour 2 max. (Gardner)	
	FARMIN O	oleyl amine	amine 95% min.; amine value 195-205; carbon chain composition C_{14} 2%, C_{16} 10%, C_{18} 8%, $C_{18'}$ 80%; iodine value 75 min.	liquid/solid; M.W. 280; colour 4 max. (Gardner)	
	FARMIN OD	distilled oleyl amine	amine 98% min.; amine value 200-215; carbon chain composition C_{14} 2%, C_{16} 10%, C_{18} 8%, $C_{18'}$ 80%; iodine value 75 min.	liquid/solid; M.W. 270; colour 150 max. (APHA)	
	FARMIN T	tallow amine	amine 95% min.; amine value 204-215; carbon chain composition C_{12} 1%, C_{14} 3%, C_{16} 35%, C_{18} 18%, $C_{18'}$ 42%, $C_{20/22}$ 1%; iodine value 40-50	paste/solid; M.W. 260; colour 4 max. (Gardner)	
	FARMIN TD	distilled tallow amine	amine 98% min.; amine value 210-220; carbon chain composition C_{14} 2%, C_{16} 35%, C_{18} 19%, $C_{18'}$ 43%, $C_{20/22}$ 1%; iodine value 40-50	paste/solid; M.W. 260; colour 150 max. (APHA)	
	FARMIN TH	hydrogenated tallow amine	amine 95% min.; amine value 209-216; carbon chain composition C_{14} 4%, C_{16} 32%, C_{18} 60%, $C_{18'}$ 3%, $C_{20/22}$ 1%; iodine value 4 max.	solid; M.W. 265; colour 3 max. (Gardner)	
	FARMIN THD	distilled hydrogenated tallow amine	amine 98% min.; amine value 208-223; carbon chain composition C_{14} 4%, C_{16} 32%, C_{18} 60%, $C_{18'}$ 3%, $C_{20/22}$ 1%; iodine value 4 max.	solid; M.W. 265; colour 100 max. (APHA)	
Lonza	BARLENE 12	alkyl dimethyl amine	95% active	liquid	chemical intermediate
	BARLENE 12S	lauryl dimethyl amine	95% active	liquid	
	BARLENE 14S	myristyl dimethyl amine	95% active	liquid	
	BARLENE 16S	cetyl dimethyl amine	95% active	liquid/solid	
	BARLENE 18S	stearyl dimethyl amine	95% active	solid	
	UNAMINE C	tertiary amine	92% active	liquid/solid	emulsifier; detergent
	UNAMINE O	tertiary amine	92% active	liquid/solid	emulsifier; detergent; anti-corrosion
Millchem	ASTON AP CONCENTRATE	cationic polyamine	solids 100%	liquid	antistatic agent; softener *industrial applications; textile industry*

Supplier	Trade name	Chemical description	Composition	General properties	Functionality / *Application*
Sandoz	DERMAGEN E	aliphatic polyamine		liquid	fixing agent / *textile industry; dyes*
	SANDOFIX TPS	aliphatic polyamine derivative		liquid	fixative / *dyeing; textile industry*
	SANDOPUR DK	polyamine in aq. soln.		liquid	levelling agent
	SANDOQUAD T2B	fatty amine derivative in aq. soln.		liquid	thickener / *industrial detergent*
Tomah Products	TOMAH ACID THICKENER	amine blend		amber paste	thickener / *strong acids*
	TOMAH ACRA 2000	amine blend	100% active	dark viscous liquid	antistripping agent / *asphalt*
	TOMAH ANTI-CAKE 17	primary ether amine	99-100% active	yellow homogenous liquid	anti-caking agent / *fertilisers*
	TOMAH ASPHALT EMULSIFIER DIAMINE	amine surfactant	100% active; amine value 320-350	pasty	emulsifier / *asphalt*
	TOMAH CLAYGELL	amine surfactant	amine 76-82%; amine value 190-200	light amber liquid	*non-asbestos roof coating bitumen*
	TOMAH PA-10	hexyl oxy propylamine	amine 95% min.; amine value 325-340	light amber liquid; M.W. 165	anti-corrosion; antistatic agent
	TOMAH PA-14	isodecyl oxy propylamine	amine 95% min.; amine value 240-255	colourless/light yellow liquid; M.W. 229	
	TOMAH PA-16	isododecyl oxy propylamine	amine 95% min.; amine value 215-230	light amber liquid; M.W. 253	anti-corrosion; flotation agent; antistatic agent; emulsifier
	TOMAH PA-17	isotridecyl oxy propylamine	amine 95% min.; amine value 195-215	colourless/light yellow liquid; M.W. 274	
	TOMAH PA-19	linear alkoxypropylamine	amine 95% min.; amine value 195-215	light to amber liquid; M.W. 272	
	TOMAH TAA 3000	amine blend	100% active	amber liquid	antistripping agent / *asphalt*
	TOMAH TAE-P	polyamine	100% active	dark paste	stabiliser; antistripping agent / *asphalt*
Witco	WITBREAK RTC-326	polymeric amine		liquid; pour point − 6.7°C; pH 8.5	reverse demulsifier / *petroleum industry*
	WITBREAK RTC-330	polymeric amine		liquid; pH 5.5	
	WITCAMINE 6606	PEG-15 tallow amine	99% solids	light amber liquid; pH 9.5	antistatic agent; dispersant; o/w emulsifier; lubricant; substantivity agent; wetting agent / *personal care products; household and industrial applications*
	WITCAMINE 6622	PEG-30 oleyl amine	80% solids	light amber liquid; pH 9.5	

Diamines

Akzo Nobel	DUOMEEN C	N-coco-1,3-diaminopropane	diamine content > 89%	liquid/paste	
	DUOMEEN CD	N-coco-1,3-diaminopropane, distilled	diamine content > 89%	liquid/paste	
	DUOMEEN HT	N-tallow-1,3-diaminopropane (hydrogenated tallow-alkyl)	diamine content > 89%	solid	
	DUOMEEN O	N-oleyl-1,3-diaminopropane	diamine content > 89%	liquid/paste	
	DUOMEEN OX	N-oleyl-1,3-diaminopropane	diamine content > 92%	liquid/paste	

Supplier	Trade name	Chemical description	Composition	General properties	Functionality / *Application*
Akzo Nobel	DUOMEEN T PASTILLES	N-tallow-1,3-diaminopropane (hydrogenated tallow-alkyl)	diamine content > 89%	pastilles	
	DUOMEEN T	N-tallow-1,3-diaminopropane	diamine content > 92%	paste	
	DUOMEEN TX	N-tallow-1,3-diaminopropane	diamine content > 92%	paste	
Elf Atochem	DINORAM 42	arachidyl behenyl propylene diamine	≥80% active	solid	
	DINORAM C	cocopropylene diamine	≥85% active	pasty liquid	
	DINORAM CD	distilled cocopropylene diamine	≥85% active	pasty liquid	
	DINORAM O	oleyl propylene diamine	≥85% active	pasty liquid	
	DINORAM OD	distilled oleyl propylene diamine	≥85% active	pasty liquid	
	DINORAM S	tallow propylene diamine	≥85% active	paste	emulsifier; anti-stripping agent *bitumen; road making*
	DINORAM SD	distilled tallow propylene diamine	≥85% active	paste	
	DINORAM SH	hydrogenated tallow propylene diamine	≥85% active	solid	chemical intermediate
	DINORAM SHD	distilled hydrogenated tallow propylene diamine	≥85% active	solid	
	DINORAM SL	derivative of N-tallow propylene diamine	≥85% active	liquid	
Fina Chemicals	RADIAMINE 6540	hydrogenated tallow propane diamine	amine 98% min.; total amine value 330; iodine value < 3; carbon chain composition C_{14} 5%, C_{16} 30%, C_{18} 65%	solid; M.W. 340; colour 3 max. (Gardner); softening point 45°C	
	RADIAMINE 6560	coconut oil propane diamine	amine 98% min.; total amine value 400; iodine value < 10; carbon chain composition C_8 5%, C_{10} 7%, C_{12} 48%, C_{14} 18%, C_{16} 12%, C_{18} 6%, $C_{18'}$ 4%	paste; M.W. 280; colour 6 max. (Gardner)	emulsifier; adhesion agent; wetting agent *road construction*
	RADIAMINE 6570	tallow propane diamine	amine 98% min.; total amine value 330; iodine value > 35; carbon chain composition C_{14} 5%, C_{16} 30%, C_{18} 20%, $C_{18'}$ 45%	paste; M.W. 340; colour 5 max. (Gardner)	
	RADIAMINE 6572	oleyl propane diamine	amine 98% min.; total amine value 325; iodine value > 60; carbon chain composition C_{14} 3%, C_{16} 9%, C_{18} 10%, $C_{18'}$ 70%, $C_{18''}$ 8%	liquid; M.W. 345; colour 6 max. (Gardner)	
Hoechst	GENAMIN LAP100D	distilled lauryl propylene diamine	95% active	paste	intermediate
	GENAMIN TAP100D	distilled tallow propylene diamine	80% active	paste	intermediate
Kao Corporation	DIAMIN AB	arachidyl behenyl 1,3-propylene diamine	amine 85% min.; amine value 300-320; carbon chain composition C_{16} 10%, C_{18} 30%, $C_{20/22}$ 60%; iodine value 3 max.	solid; M.W. 360; colour 4 max. (Gardner)	
	DIAMIN C	coco 1,3-propylene diamine	amine 85% min.; amine value 365-385; carbon chain composition C_8 3%, C_{10} 5%, C_{12} 50%, C_{14} 20%, C_{16} 10%, C_{18} 3%, $C_{18'}$ 9%; iodine value 5-15	liquid/solid; M.W. 300; colour 5 max. (Gardner)	
	DIAMIN O	oleyl 1,3-propylene diamine	amine 85% min.; amine value 310-330; carbon chain composition C_{14} 2%, C_{16} 10%, C_{18} 8%, $C_{18'}$ 80%; iodine value 60-70	liquid/solid; M.W. 350; colour 5 max. (Gardner)	

Supplier	Trade name	Chemical description	Composition	General properties	Functionality / Application
Kao Corporation	DIAMIN T	tallow 1,3-propylene diamine	amine 85% min.; amine value 320-340; carbon chain composition C_{12} 1%, C_{14} 3%, C_{16} 35%, C_{18} 18%, C_{18}' 42%, $C_{20/22}$ 1%; iodine value 35-45	solid/paste; M.W. 340; colour 5 max. (Gardner)	
	DIAMIN TH	hydrogenated tallow 1,3-propylene diamine	amine 85% min.; amine value 315-335; carbon chain composition C_{14} 4%, C_{16} 32%, C_{18} 60%, C_{18}' 3%, $C_{20/22}$ 1%; iodine value 3 max.	solid; M.W. 340; colour 3 max. (Gardner)	
Tomah Products	TOMAH DA-14	isodecyl oxy propyl-1,3-diamine propane	amine 95% min.; amine value 375-395	light amber liquid; M.W. 295	anti-corrosion; chemical intermediate
	TOMAH DA-17	isotridecyl oxy propyl-1,3-diamine propane	amine 95% min.; amine value 325-350	amber liquid; M.W. 340	anti-corrosion; chemical intermediate
	TOMAH TAE-A	diamine	100% active	light amber liquid	emulsifier
	TOMAH TAE-RI	diamine	100% active	light amber liquid	*asphalt emulsification*

Imidazolines

Supplier	Trade name	Chemical description	Composition	General properties	Functionality / Application
Akzo Nobel	BEROL 594	hydroxyethylalkylimidazoline	90% active		*corrosion inhibitors; cleaners; rust removers*
Hüls	MARLOWET 5440	substituted imidazoline	active detergent 100%	liquid	emulsifier; anti-corrosion *mineral oils; car rinsing agents*
Lakeland Laboratories	IMIDAZOLINE 12OH	hydroxyethylimidazoline	100% active	cream coloured waxy solid	anti-corrosion; dispersant; emulsifier *metal working; lubricants; chemical intermediate*
	IMIDAZOLINE 18DA	amidoethylimidazoline	100% active	clear dark amber liquid	anti-corrosion; dispersant; emulsifier *oil industry; metal working; paper industry; lubricants; road making; paint industry; inks; agriculture*
	IMIDAZOLINE 18NH	aminoethylimidazoline	100% active	clear dark amber liquid	anti-corrosion; dispersant; emulsifier; adhesion agent; flocculant *oil industry; metal working; paint industry; inks*
	IMIDAZOLINE 18OH	hydroxyethylimidazoline	100% active	clear dark amber liquid	anti-corrosion; dispersant; emulsifier; flocculant; softener *oil industry; metal working; textile industry; paper industry; lubricants; road making; paint industry; inks; agriculture*
	IMIDAZOLINE SOH	hydroxyethylimidazoline	100% active	cream coloured waxy solid	dispersant; emulsifier; flocculant *oil industry; metal working; textile industry; paper industry; lubricants; paint industry; inks; agriculture*
McIntyre Group	MACKAZOLINE C	cocoyl hydroxyethyl imidazoline	100% active	liquid; pH 11.5	emulsifier; antistatic agent; anti-corrosion
	MACKAZOLINE CY	capryl hydroxyethyl imidazoline	100% active	liquid; pH 11.5	
	MACKAZOLINE L	lauryl hydroxyethyl imidazoline	100% active	paste; pH 11.5	
	MACKAZOLINE O	oleyl hydroxyethyl imidazoline	100% active	liquid; pH 11.5	
	MACKAZOLINE T	tall oil hydroxyethyl imidazoline	100% active	liquid; pH 11.5	
Millchem	AMMONYX 4080 CG	tallow imidazoline methosulfate	74-76% active	liquid	conditioner *skin care products; hair care products*
	AMMONYX 4080	imidazolinium sulfate	75% active	liquid	softener *household and industrial applications*

Supplier	Trade name	Chemical description	Composition	General properties	Functionality Application
Millchem	CATIONIC SOFTENER CONCENTRATE	alkyl imidazoline derivative	solids 60-66%	paste	softener; lubricant; glass fibre mordant
	CATIONIC SOFTENER FLAKE	alkyl imidazoline derivative	solids 88-92%	flakes	
	CATIONIC SOFTENER VERSION C	alkyl imidazoline derivative	solids 33-36%	paste	
Scher Chemicals	SCHERCOZOLINE C	cocoyl hydroxyethyl imidazoline	imidazoline 90% min.; free amine 3% min.; alkali value 200-214	tan semi-solid; M.W. 278	
	SCHERCOZOLINE I	isostearyl hydroxyethyl imidazoline	imidazoline 90% min.; free amine 3% max.; alkali value 150-160	clear amber liquid; M.W. 378	detergent; wetting agent; antistatic agent intermediate for quaternaries
	SCHERCOZOLINE L	lauryl hydroxyethyl imidazoline	imidazoline 90% min.; free amine 3% max.; alkali value 204-214	cream-coloured solid; M.W. 268; m.p. 38-42°C	
	SCHERCOZOLINE O	oleyl hydroxyethyl imidazoline	imidazoline 90% min.; free amine 3% max.; alkali value 160-170	clear dark amber liquid; M.W. 350	w/o emulsifier; anti-corrosion intermediate for quaternaries
Servo Delden	SERVAMINE KOD 306	cocofatty acid + aminoethylethanolamine		liquid	
	SERVAMINE KOO 330	oleic acid + diethylenetriamine		liquid	
	SERVAMINE KOO 330 B	oleic acid + diethylenetriamine		liquid	
	SERVAMINE KOO 340 B	oleic acid + tetraethylene pentamine		liquid	
	SERVAMINE KOO 360	oleic acid + aminoethylethanolamine		liquid	
	SERVAMINE KOX 360	C_{10} fatty acid + aminoethylethanolamine		solid	
	SERVAMINE KOY 330 B	tallow fatty acid + diethylenetriamine		solid	
Thomas Swan & Co.	CASAMINE C	1-aminoethyl-2-alkylimidazoline	imidazoline content 90% min.	solid; M.W. 267; m.p. 43°C	detergent; wetting agent; foaming agent; antistatic agent; anti-corrosion; softener; emulsifier; thickener agriculture; plastics industry; textile industry; bitumen; paint industry; shampoos
	CASAMINE CH	1-hydroxyethyl-2-alkylimidazoline	imidazoline content 90% min.	solid; M.W. 268; m.p. 38°C	
	CASAMINE O	1-aminoethyl-2-alkylimidazoline	imidazoline content 90% min.	liquid; m.p. 43°C	
	CASAMINE OH	1-hydroxyethyl-2-alkylimidazoline	imidazoline content 90% min.	liquid; M.W. 350; pour point −17°C; m.p. 43°C	
	CASAMINE R	1-aminoethyl-2-alkylimidazoline	imidazoline content 90% min.	liquid; M.W. 349	
	CASAMINE RH	1-hydroxyethyl-2-alkylimidazoline	imidazoline content 90% min.	liquid; M.W. 350	
Thor Chemicals	SOVATEX IM12H	hydroxyethyl imidazoline of coconut fatty acid	100% active	semi-liquid	anti-corrosion; lubricant; antistatic agent
	SOVATEX IM12N	aminoethyl imidazoline of coconut fatty acid	100% active	semi-liquid	
	SOVATEX IM17H	hydroxyethyl imidazoline of oleic acid	100% active	liquid	
	SOVATEX IM17N	aminoethyl imidazoline of oleic acid	100% active	liquid	
Tomah Products	TOMAH TAA-2500	imidazoline	100% active; amine value 211	dark viscous liquid	antistripping agent asphalt
Witco	SOCHAMINE 35	imidazoline	100% active	liquid	emulsifier; anti-corrosion; wetting agent industrial applications

Supplier	Trade name	Chemical description	Composition	General properties	Functionality *Application*
Witco	WITCAMIDE 204	imidazoline		liquid; pour point −5°C; pH 11	
	WITCAMIDE 209	imidazoline		liquid; pour point −1.1°C; pH 11	
	WITCAMIDE 211	imidazoline		liquid; pour point −6.7°C; pH 11	
	WITCAMIDE 240	polyamido imidazoline		liquid; pour point 0°C; pH 9	
	WITCAMIDE 760	imidazoline		liquid; pour point −12.2°C; pH 101	anti-corrosion *petroleum industry*
	WITCAMIDE AL42-12	fatty imidazoline		liquid; pH 11	
	WITCAMIDE PA-60B	fatty imidazoline salt		liquid; pour point −7.2°C; pH 6.5	
	WITCAMIDE TI-60	amido imidazoline		liquid; pour point −5°C; pH 10	
	WITCOR 3630	imidazoline		liquid; pour point −6.7°C; pH 5.3	
	WITCOR 3635	imidazoline		liquid; pour point −1.7°C; pH 6.3	

Quaternaries

Supplier	Trade name	Chemical description	Composition	General properties	Functionality *Application*
Akcros Chemicals	AGRILAN TKA103	quaternary ammonium compound	100% active	light brown viscous liquid; pour point <0°C; viscosity 400 cSt; pH 6.0-9.0 (1% aq.)	wetting agent; spreading agent *agrochemicals*
	QUADRILAN AT	quaternised fatty amine ethoxylate	100% active	light brown viscous liquid; pour point <0°C; viscosity 400 cSt; pH 6-9 (1% aq.)	antistatic agent; additive; detergent *polymers; textile industry; detergents; rinse aids*
	QUADRILAN BC	alkyl dimethyl benzyl ammonium chloride	50% active	clear pale yellow liquid; pour point <0°C; viscosity 150 cSt; pH 7.0 (1% aq.)	bactericide; fungicide *disinfectants; detergent sanitisers*
Akzo Nobel	ARQUAD 1214B-50	alkyl (C12/14) benzyldimethylammonium chloride	50-51% active	liquid	
	ARQUAD 16-29	hexadecyltrimethylammonium chloride	27-30% active	liquid	
	ARQUAD 16-50	hexadecyltrimethylammonium chloride	49-52% active	liquid	
	ARQUAD 18-50	octadecyltrimethylammonium chloride	49-52% active	liquid/paste	
	ARQUAD 2.10-50	didecyldimethylammonium chloride	49-52% active	liquid	
	ARQUAD 2C-75 HFP	dicocodimethylammonium chloride	74-76% active	liquid	
	ARQUAD 2C-75	dicocodimethylammonium chloride	74-77% active	liquid	
	ARQUAD 2HT-75	ditallowdimethylammonium chloride (hydrogenated tallow-alkyl)	75-78% active	paste	
	ARQUAD 2T-70	ditallowdimethylammonium chloride	70-72% active	liquid	
	ARQUAD 88	ditallowdimethylammonium chloride (hydrogenated tallow-alkyl)	87-89% active	solid	
	ARQUAD B-50	cocobenzyldimethylammonium chloride	48-52% active	liquid	
	ARQUAD B-80	cocobenzyldimethylammonium chloride	78-81% active	liquid	
	ARQUAD B-90	cocobenzyldimethylammonium chloride	89-92% active	paste	

Supplier	Trade name	Chemical description	Composition	General properties	Functionality / Application
Akzo Nobel	ARQUAD B404	quaternary ammonium compound on a carrier	40% active	white powder	
	ARQUAD C-35	cocotrimethylammonium chloride	33-37% active	liquid	
	ARQUAD C-50	cocotrimethylammonium chloride	49-52% active	liquid	
	ARQUAD DM14B-90 PDR	tetradecylbenzyldimethylammonium chloride	90% active	white powder	
	ARQUAD DM350B-50	alkylbenzyldimethylammonium chloride	49-51% active	liquid	
	ARQUAD DM350B-80	alkylbenzyldimethylammonium chloride	78-81% active	liquid	
	ARQUAD DMHTB-75	tallowbenzyldimethylammonium chloride (hydrogenated tallow-alkyl)	75-78% active	paste	
	ARQUAD DMMCB-33	cocobenzyldimethylammonium chloride (fractionated coco-alkyl)	33-34% active	liquid	
	ARQUAD DMMCB-50	cocobenzyldimethylammonium chloride (fractionated coco-alkyl)	49-52% active	liquid	
	ARQUAD DMMCB-80	cocobenzyldimethylammonium chloride (fractionated coco-alkyl)	80-81% active	liquid	
	ARQUAD HC	ditallowdimethylammonium chloride (hydrogenated tallow-alkyl)	> 96% active	kibbles	
	ARQUAD HT-50	tallowtrimethylammonium chloride (hydrogenated tallow-alkyl)	49-52% active	liquid	
	ARQUAD M2HTB-85	ditallowbenzylmethylammonium chloride (hydrogenated tallow-alkyl)	82-85% active	paste	
	ARQUAD MC-50	cocotrimethylammonium chloride (fractionated coco-alkyl)	49-52% active	liquid	
	ARQUAD NF-50	dialkyldimethylammonium chloride	49-52% active	liquid	
	ARQUAD S-50	oleyltrimethylammonium chloride	49-52% active	liquid	
	ARQUAD S2C-50	oleyldimethylammonium chloride/dicocodimethylammonium chloride	49-52% active	liquid	
	ARQUAD T-30	tallowtrimethylammonium chloride	28-31% active	liquid	
	ARQUAD T-50	tallowtrimethylammonium chloride	49-52% active	liquid	
	DUOQUAD T-50	N,N,N',N'-pentamethyl-N-tallow-1,3-propanediammonium chloride	48-53% active	liquid	
	ETHOQUAD C12	cocobis(2-hydroxyethyl)methylammonium chloride	75-78% active	liquid	
	ETHOQUAD C25	polyoxyethylene (15) cocomethylammonium chloride	> 95% active	liquid	
	ETHOQUAD HT25	polyoxyethylene (15) tallowmethylammonium chloride (hydrogenated tallow-alkyl)	> 95% active	liquid	
	ETHOQUAD O12	oleylbis(2-hydroxyethyl)methylammonium chloride	72-75% active	liquid	
	QUERTON 14 BR-40	tetradecyltrimethylammonium bromide	40% active		bactericide; cosmetics; surgical scrubs

Supplier	Trade name	Chemical description	Composition	General properties	Functionality / Application
Akzo Nobel	QUERTON 210Cl-50	didecyldimethylammonium chloride	50% active		bactericide; fungicide / *medical care; food industry; detergent sanitizers*
	QUERTON 210Cl-70S	didecyldimethylammonium chloride	70% active		
	QUERTON 210Cl-80	didecyldimethylammonium chloride	80% active		
	QUERTON 246	alkyldimethylbenzylammonium chloride	80% active		
	QUERTON 280	alkyltrimethylammonium chloride	35% active		fungicide / *wood protection*
	QUERTON 28Cl-50	dioctyldimethylammonium chloride	50% active		low foam; bactericide; fungicide; algicide
	QUERTON 442-11	bis-(hydrogenated tallow)dimethylammonium chloride	ca. 77% active		softener; antistatic agent / *textile softeners; hair conditioners*
	QUERTON 442-82	bis-(hydrogenated tallow)dimethylammonium chloride	ca. 82% active		
	QUERTON 442	bis-(hydrogenated tallow)dimethylammonium chloride	ca. 75% active		
	QUERTON GCl-50	cocoalkylguanidinium chloride	50% active		bactericide / *detergent sanitizers*
	QUERTON KKBCl-50	cocoalkyldimethylbenzylammonium chloride	50% active		disinfectant; antistatic agent / *medical care; food industry; detergent sanitizers*
Albright & Wilson	EMPIGEN 5089	alkyl trimethyl ammonium chloride	34% active	liquid	
	EMPIGEN BAC50/ BP	alkyl dimethyl benzyl ammonium chloride; B.P. grade	50% active	liquid	antistatic agent; disinfectant / *textile industry; polymer industry; pharmaceuticals*
	EMPIGEN BAC50	alkyl dimethyl benzyl ammonium chloride	50% active	liquid	
	EMPIGEN BAC80	alkyl dimethyl benzyl ammonium chloride	80% active	paste	
	EMPIGEN BCB50	alkyl dimethyl benzyl ammonium chloride	50% active	liquid	
	EMPIGEN CM	alkyl trimethyl ammonium methosulfate	30% active	liquid	
Allied Colloids	ALCAMINE CWS	quaternary ammonium compound		flakes	softener; antistatic agent / *textile industry; all fibres*
	ALCAMINE HPM	quaternary ammonium compound		liquid	
	ALCOSIST BDR	quaternary ammonium compound		liquid	retarding aid; migration aid; anti-precipitant / *textile industry; dyeing of acrylics*
	ALCOSIST M	quaternary ammonium compound		liquid	migration aid / *textile industry; dyeing of acrylics*
	ALCOSIST NRL	quaternary ammonium compound		liquid	levelling agent / *textile industry; dyeing of polyamides and wools*
	ALCOSTAT TM	quaternary ammonium compound		liquid	antistatic agent / *textile industry*
Amerchol	UCARE POLYMER JR SERIES	polyquaternium-10			conditioner / *skin care products; hair care products*
	UCARE POLYMER LR SERIES	polyquaternium-10			
	UCARE POLYMER SR-10	polyquaternium-10			

Supplier	Trade name	Chemical description	Composition	General properties	Functionality / Application
Auschem Cesalpinia	CATIONICO SCL	coco-dimethyl benzyl ammonium chloride F.U.	50% active	liquid	detergent; sterilising agent / *zootechnical industry*
	TEQUAT BC	cetrimonium chloride	25% active	liquid	conditioner / *hair care products*
	TEQUAT PAN	alkyl-dimethyl benzyl ammonium chloride	50% active	liquid	detergent; sterilising agent
Bayer	ASTRAGAL ACM	quaternary ammonium compound			retarder / *textile industry*
	ASTRAGAL PAN NEW	quaternary ammonium compound			
	ASTRAGAL TR	quaternary ammonium compound			retarder / *textile industry; dyeing*
Dr. Th. Boehme	SYNTHESIN 4347	quaternary ammonium compound			spin finish / *textile industry; fibre production*
CHEM-Y	AKYPOQUAT 132	lauroyl PG-trimonium chloride	chloride 6.0-8.5%	hazy, almost colourless paste; M.W. ca. 352; colour 200 max. (Hazen; in water/ethanol 1:1); pH 3.5-4.8 (10% in water)	emollient; conditioner / *shampoos; shower preparations; shower gels; baby shampoos; liquid soaps*
Chemax	CHEMQUAT 12/50	lauryl trimethyl ammonium chloride	50% active	liquid	antistatic agent; anti-corrosion; softener; viscosity modifier / *plastics industry; latex foam; paper industry; textile industry*
	CHEMQUAT 16/50	cetyl trimethyl ammonium chloride	50% active	liquid	
	CHEMQUAT C/33W	coco trimethyl ammonium chloride	33% active	liquid	
Chemviron	MERQUAT 100	polyquaternium-6	solids 38-41%	colourless to pale yellow, clear viscous liquid; M.W. 4×10^5; viscosity 8,000-12,000 cPs; colour ≤ 70 (APHA); pH 6.0-7.0	conditioner; lubricant; antistatic agent; / *hair care products; skin care products; bath preparations; deodorants; shaving products*
	MERQUAT 280	polyquaternium-22	solids 38-41%; methyl paraben 0.1%; propyl paraben 0.02%	clear to slightly hazy yellow viscous liquid; M.W. 2×10^6; viscosity 3,500-12,000 cPs; pH 4.25-5.25	conditioner; lubricant; antistatic agent / *hair care products; skin care products; bath preparations; liquid soaps; shaving products*
	MERQUAT 295	polyquaternium-22	solids 38-40%	clear viscous pale yellow liquid; M.W. 5×10^5; viscosity 3,500-9,000 cPs; colour 125 max. (APHA); pH 4.0-5.5	conditioner / *hair care products; shampoos; hair conditioners*
	MERQUAT 550	polyquaternium-7	solids 8-10%; methyl paraben 0.1%; propyl paraben 0.02%	colourless to pale yellow clear viscous liquid; M.W. 5×10^6; viscosity 7,500-15,000 cPs; colour ≤ 70 (APHA); pH 6.5-7.5	foam booster; foam stabiliser; lubricant; film former / *hair care products; skin care products; shaving products; bath preparations; antiperspirants; deodorants*
	MERQUAT 2200	polyquaternium-7	solids 100%	white to slightly off white particles; M.W. 5×10^6	conditioner; foam stabiliser; film former / *hair care products; skin care products; antiperspirants; deodorants; shaving products; bath preparations*
	MERQUAT PLUS 3330	polyquaternium-39	solids 8-10%; methyl paraben 0.15%; propyl paraben 0.03%	colourless to very pale yellow, clear viscous liquid; M.W. 4×10^6; viscosity 4,400-10,400 cPs; pH 5.5-7.0	conditioner; antistatic agent; humectant / *hair care products; skin care products; liquid soaps; shower gels*
	MERQUAT PLUS 3331	polyquaternium-39	solids 8-10%	clear, colourless viscous liquid; M.W. 5×10^6; viscosity 5,000-15,000 cPs; pH 5.5-7.0	conditioner / *hair care products; skin care products; bath preparations*
	MERQUAT S	polyquaternium-7	solids 8-10%; methyl paraben 0.1%; propyl paraben 0.02%	colourless to pale yellow, clear viscous liquid; M.W. 7×10^6; viscosity 9,000-15,000 cPs; colour ≤ 70 (APHA); pH 6.5-7.5	film former; foam stabiliser; foam booster; lubricant / *shampoos; hair care products; skin care products; cosmetics; bath preparations; deodorants; shaving products*
Ciba	TINEGAL B	quaternary fatty amine		clear, yellowish, slightly viscous liquid	
	TINEGAL PAC	soln. of quaternary ammonium salt in H_2O/isopropanol		clear, yellowish low viscosity liquid	retardant / *textile industry; dyeing of acrylic fibres*

Supplier	Trade name	Chemical description	Composition	General properties	Functionality / Application
Croda	CRODEX C	cetrimonium bromide and cetearyl alcohol		white waxy solid	emulsifier; bactericide *skin care products; hair conditioners; pharmaceuticals*
	INCROQUAT BEHENYL BDQ/P	behenalkonium chloride supplied as a 25% soln./dispersion in propylene glycol	quaternary content 24-26%; free amine 2% max.	white paste; colour 2 max. (Gardner); pH 6.0-7.5 (10% aq. soln.)	substantivity agent; conditioner; softener *hair care products; hair conditioners*
	INCROQUAT BEHENYL TMC/P	behenyl trimethyl ammonium chloride, supplied as a 25% soln./dispersion in propylene glycol	quaternary content 24-26%; free amine < 0.2%	white paste; colour 2 max. (Gardner); pH 6.5-7.5 (10% aq. soln.)	conditioner; softener; substantivity agent *hair care products; skin care products*
	INCROQUAT BEHENYL TMC	behenyl trimethyl ammonium chloride supplied as a 25% soln./dispersion in cetostearyl alcohol	quaternary content 24-25%; free amine < 0.2 %	white flake; colour 2 max. (Gardner); m.p. ca. 50°C; pH 6.5-7.5 (10% aq. soln.)	conditioner; softener; substantivity agent; emulsifier; thickener *hair care products; skin care products*
	INCROQUAT S-25	stearalkonium chloride		paste	conditioner; softener; emollient *hair care products; skin care products*
Elf Atochem	NORAMIUM C80	coco dimethylbenzene ammonium chloride	78-81% active	liquid	bactericide; fungicide; demulsifier *cosmetics; paint industry; latexes*
	NORAMIUM DA50	coco dimethylbenzene ammonium chloride	49-51% active	liquid	
	NORAMIUM DSH75	hydrogenated tallow dimethylbenzene ammonium chloride	74-77% active	paste/liquid	
	NORAMIUM M210-50	N,N'-dialkyl didecyl ammonium chloride soln. in a H$_2$O and isopropanol mixture	49-52% active	clear, homogeneous liquid; M.W. 315 (free amine); 352 (amine chlorhydrate); colour 4 max. (Gardner); solidification point ca. − 18°C; pH 5-8 (10% in distilled H$_2$O)	bacteriostatic agent; biocide; algicide; fungicide *surface disinfectants; food industry; cosmetics; water treatment*
	NORAMIUM M2C	dicoco dimethyl ammonium chloride	74-77% active	liquid	softener *textile industry*
	NORAMIUM M2SH 100	dihydrogenated tallow) dimethyl ammonium chloride	95% active	solid	
	NORAMIUM M2SH	dihydrogenated tallow) dimethyl ammonium chloride	74-77% active	paste	
	NORAMIUM MC50	coco trimethyl ammonium chloride	49-51% active	liquid	additive *antibiotics manufacture*
	NORAMIUM MO50	oleyl trimethyl ammonium chloride	49-51% active	liquid	
	NORAMIUM MS50	tallow trimethyl ammonium chloride	49-51% active	liquid	
	NORAMIUM S75	tallow dimethylbenzene ammonium chloride	74-77% active	liquid	
	NORAMIUM 920	N,N-di(acyloxy-2-ethyl), N-hydroxy-2-ethyl, N-methyl ammonium methosulfate in isopropanol soln.	acid value 2.0 max.	yellow liquid at 50°C, paste at 20°C; viscosity 220 mPas (50°C); colour 3 max. (Gardner); solidification point ca. 29°C	antistatic agent; softener; conditioner *textile industry*
	NOXAMIUM C15M	N-alkyl coco pentadecahydroxyethyl methyl ammonium methosulfate		liquid; cloud point > 100°C (1% w/w in distilled H$_2$O); colour 15 max. (Gardner); solidification point ca. − 13°C; pH 7.8 (5% in distilled H$_2$O)	wetting agent; dispersant; emulsifier; antistatic agent *surface cleaners; textile industry*
	NOXAMIUM MC2-50	N-bis hydroxyethyl coco methyl ammonium chloride	49-51% active	liquid	
	NOXAMIUM MO2-50	N-bis hydroxyethyl oleyl methyl ammonium chloride	59-51% active	liquid	
	NOXAMIUM MS2-50	N-bis hydroxyethyl tallow methyl ammonium chloride	49-51% active	liquid	

Supplier	Trade name	Chemical description	Composition	General properties	Functionality / Application
Elf Atochem	NOXAMIUM S11M	n-alkyl tallow di(undecaoxyethyl) methyl ammonium methosulfate	99.5% active	homogeneous liquid; solidification point ca. −5°C; pH 7 (5% in distilled H_2O)	wetting agent; dispersant; emulsifier; detergent / *detergents; car washes*
Ellis & Everard	CAFLON 15QS	quaternary salt of amine ethoxylate	100% active		
	CAFLON BQC50	benzyl ammonium chloride	50% active		
	CAFLON BQC80	benzyl ammonium chloride	80% active		
	CAFLON CET	cetyl trimethyl ammonium chloride	30% active		
	CAFLON CL	lauryl dimethyl benzyl ammonium chloride			biocide; preservative
	CAFLON FS75	alkylimidazolinium methosulfate	75% active		
Fina Chemicals	RADIAQUAT 6410	di-n-decyl dimethyl ammonium chloride	product 50% min.; H_2O < 30%; carbon chain composition C_8 10%, C_{10} 90%	liquid; M.W. 361; colour 2 max. (Gardner)	antistatic agent; detergent; bactericide; germicide / *textile industry; leather industry; cosmetics; wood preservation; hard surface cleaners; laundry products*
	RADIAQUAT 6412	di-n-decyl dimethyl ammonium chloride	product 70% min.; H_2O 10%; carbon chain composition C_8 10%, C_{10} 90%	liquid; M.W. 361; colour 2 max. (Gardner)	
	RADIAQUAT 6442	di-hydrogenated tallow dimethyl ammonium chloride	product 74% min.; H_2O 10%; carbon chain composition C_{14} 5%, C_{16} 30%, C_{18} 65%	paste; M.W. 569; colour 3 max. (Gardner)	antistatic agent; detergent; conditioner / *textile industry; leather industry; cosmetics; hair conditioners; hair care products; laundry products*
	RADIAQUAT 6444	cetyl trimethyl ammonium chloride	product 50% min.; H_2O 15%; carbon chain composition C_{14} 2%, C_{16} 92%, C_{18} 6%	liquid; M.W. 319; colour 3 max. (Gardner)	
	RADIAQUAT 6445	cetyl trimethyl ammonium chloride	product 30% min.; H_2O 70%; carbon chain composition C_{14} 2%, C_{16} 94%, C_{18} 4%	liquid; M.W. 326; colour 3 max. (Gardner)	antistatic agent; detergent; bacteriostatic agent / *textile industry; leather industry; cosmetics; deodorants; personal care products*
	RADIAQUAT 6460	coconut oil trimethyl ammonium chloride	product 32-35% min.; H_2O 65-68%; carbon chain composition $C_{8/10}$ 3%, C_{12} 48-58%, C_{14} 18-24%, C_{16} 8-12%, C_{18} 10-14%	liquid; M.W. 288; colour 5 max. (Gardner)	
	RADIAQUAT 6462	di-coconut oil dimethyl ammonium chloride	product 75% min.; H_2O < 10%; carbon chain composition C_8 5%, C_{10} 7%, C_{12} 56%, C_{14} 18%, C_{16} 7%, C_{18} 5%, $C_{18'}$ 2%	paste; M.W. 439; colour 4 max. (Gardner)	antistatic agent; detergent; conditioner / *textile industry; leather industry; cosmetics; hair conditioners; hair care products*
	RADIAQUAT 6470	di-tallow dimethyl ammonium chloride	product 75% min.; H_2O 9%; carbon chain composition C_{14} 5%, C_{16} 30%, C_{18} 20%, $C_{18'}$ 45%	liquid; M.W. 569; colour 4 max. (Gardner)	
	RADIAQUAT 6471	tallow trimethyl ammonium chloride	product 50% min.; H_2O 15%; carbon chain composition C_{14} 5%, C_{16} 30%, C_{18} 20%, $C_{18'}$ 45%	liquid; M.W. 399; colour 5 max. (Gardner)	antistatic agent; detergent / *textile industry; leather industry; cosmetics*
	RADIAQUAT 6475	di-hydrogenated tallow dimethyl ammonium chloride	product 76% min.; H_2O 11%; carbon chain composition C_{14} 5%, C_{16} 30%, C_{18} 65%	paste; M.W. 569; colour 3 max. (Gardner)	
	RADIAQUAT 6480	di-hydrogenated tallow dimethyl ammonium chloride	product 8% min.; H_2O 80%; carbon chain composition C_{14} 5%, C_{16} 30%, C_{18} 65%	paste; M.W. 569; colour 3 max. (Gardner)	
Th. Goldschmidt	ABIL-QUAT 3270	quaternium-80		amber liquid	conditioner; re-fatting agent; cleansing agent; film former / *hair care products; skin cleansing preparations*
	ABIL-QUAT 3272	quaternium-80		amber liquid	
Hays Colours	DYSOFT B	cationic quaternary softening agent			softener / *textile industry; natural and synthetic fibres*
Henkel	BELFASIN OET	quaternary fatty acid ester			softener / *textile industry*
	BREVIOL RET-P 50	quaternary compound			retarding agent; / *textile industry; dyeing*

Supplier	Trade name	Chemical description	Composition	General properties	Functionality / Application
Henkel	COSMEDIA GUAR C261 N	guar hydroxypropyl trimonium chloride	92% active	fine white to yellowish powder	antistatic agent *hair care products*
	DEHYQUART A-FM	cetyl trimethyl ammonium chloride	24-26% active	liquid	algicide; fungicide; antistatic agent *water treatment*
	DEHYQUART A	cetrimonium chloride	24-26% active	clear pale yellow liquid	conditioner; antistatic agent; emulsifier *hair conditioners; hair care products; skin care products*
	DEHYQUART AU-35	bis(acyloxyethyl)hydroxyethyl methylammonium methosulfate	85 ± 2% active; isopropanol ca. 15%	semi-solid paste at 20°C, yellowish liquid at 30°C; colour < 3 (Gardner); pH 2-3 (5%)	
	DEHYQUART AU-36	bis(acyloxyethyl)hydroxyethyl methylammonium methosulfate	85 ± 2% active; isopropanol ca. 15%	semi-solid paste at 20°C, yellowish liquid at 45°C; M.W. 800; colour < 3 (Gardner); drop point 33-37°C; pH 2-3 (5%)	
	DEHYQUART AU-46	bis(acyloxyethyl)hydroxyethyl methylammonium methosulfate	90 ± 2% active; isopropanol ca. 10%	semi-solid paste at 20°C, yellowish liquid at 45°C; M.W. 800; colour < 3 (Gardner); drop point 37-42°C; pH 2-3 (5%)	softener; antistatic agent; rewetting agent *softening preparations*
	DEHYQUART AU-56	bis(acyloxyethyl)hydroxyethyl methylammonium methosulfate	90 ± 2% active; isopropanol ca. 10%	semi-solid paste at 20°C, yellowish liquid at 45°C; M.W. 800; colour < 3 (Gardner); drop point 37-42°C; pH 2-3 (5%)	
	DEHYQUART AU-57	bis(acyloxyethyl)hydroxyethyl methylammonium methosulfate	90 ± 2% active; isopropanol ca. 10%	semi-solid paste at 20°C, yellowish liquid at 45°C; colour < 3 (Gardner); pH 2-3 (5%)	
	DEHYQUART AU-67	bis(acyloxyethyl)hydroxyethyl methylammonium methosulfate	90 ± 2% active; isopropanol ca. 10%	semi-solid paste at 20°C, yellowish liquid at 45°C; colour < 3 (Gardner); pH 2-3 (5%)	
	DEHYQUART C CRYST.	laurylpyridinium chloride	90-94% active	powder	conditioner *hair care products; hair conditioners; skin care products*
	DEHYQUART C-4046	mixture of cetearyl alcohol, dipalmitoylethyl hydroxyethylmonium methosulfate and ceteareth-20	cationic active matter 21-27%; acid value 9 max.	flakes; m.p. 53-57°C; pH 2.0-3.5 (5%)	softener; conditioner; antistatic agent *hair care products*
	DEHYQUART DAM	distearyldimonium chloride	70-80% active	paste	conditioner; antistatic agent *hair care products; plastics industry; coatings*
	DEHYQUART E	hydroxycetyl hydroxyethyl dimonium chloride	21.5-23.5% active	clear to slightly turbid, slightly yellowish liquid	conditioner *shampoos; hair care products*
	DEHYQUART F30	cetearyl alcohol and dipalmitoylethyl hydroxyethylmonium methosulfate	cationic surfactant 28-32%; acid value 9 max.	white to yellow wax-like flakes; m.p. 50-56°C; pH 2.0-3.5 (5%)	conditioner *hair conditioners*
	DEHYQUART F75	distearoylethyl hydroxyethylmonium methosulfate and cetearyl alcohol	cationic surfactant 73-77%; acid value 5-12	white to yellow wax-like flakes; m.p. 60-64°C; pH 2.0-3.5 (5%)	conditioner *hair conditioners*
	DEHYQUART LDB-50	lauryl dimethyl benzyl ammonium chloride	49-51% active	liquid	bactericide; fungicide *disinfectants; cleaners*
	DEHYQUART LDB	lauralkonium chloride	34-36% active	liquid	conditioner; antistatic agent *hair care products*
	DEHYQUART LT	lauryltrimonium chloride	34-36% active	liquid	wetting agent; antistatic agent; conditioner; bactericide *hair care products; plastics industry; coatings*
	DEHYQUART SP	quaternium-52	49-51% active	clear viscous liquid	conditioner; antistatic agent *hair care products; skin care products*
	OSIMOL MA	quaternary ammonium compound			levelling agent; migration agent *textile industry; dyeing*

Supplier	Trade name	Chemical description	Composition	General properties	Functionality / *Application*
Henkel	OSIMOL RAC	quaternary ammonium compound			retarding agent / *textile industry; dyeing*
	OSIMOL TR	quaternary ammonium compound			temporary retarding agent / *textile industry; dyeing*
Hoechst	DODIGEN 95	speciality quaternary	30% active		anti-corrosion / *acids*
	DODIGEN 154	tallow ethoxy (2) BAC			anti-corrosion
	DODIGEN 181	speciality quaternary			anti-corrosion
	DODIGEN 213	speciality quaternary	50% active	liquid	disinfectants
	DODIGEN 226	coco dimethyl benzyl ammonium chloride	50% active	liquid	disinfectants
	DODIGEN 1383	speciality quaternary	50% active	liquid	pharmaceuticals; penicillin extraction
	DODIGEN 1490	dicoco dimethyl ammonium chloride	73-75% active	liquid	conditioners; detergents; car care products
	DODIGEN 1509	coco dimethyl dichloro benzyl ammonium chloride	50% active		disinfectants
	DODIGEN 1611	lauryl dimethyl benzyl ammonium chloride	50% active		disinfectants
	DODIGEN 1828	stearyl dimethyl ammonium chloride	75-80% active	solid	speciality products
	DODIGEN 2808	alkyl dimethyl benzyl ammonium chloride	80% active		disinfectants
	DODIGEN 5462	speciality quaternary		liquid	anti-corrosion / *mineral acids*
	DODIGEN 5594	soya trimethyl ammonium chloride		liquid	emulsifier; anti-corrosion / *bitumen*
	DP1334	imidazoline quaternary/dipropionate			
	GENAMIN CTAC	cetyl trimethyl ammonium chloride	30% active		*hair conditioners*
	GENAMIN DSAC	distearyl dimethyl ammonium chloride	95% active		*hair conditioners*
	GENAMIN EQ	ester quaternary	85% active		*fabric conditioners; hair conditioners*
	GENAMIN KDM-F	behenyl trimethyl ammonium chloride	80% active		*hair conditioners*
	GENAMIN KSE	ready made hair conditioner base	22% active		*hair conditioners*
	GENAMIN KSL	pentaoxethyl stearyl ammonium lactate	30% active		*hair conditioners; shampoos*
	GENAMIN PDAC	polyquaternium-6	40% active		*hair conditioners*
	GENAMIN STAC	stearyl trimethyl ammonium chloride	80% active		*hair conditioners*
	GENAPOL 3520	dialkyl quaternary	75% active		*bottle washing*
	GENAPOL 3725	dialkyl quaternary	75% active		*bottle washing*
	HOE S 4039	TEA ester quaternary	85% active	paste	*fabric conditioners*
	PRAEPAGEN WK	distearyl dimethyl ammonium chloride	75% active	paste	*fabric conditioners*
	PRAEPAGEN WKL	distearyl dialkyl ammonium chloride	75% active	liquid	*fabric conditioners; textile industry*
Hüls	MARLAZIN 7265	blend of fatty alkylimidazoline and fatty alkylammonium salt	active detergent 81%	liquid	*car rising agents*
	MARLAZIN KC 21/50	$C_{12/18}$-dimethylbenzyl ammonium chloride	active detergent 48%	liquid	bactericide / *disinfectants; cleaners*
	MARLAZIN KC 30/50	$C_{12/18}$-trimethyl ammonium chloride	active detergent 48%	liquid	*hair conditioners*

Supplier	Trade name	Chemical description	Composition	General properties	Functionality Application
Hüls	MARLOSOFT IQ 90	quaternised tallow fatty imidazolinium methosulfate	active detergent 90%	liquid/paste	*fabric softeners*
ICI	ATLAS G-265	quaternary ammonium compound		brown-yellow liquid	*household and industrial applications: agrochemicals; textile industry*
	ATLAS G-271	N-soya N-ethyl morpholinium ethosulfate	35% active	red-brown liquid	
	ATLAS G-3634A	quaternary ammonium derivative		red-brown liquid	
ISP	CERAPHYL 60	quaternium-22	solids 58-62%; H_2O 38-42%	clear, light amber liquid; pH 4.0-5.0	conditioner; *shampoos; hair conditioners; skin care products*
	CERAPHYL 65	quaternium-26	solids 53-63%; acid value 20 max.; alkali no. 25 max.	clear amber liquid	conditioner; *shampoos; hair conditioners*
	CERAPHYL 70	quaternium-70 and propylene glycol	total solids 48-58%; alkali no. 5.0 max.; sapon. no. 45-60	soft amber gel; m.p. 27-32°C	conditioner; *shampoos; hair conditioners; skin care products*
	CERAPHYL 85	stearamidopropyl cetearyl dimonium tosylate and propylene glycol	acid value 12 max.; alkali no. 5 max.; sapon. no. 20 max.	cream coloured waxy solid; m.p. 44-48°C	conditioner; *shampoos; hair conditioners; skin care products*
	GAFQUAT 734	polyquaternium-11	solids 50%	viscous alcoholic soln.	conditioner; *hair care products; shaving products; skin care products; deodorants; antiperspirants; soaps*
	GAFQUAT 755 N	polyquaternium-11	solids 20%	highly viscous. aq. soln.	
Kao Corporation	QUARTAMIN 60ETOH	alkyl trimethyl ammonium chloride	48-50% active; carbon chain composition C_{14} 3%, C_{16} 95%, C_{18} 2%	liquid; M.W. 320; colour 80 max. (APHA); pH 6-8.5 (5%)	
	QUARTAMIN 60L	alkyl trimethyl ammonium chloride	48-50% active; carbon chain composition C_{14} 3%, C_{16} 95%, C_{18} 2%	liquid; M.W. 320; colour 80 max. (APHA); pH 6-8.5 (5%)	
	QUARTAMIN CHFL	alkyl trimethyl ammonium chloride	33-37% active; carbon chain composition C_{10} 2%, C_{12} 56%, C_{14} 22%, C_{16} 10%, C_{18} 10%	liquid; M.W. 285; colour 1 max. (Red); pH 5.5-8.5 (5%)	
	QUARTAMIN CML	alkyl trimethyl ammonium chloride	49-51% active; carbon chain composition C_8 7%, C_{10} 5%, C_{12} 52%, C_{14} 18%, C_{16} 10%, C_{18} 9%	liquid; M.W. 275; colour 7 max. (Gardner); pH 5-7 (5%)	
	QUARTAMIN D86-PI	dialkyl dimethyl ammonium chloride	76-78% active; carbon chain composition C_{14} 5%, C_{10} 32%, C_{18} 61%, $C_{18'}$ 2%	solid/paste; M.W. 570; colour 4 max. (Gardner); pH 5-7 (5%)	
	QUARTAMIN D86-PIC	dialkyl dimethyl ammonium chloride	74-76% active; carbon chain composition C_{14} 5%, C_{10} 32%, C_{18} 61%, $C_{18'}$ 2%	solid/paste; M.W. 563; colour 4 max. (Gardner); pH 4.5-5.5 (5%)	
	QUARTAMIN D86-PL	dialkyl dimethyl ammonium chloride	74-76% active; carbon chain composition C_{14} 5%, C_{10} 32%, C_{18} 61%, $C_{18'}$ 2%	solid/paste; M.W. 563; colour 4 max. (Gardner); pH 4.5-5.5 (5%)	
	QUARTAMIN D86P-82	dialkyl dimethyl ammonium chloride	81-83% active; carbon chain composition C_{14} 5%, C_{10} 32%, C_{18} 61%, $C_{18'}$ 2%	solid/paste; M.W. 575; colour 7 max. (Gardner); pH 4.5-6 (5%)	
	QUARTAMIN T	alkyl trimethyl ammonium chloride	49-51% active; carbon chain composition C_{12} 1%, C_{14} 3%, C_{10} 30%, C_{18} 20%, $C_{18'}$ 40%, $C_{18''}$ 6%	liquid; M.W. 345; colour 7 max. (Gardner); pH 5-7 (5%)	
	QUARTAMIN THL	alkyl trimethyl ammonium chloride	49-51% active; carbon chain composition C_{14} 5%, C_{16} 32%, C_{18} 60%, $C_{18'}$ 3%	liquid/paste; M.W. 346; colour 7 max. (Gardner); pH 5-7 (5%)	
	TETRANYL AHT-1	N,N-di-(β-acyloxyethyl), N-β-hydroxyethyl, N, methylammonium methylsulfate	84-86% active; carbon chain composition C_{14} 5%, C_{16} 32%, C_{18} 60%, $C_{18'}$ 3%	solid; M.W. 800; colour 3 max. (Gardner); pH 2-3 (5%)	
	TETRANYL AT-1	N,N-di-(β-acyloxyethyl), N-β-hydroxyethyl, N, methylammonium methylsulfate	84-86% active; carbon chain composition C_{12} 1%, C_{14} 3%, C_{10} 30%, C_{18} 18%, $C_{18'}$ 42%, $C_{18''}$ 6%	paste/liquid; M.W. 800; colour 3 max. (Gardner); pH 2-3 (5%)	

Supplier	Trade name	Chemical description	Composition	General properties	Functionality / Application
Kao Corporation	TETRANYL AT-75	N,N-di-(β-acyloxyethyl), N-β-hydroxyethyl, N, methylammonium methylsulfate	84-86% active; carbon chain composition C_{12} 1%, C_{14} 4%, C_{16} 30%, C_{18} 31%, $C_{18'}$ 30%, $C_{18''}$ 4%	paste; M.W. 800; colour 3 max. (Gardner); pH 2-3 (5%)	
	TETRANYL AT-90	N,N-di-(β-acyloxyethyl), N-β-hydroxyethyl, N, methylammonium methylsulfate	89-91% active; carbon chain composition C_{12} 1%, C_{14} 3%, C_{16} 30%, C_{18} 18%, $C_{18'}$ 42%, $C_{18''}$ 6%	paste; M.W. 800; colour 3 max. (Gardner); pH 2-3 (5%)	
	TETRANYL B-S-25	alkyl dimethyl benzyl ammonium chloride	19.8-21.2% active; carbon chain composition C_{16} 4%, C_{18} 96%	paste; M.W. 426; pH 3-4 (2%)	
	TETRANYL BC-50	alkyl dimethyl benzyl ammonium chloride	49-51% active; carbon chain composition C_8 4%, C_{10} 5%, C_{12} 46%, C_{14} 24%, C_{16} 11%, C_{18} 3%, $C_{18'}$ 7%	liquid; M.W. 357; colour 120 max. (APHA); pH 7-8 (5%)	
	TETRANYL BC-80	alkyl dimethyl benzyl ammonium chloride	79-81% active; carbon chain composition C_8 4%, C_{10} 5%, C_{12} 46%, C_{14} 24%, C_{16} 11%, C_{18} 3%, $C_{18'}$ 7%	liquid/paste; M.W. 357; colour 120 max. (APHA); pH 7-8 (5%)	
	TETRANYL L-1	N,N-di-(β-acyloxyethyl), N-β-hydroxyethyl, N, methylammonium methylsulfate	84-86% active; carbon chain composition C_{14} 2%, C_{16} 47%, C_{18} 10%, $C_{18'}$ 33%, $C_{18''}$ 8%	liquid; M.W. 800; colour 2 max. (Gardner); pH 2-3 (5%)	
	TETRANYL L1/90	N,N-di-(β-acyloxyethyl), N-β-hydroxyethyl, N, methylammonium methylsulfate	89-91% active; carbon chain composition C_{12} 1%, C_{14} 4%, C_{16} 30%, C_{18} 31%, $C_{18'}$ 30%, $C_{18''}$ 4%	paste; M.W. 800; colour 3 max. (Gardner); pH 2-3 (5%)	
Lonza	BARDAC 2050	mixed dialkyl dimethyl ammonium chloride	50% active	liquid	fungicide; bacteriostatic agent; deodoriser; sanitizer; disinfectant / *household, institutional, industrial and water treatment applications*
	BARDAC 205M	alkyl dimethyl benzyl and dialkyl dimethyl ammonium chloride	50% active	liquid	disinfectant; sanitizer / *household, institutional, industrial and water treatment applications*
	BARDAC 208M	alkyl dimethyl benzyl and dialkyl dimethyl ammonium chloride	80% active	liquid	
	BARDAC 2080	mixed dialkyl dimethyl ammonium chloride	80% active	liquid	fungicide; bacteriostatic agent; deodoriser; sanitizer; disinfectant / *household, institutional, industrial and water treatment applications*
	BARDAC 2250	didecyl dimethyl ammonium chloride	50% active	liquid	disinfectant; sanitizer; germicide / *household, institutional, industrial and water treatment applications*
	BARDAC 2280	didecyl dimethyl ammonium chloride	80% active	liquid	
	BARQUAT 1552	alkyl dimethyl benzyl ammonium chloride and dialkyl methyl benzyl ammonium chloride	50% active	liquid	algicide / *household, institutional, industrial and water treatment applications*
	BARQUAT 4250Z	alkyl dimethyl benzyl ammonium chloride and alkyl dimethyl ethyl benzyl ammonium chloride	50% active	liquid	disinfectant; sanitizer; germicide; algicide; bacteriostatic agent; deodoriser / *household, institutional, industrial and water treatment applications*
	BARQUAT 4280	alkyl dimethyl benzyl ammonium chloride and alkyl dimethyl ethyl benzyl ammonium chloride	80% active	liquid	fungicide; disinfectant; germicide; sanitizer; algicide; deodoriser / *household, institutional, industrial and water treatment applications*

Supplier	Trade name	Chemical description	Composition	General properties	Functionality / Application
Lonza	BARQUAT 4280Z	alkyl dimethyl benzyl ammonium chloride and alkyl dimethyl ethyl benzyl ammonium chloride	80% active	liquid	disinfectant; sanitizer; germicide; algicide; bacteriostatic agent; deodoriser *household, institutional, industrial and water treatment applications*
	BARQUAT CME-35	cetethyl morpholinium ethosulfate	35% active	liquid	antistatic agent; lubricant; odour counteractant *textile industry*
	BARQUAT CT-29	cetrimonium chloride	29% active	liquid	coagulating agent *antibiotics manufacture*
	BARQUAT MB-50	alkyl dimethyl benzyl ammonium chloride	50% active	liquid	disinfectant; sanitizer; germicide; deodoriser; fungicide; algicide *household, institutional, industrial and water treatment applications*
	BARQUAT MB-80	alkyl dimethyl benzyl ammonium chloride	80% active	liquid	
	BARQUAT MX-50	alkyl dimethyl benzyl ammonium chloride	50% active	liquid	bacteriostatic agent; deodoriser; disinfectant; sanitizer *household, institutional, industrial and water treatment applications*
	BARQUAT MX-80	alkyl dimethyl benzyl ammonium chloride	80% active	liquid	
	BARQUAT OJ-50	alkyl dimethyl benzyl ammonium chloride	50% active	liquid	algicide; sanitizer *household, institutional, industrial and water treatment applications; oil recovery agent*
	BARQUAT OJ-80	alkyl dimethyl benzyl ammonium chloride	80% active	liquid	
	CARSOQUAT 816-C	blend of cetearyl alcohol, PEG-40 castor oil and stearalkonium chloride	96% active	flakes	*hair care products*
	CARSOQUAT 868E	dicetyl dimonium chloride	68% active	liquid	conditioner; softener *hair conditioners; hair care products; fabric softeners*
	CARSOQUAT CB	blend of cetyl alcohol, glyceryl stearate, dicetyl dimonium chloride, cetrimonium chloride, polysorbate 85 and PEG 40 castor oil	99% active	flakes	*hair care products*
	CARSOQUAT CT-429	cetrimonium chloride	29% active	liquid	conditioner *hair conditioners; hair care products*
	CARSOQUAT SDQ-25	stearalkonium chloride	25% active	paste	
	CARSOQUAT SDQ-85	stearalkonium chloride	85% active	flakes	
	CARSOSOFT CFI-90	alkyl imidazolinium methosulfate	90% active	clear liquid	softener *fabric softeners*
	CARSOSOFT S-90	quaternium-27	90% active	paste	softener; antistatic agent *fabric softeners*
	CARSOSOFT T-90	quaternium-53	90% active	paste	softener *fabric softeners*
	HYAMINE 3500 80%	alkyl dimethyl benzyl ammonium chloride	80% active	liquid	bactericide; disinfectant; sanitizer; germicide *pharmaceuticals; veterinary products*
	HYAMINE 3500	alkyl dimethyl benzyl ammonium chloride	50% active	liquid	
McIntyre Group	MACKERNIUM 006	polyquaternium 6	40% active	liquid; pH 7	conditioner *liquid soaps; shampoos*
	MACKERNIUM 007	polyquaternium 7	9% active	liquid; pH 7	
	MACKERNIUM CTC-30	cetrimonium chloride	30% active	liquid; pH 4	conditioner; lubricant; antistatic agent

Supplier	Trade name	Chemical description	Composition	General properties	Functionality / Application
McIntyre Group	MACKERNIUM KP	oleaalkonium chloride	50% active	liquid; pH 5	conditioner; lubricant; antistatic agent
	MACKERNIUM NLE	quaternium 84	100% active	liquid; pH 7	
	MACKERNIUM SDC-25	stearalkonium chloride	25% active	paste; pH 4	
	MACKERNIUM SDC-85	stearalkonium chloride	100% active	flake; pH 6	
	MACKERNIUM WLE	wheat lipid epoxide	100% active	liquid; pH 5	
Millchem	AMMONYX 4	stearyl dimethyl benzyl ammonium chloride	quaternary content 17-19%	paste	conditioner; softener; emollient; emulsifier *hair care products; skin care products*
	AMMONYX 4B	stearyl dimethyl benzyl ammonium chloride	quaternary content 16-18%	paste	
	AMMONYX 485	stearyl dimethyl benzyl ammonium chloride	quaternary content 85% min.	powder	
	AMMONYX 4002	stearyl dimethyl benzyl ammonium chloride	quaternary content 94% min.	flakes	
	AMMONYX CETAC-30	cetyl trimethyl ammonium chloride	quaternary content 29% min.	liquid	
	AMMONYX CETAC	cetyl trimethyl ammonium chloride	quaternary content 24-26%	liquid	
	AMMONYX KP	oleyl dimethyl benzyl ammonium chloride	quaternary content 50% min.	liquid	conditioner; antistatic agent *hair care products*
	BTC 2565	n-alkyl dimethyl benzyl ammonium chloride	quaternary content 50% min.	liquid	algicide; slimicide *water treatment*
	BTC 50 USP	benzalkonium chloride	quaternary content 50% min.	liquid	hard-surface disinfectants; sanitisers; deodorisers
	BTC 50	benzalkonium chloride	quaternary content 50% min.	liquid	hard-surface disinfectants; sanitisers; deodorisers
	BTC 65	benzalkonium chloride	quaternary content 50% min.	liquid	disinfectant; deodoriser; germicide
	BTC 99	n-didecyl dimethyl ammonium chloride	quaternary content 50% min.	liquid	algicide; slimicide; low foam *water treatment*
	BTC 471	quaternium 14	quaternary content 50%	liquid	disinfectant; deodoriser; germicide
	BTC 776	n-alkyl dimethyl benzyl ammonium chloride and dialkyl methyl benzyl ammonium chloride	quaternary content 50% min.	liquid	algicide; slimicide *water treatment*
	BTC 812	octyl dodecyl dimethyl ammonium chloride	quaternary content 50%	liquid	fungicide
	BTC 824 P-100	myristalkonium chloride	quaternary content 90% min.	powder	*disinfectants; sanitisers; deodorisers*
	BTC 824	myristalkonium chloride	quaternary content 50% min.	liquid	
	BTC 8248	myristalkonium chloride	quaternary content 80% min.	liquid	
	BTC 8249	myristalkonium chloride	quaternary content 90% min.	liquid	algicide *hard-surface disinfectants; sanitisers; water treatment*
	BTC 835	benzalkonium chloride	quaternary content 50% min.	liquid	
	BTC 8358	benzalkonium chloride	quaternary content 80% min.	liquid	
	BTC 1010	n-didecyl dimethyl ammonium chloride	quaternary content 50% min.	liquid	fungicide *hard-surface disinfectants; sanitisers*
	BTC 2125 M-80	myristalkonium chloride quaternium 14	quaternary content 80%	liquid	disinfectants; sanitisers
	BTC 2125 M	myristalkonium chloride quaternium 14	quaternary content 50%	liquid	

Supplier	Trade name	Chemical description	Composition	General properties	Functionality / Application
Millchem	BTC 2125 MP-40	myristalkonium chloride quaternium 14	quaternary content 40%	powder	disinfectants; sanitisers
	ONYXIDE 3300	quaternium 3	quaternary content 95% min.	powder	germicide / toiletries; pharmaceuticals
Nikko Chemicals	NIKKOL CA-101	benzalkonium chloride		colourless liquid	antimicrobial agent; disinfectant / cosmetics
	NIKKOL CA-1485	stearalkonium chloride		white flakes	conditioner; softener / cosmetics
	NIKKOL CA-2150	cocotrimonium chloride		colourless liquid	antimicrobial agent; disinfectant; softener / cosmetics
	NIKKOL CA-2330	cetrimonium chloride		pale yellow liquid	
	NIKKOL CA-2350	cetrimonium chloride		pale yellow liquid	
	NIKKOL CA-2450	steartrimonium chloride		pale yellow liquid	conditioner; antistatic agent; softener / cosmetics
	NIKKOL CA-2450T	tallowtrimonium chloride		yellow liquid	
	NIKKOL CA-2465	steartrimonium chloride		white paste	
	NIKKOL CA-2580	behentrimonium chloride		pale yellow solid	
	NIKKOL CA-3080	dioctyl trimethyl ammonium chloride		pale yellow liquid	antimicrobial agent; disinfectant
	NIKKOL CA-3080M	mixture		pale yellow liquid	conditioner
	NIKKOL CA-3475	distearyldimonium chloride		pale yellow paste	conditioner; softener / cosmetics
	NIKKOL LANOQUAT DES-50	lanolin quaternary ammonium salt		brown viscous liquid	cosmetics
Pentagon	PENTONIUM 2-80	lauryl dimethyl benzyl ammonium chloride	79-81% active	clear, pale yellow viscous liquid; pH 6.0-8.0 (1% aq. soln.)	disinfectants; detergent sanitisers
	PENTONIUM 24 BP	alkyl dimethyl benzyl ammonium chloride	49-51% active	colour 100 max. (Hazen); pH 5.5-6.5 (2% aq. soln.)	antistatic agent; levelling agent / disinfectants; detergent sanitisers
	PENTONIUM 24-80	alkyl dimethyl benzyl ammonium chloride	79-81% active; solvent isopropanol/H_2O	clear, pale yellow viscous liquid; pH 6.0-8.0 (1% aq. soln.)	disinfectants; detergent sanitisers
	PENTONIUM 24	alkyl dimethyl benzyl ammonium chloride	48-50% active	clear, pale yellow viscous liquid; colour 100 max. (Hazen); pH 4.0-8.0 (10% aq. soln.)	antistatic agent; levelling agent / disinfectants; detergent sanitisers
	PENTONIUM 26 SG	alkyl dimethyl benzyl ammonium chloride	80% active	clear, pale yellow viscous liquid; colour 200 max. (Hazen); pH 6.5-8.5 (10% aq. soln.)	
	PENTONIUM 26-80	alkyl dimethyl benzyl ammonium chloride	79-81% active	clear, pale yellow viscous liquid; pH 6.0-8.0 (1% aq. soln.)	disinfectants; detergent sanitisers
	PENTONIUM 26	alkyl dimethyl benzyl ammonium chloride	49-51% active	clear, viscous liquid; pH 6.0-8.0 (10% aq. soln.)	
	PENTONIUM 4BR40	myristyl trimethyl ammonium bromide (complies with the requirements of the British Pharmacopoeia 1988)	39-41% active; ethanol 7-8%; H_2O 52-53%	clear, colourless to pale yellow viscous liquid; colour 10Y:1.5R max. (lovibond)	disinfectants; detergent sanitisers; antiseptic creams
	PENTONIUM 6C130	cetyl trimethyl ammonium chloride	28-30% active; H_2O 70-72%	clear mobile liquid; pH 6.0-8.0	
	PENTONIUM 6C150	cetyl trimethyl ammonium chloride	49-51% active; ethanol 33-37%	clear mobile liquid; pH 6.0-8.0 (10% aq. soln.)	antistatic agent / hair conditioners; detergent sanitisers
	PENTONIUM 6E30	cetyl dimethyl ethyl ammonium ethosulfate	28-30% active	clear mobile liquid; pH 6.2-7.5 (1% aq. soln.)	

Supplier	Trade name	Chemical description	Composition	General properties	Functionality / Application
Pentagon	PENTONIUM DD-50	didecyl dimethyl ammonium chloride	50-51% active	clear, viscous liquid; colour 2 max. (Gardner); pH 6.5-8.0 (1% aq. soln.)	emulsifier; disinfectants; detergent sanitisers
	PENTONIUM DD-70 EG	didecyl dimethyl ammonium chloride in ethylene glycol and H$_2$O	69-71% active	yellow liquid; pH 6.5-8.0 (1% aq. soln.)	disinfectants; detergent sanitisers
	PENTONIUM DD-80	didecyl dimethyl ammonium chloride	79-81% active	clear viscous liquid; colour 3 max. (Gardner); pH 6.0-8.0 (1% aq. soln.)	disinfectants; detergent sanitisers
	PENTONIUM DO-50	dioctyl dimethyl ammonium chloride	50-51% active	clear liquid; colour 2 max. (Gardner); pH 6.5-8.5	low foam; disinfectants; detergent sanitisers
	PENTONIUM DQ-50	n-alkyl dimethyl benzyl ammonium chloride and n-alkyl dimethyl ethylbenzyl ammonium chloride	50-51% active	soln.; colour 200 max. (Hazen); pH 7.0-8.0 (10% aq. soln.)	germicide; algicide; slimicide; hard-surface disinfectants: household, institutional and industrial disinfectants; toilet bowl cleaners; swimming pool and cooling tower treatment
	PENTONIUM DQ-80	n-alkyl dimethyl benzyl ammonium chloride and n-alkyl dimethyl ethylbenzyl ammonium chloride	80-82% active	soln.; colour 200 max. (Hazen); pH 6.5-8.5 (10% aq. soln.)	
	PENTONIUM DS-75	dihydrogenated tallow) dimethyl ammonium chloride	quaternary content 74-77%	paste; colour 3 max. (Gardner)	fabric softeners; hair conditioners
	PENTONIUM SD-75	stearyl dimethyl benzyl ammonium chloride	74-80% active	soft cream paste; pH 6.0-9.0 (10% in 1:1 H$_2$O/ethanol)	hair conditioners
	PENTONIUM S65	3-chloro-2-hydroxy propyl trimethyl ammonium chloride	65% active min.	clear colourless liquid; pH 3-6	manufacture of cationic starch
	PENTONIUM S70	3-chloro-2-hydroxy propyl trimethyl ammonium chloride	69% active min.	clear colourless liquid; pH 3-6	manufacture of cationic starch
	PENTONIUM T2E	tallow bis (hydroxy ethyl) benzyl ammonium chloride	45-47% active	pale amber viscous liquid; pH 6.0-8.0	acid thickener; demulsifying agent
PPG	LAROSTAT 1084	ethoxylated amine quat	100% active	liquid	antistatic agent; textile industry; fibre production
	LAROSTAT 143	quaternary based on fatty amines	100% active	liquid	antistatic agent
	LAROSTAT 264A	alkyl dimethyl ethyl ammonium ethosulfate	35% active product in H$_2$O		antistatic agent; fibreglass; PVC; textile industry; dust control; floor finishes; polyolefins
	LAROSTAT 519	alkyl dimethyl ammonium ethosulfate	60% active product on silica		antistatic agent; polyolefins
	LAROSTAT 88	alkyl dimethyl ethyl ammonium ethosulfate quat.	10% active product in H$_2$O		antistatic agent
	M-QUAT 1033	soya ethyldimonium ethosufate	58% active	liquid	conditioner
	M-QUAT 522	isostearamidopropyl ethyldimonium ethosulfate	85% active	liquid	shampoos; hair conditioners; hair care products
	M-QUAT DIMER 18 PG	hydroxypropyl bis-stearyldimonium chloride	50% active	paste	conditioner; hair conditioners
	M-QUAT DIMER S-50 PG	hydroxypropyl bis-oleyldimonium chloride	50% active	liquid	conditioner; hair conditioners; hair care products
	M-QUAT JN	ricinoleamidopropyl ethyldimonium ethosulfate	93% active	liquid	conditioner; shampoos; hair conditioners; hair care products
	M-QUAT JS-25 SP	stearalkonium chloride	25% active	paste	conditioner
	M-QUAT JS-25	stearalkonium chloride	25% active	paste	hair conditioners
Rhone-Poulenc	CETRIMIDE BP	alkyl trimethyl ammonium bromide		powder	foaming agent (moderate); emulsifier; biocides

Supplier	Trade name	Chemical description	Composition	General properties	Functionality / *Application*
Rhone-Poulenc	GLOKILL ELC	polymeric quaternary		liquid	dispersant
	GLOKILL PQ	polymeric quaternary		liquid	*biocides*
	RHODAQUAT RP 50	benzalkonium chloride		liquid	foaming agent (moderate); emulsifier / *biocides*
	RHODAQUAT RP 80	benzalkonium chloride		liquid	foaming agent (moderate); emulsifier / *biocides*
	RHODAQUAT TFR	ethoxy quaternary		liquid	foaming agent (moderate); emulsifier / *biocides*
	RHODAQUAT WR	speciality quaternary		liquid	
Sandoz	DERMAGEN DM	based on quaternary ammonium compound		liquid	dyeing assistant / *textile industry; dyes*
	DERMAGEN K	based on a quaternary ammonium compound in aq. medium		liquid	levelling agent / *dyes; leather industry*
	LYOCOL DORM	quaternary ammonium compound in aq. soln.		liquid	bactericide / *detergents; cleaners*
	LYOGEN PAL	quaternary ammonium compound in aq. soln.		liquid	levelling agent / *textile industry; dyeing of nylon*
	RETARGAL AN	quaternary ammonium compound		liquid	retarding agent / *textile industry; dyeing*
	SANDOCLEAN FWD	quaternary amine compound in aq. soln.		liquid	softener; lubricant / *textile industry; fibres*
	SANDOLUBE SFL	quaternary ammonium compound and waxes in aq. dispersion		pale amber liquid	*special cleaners*
	SANDOQUAD C-15M LIQUID	quaternary ammonium salt of ethoxylated alkylamine		pale amber liquid	biocide
	SANDOQUAD DOR	quaternary ammonium compound		liquid	*industrial cleaners*
	SANDOQUAD KNA	quaternary ammonium compound of a fatty acid ester containing solvent		liquid	*special cleaners*
	SANDOQUAD T-2B LIQUID	quaternary ammonium salt of ethoxylated alkylamine		pale amber liquid	wetting agent; dispersant; emulsifier / *cleaners*
	SANDOQUAT C15-M	quaternary ammonium methosulfate compound of an ethoxylated monoalkylamine	99% active	pale amber free-flowing liquid	antistatic agent; emulsifier; dispersant; wetting agent / *PVC floor coverings*
	SANDOQUAT T-5M	quaternary ammonium methosulfate compound of an ethoxylated monoalkylamine	85-89% active; hexylene glycol 10%	golden yellow viscous liquid	
	SANDOSPACE DPE	quaternary ammonium compound		liquid	
Scher Chemicals	SCHERCOQUAT ALA	dilauryl acetyl dimonium chloride	dry solids 70% min.	light yellow liquid (30°C); M.W. 504	emulsifier; conditioner / *skin care products; hair care products*
	SCHERCOQUAT APAS	apricotamidopropyl ethyldimonium ethosulfate	dry solids 90% min.	amber viscous liquid; M.W. 515	conditioner
	SCHERCOQUAT BAS	behenamidopropyl ethyldimonium ethosulfate	dry solids 50% min.	amber liquid; M.W. 548	conditioner / *hair care products*
	SCHERCOQUAT CAS	cocamidopropyl ethyldimonium ethosulfate	dry solids 98% min.	amber viscous liquid; M.W. 445	antistatic agent

Supplier	Trade name	Chemical description	Composition	General properties	Functionality / Application
Scher Chemicals	SCHERCOQUAT COAS	based on canola oil fatty acid source with ethyl sulfate hydrophilic group		amber viscous liquid; M.W. 529	conditioner *hair care products; skin care products*
	SCHERCOQUAT DAS	quaternium-61	dry solids 90% min.	amber viscous liquid; M.W. 1050	conditioner *skin care products; hair care products; cosmetics*
	SCHERCOQUAT FOAS	saffloweramidopropyl ethyl dimonium ethosulfate	dry solids 90% min.	amber viscous liquid; M.W. 520	*hair conditioners*
	SCHERCOQUAT IALA	isostearamidopropyl laurylacetodimonium chloride	dry solids 80% min.	amber viscous liquid; M.W. 670	emulsifier; conditioner; lubricant *skin care products; hair care products*
	SCHERCOQUAT IAS	isostearamidopropyl ethyldimonium ethosulfate	dry solids 90% min.	amber viscous liquid; M.W. 550	antistatic agent; conditioner *shampoos; hair conditioners*
	SCHERCOQUAT IEP	quaternium-62	dry solids 80% min.	amber viscous liquid; M.W. 486	
	SCHERCOQUAT IIS	isostearyl ethyl imidonium ethosulfate	dry solids 98% min.	dark amber viscous liquid; M.W. 532	conditioner *skin care products; hair care products*
	SCHERCOQUAT ROAS	rapeseedamidopropyl ethyldimonium ethosulfate	dry solids 90% min.	amber viscous liquid; M.W. 560	conditioner; viscosity modifier *hair care products*
	SCHERCOQUAT ROEP	rapeseedamidopropyl epoxypropyl dimonium chloride	dry solids 80% min.	dark amber viscous liquid; M.W. 533	conditioner *shampoos; hair conditioners; hairspray preparations*
	SCHERCOQUAT SAS	stearamidopropyl ethyldimonium ethosulfate	dry solids 80% min.	yellow liquid; M.W. 508	conditioner *hair care products*
	SCHERCOQUAT SOAS	soyamidopropyl ethyldimonium ethosulfate	dry solids 90% min.	amber viscous liquid; M.W. 516	*hair conditioners*
	SCHERCOQUAT WOAS	wheatgermamidopropyl ethyldimonium ethosulfate	dry solids 90% min.	amber viscous liquid; M.W. 528	conditioner
Seppic	AMONYL BR 1244	lauralkonium bromide	80% active	liquid	bactericide *hair care products*
	AMONYL CL 1244 SPD	lauryldimethylammonium chloride	80% active	liquid	bactericide
	AMONYL CL 1244	lauryldimethylammonium chloride	50% active	liquid	bactericide
	AMONYL DM	quaternium-82	100% active	liquid	softener; conditioner *hair conditioners; hair care products*
Servo Delden	SERVAMINE KAC 412	N-coco-N,N,N-trimethylammonium chloride	48% active min.	liquid	
	SERVAMINE KAC 422	N-coco-N,N-dimethyl-N-benzylammonium chloride	49 ± 1.0% active	liquid	
	SERVAMINE KW 100	quaternary tallowaminepolyglycol ether (15EO)	99% active min.	liquid	
	SERVAMINE KZB 402	poly(2-hydroxypropyldimethylammonium chloride)	60 ± 1.0% active	liquid	
	SERVOSOFT XW 175	quaternary imidazoline	78 ± 0.5% active	liquid	
	SERVOSOFT XW 190	quaternary imidazoline	90 ± 1.0% active	liquid	

Supplier	Trade name	Chemical description	Composition	General properties	Functionality Application
Servo Delden	SERVOSOFT XW 445 PP	quaternary triethanolamine fatty acid ester	45 ± 1.0% active	liquid	
	SERVOSOFT XW 470 PP	quaternary triethanolamine fatty acid ester	70 ± 1.0% active	paste	
	SERVOSOFT XW 485 DPG	quaternary triethanolamine fatty acid ester	85.0% active min.	liquid	
	SERVOSOFT XW 490	quaternary triethanolamine fatty acid ester	88.0–90.0% active	paste	
	SERVOSOFT XW 490 PP	quaternary triethanolamine fatty acid ester	90 ± 1.0% active	paste	
Stepan Europe	BACTISTEP SERIES	blend of quaternary ammonium salts; made-to-order		liquid	
	CATIGENE 50 USP	alkyldimethylbenzyl ammonium chloride; made-to-order	50% active	liquid	germicide; algicide; fungicide; deodoriser; antistatic agent *surface disinfectants; water treatment*
	CATIGENE 65 USP	alkyldimethylbenzyl ammonium chloride; made-to-order	50% active	liquid	
	CATIGENE 776	alkyldimethylbenzyl ammonium chloride; made-to-order	50% active	liquid	algicide; slimicide *swimming pool and industrial water treatment*
	CATIGENE 818	dialkyl dimethyl ammonium chloride	50% active	liquid	germicide; algicide; fungicide; deodoriser; antistatic agent *surface disinfectants; water treatment*
	CATIGENE 824/ 8248	alkyl dimethyl benzyl ammonium chloride; made-to-order	50/80% active	liquid	algicide; slimicide *swimming pool and industrial water treatments; hard surface disinfectants*
	CATIGENE 1011	didecyldimethylammonium chloride	50% active	water white to pale yellow liquid	germicide; algicide; fungicide; deodoriser; antistatic agent *household, institutional and industrial cleaners; oil industry*
	CATIGENE 2125 M	alkyl dimethyl benzyl/ethylbenzyl ammonium chloride	50% active	liquid	
	CATIGENE 2125 MP40	alkyl dimethyl benzyl/ethylbenzyl ammonium chloride; made-to-order	40% active	powder	germicide; algicide; fungicide; deodoriser; antistatic agent *surface disinfectants; water treatment*
	CATIGENE 2125 P40	alkyl dimethyl benzyl/ethylbenzyl ammonium chloride; made-to-order	40% active	powder	
	CATIGENE 2565	alkyl dimethyl benzylammonium chloride; made-to-order	50% active	liquid	algicide; slimicide *swimming pool and industrial water treatments*
	CATIGENE 4513-50	alkyl dimethyl benzyl/ethylbenzyl ammonium chloride	50% active	liquid	
	CATIGENE 4513-80 M	alkyl dimethyl benzyl/ethylbenzyl ammonium chloride; made-to-order	80% active	liquid	germicide; algicide; fungicide; deodoriser; antistatic agent *surface disinfectants; water treatment*
	CATIGENE 4513-80	alkyl dimethyl benzyl/ethylbenzyl ammonium chloride	80% active	liquid	
	CATIGENE B50	alkyldimethylbenzyl ammonium chloride	50% active	liquid	
	CATIGENE B80	alkyldimethylbenzyl ammonium chloride	80% active	liquid	
	CATIGENE CA 56	alkylamidopropyltrimethyl ammonium methoxysulfate; made-to-order	56% active	pale yellow liquid	antistatic agent; additive *emulsion polymerisation*

Supplier	Trade name	Chemical description	Composition	General properties	Functionality Application
Stepan Europe	CATIGENE CETAC 30	cetyltrimethylammonium chloride; made-to-order	30% active	water white to pale yellow liquid	conditioner; lubricant; antistatic agent; emollient; softener; emulsifier; germicide; algicide; fungicide; deodoriser; *hair care products; personal care products; household and industrial cleaners; oil industry*
	CATIGENE DC 100	alkyldimethylbenzyl ammonium chloride	100% active	powder	germicide; algicide; fungicide; deodoriser; antistatic agent
	CATIGENE T50	alkyldimethylbenzyl ammonium chloride	50% active	liquid	*surface disinfectants; water treatment*
	CATIGENE T80	alkyldimethylbenzyl ammonium chloride	80% active	liquid	
	STEPANQUAT F	alkyl ammonium methoxysulfate	85% active	water white to yellow viscous liquid	fluidifying agent; dispersant; antistatic agent; viscosity modifier; stabiliser; *household, institutional and industrial cleaners*
	STEPANQUAT SERIES	dialkyl ammonium methoxysulfate	80/90% active	white to beige paste	conditioner; antistatic agent; emollient; *hair care products; personal care products*
	STEPANQUAT T	alkyl ammonium methoxysulfate	100% active	water white to yellow viscous liquid	fluidifying agent; dispersant; antistatic agent; viscosity modifier; stabiliser; *household, institutional and industrial cleaners*
	STEPANTEX DO 90	dialkyl ammonium methoxysulfate	90% active	pale yellow to beige liquid	antistatic agent; hydrophobe; softener; *metal-surface treatments; household and industrial cleaners*
	STEPANTEX SERIES	dialkyl ammonium methoxysulfate	80/90% active	white to beige paste	softener; lubricant; antistatic agent; *fabric softeners; textile industry*
Surfachem	SURFAC ARF	tallow amine ethoxy ammonium methyl sulfate			
	SURFAC BAC50	alkyl dimethyl benzyl ammonium chloride (50%)			
	SURFAC BAC80	alkyl dimethyl ethoxy ammonium chloride (80%)			
	SURFAC CAT176	cetyl trimethyl ammonium chloride			
Thor Chemicals	BAC 50	a 50% aq. blend of alkyl dimethylbenzyl ammonium chlorides	carbon chain distribution C_{12} 40%, C_{14} 50%, C_{16} 10%	pale straw-coloured slightly viscous liquid; freezing point ca. 0°C; pH 7.0	bactericide; algicide; fungicide; *disinfectants; cleaners; sterilants; water treatment*
Tomah Products	TOMAH ACID FOAMER	major component is quaternary ammonium compound	99% active	dark amber liquid	*vehicle cleaners*
	TOMAH EMULSIFIER FOUR	complex quaternary ammonium chloride	74-77% active	amber liquid	emulsifier; *car care products*
	TOMAH Q-14-2 (50%)	isodecyl oxy propyl dihydroxyethyl methyl ammonium chloride	50% active; combining weight 370	amber liquid	
	TOMAH Q-14-2	isodecyl oxy propyl dihydroxyethyl methyl ammonium chloride	75% active; combining weight 370	amber liquid	
	TOMAH Q-17-2	isotridecyl oxy propyl dihydroxyethyl methyl ammonium chloride	75% active; combining weight 410	amber liquid	detergency booster; anti-corrosion; antistatic agent; *cleaners*
	TOMAH Q-S	mono soya ammonium chloride	50% active; combining weight 410	amber liquid	
	TOMAH Q18-15	octadecyl poly (15) oxyethylene methyl ammonium chloride	100% active; combining weight 980	amber liquid	
	TOMAH QC-15	cocopoly (15) oxyethylene methyl ammonium chloride	100% active; combining weight 915	amber liquid	
	TOMAH QDT-HG	tallow diamine quaternary dichloride in hexylene glycol	70% active; combining weight 525	amber liquid	
	TOMAH QDT	tallow diamine quaternary dichloride	50% active; combining weight 525	amber liquid	
	TOMAH T-AE-DR-II	quaternary ammonium salt	75% active	amber liquid	

Supplier	Trade name	Chemical description	Composition	General properties	Functionality / Application
Unger Fabrikker	UFASOFT 75	quaternary ammonium methosulfate	75% active; isopropanol content 7.5%	white viscous liquid; pH 4-6 (1% soln.)	softener; conditioner / laundry products; fabric softener
Warwick International	MYKON 111	quaternary ammonium compound			softener / textile industry
Witco	EMCOL 4	stearalkonium chloride	25% solids	white paste; pH 4	antistatic agent; conditioner; emollient; substantivity agent / personal care products; household and industrial applications
	EMCOL CC-9	PPG-9 diethylmonium chloride	99% solids	light yellow liquid; pH 6.5	antistatic agent; conditioner; dispersant / personal care products; household and industrial applications
	EMCOL CC-42	PPG-40 diethylmonium chloride	99% solids	light yellow liquid; pH 6.5	antistatic agent; dispersant / personal care products; household and industrial applications
	EMCOL CC-55	propropoxy quaternium ammonium acetate	99% solids	light amber liquid; pH 6.5	
	REWOCID UTM 185	undecylenic acid propylamide trimethyl ammonium methosulfate	47% active	liquid	bactericide; fungicide; conditioner; antistatic agent / toiletries; shampoos
	REWOQUAT B 10	didecyl dimethyl ammonium chloride	50% active	liquid	disinfectant / cleaners; food industry
	REWOQUAT B 50	alkyl dimethyl benzyl ammonium chloride	50% active	liquid	
	REWOQUAT CPEM	coco pentaethoxy methyl ammonium methosulfate	100% active	liquid	conditioner; emulsifier / hair care products
	REWOQUAT CR 3099	difatty acid isopropylester dimethylammonium methosulfate	98% active	liquid	fabric softeners; dry cleaning
	REWOQUAT RTM 50	ricinoleic acid propylamido trimethyl ammonium methosulfate	40% active	liquid	conditioner; antistatic agent / shampoos; liquid soaps
	REWOQUAT W 222 LM	quaternary fatty diamide	88% active	paste	
	REWOQUAT W 3690 PG	dioleyl imidazoline methosulfate; free of isopropanol	75% active	liquid	fabric softeners
	REWOQUAT W 3690	dioleyl imidazoline methosulfate	75% active	liquid	
	REWOQUAT W 75 PG	quaternary imidazoline derivative; free of isopropanol	75% active	liquid	fabric softeners; hair care products
	REWOQUAT W 75 H	quaternary imidazoline derivative	75% active	paste	fabric softeners
	REWOQUAT W 75	quaternary imidazoline derivative	75% active	liquid	fabric softeners; hair care products
	REWOQUAT W 90 DPG	quaternary imidazoline derivative; free of isopropanol	90% active	soft paste	fabric softeners
	REWOQUAT W 90	quaternary imidazoline derivative	90% active	soft paste	fabric softeners
	REWOQUAT WE 18-85	quaternary dialkylester	85% active	paste	fabric softeners; hair care products
	REWOQUAT WE 18	quaternary dialkylester	90% active	paste	
	REWOQUAT WE 20	quaternary dialkylester	90% active	paste	
	SOCHAMINE 2662	quaternized imidazoline	78% active	liquid	softener / textile industry
Zschimmer & Schwarz	ADULCINOL HA	quaternary ammonium salt		liquid dispersion	softener; antistatic agent / textile industry

Supplier	Trade name	Chemical description	Composition	General properties	Functionality / *Application*
Zschimmer & Schwarz	ADULCINOL KF	quaternary ammonium compound		clear yellowish liquid	softener / *textile industry*
	ELACTIV EN	quaternary ammonium compound		straw-coloured liquid	antistatic agent / *textile industry*
	RETENTOL RM	quaternary ammonium compound	50% active	clear, light yellow, viscous liquid	levelling agent; preservative; disinfectant / *textile industry*

Miscellaneous cationics

Supplier	Trade name	Chemical description	Composition	General properties	Functionality / *Application*
Akcros Chemicals	QUADRILAN MY211	cationic surfactant	50% active	clear amber liquid; pour point <0°C; viscosity 71 cSt; pH 6.5-8.5 (1% aq.)	antistatic agent / *spray cleaners; industrial cleaners; vehicle cleaners*
Akzo Nobel	BEROL 241	cationic surfactant blend	90% active		*rinse aids; car care products*
	BEROL 556	cationic surfactant	100% active		hydrotrope; solubiliser; dispersant / *hard-surface cleaners; industrial and institutional cleaners; vehicle cleaners*
	BEROL 561	cationic surfactant	100% active		
	BEROL 563	cationic surfactant	100% active		
Allied Colloids	ALCOFIX DL			liquid	fixing agent / *textile industry; cellulosic fibres; dyes*
	ALCOFIX R			liquid	
	ALCOFIX T			liquid	
	ALCOLUBE 1375	blended softeners and lubricants		emulsion	softener; lubricant / *textile industry; synthetic and natural fibres*
	ALCOLUBE LSJ	blend of fatty amines and silicone		emulsion	softener; lubricant; brushing aid / *textile industry; cellulosic fibres*
	ALCOSTAT 1586	fibre substantive antistatic agent		liquid	antistatic agent / *textile industry; carpet industry*
Auschem Cesalpinia	ALGONINA A 25	cationic compound	25% active	paste	household and industrial softeners
	ALGONINA AC	cationic compound	100% active	flakes	household and industrial softeners
	TEQUAT BCD	special cationic blend	40% active	paste	conditioner / *hair care products*
Bayer	LEVAPON MR	alkylammonium polyglycol ether			scouring agent / *textile industry*
	LEVOGEN BF	polyammonium compound			
	LEVOGEN FWN	polyammonium compound			
	LEVOGEN HW	polyammonium compound			aftertreatment agent / *textile industry; dyeing of cellulosic fibres*
	LEVOGEN LE	polyammonium compound			
	LEVOGEN WW	polyammonium compound			
CHEM-Y	AKYPOQUAT 131 VC	mixture of behenoyl PG trimonium chloride and cetyl alcohol	H_2O 42-46%	white soft paste; pH 3.2-4.4 (1 part in 5 parts H_2O)	antistatic agent / *hair care products*
	AKYPOQUAT 131	behenoyl PG-trimonium chloride		white to yellowish, waxy solid; M.W. ca. 497; pH 3.4-4.6 (10% in H_2O)	antistatic agent / *hair care products*

Supplier	Trade name	Chemical description	Composition	General properties	Functionality / *Application*
Croda	CONDITIONER BASE CB0967	proprietary product		white waxy solid	conditioner base / *hair conditioners*
Cytec	AEROSOL C-61	alkylamine guanidine polyoxyethanol	70% active	liquid/paste; surface tension 40 dynes/cm min. (in H$_2$O)	dispersant; wetting agent; flushing agent; antistatic agent; emulsifier; softener / *plastics industry; paper industry; textile industry; adhesives; cleaners*
Elf Atochem	INIPOL 002	oleyl propylene diamine dioleate	95-105% neutralisation	pasty liquid	rust inhibitor / *paint industry*
	INIPOL OT2	tall oil propylene diamine dioleate	99-105% neutralisation	liquid	
	INIPOL SO2	tallow propylene diamine dioleate	95-105% neutralisation	pasty	
Henkel	BELFASIN 1377	combination of fatty acid amide with a quaternary compound			softener / *textile industry*
	BELFASIN LEC	combination of fatty acid ester and high molecular ethers			lubricant / *textile industry*
	BELFASIN SI	combination of a fatty acid amide and a silicone compound			softener / *textile industry*
	BELFASIN TVE	combination of hydrocarbons and fatty derivatives			lubricant / *textile industry*
	EMULGADE K	tallow alcohol, cetearalkonium bromide and ceteareth-12			ready-to-use base / *hair conditioners; hair care products*
	NONAX 1166	polyoxyethylene derivate			antistatic agent / *textile industry; finishing*
	NOPCOFLOC 222-L			emulsion	machine drying and retention aid; sludge conditioner / *paper industry*
	POLYQUART H 7102	PEG-15 cocopolyamine and stearalkonium chloride	50% active	clear amber liquid	antistatic agent / *personal care products*
	POLYQUART H	PEG-15 tallow polyamine	50% active	clear amber liquid	antistatic agent / *hair care products; skin care products*
	SELBANA 4643	polyethylene dispersion with antistatics			lubricant / *textile industry; spinning*
	STANDAMUL CONC. 1002	cetearyl alcohol, PEG-40 hydrogenated castor oil and stearalkonium chloride	99% active	off-white waxy flakes	hair care products
Hüls	MARLOWET 5401	acetic acid salt of an ethoxylated alkylamine	active detergent 100%	liquid	emulsifier / *waxes; paper industry; textile industry*
ISP	COPOLYMER VC-713	vinylcaprolactam/vinylpyrrolidone/dimethylaminoethyl methacrylate terpolymer supplied as a 37% ethanol soln.	solids 37 ± 2%	viscous liquid; cloud point 44°C (1% solids in H$_2$O)	film former / *hairspray preparations; hair care products*
	GAFQUAT HS-100	polyquaternium-28	20% active	aq. viscous soln.	conditioner / *hair care products; skin care products*
Lakeland Laboratories	LAKEWAX C37	oxidised homopolymer wax emulsion	solids 27%; wax 18%; hardness of wax 0.5 mm (ASTM D5)	off white emulsion; softening point 134°C; pH 5.0	textile finishing
	LAKEWAX C60	oxidised homopolymer wax emulsion	solids 26.5%; wax 18%; hardness of wax 5.5 mm (ASTM D5)	pale yellow translucent emulsion; softening point 104°C; pH 4.0	textile finishing
Millchem	ASTON RC (AEROSOL)	special cationic	solids 60-66%	paste	antistatic agent / *carpet industry*

Supplier	Trade name	Chemical description	Composition	General properties	Functionality / *Application*
PPG	ALUBRASOFT 116		100% active	flakes	
	ALUBRASOFT 116A		22.5% active	liquid	
	ALUBRASOFT BSF-08		100% active	paste	softener / *textile industry*
	ALUBRASOFT CSP			slightly viscous liquid emulsion	
	ALUBRASOFT ECS			slightly viscous liquid emulsion	
	ALUBRASOFT SJ-2			free flowing paste	
	ALUBRASOFT SUPER 100		100% active	flakes	
	LAROSOL ALM-1			liquid	levelling agent; migrating agent / *dyeing: textile industry*
	LAROSOL NLA-25			liquid	
	LAROSOL PDQ-2			liquid	levelling agent; retarder / *dyeing: textile industry*
	LAROSTAT 88		10% active	liquid	
	LAROSTAT 143		100% active	liquid	antistatic agent / *textile industry; fibre production*
	LAROSTAT 264-A ANHYDROUS		100% active	thick paste	
	LAROSTAT 264-A		35% active	liquid	
Protex	SURFARON C1618			liquid	dispersant / polymerisation
Rhone-Poulenc	CHDG	chlorhexidine digluconate		liquid	biocides
Sandoz	CATALIX GS	fatty acid condensation product		liquid	fat liquoring agent; finishing agent / *leather industry*
	CATALIX PNP	fatty acid condensation product		liquid	finishing agent / *leather industry*
	CATALIX U	emulsion of natural oils		paste	fat liquoring agent
	CERANINE CWD	fatty acid derivative and polyglycol ether		powder	*fabric softener*
	CERANINE HCL	fatty acid condensation product in an aq. dispersion		liquid	softener / *cellulosic and synthetic fibres*
	CERANINE RW	fatty acid derivative		paste	antistatic agent; softener; re-wetting agent
	ELFUGIN 911E	nitrogenous aliphatic fatty acid		liquid	
	ELFUGIN PF	mixture of polymeric components and ethylene glycol		liquid	antistatic agent / *textile industry; synthetic fibres*
	IMACOL S	polyglycol ether derivative in aq. soln.		liquid	lubricant; low foam / *textile industry; synthetic fibres*
	LYOGEN UL	polyglycol ether in H_2O		liquid	levelling agent / *textile industry; dyeing*
	LYOGEN WPA	fatty amide amine polyglycol ether in aq. soln.		liquid	levelling agent / *textile industry; dyeing of wools and cotton*
	SANDOFIX SWN	polyamide amine with reactive groups in aq. soln.		liquid	fixative / *dyeing: textile industry*

Supplier	Trade name	Chemical description	Composition	General properties	Functionality / Application
Sandoz	SANDOLUBE JSL	quaternary ammonium compound, nonionic surfactant and wax in aq. soln.		liquid	softener; lubricant / *textile industry; fibres*
	SANDOLUBE PSL	fatty acid derivative and cationic surfactant		liquid	
	SANDOLUBE TRA	aliphatic hydrocarbons and fatty alkyl polyamine		liquid	antistatic agent / *textile industry; synthetic fibres*
	SANDOPERM AMC	fatty amine derivative and polysiloxane		liquid	softener / *textile industry*
	SANDOPERM ME	alkylamino modified polysiloxane in aq. soln.		liquid	finishing agent / *textile industry*
	SANDOPERM MEJ	siloxane derivative in aq. organic soln.		liquid	
	SANDOPUR SW	nitrogenous condensation product and nonionic surfactants		liquid	complexing agent / *textile industry; dyes*
	SANDOQUAD RW	blend of surfactants and hydrocarbons		liquid	glossing agent; de-watering agent
Stepan Europe	CATIGENE 880	a mixture of Catigene B50 and Catigene 818; made-to-order	80% active	liquid	*surface disinfectants; water treatment*
	CATIGENE 885	a mixture of Catigene B50 and Catigene 818; made-to-order	50% active	liquid	*surface disinfectants; water treatment*
	SEVEFILM 20	cationic surfactant blend	20% active	milky liquid	film former; anti-corrosion; antiredeposition agent / *water cooling circuit treatment products*
Thor Chemicals	ATOLEX ASL/C			liquid	lubricant; antistatic agent
	ATOLEX ASL/C100		100% active	thick liquid	antistatic agent; lubricant
	ATOLEX POLYETHYLENE EMULSIONS			liquid	softener; lubricant; additive / *resin finishes*
	SUFATONE SC/L			liquid	softener / *textile industry*
	SUFATONE SCS/B			paste	softener; antistatic agent / *textile industry*
	SUFATONE SCS/B2	economy version of Sufatone SCS/B		liquid	
	SUFATONE SCS/CL			liquid	softener / *textile industry*
	SUFATONE SMC/L			liquid	softener; lubricant / *textile industry*
	SUFATONE SMC/W			semi-liquid	
Warwick International	MYKON 236S	fatty acid condensate			softener
	MYKON 338	fatty acid condensate			
	WARCOSOFT 450	anhydrous concentrate			softener / *textile industry*
	WARCOSOFT 451	fatty acid condensate			softener; lubricant / *textile industry*
	WARCOSOFT 451E	fatty acid condensate			
Witco	EMCOL E-607L	lapyrium chloride	99% solids	white to off-white powder; pH 3.9	antistatic agent; conditioner; detergent; emollient; foaming agent; substantivity agent / *personal care products; household and industrial applications*
	EMCOL E-607S	steapyrium chloride	99% solids	white to off-white powder; pH 3.4	antistatic agent; conditioner; emollient; spreading agent; substantivity agent / *personal care products; household and industrial applications*

Supplier	Trade name	Chemical description	Composition	General properties	Functionality / Application
Witco	WITCO DTA-350	dimer trimer acid		liquid; pH 4.5	anti-corrosion
	WITCOR CI-6	complex surfactant		liquid; pH 11	*petroleum industry*
Zschimmer & Schwarz	ADULCINOL ALP	fatty acid condensate with additives of paraffin and silicone		liquid dispersion	
	ADULCINOL BNA	fatty acid condensate with additives of paraffin		liquid dispersion	
	ADULCINOL BUZ	fatty acid condensate		liquid dispersion	
	ADULCINOL EBZ	fatty acid condensate		liquid dispersion	softener
	ADULCINOL KU	fatty acid condensate		scales	*textile industry*
	ADULCINOL NP	fatty acid condensate with smoothing additives		liquid dispersion	
	ADULCINOL SIV	fatty acid condensates, quaternary ammonium compounds, silicones		liquid dispersion	
	ADULCINOL SK	silicone elastomer and fatty acid condensate		liquid dispersion	
	ANTHYDRIN HSP	fatty acid condensate		white, fluid dispersion	water repellent agent *textile industry*
	AUTOPOON GK 4003	blend of cationic surfactants	92% active	liquid	
	AUTOPOON GK 4004	blend of cationic surfactants	96% active	liquid	*car washes*
	AUTOPOON NI	blend of cationic substances with solubilisers and solvents	55% active	liquid	
	ENSIMOL ML	polymerisate		colourless liquid	sizing agent *textile industry*
	SANFOROL BN	fatty acid derivatives with special additives		white emulsion	wetting agent; softener *textile industry*
	ZETESAL CR	polyglycol ethers		brown viscous liquid	reserving agent; aftertreatment agent *textile industry*
	ZETESAL CRN	polyglycol ether compound		dark yellow liquid	
	ZETESAL DN 45	cyanamide condensate		clear colourless liquid	
	ZETESAL DR	polyammonium compound		clear, weakly yellow liquid	aftertreatment agent *textile industry*
	ZETESAL OR	combination of surfactants and glycols		clear colourless liquid	

4 Nonionics

Alkoxylates

Alcohol ethoxylates

Supplier	Trade name	Chemical description	Composition	General properties	Functionality / Applications
Akcros Chemicals	ETHYLAN 172	cetyl-oleyl alcohol ethoxylate	100% active; ethylene oxide content 34%; 'n' no. 3	HLB 6.8; clear yellow liquid; pour point 16°C; viscosity 57 cSt; pH 5.5-7.5 (1% aq.)	emulsifier / *mineral oils; vegetable oils; waxes*
	ETHYLAN BAB20	end-blocked alcohol ethoxylate	100% active	opaque liquid; pour point 16°C; viscosity 38 cSt (40°C)	low foam; wetting agent / *dishwashing agents; metal cleaning*
	ETHYLAN CD103	synthetic alcohol ethoxylate; made-to-order	97% active; ethylene oxide content 47%; H_2O content 3%; 'n' no. 3.2	HLB 9.4; clear colourless liquid; pour point $-10°C$; viscosity 36 cSt; pH 5-7 (1% aq.)	emulsifier; detergent; wetting agent; solubiliser / *textile industry; metal working; leather industry; paper industry; paint industry; detergents; household products; agriculture*
	ETHYLAN CD107	synthetic alcohol ethoxylate	96% active; ethylene oxide content 65%; H_2O content 4%; 'n' no. 6.8	HLB 13.1; slightly hazy, colourless liquid; cloud point 58°C (1% aq.); pour point $-7°C$; viscosity 66 cSt; pH 5-7 (1% aq.)	
	ETHYLAN CD109	synthetic alcohol ethoxylate	97% active; ethylene oxide content 71%; H_2O content 3%; 'n' no. 9.0	HLB 14.3; opaque white liquid; cloud point 84°C (1% aq.); pour point 4°C; viscosity 87 cSt; pH 5-7 (1% aq.)	
	ETHYLAN CD122	synthetic C_{12} primary alcohol ethylene oxide condensate; made-to-order	100% active; 'n' no. 2	HLB 6.4; clear colourless liquid; cloud point 41°C (10% in 25% butyl dioxitol/H_2O); pour point $<0°C$; viscosity 40 cSt; pH 7.0 (1% aq.)	
	ETHYLAN CD123	synthetic C_{12} primary alcohol ethylene oxide condensate	100% active; 'n' no. 3	HLB 8.4; clear colourless liquid; cloud point 53°C (10% in 25% butyl dioxitol/H_2O); pour point $<0°C$; viscosity 45 cSt; pH 7.0 (1% aq.)	
	ETHYLAN CD124	synthetic C_{12} primary alcohol ethylene oxide condensate	100% active; 'n' no. 4.5	HLB 10.4; clear colourless liquid; cloud point 64°C (10% in 25% butyl dioxitol/H_2O); pour point 2°C; viscosity 55 cSt; pH 7.0 (1% aq.)	
	ETHYLAN CD127	synthetic C_{12} primary alcohol ethylene oxide condensate	95% active; H_2O content <5%; 'n' no. 7	HLB 12.4; clear colourless liquid; cloud point 42°C (1% aq.); pour point $<0°C$; viscosity 120 cSt; pH 7.0 (1% aq.)	emulsifier; detergent; wetting agent; solubiliser / *industrial and household applications*
	ETHYLAN CD128	synthetic C_{12} primary alcohol ethylene oxide condensate	95% active; H_2O content <5%; 'n' no. 8	HLB 13.1; clear colourless liquid; cloud point 55°C (1% aq.); pour point $<0°C$; viscosity 135 cSt; pH 7.0 (1% aq.)	
	ETHYLAN CD129	synthetic C_{12} primary alcohol ethylene oxide condensate; made-to-order	90% active; H_2O content 10%; 'n' no. 9	HLB 13.6; clear colourless liquid; cloud point 67°C (1% aq.); pour point $<0°C$; viscosity 125 cSt; pH 7.0 (1% aq.)	
	ETHYLAN CD1210	synthetic C_{12} primary alcohol ethylene oxide condensate	90% active; H_2O content 10%; 'n' no. 10	HLB 14.1; clear colourless liquid; cloud point 82°C (1% aq.); pour point $<0°C$; viscosity 165 cSt; pH 7.0 (1% aq.)	

Supplier	Trade name	Chemical description	Composition	General properties	Functionality Application
Akcros Chemicals	ETHYLAN CD1230	synthetic C_{12} primary alcohol ethylene oxide condensate; made-to-order	100% active; 'n' no. 30	HLB 17.5; white waxy solid; cloud point >100°C (1% aq.); pour point 42°C; viscosity 150 cSt (60°C); pH 7.0 (1% aq.)	emulsifier; detergent; wetting agent; solubiliser *household and industrial applications*
	ETHYLAN CD1260	synthetic C_{12} primary alcohol ethylene oxide condensate	100% active; 'n' no. 57	HLB 18.6; white waxy solid; cloud point >100°C (1% aq.); pour point 51°C; viscosity 245 cSt (60°C); pH 7.0 (1% aq.)	
	ETHYLAN CD4511	$C_{14/15}$ synthetic alcohol ethoxylate	100% active; 'n' no. 11	HLB 13.7; white waxy solid; cloud point 88°C (5% aq.); pour point 30°C; pH 7.0 (1% aq.)	emulsifier; detergent; solubiliser; stabiliser; levelling agent *fatty acids; alcohols; oils; waxes; essential oils; latexes; polishes; emulsion polymerisation; dyes*
	ETHYLAN CD802	synthetic branched chain alcohol ethoxylate	100% active; 'n' no. 2	HLB 8.0; clear colourless liquid; cloud point 58°C (5% in 25% BDG/H_2O); pour point −10°C; viscosity 20 cSt; pH 7.0 (1% aq.)	
	ETHYLAN CD913	synthetic lower fraction primary alcohol ethylene oxide condensate	97% active; ethylene oxide content 44%; 'n' no. 2.9	HLB 8.8; clear colourless liquid; pour point <0°C; viscosity 30 cSt; pH 5-7 (1% aq.)	emulsifier; detergent; wetting agent; solubiliser *textile industry; metal working; leather industry; paper industry; paint industry; detergents; household products; agriculture*
	ETHYLAN CD916	synthetic lower fraction primary alcohol ethylene oxide condensate	96% active; ethylene oxide content 64%; 'n' no. 6.5	HLB 12.8; clear colourless liquid; pour point 0°C; viscosity 53 cSt; pH 5-7 (1% aq.)	
	ETHYLAN CD919	synthetic lower fraction primary alcohol ethylene oxide condensate	97% active; ethylene oxide content 72%; 'n' no. 9.0	HLB 14.4; clear colourless liquid; cloud point 82°C (1% aq.); pour point 7°C; viscosity 67 cSt; pH 5-7 (1% aq.)	
	ETHYLAN CD9112	synthetic lower fraction primary alcohol ethylene oxide condensate	100% active; ethylene oxide content 77.5%; 'n' no. 12.0	HLB 15.5; white waxy solid; cloud point 100°C (1% aq.); pour point 22°C; viscosity 53 cSt (40°C); pH 5-7 (1% aq.)	
	ETHYLAN CDP2	$C_{12/14}$ alcohol ethoxylate	100% active; ethylene oxide content 30.5%; 'n' no. 2	HLB 6.1; clear colourless liquid; pour point 4°C; pH 5.5-7.5 (1% aq.)	co-emulsifier; intermediate *mineral oils; waxes; alkyd resins; paraffinic hydrocarbons; speciality surfactants; toiletry grade sulfates*
	ETHYLAN CDP3	$C_{12/14}$ alcohol ethoxylate	100% active; ethylene oxide content 40%; 'n' no. 3	HLB 8; clear colourless liquid; pour point 3°C; pH 5.5-7.5 (1% aq.)	
	ETHYLAN CDP16	natural $C_{12/14}$ alcohol ethoxylate	100% active; 'n' no. 16	HLB 15.6; white waxy solid; cloud point 77°C (1% in 10% NaCl soln.); pH 7.0 (1% aq.)	emulsifier; detergent; solubiliser; stabiliser; levelling agent *fatty acids; alcohols; oils; waxes; essential oils; latexes; polishes; emulsion polymerisation; dyes*
	ETHYLAN CO35	cetyl-oleyl alcohol ethoxylate	100% active; ethylene oxide content 85%; 'n' no. 35	HLB 17; cream-coloured waxy solid; cloud point 78°C (1% in 10% aq. NaCl); pour point 48°C; viscosity 130 cP's (60°C); pH 5.5-7.5 (1% aq.)	emulsifier; stabiliser; co-emulsifier *emulsion polymerisation; waxes; fatty acids*
	ETHYLAN CPG540	modified alcohol ethoxylate	99% active	clear colourless liquid; cloud point 29.5°C (1% aq.); pour point −10°C; viscosity 300 cSt; pH 5-7 (1% aq.)	wetting agent; detergent; low foam *dishwashing agents; rinse aids; metal cleaning; spray cleaning*
	ETHYLAN CPG630	modified alcohol ethoxylate	99% active	clear colourless liquid; cloud point 38°C (1% aq.); pour point −1°C; viscosity 64 cSt; pH 5-7 (1% aq.)	
	ETHYLAN CPG660	modified alcohol ethoxylate	99% active	clear colourless liquid; cloud point 29°C (1% aq.); pour point −10°C; viscosity 114 cSt; pH 5-7 (1% aq.)	
	ETHYLAN CPG7545	modified alcohol ethoxylate	99% active	clear colourless liquid; cloud point 37°C (1% aq.); pour point −6°C; viscosity 114 cSt; pH 5-7 (1% aq.)	
	ETHYLAN CPG945	modified alcohol ethoxylate	99% active	clear colourless liquid; cloud point 47°C (1% aq.); pour point 2°C; viscosity 108 cSt; pH 5-7 (1% aq.)	

Supplier	Trade name	Chemical description	Composition	General properties	Functionality / Application
Akcros Chemicals	ETHYLAN CS20	natural cetyl-stearyl alcohol ethoxylate; made-to-order	100% active; 'n' no. 20	HLB 15.5; off-white waxy solid; cloud point 76°C (1% in 10% NaCl soln.); pour point 40°C; pH 7.0 (1% aq.)	emulsifier; detergent; solubiliser; stabiliser; levelling agent / *fatty acids; alcohols; oils; waxes; essential oils; latexes; polishes; emulsion polymerisation; dyes*
	ETHYLAN CX138	branched chain fatty alcohol ethoxylate	100% active	soft paste; cloud point 56°C (1% aq.); pour point 12°C; viscosity 115 cSt; pH 5.5-7.5 (1% aq.)	co-emulsifier; additive / *agrochemicals*
	ETHYLAN CX308	branched chain alcohol ethoxylate	100% active; 'n' no. 8	HLB 13.0; clear colourless liquid; cloud point 52°C (5% aq.); pour point 10°C; viscosity 503 cSt; pH 7.0 (1% aq.)	emulsifier; detergent; solubiliser; stabiliser; levelling agent / *fatty acids; alcohols; oils; waxes; essential oils; latexes; polishes; emulsion polymerisation; dyes*
	ETHYLAN D253	detergent grade primary alcohol ethylene oxide condensate	100% active; ethylene oxide content 39%; 'n' no. 3	HLB 7.8; clear colourless liquid; pour point 3°C; viscosity 35 cSt; pH 5-7 (1% aq.)	emulsifier; wetting agent; detergent / *textile industry; metal working; leather industry; paper industry; paint industry; detergents; household products; agriculture*
	ETHYLAN D254	detergent grade primary alcohol ethylene oxide condensate	97% active; ethylene oxide content 49%; 'n' no. 4.5	HLB 9.8; clear colourless liquid; pour point 5°C; viscosity 45 cSt; pH 5-7 (1% aq.)	
	ETHYLAN D256	detergent grade primary alcohol ethylene oxide condensate; made-to-order	100% active; ethylene oxide content 57%; 'n' no. 6	HLB 11.4; hazy liquid; cloud point 43°C (1% aq.); pour point 15°C; viscosity 105 cSt; pH 5-7 (1% aq.)	
	ETHYLAN D2512	detergent grade primary alcohol ethylene oxide condensate	100% active; ethylene oxide content 72%; 'n' no. 12	HLB 14.4; white waxy solid; cloud point 92°C (1% aq.); pour point 29°C; viscosity 82 cSt (40°C); pH 5-7 (1% aq.)	
	ETHYLAN ME	natural cetyl-oleyl alcohol ethoxylate; made-to-order	100% active; 'n' no. 6	HLB 10.0; clear pale yellow liquid; cloud point 82°C (5% in 25% BDG/H_2O); pour point 18°C; pH 7.0 (1% aq.)	emulsifier; detergent; solubiliser; stabiliser; levelling agent / *fatty acids; alcohols; oils; waxes; essential oils; latexes; polishes; emulsion polymerisation; dyes*
	ETHYLAN OE	cetyl-oleyl alcohol ethoxylate; made-to-order	100% active; ethylene oxide content 70%; 'n' no. 13	HLB 14; cream-coloured waxy solid; cloud point 90°C (1% aq.); pour point 31°C; viscosity 55 cSt (40°C); pH 5.5-7.5 (1% aq.)	emulsifier; levelling agent; solubiliser; stabiliser / *fatty acids; alcohols; waxes; oils; latex; polishes; emulsion polymerisation; dyes*
	ETHYLAN R	cetyl-oleyl alcohol ethoxylate	100% active; ethylene oxide content 77%; 'n' no. 19	HLB 15.4; cream-coloured waxy solid; cloud point > 100°C (1% aq.); pour point 36°C; viscosity 141 cSt (40°C); pH 5.5-7.5 (1% aq.)	emulsifier; levelling agent; solubiliser; stabiliser / *fatty acids; alcohols; waxes; oils; latex; polishes; emulsion polymerisation; dyes*
Akzo Nobel	BEROL 047	tridecanol 6EO	100% active; 'n' no. 6	HLB 11.4; cloud point 64°C (5 g in 25 ml 25% BDG soln.)	wetting agent / *alkaline cleaners*
	BEROL 048	tridecanol 10EO	85% active; 'n' no. 10	HLB 13.5; cloud point 70°C (1% in H_2O)	wetting agent / *industrial cleaners*
	BEROL 050	$C_{12/14}$ fatty alcohol 3EO	100% active; 'n' no. 3	HLB 8.4; cloud point 50°C (5 g in 25 ml 25% BDG soln.)	emulsifier / *toiletries*
	BEROL 055	$C_{12/14}$ fatty alcohol 2EO	100% active; 'n' no. 2	HLB 6.5; cloud point 37°C (5 g in 25 ml 25% BDG soln.)	emulsifier / *toiletries*
	BEROL 058	tridecanol 8EO	100% active; 'n' no. 8	HLB 12.8; cloud point 44°C (1% in H_2O)	wetting agent / *industrial cleaners*
	BEROL 08 P	$C_{16/18}$ fatty alcohol 80EO	100% active; 'n' no. 80	HLB 18.7; cloud point 87°C (1 g in 100 ml 5% NaCl soln.)	dispersant / *detergents*
	BEROL 087	fatty alcohol ethoxylate	100% active	HLB 11.5; cloud point 41°C (1% in H_2O)	low foam / *dishwashing agents; industrial cleaners*
	BEROL 173	fatty alcohol ethoxylate	100% active	HLB 12.5; cloud point 68°C (1% in H_2O)	household detergents
	BEROL 185	fatty alcohol ethoxylate	90% active	HLB 13.5; cloud point 67°C (1% in H_2O)	wetting agent
	BEROL 260	ethoxylate based on SHOP alcohol	100% active; 'n' no. 4	HLB 10.5; cloud point 58°C (5 g in 25 ml 25% BDG soln.)	hard-surface cleaners; *industrial, institutional and vehicle cleaners*

Supplier	Trade name	Chemical description	Composition	General properties	Functionality Application
Akzo Nobel	BEROL 266	narrow range ethoxylate based on SHOP alcohol	100% active; 'n' no. 5.5	HLB 12.1	liquid cleaners; powder detergents
	BEROL 532	ethoxylate based on C_{11} SHOP alcohol	100% active; 'n' no. 2	HLB 6.8; cloud point 36°C (5 g in 25 ml 25% BDG soln.)	wetting agent; emulsifier detergents; hard-surface cleaners; industrial cleaners
	BEROL 533	ethoxylate based on C_{11} SHOP alcohol	100% active; 'n' no. 3	HLB 8.7; cloud point 51°C (5 g in 25 ml 25% BDG soln.)	
	BEROL 535	ethoxylate based on C_{11} SHOP alcohol	100% active; 'n' no. 5	HLB 11.2; cloud point 27°C (1% in H_2O)	wetting agent; emulsifier detergents; hard-surface cleaners; industrial cleaners; vehicle cleaners
	BEROL 537	ethoxylate based on C_{11} SHOP alcohol	100% active; 'n' no. 7	HLB 12.8; cloud point 64°C (1% in H_2O)	
	BEROL 542	ethoxylate based on C_{11} SHOP alcohol	85% active; 'n' no. 10	HLB 14.4; cloud point 62°C (1 g in 100 ml 10% NaCl soln.)	
	BEROL 543	ethoxylate based on C_{11} SHOP alcohol	85% active; 'n' no. 8	HLB 13.4; cloud point 75°C (1% in H_2O)	emulsifier household detergents; all-purpose cleaners; industrial and institutional cleaners; vehicle cleaners
	BEROL OX 25-3	ethoxylate based on $C_{12/15}$ SHOP alcohol	100% active; 'n' no. 3	HLB 8.1; cloud point 49°C (5 g in 25 ml 50% BDG soln.)	
	BEROL OX 25-7	ethoxylate based on $C_{12/15}$ SHOP alcohol	100% active; 'n' no. 7	HLB 12.0; cloud point 49°C (1% in H_2O)	wetting agent; emulsifier; dispersant laundry products
	BEROL OX 45-7	ethoxylate based on $C_{14/15}$ SHOP alcohol	100% active; 'n' no. 7	HLB 11.6; cloud point 48°C (1% in H_2O)	
	BEROL OX 45-11	ethoxylate based on $C_{14/15}$ SHOP alcohol	100% active; 'n' no. 11	HLB 13.7; cloud point 83°C (1% in H_2O)	emulsifier; dispersant; foaming agent (moderate) laundry products
	BEROL OX 91-4	ethoxylate based on $C_{9/11}$ SHOP alcohol	100% active; 'n' no. 4	HLB 10.5; cloud point 62°C (5 g in 25 ml 25% BDG soln.)	wetting agent; foaming agent (moderate) industrial, institutional and vehicle cleaners; liquid detergents; washing-up liquids
	BEROL OX 91-6	ethoxylate based on $C_{9/11}$ SHOP alcohol	100% active; 'n' no. 6	HLB 12.5; cloud point 56°C (1% in H_2O)	
	BEROL OX 91-8	ethoxylate based on $C_{9/11}$ SHOP alcohol	100% active; 'n' no. 8	HLB 13.7; cloud point 78°C (1% in H_2O)	
	ELFAPUR LM 20	lauryl myristyl alcohol ethoxylate (2)	100% active	liquid	
	ELFAPUR LM 30 S	lauryl myristyl alcohol ethoxylate (3)	100% active	liquid	
	ELFAPUR LM 75 S	lauryl myristyl alcohol ethoxylate (7.5)	90% active	liquid	
	ELFAPUR LP 25 S	$C_{12/13}$ alcohol ethoxylate (2.5)	100% active	liquid	
	ELFAPUR LP 25 SL	$C_{12/13}$ alcohol ethoxylate (2.5)	100% active	liquid	
	ELFAPUR LP 110 SLN	$C_{12/13}$ alcohol ethoxylate (11)	100% active	paste	
	ELFAPUR LT 30 SL	$C_{12/13}$ alcohol ethoxylate (3)	100% active	liquid	
	ELFAPUR LT 65 SLN	$C_{12/13}$ alcohol ethoxylate (6.5)	100% active	liquid	
	ELFAPUR LT 85/9 SLN	$C_{12/13}$ alcohol ethoxylate (8.5)	90% active	liquid	
	ELFAPUR LT 150 SLN	$C_{12/13}$ alcohol ethoxylate (15)	100% active	solid	
	ELFAPUR T 130 S	cetyl stearyl alcohol ethoxylate	100% active	solid	
	ELFAPUR T 250 PELLETS	hydrogenated tallow alcohol ethoxylate (2.5)	100% active	pellets	
Albright & Wilson	EMPILAN KA SERIES	$C_{10/11}$ alcohol ethoxylates	concentration 100%	liquids to solids	emulsifier; dispersant; wetting agent; antistatic agent; compounding aid
	EMPILAN KB SERIES	$C_{12/14}$ alcohol ethoxylates	concentration 100%	liquids to solids	agrochemicals; textile industry; emulsion polymerisation; polymer industry

Supplier	Trade name	Chemical description	Composition	General properties	Functionality Application
Albright & Wilson	EMPILAN KC SERIES	$C_{12/16}$ alcohol ethoxylates	concentration 100%	liquids	
	EMPILAN KCA SERIES	$C_{12/13}$ alcohol ethoxylates	concentration 100%	liquids	
	EMPILAN KCB SERIES	C_{11} alcohol ethoxylates	concentration 100%	liquids	
	EMPILAN KCL SERIES	$C_{12/15}$ alcohol ethoxylates	concentration 100%	liquids to pastes	emulsifier; dispersant; wetting agent; antistatic agent; compounding aid *agrochemicals; textile industry; emulsion polymerisation; polymer industry*
	EMPILAN KCP SERIES	$C_{14/15}$ alcohol ethoxylates	concentration 100%	liquids to pastes	
	EMPILAN KCX SERIES	$C_{13/15}$ alcohol ethoxylates	concentration 100%	liquids to pastes	
	EMPILAN KH SERIES	$C_{8/10}$ alcohol ethoxylates	concentration 100%	liquids	
	EMPILAN KI SERIES	C_{13} alcohol ethoxylates	concentration 100%	liquids	
	EMPILAN KLA SERIES	$C_{16/18}$ alcohol ethoxylates	concentration 100%	liquids to pastes	
	EMPILAN KM SERIES	$C_{16/18}$ alcohol ethoxylates	concentration 100%	solids, flakes and powders	
	EMPILAN KS SERIES	$C_{9/11}$ alcohol ethoxylates	concentration 100%	liquids	
Allied Colloids	DISPEX AL CONC 40	ethoxylated fatty alcohol		liquid	wetting agent; scouring agent; dispersant; emulsifier; antistatic agent *textile industry*
Amerchol	AMEROXOL OE-2	oleth-2	'n' no. 2	clear liquid	w/o emulsifier; spreading agent; solubiliser *cosmetics; toiletries; bath preparations*
	AMEROXOL OE-10	oleth-10	'n' no. 10	white semi-solid; cloud point 47-55°C (1% in 5% salt soln.)	o/w emulsifier; spreading agent; solubiliser *cosmetics; bath preparations*
	AMEROXOL OE-20	oleth-20	'n' no. 20	white waxy solid; cloud point 87-93°C (1% in 5% salt soln.)	emulsifier; solubiliser *personal care products; bath preparations*
	PROMULGEN D	cetearyl alcohol and ceteareth-20			emulsifier *personal care products*
	PROMULGEN G	stearyl alcohol and ceteareth-20			
Auschem Cesalpinia	AIONICO AF 10-12	short chain ethoxylated alcohol	100% active	liquid	degreaser; wetting agent
	BRITEX C 20	ceteth-2		HLB 5.3; solid	
	BRITEX C 100	ceteth-10		HLB 12.9; solid	
	BRITEX C 200	ceteth-20		HLB 15.7; solid	
	BRITEX CO 110	cetoleth-11		HLB 13.0; solid	
	BRITEX CO 220	cetoleth-22		HLB 15.9; solid	emulsifier *cosmetics*
	BRITEX CS 25	ceteareth-2.5		HLB 5.5; solid	
	BRITEX CS 110	ceteareth-11		HLB 13.1; solid	
	BRITEX CS 250	ceteareth-25		HLB 16.5; flakes	
	BRITEX CS 300	ceteareth-30		HLB 17.0; flakes	

Supplier	Trade name	Chemical description	Composition	General properties	Functionality Application
Auschem	BRITEX CS 1000	ceteareth-100		HLB 18.8; flakes	
Cesalpinia	BRITEX CS/BP	cetomacrogol 1000		HLB 15.5; flakes	
	BRITEX EMB	laureth-9		HLB 14.3; paste	
	BRITEX EW/BP	cetomacrogol emuls. wax		flakes	
	BRITEX L 20	laureth-2		HLB 6.5; liquid	
	BRITEX L 40	laureth-4		HLB 9.5; liquid	
	BRITEX L 100	laureth-10		HLB 13.8; paste/solid	emulsifier
	BRITEX L 230	laureth-23		HLB 16.9; solid	*cosmetics*
	BRITEX O 20	oleth-2		HLB 4.9; liquid	
	BRITEX O 100	oleth-10		HLB 12.4; paste	
	BRITEX O 200	oleth-20		HLB 15.3; solid	
	BRITEX S 20	steareth-2		HLB 4.9; solid	
	BRITEX S 100	steareth-10		HLB 12.4; solid	
	BRITEX S 200	steareth-20		HLB 15.3; flakes	
	CHIMIPAL AC 6	ethoxylated oxo-alcohol	100% active	paste	detergent; wetting agent; dispersant
	CHIMIPAL AC 7.5	ethoxylated oxo-alcohol	100% active	paste	
	CHIMIPAL AC 7	ethoxylated oxo-alcohol	100% active	paste	
	CHIMIPAL AC 9 F	ethoxylated oxo-alcohol	100% active	liquid	degreaser; wetting agent
	CHIMIPAL AC 9	ethoxylated oxo-alcohol	100% active	liquid	degreaser; wetting agent
	CHIMIPAL S 85	ethoxylated oxo-alcohol	85% active	liquid	detergent; solubiliser; emulsifier
	EMULSON CO 5	ethoxylated cetyloleil alcohol		liquid	emulsifier
	EMULSON CO 7	ethoxylated cetyloleil alcohol		liquid	*pesticides*
	EMULSON HT 72	ethoxylated tallow alcohol		flakes	emulsifier; wetting agent *pesticides*
	EMULSON O 7	ethoxylated cetyloleil alcohol		liquid	emulsifier *pesticides*
	MADEOL TR 8 L	ethoxylated iso-tridecyl alcohol		liquid	wetting agent *pesticides*
	MADEOL TR 8	ethoxylated iso-tridecyl alcohol		powder	
	POLIROL AL 435	ethoxylated fatty alcohol	82% active	liquid	viscosity modifier *urea resins; polymerisation*
	POLIROL O 55	ethoxylated fatty alcohol	100% active	flakes	stabiliser *latices; polymerisation*
	ROLFOR CO 7	POE (7) cetyloleic alcohol	100% active; 'n' no. 7	paste	
	ROLFOR CO 9	POE (9) cetyloleic alcohol	100% active; 'n' no. 9	paste	degreaser; wetting agent
	ROLFOR CO 11	POE (11) cetyloleic alcohol	100% active; 'n' no. 11	paste	
	ROLFOR CO 20	POE (20) cetyloleic alcohol	100% active; 'n' no. 20	paste	dispersant
	ROLFOR CO 30	POE (30) cetyloleic alcohol	100% active; 'n' no. 30	solid	dispersant
	ROLFOR E 270	short chain ethoxylated alcohol	100% active	paste	low foam
	ROLFOR E 527	short chain ethoxylated alcohol	100% active	liquid	*metal cleaning*

Supplier	Trade name	Chemical description	Composition	General properties	Functionality Application
Auschem Cesalpinia	ROLFOR HT 11	POE (11) cetylstearyl alcohol	100% active; 'n' no. 11	waxy	degreaser; wetting agent
	ROLFOR HT 25	POE (25) cetylstearyl alcohol	100% active; 'n' no. 25	flakes	dispersant
	ROLFOR LA 3	POE (3) lauryl alcohol	100% active; 'n' no. 3	liquid	co-detergent; co-emulsifier
	ROLFOR LA 4	POE (4) lauryl alcohol	100% active; 'n' no. 4	liquid	co-detergent; co-emulsifier
	ROLFOR LA 9	POE (9) lauryl alcohol	100% active; 'n' no. 9	paste	degreaser; wetting agent
	ROLFOR LA 23	POE (23) lauryl alcohol	100% active; 'n' no. 23	solid	dispersant
	ROLFOR TR 6	POE (6) iso tridecyl alcohol	100% active; 'n' no. 6	liquid	degreaser; wetting agent
	ROLFOR TR 9	POE (9) iso tridecyl alcohol	100% active; 'n' no. 9	liquid	degreaser; wetting agent
	ROLFOR TR 12	POE (12) iso tridecyl alcohol	85% active; 'n' no. 12	liquid	dispersant
	TEWAX TC 10	POE (9) cetearyl alcohol	'n' no. 9	HLB 12.0; flakes	emulsifier
	TEWAX TC 72	POE (8) stearyl alcohol	'n' no. 8	HLB 11.5; wax	*cosmetics*
BASF	EMULAN AF	fatty alcohol ethoxylate	100% active	cloud point 65°C (in butyl digol) — HLB=11	emulsifier
	EMULAN AT 9	fatty alcohol ethoxylate	100% active	cloud point 68°C (in H_2O)	
	EMULAN OC	fatty alcohol ethoxylate	100% active	cloud point 90°C (in NaCl soln.) — HLB=17	
	EMULAN OG	fatty alcohol ethoxylate	100% active	cloud point 92°C (in NaCl soln.)	
	EMULAN OK 5	fatty alcohol ethoxylate	100% active	cloud point 62°C (in butyl digol)	
	EMULAN OU	fatty alcohol ethoxylate	100% active	cloud point 90°C (in NaCl soln.)	
	EMULAN P	fatty alcohol ethoxylate	100% active	cloud point 52°C (in butyl digol)	
	LUTENSOL A 3	$C_{12/18}$ fatty alcohol 3EO	100% active; 'n' no. 3	cloud point 45°C (in butyl digol)	
	LUTENSOL A 4	$C_{12/18}$ fatty alcohol 4EO	100% active; 'n' no. 4	cloud point 57°C (in butyl digol)	
	LUTENSOL A 7	$C_{12/18}$ fatty alcohol 7EO	100% active; 'n' no. 7	cloud point 43°C (in H_2O)	
	LUTENSOL A 8	$C_{12/14}$ fatty alcohol 8EO	90% active; 'n' no. 8	cloud point 53°C (in H_2O)	
	LUTENSOL AO 3	$C_{13/15}$ oxo alcohol 3EO	100% active; 'n' no. 3	cloud point 45°C (in butyl digol)	
	LUTENSOL AO 4	$C_{13/15}$ oxo alcohol 4EO	100% active; 'n' no. 4	cloud point 57°C (in butyl digol)	
	LUTENSOL AO 5	$C_{13/15}$ oxo alcohol 5EO	100% active; 'n' no. 5	cloud point 62°C (in butyl digol)	
	LUTENSOL AO 7	$C_{13/15}$ oxo alcohol 7EO	100% active; 'n' no. 7	cloud point 43°C (in H_2O)	
	LUTENSOL AO 8	$C_{13/15}$ oxo alcohol 8EO	100% active; 'n' no. 8	cloud point 52°C (in H_2O)	
	LUTENSOL AO 89	$C_{13/15}$ oxo alcohol 8EO	90% active; 'n' no. 8	cloud point 52°C (in H_2O)	
	LUTENSOL AO 10	$C_{13/15}$ oxo alcohol 10EO	100% active; 'n' no. 10	cloud point 80°C (in H_2O)	
	LUTENSOL AO 109	$C_{13/15}$ oxo alcohol 10EO	90% active; 'n' no. 10	cloud point 80°C (in H_2O)	
	LUTENSOL AO 11	$C_{13/15}$ oxo alcohol 11EO	100% active; 'n' no. 11	cloud point 86°C (in H_2O)	
	LUTENSOL AO 30	$C_{13/15}$ oxo alcohol 30EO	100% active; 'n' no. 30	cloud point 91°C (in NaCl soln.)	
	LUTENSOL AO 3109	$C_{13/15}$ oxo alcohol 30EO	90% active; 'n' no. 30	cloud point 73°C (in butyl digol)	
	LUTENSOL AT 11	$C_{16/18}$ fatty alcohol 11EO	100% active; 'n' no. 11	cloud point 87°C (in H_2O)	
	LUTENSOL AT 25	$C_{16/18}$ fatty alcohol 25EO	100% active; 'n' no. 25	cloud point 95°C (in NaCl soln.)	
	LUTENSOL AT 50	$C_{16/18}$ fatty alcohol 50EO	100% active; 'n' no. 50	cloud point 92°C (in NaCl soln.)	
	LUTENSOL AT 80 E	$C_{16/18}$ fatty alcohol 80EO	100% active; 'n' no. 80	cloud point 87°C (in NaCl soln.)	

Supplier	Trade name	Chemical description	Composition	General properties	Functionality Application
BASF	LUTENSOL AT 80	$C_{16/18}$ fatty alcohol 80EO	100% active; 'n' no. 80	cloud point 87°C (in NaCl soln.)	
	LUTENSOL ON 30	C_{10} oxo alcohol 3EO	100% active; 'n' no. 3	cloud point 53°C (in butyl digol)	
	LUTENSOL ON 50	C_{10} oxo alcohol 5EO	100% active; 'n' no. 5	cloud point 67°C (in butyl digol)	
	LUTENSOL ON 60	C_{10} oxo alcohol 6EO	100% active; 'n' no. 6	cloud point 36°C (in H_2O)	
	LUTENSOL ON 70	C_{10} oxo alcohol 7EO	100% active; 'n' no. 7	cloud point 60°C (in H_2O)	
	LUTENSOL ON 80	C_{10} oxo alcohol 8EO	100% active; 'n' no. 8	cloud point 80°C (in H_2O)	
	LUTENSOL ON 110	C_{10} oxo alcohol 11EO	100% active; 'n' no. 11	cloud point 78°C (in NaCl soln.)	
	LUTENSOL TO 3	C_{13} oxo alcohol 3EO	100% active; 'n' no. 3	cloud point 40°C (in butyl digol)	
	LUTENSOL TO 5	C_{13} oxo alcohol 5EO	100% active; 'n' no. 5	cloud point 62°C (in butyl digol)	
	LUTENSOL TO 7	C_{13} oxo alcohol 7EO	100% active; 'n' no. 7	cloud point 70°C (in butyl digol)	
	LUTENSOL TO 8	C_{13} oxo alcohol 8EO	100% active; 'n' no. 8	cloud point 60°C (in H_2O)	
	LUTENSOL TO 89	C_{13} oxo alcohol 8EO	90% active; 'n' no. 8	cloud point 60°C (in H_2O)	
	LUTENSOL TO 10	C_{13} oxo alcohol 10EO	100% active; 'n' no. 10	cloud point 70°C (in H_2O)	
	LUTENSOL TO 109	C_{13} oxo alcohol 10EO	85% active; 'n' no. 10	cloud point 70°C (in H_2O)	
	LUTENSOL TO 12	C_{13} oxo alcohol 12EO	100% active; 'n' no. 12	cloud point 75°C (in NaCl soln.)	
	LUTENSOL TO 129	C_{13} oxo alcohol 12EO	85% active; 'n' no. 12	cloud point 75°C (in NaCl soln.)	
	LUTENSOL TO 15	C_{13} oxo alcohol 15EO	100% active; 'n' no. 15	cloud point 80°C (in NaCl soln.)	
	LUTENSOL TO 20	C_{13} oxo alcohol 20EO	100% active; 'n' no. 20	cloud point 86°C (in NaCl soln.)	
	LUTENSOL TO 389	C_{13} oxo alcohol ethoxylate	90% active	cloud point 70°C (in butyl digol)	
Bayer	AVOLAN AC	alkyl polyglycol ether			dispersant *textile industry; dyeing of acrylics*
	AVOLAN IW LIQUID	alkyl polyglycol ether			levelling agent; dispersant *textile industry; dyeing of wools*
	AVOLAN IW	alkyl polyglycol ether			levelling agent *textile industry; dyeing of wools*
	AVOLAN REN	modified alkylaryl polyglycol ether			dispersant *textile industry; dyeing*
	NAPHTHOPON E	alkyl polyglycol ether			
Chemax	CHEMAL 2EH-2	POE (2) 2-ethylhexyl ether	'n' no. 2	HLB 8.0; liquid; cloud point <25°C (1% aq.)	
	CHEMAL 2EH-5	POE (5) 2-ethylhexyl ether	'n' no. 5	HLB 12.6; liquid; cloud point <25°C (1% aq.)	detergent; wetting agent; emulsifier; dispersant; solubiliser; defoaming agent *textile industry; metal cleaning; household, industrial and institutional cleaners; hand cleaners*
	CHEMAL DA-4	POE (4) decyl ether	'n' no. 4	HLB 10.5; liquid; cloud point <25°C (1% aq.)	
	CHEMAL DA-6	POE (6) decyl ether	'n' no. 6	HLB 12.4; liquid; cloud point <41°C (1% aq.)	
	CHEMAL DA-9	POE (9) decyl ether	'n' no. 9	HLB 14.3; liquid; cloud point 80°C (1% aq.)	
	CHEMAL LA-4	POE (4) lauryl ether	'n' no. 4	HLB 9.2; liquid; cloud point <25°C (1% aq.)	
	CHEMAL LA-9	POE (9) lauryl ether	'n' no. 9	HLB 13.3; liquid; cloud point 76°C (1% aq.)	
	CHEMAL LA-23	POE (23) 2-ethylhexyl ether	'n' no. 23	HLB 16.7; solid; cloud point 100°C (1% aq.)	
	CHEMAL OA-4	POE (4) oleyl ether	'n' no. 4	HLB 7.9; liquid; cloud point <25°C (1% aq.)	
	CHEMAL OA-9	POE (9) oleyl ether	'n' no. 9	HLB 11.9; liquid; cloud point 52°C (1% aq.)	
	CHEMAL OA-20G	POE (20) oleyl ether	'n' no. 20	HLB 15.3; solid; cloud point >100°C (1% aq.)	

Supplier	Trade name	Chemical description	Composition	General properties	Functionality *Application*
Chemax	CHEMAL OA-23/70	POE (20) oleyl ether 70%	'n' no. 20	HLB 15.3; solid; cloud point > 100°C (1% aq.)	
	CHEMAL TDA-3	POE (3) tridecyl ether	'n' no. 3	HLB 7.9; liquid; cloud point <25°C (1% aq.)	detergent; wetting agent; emulsifier; dispersant; solubiliser; defoaming agent
	CHEMAL TDA-6	POE (6) tridecyl ether	'n' no. 6	HLB 11.4; liquid; cloud point <25°C (1% aq.)	*textile industry; metal cleaning; household, industrial and*
	CHEMAL TDA-9	POE (9) tridecyl ether	'n' no. 9	HLB 13.0; liquid; cloud point 54°C (1% aq.)	*institutional cleaners; hand cleaners*
	CHEMAL TDA-12	POE (12) tridecyl ether	'n' no. 12	HLB 14.5; paste; cloud point 70°C (1% aq.)	
	CHEMAL TDA-15	POE (15) tridecyl ether	'n' no. 15	HLB 15.4; solid; cloud point > 100°C (1% aq.)	
Ciba	IRGASOL NA	ethoxylated fatty alcohol		almost colourless low viscosity liquid	dispersant; levelling agent; fixation accelerant *textile industry*
	IRGASOL P	fatty alcohol polyglycol ether		colourless to faintly yellow low viscosity liquid	dispersant; fixation accelerant *textile industry; fluorescent whitening and dyeing of synthetic fibres*
	ULTRAVON EL	based on fatty alcohol ethoxylates		clear amber liquid	low foam; detergent *exhaust pretreatment of cotton in long liquors*
Condea	NACOLOX	fatty alcohol ethoxylates			
CPB-Companhia Petroquimica do Barreiro	QUIMIPOL EA2503	alcohol ethoxylate	100% active	pH 6.0-8.0 (1% soln.)	
	QUIMIPOL EA2504	alcohol ethoxylate	100% active	pH 6.0-8.0 (1% soln.)	
	QUIMIPOL EA2506	alcohol ethoxylate	100% active	pH 6.0-8.0 (1% soln.)	
	QUIMIPOL EA2507	alcohol ethoxylate	100% active	pH 6.0-8.0 (1% soln.)	
	QUIMIPOL EA2508	alcohol ethoxylate	100% active	pH 6.0-8.0 (1% soln.)	
	QUIMIPOL EA2509	alcohol ethoxylate	100% active	pH 6.0-8.0 (1% soln.)	
	QUIMIPOL EA2512	alcohol ethoxylate	100% active	pH 6.0-8.0 (1% soln.)	
	QUIMIPOL EA4505	alcohol ethoxylate	100% active	pH 6.0-8.0 (1% soln.)	
	QUIMIPOL EA4508	alcohol ethoxylate	100% active	pH 6.0-8.0 (1% soln.)	
	QUIMIPOL EA6801	alcohol ethoxylate	100% active	pH 6.0-8.0 (1% soln.)	
	QUIMIPOL EA6802	alcohol ethoxylate	100% active	pH 6.0-8.0 (1% soln.)	
	QUIMIPOL EA6803	alcohol ethoxylate	100% active	pH 6.0-8.0 (1% soln.)	
	QUIMIPOL EA6804	alcohol ethoxylate	100% active	pH 6.0-8.0 (1% soln.)	emulsifier *mineral oils; waxes; herbicides; aliphatic solvents*
	QUIMIPOL EA6806	alcohol ethoxylate	100% active	pH 6.0-8.0 (1% soln.)	
	QUIMIPOL EA6807	alcohol ethoxylate	100% active	pH 6.0-8.0 (1% soln.)	
	QUIMIPOL EA6808	alcohol ethoxylate	100% active	pH 6.0-8.0 (1% soln.)	
	QUIMIPOL EA6810	alcohol ethoxylate	100% active	pH 6.0-8.0 (1% soln.)	
	QUIMIPOL EA6812	alcohol ethoxylate	100% active	pH 6.0-8.0 (1% soln.)	
	QUIMIPOL EA6814	alcohol ethoxylate	100% active	pH 6.0-8.0 (1% soln.)	
	QUIMIPOL EA6818	alcohol ethoxylate	100% active	pH 6.0-8.0 (1% soln.)	
	QUIMIPOL EA6820	alcohol ethoxylate	100% active	pH 6.0-8.0 (1% soln.)	
	QUIMIPOL EA6823	alcohol ethoxylate	100% active	pH 6.0-8.0 (1% soln.)	
	QUIMIPOL EA6825	alcohol ethoxylate	100% active	pH 6.0-8.0 (1% soln.)	
	QUIMIPOL EA6850	alcohol ethoxylate	100% active	pH 6.0-8.0 (1% soln.)	
	QUIMIPOL EA9105	alcohol ethoxylate	100% active	pH 6.0-8.0 (1% soln.)	

Supplier	Trade name	Chemical description	Composition	General properties	Functionality Application
CPB-Companhia Petroquimica do Barreiro	QUIMIPOL EA9106	alcohol ethoxylate	100% active	pH 6.0-8.0 (1% soln.)	emulsifier; *mineral oils; waxes; herbicides; aliphatic solvents*
	QUIMIPOL EA9106B	alcohol ethoxylate	100% active	pH 6.0-8.0 (1% soln.)	
	QUIMIPOL EA9108	alcohol ethoxylate	100% active	pH 6.0-8.0 (1% soln.)	
Croda	CROMUL EM1207	steareth-21		HLB 15.5; white waxy solid	o/w emulsifier; *creams; lotions*
	VOLPO CS2	ceteareth-2		HLB 5.0; white waxy solid	emulsifier; wetting agent; solubiliser; dispersant; *skin care products; perfumes and essential oils; hard-surface and heavy-duty cleaners*
	VOLPO CS5	ceteareth-5		HLB 9.2; white soft solid	
	VOLPO CS10	ceteareth-10		HLB 12.6; white soft solid	
	VOLPO CS12	ceteareth-12		white waxy solid	
	VOLPO CS15	ceteareth-15		white solid	emulsifier; wetting agent; solubiliser; *skin care products; perfumes and essential oils*
	VOLPO CS20	ceteareth-20		HLB 15.7; white waxy solid	emulsifier; wetting agent; solubiliser; dispersant; *skin care products; perfumes and essential oils; liquid cleaners; degreasers*
	VOLPO CS50	POE(50) cetostearyl ether		HLB 17.8; off-white hard waxy solid	wetting agent; solubiliser; emulsifier; dispersant; *liquid cleaners; degreasers; polishes*
	VOLPO G26	glycereth-26		clear, essentially colourless liquid	humectant; viscosity modifier; solvent; *hair care products*
	VOLPO L3 SPECIAL	C$_{12/13}$ pareth-3		HLB 8.0; clear, essentially colourless liquid	emulsifier; dispersant; *bath preparations*
	VOLPO L4	POE(4) lauryl ether		HLB 9.4; colourless liquid	wetting agent; solubiliser; emulsifier; dispersant; *liquid cleaners; degreasers; polishes*
	VOLPO N3	oleth-3		HLB 6.6; clear to slightly hazy colourless to pale yellow liquid	emulsifier; wetting agent; solubiliser; dispersant; *skin care products; perfumes and essential oils; hard-surface and heavy-duty cleaners*
	VOLPO N5	oleth-5		HLB 9.0; opaque, white to off-white liquid	
	VOLPO N10	oleth-10		HLB 12.4; white turbid liquid/soft paste	
	VOLPO N15	oleth-15		HLB 14.2; white soft solid	
	VOLPO N20	POE(20) distilled oleyl ether		HLB 15.4; pale yellow soft solid	emulsifier; solubiliser; wetting agent; dispersant; *hard-surface and heavy-duty cleaners*
	VOLPO T5	POE(5) tridecyl ether		HLB 10.5; colourless liquid	wetting agent; solubiliser; emulsifier; dispersant; *liquid cleaners; degreasers; polishes*
	VOLPO T7	POE(7) tridecyl ether		HLB 12.6; colourless liquid	
	VOLPO T10	POE(10) tridecyl ether		HLB 13.7; white soft paste	
	VOLPO T15	POE(15) tridecyl ether		HLB 15.4; white soft paste	
Dac International Surfactants	BIODAC 510	fatty alcohol ethoxylate			detergent; wetting agent; *textile industry; paper industry; paint industry*
	BIODAC LS S52	mixture of alcohol ethoxylates			*low foam*
	EMULDAC AS 11	C$_{10/18}$ tallow alcohol ethoxylate			*liquid and powder detergents*
	EMULGANTE CO	alcohol cetyloleic ethoxylate			levelling agent; emulsifier; *textile industry*
	LORODAC LS 2L	synthetic fatty alcohol ethoxylate			*dishwashing agents; liquid detergents; intermediate for anionics*
	LORODAC LS 32L	synthetic fatty alcohol ethoxylate			
	LORODAC LS 33L	synthetic fatty alcohol ethoxylate			

Supplier	Trade name	Chemical description	Composition	General properties	Functionality Application
Dac International Surfactants	LORODAC LS 3L	synthetic fatty alcohol ethoxylate			dishwashing agents; liquid detergents; intermediate for anionics
	LORODAC LS 8L	synthetic alcohol ethoxylate			liquid and powder detergents
	LORODAC LS N7	synthetic alcohol ethoxylate			liquid and powder detergents
	LORODAC N2	natural $C_{12/14}$ fatty alcohol ethoxylate			shampoos; bubble baths; intermediate for anionics
	LORODAC N3	natural $C_{12/14}$ fatty alcohol ethoxylate			emulsifier; antistatic agent shampoos; bubble baths; textile industry; intermediate for anionics
	LORODAC N7	natural $C_{12/14}$ alcohol ethoxylate			liquid and powder detergents
	POLIDAC 200-300	polyethylene glycol alcohol ethoxylate			paper industry
	TRIDAC ISO 8	isotridecyl alcohol ethoxylate			wetting agent textile industry
Elf Atochem	EMULSIPAR H	ethoxylated fatty alcohol containing more than two ethylene oxide groups	acid value 3 max.	HLB ca. 12; wax; cloud point 68-70°C; colour 1 max. (Gardner); pH 5-7 (1% soln.)	emulsifier; aq. emulsions
	REMCOPAL 4	lauryl alcohol 4EO; in course of development	100% active; 'n' no. 4	HLB 9.7; liquid; cloud point 67°C (aq. soln. with 25% BDG)	
	REMCOPAL 10	oleo cetyl alcohol 9EO	100% active; carbon chain composition C_{12} <2%, C_{14} ca. 4%, C_{16} ca. 30%, C_{18} ca. 65%, C_{20} <2%; 'n' no. 9	HLB 12.2; soft paste; cloud point 84°C (aq. soln. with 25% BDG)	
	REMCOPAL 18	oleo cetyl alcohol 18EO; in course of development	100% active; carbon chain composition C_{12} <2%, C_{14} ca. 4%, C_{16} ca. 30%, C_{18} ca. 65%, C_{20} <2%; 'n' no. 18	HLB 14.7; paste; cloud point 80°C (aq. soln. of 50 g/l NaCl)	
	REMCOPAL 20	lauryl alcohol 19EO	100% active; 'n' no. 19	HLB 16.2; solid; cloud point 92°C (aq. soln. of 50 g/l NaCl)	
	REMCOPAL 25	oleo cetyl alcohol 23EO	100% active; carbon chain composition C_{12} <2%, C_{14} ca. 4%, C_{16} ca. 30%, C_{18} ca. 65%, C_{20} <2%; 'n' no. 23	HLB 16; solid; cloud point 92°C (aq. soln. of 50 g/l NaCl)	
	REMCOPAL 121	lauryl alcohol 3EO	100% active; 'n' no. 3	HLB 8.3; liquid; cloud point 58°C (aq. soln. with 25% BDG)	
	REMCOPAL 220	oleo cetyl alcohol 23EO	80% active; carbon chain composition C_{12} <2%, C_{14} ca. 4%, C_{16} ca. 30%, C_{18} ca. 65%, C_{20} <2%; 'n' no. 23	HLB 16; paste; cloud point 92°C (aq. soln. of 50 g/l NaCl)	
	REMCOPAL 229	cetostearyl alcohol 25EO; in course of development	100% active; carbon chain composition C_{14} <4%, C_{16} ca. 27%, C_{18} ca. 70%, C_{20} <2%; 'n' no. 25	HLB 15.2; solid; cloud point 91°C (aq. soln. of 50 g/l NaCl)	
	REMCOPAL 234	oleo cetyl alcohol 4EO	100% active; carbon chain composition C_{12} <2%, C_{14} ca. 4%, C_{16} ca. 30%, C_{18} ca. 65%, C_{20} <2%; 'n' no. 4	HLB 8; soft paste; cloud point 67°C (aq. soln. with 25% BDG)	
	REMCOPAL 238	cetostearyl alcohol 18EO	100% active; carbon chain composition C_{14} <4%, C_{16} ca. 27%, C_{18} ca. 70%, C_{20} <2%; 'n' no. 18	HLB 15.1; paste; cloud point 88°C (aq. soln. of 50 g/l NaCl)	
	REMCOPAL 258	isotridecanol 9EO; C_{13} carbon chain; in course of development	100% active; 'n' no. 9	HLB 13.3; liquid; cloud point 53°C	
	REMCOPAL 273	isodecanol 3EO; C_{10} carbon chain	100% active; 'n' no. 3	HLB 8.7; liquid; cloud point 51°C (aq. soln. with 25% BDG)	
	REMCOPAL 21411	lauryl alcohol 9.5EO	100% active; 'n' no. 9.5	HLB 13.5; paste; cloud point 85°C	

Supplier	Trade name	Chemical description	Composition	General properties	Functionality Application
Elf Atochem	REMCOPAL 21912AL	lauryl alcohol 12EO; in course of development	100% active; 'n' no. 12	HLB 18.8; paste; cloud point 90°C	
	REMCOPAL D	oleo cetyl alcohol 23EO	33% active; carbon chain composition C_{14} < 2%, C_{14} ca. 4% C_{16} ca. 30%, C_{18} ca. 65%, C_{20} < 2%; 'n' no. 23	HLB 16; gel; cloud point 92°C (aq. soln. of 50 g/l NaCl)	
	REMCOPAL L 9	lauryl alcohol 8EO	100% active; 'n' no. 8	HLB 12.8; solid; cloud point 66°C	
	REMCOPAL L 12	lauryl alcohol 10.5EO	100% active; 'n' no. 10.5	HLB 14; solid; cloud point 90°C	
	REMCOPAL LC	lauryl alcohol 8EO; in course of development	50% active; 'n' no. 8	HLB 12.8; gel; cloud point 66°C	
	REMCOPAL LO 2B	isodecanol 3EO; C_{10} carbon chain	65% active; 'n' no. 3	HLB 8; liquid; cloud point 80°C (aq. soln. with 25% BDG)	
	REMCOPAL LO	isotridecanol 9EO; C_{13} carbon chain; in course of development	25% active; 'n' no. 9	HLB 13.3; liquid; cloud point 53°C	
	REMCOPAL LP	lauryl alcohol 9EO; in course of development	32% active; 'n' no. 9	HLB 12.8; liquid; cloud point 66°C	
	SELLIG LA 11 100	lauryl alcohol 20EO	100% active; 'n' no. 20	HLB 16; liquid at 37°C; cloud point 92°C	
	SELLIG LA 11 100 M8	lauryl alcohol 11EO blend	100% active; 'n' no. 11	HLB 14.4; paste; cloud point 92°C	
	SELLIG LA 1150	lauryl alcohol 20EO	50% active; 'n' no. 20	HLB 16; liquid; cloud point 92°C	
	SELLIG LA 9 100	lauryl alcohol 9.5EO	100% active; 'n' no. 9.5	HLB 13.5; paste; cloud point 85°C	
	SELLIG OX 5 100	isotridecanol 5EO; C_{13} carbon chain	100% active; 'n' no. 5	HLB 10.7; liquid; cloud point 49°C	
	SELLIG OX 7.5 85	isotridecanol 8EO; C_{13} carbon chain	85% active; 'n' no. 8	HLB 13; liquid; cloud point 50°C	
	SELLIG OX 9 100	isotridecanol 9EO; C_{13} carbon chain	100% active; 'n' no. 9	HLB 13.4; paste; cloud point 60°C	
	SELLIG OX 11 100	isotridecanol 12EO; C_{13} carbon chain	100% active; 'n' no. 12	HLB 14.7; paste; cloud point 82°C	
	SELLIG SP 8 100	oleo cetyl alcohol 8EO; in course of development	100% active; carbon chain composition C_{12} < 2%, C_{14} ca. 4%, C_{16} ca. 30%, C_{18} ca. 65%, C_{20} < 2%; 'n' no. 8	HLB 11; paste; cloud point 42°C	
	SELLIG SP 16 100	oleo cetyl alcohol 16EO; in course of development	100% active; carbon chain composition C_{12} < 2%, C_{14} ca. 4% C_{16} ca. 30%, C_{18} ca. 65%, C_{20} < 2%; 'n' no. 16	HLB 14.2; liquid; cloud point 82°C (aq. soln. of 50 g/l NaCl)	
	SELLIG SP 20 100	oleo cetyl alcohol 18EO	100% active; carbon chain composition C_{12} < 2%, C_{14} ca. 4%, C_{16} ca. 30%, C_{18} ca. 65%, C_{20} < 2%; 'n' no. 18	HLB 15.2; solid; cloud point 86°C (aq. soln. of 50 g/l NaCl)	
	SELLIG SP 25 50	oleo cetyl alcohol 27EO	50% active; carbon chain composition C_{12} < 2%, C_{14} ca. 4%, C_{16} ca. 30%, C_{18} ca. 65%, C_{20} < 2%; 'n' no. 27	HLB 16.4; liquid; cloud point 90°C (aq. soln. of 50 g/l NaCl)	
	SELLIG SP 25 100	oleo cetyl alcohol 27EO	100% active; carbon chain composition C_{12} < 2%, C_{14} ca. 4%, C_{16} ca. 30%, C_{18} ca. 65%, C_{20} < 2%; 'n' no. 27	HLB 16.4; solid; cloud point 90°C (aq. soln. of 50 g/l NaCl)	
	SELLIG SP 30 100	oleo cetyl alcohol 30EO	100% active; carbon chain composition C_{12} < 2%, C_{14} ca. 4%, C_{16} ca. 30%, C_{18} ca. 65%, C_{20} < 2%; 'n' no. 30	HLB 16.7; solid; cloud point 94°C (aq. soln. of 50 g/l NaCl)	
	SELLIG SP 30 20	oleo cetyl alcohol 30EO	20% active; carbon chain composition C_{12} < 2%, C_{14} ca. 4%, C_{16} ca. 30%, C_{18} ca. 65%, C_{20} < 2%; 'n' no. 30	HLB 16.7; solid; cloud point 94°C (aq. soln. of 50 g/l NaCl)	

Supplier	Trade name	Chemical description	Composition	General properties	Functionality / *Application*
Elf Atochem	SELLIG SU 18 100	cetostearyl alcohol 18EO; in course of development	100% active; carbon chain composition C_{14} <4%, C_{16} ca. 27%, C_{18} ca. 70%, C_{20} <2%; 'n' no. 18	HLB 14.9; paste; cloud point 86°C	
	SELLIG SU 25 100	cetostearyl alcohol 20EO; in course of development	100% active; carbon chain composition C_{14} <4%, C_{16} ca. 27%, C_{18} ca. 70%, C_{20} <2%; 'n' no. 20	paste; cloud point 88°C (aq. soln. of 50 g/l NaCl)	
	SELLIG SU 30 100	cetostearyl alcohol 32EO	100% active; carbon chain composition C_{14} <4%, C_{16} ca. 27%, C_{18} ca. 70%, C_{20} <2%; 'n' no. 32	HLB 16.9; solid; cloud point 91°C (aq. soln. of 50 g/l NaCl)	
	SELLIG SU 50 100	cetostearyl alcohol 46EO	100% active; carbon chain composition C_{14} <4%, C_{16} ca. 27%, C_{18} ca. 70%, C_{20} <2%; 'n' no. 46	HLB 17.6; solid; cloud point 96°C (aq. soln. of 50 g/l NaCl)	
EniChem Augusta	ALCHEM 125/N EO	synthetic alcohol ethoxylates based on Alchem 125	'n' no. 2-3		*liquid and powder detergents; industrial applications*
	ISALCHEM 123/N EO	synthetic alcohol ethoxylates based on Isalchem 123	'n' no. 5-9		
	LIALET 111/N EO	synthetic alcohol ethoxylates based on Lial 111	'n' no. 3-7		
	LIALET 123/N EO	synthetic alcohol ethoxylates based on Lial 123	'n' no. 2-7		
	LIALET 125/N EO	synthetic alcohol ethoxylates based on Lial 125	'n' no. 2-12		
	LIALET 145/N EO	synthetic alcohol ethoxylates based on Lial 145	'n' no. 4-12		
	LIALET 158/N EO	synthetic alcohol ethoxylates based on Lial 158	'n' no. 8-12		
Gattefosse	EMULCIRE 61 WL 2659	cetyl alcohol, ceteth-20 and steareth-20		waxy solid	emulsifier / *hair care products; personal care products*
Henkel	ARLYPON F	laureth-2	99% active	liquid	thickener / *shampoos; bath preparations; shower preparations*
	BREVIOL DE	alkylpolyglycolether			levelling agent; dispersant / *textile industry; dyeing*
	BREVIOL PES	alkyl ethoxylates with additives			levelling agent / *textile industry; dyeing*
	DEFINDOL CONC.	alkylpolyglycolether			wetting agent / *textile industry; pretreatment*
	DEHYDOL 04 DEO	octyl alcohol 4EO; low odour version of Dehydol 04	99.7-100% active; 'n' no. 4	liquid	solubiliser; wetting agent; emulsifier / *solvents; oils; cleaners*
	DEHYDOL 04	octyl alcohol 4EO	99.7-100% active; 'n' no. 4	liquid	solubiliser; wetting agent; emulsifier / *solvents; oils; cleaners*
	DEHYDOL 100	fatty alcohol $C_{12/18}$ 9EO	99.5-100% active; 'n' no. 9	paste	wetting agent / *detergents; dishwashing agents*
	DEHYDOL 980	fatty alcohol $C_{10/14}$ 6EO	99.7-100% active; 'n' no. 6	liquid	
	DEHYDOL 2144	blend of fatty alcohol ethoxylates	99.5-100% active	liquid	
	DEHYDOL D 3	decyl alcohol 3EO	99.7-100% active; 'n' no. 3	liquid	wetting agent / *cleaners*
	DEHYDOL G 162	guerbet alcohol C_{16} 2EO	99.5-100% active; 'n' no. 2		defoaming agent
	DEHYDOL G 205	guerbet alcohol C_{20} 5EO	99.5-100% active; 'n' no. 5	liquid	defoaming agent; emulsifier

Supplier	Trade name	Chemical description	Composition	General properties	Functionality / Application
Henkel	DEHYDOL HD-FC 1	blend of fatty alcohol polyglycol ethers	99-100% active	paste	low foam / powder detergents; heavy-duty liquid detergents
	DEHYDOL HD-FC 2	blend of fatty alcohol polyglycol ethers	ca. 96% active	liquid	
	DEHYDOL LS 2 DEO	laureth-2	99-100% active	liquid	emulsifier; solubiliser; wetting agent / bath preparations; essential oils
	DEHYDOL LS 2	fatty alcohol $C_{12/14}$ 2EO	99.7-100% active; 'n' no. 2		solubiliser; emulsifier / solvents; oils
	DEHYDOL LS 3 DEO	laureth-3	99-100% active	liquid	viscosity modifier; solubiliser; thickener / shampoos; shower and bath preparations
	DEHYDOL LS 3	fatty alcohol $C_{12/14}$ 3EO	99.7-100% active; 'n' no. 3		solubiliser; emulsifier / solvents; oils
	DEHYDOL LS 4 DEO	laureth-4	99-100% active	liquid	solubiliser; wetting agent / bath preparations; essential oils
	DEHYDOL LS 4	fatty alcohol $C_{12/14}$ 4EO	99.7-100% active; 'n' no. 4	liquid	solubiliser; emulsifier / solvents; oils
	DEHYDOL LS 6	fatty alcohol $C_{12/14}$ 6EO	99.7-100% active; 'n' no. 6	liquid	
	DEHYDOL LT 5	fatty alcohol $C_{12/18}$ 5EO	99.7-100% active; 'n' no. 5	liquid	
	DEHYDOL LT 6	fatty alcohol $C_{12/18}$ 6EO	99.7-100% active; 'n' no. 6	liquid	cleaners; laundry products; dishwashing agents
	DEHYDOL LT 7	fatty alcohol $C_{12/18}$ 7EO	99.7-100% active; 'n' no. 7	liquid/paste	
	DEHYDOL LT 8	fatty alcohol $C_{12/18}$ 8EO	99.7-100% active; 'n' no. 8	paste	
	DEHYDOL TA 5	fatty alcohol $C_{16/18}$ 5EO	99.5-100% active; 'n' no. 5	solid	cleaners; laundry products
	DEHYDOL TA 14	fatty alcohol $C_{16/18}$ 14EO	99.5-100% active; 'n' no. 14	solid	emulsifier / laundry products; cleaners; WC sticks; paraffins; carnauba wax
	DEHYDOL WM 90	fatty alcohol $C_{12/18}$ 9EO	ca. 90% active; 'n' no. 9	solid	cleaners; laundry products
	DEHYPON G 2084	guerbet alcohol $C_{16/20}$ 8EO, n-butyl capped	99.5-100% active	liquid	defoaming agent / industrial cleaning; metal degreasers; bottle cleaners; dishwashing agents; institutional cleaning
	DEHYPON KE 2555	modified fatty alcohol polyglycol ether based on end-capped surfactants	ca. 86% active	liquid	defoaming agent / CIP cleaning
	DEHYPON KE 2619	modified fatty alcohol polyglycol ether based on end-capped surfactants	99-100% active	liquid	defoaming agent / CIP cleaning
	DEHYPON LS 104 L	fatty alcohol $C_{12/14}$ 10EO, n-butyl capped	ca. 85% active	liquid	defoaming agent / industrial cleaning; metal degreasers; bottle cleaners; dishwashing agents; institutional cleaning
	DEHYPON LS 104	fatty alcohol $C_{12/14}$ 10EO n-butyl capped	99-100% active	paste	
	DEHYPON LT 054	fatty alcohol $C_{12/18}$ 5EO, n-butyl capped	99-100% active	liquid	
	DEHYPON LT 104 L	fatty alcohol $C_{12/18}$ 10EO, n-butylcapped	ca. 85% active	liquid	
	DEHYPON LT 104	fatty alcohol $C_{12/18}$ 10EO, n-butyl capped	99-100% active	paste	
	DEHYPON MMD 2	modified fatty alcohol polyglycol ether	99-100% active	liquid	low foam / metal degreasers
	DISPONIL APE 256	mixture of ethoxylated linear fatty alcohols	ca. 80% active	HLB ca. 13.5; liquid	
	DISPONIL APE 257	mixture of ethoxylated linear fatty alcohols	ca. 65% active	HLB ca. 17; liquid	emulsifier / plastics industry; coatings
	DISPONIL APG 110	alkyl diol polyglycol ether	ca. 100% active	HLB 14.1; liquid	

Supplier	Trade name	Chemical description	Composition	General properties	Functionality / Application
Henkel	DISPONIL B 3 FLAKED	cetyl stearyl alcohol 30EO	100% active; 'n' no. 30	HLB 16.8; solid	
	DISPONIL LS 3	polyglycol ether of linear, saturated fatty alcohols	99-100% active	HLB 8.1; liquid	
	DISPONIL LS 4	polyglycol ether of linear, saturated fatty alcohols	99-100% active	HLB 9.5; liquid	
	DISPONIL LS 7	polyglycol ether of linear, saturated fatty alcohols	99-100% active	HLB 12.3; liquid	emulsifier; co-emulsifier; wetting agent; solubiliser; dispersant; detergent; levelling agent *emulsion polymerisation; plastics industry; coatings*
	DISPONIL O 5	oleyl cetyl alcohol polyglycol ether	100% active	HLB 9.5; liquid	
	DISPONIL O 10	oleyl cetyl alcohol polyglycol ether	100% active	HLB 12.5; pasty	
	DISPONIL O 250	oleyl cetyl alcohol polyglycol ether	100% active	HLB 16.5; solid	
	DISPONIL TA 11	polyglycol ether of linear, saturated fatty alcohols	99-100% active	HLB 13.0; solid	
	DISPONIL TA 430	polyglycol ether of linear, saturated fatty alcohols	99-100% active	HLB 17.4; flakes	
	EMTHOX 5882	laureth-4		colourless liquid	emulsifier; wetting agent; dispersant *hair care products; cosmetics; deodorants*
	EMTHOX 5885	ceteareth-20		white waxy solid	emulsifier; solubiliser *cosmetics; hair care products; sunscreen preparations*
	EUMULGIN 02	oleyl cetyl alcohol 2EO	99.7-100% active; 'n' no. 2	liquid	solubiliser; low foam; emulsifier *mineral oils; natural oils; solvents*
	EUMULGIN B 1	ceteareth-12		waxy solid	o/w emulsifier; emollient; conditioner *creams; ointments; lotions*
	EUMULGIN B 2	ceteareth-20		waxy solid	o/w emulsifier; emollient; conditioner *creams; ointments; lotions; laundry products; cleaners; WC sticks*
	EUMULGIN B 3	ceteareth-30		waxy solid	
	EUMULGIN EP 2	oleyl cetyl alcohol 2EO	99.5-100% active; 'n' no. 2	liquid	
	EUMULGIN EP 5L	oleyl cetyl alcohol 5EO	99.5-100% active; 'n' no. 5	liquid	low foam; emulsifier *mineral oils; natural oils; solvents*
	EUMULGIN ET 5	oleyl cetyl alcohol 5EO	99.5-100% active; 'n' no. 5	liquid	
	EUMULGIN ET 10	oleyl cetyl alcohol 10EO	99.5-100% active; 'n' no. 10	paste	o/w emulsifier *cold waves; thin liquid emulsions; transparent creams*
	EUMULGIN M 8	oleth-10 and oleth-5		gel-like substance	
	EUMULGIN O 5	oleth-5	'n' no. 5	solid/liquid	o/w emulsifier; solubiliser *cold waves; bath preparations; hand washing preparations*
	EUMULGIN O 10	oleth-10	'n' no. 10	soft waxy solid	
	EUMULGIN WM 7	oleyl cetyl alcohol 7EO	99.5-100% active; 'n' no. 7	paste	low foam *cleaners; laundry products*
	EUMULGIN WO 7	fatty alcohol $C_{12/18}$ unsaturated 7EO	99.5-100% active; 'n' no. 7	liquid	
	FORYL 100	fatty alcohol polyglycolether			scouring agent; wetting agent; dispersant; dyeing agent *textile industry; pretreatment*
	FORYL 197	fatty alcohol polyglycolether			scouring agent; aftersoaping agent *textile industry; pretreatment*
	FORYL OV	fatty alcohol polyglycolether			

Supplier	Trade name	Chemical description	Composition	General properties	Functionality / Application
Henkel	SELBANA 4236	high molecular alkylpolyglycolether			lubricant
	SELBANA UN	alkylpolyglycolether			*textile industry; spinning*
Hoechst	EMULSOGEN EPN073	speciality alcohol ethoxylate	28% active		
	EMULSOGEN EPN118	speciality alcohol ethoxylate	80% active		
	EMULSOGEN EPN207	speciality alcohol ethoxylate	70% active		*emulsion polymerisation*
	EMULSOGEN EPN287	speciality alcohol ethoxylate	70% active		
	EMULSOGEN EPN407	speciality alcohol ethoxylate	70% active		
	EMULSOGEN M	oleyl alcohol ethoxylate	100% active	clear liquid	emulsifier; wetting agent metal working; specialities
	EMULSOGEN MS-12	oleyl alcohol ethoxylate	100% active		emulsifier specialities
	GENAPOL C050	coco alcohol 5EO	100% active; 'n' no. 5		*detergents; toiletries*
	GENAPOL C080	coco alcohol 8EO	100% active; 'n' no. 8		
	GENAPOL C100	coco alcohol 10EO	100% active; 'n' no. 10		
	GENAPOL LA-030	coco alcohol 3EO; renewable source coco alcohol	100% active; 'n' no. 3		
	GENAPOL LA-040	coco alcohol 4EO; renewable source coco alcohol	100% active; 'n' no. 4		
	GENAPOL LA-050	coco alcohol 5EO; renewable source coco alcohol	100% active; 'n' no. 5		*detergents*
	GENAPOL LA-070	coco alcohol 7EO; renewable source coco alcohol	100% active; 'n' no. 7		
	GENAPOL LA-079	coco alcohol 7EO; renewable source coco alcohol	90% active; 'n' no. 7		
	GENAPOL O-020	oleic alcohol 2EO	100% active; 'n' no. 2	liquid	
	GENAPOL O-050	oleic alcohol 5EO	100% active; 'n' no. 5	turbid liquid	
	GENAPOL O-080	oleic alcohol 8EO	100% active; 'n' no. 8	turbid liquid	emulsifier; solubiliser *speciality blends; agrochemicals; metal working*
	GENAPOL O-100	oleic alcohol 10EO	100% active; 'n' no. 10	paste	
	GENAPOL O-120	oleic alcohol 12EO	100% active; 'n' no. 12	paste	
	GENAPOL O-200	oleic alcohol 20EO	100% active; 'n' no. 20	wax	
	GENAPOL OA-040	$C_{14/15}$ alcohol 4EO	100% active; 'n' no. 4		
	GENAPOL OA-070	$C_{14/15}$ alcohol 7EO	100% active; 'n' no. 7		
	GENAPOL OA-080	$C_{14/15}$ alcohol 8EO	100% active; 'n' no. 8		*detergents*
	GENAPOL OA-089	$C_{14/15}$ alcohol 8EO	90% active; 'n' no. 8		
	GENAPOL OX-030	$C_{12/15}$ alcohol 3EO	100% active; 'n' no. 3		

Supplier	Trade name	Chemical description	Composition	General properties	Functionality Application
Hoechst	GENAPOL OX-050	$C_{12/15}$ alcohol 5EO	100% active; 'n' no. 5		
	GENAPOL OX-060	$C_{12/15}$ alcohol 6EO	100% active; 'n' no. 6		
	GENAPOL OX-080	$C_{12/15}$ alcohol 8EO	100% active; 'n' no. 8		*detergents*
	GENAPOL OX-100	$C_{12/15}$ alcohol 10EO	100% active; 'n' no. 10		
	GENAPOL OX-109	$C_{12/15}$ alcohol 10EO	90% active; 'n' no. 10		
	GENAPOL OX-130	$C_{12/15}$ alcohol 13EO	100% active; 'n' no. 13		
	GENAPOL T-080	tallow alcohol 8EO	100% active; 'n' no. 8		
	GENAPOL T-110	tallow alcohol 11EO	100% active; 'n' no. 11		
	GENAPOL T-150	tallow alcohol 15EO	100% active; 'n' no. 15		*emulsifier*
	GENAPOL T-200	tallow alcohol 20EO	100% active; 'n' no. 20		*coatings; fabric conditioners*
	GENAPOL T-250	tallow alcohol 25EO	100% active; 'n' no. 25	white powder	
	GENAPOL T-250 PDR	tallow alcohol 25EO	100% active; 'n' no. 25		
	GENAPOL T-500 PDR	tallow alcohol 50EO	100% active; 'n' no. 50	powder	
	GENAPOL T-500	tallow alcohol 50EO	100% active; 'n' no. 50		
	GENAPOL T-800	tallow alcohol 80EO	100% active; 'n' no. 80		
	GENAPOL UD-030	C_{11} alcohol 3EO	100% active; 'n' no. 3		
	GENAPOL UD-050	C_{11} alcohol 5EO	100% active; 'n' no. 5		
	GENAPOL UD-080	C_{11} alcohol 8EO	100% active; 'n' no. 8		
	GENAPOL UD-088	C_{11} alcohol 8EO	80% active; 'n' no. 8		
	GENAPOL UD-110	C_{11} alcohol 11EO	100% active; 'n' no. 11		
	GENAPOL X-020	C_{13} alcohol 2EO	100% active; 'n' no. 2		
	GENAPOL X-050	C_{13} alcohol 5EO	100% active; 'n' no. 5		
	GENAPOL X-060	C_{13} alcohol 6EO	100% active; 'n' no. 6		*detergents*
	GENAPOL X-080	C_{13} alcohol 8EO	100% active; 'n' no. 8		
	GENAPOL X-090	C_{13} alcohol 9EO	100% active; 'n' no. 9		
	GENAPOL X-150	C_{13} alcohol 15EO	100% active; 'n' no. 15		
	GENAPOL Z030X	$C_{9/11}$ alcohol 3EO	100% active; 'n' no. 3		
	GENAPOL Z050X	$C_{9/11}$ alcohol 5EO	100% active; 'n' no. 5		
	GENAPOL Z060X	$C_{9/11}$ alcohol 6EO	100% active; 'n' no. 6		
	GENAPOL Z090X	$C_{9/11}$ alcohol 9EO	100% active; 'n' no. 9		
	GENAPOL Z120X	$C_{9/11}$ alcohol 12EO	100% active; 'n' no. 12		
Hüls	MARLIPAL 1/12	polyethylene glycol monomethylether	active detergent 100%	liquid	*production of methyl-terminated fatty acid polyethylene glycol esters*
	MARLIPAL 124	C_{12}-alcohol 4EO	active detergent 100%; 'n' no. 4	liquid; cloud point 68°C (10% in 25% BDG soln.)	*solubiliser*
	MARLIPAL 129	C_{12}-alcohol 9EO	active detergent 100%; 'n' no. 9	paste; cloud point 71°C (2% in H_2O)	*perfumes*

Supplier	Trade name	Chemical description	Composition	General properties	Functionality / Application
Hüls	MARLIPAL 1012/4	C$_{10/12}$ alcohol 4EO	active detergent 100%; 'n' no. 4	liquid; cloud point 66°C (10% in 25% BDG soln.)	wetting agent *hard-surface cleaners*
	MARLIPAL 1012/6	C$_{10/12}$ alcohol 6EO	active detergent 100%; 'n' no. 6	liquid; cloud point 53°C (2% in H$_2$O)	
	MARLIPAL 104	C$_{10}$-alcohol ethoxylate	active detergent 100%	liquid; cloud point 71°C (5% in 25% BDG soln.)	
	MARLIPAL 1618/6	C$_{16/18}$-alcohol 6EO	active detergent 100%; 'n' no. 6	solid; cloud point 75°C (10% in 25% BDG soln.)	
	MARLIPAL 1618/8	C$_{16/18}$-alcohol 8EO	active detergent 100%; 'n' no. 8	solid; cloud point 83°C (10% in 25% BDG soln.)	
	MARLIPAL 1618/10	C$_{16/18}$-alcohol 10EO	active detergent 100%; 'n' no. 10	solid; cloud point 70°C (2% in H$_2$O)	
	MARLIPAL 1618/11	C$_{16/18}$-alcohol 11EO	active detergent 100%; 'n' no. 11	solid; cloud point 87°C (2% in H$_2$O)	
	MARLIPAL 1618/18	C$_{16/18}$-alcohol 18EO	active detergent 100%; 'n' no. 18	solid; cloud point 74°C (2% in 10% NaCl soln.)	
	MARLIPAL 1618/25 P 6000 POWDER	C$_{16/18}$-alcohol 25EO	active detergent 100%; 'n' no. 25	spray-dried powder; cloud point 77°C (2% in 10% NaCl soln.)	dispersant; binding agent *solid cleaners; dyeing auxiliaries*
	MARLIPAL 1618/25 POWDER	C$_{16/18}$-alcohol 25EO	active detergent 100%; 'n' no. 25	spray-dried powder; cloud point 77°C (2% in 10% NaCl soln.)	
	MARLIPAL 1618/25	C$_{16/18}$-alcohol 25EO	active detergent 100%; 'n' no. 25	flakes; cloud point 77°C (2% in 10% NaCl soln.)	
	MARLIPAL 1618/40	C$_{16/18}$-alcohol 40EO	active detergent 100%; 'n' no. 40	flakes; cloud point 77°C (2% in 10% NaCl soln.)	
	MARLIPAL 1618/80	C$_{16/18}$-alcohol 80EO	active detergent 100%; 'n' no. 80	flakes; cloud point 77°C (2% in 10% NaCl soln.)	
	MARLIPAL 1850/5	oleyl alcohol 5EO	active detergent 100%; 'n' no. 5	paste; cloud point 74°C (5% in 25% BDG soln.)	
	MARLIPAL 1850/10	oleyl alcohol 10EO	active detergent 100%; 'n' no. 10	paste; cloud point 70°C (2% in H$_2$O)	
	MARLIPAL 1850/30	oleyl alcohol 30EO	active detergent 100%; 'n' no. 30	flakes; cloud point 75°C (2% in 10% NaCl soln.)	
	MARLIPAL 1850/40	oleyl alcohol 40EO	active detergent 100%; 'n' no. 40	flakes; cloud point 74°C (2% in 10% NaCl soln.)	
	MARLIPAL 1850/80	oleyl alcohol 80EO	active detergent 100%; 'n' no. 80	flakes	*laundry products; textile industry*
	MARLIPAL 24/20	C$_{12/14}$-alcohol 2EO	active detergent 100%; 'n' no. 2	liquid; cloud point 50°C (10% in 25% BDG soln.)	
	MARLIPAL 24/30	C$_{12/14}$-alcohol 3EO	active detergent 100%; 'n' no. 3	liquid; cloud point 60°C (10% in 25% BDG soln.)	
	MARLIPAL 24/40	C$_{12/14}$-alcohol 4EO	active detergent 100%; 'n' no. 4	liquid; cloud point 67°C (10% in 25% BDG soln.)	
	MARLIPAL 24/50	C$_{12/14}$-alcohol 5EO	active detergent 100%; 'n' no. 5	liquid; cloud point 73°C (10% in 25% BDG soln.)	
	MARLIPAL 24/60	C$_{12/14}$-alcohol 6EO	active detergent 100%; 'n' no. 6	liquid; cloud point 77°C (10% in 25% BDG soln.)	
	MARLIPAL 24/69	C$_{12/14}$-alcohol 6EO	active detergent 90%; 'n' no. 6	liquid; cloud point 77°C (10% in 25% BDG soln.)	dispersant; wetting agent; detergent; cleaning agent; emulsifier *textile industry*
	MARLIPAL 24/70	C$_{12/14}$-alcohol 7EO	active detergent 100%; 'n' no. 7	liquid; cloud point 54°C (2% in H$_2$O)	
	MARLIPAL 24/79	C$_{12/14}$-alcohol 7EO	active detergent 90%; 'n' no. 7	liquid; cloud point 54°C (2% in H$_2$O)	
	MARLIPAL 24/80	C$_{12/14}$-alcohol 8EO	active detergent 100%; 'n' no. 8	liquid; cloud point 67°C (2% in H$_2$O)	
	MARLIPAL 24/89	C$_{12/14}$-alcohol 8EO	active detergent 90%; 'n' no. 8	liquid/paste; cloud point 67°C (2% in H$_2$O)	
	MARLIPAL 24/90	C$_{12/14}$-alcohol 9EO	active detergent 100%; 'n' no. 9	liquid/paste; cloud point 82°C (2% in H$_2$O)	
	MARLIPAL 24/99	C$_{12/14}$-alcohol 9EO	active detergent 90%; 'n' no. 9	liquid/paste; cloud point 82°C (2% in H$_2$O)	
	MARLIPAL 24/100	C$_{12/14}$-alcohol 10EO	active detergent 100%; 'n' no. 10	paste; cloud point 56°C (2% in 10% NaCl soln.)	
	MARLIPAL 24/109	C$_{12/14}$-alcohol 10EO	active detergent 90%; 'n' no. 10	paste; cloud point 56°C (2% in 10% NaCl soln.)	

Supplier	Trade name	Chemical description	Composition	General properties	Functionality / Application
Hüls	MARLIPAL 24/150	$C_{12/14}$-alcohol 15EO	active detergent 100%; 'n' no. 15	paste; cloud point 73°C (2% in 10% NaCl soln.)	
	MARLIPAL 24/159	$C_{12/14}$-alcohol 15EO	active detergent 90%; 'n' no. 15	cloud point 73°C (2% in 10% NaCl soln.)	dispersant; wetting agent; detergent; cleaning agent; emulsifier / *textile industry*
	MARLIPAL 24/200	$C_{12/14}$-alcohol 20EO	active detergent 100%; 'n' no. 20	solid; cloud point 76°C (2% in 10% NaCl soln.)	
	MARLIPAL 24/300	$C_{12/14}$-alcohol 30EO	active detergent 100%; 'n' no. 30	solid; cloud point 76°C (2% in 10% NaCl soln.)	
	MARLIPAL 24/939	$C_{12/14}$-alcohol ethoxylate blend	active detergent 90%	liquid; cloud point 75°C (10% in 25% BDG soln.)	
	MARLIPAL KF	$C_{10/12}$-alcohol ethoxylate	active detergent 100%	liquid; cloud point 53°C (2% in H_2O)	wetting agent; degreaser / *hard-surface cleaners; textile industry; leather industry*
	MARLIPAL MG	C_{12}-alcohol ethoxylate	active detergent 100%	liquid; cloud point 62°C (2% in H_2O)	solubiliser / *cosmetics*
	MARLIPAL NE	oxo alcohol 8.3EO	active detergent 100%; 'n' no. 8.3	liquid; cloud point 53°C (2% in H_2O)	wetting agent / *textile industry*
	MARLIPAL O11/30	C_{11}-oxo alcohol 3EO	active detergent 100%; 'n' no. 3	liquid; cloud point 52°C (10% in 25% BDG soln.)	
	MARLIPAL O11/50	C_{11}-oxo alcohol 5EO	active detergent 100%; 'n' no. 5	liquid; cloud point 72°C (10% in 25% BDG soln.)	
	MARLIPAL O11/79	C_{11}-oxo alcohol 7EO	active detergent 90%; 'n' no. 7	liquid; cloud point 53°C (2% in H_2O)	wetting agent / *hard-surface cleaners; textile industry*
	MARLIPAL O11/88	C_{11}-oxo alcohol 8EO	active detergent 80%; 'n' no. 8	liquid; cloud point 61°C (2% in H_2O)	
	MARLIPAL O11/110	C_{11}-oxo alcohol 11EO	active detergent 100%; 'n' no. 11	solid; cloud point 63°C (2% in 10% NaCl soln.)	
	MARLIPAL O13/20	C_{13}-oxo alcohol 2EO	active detergent 100%; 'n' no. 2	liquid; cloud point 30°C (10% in 25% BDG soln.)	
	MARLIPAL O13/30	C_{13}-oxo alcohol 3EO	active detergent 100%; 'n' no. 3	liquid; cloud point 50°C (10% in 25% BDG soln.)	dispersant; wetting agent; detergent; cleaning agent; emulsifier / *textile industry*
	MARLIPAL O13/40	C_{13}-oxo alcohol 4EO	active detergent 100%; 'n' no. 4	liquid; cloud point 60°C (10% in 25% BDG soln.)	
	MARLIPAL O13/50	C_{13}-oxo alcohol 5EO	active detergent 100%; 'n' no. 5	liquid; cloud point 65°C (10% in 25% BDG soln.)	
	MARLIPAL O13/59	C_{13}-oxo alcohol 5EO	active detergent 90%; 'n' no. 5	liquid; cloud point 65°C (10% in 25% BDG soln.)	
	MARLIPAL O13/60	C_{13}-oxo alcohol 6EO	active detergent 100%; 'n' no. 6	liquid; cloud point 70°C (10% in 25% BDG soln.)	
	MARLIPAL O13/69	C_{13}-oxo alcohol 6EO	active detergent 90%; 'n' no. 6	liquid; cloud point 70°C (10% in 25% BDG soln.)	
	MARLIPAL O13/70	C_{13}-oxo alcohol 7EO	active detergent 100%; 'n' no. 7	liquid; cloud point 73°C (10% in 25% BDG soln.)	
	MARLIPAL O13/79	C_{13}-oxo alcohol 7EO	active detergent 90%; 'n' no. 7	liquid; cloud point 73°C (10% in 25% BDG soln.)	
	MARLIPAL O13/80	C_{13}-oxo alcohol 8EO	active detergent 100%; 'n' no. 8	liquid; cloud point 47°C (2% in H_2O)	
	MARLIPAL O13/89	C_{13}-oxo alcohol 8EO	active detergent 90%; 'n' no. 8	liquid; cloud point 47°C (2% in H_2O)	
	MARLIPAL O13/90	C_{13}-oxo alcohol 9EO	active detergent 100%; 'n' no. 9	liquid; cloud point 57°C (2% in H_2O)	
	MARLIPAL O13/99	C_{13}-oxo alcohol 9EO	active detergent 90%; 'n' no. 9	liquid; cloud point 57°C (2% in H_2O)	
	MARLIPAL O13/100	C_{13}-oxo alcohol 10EO	active detergent 100%; 'n' no. 10	liquid/paste; cloud point 75°C (2% in H_2O)	
	MARLIPAL O13/109	C_{13}-oxo alcohol 10EO	active detergent 90%; 'n' no. 10	liquid/paste; cloud point 75°C (2% in H_2O)	

Supplier	Trade name	Chemical description	Composition	General properties	Functionality Application
Hüls	MARLIPAL O13/120	C$_{13}$-oxo alcohol 12EO	active detergent 100%; 'n' no. 12	liquid/paste; cloud point 55°C (2% in 10% NaCl soln.)	
	MARLIPAL O13/129	C$_{13}$-oxo alcohol 12EO	active detergent 90%; 'n' no. 12	liquid/paste; cloud point 55°C (2% in 10% NaCl soln.)	
	MARLIPAL O13/150	C$_{13}$-oxo alcohol 15EO	active detergent 100%; 'n' no. 15	paste; cloud point 67°C (2% in 10% NaCl soln.)	dispersant; wetting agent; detergent; cleaning agent;
	MARLIPAL O13/159	C$_{13}$-oxo alcohol 15EO	active detergent 90%; 'n' no. 15	paste; cloud point 67°C (2% in 10% NaCl soln.)	emulsifier *textile industry*
	MARLIPAL O13/170	C$_{13}$-oxo alcohol 17EO	active detergent 100%; 'n' no. 17	solid; cloud point 72°C (2% in 10% NaCl soln.)	
	MARLIPAL O13/179	C$_{13}$-oxo alcohol 17EO	active detergent 90%; 'n' no. 17	solid; cloud point 72°C (2% in 10% NaCl soln.)	
	MARLIPAL O13/200	C$_{13}$-oxo alcohol 20EO	active detergent 100%; 'n' no. 20	solid; cloud point 73°C (2% in 10% NaCl soln.)	
	MARLIPAL O13/400	C$_{13}$-oxo alcohol 40EO	active detergent 100%; 'n' no. 40	flakes; cloud point 74°C (2% in 10% NaCl soln.)	
	MARLIPAL O13/500	C$_{13}$-oxo alcohol 50EO	active detergent 100%; 'n' no. 50	flakes; cloud point 74°C (2% in 10% NaCl soln.)	
	MARLIPAL O13/939	C$_{13}$-oxo alcohol ethoxylate blend	active detergent 90%	liquid; cloud point 71°C (10% in 25% BDG soln.)	
	MARLIPAL SU	oleyl alcohol ethoxylate	active detergent 100%	solid; cloud point 76°C (2% in 10% NaCl soln.)	plasticiser *soaps; textile industry*
	MARLOWET 4800	C$_{16/18}$-alcohol polyethylene glycol ether	active detergent 100%	flakes; cloud point 77°C (2% in 10% NaCl soln.)	emulsifier *textile industry; polishes; waxes*
	MARLOWET 4857	C$_{16/18}$-alcohol polyethylene glycol ether	active detergent 100%	solid; cloud point 76°C (10% in 25% BDG soln.)	emulsifier; mould release agent *paraffins; waxes; silicone oils; textile industry; paper industry; wood treatment*
	MARLOWET BL	C$_{12}$-alcohol polyethylene glycol ether	active detergent 100%	liquid; cloud point 68°C (10% in 25% BDG soln.)	emulsifier *spindle oils; white oils; textile industry*
	MARLOWET FOX	C$_{16/18}$-alcohol polyethylene glycol ether	active detergent 100%	solid; cloud point 76°C (2% in 10% NaCl soln.)	emulsifier *textile industry; polishes; waxes*
	MARLOWET GFN	C$_{18}$-alcohol polyethylene glycol ether	active detergent 100%	paste; cloud point 70°C (2% in 10% NaCl soln.)	emulsifier *floor cleaners; waxes*
	MARLOWET GFW	C$_{16/18}$-alcohol polyethylene glycol ether	active detergent 100%	flakes; cloud point 74°C (2% in 10% NaCl soln.)	emulsifier *textile industry; polishes; waxes*
	MARLOWET PW	C$_{16/18}$-alcohol polyethylene glycol ether	active detergent 100%	solid; cloud point 78°C (10% in 25% BDG soln.)	emulsifier *paraffins; waxes; textile industry; paper industry*
	MARLOX B 24/50	C$_{12/14}$-alcohol 5EO; t-butyl blocked	active detergent 100%; 'n' no. 5	liquid; cloud point 36°C (10% in 25% BDG soln.)	
	MARLOX B 24/60	C$_{12/14}$ alcohol 6EO; t-butyl blocked	active detergent 100%; 'n' no. 6	liquid; cloud point 40°C (10% in 25% BDG soln.)	low foam; wetting agent *cleaners; textile industry*
	MARLOX B 24/80	C$_{12/14}$-alcohol 8EO; t-butyl blocked	active detergent 100%; 'n' no. 8	liquid; cloud point 36°C (2% in H$_2$O)	
ICI	ARLASOLVE 200	POE-(20)-isohexadecyl alcohol		HLB 15.7; white solid	solubiliser *personal care products*
	ATLAS G-1875	POE alkyl ether		white solid	
	ATLAS G-3707	POE lauryl alcohol		HLB 12.8; colourless to pale yellow suspension	
	ATLAS G-3816	POE-(16)-cetyl alcohol		HLB 14.9; white solid	
	ATLAS G-3820	POE-(20)-cetyl alcohol		HLB 15.7; white waxy solid	*household and industrial applications; agrochemicals; textile industry*
	ATLAS G-3904	POE-(4)-oleyl cetyl alcohol		HLB 8.0; straw-coloured waxy liquid	
	ATLAS G-3910	POE-(10)-oleyl ether		HLB 12.4; yellow oily suspension	

Supplier	Trade name	Chemical description	Composition	General properties	Functionality Application
ICI	ATLAS G-3998	POE fatty alcohol ether		HLB 15.0; cream soft waxy solid	*household and industrial applications; agrochemicals; textile industry*
	ATLAS G-4822	POE-(12)-cetyl stearyl alcohol		HLB 13.9; white waxy solid	
	ATLAS G-4829	POE lauryl alcohol		HLB 14.3; colourless to pale yellow liquid	
	ATLAS G-4936	POE-(10)-cetostearyl alcohol		HLB 12.5; white waxy solid	
	ATLAS G-4938	POE-(20)-cetostearyl alcohol		HLB 15.3; white waxy solid	
	ATLAS G-4940	POE-(30)-cetostearyl alcohol		HLB 16.6; white waxy solid	
	ATLAS G-70140	POE-(3)-cetyl oleyl alcohol		HLB 6.0; pale yellow liquid	
	ATLAS G-70141	POE-(6)-cetyl oleyl alcohol		HLB 10.2; colourless liquid	
	ATLAS G-70142	POE-(13)-cetyl oleyl alcohol		HLB 13.8; white semi-solid	
	ATLAS G-70147	POE-(17)-cetostearyl alcohol		HLB 14.9; white solid	
	ATLOX 804	POE alkyl ether		HLB 14.0; white liquid	*emulsifier; dispersant agrochemicals*
	ATLOX 4881	POE alkyl ether		HLB 15.9; white suspension	
	ATLOX 4896	POE alkyl ether		HLB 15.9; white liquid	
	ATLOX 4901	POE alkyl ether		HLB 11.6; white powder	
	ATLOX 4991	POE alkyl ether		HLB 13.6; colourless cloudy liquid	
	ATLOX 4995	POE alkyl ether		HLB 13.6; white powder	
	ATPOL HD722	44 mole ethoxylate of allyl alcohol		yellow to white solid	*polymer industry*
	ATPOL HD745	23.4 mole ethoxylate of allyl alcohol		HLB 18.9; cream solid	
	ATPOL HD863	7 mole ethoxylate of allyl alcohol		HLB 16.8; light brown waxy liquid	
	ATPOL HD975	4 mole ethoxylate of allyl alcohol		HLB 15.0; yellow liquid	
	BRIJ 30	POE-(4)-lauryl alcohol	'n' no. 4	HLB 9.7; colourless suspension	*emulsifier; lubricant personal care products; industrial applications*
	BRIJ 35	POE-(23)-lauryl alcohol	'n' no. 23	HLB 16.9; white solid or sprayed	
	BRIJ 52	POE-(2)-cetyl alcohol	'n' no. 2	HLB 5.3; white solid	
	BRIJ 56	POE-(10)-cetyl alcohol	'n' no. 10	HLB 12.9; white solid	
	BRIJ 58	POE-(20)-cetyl alcohol	'n' no. 20	HLB 15.7; white solid, flaked or sprayed	
	BRIJ 72	POE-(2)-stearyl alcohol	'n' no. 2	HLB 4.9; white solid	
	BRIJ 76	POE-(10)-stearyl alcohol	'n' no. 10	HLB 12.4; white solid	
	BRIJ 78	POE-(20)-stearyl alcohol	'n' no. 20	HLB 15.3; white solid	
	BRIJ 92	POE-(2)-oleyl alcohol	'n' no. 2	HLB 4.9; pale yellow liquid	
	BRIJ 93	POE-(2)-oleyl alcohol, special quality	'n' no. 2	HLB 4.9; pale yellow liquid	
	BRIJ 96	POE-(10)-oleyl alcohol	'n' no. 10	HLB 12.4; pale yellow suspension	
	BRIJ 97	POE-(10)-oleyl alcohol, special quality	'n' no. 10	HLB 12.4; pale yellow suspension	
	BRIJ 98	POE-(20)-oleyl alcohol	'n' no. 20	HLB 15.3; cream solid	
	BRIJ 99	POE-(20)-oleyl alcohol, special quality	'n' no. 20	HLB 15.3; cream solid	
	BRIJ 700	POE-(100)-stearyl alcohol	'n' no. 100	HLB 18.8; pale yellow solid	
	BRIJ 721	POE-(21)-stearyl alcohol	'n' no. 21	HLB 15.5; colourless flakes	
	CIRRASOL AEN-XB	POE fatty alcohol		HLB 10.2; colourless liquid	*textile industry*
	CIRRASOL AEN-XF	POE fatty alcohol		HLB 13.8; white semi-solid	

Supplier	Trade name	Chemical description	Composition	General properties	Functionality / Application
ICI	CIRRASOL ALN-WF	POE fatty alcohol		HLB 14.9; white solid	
	CIRRASOL EN-MB	POE fatty alcohol		HLB 6.0; yellow liquid	
	CIRRASOL EN-MP	POE fatty alcohol		HLB 6.0; pale yellow liquid	textile industry
	CIRRASOL LAN-SF	POE fatty alcohol		HLB 7.5; pale yellow liquid	
	RENEX 30	POE-(12)-tridecyl alcohol		HLB 14.5; colourless suspension	
	RENEX 36	POE-(6)-tridecyl alcohol		HLB 11.4; colourless suspension	wetting agent; emulsifier
	RENEX 702	POE-(2)-synthetic primary $C_{13/15}$ alcohol		HLB 5.9; colourless suspension	personal care products; agrochemicals; industrial applications
	RENEX 703	POE-(3)-synthetic primary $C_{13/15}$ alcohol		HLB 7.8; colourless suspension	
	RENEX 707	POE-(7)-synthetic primary $C_{13/15}$ alcohol		HLB 12.2; colourless suspension	
	RENEX 709	POE-(9)-synthetic primary $C_{13/15}$ alcohol		HLB 12.5; white paste	
	RENEX 711	POE-(11)-synthetic primary $C_{13/15}$ alcohol		HLB 13.9; white paste	
	RENEX 720	POE-(20)-synthetic primary $C_{13/15}$ alcohol		HLB 16.2; white solid	
	SYNPERONIC 10/3	POE-(3)-isodecanol		HLB 10.0; colourless suspension	
	SYNPERONIC 10/5	POE-(5)-isodecanol		HLB 11.9; colourless suspension	
	SYNPERONIC 10/6	POE-(6)-isodecanol		HLB 12.6; colourless suspension	
	SYNPERONIC 10/7	POE-(7)-isodecanol		HLB 13.4; colourless suspension	
	SYNPERONIC 10/8	POE-(8)-isodecanol		HLB 14.2; colourless suspension	
	SYNPERONIC 10/11	POE-(11)-isodecanol		HLB 15.4; white paste	
	SYNPERONIC 13/3	POE-(3)-tridecanol		HLB 8.6; pale yellow liquid	
	SYNPERONIC 13/5	POE-(5)-tridecanol		HLB 11.2; pale yellow liquid	
	SYNPERONIC 13/6.5	POE-(6.5)-tridecanol		HLB 12.5; pale yellow liquid	
	SYNPERONIC 13/8	POE-(8)-tridecanol		HLB 13.3; pale yellow liquid	
	SYNPERONIC 13/9	POE-(9)-tridecanol		HLB 13.6; pale yellow liquid	
	SYNPERONIC 13/10	POE-(10)-tridecanol		HLB 14.1; pale yellow liquid	
	SYNPERONIC 13/12	POE-(12)-tridecanol		HLB 14.8; pale yellow liquid	
	SYNPERONIC 13/15	POE-(15)-tridecanol		HLB 14.8; pale yellow paste	
	SYNPERONIC 13/18	POE-(18)-tridecanol		HLB 16.2; pale yellow paste	
	SYNPERONIC 13/20	POE-(20)-tridecanol		HLB 16.4; pale yellow paste	
	SYNPERONIC 91/2.5	POE-(2.5)-synthetic primary $C_{9/11}$ alcohol		HLB 8.2; colourless liquid	wetting agent; emulsifier
	SYNPERONIC 91/4	POE-(4)-synthetic primary $C_{9/11}$ alcohol		HLB 10.8; colourless liquid	textile industry; industrial and institutional cleaning
	SYNPERONIC 91/5	POE-(5)-synthetic primary $C_{9/11}$ alcohol		HLB 11.8; colourless liquid	
	SYNPERONIC 91/6	POE-(6)-synthetic primary $C_{9/11}$ alcohol		HLB 12.5; colourless liquid	
	SYNPERONIC 91/7	POE-(7)-synthetic primary $C_{9/11}$ alcohol		HLB 13.4; colourless liquid	
	SYNPERONIC 91/8	POE-(8)-synthetic primary $C_{9/11}$ alcohol		HLB 13.9; colourless hazy liquid	
	SYNPERONIC 91/10	POE-(10)-synthetic primary $C_{9/11}$ alcohol		HLB 14.7; white paste	
	SYNPERONIC 91/12	POE-(12)-synthetic primary $C_{9/11}$ alcohol		HLB 15.0; white paste	
	SYNPERONIC 91/20	POE-(20)-synthetic primary $C_{9/11}$ alcohol		HLB 16.9; white solid	

Supplier	Trade name	Chemical description	Composition	General properties	Functionality / Application
ICI	SYNPERONIC A1	POE-(1)-synthetic primary $C_{13/15}$ alcohol		HLB 4.1; colourless liquid	detergent; wetting agent; emulsifier; *household, industrial and institutional cleaning; agrochemicals*
	SYNPERONIC A2	POE-(2)-synthetic primary $C_{13/15}$ alcohol		HLB 5.9; colourless liquid	
	SYNPERONIC A3	POE-(3)-synthetic primary $C_{13/15}$ alcohol		HLB 7.9; colourless liquid	
	SYNPERONIC A4	POE-(4)-synthetic primary $C_{13/15}$ alcohol		HLB 9.1; colourless liquid	
	SYNPERONIC A7	POE-(7)-synthetic primary $C_{13/15}$ alcohol		HLB 12.2; white paste	
	SYNPERONIC A9	POE-(9)-synthetic primary $C_{13/15}$ alcohol		HLB 12.5; white paste	
	SYNPERONIC A11	POE-(11)-synthetic primary $C_{13/15}$ alcohol		HLB 13.9; white solid	
	SYNPERONIC A20	POE-(20)-synthetic primary $C_{13/15}$ alcohol		HLB 16.2; white solid	
	SYNPERONIC A50	POE-(50)-synthetic primary $C_{13/15}$ alcohol		HLB 18.3; white solid	
	SYNPERONIC BD100	ethoxylated primary $C_{13/15}$ alcohol		HLB 12.0; colourless liquid	
	SYNPERONIC BD30	modified alcohol ethoxylate			detergent; wetting agent; emulsifier; *household, industrial and institutional cleaning*
	SYNPERONIC L2	POE-(2)-lauryl alcohol		HLB 6.4; colourless liquid	
	SYNPERONIC L3	POE-(3)-lauryl alcohol		HLB 8.2; colourless liquid	
	SYNPERONIC L7	POE-(7)-lauryl alcohol		HLB 12.6; colourless liquid	
	SYNPERONIC M2	POE-(2)-primary alcohol blend		HLB 6.1; colourless liquid	
	SYNPERONIC M3	POE-(3)-primary alcohol blend		HLB 7.9; colourless liquid	
	SYNPERONIC TAE 8	POE-(8)-tallow alcohol		HLB 10.2; pale yellow solid	antifoam; dispersant; *industrial applications*
	SYNPERONIC TAE 11	POE-(11)-tallow alcohol		HLB 12.1; pale yellow solid	
	SYNPERONIC TAE 25	POE-(25)-tallow alcohol		HLB 16.9; pale yellow solid	
	SYNPERONIC TAE 65	POE-(65)-tallow alcohol		HLB 18.6; pale yellow solid	
Inolex	LEXEMUL CS-20	cetearyl alcohol and cetereth-20			
Kao Corporation	FINDET 10/15	polyethoxylated decyl alcohol	OH value 192-202; 'n' no. 3	liquid; cloud point 46-50°C	
	FINDET 10/18	polyethoxylated decyl alcohol	OH value 125-140; 'n' no. 6	liquid; cloud point 51-53°C	
	FINDET 1214/14	polyethoxylated linear lauric-myristic alcohol	'n' no. 2	liquid; cloud point 31-35°C	
	FINDET 1214/15	polyethoxylated linear lauric-myristic alcohol	OH value 163-173; 'n' no. 3	liquid; cloud point 47-49°C	
	FINDET 1214/21	polyethoxylated linear lauric-myristic alcohol	'n' no. 9	solid; cloud point 74-78°C	
	FINDET 13/18.5	polyethoxylated isotridecyl alcohol	'n' no. 6.5	liquid; cloud point 68-70°C	
	FINDET 13/21	polyethoxylated isotridecyl alcohol	'n' no. 9	liquid; cloud point 43-44°C	
	FINDET 1315/15	polyethoxylated $C_{13/15}$ oxo alcohol	OH value 162-172; 'n' no. 3	paste; cloud point 41-45°C	
	FINDET 1315/19.5	polyethoxylated $C_{13/15}$ oxo alcohol	OH value 102-110; 'n' no. 7.5	liquid; cloud point 49-52°C	
	FINDET 1618/15	polyethoxylated cetyl-stearyl alcohol	'n' no. 3	solid; cloud point 51-53°C	
	FINDET 1618/18	polyethoxylated cetyl-stearyl alcohol	'n' no. 6	solid; cloud point 71-75°C	
	FINDET 1618/20	polyethoxylated cetyl-stearyl alcohol	'n' no. 8	solid; cloud point 82-86°C	
	FINDET 1618/23	polyethoxylated cetyl-stearyl alcohol	'n' no. 11	solid; cloud point 78-82°C	

Supplier	Trade name	Chemical description	Composition	General properties	Functionality / Application
Kao Corporation	FINDET 1618/35 E	polyethoxylated cetyl-stearyl alcohol	OH value 42-46; 'n' no. 23	solid; cloud point 77-80°C	
	FINDET 1618/72 E	polyethoxylated cetyl-stearyl alcohol	'n' no. 60	solid; cloud point 75-77°C	
	FINDET 1816/14	polyethoxylated oleyl-cetyl alcohol	'n' no. 2	liquid/paste; cloud point 30-34°C	
	FINDET 1816/17 AR	polyethoxylated oleyl-cetyl alcohol	'n' no. 5	liquid/paste; cloud point 62-64°C	
	FINDET 1816/32 E	polyethoxylated oleyl-cetyl alcohol	'n' no. 20	solid; cloud point 76-78°C	
	FINDET 810 D/17	polyethoxylated octo-decyl alcohol	OH value 150-160; 'n' no. 5	liquid; cloud point 43-45°C	
	FINDET FF/8750	polyethoxylated lanolin alcohol	50%; 'n' no. 75	liquid; cloud point 76-80°C	
	FINDET SE-2249	polyethoxylated decyl alcohol	OH value 140-160; 'n' no. 5	liquid; cloud point 36-38°C	
Dr. W. Kolb	IMBENTIN-AG/100/020	ethoxylated C_{10} Ziegler fatty alcohol	'n' no. ca. 2.0	HLB ca. 7.2; liquid; cloud point 38°C (5 g in 25 g BDG 25%)	*personal care products; household detergents; industrial and institutional cleaning; pharmaceuticals*
	IMBENTIN-AG/100/030	ethoxylated C_{10} Ziegler fatty alcohol	'n' no. ca. 3.0	HLB ca. 9.1; liquid; cloud point 54°C (5 g in 25 g BDG 25%)	
	IMBENTIN-AG/100/040	ethoxylated C_{10} Ziegler fatty alcohol	'n' no. ca. 4.0	HLB ca. 10.5; liquid cloud point 61°C (5 g in 25 g BDG 25%)	
	IMBENTIN-AG/100/050	ethoxylated C_{10} Ziegler fatty alcohol	'n' no. ca. 5.0	HLB ca. 11.6; liquid; cloud point 30°C (1% in deionised water)	
	IMBENTIN-AG/100/52	ethoxylated C_{10} Ziegler fatty alcohol	'n' no. ca. 6.0	HLB ca. 12.5; liquid; cloud point 52°C (1% in deionised water)	
	IMBENTIN-AG/100/065	ethoxylated C_{10} Ziegler fatty alcohol	'n' no. ca. 6.5	HLB ca. 12.9; liquid; cloud point 60°C (1% in deionised water)	
	IMBENTIN-AG/100/080	ethoxylated C_{10} Ziegler fatty alcohol	'n' no. ca. 8.0	HLB ca. 13.8; liquid; cloud point 84°C (1% in deionised water)	
	IMBENTIN-AG/100/100	ethoxylated C_{10} Ziegler fatty alcohol	'n' no. ca. 10.0	HLB ca. 14.7; liquid/paste; cloud point 63°C (1% in 10% NaCl)	
	IMBENTIN-AG/100/120	ethoxylated C_{10} Ziegler fatty alcohol	'n' no. ca. 12.0	HLB ca. 15.4; paste; cloud point 68°C (1% in 10% NaCl)	
	IMBENTIN-AG/100/150	ethoxylated C_{10} Ziegler fatty alcohol	'n' no. ca. 15.0	HLB ca. 16.1; solid; cloud point 73°C (1% in 10% NaCl)	
	IMBENTIN-AG/100/200	ethoxylated C_{10} Ziegler fatty alcohol	'n' no. ca. 20.0	HLB ca. 17.0; solid; cloud point 75°C (1% in 10% NaCl)	
	IMBENTIN-AG/102/025	ethoxylated $C_{10/12}$ Ziegler fatty alcohol	'n' no. ca. 2.5	HLB ca. 8.0; liquid; cloud point 45°C (5 g in 25 g BDG 25%)	
	IMBENTIN-AG/102/040	ethoxylated $C_{10/12}$ Ziegler fatty alcohol	'n' no. ca. 4.0	HLB ca. 10.3; liquid; cloud point 64°C (5 g in 25 g BDG 25%)	
	IMBENTIN-AG/102/35	ethoxylated $C_{10/12}$ Ziegler fatty alcohol	'n' no. ca. 5.0	HLB ca. 11.4; liquid; cloud point 36°C (1% in deionised water)	
	IMBENTIN-AG/102/55	ethoxylated $C_{10/12}$ Ziegler fatty alcohol	'n' no. ca. 6.0	HLB ca. 12.3; liquid; cloud point 55°C (1% in deionised water)	
	IMBENTIN-AG/102/070	ethoxylated $C_{10/12}$ Ziegler fatty alcohol	'n' no. ca. 7.0	HLB ca. 13.0; liquid; cloud point 68°C (1% in deionised water)	
	IMBENTIN-AG/102/090	ethoxylated $C_{10/12}$ Ziegler fatty alcohol	'n' no. ca. 9.0	HLB ca. 14.1; liquid; cloud point 91°C (1% in tap water)	
	IMBENTIN-AG/124/020	ethoxylated $C_{12/14}$ Ziegler fatty alcohol	'n' no. ca. 2.0	HLB ca. 6.1; liquid; cloud point 38°C (5 g in 25 g BDG 25%)	

Supplier	Trade name	Chemical description	Composition	General properties	Functionality / Application
Dr W. Kolb	IMBENTIN-AG/124/030	ethoxylated $C_{12/14}$ Ziegler fatty alcohol	'n' no. ca. 3.0	HLB ca. 8.0; liquid; cloud point 54°C (5 g in 25 g BDG 25%)	*personal care products; household detergents; industrial and institutional cleaning; pharmaceuticals*
	IMBENTIN-AG/124/35	ethoxylated $C_{12/14}$ Ziegler fatty alcohol	'n' no. ca. 6.5	HLB ca. 11.8; liquid; cloud point 35°C (1% in deionised water)	
	IMBENTIN-AG/124/43	ethoxylated $C_{12/14}$ Ziegler fatty alcohol	'n' no. ca. 7.0	HLB ca. 12.1; liquid; cloud point 44°C (1% in deionised water)	
	IMBENTIN-AG/124/55	ethoxylated $C_{12/14}$ Ziegler fatty alcohol	'n' no. ca. 7.5	HLB ca. 12.5; liquid; cloud point 65°C (1% in deionised water)	
	IMBENTIN-AG/124/120	ethoxylated $C_{12/14}$ Ziegler fatty alcohol	'n' no. ca. 12.0	HLB ca. 14.5; solid; cloud point 68°C (1% in 10% NaCl)	
	IMBENTIN-AG/124/150	ethoxylated $C_{12/14}$ Ziegler fatty alcohol	'n' no. ca. 15.0	HLB ca. 15.3; solid; cloud point 73°C (1% in 10% NaCl)	
	IMBENTIN-AG/124PG/020	ethoxylated lauryl-myristyl alcohol; $C_{12/14/16}$ carbon chain	'n' no. ca. 2.0	HLB ca. 6.1; liquid; cloud point 37°C (5 g in 25 g BDG 25%)	
	IMBENTIN-AG/124PG/030	ethoxylated lauryl-myristyl alcohol; $C_{12/14/16}$ carbon chain	'n' no. ca. 3.0	HLB ca. 8.0; liquid; cloud point 52°C (5 g in 25 g BDG 25%)	
	IMBENTIN-AG/124S/020	ethoxylated lauryl-myristyl alcohol; $C_{12/14}$ carbon chain	'n' no. ca. 2.0	HLB ca. 6.1; liquid; cloud point 37°C (5 g in 25 g BDG 25%)	
	IMBENTIN-AG/124S/025	ethoxylated lauryl-myristyl alcohol; $C_{12/14}$ carbon chain	'n' no. ca. 2.5	HLB ca. 7.1; liquid; cloud point 47°C (5 g in 25 g BDG 25%)	
	IMBENTIN-AG/124S/030	ethoxylated lauryl-myristyl alcohol; $C_{12/14}$ carbon chain	'n' no. ca. 3.0	HLB ca. 8.0; liquid; cloud point 52°C (5 g in 25 g BDG 25%)	
	IMBENTIN-AG/124S/040	ethoxylated lauryl-myristyl alcohol; $C_{12/14}$ carbon chain	'n' no. ca. 4.0	HLB ca. 9.4; liquid; cloud point 62°C (5 g in 25 g BDG 25%)	
	IMBENTIN-AG/124S/045	ethoxylated lauryl-myristyl alcohol; $C_{12/14}$ carbon chain	'n' no. ca. 4.5	HLB ca. 10.0; liquid; cloud point 66°C (5 g in 25 g BDG 25%)	
	IMBENTIN-AG/124S/060	ethoxylated lauryl-myristyl alcohol; $C_{12/14}$ carbon chain	'n' no. ca. 6.0	HLB ca. 11.4; liquid; cloud point 44°C (1% in deionised water)	
	IMBENTIN-AG/124S/065	ethoxylated lauryl-myristyl alcohol; $C_{12/14}$ carbon chain	'n' no. ca. 6.5	HLB ca. 11.8; liquid; cloud point 48°C (1% in deionised water)	
	IMBENTIN-AG/124S/070	ethoxylated lauryl-myristyl alcohol; $C_{12/14}$ carbon chain	'n' no. ca. 7.0	HLB ca. 12.1; liquid; cloud point 54°C (1% in deionised water)	
	IMBENTIN-AG/124S/080	ethoxylated lauryl-myristyl alcohol; $C_{12/14}$ carbon chain	'n' no. ca. 8.0	HLB ca. 12.8; liquid/paste; cloud point 71°C (1% in deionised water)	
	IMBENTIN-AG/124S/230	ethoxylated lauryl-myristyl alcohol; $C_{12/14}$ carbon chain	'n' no. ca. 23.0	HLB ca. 16.7; solid/flakes; cloud point 77°C (1% in 10% NaCl)	
	IMBENTIN-AG/128/050	ethoxylated $C_{12/18}$ Ziegler fatty alcohol	'n' no. ca. 5.0	HLB ca. 10.2; liquid; cloud point 73°C (5 g in 25 g BDG 25%)	
	IMBENTIN-AG/128/080	ethoxylated $C_{12/18}$ Ziegler fatty alcohol	'n' no. ca. 8.0	HLB ca. 12.5; liquid/paste; cloud point 60°C (1% in deionised water)	
	IMBENTIN-AG/128/110	ethoxylated $C_{12/18}$ Ziegler fatty alcohol	'n' no. ca. 11.0	HLB ca. 13.9; paste; cloud point 85°C (1% in deionised water)	
	IMBENTIN-AG/168/080	ethoxylated $C_{16/18}$ Ziegler fatty alcohol	'n' no. ca. 8.0	HLB ca. 11.5; paste; cloud point 83°C (5 g in 25 g BDG 25%)	

Supplier	Trade name	Chemical description	Composition	General properties	Functionality Application
Dr. W. Kolb	IMBENTIN-AG/168/110	ethoxylated $C_{16/18}$ Ziegler fatty alcohol	'n' no. ca. 11.0	HLB ca. 13.0; solid; cloud point 75°C (1% in deionised water)	personal care products; household detergents; industrial and institutional cleaning; pharmaceuticals
	IMBENTIN-AG/168/150	ethoxylated $C_{16/18}$ Ziegler fatty alcohol	'n' no. ca. 15.0	HLB ca. 14.3; solid; cloud point 72°C (1% in 10% NaCl)	
	IMBENTIN-AG/168/250	ethoxylated $C_{16/18}$ Ziegler fatty alcohol	'n' no. ca. 25.0	HLB ca. 16.2; solid/flakes; cloud point 78°C (1% in 10% NaCl)	
	IMBENTIN-AG/168/400	ethoxylated $C_{16/18}$ Ziegler fatty alcohol	'n' no. ca. 40.0	HLB ca. 17.4; solid/flakes	
	IMBENTIN-AG/168/470	ethoxylated $C_{16/18}$ Ziegler fatty alcohol	'n' no. ca. 47.0	HLB ca. 17.8; solid/flakes	
	IMBENTIN-AG/168S/020	ethoxylated cetyl-stearyl alcohol; $C_{16/18}$ carbon chain	'n' no. ca. 2.0	HLB ca. 5.0; paste/solid; cloud point 66°C (5% in BDG 25%)	
	IMBENTIN-AG/168S/025	ethoxylated cetyl-stearyl alcohol; $C_{16/18}$ carbon chain	'n' no. ca. 2.5	HLB ca. 5.9; paste/solid; cloud point 48°C (5 g in 25 g BDG 25%)	
	IMBENTIN-AG/168S/055	ethoxylated cetyl-stearyl alcohol; $C_{16/18}$ carbon chain	'n' no. ca. 5.5	HLB ca. 9.6; paste/solid; cloud point 75°C (5 g in 25 g BDG 25%)	
	IMBENTIN-AG/168S/060	ethoxylated cetyl-stearyl alcohol; $C_{16/18}$ carbon chain	'n' no. ca. 6.0	HLB ca. 10.0; paste/solid; cloud point 77°C (5 g in 25 g BDG 25%)	
	IMBENTIN-AG/168S/080	ethoxylated cetyl-stearyl alcohol; $C_{16/18}$ carbon chain	'n' no. ca. 8.0	HLB ca. 11.5; solid; cloud point 86°C (5 g in 25 g BDG 25%)	
	IMBENTIN-AG/168S/100	ethoxylated cetyl-stearyl alcohol; $C_{16/18}$ carbon chain	'n' no. ca. 10.0	HLB ca. 12.5; solid; cloud point 88°C (5 g in 25 g BDG 25%)	
	IMBENTIN-AG/168S/110	ethoxylated cetyl-stearyl alcohol; $C_{16/18}$ carbon chain	'n' no. ca. 11.0	HLB ca. 13.0; solid; cloud point 89°C (5 g in 25 g BDG 25%)	
	IMBENTIN-AG/168S/120	ethoxylated cetyl-stearyl alcohol; $C_{16/18}$ carbon chain	'n' no. ca. 12.0	HLB ca. 13.4; solid; cloud point 90°C (5 g in 25 g BDG 25%)	
	IMBENTIN-AG/168S/140	ethoxylated cetyl-stearyl alcohol; $C_{16/18}$ carbon chain	'n' no. ca. 14.0	HLB ca. 14.0; solid; cloud point 66°C (1% in 10% NaCl)	
	IMBENTIN-AG/168S/150	ethoxylated cetyl-stearyl alcohol; $C_{16/18}$ carbon chain	'n' no. ca. 15.0	HLB ca. 14.3; solid; cloud point 72°C (1% in 10% NaCl)	
	IMBENTIN-AG/168S/180	ethoxylated cetyl-stearyl alcohol; $C_{16/18}$ carbon chain	'n' no. ca. 18.0	HLB ca. 15.0; solid/flakes; cloud point 75°C (1% in 10% NaCl)	
	IMBENTIN-AG/168S/250	ethoxylated cetyl-stearyl alcohol; $C_{16/18}$ carbon chain	'n' no. ca. 25.0	HLB ca. 16.2; solid/flakes; cloud point 78°C (1% in 10% NaCl)	
	IMBENTIN-AG/168S/300	ethoxylated cetyl-stearyl alcohol; $C_{16/18}$ carbon chain	'n' no. ca. 30.0	HLB ca. 16.7; solid/flakes	
	IMBENTIN-AG/168S/360	ethoxylated cetyl-stearyl alcohol; $C_{16/18}$ carbon chain	'n' no. ca. 36.0	HLB ca. 17.2; solid/flakes	
	IMBENTIN-AG/168S/500	ethoxylated cetyl-stearyl alcohol; $C_{16/18}$ carbon chain	'n' no. ca. 50.0	HLB ca. 17.9; solid/flakes	
	IMBENTIN-AG/168S/600	ethoxylated cetyl-stearyl alcohol; $C_{16/18}$ carbon chain	'n' no. ca. 60.0	HLB ca. 18.2; solid/flakes	
	IMBENTIN-AG/168S/800	ethoxylated cetyl-stearyl alcohol; $C_{16/18}$ carbon chain	'n' no. ca. 80.0	HLB ca. 18.6; solid/flakes	

personal care products; household detergents; industrial and institutional cleaning; pharmaceuticals

Supplier	Trade name	Chemical description	Composition	General properties
Dr. W. Kolb	IMBENTIN-AG/1685/950	ethoxylated cetyl-stearyl alcohol; $C_{16/18}$ carbon chain	'n' no. ca. 95.0	HLB ca. 18.8; solid/flakes
	IMBENTIN-AG/200/025	ethoxylated C_{20} Ziegler fatty alcohol	'n' no. ca. 2.5	HLB ca. 5.2; solid; cloud p... water)
	IMBENTIN-AG/810/022	ethoxylated $C_{8/10}$ Ziegler fatty alcohol	'n' no. ca. 2.2	HLB ca. 8.0; liquid; cloud [g BDG 25%)
	IMBENTIN-AG/810/050	ethoxylated $C_{8/10}$ Ziegler fatty alcohol	'n' no. ca. 5.0	HLB ca. 12.0; liquid; cloud deionised water)
	IMBENTIN-AG/810/055	ethoxylated $C_{8/10}$ Ziegler fatty alcohol	'n' no. ca. 5.5	HLB ca. 12.5; liquid; cloud deionised water)
	IMBENTIN-AG/810/080	ethoxylated $C_{8/10}$ Ziegler fatty alcohol	'n' no. ca. 8.0	HLB ca. 14.0; liquid; cloud deionised water)
	IMBENTIN-C/125/025	ethoxylated $C_{12/15}$ oxo alcohol	'n' no. ca. 2.5	HLB ca. 7.0; liquid; cloud] ml alcohol (30°C))
	IMBENTIN-C/125/17 A	ethoxylated $C_{12/15}$ oxo alcohol	'n' no. ca. 3.0	HLB ca. 7.5; liquid; cloud point 17°C (3 g in 10 ml alcohol (30°C))
	IMBENTIN-C/125/050	ethoxylated $C_{12/15}$ oxo alcohol	'n' no. ca. 5.0	HLB ca. 10.0; liquid; cloud point 68°C (5 g in 25 g BDG 25%)
	IMBENTIN-C/125/060	ethoxylated $C_{12/15}$ oxo alcohol	'n' no. ca. 6.0	HLB ca. 11.0; liquid; cloud point 79°C (10 g in 50 ml BDG 25%)
	IMBENTIN-C/125/55	ethoxylated $C_{12/15}$ oxo alcohol	'n' no. ca. 8.0	HLB ca. 12.5; liquid/paste; cloud point 55°C (1% in deionised water)
	IMBENTIN-C/125/85	ethoxylated $C_{12/15}$ oxo alcohol	'n' no. ca. 10.0	HLB ca. 13.5; paste; cloud point 82°C (1% in deionised water)
	IMBENTIN-C/135/020	ethoxylated $C_{13/15}$ oxo alcohol	'n' no. ca. 2.0	HLB ca. 6.0; liquid; cloud point 45°C (10% in BDG 25%)
	IMBENTIN-C/135/030	ethoxylated $C_{13/15}$ oxo alcohol	'n' no. ca. 3.0	HLB ca. 7.5; liquid; cloud point 57°C (10% in BDG 25%)
	IMBENTIN-C/135/070	ethoxylated $C_{13/15}$ oxo alcohol	'n' no. ca. 7.0	HLB ca. 12.0; liquid; cloud point 47°C (1% in deionised water)
	IMBENTIN-C/135/093	ethoxylated $C_{13/15}$ oxo alcohol	'n' no. ca. 9.3	HLB ca. 13.5; liquid/paste; cloud point 64°C (1% in deionised water)
	IMBENTIN-C/135/110	ethoxylated $C_{13/15}$ oxo alcohol	'n' no. ca. 11.0	HLB ca. 14.0; paste; cloud point 87°C (1% in deionised water)
	IMBENTIN-C/145/050	ethoxylated $C_{14/15}$ oxo alcohol	'n' no. ca. 5.0	HLB ca. 10.0; paste; cloud point 71°C (10% in BDG 25%)
	IMBENTIN-C/145/180	ethoxylated $C_{14/15}$ oxo alcohol	'n' no. ca. 18.0	HLB ca. 15.5; solid; cloud point 73°C (1% in 10% NaCl)
	IMBENTIN-C/91/17 A	ethoxylated $C_{9/11}$ oxo alcohol	'n' no. ca. 2.0	HLB ca. 7.2; liquid; cloud point 17°C (3 g in 10 ml alcohol (30°C))
	IMBENTIN-C/91/040	ethoxylated $C_{9/11}$ oxo alcohol	'n' no. ca. 4.0	HLB ca. 10.5; liquid; cloud point 63°C (5 g in 25 g BDG 25%)
	IMBENTIN-C/91/35	ethoxylated $C_{9/11}$ oxo alcohol	'n' no. ca. 5.0	HLB ca. 11.5; liquid; cloud point 36°C (1% in deionised water)

Supplier	Trade name	Chemical description	Composition	General properties	Functionality Application
Dr. W. Kolb	IMBENTIN-C/91/52	ethoxylated $C_{9/11}$ oxo alcohol	'n' no. ca. 6.0	HLB ca. 12.5; liquid; cloud point 52°C (1% in deionised water)	
	IMBENTIN-C/91/080	ethoxylated $C_{9/11}$ oxo alcohol	'n' no. ca. 8.0	HLB ca. 13.5; liquid; cloud point 80°C (1% in deionised water)	
	IMBENTIN-C/91/85	ethoxylated $C_{9/11}$ oxo alcohol	'n' no. ca. 9.0	HLB ca. 14.0; liquid; cloud point 85°C (1% in deionised water)	
	IMBENTIN-C/91/120	ethoxylated $C_{9/11}$ oxo alcohol	'n' no. ca. 12.0	HLB ca. 15.5; paste; cloud point 68°C (1% in 10% NaCl)	
	IMBENTIN-L/111/020	ethoxylated C_{11} branched oxo alcohol	'n' no. ca. 2.0	HLB ca. 6.8; liquid; cloud point 32°C (5 g in 25 g BDG 25%)	
	IMBENTIN-L/111/030	ethoxylated C_{11} branched oxo alcohol	'n' no. ca. 3.0	HLB ca. 8.7; liquid; cloud point 40°C (5 g in 25 g BDG 25%)	
	IMBENTIN-L/111/050	ethoxylated C_{11} branched oxo alcohol	'n' no. ca. 5.0	HLB ca. 11.3; liquid; cloud point 68°C (5 g in 25 g BDG 25%)	
	IMBENTIN-L/111/060	ethoxylated C_{11} branched oxo alcohol	'n' no. ca. 6.0	HLB ca. 12.5; liquid; cloud point 41°C (1% in deionised water)	
	IMBENTIN-L/111/070	ethoxylated C_{11} branched oxo alcohol	'n' no. ca. 7.0	HLB ca. 12.9; liquid; cloud point 51°C (1% in deionised water)	
	IMBENTIN-L/111/080	ethoxylated C_{11} branched oxo alcohol	'n' no. ca. 8.0	HLB ca. 13.5; liquid; cloud point 63°C (1% in deionised water)	
	IMBENTIN-L/111/090	ethoxylated C_{11} branched oxo alcohol	'n' no. ca. 9.0	HLB ca. 14.0; liquid/paste; cloud point 78°C (1% in deionised water)	personal care products; household detergents; industrial and institutional cleaning; pharmaceuticals
	IMBENTIN-L/111/100	ethoxylated C_{11} branched oxo alcohol	'n' no. ca. 10.0	HLB ca. 14.4; liquid/paste; cloud point 89°C (1% in deionised water)	
	IMBENTIN-L/111/110	ethoxylated C_{11} branched oxo alcohol	'n' no. ca. 11.0	HLB ca. 14.8; paste; cloud point 63°C (1% in 10% NaCl)	
	IMBENTIN-L/111/150	ethoxylated C_{11} branched oxo alcohol	'n' no. ca. 15.0	HLB ca. 15.9; solid; cloud point 73°C (1% in 10% NaCl)	
	IMBENTIN-L/111/200	ethoxylated C_{11} branched oxo alcohol	'n' no. ca. 20.0	HLB ca. 16.8; solid; cloud point 77°C (1% in 10% NaCl)	
	IMBENTIN-L/123/020	ethoxylated $C_{12/13}$ branched oxo alcohol	'n' no. ca. 2.0	HLB ca. 6.5; liquid; cloud point 25°C (5 g in 25 g BDG 25%)	
	IMBENTIN-L/123/030	ethoxylated $C_{12/13}$ branched oxo alcohol	'n' no. ca. 3.0	HLB ca. 8.1; liquid; cloud point 54°C (5 g in 25 g BDG 25%)	
	IMBENTIN-L/125/030	ethoxylated $C_{12/15}$ branched oxo alcohol	'n' no. ca. 3.0	HLB ca. 7.8; liquid; cloud point 43°C (5 g in 25 g BDG 25%)	
	IMBENTIN-L/125/050	ethoxylated $C_{12/15}$ branched oxo alcohol	'n' no. ca. 5.0	HLB ca. 10.4; liquid; cloud point 70°C (5 g in 25 g BDG 25%)	
	IMBENTIN-L/125/080	ethoxylated $C_{12/15}$ branched oxo alcohol	'n' no. ca. 8.0	HLB ca. 12.6; liquid; cloud point 78°C (5 g in 25 g BDG 25%)	
	IMBENTIN-L/125/110	ethoxylated $C_{12/15}$ branched oxo alcohol	'n' no. ca. 11.0	HLB ca. 14.0; liquid; cloud point 82°C (1% in deionised water)	
	IMBENTIN-L/145/040	ethoxylated $C_{14/15}$ branched oxo alcohol	'n' no. ca. 4.0	HLB ca. 8.9; liquid; cloud point 51°C (5 g in 25 g BDG 25%)	

IMBENTIN POA 024
Cetyl/Oleyl +2.4

Supplier	Trade name	Chemical description	Composition	General properties	Functionality / Application
Dr. W. Kolb	IMBENTIN-L/145/050	ethoxylated $C_{14/15}$ branched oxo alcohol	'n' no. ca. 5.0	HLB ca. 10.0; liquid; cloud point 60°C (5 g in 25 g BDG 25%)	
	IMBENTIN-L/145/080	ethoxylated $C_{14/15}$ branched oxo alcohol	'n' no. ca. 8.0	HLB ca. 12.3; paste; cloud point 78°C (5 g in 25 g BDG 25%)	
	IMBENTIN-L/145/130	ethoxylated $C_{14/15}$ branched oxo alcohol	'n' no. ca. 13.0	HLB ca. 14.4; solid; cloud point 59°C (1% in 10% NaCl)	
	IMBENTIN-L/145/180	ethoxylated $C_{14/15}$ branched oxo alcohol	'n' no. ca. 18.0	HLB ca. 15.6; solid; cloud point 71°C (1% in 10% NaCl)	
	IMBENTIN-OA/020	ethoxylated 2-ethylhexanol; C_8 carbon chain	'n' no. ca. 2.0	HLB ca. 7.3; liquid; cloud point 50°C (5% in BDG 25%)	
	IMBENTIN-OA/030	ethoxylated 2-ethylhexanol; C_8 carbon chain	'n' no. ca. 3.0	HLB ca. 10.0; liquid; cloud point 42°C (5 g in 25 g BDG 25%)	
	IMBENTIN-OA/050	ethoxylated 2-ethylhexanol; C_8 carbon chain	'n' no. ca. 5.0	HLB ca. 12.5; liquid; cloud point 45°C (5 g in 25 g BDG 25%)	
	IMBENTIN-OA/080	ethoxylated 2-ethylhexanol; C_8 carbon chain	'n' no. ca. 8.0	HLB ca. 14.6; liquid; cloud point 60°C (5 g in 25 g BDG 25%)	
	IMBENTIN-OA/100	ethoxylated 2-ethylhexanol; C_8 carbon chain	'n' no. ca. 10.0	HLB ca. 15.4; liquid; cloud point 67°C (5 g in 25 g BDG 25%)	
	IMBENTIN-OA/110	ethoxylated 2-ethylhexanol; C_8 carbon chain	'n' no. ca. 11.0	HLB ca. 17.0; liquid/paste; cloud point 85°C (5 g in 25 g BDG 25%)	*personal care products; household detergents; industrial and institutional cleaning; pharmaceuticals*
	IMBENTIN-POA/020	ethoxylated cetyl-oleyl alcohol; $C_{16/18}$ carbon chain	'n' no. ca. 2.0	HLB ca. 5.1; liquid; cloud point 32°C (5 g in 25 g BDG 25%)	
	IMBENTIN-POA/024	ethoxylated cetyl-oleyl alcohol; $C_{16/18}$ carbon chain	'n' no. ca. 2.4	HLB ca. 5.9; liquid; cloud point 39°C (5 g in 25 g BDG 25%)	
	IMBENTIN-POA/050	ethoxylated cetyl-oleyl alcohol; $C_{16/18}$ carbon chain	'n' no. ca. 5.0	HLB ca. 9.3; liquid; cloud point 67°C (5 g in 25 g BDG 25%)	
	IMBENTIN-POA/060	ethoxylated cetyl-oleyl alcohol; $C_{16/18}$ carbon chain	'n' no. ca. 6.0	HLB ca. 10.2; liquid; cloud point 73°C (5 g in 25 g BDG 25%)	
	IMBENTIN-POA/080	ethoxylated cetyl-oleyl alcohol; $C_{16/18}$ carbon chain	'n' no. ca. 8.0	HLB ca. 11.6; liquid; cloud point 83°C (5 g in 25 g BDG 25%)	
	IMBENTIN-POA/100	ethoxylated cetyl-oleyl alcohol; $C_{16/18}$ carbon chain	'n' no. ca. 10.0	HLB ca. 12.7; paste/solid; cloud point 46°C (1% in 10% NaCl)	
	IMBENTIN-POA/140	ethoxylated cetyl-oleyl alcohol; $C_{16/18}$ carbon chain	'n' no. ca. 14.0	HLB ca. 14.2; solid; cloud point 66°C (1% in 10% NaCl)	
	IMBENTIN-POA/180	ethoxylated cetyl-oleyl alcohol; $C_{16/18}$ carbon chain	'n' no. ca. 18.0	HLB ca. 15.1; solid; cloud point 73°C (1% in 10% NaCl)	
	IMBENTIN-POA/220	ethoxylated cetyl-oleyl alcohol; $C_{16/18}$ carbon chain	'n' no. ca. 22.0	HLB ca. 15.8; solid/flakes; cloud point 75°C (1% in 10% NaCl)	
	IMBENTIN-POA/310	ethoxylated cetyl-oleyl alcohol; $C_{16/18}$ carbon chain	'n' no. ca. 31.0	HLB ca. 16.9; solid/flakes	
	IMBENTIN-POA/450	ethoxylated cetyl-oleyl alcohol; $C_{16/18}$ carbon chain	'n' no. ca. 45.0	HLB ca. 17.7; solid/flakes	
	IMBENTIN-POA/800	ethoxylated cetyl-oleyl alcohol; $C_{16/18}$ carbon chain	'n' no. ca. 80.0	HLB ca. 18.7; solid/flakes	

Supplier	Trade name	Chemical description	Composition	General properties	Functionality Application
Dr. W. Kolb	IMBENTIN-T/020	ethoxylated isotridecyl alcohol; C_{13} carbon chain	'n' no. ca. 2.0	HLB ca. 6.1: liquid; cloud point 55°C (5% in BDG 25%)	
	IMBENTIN-T/030	ethoxylated isotridecyl alcohol; C_{13} carbon chain	'n' no. ca. 3.0	HLB ca. 8.0: liquid; cloud point 36°C (5 g in 25 g BDG 25%)	
	IMBENTIN-T/035	ethoxylated isotridecyl alcohol; C_{13} carbon chain	'n' no. ca. 3.5	HLB ca. 8.7: liquid; cloud point 41°C (5 g in 25 g BDG 25%)	
	IMBENTIN-T/040	ethoxylated isotridecyl alcohol; C_{13} carbon chain	'n' no. ca. 4.0	HLB ca. 9.4: liquid; cloud point 48°C (5 g in 25 g BDG 25%)	
	IMBENTIN-T/050	ethoxylated isotridecyl alcohol; C_{13} carbon chain	'n' no. ca. 5.0	HLB ca. 10.5: liquid; cloud point 58°C (5 g in 25 g BDG 25%)	
	IMBENTIN-T/060	ethoxylated isotridecyl alcohol; C_{13} carbon chain	'n' no. ca. 6.0	HLB ca. 11.4: liquid; cloud point 66°C (5 g in 25 g BDG 25%)	*personal care products; household detergents; industrial and institutional cleaning; pharmaceuticals*
	IMBENTIN-T/070	ethoxylated isotridecyl alcohol; C_{13} carbon chain	'n' no. ca. 7.0	HLB ca. 12.1: liquid; cloud point 72°C (5 g in 25 g BDG 25%)	
	IMBENTIN-T/080	ethoxylated isotridecyl alcohol; C_{13} carbon chain	'n' no. ca. 8.0	HLB ca. 12.8: liquid; cloud point 76°C (5 g in 25 g BDG 25%)	
	IMBENTIN-T/090	ethoxylated isotridecyl alcohol; C_{13} carbon chain	'n' no. ca. 9.0	HLB ca. 13.3: liquid; cloud point 80°C (5 g in 25 g BDG 25%)	
	IMBENTIN-T/100	ethoxylated isotridecyl alcohol; C_{13} carbon chain	'n' no. ca. 10.0	HLB ca. 13.8: liquid/paste; cloud point 77°C (1% in deionised water)	
	IMBENTIN-T/120	ethoxylated isotridecyl alcohol; C_{13} carbon chain	'n' no. ca. 12.0	HLB ca. 14.5: paste; cloud point 86°C (1% in deionised water)	
	IMBENTIN-T/150	ethoxylated isotridecyl alcohol; C_{13} carbon chain	'n' no. ca. 15.0	HLB ca. 15.4: solid; cloud point 68°C (1% in 10% NaCl)	
	IMBENTIN-T/250	ethoxylated isotridecyl alcohol; C_{13} carbon chain	'n' no. ca. 25.0	HLB ca. 16.9: solid; cloud point 76°C (1% in 10% NaCl)	
	IMBENTIN-T/400	ethoxylated isotridecyl alcohol; C_{13} carbon chain	'n' no. ca. 40.0	HLB ca. 18.0: solid/flakes; cloud point 74°C (2% in 10% NaCl)	
	IMBENTIN-T/900	ethoxylated isotridecyl alcohol; C_{13} carbon chain	'n' no. ca. 90.0	HLB ca. 19.0: solid/flakes	
	SYMPATENS-AC/020	ceteth-2	OH value 170; 'n' no. 2	HLB 5.3; white solid; viscosity 10 mPas (50°C); pH 5-7 (1% aq.)	solubiliser *essential oils; perfumes*
	SYMPATENS-AC/100	ceteth-10	OH value 80; 'n' no. 10	HLB 12.9; white solid; viscosity 30 mPas; pH 5-7 (1% aq.)	solubiliser; o/w co-emulsifier *essential oils; perfumes*
	SYMPATENS-AC/200	ceteth-20	OH value 50; 'n' no. 20	HLB 15.7: white solid; viscosity 75 mPas (50°C); pH 5-7 (1% aq.)	solubiliser; o/w co-emulsifier *essential oils; perfumes*
	SYMPATENS-ACS/060	ceteareth-6	OH value 110; 'n' no. 6	HLB 10.0: white solid; viscosity 25 mPas (50°C); pH 5-7 (1% aq.)	o/w emulsifier; stabiliser; viscosity modifier
	SYMPATENS-ACS/080	ceteareth-8	OH value 90; 'n' no. 8	HLB 11.5: white solid; viscosity 25 mPas (50°C); pH 5-7 (1% aq.)	o/w emulsifier; stabiliser; viscosity modifier *cosmetics*
	SYMPATENS-ACS/100	ceteareth-10	OH value 80; 'n' no. 10	HLB 12.5: white solid; viscosity 35 mPas (50°C); pH 5-7 (1% aq.)	o/w emulsifier; stabiliser; viscosity modifier *cosmetics*
	SYMPATENS-ACS/120	ceteareth-12	OH value 70; 'n' no. 12	HLB 13.4: white solid; viscosity 40 mPas (50°C); pH 5-7 (1% aq.)	o/w emulsifier; stabiliser *cosmetics*
	SYMPATENS-ACS/200	ceteareth-20	OH value 50; 'n' no. 20	HLB 15.4: white solid; viscosity 60 mPas (50°C); pH 5-7 (1% aq.)	o/w emulsifier; stabiliser *cosmetics*

Supplier	Trade name	Chemical description	Composition	General properties	Functionality Application
Dr. W. Kolb	SYMPATENS-ACS/250	ceteareth-25	OH value 40; 'n' no. 25	HLB 16.1; white solid; viscosity 75 mPas (50°C); pH 5-7 (1% aq.)	o/w emulsifier; stabiliser; *cosmetics*
	SYMPATENS-AIC/200	isoceteth-20	OH value 50; 'n' no. 20	HLB 15.7; white solid; viscosity 65 mPas (50°C); pH 5-7 (1% aq.)	solubiliser; foam stabiliser; *perfumes*
	SYMPATENS-ALM/020	laureth-2	OH value 195; 'n' no. 2	HLB 6.1; cloudy paste; viscosity 20 mPas; pH 5-7 (1% aq.)	w/o co-emulsifier; solubiliser; *cosmetics; bath preparations; essential oils*
	SYMPATENS-ALM/030	laureth-3	OH value 170; 'n' no. 3	HLB 8.0; cloudy paste; viscosity 20 mPas; pH 5-7 (1% aq.)	w/o co-emulsifier; solubiliser; dispersant; *cosmetics; bath preparations; essential oils*
	SYMPATENS-ALM/040	laureth-4	OH value 150; 'n' no. 4	HLB 9.4; clear liquid/paste; viscosity 30 mPas; pH 5-7 (1% aq.)	o/w co-emulsifier; solubiliser; *cosmetics; bath preparations; essential oils*
	SYMPATENS-ALM/060	laureth-6	OH value 120; 'n' no. 6	HLB 11.4; clear liquid/paste; viscosity 40 mPas; pH 5-7 (1% aq.)	o/w co-emulsifier; solubiliser; *cosmetics; bath preparations; essential oils*
	SYMPATENS-ALM/080	laureth-8	OH value 100; 'n' no. 8	HLB 12.7; clear liquid; viscosity 20 mPas (50°C); pH 5-7 (1% aq.)	o/w co-emulsifier; solubiliser; *cosmetics; essential oils*
	SYMPATENS-ALM/090	laureth-9	OH value 95; 'n' no. 9	HLB 13.6; colourless liquid; viscosity 50 mPas; pH 5-7 (1% aq.)	anaesthetising agent; *pharmaceuticals*
	SYMPATENS-ALM/230	laureth-23	OH value 45; 'n' no. 23	HLB 16.7; white solid; viscosity 55 mPas (50°C); pH 5-7 (1% aq.)	o/w co-emulsifier; solubiliser; wetting agent; *cosmetics; essential oils; hair care products*
	SYMPATENS-AO/100	oleth-10	OH value 80; 'n' no. 10	HLB 12.5; white-yellow solid; viscosity 35 mPas (50°C); pH 5-7 (1% aq.)	
	SYMPATENS-AO/200	oleth-20	OH value 50; 'n' no. 20	HLB 15.3; white-yellow solid; viscosity 70 mPas (50°C); pH 5-7 (1% aq.)	
	SYMPATENS-AOC/020	oleth-2	OH value 155; 'n' no. 2	HLB 4.9; white-yellow liquid; viscosity 35 mPas; pH 5-7 (1% aq.)	o/w emulsifier; *cosmetics*
	SYMPATENS-AOC/050	oleth-5	OH value 115; 'n' no. 5	HLB 9.0; white-yellow liquid; viscosity 25 mPas; pH 5-7 (1% aq.)	
	SYMPATENS-AOC/100	oleth-10	OH value 80; 'n' no. 10	HLB 12.4; white-yellow solid; viscosity 35 mPas (50°C); pH 5-7 (1% aq.)	
	SYMPATENS-AS/020	steareth-2	OH value 155; 'n' no. 2	HLB 4.9; white solid; viscosity 15 mPas (50°C); pH 5-7 (1% aq.)	
	SYMPATENS-AS/100	steareth-10	OH value 80; 'n' no. 10	HLB 12.4; white solid; viscosity 35 mPas (50°C); pH 5-7 (1% aq.)	
	SYMPATENS-AS/200	steareth-20	OH value 50; 'n' no. 20	HLB 15.3; white solid; viscosity 70 mPas (50°C); pH 5-7 (1% aq.)	o/w emulsifier; viscosity modifier; *cosmetics*
	SYMPATENS-AS/1000	steareth-100	OH value 10; 'n' no. 100	HLB 18.8; white-yellow solid; viscosity 520 mPas (60°C); pH 5-7 (1% aq.)	
Lonza	ETHOSPERSE CA-2	ceteth-2	100% active; OH value 170	HLB 6; solid; colour 1 (Gardner)	*hair care products*
	ETHOSPERSE CA-20	ceteth-20	100% active; OH value 55	HLB 16; soft solid; colour 1 (Gardner)	*hair care products*
	ETHOSPERSE G-26	glycereth-26	100% active; OH value 137	HLB 18; clear liquid; colour 1 (Gardner)	humectant; viscosity modifier
Nikko Chemicals	NIKKOL BB-5	beheneth-5		white solid	
	NIKKOL BB-10	beheneth-10		white solid	
	NIKKOL BB-20	beheneth-20		white solid	emulsifier; *cosmetics*
	NIKKOL BB-30	beheneth-30		white solid	
	NIKKOL BC-2	ceteth-2		colourless liquid	

Supplier	Trade name	Chemical description	Composition	General properties	Functionality Application
Nikko Chemicals	NIKKOL BC-5.5	ceteth-6		white solid	emulsifier / cosmetics
	NIKKOL BC-7	POE (7) cetyl ether		white solid	
	NIKKOL BC-10TX	ceteth-10		white solid	
	NIKKOL BC-15TX	POE (15) cetyl ether		white solid	
	NIKKOL BC-20TX	ceteth-20		white solid	
	NIKKOL BC-23	POE (23) cetyl ether		white solid	emulsifier; dispersant; solubiliser / cosmetics
	NIKKOL BC-25TX	ceteth-25		white solid	
	NIKKOL BC-30TX	ceteth-30		white solid	
	NIKKOL BC-40TX	POE (40) cetyl ether		white solid	
	NIKKOL BD-2	POE (2) synthetic alcohol		colourless liquid	emulsifier; dispersant / cosmetics
	NIKKOL BD-4	POE (4) synthetic alcohol		colourless liquid	
	NIKKOL BD-10	POE (10) synthetic alcohol		white solid	
	NIKKOL BL-2	laureth-2		colourless liquid	cosmetics
	NIKKOL BL-4.2	laureth-4		colourless liquid	cosmetics
	NIKKOL BL-9EX	laureth-9		colourless liquid	analgesic; antipruritic / cosmetics
	NIKKOL BL-21	POE (21) lauryl ether		white solid	solubiliser / cosmetics
	NIKKOL BL-25	laureth-25		white solid	
	NIKKOL BO-2	oleth-2		pale yellow liquid	super-fatting agent / hair care products
	NIKKOL BO-7	oleth-7		pale yellow liquid	
	NIKKOL BO-10TX	oleth-10		pale yellow liquid	emulsifier; dispersant / cosmetics
	NIKKOL BO-15TX	oleth-15		pale yellow liquid	
	NIKKOL BO-20	oleth-20		pale yellow solid	solubiliser / cosmetics
	NIKKOL BO-50	oleth-50		white flakes	
	NIKKOL BS-2	steareth-2		white solid	emulsifier / cosmetics
	NIKKOL BS-4	steareth-4		white solid	
	NIKKOL BS-20	steareth-20		white solid	emulsifier; dispersant / cosmetics
	NIKKOL BT-3	POE (3) secondary alkyl ether		colourless liquid	detergent; antistatic agent / cosmetics
	NIKKOL BT-5	POE (5) secondary alkyl ether		colourless liquid	
	NIKKOL BT-7	POE (7) secondary alkyl ether		colourless liquid	
	NIKKOL BT-9	POE (9) secondary alkyl ether		colourless liquid	
	NIKKOL BT-12	POE (12) secondary alkyl ether		colourless liquid	
Nippon Shokubai	SOFTANOL 30	$C_{12/14}$ secondary alcohols reacted with ethylene oxide	'n' no. 3	HLB 7.9; clear liquid; cloud point <0°C (1% aq. soln.); pour point −56°C; viscosity 25 cPs; surface tension 28 dynes/cm (0.1% aq. soln.)	foaming agent; detergent; wetting agent; scouring agent; emulsifier; dispersant; solubiliser / laundry products; dishwashing agents; hard-surface cleaning; textile industry; paper industry; paint industry; fertilisers; leather industry; metal working

Supplier	Trade name	Chemical description	Composition	General properties	Functionality Application
Nippon Shokubai	SOFTANOL 50	$C_{12/14}$ secondary alcohols reacted with ethylene oxide	'n' no. 5	HLB 10.5; clear liquid; cloud point < 0°C (1% aq. soln.); pour point −24°C; viscosity 33 cPs; surface tension 28 dynes/cm (0.1% aq. soln.)	foaming agent; detergent; wetting agent; scouring agent; emulsifier; dispersant; solubiliser
	SOFTANOL 70	$C_{12/14}$ secondary alcohols reacted with ethylene oxide	'n' no. 7	HLB 12.1; clear liquid; cloud point 33°C (1% aq. soln.); pour point −3°C; viscosity 52 cPs; surface tension 28 dynes/cm (0.1% aq. soln.)	laundry products; dishwashing agents; hard-surface cleaning; textile industry; paper industry; paint industry; fertilisers; leather industry; metal working
	SOFTANOL 90	$C_{12/14}$ secondary alcohols reacted with ethylene oxide	'n' no. 9	HLB 13.3; clear liquid; cloud point 56°C (1% aq. soln.); pour point 8°C; viscosity 65 cPs; surface tension 29 dynes/cm (0.1% aq. soln.)	
	SOFTANOL 120	$C_{12/14}$ secondary alcohols reacted with ethylene oxide	'n' no. 12	HLB 14.5; clear liquid; cloud point 83°C (1% aq. soln.); pour point 17°C; viscosity 87 cPs; surface tension 32 dynes/cm (0.1% aq. soln.)	
PPG	MACOL CSA 20	ceteareth 20		HLB 15.2; solid	emulsifier; solubiliser cosmetics; hair care products; skin care products
	MACOL LA-4	laureth 4		HLB 9.5; liquid; cloud point insoluble in water; viscosity 60 cPs	lubricant; w/o emulsifier; detergent; coupling agent; wetting agent; dispersant metal working; cosmetics; deodorants; antiperspirants; hair care products
	MACOL LA-12	laureth 12		HLB 14.6; solid; cloud point 95°C (1% aq.)	lubricant; o/w emulsifier; detergent; coupling agent; wetting agent; solubiliser metal working; cosmetics; hair care products; deodorants; personal care products
	MACOL LA-23	laureth 23		solid	solubiliser; emulsifier cosmetics; deodorants; hair care products; personal care products
	MACOL LA-790	laureth 7	90% active	HLB 10.8; liquid	wetting agent; dispersant; emulsifier bubble baths; shower gels
	MACOL SA 2	steareth 2		HLB 4.7; solid	emulsifier personal care products; cosmetics; skin care products; hair care products
	MACOL TD-3	tridecyl alcohol 3EO		HLB 8.0; liquid; viscosity 65 cPs	lubricant; emulsifier; detergent; coupling agent; wetting agent metal working
	MACOL TD-12	tridecyl alcohol 12EO		HLB 14.2; gel; cloud point 91°C (1% aq.); viscosity 540 cPs	
	MACOL TD-610	tridecyl alcohol 6EO		HLB 11.2; liquid; cloud point 41°C (1% aq.); viscosity 115 cPs	
Protex	PROX-ONIC 2EHA-1/02	POE (2) 2-ethylhexyl ether	'n' no. 2	HLB 8.0; liquid	detergent; wetting agent; emulsifier; dispersant; solubiliser; defoaming agent
	PROX-ONIC 2EHA-1/05	POE (5) 2-ethylhexyl ether	'n' no. 5	HLB 12.6; liquid; cloud point < 25°C (1% aq.)	textile industry; metal cleaning; household, industrial and institutional cleaners
	PROX-ONIC CSA-1/04	ceteareth 4	'n' no. 4	HLB 8.0; solid	
	PROX-ONIC CSA-1/06	ceteareth 6	'n' no. 6	HLB 10.1; solid	
	PROX-ONIC CSA-1/010	ceteareth 10	'n' no. 10	HLB 12.4; solid; cloud point 73°C (1% aq.)	
	PROX-ONIC CSA-1/015	ceteareth 15	'n' no. 15	HLB 14.3; paste; cloud point 95-100°C (1% aq.)	

Supplier	Trade name	Chemical description	Composition	General properties	Functionality Application
Protex	PROX-ONIC CSA-1/020	ceteareth 20	'n' no. 20	HLB 15.3; solid; cloud point > 100°C (1% aq.)	
	PROX-ONIC CSA-1/030	ceteareth 30	'n' no. 30	HLB 15.3; solid; cloud point > 100°C (1% aq.)	
	PROX-ONIC CSA-1/050	ceteareth 50	'n' no. 50	HLB 16.9; solid; cloud point < 100°C (1% aq.)	
	PROX-ONIC CSA-2/04	ceteareth 4	'n' no. 4	HLB 8.0; solid	
	PROX-ONIC CSA-2/06	ceteareth 6	'n' no. 6	HLB 10.1; solid	
	PROX-ONIC CSA-2/010	ceteareth 10	'n' no. 10	HLB 12.4; solid; cloud point 73°C (1% aq.)	detergent; wetting agent; emulsifier; dispersant; solubiliser; defoaming agent
	PROX-ONIC CSA-2/015	ceteareth 15	'n' no. 15	HLB 14.3; paste; cloud point 95-100°C (1% aq.)	*textile industry; metal cleaning; household, industrial and institutional cleaners*
	PROX-ONIC CSA-2/020	ceteareth 20	'n' no. 20	HLB 15.3; solid; cloud point > 100°C (1% aq.)	
	PROX-ONIC CSA-2/030	ceteareth 30	'n' no. 30	HLB 15.3; solid; cloud point > 100°C (1% aq.)	
	PROX-ONIC CSA-2/050	ceteareth 50	'n' no. 50	HLB 16.9; solid; cloud point < 100°C (1% aq.)	
	PROX-ONIC DA-1/04	deceth 4	'n' no. 4	HLB 10.5; liquid; cloud point < 25°C (1% aq.)	
	PROX-ONIC DA-1/06	deceth 6	'n' no. 6	HLB 12.4; liquid; cloud point 41°C (1% aq.)	
	PROX-ONIC DA-1/09	deceth 9	'n' no. 9	HLB 14.3; liquid; cloud point 80°C (1% aq.)	
	PROX-ONIC DA-2/04	deceth 4	'n' no. 4	HLB 10.5; liquid; cloud point < 25°C (1% aq.)	
	PROX-ONIC DA-2/06	deceth 6	'n' no. 6	HLB 12.4; liquid; cloud point 41°C (1% aq.)	
	PROX-ONIC DA-2/09	deceth 9	'n' no. 9	HLB 14.3; liquid; cloud point 80°C (1% aq.)	
	PROX-ONIC L 081-05	POE (5) linear alcohol ether	'n' no. 5		low foam; detergent; wetting agent; defoaming agent
	PROX-ONIC L 101-05	POE (5) linear alcohol ether	'n' no. 5		*dishwashing agents; metal cleaning; detergents*
	PROX-ONIC L 102-02	POE (2) linear alcohol ether	'n' no. 2		
	PROX-ONIC L 121-09	POE (9) linear alcohol ether	'n' no. 9		
	PROX-ONIC L 161-05	POE (5) linear alcohol ether	'n' no. 5		
	PROX-ONIC L 181-05	POE (5) linear alcohol ether	'n' no. 5		
	PROX-ONIC L 201-02	POE (2.5) linear alcohol ether	'n' no. 2.5		

Supplier	Trade name	Chemical description	Composition	General properties	Functionality/Application
Protex	PROX-ONIC LA-1/02	laureth 2	'n' no. 2	HLB 6.4; liquid	detergent; wetting agent; emulsifier; dispersant; solubiliser; defoaming agent; *textile industry; metal cleaning; household, industrial and institutional cleaners*
	PROX-ONIC LA-1/04	laureth 4	'n' no. 4	HLB 9.2; liquid; cloud point 52°C (1% aq.)	
	PROX-ONIC LA-1/09	laureth 9	'n' no. 9	HLB 13.3; liquid; cloud point 73-76°C (1% aq.)	
	PROX-ONIC LA-1/012	laureth 12	'n' no. 12	HLB 14.5; solid; cloud point < 100°C (1% aq.)	
	PROX-ONIC LA-1/023	laureth 23	'n' no. 23	HLB 16.7; solid	
	PROX-ONIC LA-2/02	laureth 2	'n' no. 2	HLB 6.4; liquid	
	PROX-ONIC LA-2/04	laureth 4	'n' no. 4	HLB 9.2; liquid; cloud point 52°C (1% aq.)	
	PROX-ONIC LA-2/09	laureth 9	'n' no. 9	HLB 13.3; liquid; cloud point 73-76°C (1% aq.)	
	PROX-ONIC LA-2/012	laureth 12	'n' no. 12	HLB 14.5; solid; cloud point < 100°C (1% aq.)	
	PROX-ONIC LA-2/023	laureth 23	'n' no. 23	HLB 16.7; solid	
	PROX-ONIC OA-1/04	oleth 4	'n' no. 4	HLB 7.9; liquid	
	PROX-ONIC OA-1/09	oleth 9	'n' no. 9	HLB 11.9; liquid; cloud point 52°C (1% aq.)	
	PROX-ONIC OA-1/020	oleth 20	'n' no. 20	HLB 15.3; solid; cloud point > 100°C (1% aq.)	
	PROX-ONIC OA-2/04	oleth 4	'n' no. 4	HLB 7.9; liquid	
	PROX-ONIC OA-2/09	oleth 9	'n' no. 9	HLB 11.9; liquid; cloud point 52°C (1% aq.)	
	PROX-ONIC OA-2/020	oleth 20	'n' no. 20	HLB 15.3; cloud point > 100°C (1% aq.)	
	PROX-ONIC OCA-1/06	cetoeth 6	'n' no. 6	HLB 10.7; paste	
	PROX-ONIC OCA-1/030	cetoeth 30	'n' no. 30	liquid; cloud point > 100°C (1% aq.)	
	PROX-ONIC OCA-2/06	cetoeth 6	'n' no. 6	HLB 10.7; paste	
	PROX-ONIC OCA-2/030	cetoeth 30	'n' no. 30	liquid; cloud point > 100°C (1% aq.)	
	PROX-ONIC SA-1/02	steareth 2	'n' no. 2	HLB 4.9; solid; cloud point 57-61°C (1% aq.)	
	PROX-ONIC SA-1/010	steareth 10	'n' no. 10	HLB 12.4; solid; cloud point 60-64°C (1% aq.)	
	PROX-ONIC SA-1/020	steareth 20	'n' no. 20	HLB 15.3; solid; cloud point 73-77°C (1% aq.)	

Supplier	Trade name	Chemical description	Composition	General properties	Functionality Application
Protex	PROX-ONIC SA-2/02	stearth 2	'n' no. 2	HLB 4.9; solid; cloud point 57-61°C (1% aq.)	detergent; wetting agent; emulsifier; dispersant; solubiliser; defoaming agent / textile industry; metal cleaning; household, industrial and institutional cleaners
	PROX-ONIC SA-2/010	stearth 10	'n' no. 10	HLB 12.4; solid; cloud point 60-64°C (1% aq.)	
	PROX-ONIC SA-2/020	stearth 20	'n' no. 20	HLB 15.3; solid; cloud point 73-77°C (1% aq.)	
	PROX-ONIC TD-1/03	trideceth 3	'n' no. 3	HLB 7.9; liquid	
	PROX-ONIC TD-1/06	trideceth 6	'n' no. 6	HLB 11.4; liquid	
	PROX-ONIC TD-1/09	trideceth 9	'n' no. 9	HLB 13.0; liquid; cloud point 54°C (1% aq.)	
	PROX-ONIC TD-1/012	trideceth 12	'n' no. 12	HLB 14.5; paste; cloud point 70°C (1% aq.)	
	PROX-ONIC UA-03	isoundeceth 3	'n' no. 3	liquid; cloud point < 100°C (1% aq.)	
	PROX-ONIC UA-06	isoundeceth 6	'n' no. 6	liquid	
	PROX-ONIC UA-09	isoundeceth 9	'n' no. 9	liquid	
	PROX-ONIC UA-012	isoundeceth 12	'n' no. 12	liquid	
Raschig	SURFAROX 10AX 0645	6 mole ethoxylated fatty alcohol			emulsifier
	SURFAROX 8A06	polyoxyethylene fatty alcohol			low foam
	RALUFON EN 16-80	ethyl hexanol ethoxylate			
Rhone-Poulenc	RHODASURF ID SERIE	ethoxylated isodecyl alcohols		liquid	foaming agents; emulsifiers; wetting agents; dispersants
	RHODASURF LA SERIE	ethoxylated lauryl alcohols		liquid	foaming agents; emulsifiers; wetting agents; dispersants
	RHODASURF ON 870	oleyl alcohol ethoxylate		HLB 15.4; wax	foaming agent (high); emulsifier; dispersant
Sandoz	LYOGEN DFT	alkyl polyglycol ether		liquid	levelling agent; stripping agent / textile industry; dyeing
	SANDOCLEAN PC-LF	alkyl polyglycol ether in aq. glycol soln.		liquid	detergent; low foam / industrial cleaning; textile industry
	SANDOCLEAN PC	alkyl polyglycol ether in aq. glycol soln.		liquid	detergent / industrial cleaning; textile industry
	SANDOPAN LFW LIQUID	alkylpolyglycol ether		clear, colourless liquid	wetting agent; detergent; non-foaming / detergents; cleaners
	SANDOPAN LFW	alkyl polyglycol ether		liquid	wetting agent; low foam / textile industry; industrial applications
	SANDOXYLATE A20	ethoxylated fatty alcohol		liquid	wetting agent / industrial detergents
	SANDOXYLATE AC-20	ethoxylated fatty alcohol		solid	
	SANDOXYLATE AXT-6	alkyl polyglycol ether		liquid	

Supplier	Trade name	Chemical description	Composition	General properties	Functionality Application
Sandoz	SANDOZIN BFN	ethoxylated fatty alcohol derivative		liquid	emollient
	VELSAN P8-16	fatty alcohol ethoxylate		paste	*cosmetics; toiletries*
Seppic	CIRE DE LANOL CTO	cetearyl alcohol and ceteareth-33	100% active	flakes	self emulsifying base *personal care products*
	SIMULSOL 52	cetyl alcohol 2EO	100% active; 'n' no. 2	HLB 5.3; wax	emulsifier; wetting agent
	SIMULSOL 56	cetyl alcohol 10EO	100% active; 'n' no. 10	HLB 12.9; wax	emulsifier; wetting agent
	SIMULSOL 58	ceteth-20	100% active; 'n' no. 20	HLB 15.7; flakes	emulsifier; wetting agent; solubiliser *personal care products*
	SIMULSOL 68	cetyl-stearyl alcohol 20EO	100% active; 'n' no. 20	HLB 15.5; flakes	⎱ emulsifier; wetting agent
	SIMULSOL 72	stearyl alcohol 2EO	100% active; 'n' no. 2	HLB 4.9; wax	
	SIMULSOL 76	stearyl alcohol 10EO	100% active; 'n' no. 10	HLB 12.4; wax	
	SIMULSOL 760	stearyl alcohol 75EO	100% active; 'n' no. 75	HLB 19.0; flakes	
	SIMULSOL 78	steareth-20	100% active; 'n' no. 20	HLB 15.3; wax	emulsifier; wetting agent *personal care products*
	SIMULSOL 92	oleyl alcohol 2EO	100% active; 'n' no. 2	HLB 4.9; liquid	⎱ emulsifier; wetting agent
	SIMULSOL 96	oleyl alcohol 10EO	100% active; 'n' no. 10	HLB 12.4; liquid	
	SIMULSOL 967 AP	oxo alcohol 16EO	100% active; 'n' no. 16	HLB 16.1; soft wax	
	SIMULSOL 98	oleth-20	100% active; 'n' no. 21	HLB 15.6; wax	emulsifier; wetting agent; solubiliser *personal care products*
	SIMULSOL CS	ceteareth-33	100% active; 'n' no. 33	HLB 16.2; flakes	emulsifier; wetting agent *personal care products*
	SIMULSOL P 2	lauryl alcohol 2EO	100% active; 'n' no. 2	HLB 6.6; liquid	emulsifier; wetting agent
	SIMULSOL P 4	laureth-4	100% active; 'n' no. 4	HLB 9.7; liquid	emulsifier; wetting agent *personal care products*
	SIMULSOL P 23	laureth-23	100% active; 'n' no. 23	HLB 16.9; paste	⎱ emulsifier; wetting agent
	SIMULSOL T 300	oxo alcohol 4.5EO	100% active; 'n' no. 4.5	HLB 11.1; viscous liquid	emulsifier; wetting agent
Servo Delden	SERDOX NAS 5	C_{11} oxoalcohol 5EO		liquid; cloud point 70-73°C (10% in 25% BDG)	
	SERDOX NAS 6.6/90	C_{11} oxoalcohol 6.6EO		liquid; cloud point 41-44°C (2% in water)	
	SERDOX NBS 5.5	$C_{9/11}$ oxoalcohol 5.5EO		liquid; cloud point 38-40°C (1% in water)	
	SERDOX NBS 6.6	$C_{9/11}$ oxoalcohol 6.6EO		liquid; cloud point 60-64°C (1% in water)	
	SERDOX NBS 6.6/90	$C_{9/11}$ oxoalcohol 6.6EO		liquid	
	SERDOX NOL 2	oleylalcohol 2EO		liquid	
	SERDOX NKL 6	tallow alcohol 6EO		solid; cloud point 74-77°C (10% in 25% BDG)	
	SERDOX NKL 25	tallow alcohol 25EO		solid; cloud point 76-78°C (2% in 10% NaCl soln.)	
	SERDOX NKZ 650	blend of tallow fatty alcohol 6 & 50EO		solid	
	SERDOX NSL 30	stearylalcohol 30EO		solid; cloud point 77-81°C (1% in 10% NaCl soln.)	
	SERDOX NSL 50	stearylalcohol 50EO		solid; cloud point 76-78°C (1% in 10% NaCl soln.)	
	SERDOX NSL 100	stearylalcohol 100EO		solid; cloud point 76-78°C (1% in 10% NaCl soln.)	

Supplier	Trade name	Chemical description	Composition	General properties	Functionality Application
Shell Chemicals	DOBANOL 1-5	ethoxylated alcohol	100% active; acid no. 0.10; OH no. 143	HLB 11.2; cloud point 27°C; M.W. 392; pour point 6°C; kin viscosity 18 mm²/s (40°C); colour < 50 (Pt-Co); pH 6.8 (0.5% aq. soln.)	
	DOBANOL 1-9	ethoxylated alcohol	100% active; acid no. 0.10; OH no. 99	HLB 13.9; cloud point 84°C; M.W. 568; pour point 17°C; kin viscosity 3 mm²/s (40°C); colour < 50 (Pt-Co); pH 6.8 (0.5% aq. soln.)	
	DOBANOL 23-0.8	ethoxylated alcohol	100% active; acid no. 0.05-0.20; OH no. 241-251	HLB 3.1; M.W. 228; pour point 3°C; kin viscosity 13 mm²/s (40°C); colour 30 max. (Pt-Co); pH 6.8 (0.5% aq. soln.)	
	DOBANOL 23-2	ethoxylated alcohol	100% active; acid no. 0.05-0.20; OH no. 194-204	HLB 6.2; M.W. 282; pour point 4°C; kin viscosity 15 mm²/s; colour 50 max. (Pt-Co); pH 6.8 (0.5% aq. soln.)	
	DOBANOL 23-3	ethoxylated alcohol	100% active; acid no. 0.05-0.20; OH no. 168-176	HLB 8.1; M.W. 325; pour point 4°C; kin viscosity 16 mm²/s (40°C); colour 50 max. (Pt-Co); pH 6.8 (0.5% aq. soln.)	
	DOBANOL 23-6.5	ethoxylated alcohol	100% active; acid no. 0.05-0.20; OH no. 112-122	HLB 11.9; cloud point 38-44°C; M.W. 480; pour point 18°C; kin viscosity 26 mm²/s (40°C); colour 50 max. (Pt-Co); pH 6.8 (0.5% aq. soln.)	
	DOBANOL 25-2.5	ethoxylated alcohol	100% active; acid no. 0.05-0.20; OH no. 171-183	HLB 6.9; M.W. 317; pour point 6°C; kin viscosity 16 mm²/s (40°C); colour 50 max. (Pt-Co); pH 6.8 (0.5% aq. soln.)	
	DOBANOL 25-2.75	ethoxylated alcohol	100% active; acid no. 0.05-0.20; OH no. 165-177	HLB 7.3; M.W. 328; pour point 7°C; kin viscosity 17 mm²/s (40°C); colour 50 max. (Pt-Co); pH 6.8 (0.5% aq. soln.)	
	DOBANOL 25-3	ethoxylated alcohol	100% active; acid no. 0.05-0.20; OH no. 160-172	HLB 7.8; M.W. 339; pour point 5°C; kin viscosity 17 mm²/s (40°C); colour 50 max. (Pt-Co); pH 6.8 (0.5% aq. soln.)	washing-up liquids; laundry products; hard-surface cleaners; textile industry; industrial applications
	DOBANOL 25-7	ethoxylated alcohol	100% active; acid no. 0.05-0.20; OH no. 104-114	HLB 12.0; cloud point 46-52°C; M.W. 515; pour point 22°C; kin viscosity 32 mm²/s (40°C); colour 50 max. (Pt-Co); pH 6.8 (0.5% aq. soln.)	
	DOBANOL 25-9	ethoxylated alcohol	100% active (also available as a 90% aq. soln.); acid no. 0.05-0.20; OH no. 88-96	HLB 13.2; cloud point 74-82°C; M.W. 609; pour point 25°C; kin viscosity 41 mm²/s (40°C); colour 50 max. (Pt-Co); pH 6.8 (0.5% aq. soln.)	
	DOBANOL 45-4	ethoxylated alcohol	100% active; acid no. 0.05-0.20; OH no. 136-146	HLB 8.9; M.W. 397; pour point 13°C; kin viscosity 25 mm²/s (40°C); colour 50 max. (Pt-Co); pH 6.8 (0.5% aq. soln.)	
	DOBANOL 45-7	ethoxylated alcohol	100% active; acid no. 0.05-0.20; OH no. 98-106	HLB 11.6; cloud point 43-49°C; M.W. 529; pour point 22°C; kin viscosity 34 mm²/s (40°C); colour 50 max. (Pt-Co); pH 6.8 (0.5% aq. soln.)	
	DOBANOL 45-11	ethoxylated alcohol	100% active; acid no. 0.10; OH no. 80	HLB 13.7; cloud point 86°C; M.W. 702; pour point 30°C; kin viscosity 47 mm²/s (40°C); colour < 50 (Pt-Co); pH 6.8 (0.5% aq. soln.)	
	DOBANOL 91-2.5	ethoxylated alcohol	100% active; acid no. 0.05-0.20; OH no. 202-214	HLB 8.1; M.W. 270; pour point − 20°C; kin viscosity 11 mm²/s (40°C); colour 50 max. (Pt-Co); pH 6.8 (0.5% aq. soln.)	
	DOBANOL 91-5	ethoxylated alcohol	100% active; acid no. 0.05-0.20; OH no. 140-150	HLB 11.6; cloud point 33-39°C; M.W. 380; pour point 4°C; kin viscosity 18 mm²/s (40°C); colour 50 max. (Pt-Co); pH 6.8 (0.5% aq. soln.)	

Supplier	Trade name	Chemical description	Composition	General properties	Functionality / Application
Shell Chemicals	DOBANOL 91-6	ethoxylated alcohol	100% active; acid no. 0.05-0.20; OH no. 127-139	HLB 12.5; cloud point 51-57°C; M.W. 424; pour point 6°C; kin viscosity 21 mm²/s (40°C); colour 50 max. (Pt-Co); pH 6.8 (0.5% aq. soln.)	*washing-up liquids; laundry products; hard-surface cleaners; textile industry; industrial applications*
	DOBANOL 91-8	ethoxylated alcohol	100% active; acid no. 0.05-0.20; OH no. 105-115	HLB 13.7; cloud point 78-85°C; M.W. 512; pour point 16°C; kin viscosity 27 mm²/s (40°C); colour 50 max. (Pt-Co); pH 6.8 (0.5% aq. soln.)	
	DOBANOL 91-10	ethoxylated alcohol	100% active; acid no. 0.05-0.20; OH no. 89-99	HLB 14.7; M.W. 600; pour point 21°C; kin viscosity 27 mm²/s (40°C); colour 50 max. (Pt-Co); pH 6.8 (0.5% aq. soln.)	
	DOBANOL 91-12	ethoxylated alcohol	100% active; acid no. 0.05-0.20; OH no. 76.5-86.5	HLB 15.4; cloud point 27°C; M.W. 690; pour point 25°C; kin viscosity 39 mm²/s (40°C); colour 50 max. (Pt-Co); pH 6.8 (0.5% aq. soln.)	
	NONIDET LE	alcohol ethoxylate	100% active	clear or slightly turbid liquid; cloud point 54°C (1% aq.); pour point 7°C; kin viscosity 50 mm²/s; colour <50 (Pt-Co); pH 6.8 (0.5% aq. soln.)	*detergent; textile industry*
	NONIDET LG	alcohol ethoxylate	100% active	turbid liquid or slurry; cloud point 81°C (1% aq.); pour point 16°C; kin viscosity 28 mm²/s (40°C); colour <50 (Pt-Co); pH 6.8 (0.5% aq. soln.)	
	NONIDET SH30	alcohol ethoxylate	30% active	clear, very pale, straw coloured liquid; cloud point 79°C (1% aq.); pour point 0°C; kin viscosity 46 mm²/s; colour <70 (Pt-Co); pH 7.0 (0.5% aq. soln.)	
	OXITEX 60	alcohol ethoxylate		straw-coloured liquid; kin viscosity 32-37 mm²/s (40°C); colour 60 max. (10% in methanol; Pt-Co)	
	OXITEX 70	alcohol ethoxylate		straw-coloured liquid; kin viscosity 15-21 mm²/s (40°C); colour 60 max. (10% in methanol; Pt-Co)	
	OXITEX 80	alcohol ethoxylate		straw-coloured liquid; kin viscosity 110-124 mm²/s (40°C); colour 80 max. (Pt-Co)	
Stepan Europe	BIO-SOFT CS 50	fatty alcohol ethoxylate; made-to-order	100% active; 'n' no. 50	white solid or flakes	emulsifier; dispersant; detergent; levelling agent *laundry products; textile industry*
Surfachem	SURFAC BL6	cetyl oleyl alcohol 6EO	'n' no. 6		
	SURFAC BL10	cetyl oleyl alcohol 10EO	'n' no. 10		
	SURFAC GM590	synthetic fatty alcohol 5EO (90%)	'n' no. 5		
	SURFAC GM880	synthetic fatty alcohol 8EO (80%)	'n' no. 8		
	SURFAC JH60	cetyl stearyl alcohol 6EO	'n' no. 6		
	SURFAC JH200	cetyl stearyl alcohol 20EO	'n' no. 20		
	SURFAC JH500	cetyl stearyl alcohol 50EO	'n' no. 50		
	SURFAC LM30	detergent alcohol 3EO	'n' no. 3		
	SURFAC LM70/90	detergent alcohol 7EO	90% active; 'n' no. 7		
	SURFAC MB2	lauryl alcohol 2EO	'n' no. 2		
	SURFAC MB3	lauryl alcohol 3EO	'n' no. 3		
	SURFAC MB12	lauryl alcohol 12EO	'n' no. 12		

Supplier	Trade name	Chemical description	Composition	General properties	Functionality / *Application*
Surfachem	SURFAC MK70	C$_{10}$ alcohol ethoxylate 7EO	'n' no. 7		
	SURFAC UN65/95	alcohol ethoxylate 6.5EO (95%)	'n' no. 6.5		
	SURFAC UN65	alcohol ethoxylate 6.5EO	'n' no. 6.5		
	SURFAC UN70	alcohol ethoxylate 7EO	'n' no. 7		
	SURFAC UN90	alcohol ethoxylate 9EO	'n' no. 9		
	SURFAC UN120	alcohol ethoxylate 12EO	'n' no. 12		
Thor Chemicals	SOVATEX NI/100B (B)	alcohol ethoxylate	97% active	liquid	detergent; wetting agent
	SOVATEX NI/B (B)	alcohol ethoxylate	25% active	liquid	detergent; wetting agent
Union Carbide	TERGITOL 15-S-3	mixture of C$_{11/15}$ secondary alcohols reacted with ethylene oxide	100% active; OH no. 167; 'n' no. 3.2	HLB 8.3; clear liquid; cloud point 17°C (1% in 1:1 isopropanol/H$_2$O); M.W. 336; pour point −46°C; viscosity 26 cP; colour 30 (Pt-Co; APHA); pH 4.0 (1% in 10:6 isopropanol/H$_2$O)	emulsifier; wetting agent; coupling agent; antistatic agent *textile industry; laundry products; agriculture; plastics industry; paper industry; oil industry; water treatment; metal working; dry cleaning*
	TERGITOL 15-S-5	mixture of C$_{11/15}$ secondary alcohols reacted with ethylene oxide	100% active; OH no. 135; 'n' no. 5.0	HLB 10.6; clear liquid; cloud point 100°C (1% in 1:1 isopropanol/H$_2$O); M.W. 415; pour point −24°C; viscosity 35 cP; colour 45 (Pt-Co; APHA); pH 7.4 (1% in 10:6 isopropanol/H$_2$O)	
	TERGITOL 15-S-7	mixture of C$_{11/15}$ secondary alcohols reacted with ethylene oxide	100% active; OH no. 109; 'n' no. 7.3	HLB 12.4; clear liquid; cloud point 37°C (1% aq. soln.); M.W. 515; pour point 2°C; viscosity 51 cP; colour 40 (Pt-Co; APHA); surface tension 28 dynes/cm (0.1% aq. soln.); pH 6.8 (1% aq. soln.)	detergent; wetting agent; coupling agent; dispersant; levelling agent; emulsifier; stabiliser *laundry products; detergents; degreasers; household and industrial cleaners; textile industry; agriculture; paint industry; metal cleaning; leather industry; oil industry; water treatment*
	TERGITOL 15-S-9	mixture of C$_{11/15}$ secondary alcohols reacted with ethylene oxide	100% active; OH no. 96; 'n' no. 8.9	HLB 13.3; clear liquid; cloud point 60°C (1% aq. soln.); M.W. 584; pour point 9°C; viscosity 60 cP; colour 45 (Pt-Co; APHA); surface tension 30 dynes/cm (0.1% aq. soln.); pH 7.1 (1% aqueous soln.)	
	TERGITOL 15-S-12	mixture of C$_{11/15}$ secondary alcohols reacted with ethylene oxide	100% active; OH no. 76; 'n' no. 12.3	HLB 14.7; clear liquid; cloud point 88°C (1% aq. soln.); M.W. 738; pour point 20°C; viscosity 85 cP; surface tension 31 dynes/cm (0.1% aq. soln.); pH 6.2 (1% aq. soln.)	o/w emulsifier *textile industry; industrial cleaners and degreasers; bottle washing; oil industry latexes; paper industry*
	TERGITOL 15-S-15	mixture of C$_{11/15}$ secondary alcohols reacted with ethylene oxide	100% active; OH no. 64; 'n' no. 15.5	HLB 15.6; slightly hazy liquid at 50°C, white waxy solid at room temperature; cloud point 100°C (1% aq. soln.); M.W. 877; pour point 28°C; viscosity 43 cP (50°C); colour 40 (Pt-Co; APHA); surface tension 34 dynes/cm (0.1% aq. soln.); pH 6.4 (1% aq. soln.)	emulsifier *textile industry; industrial cleaners and degreasers; bottle washing; oil industry latexes; paper industry*
	TERGITOL 15-S-20	mixture of C$_{11/15}$ secondary alcohols reacted with ethylene oxide	100% active; OH no. 52; 'n' no. 20.1	HLB 16.4; slightly hazy liquid at 50°C, white waxy solid at room temperature; cloud point >100°C (1% aq. soln.); M.W. 1079; pour point 32°C; viscosity 49 cP (50°C); colour 35 (Pt-Co; APHA); surface tension 35 dynes/cm (0.1% aq. soln.); pH 6.6 (1% aq. soln.)	emulsifier *textile industry; industrial cleaners and degreasers; bottle washing; oil industry; latexes; solid toilet cleaners; laundry products*
	TERGITOL 15-S-30	mixture of C$_{11/15}$ secondary alcohols reacted with ethylene oxide	100% active; OH no. 36; 'n' no. 31.0	HLB 17.5; hazy liquid at 50°C, white waxy solid at room temperature; cloud point >100°C (1% aq. soln.); M.W. 1558; pour point 39°C; viscosity 92 cP (50°C); colour 35 (Pt-Co; APHA); surface tension 39 dynes/cm (0.1% aq. soln.); pH 6.5 (1% aq. soln.)	emulsifier *textile industry; industrial cleaners and degreasers; bottle washing; latexes; solid toilet cleaners; laundry products*

Supplier	Trade name	Chemical description	Composition	General properties	Functionality / Application
Union Carbide	TERGITOL 15-S-40	mixture of C$_{11/15}$ secondary alcohols reacted with ethylene oxide	100% active; OH no. 28; 'n' no. 41.1	HLB 18.0; hazy liquid at 50°C, white waxy solid at room temperature; cloud point 100°C (1% aq. soln.); M.W. 2004; pour point 44°C; viscosity 166 cP (50°C); colour 50 (Pt-Co; APHA); surface tension 42 dynes/cm (0.1% aq. soln.); pH 7.0 (1% aq. soln.)	emulsifier / textile industry; industrial cleaners and degreasers; bottle washing; latexes; solid toilet cleaners; laundry products
	TERGITOL TMN-10 (90% AQ.)	2,6,8-trimethyl-4-nonanol reacted with ethylene oxide	90% active; H$_2$O 8-10%; OH no. 82; 'n' no. 11	HLB 14.1; clear liquid; cloud point 77°C (1% aq. soln.); M.W. 683; pour point −16°C; viscosity 96 cP; colour 30 (Pt-Co; APHA); surface tension 27 dynes/cm (0.1% aq. soln.); pH 6.4 (10% aq. soln.)	wetting agent; penetrant; levelling agent; dispersant; coupling agent / textile industry; agriculture; paper industry; metal cleaning; leather industry; water treatment
Witco	CETALOX 8	C$_{16/18}$ fatty alcohol 8EO	100% active; 'n' no. 8	paste	low foam; detergent / laundry products
	CETALOX 11	C$_{16/18}$ fatty alcohol 11EO	100% active; 'n' no. 11	paste	low foam; detergent / household detergents; laundry products
	CETALOX 16	C$_{16/18}$ fatty alcohol 16EO	100% active; 'n' no. 16	wax	
	CETALOX 50	C$_{16/18}$ fatty alcohol 50EO	100% active; 'n' no. 50	flakes	
	CETALOX AT	C$_{16/18}$ fatty alcohol 50EO	100% active; 'n' no. 50	powder	
	LAUROPAL 0205	C$_{10/12}$ fatty alcohol 5EO	100% active; 'n' no. 5	liquid	industrial detergents; laundry products
	LAUROPAL 0207	C$_{10/12}$ fatty alcohol 7EO	100% active; 'n' no. 7	liquid	emulsifier; wetting agent; detergent
	LAUROPAL 2	C$_{12/14}$ fatty alcohol 2EO	100% active; 'n' no. 2	liquid	emulsifier; wetting agent / mineral oils; pigments
	LAUROPAL 4	C$_{12/14}$ fatty alcohol 4EO	100% active; 'n' no. 4	liquid	emulsifier; wetting agent / mineral oils; pigments; metal working; textile industry
	LAUROPAL 9	C$_{12/14}$ fatty alcohol 9EO	100% active; 'n' no. 9	paste	levelling agent; dispersant / paint industry; plastics industry; metal working; textile industry
	LAUROPAL 11	C$_{12/14}$ fatty alcohol 11EO	100% active; 'n' no. 11	paste	emulsifier / emulsion polymerisation; textile industry
	LAUROXAL 3	C$_{13}$ fatty alcohol 3EO	100% active; 'n' no. 3	liquid	emulsifier; wetting agent / mineral oils
	LAUROXAL 6	C$_{13}$ fatty alcohol 6EO	100% active; 'n' no. 6	liquid	
	LAUROXAL 8	C$_{13}$ fatty alcohol 8EO	100% active; 'n' no. 8	liquid/paste	wetting agent; detergent / textile industry; leather industry
	LAUROXAL 10	C$_{13}$ fatty alcohol 10EO	100% active; 'n' no. 10	liquid/paste	metal cleaning
	OCETOX 55-25	olecetylic fatty alcohol 25EO	100% active; 'n' no. 25	wax	emulsifier / detergents; textile industry
	REWOPAL LA 3	lauryl alcohol 3EO	100% active; 'n' no. 3	liquid	emulsifier / detergents; bath preparations; textile industry; metal working
	REWOPAL LA 6	lauryl alcohol 6EO	100% active; 'n' no. 6	liquid	detergents; cleaners
	REWOPAL LA 10-80	lauryl alcohol 10EO	80% active; 'n' no. 10	liquid	
	REWOPAL LA 10	lauryl alcohol 10EO	100% active; 'n' no. 10	soft wax	
	REWOPAL TA 25	tallow alcohol 25EO	100% active; 'n' no. 25	powder	dispersant / detergents; cleaners

Supplier	Trade name	Chemical description	Composition	General properties	Functionality / Application
Witco	WITCOWET 1003	alkylpolyglycolether adsorbed on silica	60% active; moisture 3-4%	white powder; pH 7.0 (3% in H_2O)	low foam; wetting agent / *wettable powders*
	WITCOWET 1207	ethoxylated fatty alcohol adsorbed on silica	50% active; moisture 4.2%	white powder; pH 6.8 (3% in H_2O)	wetting agent / *wettable powders*
Zschimmer & Schwarz	MULSIFAN CB	fatty alcohol ethoxylate	100% active	HLB 10; wax	o/w emulsifier / *cosmetics*
	MULSIFAN CPA	fatty alcohol ethoxylate	100% active	HLB 9; liquid	emulsifier / *waxes*
	MULSIFAN RT 11	fatty alcohol ethoxylate	100% active	HLB 16; wax	emulsifier / *waxes*
	MULSIFAN RT 19	fatty alcohol ethoxylate	100% active	HLB 13; paste	emulsifier
	MULSIFAN RT 23	fatty alcohol ethoxylate	100% active	HLB 11; liquid	emulsifier / *paraffin oils*
	MULSIFAN RT 24	fatty alcohol ethoxylate	100% active	HLB 9; liquid	emulsifier / *white oils; paraffin oils*
	MULSIFAN RT 110	fatty alcohol ethoxylate	100% active	HLB 10; wax	emulsifier / *paraffins*
	MULSIFAN RT 125	fatty alcohol ethoxylate	100% active	HLB 16; wax	emulsifier
	MULSIFAN RT 157	fatty alcohol ethoxylate	100% active	HLB 15; wax	emulsifier / *waxes*
	MULSIFAN RT 203/80	fatty alcohol ethoxylate	80% active	HLB 14; liquid	solubiliser; emulsifier / *perfumes; essential oils*
	MULSIFAN RT 258	fatty alcohol ethoxylate	100% active	HLB 10; wax	emulsifier / *paraffins*
	MULSIFAN RT 269	fatty alcohol ethoxylate	100% active	liquid	co-emulsifier; thickener / *liquid cleansers*
	OXETAL 500/85	fatty alcohol 5EO	85% active	liquid; cloud point 65-68°C (5 g in 20 ml of 25% BDG)	
	OXETAL 800/85	fatty alcohol 8EO	85% active	liquid; cloud point 66-70°C (1% in distilled H_2O)	
	OXETAL ID 104	isodecyl alcohol 4EO	100% active	liquid; cloud point 50-53°C (5 g in 20 ml of 25% BDG)	
	OXETAL O 108	oleyl/cetyl alcohol 8EO	100% active	liquid; cloud point 45-50°C (1% in distilled H_2O)	*detergents; cleansing preparations; all-purpose cleaners*
	OXETAL O 112	oleyl/cetyl alcohol 12EO	100% active	paste; cloud point 83-87°C (1% in distilled H_2O)	
	OXETAL TG 111	tallow fatty alcohol 11EO	100% active	wax; cloud point 70-75°C (1% in distilled H_2O)	
	OXETAL TG 118	tallow fatty alcohol 18EO	100% active	wax; cloud point 72-76°C (1% in 10% NaCl soln.)	
	OXETAL VD 20	fatty alcohol polyglycol ether	100% active	liquid	thickener
	OXETAL VD 28	fatty alcohol polyglycol ether	100% active	liquid	*cleansing preparations; all-purpose cleaners; detergents*
	OXTEL D 104	decyl alcohol 4EO	100% active	liquid; cloud point 58-61°C (5 g in 20 ml of 25% BDG)	*detergents; cleansing preparations; all-purpose cleaners*
	TISSOCYL HWB	alkyl polyglycol ether		colourless liquid or paste depending on temperature	wetting agent; washing agent; detergent
	TISSOCYL RLB	alkyl polyglycol ether		colourless viscous liquid	*textile industry*

Supplier	Trade name	Chemical description	Composition	General properties	Functionality / Application
Zschimmer & Schwarz	ZUSOLAT 1004	fatty alcohol 4EO	100% active	liquid; cloud point 57-61°C (5 g in 20 ml of 25% BDG)	wetting agent
	ZUSOLAT 1005/85	fatty alcohol 5EO	85% active	liquid; cloud point 65-68°C (5 g in 20 ml of 25% BDG)	detergents; cleansing preparations; all-purpose cleaners; floor cleaners; high-pressure cleaners
	ZUSOLAT 1008/85	fatty alcohol 8EO	85% active	liquid; cloud point 66-70°C (1% in distilled H_2O)	

Alkylphenol ethoxylates

Supplier	Trade name	Chemical description	Composition	General properties	Functionality / Application
Akcros Chemicals	ETHYLAN 20	nonyl phenol ethoxylate; made-to-order	100% active; ethylene oxide content 80%; 'n' no. 20	HLB 16.0; white waxy solid; cloud point > 100°C (1% aq.); pour point 30°C; viscosity 160 cSt (40°C); pH 5-7 (1% aq.)	foam booster; foam stabiliser; emulsifier; solubiliser; wetting agent; detergent; industrial detergents; essential oils; perfumes; emulsion polymerisation; latexes; waxes; polyester resin
	ETHYLAN 44	nonyl phenol ethoxylate	100% active; ethylene oxide content 45%; 'n' no. 4	HLB 9.0; clear pale yellow liquid; pour point < 0°C; viscosity 340 cSt; pH 5-7 (1% aq.)	emulsifier; hydrocarbons; silicones; hand cleaning gels; degreasers
	ETHYLAN 55	nonyl phenol ethoxylate	100% active; ethylene oxide content 52%; 'n' no. 5.5	HLB 10.5; clear pale yellow liquid; pour point < 0°C; viscosity 290 cSt; pH 5-7 (1% aq.)	
	ETHYLAN 77	nonyl phenol ethoxylate	100% active; ethylene oxide content 55%; 'n' no. 6	HLB 10.9; clear pale yellow liquid; cloud point < 2°C; pour point < 0°C; viscosity 360 cSt; pH 5-7 (1% aq.)	
	ETHYLAN BCP	nonyl phenol ethoxylate	100% active; ethylene oxide content 65%; 'n' no. 9	HLB 12.9; clear pale yellow liquid; cloud point 54°C (1% aq.); pour point 3°C; viscosity 320 cSt; pH 5-7 (1% aq.)	detergent; scouring agent; wetting agent; emulsifier; textile industry; oils; aromatic solvents
	ETHYLAN BV	nonyl phenol ethoxylate	100% active; ethylene oxide content 72%; 'n' no. 13	HLB 14.5; white soft paste; cloud point 90°C (1% aq.); pour point 17°C; viscosity 110 cSt (40°C); pH 5-7 (1% aq.)	foam stabiliser; foam booster; emulsifier; solubiliser; industrial detergents; essential oils; perfumes; emulsion polymerisation
	ETHYLAN DP	nonyl phenol ethoxylate	100% active; ethylene oxide content 70%; 'n' no. 12	HLB 14.0; clear pale yellow liquid; cloud point 81°C (1% aq.); pour point 14°C; viscosity 400 cSt; pH 5-7 (1% aq.)	foam stabiliser; foam booster; emulsifier; solubiliser; industrial detergents; essential oils; perfumes; emulsion polymerisation
	ETHYLAN ENT EXTRA	di-nonyl phenol ethoxylate	100% active	HLB 9.0; clear pale yellow liquid; pour point 2°C; viscosity 440 cSt; pH 5-7 (1% aq.)	emulsifier; plasticiser; mineral oils; waxes; mastic
	ETHYLAN HA FLAKE	nonyl phenol ethoxylate	100% active; ethylene oxide content 87%; 'n' no. 35	HLB 17.4; white flakes; cloud point > 100°C (1% aq.); pour point 43°C; viscosity 120 cSt (60°C); pH 5-7 (1% aq.)	foam booster; foam stabiliser; emulsifier; solubiliser; wetting agent; detergent; industrial detergents; essential oils; perfumes; emulsion polymerisation; latexes; waxes; polyester resin
	ETHYLAN N30	nonyl phenol ethoxylate; made-to-order	100% active; ethylene oxide content 85%; 'n' no. 30	HLB 17.0; white waxy solid; cloud point > 100°C (1% aq.); pour point 39°C; viscosity 90 cSt (60°C); pH 5-7 (1% aq.)	
	ETHYLAN NP1	nonyl phenol ethoxylate	100% active; ethylene oxide content 23%; 'n' no. 1.5	HLB 4.5; clear pale yellow liquid; pour point < 0°C; viscosity 650 cSt; pH 5-7 (1% aq.)	
	ETHYLAN TU	nonyl phenol ethoxylate	100% active; ethylene oxide content 62%; 'n' no. 8	HLB 12.3; clear pale yellow liquid; cloud point 32°C (1% aq.); pour point < 0°C; viscosity 366 cSt; pH 5-7 (1% aq.)	detergent; scouring agent; wetting agent; emulsifier; textile industry; oils; aromatic solvents
	LEVELAN P208	nonyl phenol ethoxylate	80% active; H_2O content 20%; 'n' no. 20	HLB 16.0; clear pale yellow liquid; cloud point > 100°C (1% aq.); pour point < 0°C; viscosity 460 cSt; pH 5-7 (1% aq.)	foam booster; foam stabiliser; emulsifier; solubiliser; wetting agent; detergent; industrial detergents; essential oils; perfumes; emulsion polymerisation; latexes; waxes; polyester resin

Supplier	Trade name	Chemical description	Composition	General properties	Functionality Application
Akzo Nobel	BEROL 26	nonylphenol 4EO	100% active; 'n' no. 4	HLB 8.9; cloud point 44°C (5 g in 25 ml 25% BDG soln.)	*emulsifier* *non-polar hydrocarbons and oils; hard-surface cleaners; antifoamers; toiletries; degreasers; vehicle cleaners*
	BEROL 259	nonylphenol 2EO	100% active; 'n' no. 2	HLB 5.7; cloud point 52°C (5 g in 25 ml 50% BDG soln.)	*non-polar hydrocarbons and oils; hard-surface cleaners; antifoamers; toiletries*
	BEROL 265	nonylphenol 13EO	100% active; 'n' no. 13	HLB 14.4; cloud point 84°C (1% in H_2O)	*wetting agent; pigment dispersant* *detergents*
	BEROL 267	nonylphenol 8EO	100% active; 'n' no. 8	HLB 12.3; cloud point 9-18°C (1% in H_2O)	*industrial cleaners; vehicle cleaners; detergents*
	BEROL 268	nonylphenol 11EO	100% active; 'n' no. 11	HLB 13.7; cloud point 64°C (1% in H_2O)	*wetting agent; pigment dispersant* *detergents*
	BEROL 271	nonylphenol 8.7EO	100% active; 'n' no. 8.7	HLB 12.7; cloud point 32°C (1% in H_2O)	*wetting agent; detergent*
	BEROL 274	nonylphenol 30EO	100% active; 'n' no. 30	HLB 17.0; cloud point 77°C (1 g in 100 ml 10% NaCl soln.)	*dispersant* *detergents; foam controllers; metal working*
	BEROL 277	nonylphenol 30EO	70% active; 'n' no. 30	HLB 17.1; cloud point 77°C (1 g in 100 ml 10% NaCl soln.)	*dispersant* *detergents; foam controllers; metal working*
	BEROL 281	nonylphenol 20EO	80% active; 'n' no. 20	HLB 16.0; cloud point 73°C (1 g in 100 ml 10% NaCl soln.)	*dispersant; solubiliser* *detergents; metal working; toiletries; perfumes*
	BEROL 291	nonylphenol 50EO	100% active; 'n' no. 50	HLB 18.2; cloud point 77°C (1 g in 100 ml 10% NaCl soln.)	*dispersant* *toiletries*
	BEROL 292	nonylphenol 20EO	100% active; 'n' no. 20	HLB 16.0; cloud point 73°C (1 g in 100 ml 10% NaCl soln.)	*dispersant; solubiliser* *detergents; metal working; toiletries; perfumes*
	BEROL 295	nonylphenol 40EO	70% active; 'n' no. 40	HLB 17.8; cloud point 77°C (1 g in 100 ml 10% NaCl soln.)	*dispersant; emulsifier* *waxes; toiletries*
	BEROL 296	nonylphenol 16EO	100% active; 'n' no. 16	HLB 15.2; cloud point 67°C (1 g in 100 ml 10% NaCl soln.)	*dispersant; solubiliser* *detergents; metal working; toiletries; perfumes*
	BEROL O2	nonylphenol 6EO	100% active; 'n' no. 6	HLB 10.9; cloud point 64°C (5 g in 25 ml 25% BDG soln.)	*industrial cleaners; vehicle cleaners; detergents*
	BEROL O9	nonylphenol 10EO	100% active; 'n' no. 10	HLB 13.3; cloud point 55°C (1% in H_2O)	*wetting agent* *hard-surface cleaners; vehicle cleaners; detergents; wool scouring*
	WASC	nonylphenol 12EO	95% active; 'n' no. 12	HLB 14.1; cloud point 77°C (1% in H_2O)	*wetting agent; pigment dispersant* *detergents*
Albright & Wilson	DEHSCOFIX 505	substituted phenol ethoxylate	concentration 80%	liquid	
	DEHSCOFIX 908	substituted phenol ethoxylate	concentration 98%	soft paste	
	DEHSCOFIX 944	substituted phenol ethoxylate	concentration 100%		
	EMPILAN NP SERIES	nonylphenol ethoxylates	concentration 100%	liquids to solids	*wetting agent; antistatic agent; dispersant; emulsifier; detergent; plasticiser* *coatings; agrochemicals; leather industry; mortar; construction; emulsion polymerisation*
	EMPILAN OPE 9.5 SERIES	octylphenol ethoxylates	concentration 100%	liquids	*emulsifier* *agrochemicals*
Auschem Cesalpinia	EMULSON NF 2	POE (2) nonylphenol	100% active; 'n' no. 2	liquid	*emulsifier* *pesticides*
	EMULSON NF 4	POE (4) nonylphenol	100% active; 'n' no. 4	liquid	*pesticides*
	EMULSON NF 7 C	POE (7) nonylphenol	100% active; 'n' no. 7	liquid	*spreading agent; sticker; wetting agent* *pesticides*

Supplier	Trade name	Chemical description	Composition	General properties	Functionality / *Application*
Auschem Cesalpinia	EMULSON NF 9	POE (9) nonylphenol	100% active; 'n' no. 9	liquid	emulsifier / *pesticides*
	EMULSON NF 10	POE (10) nonylphenol	100% active; 'n' no. 10	liquid	
	EMULSON NF 12	POE (12) nonylphenol	100% active; 'n' no. 12	liquid	wetting agents / *pesticides*
	MADEOL 10 B	POE (10) nonylphenol	50% active; 'n' no. 10	powder	
	POLIROL NF 9	POE (9) nonylphenol	100% active; 'n' no. 9	liquid	
	POLIROL NF 10	POE (10) nonylphenol	100% active; 'n' no. 10	liquid	
	POLIROL NF 15/85	POE (15) nonylphenol	85% active; 'n' no. 15	liquid	
	POLIROL NF 15	POE (15) nonylphenol	100% active; 'n' no. 15	liquid	
	POLIROL NF 20/30	POE (20) nonylphenol	30% active; 'n' no. 20	liquid	
	POLIROL NF 20/75	POE (20) nonylphenol	75% active; 'n' no. 20	liquid	
	POLIROL NF 20	POE (20) nonylphenol	100% active; 'n' no. 20	waxy	
	POLIROL NF 30 S	POE (30) nonylphenol	100% active; 'n' no. 30	flakes	emulsifier; stabiliser; homogeniser / *emulsion polymerisation; latices*
	POLIROL NF 30/70	POE (30) nonylphenol	70% active; 'n' no. 30	liquid	
	POLIROL NF 30	POE (30) nonylphenol	100% active; 'n' no. 30	waxy	
	POLIROL NF 40 S	POE (40) nonylphenol	100% active; 'n' no. 40	flakes	
	POLIROL NF 40/30	POE (40) nonylphenol	30% active; 'n' no. 40	liquid	
	POLIROL NF 40/60	POE (40) nonylphenol	60% active; 'n' no. 40	liquid	
	POLIROL NF 40	POE (40) nonylphenol	100% active; 'n' no. 40	solid	
	POLIROL NF 80 S	POE (80) nonylphenol	100% active; 'n' no. 80	flakes	
	POLIROL NF 80/30	POE (80) nonylphenol	30% active; 'n' no. 80	liquid	
	POLIROL NF 80/60	POE (80) nonylphenol	60% active; 'n' no. 80	liquid	
	POLIROL NF 80	POE (80) nonylphenol	100% active; 'n' no. 80	waxy	
	RIOKLEN NF 2	POE (2) nonylphenol	100% active; 'n' no. 2	liquid	defoaming agent; co-emulsifier
	RIOKLEN NF 4	POE (4) nonylphenol	100% active; 'n' no. 4	liquid	
	RIOKLEN NF 6	POE (6) nonylphenol	100% active; 'n' no. 6	liquid	
	RIOKLEN NF 7	POE (7) nonylphenol	100% active; 'n' no. 7	liquid	
	RIOKLEN NF 8	POE (8) nonylphenol	100% active; 'n' no. 8	liquid	
	RIOKLEN NF 9	POE (9) nonylphenol	100% active; 'n' no. 9	liquid	wetting agent; solubiliser; dispersant; demulsifier
	RIOKLEN NF 10	POE (10) nonylphenol	100% active; 'n' no. 10	liquid	
	RIOKLEN NF 12	POE (12) nonylphenol	100% active; 'n' no. 12	liquid	
	RIOKLEN NF 15	POE (15) nonylphenol	100% active; 'n' no. 15	solid	
	RIOKLEN NF 20	POE (20) nonylphenol	100% active; 'n' no. 20	solid	
	RIOKLEN NF 30	POE (30) nonylphenol	100% active; 'n' no. 30	flakes	
	RIOKLEN NF 40	POE (40) nonylphenol	100% active; 'n' no. 40	flakes	
BASF	EMULAN NP 2080	alkylphenol ethoxylate	80% active	cloud point 85°C (in NaCl soln.)	emulsifier
	EMULAN OP25	alkylphenol ethoxylate	100% active	cloud point 88°C (in NaCl soln.)	
	EMULAN PO	alkylphenol ethoxylate	100% active	cloud point 46°C (in butyl digol)	
	LUTENSOL AP 6	alkylphenol 6EO	100% active; 'n' no. 6	cloud point 61°C (in butyl digol)	

Supplier	Trade name	Chemical description	Composition	General properties	Functionality / Application
BASF	LUTENSOL AP 7	alkylphenol 7EO	100% active; 'n' no. 7	cloud point 62°C (in butyl digol)	
	LUTENSOL AP 8	alkylphenol 8EO	100% active; 'n' no. 8	cloud point 34°C (in H_2O)	
	LUTENSOL AP 9	alkylphenol 9EO	100% active; 'n' no. 9	cloud point 51°C (in H_2O)	
	LUTENSOL AP 10	alkylphenol 10EO	100% active; 'n' no. 10	cloud point 60°C (in H_2O)	
	LUTENSOL AP 14	alkylphenol 14EO	100% active; 'n' no. 14	cloud point 76°C (in NaCl soln.)	
	LUTENSOL AP 20	alkylphenol 20EO	100% active; 'n' no. 20	cloud point 85°C (in NaCl soln.)	
Bayer	LEVEGAL KNS	alkylphenol polyglycol ether			levelling agent; low foam; *textile industry; dyeing of cellulosic fibres*
Dr. Th. Boehme	SYNTHESIN BZ-K	alkyl phenol polyglycol ether			spin finish; *textile industry; fibre production*
Chemax	CHEMAX DNP-8	POE (8) dinonyl phenol	'n' no. 8	HLB 10.4; liquid; cloud point <25°C (1% aq.)	emulsifier; detergent; wetting agent; dispersant; solubiliser; coupling agent; *textile industry; metal working; household applications; industrial applications; agriculture; paper industry; paint industry*
	CHEMAX DNP-150/50	POE (150) dinonyl phenol 50%	'n' no. 150	HLB 19.0; liquid; cloud point >100°C (1% aq.)	
	CHEMAX DNP-150	POE (150) dinonyl phenol	'n' no. 150	HLB 19.0; solid; cloud point >100°C (1% aq.)	
	CHEMAX NP-1.5	POE (1.5) nonyl phenol	'n' no. 1.5	HLB 4.6; liquid; cloud point <25°C (1% aq.)	
	CHEMAX NP-4	POE (4) nonyl phenol	'n' no. 4	HLB 8.9; liquid; cloud point <25°C (1% aq.)	
	CHEMAX NP-6	POE (6) nonyl phenol	'n' no. 6	HLB 10.9; liquid; cloud point <25°C (1% aq.)	
	CHEMAX NP-9	POE (9) nonyl phenol	'n' no. 9	HLB 13.0; liquid; cloud point 54°C (1% aq.)	
	CHEMAX NP-10	POE (10) nonyl phenol	'n' no. 10	HLB 13.5; liquid; cloud point 72°C (1% aq.)	
	CHEMAX NP-15	POE (15) nonyl phenol	'n' no. 15	HLB 15.0; paste; cloud point 96°C (1% aq.)	
	CHEMAX NP-20	POE (20) nonyl phenol	'n' no. 20	HLB 16.0; solid; cloud point >100°C (1% aq.)	
	CHEMAX NP-30/70	POE (30) nonyl phenol 70%	'n' no. 30	HLB 17.1; liquid; cloud point >100°C (1% aq.)	
	CHEMAX NP-30	POE (30) nonyl phenol	'n' no. 30	HLB 17.1; solid; cloud point >100°C (1% aq.)	
	CHEMAX NP-40/70	POE (40) nonyl phenol 70%	'n' no. 40	HLB 17.8; liquid; cloud point >100°C (1% aq.)	
	CHEMAX NP-40	POE (40) nonyl phenol	'n' no. 40	HLB 17.8; solid; cloud point >100°C (1% aq.)	
	CHEMAX NP-50/70	POE (50) nonyl phenol 70%	'n' no. 50	HLB 18.2; liquid; cloud point >100°C (1% aq.)	
	CHEMAX NP-50	POE (50) nonyl phenol	'n' no. 50	HLB 18.2; solid; cloud point >100°C (1% aq.)	
	CHEMAX NP-100	POE (100) nonyl phenol	'n' no. 100	HLB 19.0; solid; cloud point >100°C (1% aq.)	
	CHEMAX NP-100/70	POE (100) nonyl phenol 70%	'n' no. 100	HLB 19.0; liquid; cloud point >100°C (1% aq.)	
	CHEMAX OP-30/70	POE (30) octyl phenol 70%	'n' no. 30	HLB 17.3; liquid; cloud point >100°C (1% aq.)	
	CHEMAX OP-40/70	POE (40) octyl phenol	'n' no. 40	HLB 17.9; liquid; cloud point >100°C (1% aq.)	
Ciba	TINOVETIN JU	alkylphenol ethoxylate derivative		clear yellowish syrupy liquid	detergent; wetting agent; *general purpose scouring and wetting*
CPB-Companhia Petroquimica do Barreiro	QUIMIPOL E 2NF-300	nonylphenol ethoxylate	100% active	pH 6.0-8.0 (1% soln.)	
	QUIMIPOL E NF 15	nonylphenol ethoxylate	100% active	pH 6.0-8.0 (1% soln.)	emulsifier; *hydrocarbons; silicones; aromatic oils; solvents*
	QUIMIPOL E NF 20	nonylphenol ethoxylate	100% active	pH 6.0-8.0 (1% soln.)	
	QUIMIPOL E NF 30	nonylphenol ethoxylate	100% active	pH 6.0-8.0 (1% soln.)	
	QUIMIPOL E NF 40	nonylphenol ethoxylate	100% active	pH 6.0-8.0 (1% soln.)	

Supplier	Trade name	Chemical description	Composition	General properties	Functionality / Application
CPB-Companhia Petroquímica do Barreiro	QUIMIPOL E NF 55	nonylphenol ethoxylate	100% active	pH 6.0-8.0 (1% soln.)	
	QUIMIPOL E NF 65	nonylphenol ethoxylate	100% active	pH 6.0-8.0 (1% soln.)	
	QUIMIPOL E NF 80	nonylphenol ethoxylate	100% active	pH 6.0-8.0 (1% soln.)	
	QUIMIPOL E NF 90	nonylphenol ethoxylate	100% active	pH 6.0-8.0 (1% soln.)	
	QUIMIPOL E NF 95	nonylphenol ethoxylate	100% active	pH 6.0-8.0 (1% soln.)	emulsifier
	QUIMIPOL E NF 100	nonylphenol ethoxylate	100% active	pH 6.0-8.0 (1% soln.)	*hydrocarbons; silicones; aromatic oils; solvents*
	QUIMIPOL E NF 110	nonylphenol ethoxylate	100% active	pH 6.0-8.0 (1% soln.)	
	QUIMIPOL E NF 120	nonylphenol ethoxylate	100% active	pH 6.0-8.0 (1% soln.)	
	QUIMIPOL E NF 140	nonylphenol ethoxylate	100% active	pH 6.0-8.0 (1% soln.)	
	QUIMIPOL E NF 170	nonylphenol ethoxylate	100% active	pH 6.0-8.0 (1% soln.)	
	QUIMIPOL E NF 200	nonylphenol ethoxylate	100% active	pH 6.0-8.0 (1% soln.)	
	QUIMIPOL E NF 230	nonylphenol ethoxylate	100% active	pH 6.0-8.0 (1% soln.)	
Dac International Surfactants	NONFIX 10	nonylphenol ethoxylate			detergent; wetting agent *textile industry; paint industry*
	NONFIX 30	nonylphenol ethoxylate			*paint industry*
	NONFIX 40	nonylphenol ethoxylate			*emulsion polymerization*
	NONFIX 11000	nonylphenol ethoxylate			*emulsion polymerization*
	OTIX 10	octylphenol ethoxylate			softener *textile industry*
Elf Atochem	REMCOPAL 011	alkylphenol blend 10EO	100% active; 'n' no. 10	HLB 13.4; liquid	
	REMCOPAL 012	alkylphenol blend 11EO	100% active; 'n' no. 11	HLB 14.1; paste; cloud point 81°C	
	REMCOPAL 306	alkylphenol blend 5EO	100% active; 'n' no. 5	HLB 10.5; liquid; cloud point 65°C (aq. soln. with 25% BDG)	
	REMCOPAL N 2 100	nonyl phenol 2EO	100% active; 'n' no. 2	HLB 5.3; liquid; cloud point 16-17°C	
	REMCOPAL N 4 100	nonyl phenol 4EO	100% active; 'n' no. 4	HLB 8.5; liquid; cloud point 55°C (aq. soln. with 25% BDG)	
	REMCOPAL N 5 100	nonyl phenol 5EO	100% active; 'n' no. 5	HLB 10.3; liquid; cloud point 66°C (aq. soln. with 25% BDG)	
	REMCOPAL N 6 100	nonyl phenol 6EO	100% active; 'n' no. 6	HLB 11.2; liquid; cloud point 70°C (aq. soln. with 25% BDG)	
	REMCOPAL N 8 100	nonyl phenol 8EO	100% active; 'n' no. 8	HLB 12.3; liquid; cloud point 33°C	
	REMCOPAL N 9 100	nonyl phenol 9EO	100% active; 'n' no. 9	HLB 12.7; liquid; cloud point 40°C	
	REMCOPAL N 9.5 100	nonyl phenol 9.5EO	100% active; 'n' no. 9.5	HLB 12.8; liquid; cloud point 50°C	
	REMCOPAL N 10 100	nonyl phenol 10EO	100% active; 'n' no. 10	HLB 13.1; liquid; cloud point 64°C	
	REMCOPAL N 11 100	nonyl phenol 11EO	100% active; 'n' no. 11	HLB 13.5; liquid; cloud point 69°C	

Supplier	Trade name	Chemical description	Composition	General properties	Functionality Application
Elf Atochem	REMCOPAL N 11.5 97	nonyl phenol 11.5EO	97% active; 'n' no. 11.5	HLB 13.9; viscous liquid; cloud point 82°C	
	REMCOPAL N 12.5 100	nonyl phenol 12.5EO	100% active; 'n' no. 12.5	HLB 14.3; viscous liquid; cloud point 89°C	
	REMCOPAL N 15 100	nonyl phenol 15EO	100% active; 'n' no. 15	HLB 15.1; paste; cloud point 98°C	
	REMCOPAL N 17 100	nonyl phenol 17EO	100% active; 'n' no. 17	HLB 15.3; paste; cloud point 83°C (aq. soln. of 50 g/l NaCl)	
	REMCOPAL N 17 80	nonyl phenol 17EO	80% active; 'n' no. 17	HLB 15.3; liquid; cloud point 83°C (aq. soln. of 50 g/l NaCl)	
	REMCOPAL N 20 80	nonyl phenol 20EO	80% active; 'n' no. 20	HLB 15.8; liquid; cloud point 86°C (aq. soln. of 50 g/l NaCl)	
	REMCOPAL N 20 100	nonyl phenol 20EO	100% active; 'n' no. 20	HLB 15.8; solid; cloud point 86°C (aq. soln. of 50 g/l NaCl)	
	REMCOPAL N 30 70	nonyl phenol 30EO	70% active; 'n' no. 30	HLB 16.6; liquid; cloud point 90°C (aq. soln. of 50 g/l NaCl)	
	REMCOPAL N 30 100	nonyl phenol 30EO	100% active; 'n' no. 30	HLB 16.6; solid; cloud point 90°C (aq. soln. of 50 g/l NaCl)	
	REMCOPAL N 40 70	nonyl phenol 40EO	70% active; 'n' no. 40	HLB 17.8; liquid; cloud point 93°C (aq. soln. of 50 g/l NaCl)	
	REMCOPAL N 50 100	nonyl phenol 50EO	100% active; 'n' no. 50	HLB 18.2; solid; cloud point 93°C (aq. soln. of 50 g/l NaCl)	
	SELLIG DN 10 100	dinonyl phenol 10EO; in course of development	100% active; 'n' no. 10	HLB 10	
	SELLIG DN 22 100	dinonyl phenol 22EO; in course of development	100% active; 'n' no. 22	HLB 13.8	
	SELLIG O 4 100	octyl phenol 4EO; in course of development	100% active; 'n' no. 4	HLB 9.3; liquid; cloud point 31°C	
	SELLIG O 5 100	octyl phenol 5EO; in course of development	100% active; 'n' no. 5	HLB 10.4; liquid; cloud point 42°C	
	SELLIG O 6 100	octyl phenol 6EO	100% active; 'n' no. 6	HLB 10.8; liquid; cloud point 51°C	
	SELLIG O 8 100	octyl phenol 8EO; in course of development	100% active; 'n' no. 8	HLB 12.6; liquid; cloud point 45°C	
	SELLIG O 9 100	octyl phenol 9EO	100% active; 'n' no. 9	HLB 13.1; liquid; cloud point 61°C	
	SELLIG O 11 100	octyl phenol 11EO	100% active; 'n' no. 11	HLB 14; liquid; cloud point 82°C	
	SELLIG O 12 100	octyl phenol 12EO	100% active; 'n' no. 12	HLB 14.6; paste; cloud point 91°C	
	SELLIG O 20 100	octyl phenol 20EO; in course of development	100% active; 'n' no. 20	HLB 16.2; cloud point 87°C (aq. soln. of 50 g/l NaCl)	
Ellis & Everard	CAFLON P1	phenol ethoxylate			
	CAFLON P4	phenol ethoxylate		viscous liquid	dispersant; solubiliser; *cosmetics; perfumes; shampoos*
Gattefosse	OXYPOL	octoxynol-11			
Henkel	DISPONIL NP 4	nonylphenol 4EO	99% active; 'n' no. 4	HLB 8.8; liquid	emulsifier
	DISPONIL NP 6	nonylphenol 6EO	99% active; 'n' no. 6	HLB 11.3; liquid	
	DISPONIL NP 10	nonylphenol 10EO	99% active; 'n' no. 10	HLB 13.3; liquid	
	DISPONIL NP 11	nonylphenol 11EO	99% active; 'n' no. 11	HLB 13.8; liquid	*emulsion polymerisation*

Supplier	Trade name	Chemical description	Composition	General properties	Functionality Application
Henkel	DISPONIL NP 208	nonylphenol 20EO	ca. 80% active; 'n' no. 20	HLB 16.0; liquid	emulsifier
	DISPONIL NP 307	nonylphenol 30EO	ca. 70% active; 'n' no. 30	HLB 17.2; liquid	emulsion polymerisation
	EUMULGIN 286	nonoxyl-10	99-100% active	liquid	o/w emulsifier; wetting agent; solubiliser perfumes ; hair care products; bath preparations
Hoechst	ARKOPAL N040	nonylphenol 4EO	100% active; 'n' no. 4	liquid	industrial detergents; textile industry; construction industry
	ARKOPAL N060	nonylphenol 6EO	100% active; 'n' no. 6	viscous, slightly yellow liquid	
	ARKOPAL N080	nonylphenol 8EO	100% active; 'n' no. 8	viscous, slightly yellow liquid	
	ARKOPAL N090	nonylphenol 9EO	100% active; 'n' no. 9	viscous, slightly yellow liquid	
	ARKOPAL N100	nonylphenol 10EO	100% active; 'n' no. 10	viscous, slightly yellow liquid	
	ARKOPAL N110	nonylphenol 11EO	100% active; 'n' no. 11	slightly yellow pourable paste	
	ARKOPAL N130	nonylphenol 13EO	100% active; 'n' no. 13	slightly yellow paste	
	ARKOPAL N150	nonylphenol 15EO	100% active; 'n' no. 15	slightly yellow wax-like product	
	ARKOPAL N230	nonylphenol 23EO	100% active; 'n' no. 23	slightly yellow wax-like product	
	ARKOPAL N300	nonylphenol 30EO	100% active; 'n' no. 30	clear liquid	emulsifier
	EMULSOGEN DG	tributylphenol EO	100% active	turbid liquid	
	SAPOGENAT T040	tributylphenol 4EO	100% active; 'n' no. 4	turbid liquid	
	SAPOGENAT T060	tributylphenol 6EO	100% active; 'n' no. 6	turbid liquid	
	SAPOGENAT T080	tributylphenol 8EO	100% active; 'n' no. 8	turbid liquid	
	SAPOGENAT T110	tributylphenol 11EO	100% active; 'n' no. 11	turbid liquid	emulsifier detergents; agrochemicals
	SAPOGENAT T130	tributylphenol 13EO	100% active; 'n' no. 13	turbid liquid	
	SAPOGENAT T139	tributylphenol 13EO	90% active; 'n' no. 13	clear liquid	
	SAPOGENAT T180	tributylphenol 18EO	100% active; 'n' no. 18	paste	
	SAPOGENAT T300	tributylphenol 30EO	100% active; 'n' no. 30	wax	
	SAPOGENAT T500	tributylphenol 50EO	100% active; 'n' no. 50	wax paste	
Hüls	MARLOPHEN 1028	octylphenol 8EO	active detergent 100%; 'n' no. 8	liquid; cloud point 54°C (2% in H_2O)	wetting agent cleaners; textile industry
	MARLOPHEN 1028 N	nonylphenol 9EO	active detergent 100%; 'n' no. 9	liquid; cloud point 54°C (2% in H_2O)	
	MARLOPHEN 81 N	nonylphenol 1EO	active detergent 100%; 'n' no. 1	liquid; cloud point 66°C (2% in 25% BDG soln.)	floor care
	MARLOPHEN 82 N	nonylphenol 2EO	active detergent 100%; 'n' no. 2	liquid; cloud point 29°C (10% in 25% BDG soln.)	
	MARLOPHEN 83 N	nonylphenol 3EO	active detergent 100%; 'n' no. 3	liquid; cloud point 44°C (10% in 25% BDG soln.)	
	MARLOPHEN 84 N	nonylphenol 4EO	active detergent 100%; 'n' no. 4	liquid; cloud point 55°C (10% in 25% BDG soln.)	detergents; cleaners
	MARLOPHEN 85	octylphenol 5EO	active detergent 100%; 'n' no. 5	liquid; cloud point 61°C (10% in 25% BDG soln.)	wetting agent cleaners
	MARLOPHEN 85 N	nonylphenol 5EO	active detergent 100%; 'n' no. 5	liquid; cloud point 61°C (10% in 25% BDG soln.)	detergents; cleaners
	MARLOPHEN 86	octylphenol 6.5EO	active detergent 100%; 'n' no. 6.5	liquid; cloud point 72°C (10% in 25% BDG soln.)	wetting agent cleaners

Supplier	Trade name	Chemical description	Composition	General properties	Functionality / Application
Hüls	MARLOPHEN 86 N	nonylphenol 6.5EO	active detergent 100%; 'n' no. 6.5	liquid; cloud point 72°C (10% in 25% BDG soln.)	detergents; cleaners
	MARLOPHEN 86 N/S	nonylphenol 6EO	active detergent 100%; 'n' no. 6	liquid; cloud point 68°C (10% in 25% BDG soln.)	
	MARLOPHEN 87	octylphenol 7EO	active detergent 100%; 'n' no. 7	liquid; cloud point 23°C (2% in H_2O)	wetting agent, cleaners; textile industry
	MARLOPHEN 87 N	nonylphenol 7EO	active detergent 100%; 'n' no. 7	liquid; cloud point 23°C (2% in H_2O)	
	MARLOPHEN 88	octylphenol 8EO	active detergent 100%; 'n' no. 8	liquid; cloud point 42°C (2% in H_2O)	
	MARLOPHEN 88 N	nonylphenol 8EO	active detergent 100%; 'n' no. 8	liquid; cloud point 42°C (2% in H_2O)	
	MARLOPHEN 89	octylphenol 9.5EO	active detergent 100%; 'n' no. 9.5	liquid; cloud point 61°C (2% in H_2O)	
	MARLOPHEN 89 N	nonylphenol 9.5EO	active detergent 100%; 'n' no. 9.5	liquid; cloud point 61°C (2% in H_2O)	
	MARLOPHEN 810	octylphenol 10EO	active detergent 100%; 'n' no. 10	liquid/paste; cloud point 72°C (2% in H_2O)	
	MARLOPHEN 810 N	nonylphenol 10EO	active detergent 100%; 'n' no. 10	liquid/paste; cloud point 72°C (2% in H_2O)	
	MARLOPHEN 812	octylphenol 12EO	active detergent 100%; 'n' no. 12	liquid/paste; cloud point 85°C (2% in H_2O)	
	MARLOPHEN 812 N	nonylphenol 9EO	active detergent 100%; 'n' no. 12	liquid/paste; cloud point 85°C (2% in H_2O)	
	MARLOPHEN 814	octylphenol 14EO	active detergent 100%; 'n' no. 14	paste; cloud point 61°C (2% in 10% NaCl soln.)	binding agent; emulsifier, cleaners; waxes
	MARLOPHEN 814 N	nonylphenol 14EO	active detergent 100%; 'n' no. 14	paste; cloud point 61°C (2% in 10% NaCl soln.)	
	MARLOPHEN 820	octylphenol 20EO	active detergent 100%; 'n' no. 20	paste; cloud point 73°C (2% in 10% NaCl soln.)	
	MARLOPHEN 820 N	nonylphenol 20EO	active detergent 100%; 'n' no. 20	paste; cloud point 73°C (2% in 10% NaCl soln.)	
	MARLOPHEN 830 N	nonylphenol 30EO	active detergent 100%; 'n' no. 30	solid; cloud point 73°C (2% in 10% NaCl soln.)	binding agent, solid cleaners
	MARLOPHEN 840 N	nonylphenol 40EO	active detergent 100%; 'n' no. 40	flakes; cloud point 73°C (2% in 10% NaCl soln.)	
	MARLOPHEN 850 N	nonylphenol 50EO	active detergent 100%; 'n' no. 50	flakes; cloud point 73°C (2% in 10% NaCl soln.)	
	MARLOPHEN DNP 16	dinonylphenol 16EO	active detergent 100%; 'n' no. 16	solid; cloud point 58°C (2% in H_2O)	dispersant, textile industry; paper industry
	MARLOPHEN DNP 18	dinonylphenol 18EO	active detergent 100%; 'n' no. 18	solid; cloud point 75°C (2% in H_2O)	
	MARLOPHEN DNP 30	dinonylphenol 30EO	active detergent 100%; 'n' no. 30	solid; cloud point 70°C (2% in 10% NaCl soln.)	
	MARLOPHEN DNP 150	dinonylphenol 150EO	active detergent 100%; 'n' no. 150	flakes; cloud point 71°C (2% in 10% NaCl soln.)	
	MARLOPHEN P 1	phenol 1EO	active detergent 100%; 'n' no. 1	liquid; cloud point 76°C (5% in 25% BDG soln.)	solvent
	MARLOPHEN P 4	phenol 4EO	active detergent 100%; 'n' no. 4	liquid; cloud point 33°C (2% in 10% NaCl soln.)	solubiliser
	MARLOPHEN P 7	phenol 7EO	active detergent 100%; 'n' no. 7	liquid; cloud point 65°C (2% in 10% NaCl soln.)	
	MARLOWET 4900	nonylphenol polyethylene glycol ether	active detergent 100%	liquid; cloud point 72°C (10% in 25% BDG soln.)	emulsifier, mineral oils; metal working; cold cleaners
	MARLOWET 4901	nonylphenol polyethylene glycol ether	active detergent 100%	liquid; cloud point 23°C (2% in H_2O)	
	MARLOWET 4902	nonylphenol polyethylene glycol ether	active detergent 100%	liquid; cloud point 51°C (2% in H_2O)	emulsifier, solvents; pesticides; cleaners

Supplier	Trade name	Chemical description	Composition	General properties	Functionality / Application
Hüls	MARLOWET 4930	nonylphenol polyethylene glycol ether	active detergent 100%	solid; cloud point 73°C (2% in 10% NaCl soln.)	dispersant
	MARLOWET 4938	nonylphenol polyethylene glycol ether	active detergent 100%	liquid; cloud point 55°C (10% in 25% BDG soln.)	emulsifier / *mineral oils; metal working*
	MARLOWET 4940	nonylphenol polyethylene glycol ether	active detergent 100%	liquid/paste; cloud point 72°C (2% in H_2O)	dispersant
	MARLOWET 4941	nonylphenol polyethylene glycol ether	active detergent 100%	paste; cloud point 73°C (2% in 10% NaCl soln.)	dispersant / *coal-water slurries*
	MARLOWET 5641	modified dinonylphenol polyethylene glycol ether	active detergent 45%	liquid	emulsifier / *solvents; pesticides; cleaners*
	MARLOWET ISM	nonylphenol polyethylene glycol ether	active detergent 100%	liquid; cloud point 42°C (2% in H_2O)	emulsifier / *mineral oils; metal working; cleaners*
	MARLOWET TM	nonylphenol polyethylene glycol ether	active detergent 100%	liquid; cloud point 61°C (10% in 25% BDG soln.)	
ICI	ATLOX 1690	POE-(10)-nonylphenol		HLB 13.3; colourless to pale yellow liquid	emulsifier; dispersant / *agrochemicals*
	CIRRASOL AEN-XZ	POE alkyl phenol		HLB 12.3; colourless liquid	*textile industry*
	RENEX 647	POE-(4)-nonylphenol	'n' no. 4	HLB 8.9; pale yellow liquid	
	RENEX 648	POE-(5)-nonylphenol	'n' no. 5	HLB 10.5; yellow liquid	
	RENEX 649	POE-(20)-nonylphenol	'n' no. 20	HLB 16.0; cream solid	
	RENEX 650	POE-(30)-nonylphenol	'n' no. 30	HLB 17.1; cream solid	
	RENEX 670	POE-(11)-nonylphenol	'n' no. 11	HLB 13.6; colourless/pale yellow liquid	
	RENEX 678	POE-(15)-nonylphenol	'n' no. 15	HLB 15.0; colourless/pale yellow paste	
	RENEX 679	POE-(13)-nonylphenol	'n' no. 13	HLB 14.4; colourless/pale yellow paste	
	RENEX 682	POE-(12)-nonylphenol	'n' no. 12	HLB 13.9; colourless/pale yellow suspension	
	RENEX 688	POE-(8)-nonylphenol	'n' no. 8	HLB 12.3; colourless/pale yellow liquid	
	RENEX 690	POE-(10)-nonylphenol	'n' no. 10	HLB 13.3; colourless/pale yellow liquid	
	RENEX 697	POE-(6)-nonylphenol	'n' no. 6	HLB 10.9; colourless/pale yellow liquid	
	RENEX 698	POE-(9)-nonylphenol	'n' no. 9	HLB 13.0; colourless/pale yellow liquid	
	RENEX 750	POE-(10)-octylphenol	'n' no. 10	HLB 13.6; pale yellow liquid	wetting agent; emulsifier / *personal care products; agrochemicals; industrial applications*
	RENEX 759	POE-(9)-octylphenol	'n' no. 9	HLB 13.0; pale yellow liquid	
	SYNPERONIC NP1	POE-(1)-nonylphenol		HLB 3.3; pale yellow liquid	
	SYNPERONIC NP2	POE-(2)-nonylphenol		HLB 5.7; pale yellow liquid	
	SYNPERONIC NP4	POE-(4)-nonylphenol		HLB 8.9; pale yellow liquid	
	SYNPERONIC NP5	POE-(5)-nonylphenol		HLB 10.5; pale yellow liquid	
	SYNPERONIC NP5.5	POE-(5.5)-nonylphenol		HLB 10.7; pale yellow liquid	
	SYNPERONIC NP6	POE-(6)-nonylphenol		HLB 10.9; pale yellow liquid	wetting agent / *textile industry; agrochemicals; industrial applications*
	SYNPERONIC NP7	POE-(7)-nonylphenol		HLB 11.7; pale yellow liquid	
	SYNPERONIC NP8	POE-(8)-nonylphenol		HLB 12.3; pale yellow liquid	
	SYNPERONIC NP8.5	POE-(8.5)-nonylphenol		HLB 12.6; pale yellow liquid	
	SYNPERONIC NP8.75	POE-(8.75)-nonylphenol		HLB 12.7; pale yellow liquid	
	SYNPERONIC NP9	POE-(9)-nonylphenol		HLB 12.8; pale yellow liquid	

Supplier	Trade name	Chemical description	Composition	General properties	Functionality / Application
ICI	SYNPERONIC NP9.5	POE-(9.5)-nonylphenol		HLB 13.0; pale yellow liquid	
	SYNPERONIC NP9.75	POE-(9.75)-nonylphenol		HLB 13.2; pale yellow liquid	
	SYNPERONIC NP10	POE-(10)-nonylphenol		HLB 13.3; pale yellow liquid	
	SYNPERONIC NP12	POE-(12)-nonylphenol		HLB 13.9; pale yellow liquid	
	SYNPERONIC NP13	POE-(13)-nonylphenol		HLB 14.4; pale yellow paste	
	SYNPERONIC NP15	POE-(15)-nonylphenol		HLB 15.0; pale yellow paste	wetting agent *textile industry; agrochemicals; industrial applications*
	SYNPERONIC NP17	POE-(17)-nonylphenol		HLB 16.0; pale yellow solid	
	SYNPERONIC NP20	POE-(20)-nonylphenol		HLB 17.1; pale yellow solid	
	SYNPERONIC NP25	POE-(25)-nonylphenol		HLB 17.2; pale yellow solid	
	SYNPERONIC NP30	POE-(30)-nonylphenol		HLB 17.9; pale yellow solid	
	SYNPERONIC NP35	POE-(35)-nonylphenol		HLB 17.5; pale yellow solid	
	SYNPERONIC NP40	POE-(40)-nonylphenol		HLB 17.8; pale yellow solid	
	SYNPERONIC NP50	POE-(50)-nonylphenol		HLB 18.2; pale yellow solid	
	SYNPERONIC OP3	POE-(3)-octylphenol		HLB 7.1; pale yellow liquid	
	SYNPERONIC OP4.5	POE-(4.5)-octylphenol		HLB 9.4; pale yellow liquid	
	SYNPERONIC OP6	POE-(6)-octylphenol		HLB 10.5; pale yellow liquid	
	SYNPERONIC OP7.5	POE-(7.5)-octylphenol		HLB 11.7; pale yellow liquid	wetting agent *agrochemicals; industrial applications*
	SYNPERONIC OP8	POE-(8)-octylphenol		HLB 12.6; pale yellow liquid	
	SYNPERONIC OP10	POE-(10)-octylphenol		HLB 13.3; pale yellow liquid	
	SYNPERONIC OP10.5	POE-(10.5)-octylphenol		HLB 13.5; pale yellow liquid	
	SYNPERONIC OP11	POE-(11)-octylphenol		HLB 13.6; pale yellow liquid	
	SYNPERONIC OP12.5	POE-(12.5)-octylphenol		HLB 14.3; pale yellow liquid	
	SYNPERONIC OP16.5	POE-(16.5)-octylphenol		HLB 15.3; pale yellow solid	
	SYNPERONIC OP20	POE-(20)-octylphenol		HLB 16.2; pale yellow solid	
	SYNPERONIC OP25	POE-(25)-octylphenol		HLB 16.7; pale yellow solid	
	SYNPERONIC OP30	POE-(30)-octylphenol		HLB 17.2; pale yellow solid	
	SYNPERONIC OP40	POE-(40)-octylphenol		HLB 17.4; pale yellow solid	
Kao Corporation	FINDET 9Q/16	polyethoxylated nonylphenol	'n' no. 4	liquid; cloud point 41-44°C	
	FINDET 9Q/17	polyethoxylated nonylphenol	'n' no. 5	liquid; cloud point 53-55°C	
	FINDET 9Q/18	polyethoxylated nonylphenol	'n' no. 6	liquid; cloud point 62-66°C	
	FINDET 9Q/19	polyethoxylated nonylphenol	'n' no. 7	liquid; cloud point 68-70°C	
	FINDET 9Q/20	polyethoxylated nonylphenol	'n' no. 8	liquid; cloud point 29-31°C	
	FINDET 9Q/21	polyethoxylated nonylphenol	'n' no. 9	liquid; cloud point 53-54°C	
	FINDET 9Q/21.5	polyethoxylated nonylphenol	'n' no. 9.5	liquid; cloud point 58-61°C	
	FINDET 9Q/22	polyethoxylated nonylphenol	'n' no. 10	liquid; cloud point 62-66°C	

Supplier	Trade name	Chemical description	Composition	General properties	Functionality / Application
Kao Corporation	FINDET 9Q/25	polyethoxylated nonylphenol	'n' no. 13	liquid/paste; cloud point 88-92°C	
	FINDET 9Q/27	polyethoxylated nonylphenol	'n' no. 15	paste; cloud point 67-69°C	
	FINDET 9Q/32	polyethoxylated nonylphenol	'n' no. 20	solid; cloud point 73-75°C	
	FINDET 9Q/42	polyethoxylated nonylphenol	'n' no. 30	solid; cloud point 75-77°C	
	FINDET 9Q/52	polyethoxylated nonylphenol	'n' no. 40	solid; cloud point 76-78°C	
	FINDET 9Q/3280	polyethoxylated nonylphenol	80% active; 'n' no. 20	liquid	
	FINDET 9Q/8750	polyethoxylated nonylphenol	50% active; 'n' no. 75	liquid	
	FINDET 58Q/19	polyethoxylated iso-octylphenol	'n' no. 7	liquid; cloud point 68-70°C	
	FINDET 58Q/20	polyethoxylated iso-octylphenol	'n' no. 8	liquid; cloud point 71-73°C	
	FINDET 58Q/21	polyethoxylated iso-octylphenol	'n' no. 9	liquid; cloud point 65-67°C	
	FINDET 58Q/4270	polyethoxylated iso-octylphenol	70% active; 'n' no. 30	liquid; cloud point 73-77°C	
Dr. W. Kolb	IMBENTIN-N/020	ethoxylated nonylphenol	'n' no. ca. 1.5	HLB ca. 4.6; liquid; cloud point 48°C (5% in BDG 25%)	*personal care products; household detergents; industrial and institutional cleaning; pharmaceuticals*
	IMBENTIN-N/060	ethoxylated nonylphenol	'n' no. ca. 6.0	HLB ca. 10.9; liquid; cloud point 63°C (5 g in 25 g BDG 25%)	
	IMBENTIN-N/7 A	ethoxylated nonylphenol	'n' no. ca. 2.0	HLB ca. 5.5; liquid; cloud point 52°C (5% in BDG 25%)	
	IMBENTIN-N/11 A	ethoxylated nonylphenol	'n' no. ca. 3.0	HLB ca. 7.5; liquid; cloud point 61°C (5% in BDG 25%)	
	IMBENTIN-N/16 A	ethoxylated nonylphenol	'n' no. ca. 4.0	HLB ca. 9.1; liquid; cloud point 68°C (5% in BDG 25%)	
	IMBENTIN-N/20 A	ethoxylated nonylphenol	'n' no. ca. 5.0	HLB ca. 10.0; liquid; cloud point 56°C (5 g in 25 g BDG 25%)	
	IMBENTIN-N/22	ethoxylated nonylphenol	'n' no. ca. 7.0	HLB ca. 11.7; liquid; cloud point 69°C (5 g in 25 g BDG 25%)	
	IMBENTIN-N/26 A	ethoxylated nonylphenol	'n' no. ca. 5.5	HLB ca. 10.5; liquid; cloud point 60°C (5 g in 25 g BDG 25%)	
	IMBENTIN-N/30	ethoxylated nonylphenol	'n' no. ca. 7.5	HLB ca. 12.0; liquid; cloud point 30°C (1% in deionised water)	
	IMBENTIN-N/35	ethoxylated nonylphenol	'n' no. ca. 8.0	HLB ca. 12.3; liquid; cloud point 35°C (1% in deionised water)	
	IMBENTIN-N/40 A	ethoxylated nonylphenol	'n' no. ca. 6.5	HLB ca. 11.3; liquid; cloud point 66°C (5 g in 25 g BDG 25%)	
	IMBENTIN-N/44	ethoxylated nonylphenol	'n' no. ca. 8.5	HLB ca. 12.6; liquid; cloud point 44°C (1% in deionised water)	
	IMBENTIN-N/52	ethoxylated nonylphenol	'n' no. ca. 9.0	HLB ca. 12.9; liquid; cloud point 52°C (1% in deionised water)	
	IMBENTIN-N/55	ethoxylated nonylphenol	'n' no. ca. 9.0	HLB ca. 12.9; liquid; cloud point 55°C (1% in deionised water)	
	IMBENTIN-N/57	ethoxylated nonylphenol	'n' no. ca. 10.0	HLB ca. 13.3; liquid; cloud point 57°C (1% in deionised water)	
	IMBENTIN-N/61	ethoxylated nonylphenol	'n' no. ca. 10.0	HLB ca. 13.3; liquid; cloud point 61°C (1% in deionised water)	

Supplier	Trade name	Chemical description	Composition	General properties	Functionality Application
Dr. W. Kolb	IMBENTIN-N/63	ethoxylated nonylphenol	'n' no. ca. 10.0	HLB ca. 13.3; liquid; cloud point 63°C (1% in deionised water)	
	IMBENTIN-N/66	ethoxylated nonylphenol	'n' no. ca. 11.0	HLB ca. 13.8; liquid; cloud point 66°C (1% in deionised water)	
	IMBENTIN-N/75	ethoxylated nonylphenol	'n' no. ca. 12.0	HLB ca. 14.1; liquid; cloud point 75°C (1% in deionised water)	
	IMBENTIN-N/85	ethoxylated nonylphenol	'n' no. ca. 12.5	HLB ca. 14.3; liquid; cloud point 85°C (1% in deionised water)	
	IMBENTIN-N/91	ethoxylated nonylphenol	'n' no. ca. 14.0	HLB ca. 14.7; liquid/paste; cloud point 91°C (1% in deionised water)	
	IMBENTIN-N/98	ethoxylated nonylphenol	'n' no. ca. 14.5	HLB ca. 14.9; liquid/paste; cloud point 98°C (1% in deionised water)	
	IMBENTIN-N/185	ethoxylated nonylphenol	'n' no. ca. 18.5	HLB ca. 15.7; solid; cloud point 71°C (1% in 10% NaCl)	personal care products; household detergents; industrial and institutional cleaning; pharmaceuticals
	IMBENTIN-N/200	ethoxylated nonylphenol	'n' no. ca. 20.0	HLB ca. 16.0; solid; cloud point 73°C (1% in 10% NaCl)	
	IMBENTIN-N/265	ethoxylated nonylphenol	'n' no. ca. 26.5	HLB ca. 16.8; solid; cloud point 76°C (1% in 10% NaCl)	
	IMBENTIN-N/300	ethoxylated nonylphenol	'n' no. ca. 30.0	HLB ca. 17.1; solid/flakes; cloud point 78°C (1% in 10% NaCl)	
	IMBENTIN-N/400	ethoxylated nonylphenol	'n' no. ca. 40.0	HLB ca. 17.8; solid/flakes	
	IMBENTIN-N/500	ethoxylated nonylphenol	'n' no. ca. 50.0	HLB ca. 18.2; solid/flakes	
	IMBENTIN-O/050	ethoxylated octylphenol	'n' no. ca. 5.0	HLB ca. 10.3; liquid; cloud point 55°C (5 g in 25 g BDG 25%)	
	IMBENTIN-O/080	ethoxylated octylphenol	'n' no. ca. 8.0	HLB ca. 12.6; liquid; cloud point 49°C (1% in deionised water)	
	IMBENTIN-O/100	ethoxylated octylphenol	'n' no. ca. 10.0	HLB ca. 13.6; liquid; cloud point 68°C (1% in deionised water)	
	IMBENTIN-O/130	ethoxylated octylphenol	'n' no. ca. 13.0	HLB ca. 14.7; paste; cloud point 89°C (1% in deionised water)	
	IMBENTIN-O/150	ethoxylated octylphenol	'n' no. ca. 15.0	HLB ca. 15.2; paste; cloud point 86°C (5 g in 25 g BDG 25%)	
	IMBENTIN-O/200	ethoxylated octylphenol	'n' no. ca. 20.0	HLB ca. 16.2; solid; cloud point 72°C (1% in 10% NaCl)	
	IMBENTIN-O/300	ethoxylated octylphenol	'n' no. ca. 30.0	HLB ca. 17.3; solid; cloud point 74°C (1% in 10% NaCl)	
	IMBENTIN-O/400	ethoxylated octylphenol	'n' no. ca. 40.0	HLB ca. 17.9; solid; cloud point 73°C (5% in 10% NaCl)	
Lonza	CARSONON N-4	nonoxynol 4	100% active; OH value 140	HLB 9; liquid; colour 100 (APHA)	emulsifier
	CARSONON N-6	nonoxynol 6	100% active; OH value 128	HLB 11; liquid; colour 100 (APHA)	household and industrial cleaners and detergents
	CARSONON N-9	nonoxynol 9	100% active; OH value 98	HLB 13; liquid; cloud point 54.4°C; colour 75 (APHA)	

Supplier	Trade name	Chemical description	Composition	General properties	Functionality / Application
Lonza	CARSONON N-12	nonoxynol 12	100% active; OH value 75	HLB 14; liquid; cloud point 87.8°C; colour 75 (APHA)	emulsifier / *household and industrial cleaners and detergents*
	CARSONON N-30	nonoxynol 30	70% active; OH value 36	HLB 17; liquid; cloud point 73.9°C (in 10% salt); colour 150 (APHA)	
	CARSONON N-50	nonoxynol 50	70% active	HLB 18; liquid; cloud point 76.7°C (in 10% salt); colour 3 (Gardner)	
Millchem	NEUTRONYX 600	nonoxynol-9	'n' no. 9.5	liquid; cloud point 54°C	detergent; dispersant; emulsifier; wetting agent / *household detergents; disinfectants; sanitisers; dishwashing agents; laundry products*
	NEUTRONYX 656	nonoxynol-10	'n' no. 11	liquid; cloud point 71°C	
Nikko Chemicals	NIKKOL NP-2	nonoxynol-2		colourless liquid	cosmetics
	NIKKOL NP-5	nonoxynol-5		colourless liquid	
	NIKKOL NP-7.5	nonoxynol-8		colourless liquid	wetting agent; penetrant; emulsifier / *cosmetics*
	NIKKOL NP-10	nonoxynol-10		colourless liquid	
	NIKKOL NP-15	nonoxynol-15		colourless liquid	emulsifier / *cosmetics*
	NIKKOL NP-18TX	nonoxynol-18		colourless liquid	solubiliser / *cosmetics*
	NIKKOL NP-20	nonoxynol-20		white paste	
	NIKKOL OP-3	octoxynol-3		colourless liquid	wetting agent; penetrant; emulsifier / *cosmetics*
	NIKKOL OP-10	octoxynol-10		colourless liquid	
	NIKKOL OP-30	POE (30) octylphenyl ether		white solid	solubiliser / *cosmetics*
PPG	MACOL DNP-10	dinonylphenol 10EO		HLB 11.3; liquid; dispersible in water; viscosity 390 cPs	
	MACOL DNP-21	dinonylphenol 21EO		HLB 14.8; solid; cloud point 91°C (1% aq.)	
	MACOL NP-4	nonylphenol 4EO		HLB 8.9; liquid; viscosity 350 cPs	
	MACOL NP-6	nonylphenol 6EO		HLB 10.9; liquid; dispersible in water; viscosity 300 cPs	
	MACOL NP-9.5	nonylphenol 9.5EO		HLB 12.9; liquid; cloud point 55°C (1% aq.); viscosity 275 cPs	
	MACOL NP-12	nonylphenol 12EO		HLB 14.0; liquid; cloud point 81°C (1% aq.); viscosity 325 cPs	
	MACOL OP-3	octylphenol 3EO		HLB 7.8; liquid; viscosity 350 cPs	lubricant; w/o emulsifier; detergent; coupling agent; wetting agent / *metal working*
	MACOL OP-5	octylphenol 5EO		HLB 10.4; liquid; dispersible in water; viscosity 300 cPs	
	MACOL OP-8	octylphenol 8EO		HLB 12.3; liquid; cloud point 23°C (1% aq.); viscosity 275 cPs	
	MACOL OP-10 SP	modified octylphenol 10EO		HLB 13.4; liquid; cloud point 65°C (1% aq.); viscosity 250 cPs	
	MACOL OP-12	octylphenol 12EO		HLB 14.6; liquid; cloud point 88°C (1% aq.); viscosity 335 cPs	

Supplier	Trade name	Chemical description	Composition	General properties	Functionality / Application
Protex	PROX-ONIC DDP-09	POE (9) dodecyl phenol	'n' no. 9	liquid; cloud point 18°C (1% aq.)	
	PROX-ONIC DDP-012	POE (12) dodecyl phenol	'n' no. 12	liquid; cloud point 76°C (1% aq.)	
	PROX-ONIC DNP-08	POE (8) dinonyl phenol	'n' no. 8	HLB 10.4; liquid; cloud point <25°C (1% aq.)	emulsifier; detergent; wetting agent; dispersant; solubiliser; coupling agent; *metal working; agriculture; paper industry; paint industry; household and industrial applications*
	PROX-ONIC DNP-0150	POE (150) dinonyl phenol	'n' no. 150	HLB 19.0; solid; cloud point >100°C (1% aq.)	
	PROX-ONIC DNP-0150/50	POE (150) dinonyl phenol 50%	'n' no. 150	HLB 19.0; liquid; cloud point >100°C (1% aq.)	
	PROX-ONIC NP-1.5	POE (1.5) nonyl phenol	'n' no. 1.5	HLB 4.6; liquid	
	PROX-ONIC NP-04	POE (4) nonyl phenol	'n' no. 4	HLB 8.9; liquid	
	PROX-ONIC NP-06	POE (6) nonyl phenol	'n' no. 6	HLB 10.9; liquid; cloud point <25°C (1% aq.)	emulsifier; detergent; wetting agent; dispersant; solubiliser; *metal working; agriculture; paper industry; paint industry; household and industrial applications*
	PROX-ONIC NP-09	POE (9) nonyl phenol	'n' no. 9	HLB 13.0; liquid; cloud point 54°C (1% aq.)	
	PROX-ONIC NP-010	POE (10) nonyl phenol	'n' no. 10	HLB 13.5; liquid; cloud point 72°C (1% aq.)	
	PROX-ONIC NP-015	POE (15) nonyl phenol	'n' no. 15	HLB 15.0; paste; cloud point 96°C (1% aq.)	
	PROX-ONIC NP-020	POE (20) nonyl phenol	'n' no. 20	HLB 16.0; solid; cloud point >100°C (1% aq.)	
	PROX-ONIC NP-030	POE (30) nonyl phenol	'n' no. 30	HLB 17.1; solid; cloud point >100°C (1% aq.)	
	PROX-ONIC NP-030/70	POE (30) nonyl phenol 70%	'n' no. 30	HLB 17.1; solid; cloud point >100°C (1% aq.)	
	PROX-ONIC NP-040	POE (40) nonyl phenol	'n' no. 40	HLB 17.8; solid; cloud point >100°C (1% aq.)	
	PROX-ONIC NP-040/70	POE (40) nonyl phenol 70%	'n' no. 40	HLB 17.8; liquid; cloud point >100°C (1% aq.)	
	PROX-ONIC NP-050	POE (50) nonyl phenol	'n' no. 50	HLB 18.2; solid; cloud point >100°C (1% aq.)	
	PROX-ONIC NP-050/70	POE (50) nonyl phenol 70%	'n' no. 50	HLB 18.2; liquid; cloud point >100°C (1% aq.)	
	PROX-ONIC NP-0100	POE (100) nonyl phenol	'n' no. 100	HLB 19.0; solid; cloud point >100°C (1% aq.)	
	PROX-ONIC NP-0100/70	POE (100) nonyl phenol 70%	'n' no. 100	HLB 19.0; liquid; cloud point >100°C (1% aq.)	emulsifier; detergent; wetting agent; dispersant; solubiliser; coupling agent; *metal working; agriculture; paper industry; paint industry; household and industrial applications*
	PROX-ONIC OP-09	POE (9) octyl phenol	'n' no. 9	HLB 13.5; liquid; cloud point 65°C (1% aq.)	
	PROX-ONIC OP-016	POE (16) octyl phenol	'n' no. 16	HLB 15.8; paste; cloud point >100°C (1% aq.)	
	PROX-ONIC OP-030/70	POE (30) octyl phenol 70%	'n' no. 30	HLB 17.3; liquid; cloud point >100°C (1% aq.)	
	PROX-ONIC OP-040/70	POE (40) octyl phenol 70%	'n' no. 40	HLB 17.9; liquid; cloud point >100°C (1% aq.)	
	PROX-ONIC TBP-08	POE (8) tertiary butyl phenol	'n' no. 8		
	PROX-ONIC TBP-030	POE (30) tertiary butyl phenol	'n' no. 30		
	SURFAROX NP 01	1 mole nonyl phenol	'n' no. 1		emulsifier
	SURFAROX NP6	6 mole nonyl phenol ethoxylate	'n' no. 6		
	SURFAROX NP10	10 mole nonyl phenol	'n' no. 10		non-foaming
	SURFAROX NP10P	modified 10 mole nonyl phenol	'n' no. 10		
	SURFAROX NP11C	11 mole nonyl phenol	'n' no. 11		*cosmetics*

Supplier	Trade name	Chemical description	Composition	General properties	Functionality Application
Protex	SURFAROX NP20	20 mole nonyl phenol	'n' no. 20		
	SURFAROX NP20B	20 mole nonyl phenol	'n' no. 20		non-foaming
	SURFAROX NP50	50 mole nonyl phenol	'n' no. 50		
Rhone-Poulenc	IGEPAL CA-890	octylphenol 40EO	100% solids; 'n' no. 40	solid; surface tension 34 dynes/cm	
	IGEPAL CA-897	octylphenol 40EO	70% solids; surface	liquid; surface tension 34 dynes/cm	
	IGEPAL CO-630	nonylphenol 9EO	100% solids; 'n' no. 9	liquid; surface tension 34 dynes/cm	
	IGEPAL CO-850	nonylphenol 20EO	100% solids; 'n' no. 20	liquid; surface tension 41 dynes/cm	
	IGEPAL CO-880	nonylphenol 30EO	100% solids; 'n' no. 30	solid; surface tension 45 dynes/cm	
	IGEPAL CO-887	nonylphenol 30EO	70% solids; 'n' no. 30	liquid; surface tension 45 dynes/cm	emulsifier; stabiliser *emulsion polymerisation*
	IGEPAL CO-890	nonylphenol 40EO	100% solids; 'n' no. 40	solid; surface tension 44 dynes/cm	
	IGEPAL CO-897	nonylphenol 40EO	70% solids; 'n' no. 40	liquid; surface tension 44 dynes/cm	
	IGEPAL CO-970	nonylphenol 50EO	100% solids; 'n' no. 50	solid; surface tension 44 dynes/cm	
	IGEPAL CO-977	nonylphenol 50EO	70% solids; 'n' no. 50	paste; surface tension 44 dynes/cm	
	IGEPAL CO-990	nonylphenol 100EO	100% solids; 'n' no. 100	solid; surface tension 50 dynes/cm	
	IGEPAL CO-997	nonylphenol 100EO	70% solids; 'n' no. 100	paste; surface tension 50 dynes/cm	stabiliser; low foam *emulsion polymerisation*
	IGEPAL DM-730	dinonylphenol 24EO	100% solids; 'n' no. 24	paste; surface tension 36 dynes/cm	stabiliser *emulsion polymerisation*
	IGEPAL DM-880	dinonylphenol 49EO	100% solids; 'n' no. 49	wax; surface tension 44 dynes/cm	
Sandoz	SANDOXYLATE PN 6	alkyl phenol polyglycol ether		liquid	
	SANDOXYLATE PN 8	alkyl phenol polyglycol ether		liquid	
	SANDOXYLATE PN 9	alkyl phenol polyglycol ether		liquid	wetting agent *industrial detergents*
	SANDOXYLATE PN 10	alkyl phenol polyglycol ether		liquid	
	SANDOXYLATE PN 25	alkyl phenol polyglycol ether		liquid	
	SANDOZIN NI	alkyl phenol ethoxylate		liquid	wetting agent; detergent
	SANDOZIN NIE	alkyl phenol ethoxylate		liquid	wetting agent; detergent
Seppic	NONAROX 575	nonylphenol 5.5EO	98-99% active; 'n' no. 5.5	HLB 10.5; liquid	emulsifier
	NONAROX 730	nonylphenol 7EO	98-99% active; 'n' no. 7	HLB 11.7; liquid	wetting agent; detergent
	NONAROX 930	nonylphenol 9EO	98-99% active; 'n' no. 9	HLB 13.0; liquid	wetting agent; detergent; dispersant
	NONAROX 1030	nonylphenol 10EO	98-99% active; 'n' no. 10	HLB 13.4; liquid	wetting agent; detergent; dispersant
	OCTAROX 330	octylphenol 3EO	100% active; 'n' no. 3	HLB 7.8; liquid	emulsifier
	OCTAROX 530	octylphenol 5EO	100% active; 'n' no. 5	HLB 10.4; liquid	emulsifier
	OCTAROX 1630	octylphenol 16EO	70% active; 'n' no. 16	HLB 15.5; liquid	detergent; wetting agent; dispersant
	OCTAROX 4030	octylphenol 40EO	70% active; 'n' no. 40	HLB 17.5; liquid	emulsifier; dispersant
	SIMULSOL 430 NP	nonylphenol 4EO	100% active; 'n' no. 4	HLB 9.0; liquid	emulsifier

Supplier	Trade name	Chemical description	Composition	General properties	Functionality / Application
Seppic	SIMUSOL 575 NP	nonylphenol 5.5EO	100% active; 'n' no. 6	HLB 10.5; liquid	emulsifier
	SIMUSOL 630 NP	nonylphenol 6EO	100% active; 'n' no. 7	HLB 10.9; liquid	wetting agent; detergent
	SIMUSOL 730 NP	nonylphenol 7EO	100% active; 'n' no. 8	HLB 11.7; liquid	
	SIMUSOL 830 NP	nonylphenol 8 EO	100% active; 'n' no. 9	HLB 12.3; liquid	
	SIMUSOL 930 NP	nonylphenol 9EO	100% active; 'n' no. 10	HLB 13.0; liquid	wetting agent; detergent; dispersant
	SIMUSOL 1030 NP	nonylphenol 10EO	100% active; 'n' no. 12	HLB 13.4; liquid	
	SIMUSOL 1230 NP	nonylphenol 12EO	100% active; 'n' no. 30	HLB 14.0; soft paste	emulsifier; dispersant
	SIMUSOL 3030 NP	nonylphenol 30EO		HLB 17.0; paste	
Stepan Europe	MAKON 4	nonylphenol ethoxylate	100% active	water white to yellow liquid	emulsifier; dispersant; wetting agent; low foam; detergent *household, institutional and industrial cleaners; agrochemicals*
	MAKON 5	nonylphenol ethoxylate	100% active	water white to yellow liquid	
	MAKON 6	nonylphenol ethoxylate	100% active	water white to yellow liquid	emulsifier; dispersant; wetting agent; low foam; detergent *household, institutional and industrial cleaners; textile industry; oil industry; agrochemicals*
	MAKON 7	nonylphenol ethoxylate	100% active	water white to yellow liquid to solid	detergent; wetting agent; emulsifier; dispersant *household, institutional and industrial cleaners; textile industry; oil industry; agrochemicals*
	MAKON 8	nonylphenol ethoxylate	100% active	water white to yellow liquid to solid	
	MAKON 9	nonylphenol ethoxylate	100% active	water white to yellow liquid to solid	
	MAKON 10	nonylphenol ethoxylate	100% active	water white to yellow liquid to solid	
	MAKON 11	nonylphenol ethoxylate	100% active	water white to yellow liquid to solid	
	MAKON 12	nonylphenol ethoxylate	100% active	water white to yellow liquid to solid	dispersant; emulsifier; detergent agrochemicals; oil industry
	MAKON 13	nonylphenol ethoxylate	100% active	water white to yellow liquid to solid	
	MAKON 14	nonylphenol ethoxylate	100% active	water white to yellow liquid to solid	
	MAKON 15	nonylphenol ethoxylate	100% active	water white to yellow liquid to solid	
	MAKON 16	nonylphenol ethoxylate	100% active	water white to yellow liquid to solid	
	MAKON 17	nonylphenol ethoxylate	100% active	water white to yellow liquid to solid	
	MAKON 18	nonylphenol ethoxylate	100% active	water white to yellow liquid to solid	
	MAKON 19	nonylphenol ethoxylate	100% active	water white to yellow liquid to solid	
	MAKON 20	nonylphenol ethoxylate	100% active	water white to yellow liquid to solid	
	MAKON 21	nonylphenol ethoxylate	100% active	water white to yellow liquid to solid	
	MAKON 22	nonylphenol ethoxylate	100% active	water white to yellow liquid to solid	
	MAKON 23	nonylphenol ethoxylate	100% active	water white to yellow liquid to solid	
	MAKON 24	nonylphenol ethoxylate	100% active	water white to yellow liquid to solid	
	MAKON 25	nonylphenol ethoxylate	100% active	water white to yellow liquid to solid	dispersant; emulsifier *agrochemicals*
	MAKON 26	nonylphenol ethoxylate	100% active	water white to yellow liquid to solid	
	MAKON 27	nonylphenol ethoxylate	100% active	water white to yellow liquid to solid	
	MAKON 28	nonylphenol ethoxylate	100% active	water white to yellow liquid to solid	
	MAKON 29	nonylphenol ethoxylate	100% active	water white to yellow liquid to solid	

Supplier	Trade name	Chemical description	Composition	General properties	Functionality *Application*
Stepan Europe	MAKON 30	nonylphenol ethoxylate	100% active	water white to yellow liquid to solid	
	MAKON 31	nonylphenol ethoxylate	100% active	water white to yellow liquid to solid	
	MAKON 32	nonylphenol ethoxylate	100% active	water white to yellow liquid to solid	
	MAKON 33	nonylphenol ethoxylate	100% active	water white to yellow liquid to solid	
	MAKON 34	nonylphenol ethoxylate	100% active	water white to yellow liquid to solid	
	MAKON 35	nonylphenol ethoxylate	100% active	water white to yellow liquid to solid	
	MAKON 36	nonylphenol ethoxylate	100% active	water white to yellow liquid to solid	
	MAKON 37	nonylphenol ethoxylate	100% active	water white to yellow liquid to solid	
	MAKON 38	nonylphenol ethoxylate	100% active	water white to yellow liquid to solid	
	MAKON 39	nonylphenol ethoxylate	100% active	water white to yellow liquid to solid	dispersant; emulsifier *agrochemicals*
	MAKON 40	nonylphenol ethoxylate	100% active	water white to yellow liquid to solid	
	MAKON 41	nonylphenol ethoxylate	100% active	water white to yellow liquid to solid	
	MAKON 42	nonylphenol ethoxylate	100% active	water white to yellow liquid to solid	
	MAKON 43	nonylphenol ethoxylate	100% active	water white to yellow liquid to solid	
	MAKON 44	nonylphenol ethoxylate	100% active	water white to yellow liquid to solid	
	MAKON 45	nonylphenol ethoxylate	100% active	water white to yellow liquid to solid	
	MAKON 46	nonylphenol ethoxylate	100% active	water white to yellow liquid to solid	
	MAKON 47	nonylphenol ethoxylate	100% active	water white to yellow liquid to solid	
	MAKON 48	nonylphenol ethoxylate	100% active	water white to yellow liquid to solid	
	MAKON 49	nonylphenol ethoxylate	100% active	water white to yellow liquid to solid	
	MAKON 50	nonylphenol ethoxylate	100% active	water white to yellow liquid to solid	
	MAKON OP 6	alkylphenol ethoxylate	100% active	water white to yellow liquid	detergent; wetting agent; emulsifier *household, institutional and industrial cleaners*
	MAKON OP 9	alkylphenol ethoxylate	100% active	water white to yellow liquid	
	POLYSTEP F 1	alkylphenol ethoxylate; made-to-order	100% active	water white to pale yellow liquid	emulsifier *emulsion polymerisation*
	POLYSTEP F 3	nonylphenol ethoxylate; made-to-order	100% active	water white to pale yellow liquid	
	POLYSTEP F 4	nonylphenol ethoxylate; made-to-order	100% active	water white to pale yellow liquid	
	POLYSTEP F 5	nonylphenol ethoxylate; made-to-order	100% active	water white to pale yellow liquid	
	POLYSTEP F 6	nonylphenol ethoxylate; made-to-order	100% active	water white to pale yellow liquid	
	POLYSTEP F 7	nonylphenol ethoxylate; made-to-order	100% active	water white to pale yellow liquid	
	POLYSTEP F 8	nonylphenol ethoxylate; made-to-order	100% active	water white to pale yellow liquid	
	POLYSTEP F 9	nonylphenol ethoxylate; made-to-order	100% active	water white to pale yellow liquid	
	POLYSTEP F 95 B	nonylphenol ethoxylate; made-to-order	70% active	clear liquid	
Surfachem	SURFAC NO40	nonylphenol condensate 4EO	'n' no. 4		
	SURFAC NO50	nonylphenol condensate 5EO	'n' no. 5		
	SURFAC NO60	nonylphenol condensate 6EO	'n' no. 6		
	SURFAC NO65	nonylphenol condensate 6.5EO	'n' no. 6.5		
	SURFAC NO80	nonylphenol condensate 8EO	'n' no. 8		
	SURFAC NO90	nonylphenol condensate 9EO	'n' no. 9		

Supplier	Trade name	Chemical description	Composition	General properties	Functionality / Application
Surfachem	SURFAC NO100	nonylphenol condensate 10EO	'n' no. 10		
	SURFAC NO120	nonylphenol condensate 12EO	'n' no. 12		
	SURFAC NO140	nonylphenol condensate 14EO	'n' no. 14		
Unger Fabrikker	EMULGATOR F-8	fatty alcohol ethoxylate	≥99% active; 'n' no. 8	HLB 13.7; colourless liquid	
	EMULGATOR U4	nonyl phenol ethoxylate	≥99% active; 'n' no. 4	HLB 8.8; colourless liquid	
	EMULGATOR U6	nonyl phenol ethoxylate	≥99% active; 'n' no. 6	HLB 10.0; colourless liquid	emulsifier; *liquid detergents; industrial preparations*
	EMULGATOR U9	nonyl phenol ethoxylate	≥99% active; 'n' no. 9	HLB 12.8; colourless liquid	
	EMULGATOR U12	nonyl phenol ethoxylate	≥99% active; 'n' no. 12	HLB 13.9; colourless liquid	
Union Carbide	TRITON N-42	nonylphenol ethoxylate	100% active	HLB 9.1; liquid; pour point $-26°C$	antifogging agent; *plastics industry*
	TRITON N-57	nonylphenol ethoxylate	100% active	HLB 10.0; liquid; pour point $-31°C$	emulsifier; *industrial cleaners*
	TRITON N-60	nonylphenol ethoxylate	100% active	HLB 10.9; liquid; pour point $-31°C$	*industrial cleaners*
	TRITON N-101	nonylphenol ethoxylate	100% active	HLB 13.4; liquid; cloud point 54°C (1% aq.); pour point 5°C	*hard-surface cleaners; laundry products*
	TRITON N-111	nonylphenol ethoxylate	100% active	HLB 13.8; liquid; cloud point 72°C (1% aq.); pour point 13°C	*hard-surface cleaners*
	TRITON X-100	octylphenol ethoxylate	100% active	HLB 13.5; liquid; cloud point 65°C (1% aq.); pour point 7°C	*metal cleaning; institutional cleaners*
	TRITON X-15	octylphenol ethoxylate	100% active	HLB 3.6; liquid; pour point $-9°C$	emulsifier; *polishes; waxes*
	TRITON X-35	octylphenol ethoxylate	100% active	HLB 7.8; liquid; pour point $-23°C$	
	TRITON X-45	octylphenol ethoxylate	100% active	HLB 10.4; liquid; cloud point <0°C (1% aq.); pour point $-26°C$	*metal cleaning; dry cleaning*
	TRITON X-100CG	octylphenol ethoxylate	100% active	HLB 13.5; liquid; cloud point 65°C (1% aq.); pour point 7°C	emulsifier; *cosmetics*
	TRITON X-102	octylphenol ethoxylate	100% active	HLB 14.6; liquid; cloud point 88°C (1% aq.); pour point 16°C	*metal cleaning; institutional cleaners*
	TRITON X-114	octylphenol ethoxylate	100% active	HLB 12.4; liquid; cloud point 22°C (1% aq.); pour point $-9°C$	*metal cleaning; textile industry; laundry products*
	TRITON X-120	octylphenol 9-10EO	40% active; 'n' no. 9-10	powder	wetting agent; *agriculture*
	TRITON X-165.70%	octylphenol ethoxylate	70% active	HLB 15.8; liquid; cloud point > 100°C (1% aq.); pour point 13°C	stabiliser; *polymerisation*
	TRITON X-305.70%	octylphenol ethoxylate	70% active	HLB 17.3; liquid; cloud point > 100°C (1% aq.); pour point 2°C	additive; *paint industry; emulsion polymerisation*
	TRITON X-405.70%	octylphenol ethoxylate	70% active	HLB 17.9; liquid; cloud point > 100°C (1% aq.); pour point $-4°C$	*emulsion polymerisation*
	TRITON X-705.70%	octylphenol ethoxylate	70% active	HLB 18.7; liquid; cloud point > 100°C (1% aq.); pour point 6°C	*emulsion polymerisation*
Witco	REWOPAL HV 4	nonylphenol 4EO	100% active; 'n' no. 4	liquid	emulsifier; *mineral oils; petroleum; aliphatic hydrocarbons*

Supplier	Trade name	Chemical description	Composition	General properties	Functionality / Application
Witco	REWOPAL HV 5	nonylphenol 5EO	100% active; 'n' no. 5	liquid	emulsifier / *mineral oils; petroleum; aliphatic hydrocarbons; emulsion polymerisation*
	REWOPAL HV 6	nonylphenol 6EO	100% active; 'n' no. 6	liquid	emulsifier / *mineral oils; petroleum; aliphatic hydrocarbons; degreasers; dishwashing agents*
	REWOPAL HV 8	nonylphenol 8EO	100% active; 'n' no. 8	liquid	emulsifier / *mineral oils; petroleum; aliphatic hydrocarbons; degreasers; wool scouring; emulsion polymerisation*
	REWOPAL HV 9	nonylphenol 9EO	100% active; 'n' no. 9	liquid	emulsifier / *mineral oils; petroleum; aliphatic hydrocarbons; degreasers; wool scouring*
	REWOPAL HV 10	nonylphenol 10EO	100% active; 'n' no. 10	liquid	emulsifier / *aromatic hydrocarbons; chlorinated hydrocarbons; cleaners; emulsion polymerisation*
	REWOPAL HV 25	nonylphenol 25EO	100% active; 'n' no. 25	wax	emulsifier / *fatty acids; waxes; detergents; electroplating; emulsion polymerisation*
	SYNTOPON 2 A	nonylphenol 2EO	100% active; 'n' no. 2	liquid	defoaming agent / *drilling muds*
	SYNTOPON 8 A	octylphenol 6EO	100% active; 'n' no. 6	liquid	emulsifier; detergent; wetting agent / *paint industry; detergents*
	SYNTOPON 8 B	octylphenol 8EO	100% active; 'n' no. 8	liquid	detergent; wetting agent / *metal cleaning*
	SYNTOPON 8 C	octylphenol 9EO	100% active; 'n' no. 9	liquid	emulsifier; dispersant / *plastics industry; industrial detergents*
	SYNTOPON 8 D 1	octylphenol 11EO	100% active; 'n' no. 11	liquid	dispersant; emulsifier; solubiliser / *flavours; fragrances*
	SYNTOPON A	nonylphenol 6EO	100% active; 'n' no. 6	liquid	emulsifier / *agrochemicals; detergents; metal working; plastics industry*
	SYNTOPON A 100	nonylphenol 4EO	100% active; 'n' no. 4	liquid	emulsifier / *agrochemicals; detergents; metal working*
	SYNTOPON B	nonylphenol 8EO	100% active; 'n' no. 8	liquid	emulsifier / *agrochemicals; metal cleaning; textile industry; dry cleaning*
	SYNTOPON C	nonylphenol 9EO	100% active; 'n' no. 9	liquid	emulsifier; wetting agent; solubiliser / *laundry products; industrial detergents; metal working; dry cleaning*
	SYNTOPON D	nonylphenol 10EO	100% active; 'n' no. 10	liquid	emulsifier; wetting agent; solubiliser / *laundry products; agrochemicals*
	SYNTOPON D 2	nonylphenol 12EO	90% active; 'n' no. 12	liquid	emulsifier; wetting agent / *agrochemicals; metal cleaning; detergents*
	SYNTOPON E	nonylphenol 13EO	100% active; 'n' no. 13	liquid/paste	emulsifier; wetting agent / *agrochemicals; metal cleaning; detergents*
	SYNTOPON F	nonylphenol 15EO	100% active; 'n' no. 15	paste	emulsifier / *waxes; oil industry; household products*
	SYNTOPON F 100	nonylphenol 17EO	100% active; 'n' no. 17	paste	emulsifier / *waxes; oil industry*

Supplier	Trade name	Chemical description	Composition	General properties	Functionality / *Application*
Witco	SYNTOPON G	nonylphenol 20EO	100% active; 'n' no. 20	wax	wetting agent; stabiliser / *plastics industry; metal cleaning*
	SYNTOPON H	nonylphenol 30EO	100% active; 'n' no. 30	wax	wetting agent
Zschimmer & Schwarz	MULSIFAN RT 18	nonyl phenol ethoxylate	100% active	HLB 13; liquid	solubiliser; emulsifier / *perfumes; essential oils*
	TISSOCYL HW	alkphenol polyglycol ether		straw-coloured, viscosity depending on temperature	washing agent; detergent / *textile industry*
	TISSOCYL RL 88	alkylphenol polyglycol ether		clear, straw-coloured viscous liquid	washing agent; wetting agent; detergent / *textile industry*
	TISSOCYL SPECIAL	alkylphenol polyglycol ether		yellowish viscous liquid	

Amide ethoxylates

Supplier	Trade name	Chemical description	Composition	General properties	Functionality / *Application*
Akcros Chemicals	ETHYLAN CH	ethoxylated alkylolamide		amber liquid/paste; pour point 14°C; viscosity 250 cSt; pH 8.0 (1% aq.)	foam stabiliser; thickener; emulsifier; antistatic agent; detergent; dispersant / *liquid detergents; hard-surface cleaners; general-purpose cleaners; shampoos; hand cleaning gels; plastics industry; industrial degreasers*
	ETHYLAN CRS	ethoxylated alkylolamide		clear amber liquid; pour point 10°C; viscosity 270 cSt; pH 8.0 (1% aq.)	
	ETHYLAN LM2	ethoxylated alkylolamide; made-to-order		off-white soft paste; pour point 2°C; viscosity 108 cSt (40°C); pH 8.0 (1% aq.)	
Akzo Nobel	ELFAPUR KA 45	coco monoethanol amide ethoxylate (4.5)	100% active	liquid	
	ETHOMID HT/15	polyoxyethylene (5) tallowamide (hydrogenated tallow-alkyl)	OH value 95-120	solid	
	ETHOMID HT/60 PASTILLES	polyoxyethylene (50) tallowamide (hydrogenated tallow-alkyl)	OH value 40-50	pastilles	
	ETHOMID O/15	polyoxyethylene (5) oleamide	OH value 100-120	liquid	
Albright & Wilson	EMPILAN LPZ	coconut monoethanolamide ethoxylate	100% active	soft paste	
	EMPILAN MAA	coconut monoethanolamide ethoxylate	active 100%	paste	
Auchem Cesalpinia	ROLAMID DE 7	alkanolamide polyglycol ether	100% active		*liquid detergents*
BASF	LUTENSOL FSA 10	fatty acid amide 10EO	100% active; 'n' no. 10	cloud point 85°C (in butyl digol)	
CHEM-Y	AMINOL N	PEG-4 rapeseedamide	H_2O 6.5-8.5%	clear, yellowish liquid; viscosity 500 mPas max.; colour 4 max. (Gardner); pH 9.2-10.2 (1% in water)	thickener; emulsifier; foam stabiliser / *shower preparations; shampoos; bath preparations; bubble baths*
	AMINOL TEC N	PEG-4 rapeseedamide	H_2O 6.5-8.5%	clear yellow liquid; viscosity 500 mPas max.; colour 4 max. (Gardner); pH 9.2-10.2 (1% in water)	emulsifier / *metal working; lubricants*
Dac International Surfactants	DAIMMIN S 15	tallow amide ethoxylate			washing agent / *textile industry*
	DAIMMIN S 25	tallow amide ethoxylate			
Henkel	EUMULGIN C4	coconut fatty acid monoethanol amide 4EO	ca. 99% active	liquid	solubiliser; emulsifier / *detergents*

Supplier	Trade name	Chemical description	Composition	General properties	Functionality Application
Kao Corporation	FINDET 118/17.4	ethoxylated coco monoethanolamide	carbon chain composition C$_8$ 5-9%, C$_{10}$ 5-8%, C$_{12}$ 46-52%, C$_{14}$ 14-20%, C$_{16}$ 8-10%, C$_{18}$ 2-4%, C$_{18'}$ 6-9%	liquid/paste; cloud point 80-83°C; colour 6 max. (Gardner)	
Dr. W. Kolb	IMBENTIN -CDEA/100	ethoxylated cocodiethanolamide	'n' no. ca. 10.0	liquid; cloud point 44°C (10% in 10% NaCl)	
	IMBENTIN -CMEA/020	ethoxylated cocomonoethanolamide	'n' no. ca. 2.0	liquid; cloud point 65°C (5 g in 25 g BDG 25%)	
	IMBENTIN -CMEA/045	ethoxylated cocomonoethanolamide	'n' no. ca. 4.5	liquid; cloud point 78°C (5 g in 25 g BDG 25%)	*personal care products; household detergents; industrial and institutional cleaning; pharmaceuticals*
	IMBENTIN -CMEA/100	ethoxylated cocomonoethanolamide	'n' no. ca. 10.0	paste; cloud point 60°C (1% in 10% NaCl)	
	IMBENTIN -CMEA/130	ethoxylated cocomonoethanolamide	'n' no. ca. 13.0	paste; cloud point 63°C (1% in 10% NaCl)	
PPG	MAZAMIDE C-2	PEG-3 cocamide	free fatty acid 0.5% max.		*solubiliser; coupling agent personal care products*
	MAZAMIDE C-5	PEG-6 cocamide	free fatty acid 0.5% max.		
	MAZAMIDE L-5	PEG-6 lauramide	free fatty acid 0.5% max.		
Scher Chemicals	SCHERCOTERGE 140	ethoxylated amide	100% active	liquid	*fulling agent; scouring agent textile industry*
Servo Delden	SERDOX NXC 3H	oleic acid monoethanolamide 3EO		liquid; cloud point 65-70°C (10% in 25% BDG)	
	SERDOX NXC 6H	oleic acid monoethanolamide 6EO		liquid; cloud point 78°C (10% in 25% BDG)	
	SERDOX NXC 14H	oleic acid monoethanolamide 14EO		liquid; cloud point 57-65°C (2% in 10% NaCl soln.)	

Amine ethoxylates

Supplier	Trade name	Chemical description	Composition	General properties	Functionality Application
Akcros Chemicals	ETHYLAN TD3	diamine ethoxylate	100% active; ethylene oxide content 29%; 'n' no. 3	brown liquid; pour point 13°C; pH 10 (5% aq.)	
	ETHYLAN TD10	diamine ethoxylate	100% active; ethylene oxide content 58%; 'n' no. 10	dark brown liquid; pour point -7°C; pH 10 (5% aq.)	
	ETHYLAN TD15	diamine ethoxylate	100% active; ethylene oxide content 67%; 'n' no. 15	dark brown liquid; pour point -10°C; pH 10 (5% aq.)	wetting agent; anti-corrosion; antistatic agent; emulsifier; detergent
	ETHYLAN TH30	fatty amine ethoxylate	100% active; ethylene oxide content 83%; 'n' no. 30	HLB 16.6; yellow waxy solid; pour point 30°C; viscosity 143 cSt (40°C); pH 9.0 (1% aq.)	*textile industry; paint industry; metal working; agriculture; polishes*
	ETHYLAN TLM	fatty amine ethoxylate	100% active; ethylene oxide content 78%; 'n' no. 15	HLB 15.6; pale brown liquid; pour point -5°C; viscosity 114 cSt; pH 10 (5% aq.)	
	ETHYLAN TN10	fatty amine ethoxylate	100% active; ethylene oxide content 70%; 'n' no. 10	HLB 14.0; pale brown liquid; pour point <0°C; viscosity 174 cSt; pH 9.5 (1% aq.)	
	ETHYLAN TT05	fatty amine ethoxylate	100% active; ethylene oxide content 44%; 'n' no. 5	HLB 8.8; pale brown liquid; pour point 10°C; viscosity 130 cSt; pH 10 (5% aq.)	
	ETHYLAN TT07	fatty amine ethoxylate; made-to-order	100% active; ethylene oxide content 53%; 'n' no. 7	HLB 10.6; pale brown liquid; pour point 1°C; viscosity 151 cSt; pH 10 (5% aq.)	

Supplier	Trade name	Chemical description	Composition	General properties	Functionality / Application
Akcros Chemicals	ETHYLAN TT15	fatty amine ethoxylate	100% active; ethylene oxide content 71%; 'n' no. 15	HLB 14.2; pale brown liquid; pour point 0°C; viscosity 252 cSt; pH 9.5 (1% aq.)	wetting agent; anti-corrosion; antistatic agent; emulsifier; detergent *textile industry; paint industry; metal working; agriculture; polishes*
	ETHYLAN TT30	fatty amine ethoxylate	100% active; ethylene oxide content 83%; 'n' no. 30	HLB 16.6; pale brown solid; pour point 27°C; pH 10 (5% aq.)	
	ETHYLAN TT40	fatty amine ethoxylate	100% active; ethylene oxide content 87%; 'n' no. 40	HLB 17.4; pale brown solid; pour point 30°C; pH 10 (5% aq.)	
	ETHYLAN TT203	fatty amine ethoxylate	100% active; ethylene oxide content 27%; 'n' no. 2	HLB 5.5; pale brown liquid; pour point 15°C; viscosity 65 cSt (40°C); pH 9.5 (1% aq.)	
Akzo Nobel	ETHODUOMEEN HT/13	N,N,N'-tris(2-hydroxyethyl)-N-tallow-1,3-diaminopropane (hydrogenated tallow-alkyl)		paste	
	ETHODUOMEEN T/13	N,N,N'-tris (2-hydroxyethyl)-N-tallow-1,3-diaminopropane		liquid	
	ETHODUOMEEN T/20	N,N,N'-polyoxyethylene (10)-N-tallow-1,3-diaminopropane		liquid	
	ETHODUOMEEN T/25	N,N,N'-polyoxyethylene (15)-N-tallow-1,3-diaminopropane		liquid	
	ETHODUOMEEN T/38	N,N,N'-polyoxyethylene (28)-N-tallow-1,3-diaminopropane		liquid/paste	
	ETHOMEEN C/12	cocobis(2-hydroxyethyl)amine		liquid	
	ETHOMEEN C/15	polyoxyethylene (5) cocoamine		liquid	
	ETHOMEEN C/25	polyoxyethylene (15) cocoamine		liquid	
	ETHOMEEN C/215	polyoxyethylene (11.5) cocoamine		liquid	
	ETHOMEEN C/2620	polyoxyethylene (15-20) cocoamine		liquid	
	ETHOMEEN HT/12	tallow bis(2-hydroxyethyl)amine (hydrogenated tallow-alkyl)		solid	
	ETHOMEEN HT/15	polyoxyethylene (5) tallow amine (hydrogenated tallow-alkyl)		solid/liquid	
	ETHOMEEN HT/25	polyoxyethylene (15) tallow amine (hydrogenated tallow-alkyl)		liquid	
	ETHOMEEN HT/60	polyoxyethylene (50) tallow amine (hydrogenated tallow-alkyl)		solid	
	ETHOMEEN S/12	oleylbis(2-hydroxyethyl)amine		liquid	
	ETHOMEEN S/15	polyoxyethylene (5) oleylamine		liquid	
	ETHOMEEN S/20	polyoxyethylene (10) oleylamine		liquid	
	ETHOMEEN S/25	polyoxyethylene (15) oleylamine		liquid	
	ETHOMEEN T/12	tallow bis(2-hydroxyethyl)amine		liquid/paste	
	ETHOMEEN T/15	polyoxyethylene (5) tallow amine		liquid/paste	
	ETHOMEEN T/25	polyoxyethylene (15) tallow amine		liquid	
	ETHOMEEN TO/12	tallow/oleylbis (2-hydroxyethyl)amine		liquid	
Albright & Wilson	EMPILAN AMT SERIES	fatty amine ethoxylates	concentration 100%	liquids, pastes and solids	emulsifier; antistatic agent *agrochemicals; polymer industry*
Allied Colloids	ALCOSIST ACP	ethoxylated amine		liquid	levelling agent *textile industry; dyeing; polyamide and wool fibres*

Supplier	Trade name	Chemical description	Composition	General properties	Functionality / Application
Allied Colloids	ALCOSIST PL	ethoxylated amine		liquid	levelling agent; migration aid; anti-precipitant *textile industry: dyeing of polyamides and wools*
Auschem Cesalpinia	CHIMIPAL NC 15	POE (15) coco amine	100% active; 'n' no. 15	fluid	
	CHIMIPAL NH 6	POE (6) oleic amine	100% active; 'n' no. 6	liquid	
	CHIMIPAL NH 10	POE (10) oleic amine	100% active; 'n' no. 10	fluid/paste	wetting agent; suspending agent
	CHIMIPAL NH 20	POE (20) oleic amine	100% active; 'n' no. 20	paste	
	CHIMIPAL NH 30	POE (30) oleic amine	100% active; 'n' no. 30	waxy	
	CHIMIPAL NS 20	POE (20) tallow amine	100% active; 'n' no. 20	waxy	
BASF	LUTENSOL FA 12	fatty amine 12EO	100% active; 'n' no. 12	cloud point 85°C (in NaCl soln.)	
Bayer	AVOLAN AV 200%	alkylarylamine polyglycol ether			levelling agent *textile industry: dyeing of wools*
Chemax	CHEMEEN 18-2	POE (2) stearyl amine	'n' no. 2	HLB 4.8: solid; M.W. 365	
	CHEMEEN 18-50	POE (50) stearyl amine	'n' no. 50	HLB 17.8: solid; M.W. 2400	
	CHEMEEN C-2	POE (2) coco amine	'n' no. 2	HLB 6.1: liquid; M.W. 290	
	CHEMEEN C-5	POE (5) coco amine	'n' no. 5	HLB 10.4: liquid; M.W. 425	
	CHEMEEN C-15	POE (15) coco amine	'n' no. 15	HLB 15.0: liquid; M.W. 890	emulsifier; additive; antistatic agent; detergent; anti-corrosion *lubricants; textile industry; plastics industry; degreaser formulations*
	CHEMEEN DT-3	POE (3) tallow diamine	'n' no. 3	HLB 4.9: liquid; M.W. 535	
	CHEMEEN DT-15	POE (15) tallow diamine	'n' no. 15	HLB 13.0: liquid; M.W. 1020	
	CHEMEEN DT-30	POE (30) tallow diamine	'n' no. 30	HLB 15.9: liquid; M.W. 1665	
	CHEMEEN HT-5	POE (5) hydrogenated tallow amine	'n' no. 5	HLB 9.0: paste; M.W. 495	
	CHEMEEN HT-15	POE (15) hydrogenated tallow amine	'n' no. 15	HLB 14.3: liquid; M.W. 925	
	CHEMEEN O-30/80	POE (30) oleyl amine 80%	'n' no. 30	HLB 16.5: liquid; M.W. 1600	
	CHEMEEN T-2	POE (2) tallow amine	'n' no. 2	HLB 5.0: paste; M.W. 350	
	CHEMEEN T-5	POE (5) tallow amine	'n' no. 5	HLB 9.0: liquid; M.W. 490	
	CHEMEEN T-15	POE (15) tallow amine	'n' no. 15	HLB 14.3: liquid; M.W. 930	
	CHEMEEN T-20	POE (20) tallow amine	'n' no. 20	HLB 15.7: liquid; M.W. 1120	
Ciba	ALBEGAL SW	alkylamine polyglycol ether		yellow, clear, low viscosity liquid	levelling agent *textile industry: dyeing*
Croda	CRODAMET C5	POE(5) coconut amine		clear amber liquid	
	CRODAMET O2	POE(2) oleyl amine		clear yellow to amber liquid	emulsifier; thickener; antistatic agent *household products*
	CRODAMET T5	POE(5) tallow amine		clear yellow to amber liquid	
	CRODAMET T8	POE(8) tallow amine		clear yellow to amber liquid	
Elf Atochem	DINORAMOX S3	based on NNN'-tri-2-hydroxy ethyl N-tallow propylene diamine		liquid/paste; cloud point 15°C	
	DINORAMOX S7	based on NNN'-polyethoxylated N-tallow propylene diamine		liquid; cloud point 5°C	
	DINORAMOX S12	based on NNN'-polyethoxylated N-tallow propylene diamine		liquid; cloud point −5°C	
	NORAMOX C2	ethoxylated coco amine	'n' no. 2	liquid; cloud point 7°C	
	NORAMOX C5	ethoxylated coco amine	'n' no. 5	liquid; cloud point <10°C	

Supplier	Trade name	Chemical description	Composition	General properties	Functionality / Application
Elf Atochem	NORAMOX C11	ethoxylated coco amine	'n' no. 11	liquid; cloud point −10°C	
	NORAMOX C15	ethoxylated coco amine	'n' no. 15	liquid; cloud point −10°C	
	NORAMOX O2	ethoxylated oleylamine	'n' no. 2	liquid; cloud point 5°C	
	NORAMOX O5	ethoxylated oleylamine	'n' no. 5	liquid; cloud point 0°C	
	NORAMOX O11	ethoxylated oleylamine	'n' no. 11	liquid; cloud point −5°C	
	NORAMOX S2	ethoxylated tallow amine	98% active; 'n' no. 2	paste; cloud point 27°C	
	NORAMOX S5	ethoxylated tallow amine	98% active; 'n' no. 5	liquid; cloud point <15°C	
	NORAMOX S7	ethoxylated tallow amine	98% active; 'n' no. 7	liquid; cloud point 10°C	
	NORAMOX S11	ethoxylated tallow amine	98% active; 'n' no. 11	liquid; cloud point 4°C	
	NORAMOX S15	ethoxylated tallow amine	98% active; 'n' no. 15	liquid	
	NORAMOX S25	tallow amine 25EO	100% active; alkali value 0.73; 'n' no. 25	solid	
	SELLIG LNS 5 100	tallow amine 5EO	100% active; alkali value 2.03; 'n' no. 5	liquid	
	SELLIG LNS 7 100	tallow amine 7EO	100% active; alkali value 1.7; 'n' no. 7	HLB 10.7; liquid	
	SELLIG LNS 11 100	tallow amine 11EO	100% active; 'n' no. 11	liquid	
	SELLIG LNS 25 100	tallow amine 25EO	100% active; alkali value 0.73; 'n' no. 25	solid	
Henkel	ARAPHEN K 100	coconut amine 12EO	99-100% active	liquid	degreaser; cleaners
	FIXEGAL N	fatty amine polyglycolether			levelling agent; *textile industry; dyeing*
	OSIMOL R	alkylamine polyglycolether			
	POLYQUART H 81	PEG-15 coco polyamine	49-51%	liquid	antistatic agent; conditioner; *shampoos; hair care preparations*
Hoechst	GENAGEN CA-050	ethoxylated cocoalkanolamine	100% active	liquid	*cosmetics; detergents*
	GENAMIN C020	coco amine 2EO	100% active; 'n' no. 2	liquid	
	GENAMIN C050	coco amine 5EO	100% active; 'n' no. 5	liquid	
	GENAMIN C100	coco amine 10EO	100% active; 'n' no. 10	liquid	antistatic agent; *detergents; textile industry; agrochemicals*
	GENAMIN C150	coco amine 15EO	100% active; 'n' no. 15	liquid	
	GENAMIN C200	coco amine 20EO	100% active; 'n' no. 20	liquid	
	GENAMIN O-020	oleyl amine 2EO	100% active; 'n' no. 2	liquid	
	GENAMIN O-050	oleyl amine 5EO	100% active; 'n' no. 5	liquid	
	GENAMIN O-080	oleyl amine 8EO	100% active; 'n' no. 8	liquid	antistatic agent; *detergents; textile industry*
	GENAMIN O-200	oleyl amine 20EO	100% active; 'n' no. 20	liquid	
	GENAMIN S-020	stearyl amine 2EO	100% active; 'n' no. 2	liquid	
	GENAMIN S-080	stearyl amine 8EO	100% active; 'n' no. 8	liquid	
	GENAMIN S-100	stearyl amine 10EO	100% active; 'n' no. 10	liquid	
	GENAMIN S-120	stearyl amine 12EO	100% active; 'n' no. 12	liquid	
	GENAMIN S-150	stearyl amine 15EO	100% active; 'n' no. 15	liquid	
	GENAMIN S-200	stearyl amine 20EO	100% active; 'n' no. 20	liquid	
	GENAMIN S-250	stearyl amine 25EO	100% active; 'n' no. 25	paste	

Supplier	Trade name	Chemical description	Composition	General properties	Functionality / Application
Hoechst	GENAMIN T-020	tallow amine 2EO	100% active; 'n' no. 2		antistatic agent
	GENAMIN T-150	tallow amine 15EO	100% active; 'n' no. 15		detergents; textile industry; agrochemicals
	GENAMIN T-200	tallow amine 20EO	100% active; 'n' no. 20		
Hüls	MARLAZIN L 10	laurylamine 10EO	active detergent 100%; 'n' no. 10	liquid; cloud point 65°C (0.1% in 6% NaOH soln.)	low-foaming cleaners; textile industry
	MARLAZIN OL 2	oleylamine 2EO	active detergent 100%; 'n' no. 2	liquid	cleaners
	MARLAZIN OL 20	oleylamine 20EO	active detergent 100%; 'n' no. 20	liquid; cloud point 70°C (2% in 10% NaCl soln.)	
	MARLAZIN S 10	stearylamine 10EO	active detergent 100%; 'n' no. 10	liquid; cloud point 72°C (2% in 10% NaCl soln.)	
	MARLAZIN S 40	stearylamine 40EO	active detergent 100%; 'n' no. 40	solid; cloud point 84°C (2% in 10% NaCl soln.)	
	MARLAZIN T 10	tallow fatty amine 10EO	active detergent 100%; 'n' no. 10	liquid; cloud point 71°C (2% in 10% NaCl soln.)	textile industry
	MARLAZIN T 15/2	tallow fatty amine 15EO	active detergent 100%; 'n' no. 15	liquid/paste	
	MARLAZIN T 16/1	tallow fatty amine 16EO	active detergent 100%; 'n' no. 16	solid	
	MARLAZIN T 50	tallow fatty amine 50EO	active detergent 70%; 'n' no. 50	liquid; cloud point 85°C (2% in 10% NaCl soln.)	dispersant
					fabric softeners
	MARLOWET 5400	alkylamine polyethylene glycol ether	active detergent 100%	liquid	emulsifier
					mineral and paraffin oils; car rinsing agents; polishes; textile industry
	MARLOWET OAM	alkylamine polyethylene glycol ether	active detergent 100%	liquid	anti-corrosion
	MARLOWET OAM SPEC.	alkylamine polyethylene glycol ether	active detergent 100%	liquid	lubricants
ICI	ATLAS G-3770	POE-(10)-tallow amine		yellow-brown waxy liquid	
	ATLAS G-3780A	POE alkyl amine		HLB 15.5; amber liquid	
	ATLAS G-3835	POE fatty amine		HLB 10.6; yellow liquid	household and industrial applications; agrochemicals; textile industry
	ATLAS G-4961	POE alkyl amine		HLB 15.5; amber liquid	
	ATLAS G-4962	POE fatty amine		HLB 10.5; red-amber liquid	
	ATLOX 4879	POE-fatty amine		HLB 9.5; red-brown liquid	emulsifier; dispersant
					agrochemicals
Kao Corporation	AMIET CD/14	polyethoxylated coconut fatty amine	amine value 187-200; 'n' no. 2	liquid	
	AMIET CD/17	polyethoxylated coconut fatty amine	amine value 129-139; 'n' no. 5	liquid; colour 12 max. (Gardner)	
	AMIET CD/27	polyethoxylated coconut fatty amine	amine value 63-67; 'n' no. 15	liquid; colour 10 max. (Gardner)	
	AMIET DO/15	polyethoxylated oleyl diamine	amine value 224-235; 'n' no. 3	liquid; colour 15 max. (Gardner)	
	AMIET DT/15	polyethoxylated tallow diamine	amine value 232-241; 'n' no. 3	liquid	
	AMIET DT/19	polyethoxylated tallow diamine	amine value 170-175; 'n' no. 7	liquid	
	AMIET OD/14.4	polyethoxylated oleyl amine	amine value 145-155; 'n' no. 2.4	liquid	
	AMIET OD/17	polyethoxylated oleyl amine	amine value 111-118; 'n' no. 5	liquid; colour 10 max. (Gardner)	
	AMIET OD/24	polyethoxylated oleyl amine	amine value 68-72; 'n' no. 12	liquid; colour 10 max. (Gardner)	
	AMIET OD/27	polyethoxylated oleyl amine	amine value 55-65; 'n' no. 15	paste	
	AMIET OD/32	polyethoxylated oleyl amine	amine value 46-51; 'n' no. 20	solid; colour 9 max. (Gardner)	
	AMIET TD/14	polyethoxylated tallow amine	amine value 156-165; 'n' no. 2	liquid; colour 9 max. (Gardner)	
	AMIET TD/17	polyethoxylated tallow amine	amine value 113-121; 'n' no. 5	liquid; colour 10 max. (Gardner)	

Supplier	Trade name	Chemical description	Composition	General properties	Functionality Application
Kao Corporation	AMIET TD/19	polyethoxylated tallow amine	amine value 92-104; 'n' no. 7	liquid	
	AMIET TD/20	polyethoxylated tallow amine	amine value 88-94; 'n' no. 8	liquid; colour 10 max. (Gardner)	
	AMIET TD/23	polyethoxylated tallow amine	amine value 73-78; 'n' no. 11	liquid; colour 10 max. (Gardner)	
	AMIET TD/27	polyethoxylated tallow amine	amine value 59-63; 'n' no. 15	liquid; colour 8 max. (Gardner)	
	AMIET TD/32	polyethoxylated tallow amine	amine value 47-52; 'n' no. 20	paste; colour 8 max. (Gardner)	
	AMIET TD/42	polyethoxylated tallow amine	amine value 34-38; 'n' no. 30	solid; colour 8 max. (Gardner)	
	AMIET THD/21	polyethoxylated hydrogenated tallow amine	amine value 81-86; 'n' no. 9	paste; colour 7 max. (Gardner)	
Dr. W. Kolb	IMBENTIN-CAM/020	ethoxylated cocoamine; $C_{12/18}$ carbon chain	amine value ca. 160; 'n' no. ca. 2.2	liquid	personal care products; household detergents; industrial and institutional cleaning; pharmaceuticals
	IMBENTIN-CAM/120	ethoxylated cocoamine; $C_{12/18}$ carbon chain	amine value ca. 75; 'n' no. ca. 12.0	liquid	
	IMBENTIN-DC/030	ethoxylated triethanolamine	amine value 200; 'n' no. ca. 3.0	liquid	
	IMBENTIN-DT/030	ethoxylated tallow propylenediamine	amine value 235; 'n' no. ca. 3.0	liquid	
	IMBENTIN-DT/050	ethoxylated tallow propylenediamine	amine value 215; 'n' no. ca. 5.0	liquid	
	IMBENTIN-DT/100	ethoxylated tallow propylenediamine	amine value 180; 'n' no. ca. 10.0	liquid	
	IMBENTIN-EDA/040	ethoxylated ethylenediamine	amine value 475; 'n' no. ca. 4.0	liquid	
	IMBENTIN-OAM/020	ethoxylated oleylamine; C_{18} carbon chain	amine value ca. 160; 'n' no. ca. 2.0	liquid	
	IMBENTIN-OAM/200	ethoxylated oleylamine; C_{18} carbon chain	amine value ca. 50; 'n' no. ca. 20.0	paste	
	IMBENTIN-SAM/085	ethoxylated stearylamine; C_{18} carbon chain	amine value ca. 91; 'n' no. ca. 8.5	liquid	
	IMBENTIN-SAM/100	ethoxylated stearylamine; C_{18} carbon chain	amine value ca. 80; 'n' no. ca. 10.0	liquid	
	IMBENTIN-SAM/350	ethoxylated stearylamine; C_{18} carbon chain	amine value ca. 32; 'n' no. ca. 35.0	solid	
	IMBENTIN-SAM/400	ethoxylated stearylamine; C_{18} carbon chain	amine value ca. 30; 'n' no. ca. 40.0	solid	
	IMBENTIN-TAM/020	ethoxylated tallow amine; $C_{16/18}$ carbon chain	amine value ca. 160; 'n' no. ca. 2.0	liquid	
	IMBENTIN-TAM/050	ethoxylated tallow amine; $C_{16/18}$ carbon chain	amine value ca. 115; 'n' no. ca. 5.0	liquid	
	IMBENTIN-TAM/080	ethoxylated tallow amine; $C_{16/18}$ carbon chain	amine value ca. 91; 'n' no. ca. 8.0	liquid	
	IMBENTIN-TAM/100	ethoxylated tallow amine; $C_{16/18}$ carbon chain	amine value ca. 80; 'n' no. ca. 10.0	liquid	
	IMBENTIN-TAM/120	ethoxylated tallow amine; $C_{16/18}$ carbon chain	amine value ca. 70; 'n' no. ca. 12.0	liquid	

Supplier	Trade name	Chemical description	Composition	General properties	Functionality Application
Dr. W. Kolb	IMBENTIN-TAM/150	ethoxylated tallow amine; $C_{16/18}$ carbon chain	amine value ca. 60; 'n' no. ca. 15.0	liquid	
	IMBENTIN-TAM/200	ethoxylated tallow amine; $C_{16/18}$ carbon chain	amine value ca. 51; 'n' no. ca. 20.0	liquid	*personal care products; household detergents; industrial and institutional cleaning; pharmaceuticals*
	IMBENTIN-TAM/300	ethoxylated tallow amine; $C_{16/18}$ carbon chain	amine value ca. 35; 'n' no. ca. 30.0	solid	
	IMBENTIN-TAM/400	ethoxylated tallow amine; $C_{16/18}$ carbon chain	amine value ca. 28; 'n' no. ca. 40.0	solid	
PPG	LAROSTAT C-2	ethoxylated cocoamine			antistatic agent
	LAROSTAT T-2	ethoxylated tallow amine			*polyolefins*
	MAZEEN C-2	coco amine 2EO		liquid	rust inhibitor; viscosity modifier *metal working*
	MAZEEN C-5	coco amine 5EO		liquid	rust inhibitor *metal working*
	MAZEEN C-15	coco amine 15EO		liquid	o/w emulsifier *metal working*
	MAZEEN T-2	tallow amine 2EO		gel	rust inhibitor; viscosity modifier *metal working*
	MAZEEN T-3.5	tallow amine 3.5EO		liquid	
	MAZEEN T-5	tallow amine 5EO		liquid	lubricant; emulsifier; rust inhibitor; viscosity modifier *metal working*
Protex	PROX-ONIC DT-03	POE (3) tallow diamine	'n' no. 3	HLB 5.5; liquid; M.W. 475; pH 8-10 (10% aq. soln.)	
	PROX-ONIC DT-015	POE (15) tallow diamine	'n' no. 15	HLB 13.0; liquid; M.W. 1020; pH 8-10 (10% aq. soln.)	
	PROX-ONIC DT-030	POE (30) tallow diamine	'n' no. 30	HLB 15.9; liquid; M.W. 1665; pH 8-10 (10% aq. soln.)	
	PROX-ONIC MC-02	POE (2) coco amine	'n' no. 2	HLB 6.1; liquid; M.W. 290; pH 8-10 (10% aq. soln.)	
	PROX-ONIC MC-05	POE (5) coco amine	'n' no. 5	HLB 10.4; liquid; M.W. 425; pH 8-10 (10% aq. soln.)	emulsifier; antistatic agent; anti-corrosion; detergent; dyeing assistant
	PROX-ONIC MC-015	POE (15) coco amine	'n' no. 15	HLB 15.0; liquid; M.W. 890; pH 8-10 (10% aq. soln.)	*textile industry; degreasing formulations; plastics industry; metal working*
	PROX-ONIC MHT-05	POE (5) hydrogenated tallow amine	'n' no. 5	HLB 9.0; paste; M.W. 495; pH 8-10 (10% aq. soln.)	
	PROX-ONIC MHT-015	POE (15) hydrogenated tallow amine	'n' no. 15	HLB 14.3; liquid; M.W. 925; pH 8-10 (10% aq. soln.)	
	PROX-ONIC MO-02	POE (2) oleyl amine	'n' no. 2	HLB 4.1; liquid; M.W. 358; pH 8-10 (10% aq. soln.)	
	PROX-ONIC MO-015	POE (15) oleyl amine	'n' no. 15	HLB 14.2; liquid; M.W. 930; pH 8-10 (10% aq. soln.)	
	PROX-ONIC MO-030	POE (30) oleyl amine	'n' no. 30	HLB 16.5; liquid; M.W. 1600; pH 8-10 (10% aq. soln.)	
	PROX-ONIC MO-030-80	POE (30) oleyl amine	'n' no. 30	HLB 16.5; pH 8-10 (10% aq. soln.)	

Supplier	Trade name	Chemical description	Composition	General properties	Functionality Application
Protex	PROX-ONIC MS-02	POE (2) stearyl amine	'n' no. 2	HLB 6.9; solid; M.W. 353; pH 8-10 (10% aq. soln.)	
	PROX-ONIC MS-05	POE (5) stearyl amine	'n' no. 5	HLB 9.0; solid; M.W. 485; pH 8-10 (10% aq. soln.)	
	PROX-ONIC MS-011	POE (11) stearyl amine	'n' no. 11	HLB 12.9; liquid; M.W. 750; pH 8-10 (10% aq. soln.)	
	PROX-ONIC MS-050	POE (50) stearyl amine	'n' no. 50	HLB 17.8; solid; M.W. 2465; pH 8-10 (10% aq. soln.)	emulsifier; antistatic agent; anti-corrosion; detergent; dyeing assistant
	PROX-ONIC MT-02	POE (2) tallow amine	'n' no. 2	HLB 5.0; paste; M.W. 350; pH 8-10 (10% aq. soln.)	textile industry; degreasing formulations; plastics industry; metal working
	PROX-ONIC MT-05	POE (5) tallow amine	'n' no. 5	HLB 9.0; liquid; M.W. 490; pH 8-10 (10% aq. soln.)	
	PROX-ONIC MT-015	POE (15) tallow amine	'n' no. 15	HLB 14.3; liquid; M.W. 930; pH 8-10 (10% aq. soln.)	
	PROX-ONIC MT-020	POE (20) tallow amine	'n' no. 20	HLB 15.7; liquid; M.W. 1150; pH 8-10 (10% aq. soln.)	
Sandoz	DERMAGEN PR	fatty amine polyglycol ether		liquid	levelling agent textile industry; dyes; polyamide fibres
	ELFUGIN BM	fatty amine polyglycol ether		liquid	antistatic agent
	LYOGEN CFA	fatty amine polyglycol ether in aq. soln.		liquid	levelling agent textile industry; dyeing
	LYOGEN MS	fatty amine polyglycol ether in aq. medium		liquid	
	LYOGEN NLM	fatty amine polyglycol ether in aq. medium		liquid	
	LYOGEN WD	fatty amine polyglycol ether in aq. soln.		liquid	levelling agent dyeing; textile industry
	SANDOGEN NH	fatty amine polyglycol ether		liquid	levelling agent textile industry; dyeing of polyamides; dyes
	SANDOQUAD C-ISM	aliphatic amine polyglycol ether		liquid	industrial detergents; vehicle cleaning
	SANDOQUAD T5M	fatty amine polyglycol ether and solvent in aq. soln.		liquid	industrial detergents; antistatic products
	SANDOXYLATE NT 2	tallow amine ethoxylate		paste	wetting agent industrial detergents
	SANDOXYLATE NT 5	fatty amine polyglycol ether		liquid	wetting agent industrial detergents
Scher Chemicals	THIOTAN TR	fatty amine polyglycol ether in aq. soln.		liquid	
	SCHERCOPOL DS-120	modified ethoxylated alkylamine	20% active	liquid	dispersant textile industry
	SCHERCOPOL DS-140	modified ethoxylated alkylamine	40% active	liquid	dispersant textile industry
Seppic	MONTEGAL SH 25 150	stearylamine 25EO	42% active; 'n' no. 25	liquid	retardant for dyes; levelling agent
Servo Delden	NEOLISAL HCN	oleylamine 10EO	100% active; 'n' no. 10	liquid	retardant for dyes; levelling agent
	SERDOX NCA 5	cocoamine 5EO		liquid	retardant for dyes; stripping agent

Supplier	Trade name	Chemical description	Composition	General properties	Functionality / *Application*
Servo Delden	SERDOX NCA 15	cocoamine 15EO		liquid; cloud point 70-72°C (2% in 10% NaCl soln.)	emulsifier; wetting agent; anti-corrosion *household, institutional and industrial cleaners*
	SERDOX NJAD 10 S	tallow amine 10EO		liquid; cloud point 70-72°C (2% in 10% NaCl soln.)	
	SERDOX NJAD 20	tallow amine 20EO		solid; cloud point 86-89°C (1% in 10% NaCl soln.)	
	SERDOX NJAD 33 S	tallow amine 33EO		solid; cloud point 83-85°C (2% in 10% NaCl soln.)	
	SERDOX NJAD 45	tallow amine 45EO		solid; cloud point 83-87°C (1% in 10% NaCl soln.)	
	SERDOX NRA 4	oleylamine 4EO		liquid; cloud point 72-75°C (10% in 25% BDG)	
	SERDOX NRA 10	oleylamine 10EO		liquid; cloud point 64°C (1% in 10% NaCl soln.)	
Stepan Europe	SECOMINE TA SERIES	tallow amine ethoxylates	100% active	beige liquid to paste	
Surfachem	SURFAC TDA100	tallow diamine condensate 10EO	'n' no. 10		
	SURFAC TFA150	tallow amine condensate 15EO	'n' no. 15		
Tomah Products	TOMAH E-14-2	ethoxylated isodecyloxypropyl amine	95-100% active; amine value 170-190; typical combining weight 310; 'n' no. 2	amber liquid	emulsifier; anti-corrosion *textile industry*
	TOMAH E-14-5	ethoxylated isodecyloxypropyl amine	95-100% active; amine value 124-128; typical combining weight 445; 'n' no. 5	amber liquid	
	TOMAH E-14-15	ethoxylated isodecyloxypropyl amine	95-100% active; amine value 62-64; typical combining weight 885; 'n' no. 15	amber liquid	
	TOMAH E-17-2	ethoxylated isotridecyloxypropyl amine	95-100% active; amine value 150-165; typical combining weight 335; 'n' no. 2	amber liquid	
	TOMAH E-18-2	ethoxylated C_{18} fatty amine	95-100% active; amine value 153-163; typical combining weight 360; 'n' no. 2	wax	
	TOMAH E-18-5	ethoxylated C_{18} fatty amine	95-100% active; amine value 112-117; typical combining weight 490; 'n' no. 5	paste	
	TOMAH E-18-8	ethoxylated C_{18} fatty amine	95-100% active; amine value 88-95; typical combining weight 610; 'n' no. 8	amber liquid	
	TOMAH E-18-10	ethoxylated C_{18} fatty amine	95-100% active; amine value 77-81; typical combining weight 710; 'n' no. 10	amber liquid	
	TOMAH E-18-15	ethoxylated C_{18} fatty amine	95-100% active; amine value 58-62; typical combining weight 930; 'n' no. 15	amber liquid	
	TOMAH E-DT-3	ethoxylated tallow fatty amine	95-100% active; amine value 230-245; typical combining weight 240; 'n' no. 3	amber liquid	
	TOMAH E-S-2	ethoxylated soya amine	95-100% active; amine value 145-160; typical combining weight 358; 'n' no. 2	amber liquid	
	TOMAH E-S-5	ethoxylated soya amine	95-100% active; amine value 113-119; typical combining weight 485; 'n' no. 5	amber liquid	
	TOMAH E-S-15	ethoxylated soya amine	95-100% active; amine value 57-63; typical combining weight 925; 'n' no. 15	amber liquid	

Supplier	Trade name	Chemical description	Composition	General properties	Functionality *Application*
Tomah Products	TOMAH E-T-2	ethoxylated tallow fatty amine	95-100% active; amine value 151-161; typical combining weight 360; 'n' no. 5	paste	
	TOMAH E-T-5	ethoxylated tallow fatty amine	95-100% active; amine value 112-120; typical combining weight 485; 'n' no. 5	amber liquid	emulsifier; anti-corrosion *textile industry*
	TOMAH E-T-15	ethoxylated tallow fatty amine	95-100% active; amine value 59-63; typical combining weight 925; 'n' no. 15	amber liquid	
Union Carbide	TRITON RW-20	tertiary alkylamine ethoxylate	100% active; 'n' no. 2	HLB 6-8; liquid; pour point − 20°C	o/w emulsifier; degreaser
	TRITON RW-50	tertiary alkylamine ethoxylate	100% active; 'n' no. 5	HLB 12-14; liquid; cloud point < 0°C (1% aq.); pour point − 32°C	o/w emulsifier; detergent *laundry products*
	TRITON RW-75	tertiary alkylamine ethoxylate	100% active; 'n' no. 7.5	HLB 14-16; liquid; cloud point 32°C (1% aq.); pour point − 5°C	*metal cleaning; laundry products*
	TRITON RW-100	tertiary alkylamine ethoxylate	100% active; 'n' no. 10	HLB 16; liquid; cloud point 67°C (1% aq.); pour point 0°C	*metal cleaning; laundry products*
	TRITON RW-150	tertiary alkylamine ethoxylate	100% active; 'n' no. 15	HLB >16; liquid; cloud point 96°C (1% aq.); pour point 19°C	*metal cleaning*
Witco	ETHOXAMINE SF 11	modified tallow amine 11EO	100% active; 'n' no. 11	liquid	emulsifier; anti-corrosion; wetting agent *printing inks; metal cleaning*
	ETHOXAMINE SF 15	modified tallow amine 15EO	100% active; 'n' no. 15	liquid	emulsifier; anti-corrosion; wetting agent *printing inks; metal cleaning*
	REWOPAL C 6	coconut fatty acid monoethanolamide 6EO	100% active	liquid	emulsifier; foam booster; solubiliser *perfumes*
	REWOPAL TPD 30	fatty amine ethoxylate	100% active	liquid	emulsifier
	WITCAMIDE RAD 0500	polyoxyethylated rosin amine		liquid; pour point 16.1°C; pH 10	
	WITCAMIDE RAD 0515	polyoxyethylated rosin amine		liquid; pour point 16.7°C; pH 10	
	WITCAMIDE RAD 1100	polyoxyethylated rosin amine		liquid; pH 10	anti-corrosion *petroleum industry*
	WITCAMIDE RAD 1110	polyoxyethylated rosin amine		liquid; pH 10	
Zschimmer & Schwarz	SETAVIN RE	alkylamine polyglycol ether		clear, straw-coloured to brown liquid	levelling agent; dispersant *textile industry*

Block polymers

Akcros Chemicals	MONOLAN 2000/ E12	condensate of ethylene oxide with propylene oxide, having two terminal hydroxyl groups	100% active; ethylene oxide content 12%	clear colourless liquid; M.W. 2000; pour point < 0°C; viscosity 390 cSt; pH 7.0 (1% aq.)	defoaming agent; low foam; wetting agent *agrochemicals; emulsion polymerisation; resins; latexes; textile industry*
	MONOLAN 2500/ E30	condensate of ethylene oxide with propylene oxide, having two terminal hydroxyl groups	100% active; ethylene oxide content 30%	clear colourless liquid; cloud point 35°C (1% aq.); M.W. 2500; pour point < 0°C; viscosity 650 cSt; pH 7.0 (1% aq.)	

Supplier	Trade name	Chemical description	Composition	General properties	Functionality Application
Akcros Chemicals	MONOLAN 8000/E80	block polymer	100% active	white flakes; cloud point 100°C (1% aq.); pour point 49°C; viscosity 1100 cSt (60°C); pH 5.5-7.5 (1% aq.)	co-dispersant; *aq. pigment systems*
	MONOLAN 8000/E80 FLAKE	condensate of ethylene oxide with propylene oxide, having two terminal hydroxyl groups	100% active; ethylene oxide content 80%	white flake; cloud point 100°C (1% aq.); pour point 49°C, viscosity 1100 cSt (60°C); pH 7.0 (1% aq.)	defoaming agent; low foam; wetting agent; *agrochemicals; emulsion polymerisation; resins; latexes; textile industry*
	MONOLAN P165	complex ethylene oxide-propylene oxide copolymer having three terminal hydroxyl groups	100% active	clear amber liquid; pour point <0°C; viscosity 600 cSt	*hydraulic fluids*
	MONOLAN P222	condensate of ethylene oxide with propylene oxide, having two terminal hydroxyl groups	100% active	clear colourless liquid; pour point <0°C; viscosity 440 cSt; pH 7.0 (1% aq.)	defoaming agent; low foam; wetting agent; *agrochemicals; emulsion polymerisation; resins; latexes; textile industry*
	MONOLAN PB	complex ethylene oxide-propylene oxide copolymer having three terminal hydroxyl groups	100% active	colourless liquid; cloud point 19°C (1% aq.); pour point <0°C; viscosity 680 cSt	
	MONOLAN PC	complex ethylene oxide-propylene oxide copolymer having three terminal hydroxyl groups	100% active	colourless liquid; cloud point 44°C (1% aq.); pour point 8°C; viscosity 800 cSt	defoaming agent; low foam; wetting agent; *agrochemicals; textile industry*
	MONOLAN PK	high molecular weight trifunctional ethylene oxide-propylene oxide copolymer	100% active	clear pale yellow liquid; pour point <0°C; viscosity 500 cSt	defoaming agent; *sugar beet extraction; fermentation industry*
	MONOLAN PL	high molecular weight trifunctional ethylene oxide-propylene oxide copolymer	100% active	clear pale yellow liquid; pour point <0°C; viscosity 930 cSt	
	MONOLAN PT	complex ethylene oxide-propylene oxide copolymer having three terminal hydroxyl groups	100% active	colourless liquid; cloud point 28°C (1% aq.); pour point <0°C; viscosity 750 cSt	defoaming agent; low foam; wetting agent; *agrochemicals; textile industry*
Akzo Nobel	BEROL 370	block polymer	100% active	cloud point 32°C (1% in H_2O); M.W. 1400	low foam; *rinse aids*
	BEROL 374	block polymer	100% active	cloud point 25°C (1% in H_2O); M.W. 2200	low foam; *dishwashing agents*
BASF	PLURONIC PE 3100	PO-EO block copolymer, 10% EO	100% active	cloud point 40°C (in H_2O)	
	PLURONIC PE 4300	PO-EO block copolymer, 30% EO	100% active	cloud point 60°C (in butyl digol)	
	PLURONIC PE 6100	PO-EO block copolymer, 10% EO	100% active	cloud point 23°C (in H_2O)	
	PLURONIC PE 6200	PO-EO block copolymer, 20% EO	100% active	cloud point 33°C (in H_2O)	
	PLURONIC PE 6400	PO-EO block copolymer, 40% EO	100% active	cloud point 59°C (in H_2O)	
	PLURONIC PE 6800	PO-EO block copolymer, 80% EO	100% active	cloud point 88°C (in NaCl soln.)	
	PLURONIC PE 8100	PO-EO block copolymer, 10% EO	100% active	cloud point 36°C (in butyl digol)	low foam
	PLURONIC PE 9200	PO-EO block copolymer, 20% EO	100% active	cloud point 49°C (in butyl digol)	
	PLURONIC PE 9400	PO-EO block copolymer, 40% EO	100% active	cloud point 80°C (in butyl digol)	
	PLURONIC PE 10100	PO-EO block copolymer, 10% EO	100% active	cloud point 35°C (in butyl digol)	
	PLURONIC PE 10500	PO-EO block copolymer, 50% EO	100% active	cloud point 75°C (in NaCl soln.)	
	PLURONIC RPE 2520	EO-PO block copolymer, 20% EO	100% active	cloud point 31°C (in butyl digol)	
	PLURONIC RPE 3110	EO-PO block copolymer, 10% EO	100% active	cloud point 25°C (in butyl digol)	

Supplier	Trade name	Chemical description	Composition	General properties	Functionality Application
Dr. Th. Boehme	FILIPAN 7521	ethylene oxide-propylene oxide block polymer			spin finish *textile industry; fibre production*
Chemax	CHEMAL BP-261	difunctional block polymer ending in primary hydroxyl group	100% active	HLB 3.0: liquid; cloud point 24°C (1% aq.)	defoaming agent *metal working; cosmetics; paper industry; textile industry; dishwashing agents; lubricants*
	CHEMAL BP-262	difunctional block polymer ending in primary hydroxyl group	100% active	HLB 7.0: liquid; cloud point 30°C (1% aq.)	
	CHEMAL BP-262LF	difunctional block polymer ending in primary hydroxyl group	100% active	HLB 6.5: liquid; cloud point 28°C (1% aq.)	
	CHEMAL BP-2101	difunctional block polymer ending in primary hydroxyl group	100% active	HLB 1.0: liquid; cloud point 16°C (1% aq)	
Henkel	TEXADRIL 2010	EO/PO block copolymer	ca. 100% active	liquid	low foam; wetting agent; co-emulsifier *plastics industry; coatings*
Hoechst	GENAPOL PF10	block polymer (10% EO)	100% active	clear liquid	low foam: dispersant *detergents; dishwashing agents*
	GENAPOL PF20	block polymer (20% EO)	100% active	clear liquid	
	GENAPOL PF40	block polymer (40% EO)	100% active	clear liquid	
	GENAPOL PF80	block polymer (80% EO)	100% active	white flakes	
	GENAPOL PN-73	diamine-based block polymer	100% active		
Hüls	MARLOX LM 75/30	polyethylene-polypropylene glycol ether	active detergent 100%	liquid; cloud point 42°C (2% in H_2O)	foam regulator *low foam industrial cleaners*
	MARLOX LP 90/20	polyethylene-polypropylene glycol ether	active detergent 100%	liquid; cloud point 19°C (2% in H_2O)	
	MARLOX LP 90/20 E	polyethylene-polypropylene glycol ether; low electrolyte form	active detergent 100%	liquid; cloud point 19°C (2% in H_2O)	
ICI	ATLAS G-3888	capped EO/PO block polymer		white liquid	household and industrial applications; agrochemicals; textile industry
	HYPERMER B246	nonionic block copolymer surfactant		HLB 6.0: red-brown waxy solid	emulsifier *polymer industry*
	HYPERMER B259	nonionic block copolymer surfactant		HLB 12.0: yellow-brown waxy solid	
	HYPERMER B261	nonionic block copolymer surfactant		HLB 8.0: red-brown waxy solid	
	SYNPERONIC PE/25R2	POE/POP block copolymers		colourless liquid; cloud point 26°C (10% aq. soln.)	emulsifier; foam control; wetting agent *agrochemicals; household and industrial applications*
	SYNPERONIC PE/F38	POE/POP block copolymers		white flake; cloud point > 100°C (10% aq. soln.)	
	SYNPERONIC PE/F68	POE/POP block copolymers		white flake; cloud point > 100°C (10% aq. soln.)	
	SYNPERONIC PE/F77	POE/POP block copolymers		white flake; cloud point > 100°C (10% aq. soln.)	
	SYNPERONIC PE/F87	POE/POP block copolymers		white flake; cloud point > 100°C (10% aq. soln.)	
	SYNPERONIC PE/F88	POE/POP block copolymers		white flake; cloud point > 100°C (10% aq. soln.)	
	SYNPERONIC PE/F108	POE/POP block copolymers		white flake; cloud point > 100°C (10% aq. soln.)	
	SYNPERONIC PE/F127	POE/POP bluuk copolymers		white flake; cloud point > 100°C (10% aq. soln.)	
	SYNPERONIC PE/L31	POE/POP block copolymers		colourless liquid; cloud point 29°C (10% aq. soln.)	

Supplier	Trade name	Chemical description	Composition	General properties	Functionality / Application
ICI	SYNPERONIC PE/L35	POE/POP block copolymers		colourless liquid; cloud point 80°C (10% aq. soln.)	emulsifier; foam control; wetting agent agrochemicals; household and industrial applications
	SYNPERONIC PE/L42	POE/POP block copolymers		colourless liquid; cloud point 28°C (10% aq. soln.)	
	SYNPERONIC PE/L43	POE/POP block copolymers		colourless liquid; cloud point 33°C (10% aq. soln.)	
	SYNPERONIC PE/L44	POE/POP block copolymers		colourless liquid; cloud point 71°C (10% aq. soln.)	
	SYNPERONIC PE/L61	POE/POP block copolymers		colourless liquid; cloud point 17°C (10% aq. soln.)	
	SYNPERONIC PE/L61N	POE/POP block copolymers		colourless liquid; cloud point 17°C (10% aq. soln.)	
	SYNPERONIC PE/L62	POE/POP block copolymers		colourless liquid; cloud point 24°C (10% aq. soln.)	
	SYNPERONIC PE/L62LF	POE/POP block copolymers		colourless liquid; cloud point 22°C (10% aq. soln.)	
	SYNPERONIC PE/L64	POE/POP block copolymers		colourless liquid; cloud point 59°C (10% aq. soln.)	
	SYNPERONIC PE/L81	POE/POP block copolymers		colourless liquid; cloud point 16°C (10% aq. soln.)	
	SYNPERONIC PE/L92	POE/POP block copolymers		colourless liquid; cloud point 16°C (10% aq. soln.)	
	SYNPERONIC PE/L101	POE/POP block copolymers		colourless liquid; cloud point 11°C (10% aq. soln.)	
	SYNPERONIC PE/L121	POE/POP block copolymers		colourless liquid; cloud point 10°C (10% aq. soln.)	
	SYNPERONIC PE/P75	POE/POP block copolymers		white paste; cloud point 87°C (10% aq. soln.)	
	SYNPERONIC PE/P84	POE/POP block copolymers		white paste; cloud point 71°C (10% aq. soln.)	
	SYNPERONIC PE/P85	POE/POP block copolymers		white paste; cloud point 86°C (10% aq. soln.)	
	SYNPERONIC PE/P94	POE/POP block copolymers		white paste; cloud point 73°C (10% aq. soln.)	
	SYNPERONIC PE/P103	POE/POP block copolymers		white paste; cloud point 51°C (10% aq. soln.)	
	SYNPERONIC PE/P105	POE/POP block copolymers		white paste; cloud point 90°C (10% aq. soln.)	
	SYNPERONIC T/110R2	EDA POE/POP copolymers		yellow liquid	antifoam; dispersant industrial applications
	SYNPERONIC T/304	EDA POE/POP copolymers		yellow liquid; cloud point 73°C (10% aq. soln.)	
	SYNPERONIC T/701	EDA POE/POP copolymers		yellow liquid; cloud point 16.5°C (10% aq. soln.)	

Supplier	Trade name	Chemical description	Composition	General properties	Functionality / Application
ICI	SYNPERONIC T/707	EDA POE/POP copolymers		yellow/amber solid flakes; cloud point >100°C (10% aq. soln.)	
	SYNPERONIC T/904	EDA POE/POP copolymers		yellow liquid paste; cloud point 69°C (10% aq. soln.)	
	SYNPERONIC T/908	EDA POE/POP copolymers		yellow/amber solid flakes; cloud point >100°C (10% aq. soln.)	antifoam; dispersant *industrial applications*
	SYNPERONIC T/1301	EDA POE/POP copolymers		yellow liquid; cloud point 12.5°C (10% aq. soln.)	
	SYNPERONIC T/1302	EDA POE/POP copolymers		yellow liquid; cloud point 29°C (10% aq. soln.)	
	UKANIL 2239	POE/POP copolymer		colourless paste; cloud point 32.0°C (1% aq. soln.)	
	UKANIL 2252	copolymer of ethylene oxide and propylene oxide polyamine based		colourless liquid; cloud point <10°C (1% aq. soln.)	
	UKANIL 2262	copolymer of propylene oxide and ethylene oxide polyamine based		amber liquid; cloud point <10°C (1% aq. soln.)	defoaming agent; wetting agent *agrochemicals; textile industry; industrial and institutional cleaners*
	UKANIL 3500	copolymer of ethylene oxide and propylene oxide on a polyfunctional alcohol		pale yellow liquid; cloud point <10°C (1% aq. soln.)	
	UKANIL 3501	copolymer of ethylene oxide and propylene oxide on a polyfunctional alcohol		colourless liquid; cloud point <10°C (1% aq. soln.)	
	UKANIL FD42	copolymer of ethylene oxide and propylene oxide on a polyfunctional alcohol		pale yellow liquid; cloud point <10°C (1% aq. soln.)	
Kao Corporation	DANOX LF-102	ethylene oxide-propylene oxide block copolymers		liquid; cloud point 45-47°C	
	FINDET SE-1851	ethylene oxide-propylene oxide block copolymers		liquid; cloud point 41-43°C	
	FINDET SE-2411	ethylene oxide-propylene oxide block copolymers		liquid; cloud point 40-42°C	
Dr. W. Kolb	IMBENTIN-PAP/3100	EO/PO copolymer		liquid; cloud point 51°C (10% in BDG 25%)	*personal care products; household detergents; industrial and institutional cleaning; pharmaceuticals*
	IMBENTIN-PAP/3500	EO/PO copolymer		liquid; cloud point 80°C (10% in BDG 25%)	
	IMBENTIN-PAP/6100	EO/PO copolymer		liquid; cloud point 24°C (1% in deionised water)	
	IMBENTIN-PAP/6200	EO/PO copolymer		liquid; cloud point 34°C (1% in deionised water)	
	IMBENTIN-PAP/6400	EO/PO copolymer		liquid; cloud point 77°C (5% in BDG 25%)	
	IMBENTIN-PAP/6800	EO/PO copolymer		solid; cloud point 71°C (1% in 10% NaCl)	
	IMBENTIN-PAP/8100	EO/PO copolymer		liquid; cloud point 53°C (5% in BDG 25%)	
	IMBENTIN-PAP/10200	EO/PO copolymer		liquid; cloud point 17°C (3 g in 10 ml alcohol (30°C))	

Supplier	Trade name	Chemical description	Composition	General properties	Functionality / *Application*
Olin Chemicals	POLY-TERGENT E-17A	EO/PO block polymer		cloud point 25°C (1% aq. soln.); surface tension 42 dynes/cm (0.1%); Draves wetting 35 sec (0.25%); Ross-Miles foam (0.25%): initial 20 mm, 5 min 0 mm; freezing point −30°C	
	POLY-TERGENT E-17B	EO/PO block polymer		cloud point 35°C (1% aq. soln.); surface tension 44 dynes/cm (0.1%); Draves wetting >300 sec (0.25%); Ross-Miles foam (0.25%): initial 30 mm, 5 min 0 mm; freezing point −35°C	
	POLY-TERGENT E-25B	EO/PO block polymer		cloud point 29°C (1% aq. soln.); surface tension 42 dynes/cm (0.1%); Draves wetting >300 sec (0.25%); Ross-Miles foam (0.25%): initial 30 mm, 5 min 0 mm; freezing point −10°C	
	POLY-TERGENT P-9E	EO/PO block polymer		cloud point 75°C (1% aq. soln.); surface tension 42 dynes/cm (0.1%); Draves wetting >300 sec (0.25%); Ross-Miles foam (0.25%): initial 45 mm, 5 min 10 mm; freezing point 12°C	
	POLY-TERGENT P-17A	EO/PO block polymer		cloud point 28°C (1% aq. soln.); surface tension 37 dynes/cm (0.1%); Draves wetting 35 sec (0.25%); Ross-Miles foam (0.25%): initial 20 mm, 5 min 0 mm; freezing point −47°C	
	POLY-TERGENT P-17B	EO/PO block polymer		cloud point 31°C (1% aq. soln.); surface tension 40 dynes/cm (0.1%); Draves wetting 137 sec (0.25%); Ross-Miles foam (0.25%): initial 30 mm, 5 min 10 mm; freezing point −3°C	
	POLY-TERGENT P-17BX	EO/PO block polymer		cloud point 32°C (1% aq. soln.); surface tension 39 dynes/cm (0.1%); Draves wetting 180 sec (0.25%); Ross-Miles foam (0.25%): initial 35 mm, 5 min 0 mm; freezing point −12°C	low foam
	POLY-TERGENT P-17D	EO/PO block polymer		cloud point 59°C (1% aq. soln.); surface tension 41 dynes/cm (0.1%); Draves wetting >180 sec (0.25%); Ross-Miles foam (0.25%): initial 60 mm, 5 min 15 mm; freezing point 9°C	
	POLY-TERGENT P-17LF	EO/PO block polymer		cloud point 31°C (1% aq. soln.); surface tension 40 dynes/cm (0.1%); Draves wetting 90 sec (0.25%); Ross-Miles foam (0.25%): initial 20 mm, 5 min 0 mm; freezing point −10°C	
	POLY-TERGENT P-22A	EO/PO block polymer		cloud point 20°C (1% aq. soln.); surface tension 41 dynes/cm (0.1%); Draves wetting 20 sec (0.25%); Ross-Miles foam (0.25%): initial 12 mm, 5 min 0 mm; freezing point −25°C	
	POLY-TERGENT P-32A	EO/PO block polymer		cloud point 15°C (1% aq. soln.); surface tension 35 dynes/cm (0.01%); Ross-Miles foam (0.25%): initial 0 mm, 5 min 0 mm; freezing point −12°C	
	POLY-TERGENT P-32D	EO/PO block polymer		cloud point 81°C (1% aq. soln.); surface tension 34 dynes/cm (0.1%); Draves wetting 13 sec (0.25%); Ross-Miles foam (0.25%): initial 110 mm, 5 min 100 mm; freezing point 15°C	
PPG	MACOL 1	polymers of propylene oxide and ethylene oxide arranged in blocks		liquid; cloud point 24°C (1% aq.); viscosity 335 cPs	lubricant; wetting agent; w/o emulsifier; solubiliser; defoaming agent *metal working*

Supplier	Trade name	Chemical description	Composition	General properties	Functionality Application
PPG	MACOL 2	polymers of propylene oxide and ethylene oxide arranged in blocks		liquid; cloud point 32°C (1% aq.); viscosity 415 cPs	lubricating; wetting agent; w/o emulsifier; solubiliser; defoaming agent; *metal working*
	MACOL 2D	polymers of propylene oxide and ethylene oxide arranged in blocks		liquid; cloud point 35°C (1% aq.); viscosity 400 cPs	
	MACOL 2 LF	polymers of propylene oxide and ethylene oxide arranged in blocks		liquid; cloud point 28°C (1% aq.); viscosity 400 cPs	
	MACOL 16	meroxapol 108		flakes; mp. 46°C	solubiliser; humectant; *skin care products; toiletries*
	MACOL 18	polymers of propylene oxide and ethylene oxide arranged in blocks		liquid; cloud point 32°C (1% aq.); viscosity 300 cPs	lubricant; wetting agent; emulsifier; solubiliser; defoaming agent; *metal working*
	MACOL 19	polymers of propylene oxide and ethylene oxide arranged in blocks		liquid; cloud point 36°C (1% aq.); viscosity 425 cPs	
	MACOL 22	polymers of propylene oxide and ethylene oxide arranged in blocks		liquid; cloud point 17°C (1% aq.); viscosity 525 cPs	
	MACOL 27	poloxamer 407		flakes; mp. 56°C	wetting agent; foaming agent; *mouthwashes; baby shampoos; skin cleansing preparations*
	MACOL 33	polymers of propylene oxide and ethylene oxide arranged in blocks		liquid; cloud point 25°C (1% aq.); viscosity 650 cPs	lubricant; wetting agent; emulsifier; solubiliser; defoaming agent; *metal working*
	MACOL 40	polymers of propylene oxide and ethylene oxide arranged in blocks		liquid; cloud point 29°C (1% aq.); viscosity 675 cPs	
Protex	PROX-ONIC EP 1090-1	difunctional block polymer ending in primary hydroxyl group	100% active	HLB 3.0; liquid; cloud point 24°C (1% aq.); M.W. 2000	defoaming agent; *dishwashing agents; metal working; cosmetics; paper industry; textile industry*
	PROX-ONIC EP 1090-2	difunctional block polymer ending in primary hydroxyl group	100% active	HLB 6.5; liquid; cloud point 28°C (1% aq.); M.W. 2600	
	PROX-ONIC EP 2080-1	difunctional block polymer ending in primary hydroxyl group	100% active	HLB 7.0; liquid; cloud point 30°C (1% aq.); M.W. 2500	
	PROX-ONIC EP 4060-1	difunctional block polymer ending in primary hydroxyl group	100% active	HLB 1.0; liquid; cloud point 16°C (1% aq.); M.W. 3000	
Surfachem	SURFAC HT10	block copolymer of propylene and ethylene oxides			
	SURFAC HT20	block copolymer of propylene and ethylene oxides			
Union Carbide	TERGITOL XD	high molecular weight copolymer of ethylene oxide and propylene oxide	100% active; OH no. 18	white waxy solid at room temperature, hazy liquid at 50°C; cloud point 76°C (1% aq. soln.); M.W. 3117; pour point 35°C; viscosity 251 cP (50°C); colour 30 (Pt-Co; APHA); surface tension 38 dynes/cm (0.1% aq. soln.); pH 6.5 (20% aq. soln.)	emulsifier; dispersant; *insecticides; herbicides; silicone oils; leather industry; emulsion polymerisation; paint industry*
	TERGITOL XH	high molecular weight copolymer of ethylene oxide and propylene oxide	100% active; OH no. 15	white waxy solid at room temperature, hazy liquid at 50°C; cloud point 99°C (1% aq. soln.); M.W. 3740; pour point 44°C; viscosity 319 cP (50°C); colour 50 (Pt-Co; APHA); surface tension 39 dynes/cm (0.1% aq. soln.); pH 5.4 (10% aq. soln.)	
Witco	REWOPAL PO	ethylenoxide/propyleneoxide block polymer	100% active	liquid	wetting agent; *non-foaming cleaners*

Supplier	Trade name	Chemical description	Composition	General properties	Functionality *Application*
Witco	SYNTHIONIC D 9500	PO/EO condensate	100% active	liquid	detergent; wetting agent; low foam *dishwashing agents*
	SYNTHIONIC E 7525	PO/EO condensate	100% active	paste	low foam; detergent *industrial cleaners; metal cleaning*
	SYNTHIONIC P 8020	PO/EO condensate	100% active	liquid	detergent; wetting agent; low foam *dishwashing agents*

Ester ethoxylates

Supplier	Trade name	Chemical description	Composition	General properties	Functionality *Application*
Akcros Chemicals	ETHYLAN GEL2	POE (20) sorbitan monolaurate	97% active; 'n' no. 20	HLB 16.5; clear amber liquid; pour point −10°C; viscosity 350 cSt	
	ETHYLAN GEO8	POE (20) sorbitan mono-oleate	97% active; 'n' no. 20	HLB 15.0; clear amber liquid; pour point −20°C; viscosity 720 cSt	w/o emulsifier; solubiliser; adjuvant; lubricant; antistatic agent; additive; antifogging agent; softener *cosmetics; pharmaceuticals; agrochemicals; toiletries; textile industry; oil slick dispersants; perfumes; essential oils; plastic films; emulsion and suspension polymerisation*
	ETHYLAN GES6	POE (20) sorbitan monostearate	97% active; 'n' no. 20	HLB 15.0; pale brown liquid/paste; pour point 22°C; viscosity 190 cSt (40°C)	
	ETHYLAN GOE81	POE (5) sorbitan mono-oleate	100% active; 'n' no. 5	HLB 10.0; clear amber liquid; pour point −10°C; viscosity 465 cSt	
	ETHYLAN GPS85	POE (20) sorbitan trioleate	100% active; 'n' no. 20	HLB 11.0; viscous amber liquid; pour point −20°C; viscosity 270 cSt	
Akzo Nobel	ARMOTAN PML 20	polyoxyethylene (20) sorbitan monolaurate	100% active	liquid	
	ARMOTAN PMO 20	polyoxyethylene (20) sorbitan monooleate	100% active	liquid	
	ARMOTAN PMS 20	polyoxyethylene (20) sorbitan monostearate	100% active	paste	
Allied Colloids	DRAFCOL A108	ethoxylated fatty acid ester		liquid	processing aid; antistatic agent; lubricant *textile industry; natural and synthetic fibres*
	DRAFCOL PPL	ethoxylated fatty acid ester plus emulsifiers and antistats		liquid	processing aid; antistatic agent; lubricant *textile industry; polyolefine fibres*
Auschem Cesalpinia	SORBILENE ISM	PEG-20 sorbitan isostearate		HLB 14.8; liquid	
	SORBILENE L	polysorbate 20		HLB 16.7; liquid	o/w emulsifier; dispersant; solubiliser *cosmetics; pesticides; zootechnical industry*
	SORBILENE L/4	polysorbate 21		HLB 13.3; liquid	
	SORBILENE O	polysorbate 80		HLB 15.0; liquid	
	SORBILENE O/5	polysorbate 81		HLB 10.0; liquid	
	SORBILENE P	polysorbate 40		HLB 15.7; liquid/gel	
	SORBILENE S	polysorbate 60		HLB 14.8; semi solid	
	SORBILENE S/4	polysorbate 61		HLB 9.6; solid	
	SORBILENE TO	polysorbate 85		HLB 11.0; liquid	
	SORBILENE TS	polysorbate 65		HLB 10.5; solid	
	STEROL GO	PEG-10 glyceryl oleate		HLB 11.0; liquid	emulsifier; stabiliser *cosmetics; zootechnical industry*
	STEROL GS 5	PEG-5 glyceryl stearate		HLB 8.5; paste	
	STEROL SC 20	POE (2) sorbitol cocoate	'n' no. 2	HLB 5.5; liquid	emulsifier; co-emulsifier; stabiliser; thickener *cosmetics*

Supplier	Trade name	Chemical description	Composition	General properties	Functionality Application
Auschem Cesalpinia	STEROL SS 70	POE (7) stearylstearate	'n' no. 7	HLB 9.0; solid	emulsifier; co-emulsifier; stabiliser; thickener; cosmetics
	STEROL TE 200	POE (6) stearate palmitate	'n' no. 6	HLB 9.7; paste	cosmetics
	TEWAX TC 70	POE (10) stearyl stearate and mineral oil		paste	emulsifier; cosmetics
	TEWAX TC 83	POE (5) glycerol stearate		HLB 8.7; flakes	emulsifier; cosmetics
Chemax	SORBAX HO-40	POE-sorbitol ester	sapon. no. 97	HLB 10.4; liquid	emulsifier; agrochemicals; emulsion polymerisation; metal working; lubricants
	SORBAX HO-50	POE-sorbitol ester	sapon. no. 85	HLB 11.4; liquid	emulsifier; lubricants
	SORBAX PML-20	POE (20) sorbitan monolaurate	'n' no. 20; sapon. no. 45	HLB 16.7; liquid	emulsifier; solubiliser; agriculture; cosmetics; leather industry; metal working; textile industry
	SORBAX PMO-5	POE (5) sorbitan monooleate	'n' no. 5; sapon. no. 100	HLB 10.0; liquid	
	SORBAX PMO-20	POE (20) sorbitan monooleate	'n' no. 20; sapon. no. 50	HLB 15.0; liquid	
	SORBAX PMP-20	POE (20) sorbitan monopalmitate	'n' no. 20; sapon. no. 46	HLB 15.6; liquid	
	SORBAX PMS-20	POE (20) sorbitan monostearate	'n' no. 20; sapon. no. 50	HLB 14.9; soft paste	
	SORBAX PTO-20	POE (20) sorbitan trioleate	'n' no. 20; sapon. no. 88	HLB 11.0; soft paste	
	SORBAX PTS-20	POE (20) sorbitan tristearate	'n' no. 20; sapon. no. 93	HLB 10.5; solid	
Croda	CRILLET 1	polysorbate 20	acid value 2 max.; OH value 96-108; H_2O content 3% max.; iodine value 5 max.; 'n' no. 20; sapon. no. 40-50	HLB 16.7; clear yellow liquid; colour 4.5 max. (Gardner)	solubiliser; emulsifier; dispersant; wetting agent; cosmetics; pharmaceuticals; food industry
	CRILLET 2	polysorbate 40	acid value 2 max.; OH value 89-105; H_2O content 3% max.; 'n' no. 20; sapon. no. 43-49	HLB 15.6; yellow pasty liquid; colour 8 max. (Gardner)	emulsifier; dispersant; penetrant; levelling agent; lubricant; antistatic agent; wetting agent; solubiliser; cosmetics; pesticides; pharmaceuticals; polishes; cleaners; textile industry; food industry
	CRILLET 3	polysorbate 60	acid value 2 max.; OH value 81-96; H_2O content 3% max.; 'n' no. 20; sapon. no. 45-55	HLB 14.9; yellow soft solid; colour 8 max. (Gardner)	emulsifier; dispersant; penetrant; levelling agent; lubricant; antistatic agent; wetting agent; solubiliser; cosmetics; pesticides; pharmaceuticals; plastics industry; polishes; cleaners; textile industry; food industry
	CRILLET 4	polysorbate 80	acid value 2 max.; OH value 65-80; H_2O content 3% max.; 'n' no. 20; sapon. no. 45-55	HLB 15.0; clear yellow liquid; colour 10 max. (Gardner); viscosity 300-500 cS	emulsifier; dispersant; solubiliser; lubricant; detergent; antistatic agent; wetting agent; cosmetics; pesticides; leather industry; metal washing; oil slick dispersants; paint industry; inks; pharmaceuticals; plastics industry; polishes; cleaners; textile industry; household products; paper industry; food industry
	CRILLET 6	PEG-20 sorbitan isostearate	acid value 2 max.; OH value 65-85; H_2O content 3% max.; 'n' no. 20; sapon. no. 40-50	HLB 14.9; clear yellow liquid; colour 5 max. (Gardner)	emulsifier; solubiliser; wetting agent; dispersant; cosmetics; pesticides; pharmaceuticals; polishes; cleaners; textile industry; pigments
	CRILLET 11	polysorbate 21	acid value 2 max.; OH value 225-265; H_2O content 3% max.; 'n' no. 4; sapon. no. 100-115	HLB 13.3; yellow liquid; colour 7 max. (Gardner)	solubiliser; emulsifier; dispersant; wetting agent; cosmetics; pharmaceuticals; food industry
	CRILLET 31	polysorbate 61	acid value 2 max.; OH value 170-200; H_2O content 3% max.; 'n' no. 4; sapon. no. 98-113	HLB 9.6; yellow/amber solid; colour 9 max. (Gardner)	emulsifier; dispersant; penetrant; levelling agent; lubricant; antistatic agent; wetting agent; cosmetics; pesticides; pharmaceuticals; polishes; cleaners; textile industry
	CRILLET 35	polysorbate 65	acid value 2 max.; OH value 44-60; H_2O content 3% max.; 'n' no. 20; sapon. no. 88-98	HLB 10.5; yellow waxy solid; colour 11 max. (Gardner)	emulsifier; dispersant; penetrant; levelling agent; lubricant; antistatic agent; wetting agent; solubiliser; cosmetics; pesticides; pharmaceuticals; polishes; cleaners; textile industry; food industry

Supplier	Trade name	Chemical description	Composition	General properties	Functionality / Application
Croda	CRILLET 41	polysorbate 81	acid value 2 max.; OH value 134-150; H$_2$O content 3% max.; 'n' no. 5; sapon. no. 96-104	HLB 10.0: amber liquid; colour 8 max. (Gardner)	emulsifier; dispersant; wetting agent; solubiliser *cosmetics; pesticides; leather industry; metal working; paint industry; inks; pharmaceuticals; polishes; cleaners; textile industry; household products; paper industry*
	CRILLET 45	polysorbate 85	acid value 2 max.; OH value 39-52; H$_2$O content 5% max.; 'n' no. 20; sapon. no. 82-95	HLB 11.0: viscous amber liquid; colour 10 max. (Gardner)	emulsifier; dispersant; wetting agent; solubiliser *cosmetics; pesticides; leather industry; metal working; paint industry; inks; pharmaceuticals; polishes; cleaners; textile industry; household products; paper industry; food industry*
	CRILLET 50	sorbitan monooleate 20EO, technical grade	acid value 2 max.; 'n' no. 20		emulsifier; wetting agent; solubiliser
	CRILLET 65	PEG-20 sorbitan tri-isostearate		HLB 11.0: slightly hazy amber liquid	o/w emulsifier; solubiliser; wetting agent; dispersant *cosmetics; toiletries; bath preparations; hair care products*
	CROTHIX	PEG-150 pentaerythrityl tetrastearate		white waxy solid	viscosity modifier *detergents*
	GLYCEROX HE	PEG-7 glyceryl cocoate	'n' no. 7	HLB 10.6: clear, colourless to pale yellow liquid	o/w emulsifier; solubiliser; emollient *shampoos; foam baths; fragrances; skin care products; antiperspirants*
	GLYCEROX L8	PEG-8 glyceryl laurate	'n' no. 8	HLB 11.0: clear, colourless to pale yellow liquid	
	GLYCEROX L15	PEG-15 glyceryl laurate	'n' no. 15	HLB 14.0: clear, colourless to pale yellow liquid	
	GLYCEROX L40	PEG-40 glyceryl laurate	'n' no. 40	HLB 17.0: white paste	
Th. Goldschmidt	TAGAT I	PEG-30 glyceryl isostearate	OH value 90-105; acid value 2 max.; iodine value 4 max.; 'n' no. 30; sapon. no. 25-40	HLB 15.6: pale yellow liquid	solubiliser
	TAGAT I 2	PEG-20 glyceryl isostearate	OH value 110-130; acid value 2 max.; iodine value 4 max.; 'n' no. 20; sapon. no. 45-65	HLB 14.2: pale yellow liquid	
	TAGAT L	PEG-30 glyceryl laurate	OH value 52-68; acid value 2 max.; iodine value 4 max.; 'n' no. 30; sapon. no. 35-55	HLB 17.0: pale yellow liquid	
	TAGAT L 2	PEG-20 glyceryl laurate	OH value 60-80; acid value 2 max.; iodine value 4 max.; 'n' no. 20; sapon. no. 50-70	HLB 15.7: ivory liquid	*baby care products; baby shampoos; hair care products; shampoos; hair conditioners; skin cleansing preparations; bath preparations; shower gels; bubble baths; liquid soaps; skin care products; shaving products; sunscreen products; deodorants; antiperspirants; cosmetics; perfumes*
	TAGAT O	PEG-30 glyceryl oleate	OH value 50-65; acid value 2 max.; iodine value 15-19; 'n' no. 30; sapon. no. 30-45	HLB 16.4: yellow liquid	
	TAGAT O 2	PEG-20 glyceryl oleate	OH value 70-85; acid value 2 max.; iodine value 21-27; 'n' no. 20; sapon. no. 40-55	HLB 15.0: yellow liquid	
	TAGAT S	PEG-30 glyceryl stearate	OH value 53-70; acid value 2 max.; iodine value 2 max.; 'n' no. 30; sapon. no. 30-47	HLB 16.4: ivory liquid/solid	
	TAGAT S 2	PEG-20 glyceryl stearate	OH value 65-85; acid value 2 max.; iodine value 2 max.; 'n' no. 20; sapon. no. 40-60	HLB 15.0: ivory solid/liquid	
	TAGAT TO	PEG-25 glyceryl trioleate	OH value 18-33; acid value 12 max.; iodine value 34-40; 'n' no. 25; sapon. no. 75-90	HLB 11.3; amber liquid	
Henkel	CETIOL HE	PEG-7 glyceryl cocoate	iodine value 5; 'n' no. 7	clear, low viscosity oil	emollient; superfatting agent *shampoos; foam baths; shower preparations; cosmetics*
	CUTINA E 24	PEG-20 glyceryl stearate	'n' no. 20	liquid/pasty	o/w emulsifier *sunscreen preparations; baby care products*
	DISPONIL SML 120 SPEC.	sorbitan-20EO-monolaurate	100% active; 'n' no. 20	HLB 16.5; liquid	*plastics industry; coatings*
	DISPONIL SMO 120 SPEC.	sorbitan-20EO-monooleate	100% active; 'n' no. 20	HLB 15.0; liquid	
	DISPONIL SMS 120 SPEC.	sorbitan-20EO-monostearate	100% active; 'n' no. 20	HLB 15.0; liquid	

Supplier	Trade name	Chemical description	Composition	General properties	Functionality / Application
Henkel	EMSORB 2720	polysorbate 20		liquid	o/w emulsifier; stabiliser; viscosity modifier shampoos; hair conditioners; liquid soaps; personal care products
	EMSORB 2721	PEG-80 sorbitan laurate	72% active	liquid	anti-irritant baby care products
	EMSORB 2722	polysorbate 80		yellow clear liquid	emulsifier; dispersant; solubiliser personal care products; fragrances; cosmetics
	EMSORB 2726	PEG-40 sorbitan diisostearate		yellow liquid	emulsifier; solubiliser fragrances; mouthwashes; cosmetics
	EMSORB 2728	polysorbate 60		waxy semi-solid	o/w emulsifier cosmetics; sunscreen preparations; personal care products
	EUMULGIN SML 20	polysorbate 20	99-100% active	liquid	o/w emulsifier; solubiliser creams; emulsions; essential oils
	EUMULGIN SMO 20	polysorbate 80	99-100% active	liquid	
	EUMULGIN SMS 20	polysorbate 60	99-100% active	viscous liquid to pasty	
	LAMACIT GML 20	PEG-20 glyceryl laurate	'n' no. 20	liquid	solubiliser; o/w emulsifier cosmetics
ICI	ARLATONE 970	POE sorbitan fatty acid ester, saturated		HLB 14.3; yellow amber liquid	o/w emulsifier; solubiliser personal care products
	ARLATONE 983	POE fatty acid ester		HLB 8.7; pale cream solid or sprayed	
	ARLATONE 985	POE-(5)-stearyl stearate		HLB 7.5; pale cream pellets	
	ARLATONE T	POE-(40)-sorbitol septaoleate		HLB 9.5; yellow liquid	
	ATLAS G-310	POE ester		yellow-brown liquid	
	ATLAS G-1049	POE-sorbitol septa-isostearate		HLB 9.0; yellow liquid	
	ATLAS G-1086	POE sorbitol hexaoleate		HLB 10.2; pale yellow liquid	
	ATLAS G-1087	POE sorbitol oleate		HLB 9.5; yellow liquid	
	ATLAS G-1096	POE sorbitol hexaoleate		HLB 11.4; pale yellow liquid	
	ATLAS G-1256	POE sorbitol tall oil ester		HLB 9.7; red-brown liquid	
	ATLAS G-1821	POE-(150)-distearate		white solid	
	ATLAS G-1822	solubilised POE-(150)-distearate		amber suspension	household and industrial applications; agrochemicals; textile industry
	ATLAS G-1823	solubilised POE-(150)-distearate		amber suspension	
	ATLAS G-4252	POE sorbitan palmitate		HLB 18.9; yellow liquid	
	ATLAS G-4280	POE sorbitan laurate		HLB 19.1; yellow liquid	
	ATLAS G-4897	ethoxylated fatty acid ester		HLB 12.4; amber liquid	
	ATLAS G-4905	technical POE-(20)-sorbitan monooleate		HLB 15.0; brown liquid	
	ATLAS G-4964	POE-(2)-myristyl myristate		HLB 3.5; white solid	
	ATLAS G-7074	POE glycerol monolaurate		HLB 15.7; clear liquid	
	ATLAS G-8916PF	POE sorbitan esters of mixed fatty and resin acids		HLB 14.6; red-brown liquid	
	ATLOX 1045A	POE sorbitol oleate laurate		HLB 11.4; yellow liquid	emulsifier; dispersant agrochemicals
	ATLOX 1086	POE sorbitol hexaoleate		HLB 10.2; pale yellow liquid	
	ATLOX 1096	POE sorbitol hexaoleate		HLB 11.4; pale yellow liquid	

Supplier	Trade name	Chemical description	Composition	General properties	Functionality Application
ICI	ATLOX 4875	POE ester		HLB 18.2; yellow liquid	emulsifier; dispersant
	ATLOX 8916PF	POE sorbitan esters of mixed fatty and resin acids		HLB 14.6; red-brown liquid	agrochemicals
	CIRRASOL ALN-GM	POE polyol ester		HLB 15.7; yellow liquid	textile industry
	RENEX 20	POE ester of mixed fatty and rosin acids		HLB 13.8; yellow-amber suspension	wetting agent; emulsifier
	RENEX 22	POE ester of mixed fatty and rosin acids		HLB 12.2; amber suspension	personal care products; agrochemical; industrial applications
	TWEEN 20	POE-(20)-sorbitan monolaurate	'n' no. 20	HLB 16.7; yellow liquid	
	TWEEN 21	POE-(4)-sorbitan monolaurate	'n' no. 4	HLB 13.3; yellow liquid	
	TWEEN 40	POE-(20)-sorbitan monopalmitate	'n' no. 20	HLB 16.7; pale yellow liquid gel with suspended solids	
	TWEEN 60	POE-(20)-sorbitan monostearate	'n' no. 20	HLB 14.9; pale yellow semi-solid	emulsifier; solubiliser
	TWEEN 61	POE-(4)-sorbitan monostearate	'n' no. 4	HLB 9.6; ivory solid	personal care products; industrial applications; textile industry
	TWEEN 65	POE-(20)-sorbitan tristearate	'n' no. 20	HLB 10.5; pale yellow solid	
	TWEEN 80	POE-(20)-sorbitan monooleate	'n' no. 20	HLB 15.0; yellow-brown liquid	
	TWEEN 81	POE-(5)-sorbitan monooleate	'n' no. 5	HLB 10.0; yellow-brown liquid	
	TWEEN 85	POE-(20)-sorbitan trioleate	'n' no. 20	HLB 11.0; yellow-brown liquid	
Kao Corporation	KAOPAN TW-L-120	POE (20) sorbitan monolaurate	OH value 96-114; sapon. no. 40-50	liquid	
	KAOPAN TW-O-120	POE (20) sorbitan monooleate	OH value 65-80; sapon. no. 45-55	liquid	
	KAOPAN TW-S-120	POE (20) sorbitan monostearate	OH value 80-96; sapon. no. 45-55	paste	
Dr. W. Kolb	KOTILEN-L/1	polysorbate 20	acid value < 2; OH value 100; 'n' no. 20; sapon. no. 45	HLB 16.7; yellow liquid; viscosity 100 mPas (50°C); pH ca. 7 (1% aq.)	
	KOTILEN-O/1	polysorbate 80	acid value < 2; OH value 70; 'n' no. 20; sapon. no. 50	HLB 15.0; yellow-brown liquid; viscosity 110 mPas (50°C); pH ca. 7 (1% aq.)	o/w emulsifier; wetting agent; solubiliser
	KOTILEN-O/1/050	polysorbate 81	acid value < 2; OH value 130; 'n' no. 5; sapon. no. 100	HLB 10.0; yellow-brown liquid; viscosity 100 mPas (50°C)	cosmetics; personal care products; household detergents; industrial and institutional cleaning; pharmaceuticals
	KOTILEN-O/3	polysorbate 85	acid value < 2; OH value 45; 'n' no. 20; sapon. no. 90	HLB 11.0; yellow-brown liquid; viscosity 80 mPas (50°C)	
	KOTILEN-O/3/020	polysorbate	acid value < 2; OH value 55; 'n' no. 2; sapon. no. 165	HLB 5.0; yellow-brown liquid; viscosity 45 mPas (50°C)	
	KOTILEN-P/1	polysorbate 40	acid value < 2; OH value 100; 'n' no. 20; sapon. no. 45	HLB 15.6; yellow liquid/paste; viscosity 200 mPas (50°C); pH ca. 7 (1% aq.)	
	KOTILEN-S/1	polysorbate 60	acid value < 2; OH value 90; 'n' no. 20; sapon. no. 50	HLB 14.9; yellow paste; viscosity 700 mPas (50°C); pH ca. 7 (1% aq.)	
	KOTILEN-S/1/040	ethoxylated sorbitan monostearate	'n' no. ca. 4.0	HLB ca. 5.5; liquid	personal care products; household detergents; industrial and institutional cleaning; pharmaceuticals
	KOTILEN-S/3	polysorbate 65	acid value < 2; OH value 50; 'n' no. 20; sapon. no. 95	HLB 10.5; white-yellow solid; viscosity 77 mPas (50°C)	o/w emulsifier; wetting agent; solubiliser; cosmetics; personal care products; household detergents; industrial and institutional cleaning; pharmaceuticals
	SYMPATENS-GMIS/150	PEG-15 glyceryl isostearate	acid value < 3; 'n' no. 15; sapon. no. 60	HLB ca. 12; yellow liquid; viscosity 290 mPas; pH ca. 7 (1% aq.)	
	SYMPATENS-GMIS/200	PEG-20 glyceryl isostearate	acid value < 3; 'n' no. 20; sapon. no. 55	HLB ca. 15; yellow liquid; viscosity 360 mPas; pH ca. 7 (1% aq.)	o/w emulsifier; solubiliser; refatting agent; cosmetics

Supplier	Trade name	Chemical description	Composition	General properties	Functionality Application
Dr. W. Kolb	SYMPATENS-GMS/200	PEG-20 glyceryl stearate	acid value < 3; 'n' no. 20; sapon. no. 50	HLB ca. 15; white-yellow liquid; viscosity 250 mPas (30°C); pH ca. 7 (1% aq.)	o/w emulsifier; stabiliser; refatting agent *cosmetics*
	SYMPATENS-SHO/400	PEG-40 sorbitan hexaoleate	acid value < 12; OH value 40; 'n' no. 40; sapon. no. 100	HLB 10.0; yellow liquid; viscosity 180 mPas	o/w emulsifier; w/o co-emulsifier; solubiliser *cosmetics*
	SYMPATENS-SPO/400	PEG-40 sorbitan peroleate	acid value < 12; OH value 40; 'n' no. 40; sapon. no. 105	HLB 9.5; yellow liquid; viscosity 175 mPas	
Lonza	ALDOSPERSE 40/60 FG	40% Aldosperse MS-20, 60% Aldo HMS; food grade	acid value 2; sapon. no. 128	HLB 7; colourless beads; colour 1 (Gardner); m.p. 57°C	strengthening agent *food industry: bakery goods*
	ALDOSPERSE 40/60 KFG	40% Aldosperse MS-20, 60% Aldo HMS; Kosher food grade	acid value 2; sapon. no. 128	HLB 7; colourless beads; colour 2 (Gardner); m.p. 57°C	
	ALDOSPERSE ML-23	POE 23 glyceryl monolaurate	acid value 3; sapon. no. 46	HLB 17; colourless liquid; colour 1 (Gardner)	emulsifier *textile industry; cosmetics*
	ALDOSPERSE MS-20 FG	POE 20 glyceryl monostearate; food grade	acid value 2; sapon. no. 70	HLB 13; colourless soft solid; colour 2 (Gardner); m.p. 27°C	strengthening agent *food industry: bakery goods*
	ALDOSPERSE MS-20 KFG	POE 20 glyceryl monostearate; Kosher food grade	acid value 2; sapon. no. 70	HLB 13; colourless soft solid; colour 2 (Gardner); m.p. 27°C	strengthening agent *food industry: bakery goods*
	ALDOSPERSE O-20 KFG	20% Glycosperse O-20, 80% Aldo MS LG; Kosher food grade	acid value 2; sapon. no. 148	HLB 5; colourless beads; colour 5 (Gardner); m.p. 60°C	emulsifier *food industry: ice cream*
	ALDOSPERSE TS-40 KFG	40% Glycosperse TS-20, 60% Aldo MS LG; Kosher food grade	acid value 2; sapon. no. 140	HLB 6; colourless beads; colour 5 (Gardner); m.p. 55°C	emulsifier *food industry: desserts*
	GLYCOSPERSE HTO-40	POE 40 sorbitan hexatallate	acid value 10	HLB 10; liquid; colour 7 (Gardner)	o/w emulsifier *food industry; cosmetics; household and industrial applications*
	GLYCOSPERSE L-10	POE 10 sorbitan monolaurate	acid value 2	HLB 8; liquid; colour 5 (Gardner)	
	GLYCOSPERSE L-20	POE 20 sorbitan monolaurate	acid value 2	HLB 17; liquid; colour 4 (Gardner)	
	GLYCOSPERSE O-5	POE 5 sorbitan monooleate	acid value 2	HLB 10; liquid; colour 6 (Gardner)	
	GLYCOSPERSE O-20 FG	POE 20 sorbitan monooleate; food grade	acid value 2	HLB 15; liquid; colour 5 (Gardner)	
	GLYCOSPERSE O-20 KFG	POE 20 sorbitan monooleate; Kosher food grade	acid value 2	HLB 15; liquid; colour 5 (Gardner)	
	GLYCOSPERSE S-20 FG	POE 20 sorbitan monostearate; food grade	acid value 2	HLB 15; soft solid; colour 7 (Gardner); m.p. 28°C	
	GLYCOSPERSE S-20 KFG	POE 20 sorbitan monostearate; Kosher food grade	acid value 2	HLB 15; soft solid; colour 7 (Gardner); m.p. 28°C	
	GLYCOSPERSE TS-20 FG	POE 20 sorbitan tristearate; food grade	acid value 2	HLB 11; soft solid; colour 6 (Gardner); m.p. 31°C	
	GLYCOSPERSE TS-20 KFG	POE 20 sorbitan tristearate; Kosher food grade	acid value 2	HLB 11; soft solid; colour 6 (Gardner); m.p. 31°C	
	PEGOSPERSE 100 L	diethylene glycol laurate	acid value 4; sapon. no. 165	HLB 7; liquid; colour 2 (Gardner)	emulsifier; dispersant; opacifier; defoaming agent; viscosity modifier *cosmetics; household applications; textile industry; plastics industry; water treatment*
	PEGOSPERSE 100 O	diethylene glycol oleate	acid value 86; sapon. no. 160	HLB 4; liquid; colour 4 (Gardner); m.p. 52°C	
	PEGOSPERSE 100 S	diethylene glycol stearate	acid value 100; sapon. no. 170	HLB 4; beads; colour 2 (Gardner); m.p. 52°C	
	PEGOSPERSE 200 DL	PEG 200 dilaurate	acid value 5; sapon. no. 178	HLB 7; liquid; colour 2 (Gardner)	
	PEGOSPERSE 200 ML	PEG 200 monolaurate	acid value 5; sapon. no. 154	HLB 9; liquid; colour 1 (Gardner)	

Supplier	Trade name	Chemical description	Composition	General properties	Functionality / *Application*
Lonza	PEGOSPERSE 400 DL	PEG 400 dilaurate	acid value 5; sapon. no. 135	HLB 10; liquid; colour 2 (Gardner)	emulsifier; dispersant; opacifier; defoaming agent; viscosity modifier *cosmetics; household applications; textile industry; plastics industry; water treatment*
	PEGOSPERSE 400 DOT	PEG 400 ditallate	acid value 15; sapon. no. 121	HLB 8; liquid; colour 10 (Gardner)	
	PEGOSPERSE 400 DS	PEG 400 distearate	acid value 10; sapon. no. 120	HLB 8; solid; colour 1 (Gardner); m.p. 33°C	
	PEGOSPERSE 400 ML	PEG 400 monolaurate	acid value 3; sapon. no. 95	HLB 14; liquid; colour 1 (Gardner)	
	PEGOSPERSE 400 MO	PEG 400 monooleate	acid value 5; sapon. no. 84	HLB 11; liquid; colour 3 (Gardner)	
	PEGOSPERSE 400 MS	PEG 400 monostearate	acid value 3; sapon. no. 89	HLB 11; solid; colour 1 (Gardner); m.p. 31°C	
	PEGOSPERSE 600 DOT	PEG 600 ditallate	acid value 15; sapon. no. 101	HLB 9; liquid; colour 10 (Gardner)	
	PEGOSPERSE 1750 MS	PEG 1750 monostearate	acid value 1; sapon. no. 30	HLB 18; flakes; colour 3 (Gardner); m.p. 46°C	
Nikko Chemicals	NIKKOL CDIS-400	PEG disostearate		pale yellow liquid	emulsifier *cosmetics*
	NIKKOL CDO-600	PEG dioleate		pale yellow liquid	
	NIKKOL CDS-400	PEG distearate		pale yellow solid	emulsifier; thickener *cosmetics*
	NIKKOL CDS-6000P	PEG distearate		pale yellow flakes	
	NIKKOL DEGS	PEG-2 stearate		pale yellow solid	emulsifier *cosmetics*
	NIKKOL GL-1	PEG (6) sorbit monolaurate		colourless liquid	*cosmetics*
	NIKKOL GO-4	sorbeth-6 tetraoleate		pale yellow liquid	
	NIKKOL GO-430	sorbeth-30 tetraoleate		pale yellow liquid	
	NIKKOL GO-440	sorbeth-40 tetraoleate		pale yellow liquid	
	NIKKOL GO-460	sorbeth-60 tetraoleate		pale yellow liquid	
	NIKKOL GS-6	sorbeth-6 hexastearate		white solid	emulsifier *cosmetics*
	NIKKOL GS-460	sorbeth-60 tetrastearate		pale yellow paste	
	NIKKOL MYL-10	PEG-10 laurate		pale yellow liquid	
	NIKKOL MYO-2	PEG (2EO) monooleate		pale yellow liquid	
	NIKKOL MYO-6	PEG-6 oleate		pale yellow liquid	
	NIKKOL MYO-10	PEG-10 oleate		pale yellow liquid	
	NIKKOL MYS-1EX	PEG (1EO) monostearate		pale yellow solid	pearlescent agent *cosmetics*
	NIKKOL MYS-2	PEG-2 stearate		pale yellow solid	emulsifier *cosmetics*
	NIKKOL MYS-4	PEG-4 stearate		pale yellow solid	
	NIKKOL MYS-10	PEG-10 stearate		pale yellow solid	
	NIKKOL MYS-25	PEG-25 stearate		pale yellow solid	
	NIKKOL MYS-40	PEG-40 stearate		pale yellow solid	

Supplier	Trade name	Chemical description	Composition	General properties	Functionality Application
Nikko Chemicals	NIKKOL MYS-45	PEG-45 stearate		pale yellow flakes	emulsifier cosmetics
	NIKKOL MYS-55	PEG-55 stearate		pale yellow flakes	
	NIKKOL TI-10	PEG-20 sorbitan isostearate		pale yellow liquid	
	NIKKOL TL-10	polysorbate 20		pale yellow liquid	emulsifier; solubiliser; dispersant cosmetics
	NIKKOL TO-10	polysorbate 80		yellow liquid	
	NIKKOL TO-10M	polysorbate 80		yellow liquid	
	NIKKOL TO-30	polysorbate 85		yellow liquid	emulsifier cosmetics
	NIKKOL TO-106	PEG-6 sorbitan oleate		yellow liquid	emulsifier; solubiliser; dispersant cosmetics
	NIKKOL TP-10	polysorbate 40		yellow liquid	emulsifier cosmetics
	NIKKOL TS-10	polysorbate 60		yellow viscous liquid	
	NIKKOL TS-30	polysorbate 65		yellow semi-solid	
	NIKKOL TS-106	PEG-6 sorbitan stearate		yellow paste	
PPG	MAZOL 80 MGK	PEG-20 glyceryl stearate			emulsifier personal care products
	MAZOL 159	PEG-7 glyceryl cocoate			solubiliser; emulsifier skin cleansing preparations; bath preparations; shower gels
	MAZON 1045 A	ethoxylated sorbitol ester		HLB 13.0; liquid; viscosity 260 cPs	emulsifier; coupling agent; lubricant metal working; agrochemicals
	MAZON 1086	ethoxylated sorbitol ester		HLB 10.4; liquid; viscosity 200 cPs	
	MAZON 1096	ethoxylated sorbitol ester		HLB 11.2; liquid; viscosity 240 cPs	
	T-MAZ 20	polysorbate 20		HLB 16.7; yellow liquid	solubiliser; emulsifier baby shampoos; skin cleansing preparations; skin care products; bath preparations
	T-MAZ 28	PEG 80 sorbitan laurate		HLB 19.2; pale yellow liquid	
	T-MAZ 60K	polysorbate 60		HLB 14.9; amber semi-solid	emulsifier; binder personal care products; skin cleansing preparations; cosmetics
	T-MAZ 80	polysorbate 80		HLB 15.0; light amber liquid; viscosity 420 cPs	lubricant; dispersant; solubiliser; viscosity modifier; emulsifier; emollient metal working; cosmetics; skin care products; hair conditioners; toiletries
	T-MAZ 81	sorbitan monooleate 5EO		HLB 10.0; liquid; viscosity 550 cPs	lubricant; dispersant; solubiliser; viscosity modifier metal working
	T-MAZ 85	polysorbate 85		HLB 11.1; light amber liquid	dispersant; emulsifier; solubiliser cosmetics; skin care products; hair conditioners; toiletries
Protex	PROX-ONIC SML-020	POE (20) sorbitan monolaurate	'n' no. 20; sapon. no. 45	HLB 16.7; liquid	
	PROX-ONIC SMO-05	POE (5) sorbitan monooleate	'n' no. 5; sapon. no. 100	HLB 10.0; liquid	
	PROX-ONIC SMO-020	POE (20) sorbitan monooleate	'n' no. 20; sapon. no. 50	HLB 15.0; liquid	o/w emulsifier; solubiliser agriculture; cosmetics; leather industry; metal working; textile industry
	PROX-ONIC SMP-020	POE (20) sorbitan monopalmitate	'n' no. 20; sapon. no. 46	HLB 15.6; liquid	
	PROX-ONIC SMS-020	POE (20) sorbitan monostearate	'n' no. 20; sapon. no. 50	HLB 14.9; soft paste	

Supplier	Trade name	Chemical description	Composition	General properties	Functionality / Application
Protex	PROX-ONIC STO-020	POE (20) sorbitan trioleate	'n' no. 20; sapon. no. 88	HLB 11.0; soft paste	o/w emulsifier; solubiliser *agriculture; cosmetics; leather industry; metal working; textile industry*
	PROX-ONIC STS-020	POE (20) sorbitan tristearate	'n' no. 20; sapon. no. 93	HLB 10.5; solid	
Seppic	MONTANOX 20	sorbitan monolaurate 20EO	100% active; 'n' no. 20	HLB 16.7; liquid	antistatic agent
	MONTANOX 20 DF	polysorbate 20 (dioxane free)	100% active; 'n' no. 20	HLB 16.7; liquid	antistatic agent *personal care products*
	MONTANOX 40	sorbitan monopalmitate 20EO	100% active; 'n' no. 20	HLB 15.6; viscous liquid	emulsifier; wetting agent; dispersant
	MONTANOX 40 DF	polysorbate 40 (dioxane free)	100% active	paste	emulsifier *personal care products*
	MONTANOX 60	sorbitan monostearate 20EO	100% active; 'n' no. 20	HLB 14.9; viscous liquid	emulsifier; wetting agent; dispersant
	MONTANOX 60 DF	polysorbate 60 (dioxane free)	100% active; 'n' no. 20	HLB 14.9; viscous liquid	emulsifier; wetting agent; dispersant *personal care products*
	MONTANOX 70	sorbitan monoisostearate 20EO	100% active; 'n' no. 20	HLB 15.6; liquid	emulsifier; wetting agent; dispersant
	MONTANOX 80	sorbitan monooleate 20EO	100% active; 'n' no. 20	HLB 15; liquid	emulsifier; wetting agent; dispersant
	MONTANOX 80 DF	polysorbate 80 (dioxane free)	100% active; 'n' no. 20	HLB 15; liquid	emulsifier; wetting agent; dispersant *personal care products*
	MONTANOX 85	sorbitan trioleate 20EO	100% active; 'n' no. 20	HLB 11; liquid	emulsifier; wetting agent; dispersant
	MONTANOX 85 DF	polysorbate 85 (dioxane free)	100% active; 'n' no. 20	liquid	emulsifier *personal care products*
	SIMULSOL 220 TM	PEG-200 glyceryl stearate	100% active; 'n' no. 200	wax	emollient; thickener *cosmetics; hair care products*
	SIMULSOL CG	glycerol cocoate 40EO	75% active; 'n' no. 40	liquid	emollient *cosmetics*
Witco	SORBANOX AL	sorbitan monolaurate 20EO	100% active; 'n' no. 20	liquid	antistatic agent; defoaming agent; solubiliser *food industry*
	SORBANOX AO	sorbitan monooleate 20EO	100% active; 'n' no. 20	liquid	solubiliser; emulsifier *metal working; pharmaceuticals; vitamin preparations*
	SORBANOX AST	sorbitan monostearate 20EO	100% active; 'n' no. 20	paste	solubiliser; emulsifier *paraffins*
	SORBANOX CO	sorbitan trioleate 20EO	100% active; 'n' no. 20	liquid	emulsifier *agrochemicals*
	WITCONOL 2720	polysorbate 20		HLB 16.7; liquid; pour point −10°C; colour 6 (Gardner)	coupling agent; dispersant; o/w emulsifier; viscosity modifier *personal care products; household and industrial applications*
	WITCONOL 2722	polysorbate 80		HLB 15.0; liquid; pour point −12°C; colour 6 (Gardner)	
	WITCONOL 6903	polysorbate 85		HLB 11.1; liquid; pour point −15°C	coupling agent; dispersant; o/w emulsifier; lubricant *personal care products; household and industrial applications*
Zschimmer & Schwarz	MULSIFAN RT 141	sorbitan monolaurate ethoxylate	100% active	HLB 15-16; liquid	solubiliser; emulsifier *perfumes; essential oils*
	MULSIFAN RT 146	sorbitan monooleate ethoxylate	100% active	HLB 15-16; liquid	

Fatty acid ethoxylates

Supplier	Trade name	Chemical description	Composition	General properties	Functionality Application
Albright & Wilson	EMPILAN BQ SERIES	oleic acid ethoxylates	concentration 100%	liquid/paste	antistatic agent *polymer industry*
Auschem Cesalpinia	ANTISCHIUMA G8	ethoxylated fatty acid	15% active	emulsion	antifoaming agent *general purpose; textile, paper and food industries; water purification*
	POLIROL 400 SA	ethoxylated fatty acids in aq. alcohol soln.	80% active	liquid	viscosity modifier
	POLIROL C 5	ethoxylated fatty acids	100% active	liquid	*PVC emulsions; polymerisation*
	POLIROL L 400	ethoxylated fatty acids	100% active	liquid	
	ROLFAT CC 9	POE (9) coco acid	100% active; 'n' no. 9	liquid	
	ROLFAT OL 6	POE (6) oleic acid	100% active; 'n' no. 6	liquid	
	ROLFAT OL 7	POE (7) oleic acid	100% active; 'n' no. 7	liquid	
	ROLFAT OL 9	POE (9) oleic acid	100% active; 'n' no. 9	liquid	
	ROLFAT OL 12	POE (12) oleic acid	100% active; 'n' no. 12	liquid	lubricant
	ROLFAT SO 6	POE (6) fatty acids from vegetable oil	100% active; 'n' no. 6	liquid	*special purpose formulations*
	ROLFAT ST 6	POE (6) stearic acid	100% active; 'n' no. 6	waxy	
	ROLFAT ST 20	POE (20) stearic acid	100% active; 'n' no. 20	waxy	
	ROLFAT ST 40	POE (40) stearic acid	100% active; 'n' no. 40	flakes	
	ROLFAT ST 100	POE (100) stearic acid	100% active; 'n' no. 100	flakes	
BASF	EMULAN A	oleic acid ethoxylate	100% active	cloud point 50°C (in butyl digol)	emulsifier
Dr. Th. Boehme	SYNTHESIN 32	fatty acid polyglycol ester			spin finish
	SYNTHESIN PEF 100	fatty acid polyglycol ester			*textile industry; fibre production*
Chemax	CHEMAX AR-497	POE (15) rosin acid	'n' no. 15	HLB 13.0; paste; cloud point 75°C (1% aq.)	emulsifier; detergent *degreasers; metal cleaning*
	CHEMAX E-200ML	POE (5) coco fatty acid	'n' no. 5; sapon. no. 135	HLB 9.3; liquid	
	CHEMAX E-200MO	POE (5) oleic fatty acid	'n' no. 5; sapon. no. 118	HLB 8.3; liquid	
	CHEMAX E-200MS	POE (5) stearic fatty acid	'n' no. 5; sapon. no. 112	HLB 8.5; paste	
	CHEMAX E-400ML	POE (9) coco fatty acid	'n' no. 9; sapon. no. 90	HLB 13.2; liquid	
	CHEMAX E-400MO	POE (9) oleic fatty acid	'n' no. 9; sapon. no. 85	HLB 11.8; liquid	emulsifier; additive; viscosity modifier
	CHEMAX E-400MS	POE (9) stearic fatty acid	'n' no. 9; sapon. no. 87	HLB 12.0; paste	*lubricants; cosmetics*
	CHEMAX E-600ML	POE (14) coco fatty acid	'n' no. 14; sapon. no. 68	HLB 14.8; liquid	
	CHEMAX E-600MO	POE (14) oleic fatty acid	'n' no. 14; sapon. no. 65	HLB 13.6; liquid	
	CHEMAX E-600MS	POE (14) stearic fatty acid	'n' no. 14; sapon. no. 62	HLB 13.8; paste	
	CHEMAX E-1000MS	POE (23) stearic fatty acid	'n' no. 23; sapon. no. 43	HLB 15.7; solid	
	CHEMAX TO-8	POE (8) tall oil acid	'n' no. 8	HLB 10.5; liquid; cloud point <25°C (1% aq.)	emulsifier; detergent *degreasers*

Supplier	Trade name	Chemical description	Composition	General properties	Functionality Application
Chemax	CHEMAX TO-10	POE (10) tall oil acid	'n' no. 10	HLB 11.5; liquid; cloud point <25°C (1% aq.)	emulsifier; detergent *degreaser*
	CHEMAX TO-16	POE (16) tall oil acid	'n' no. 16	HLB 13.4; liquid; cloud point 70°C (1% aq.)	
Ciba	ERIOPON OL	based on ethoxylated fatty acid derivatives		clear, yellow brown low viscosity liquid	afterclearing agent; emulsifier *textile industry; dyeing*
Elf Atochem	REMCOPAL 6	oleic acid 7.4EO; C_{18} carbon chain	100% active; 'n' no. 7.4	HLB 10.7; liquid; cloud point 65°C (aq. soln. with 25% BDG)	
	REMCOPAL 207	oleic acid 4.5EO; C_{18} carbon chain	100% active; 'n' no. 4.5	HLB 8; liquid; cloud point 53°C (aq. soln. with 25% BDG)	
	SELLIG AO 6 100	oleic acid 6EO; C_{18} carbon chain	100% active; 'n' no. 6	HLB 9.6; liquid; cloud point 61°C (aq. soln. with 25% BDG)	
	SELLIG AO 9 100	oleic acid 8.5EO; C_{18} carbon chain	100% active; 'n' no. 8.5	HLB 11.4; liquid; cloud point 68°C (aq. soln. with 25% BDG)	
	SELLIG AO 15 100	oleic acid 15EO; C_{18} carbon chain; in course of development	100% active; 'n' no. 15	HLB 14; liquid; cloud point 42°C (aq. soln. of 50 g/l NaCl)	
	SELLIG S 30 100	stearic acid 30EO; C_{18} carbon chain; in course of development	100% active; 'n' no. 30	HLB 11.9; solid; cloud point 75°C (aq. soln. of 50 g/l NaCl)	
	SELLIG STEARO 6	stearic acid 6EO; C_{18} carbon chain; in course of development	100% active; 'n' no. 6	HLB 11.9; solid; cloud point 60°C (aq. soln. with 25% BDG)	
	SELLIG T 3 100	tall oil 3EO	100% active; 'n' no. 3	HLB 6; liquid; cloud point 3.2°C	
	SELLIG T 14 100	tall oil 14EO	100% active; 'n' no. 14	HLB 13.1; liquid; cloud point 70°C (aq. soln. with 25% BDG)	
	SELLIG T 1790	tall oil 17EO	90% active; 'n' no. 17	HLB 14.2; viscous liquid	
	STEARATE PEG 1000	stearic acid 21EO; C_{18} carbon chain	100% active; 'n' no. 21	HLB 15.3; solid; cloud point 85°C	
Henkel	OSIMOL OV	fatty acid ethoxylate			levelling agent; dispersant *textile industry; dyeing*
	OSIMOL PHT	fatty acid ethoxylate			lubricant *textile industry; spinning*
	SELBANA WD	fatty acid polyglycolester			emulsifier; wetting agent *metal working; specialities*
Hoechst	EMULSOGEN A	oleic acid EO	100% active		
Hüls	MARLOSOL 183	stearic acid 3EO	active detergent 100%; 'n' no. 3	solid; cloud point 65°C (2% in 60% BDG soln.)	
	MARLOSOL 186	stearic acid 6EO	active detergent 100%; 'n' no. 6	solid; cloud point 87°C (5 g in 25 g BDG plus 25 g H_2O)	emulsifier *textile industry*
	MARLOSOL 188	stearic acid 8EO	active detergent 100%; 'n' no. 8	solid; cloud point 65°C (10% in 25% BDG soln.)	
	MARLOSOL 1820	stearic acid 20EO	active detergent 100%; 'n' no. 20	solid; cloud point 80°C (2% in H_2O)	
	MARLOSOL OL 2	oleic acid 2EO	active detergent 100%; 'n' no. 2	liquid	*fabric softeners*
	MARLOSOL OL 7	oleic acid 7EO	active detergent 100%; 'n' no. 7	liquid; cloud point 62°C (10% in 25% BDG soln.)	
	MARLOSOL OL 10	oleic acid 10EO	active detergent 100%; 'n' no. 10	liquid/paste; cloud point 70°C (10% in 25% BDG soln.)	*textile industry*
	MARLOSOL OL 15	oleic acid 15EO	active detergent 100%; 'n' no. 15	paste; cloud point 46°C (2% in 5% NaCl soln.)	
	MARLOSOL OL 20	oleic acid 20EO	active detergent 100%; 'n' no. 20	paste; cloud point 62°C (2% in 5% NaCl soln.)	
	MARLOSOL RF 3	rape seed oil fatty acid 3EO	active detergent 100%; 'n' no. 3	liquid	

Supplier	Trade name	Chemical description	Composition	General properties	Functionality / Application
Hüls	MARLOSOL TF 3	tall oil fatty acid 3EO	active detergent 100%; 'n' no. 3	liquid	*textile industry*
	MARLOSOL TF 4	tall oil fatty acid 4EO	active detergent 100%; 'n' no. 4	liquid	
ICI	ATLAS G-2079	POE-(20)-palmitate		HLB 15.5; pale cream solid	*household and industrial applications; agrochemicals; textile industry*
	ATLAS G-2109	POE laurate		HLB 13.3; pale yellow liquid	
	ATLAS G-2127	POE laurate		HLB 12.8; pale yellow liquid	
	ATLAS G-2143	POE-(10)-monooleate		light yellow liquid	
	ATLAS G-5507	POE oleate		HLB 10.4; red-brown liquid	
	CIRRASOL ALN-FP	POE fatty acid		HLB 11.6; white semi-solid	*textile industry*
	CIRRASOL ALN-TF	POE fatty acid		HLB 13.6; pale yellow liquid	
	CIRRASOL ALN-TS	POE fatty acid		HLB 13.6; yellow-brown liquid	
	MYRJ 45	POE-(8)-stearate		HLB 11.1; creamy white solid	emulsifier; dispersant / *personal care products; industrial applications*
	MYRJ 49	POE-(20)-stearate		HLB 15.0; creamy white solid	
	MYRJ 51	POE-(30)-stearate		HLB 16.0; creamy white solid	
	MYRJ 52	POE-(40)-stearate		HLB 16.9; creamy white solid, flaked or sprayed	
	MYRJ 53	POE-(50)-stearate		HLB 17.9; creamy white solid or sprayed	
	MYRJ 59	POE-(100)-stearate		HLB 18.8; creamy white solid or sprayed	
Kao Corporation	FINDET OR/16	polyethoxylated unsaturated fatty acids	'n' no. 4; sapon. no. 105-115	liquid	
	FINDET OR/17.4	polyethoxylated unsaturated fatty acids	'n' no. 5.4; sapon. no. 100-110	liquid	
	FINDET OR/22	polyethoxylated unsaturated fatty acids	'n' no. 10; sapon. no. 68-80	liquid	
	FINDET OR/25	polyethoxylated unsaturated fatty acids	'n' no. 13; sapon. no. 60-68	liquid	
	FINDET OR/32	polyethoxylated unsaturated fatty acids	'n' no. 20; sapon. no. 42-50	paste/solid	
Dr. W. Kolb	HEDIPIN-CFA/090	ethoxylated coco fatty acid; $C_{12/14}$ carbon chain	'n' no. ca. 9.0	liquid; cloud point 63°C (5 g in 25 g BDG 25%)	*personal care products; household detergents; industrial and institutional cleaning; pharmaceuticals*
	HEDIPIN-CFA/100	ethoxylated coco fatty acid; $C_{12/14}$ carbon chain	'n' no. ca. 10.0	liquid; cloud point 73°C (5% in BDG 25%)	
	HEDIPIN-PO/060	ethoxylated oleic acid; C_{18} carbon chain	'n' no. ca. 6.0	liquid	
	HEDIPIN-PO/080	ethoxylated oleic acid; C_{18} carbon chain	'n' no. ca. 8.0	liquid; cloud point 70°C (5% in BDG 25%)	
	HEDIPIN-PO/100	ethoxylated oleic acid; C_{18} carbon chain	'n' no. ca. 10.0	liquid; cloud point 77°C (5% in BDG 25%)	
	HEDIPIN-PO/120	ethoxylated oleic acid; C_{18} carbon chain	'n' no. ca. 12.0	liquid; cloud point 70°C (5 g in 25 g BDG 25%)	
	HEDIPIN-PO/140	ethoxylated oleic acid; C_{18} carbon chain	'n' no. ca. 14.0	liquid; cloud point 52°C (5 g in 25 g BDG 25%)	
	HEDIPIN-PS/060	ethoxylated stearic acid; C_{18} carbon chain	'n' no. ca. 6.0	paste; cloud point 70°C (5% in BDG 25%)	
	HEDIPIN-PS/090	ethoxylated stearic acid; C_{18} carbon chain	'n' no. ca. 9.0	solid; cloud point 75°C (5% in BDG 25%)	
	HEDIPIN-PS/130	ethoxylated stearic acid; C_{18} carbon chain	'n' no. ca. 13.0	solid; cloud point 58°C (1% in deionised water)	
	HEDIPIN-PS/230	ethoxylated stearic acid; C_{18} carbon chain	'n' no. ca. 23.0	solid/flakes; cloud point 62°C (1% in 10% NaCl)	
	HEDIPIN-PS/300	ethoxylated stearic acid; C_{18} carbon chain	'n' no. ca. 30.0	solid/flakes	
	HEDIPIN-PS/400	ethoxylated stearic acid; C_{18} carbon chain	'n' no. ca. 40.0	solid/flakes	
	HEDIPIN-PT/045	ethoxylated tall oil fatty acid; $C_{18/18'}$ carbon chain	'n' no. ca. 4.5	liquid; cloud point 43°C (5 g in 25 g BDG 25%)	
	HEDIPIN-PT/100	ethoxylated tall oil fatty acid; $C_{18/18''}$ carbon chain	'n' no. ca. 10.0	liquid; cloud point 64°C (50% in BDG 25%)	

Supplier	Trade name	Chemical description	Composition	General properties	Functionality Application
Dr. W. Kolb	SYMPATENS-BCO/050	PEG-5 cocoate	'n' no. 5; sapon. no. 135	HLB 7.2; yellow liquid; viscosity 50 mPas; pH 5-7 (1% aq.)	
	SYMPATENS-BCO/100	PEG-10 cocoate	'n' no. 10; sapon. no. 70	HLB 13.5; yellow liquid; viscosity 75 mPas; pH 5-7 (1% aq.)	refatting agent; w/o co-emulsifier *cosmetics*
	SYMPATENS-BCO/150	PEG-15 cocoate	'n' no. 15; sapon. no. 65	HLB 15.0; yellow liquid; viscosity 110 mPas; pH 5-7 (1% aq.)	
	SYMPATENS-BO/080	PEG-8 oleate	'n' no. 8; sapon. no. 90	HLB 11.2; white-yellow liquid; viscosity 90 mPas; pH 5-7 (1% aq.)	
	SYMPATENS-BO/100	PEG-10 oleate	'n' no. 10; sapon. no. 80	HLB 12.2; white-yellow liquid; viscosity 105 mPas; pH 5-7 (1% aq.)	
	SYMPATENS-BO/120	PEG-12 oleate	'n' no. 12; sapon. no. 70	HLB 13.0; white-yellow liquid; viscosity 120 mPas; pH 5-7 (1% aq.)	
	SYMPATENS-BS/060	PEG-6 stearate	'n' no. 6; sapon. no. 105	HLB 9.6; white-yellow paste; viscosity 25 mPas (50°C); pH 5-7 (1% aq.)	o/w co-emulsifier; refatting agent *personal care products; cosmetics*
	SYMPATENS-BS/080	PEG-8 stearate	'n' no. 8; sapon. no. 90	HLB 11.1; white-yellow solid; viscosity 30 mPas (50°C); pH 5-7 (1% aq.)	
	SYMPATENS-BS/090	PEG-9 stearate	'n' no. 9; sapon. no. 85	HLB 11.7; white-yellow solid; viscosity 30 mPas (50°C); pH 5-7 (1% aq.)	
	SYMPATENS-BS/120	PEG-12 stearate	'n' no. 12; sapon. no. 70	HLB 13.0; white-yellow solid; viscosity 40 mPas (50°C); pH 5-7 (1% aq.)	
	SYMPATENS-BS/200	PEG-20 stearate	'n' no. 20; sapon. no. 50	HLB 15.1; white-yellow solid; viscosity 50 mPas (60°C); pH 5-7 (1% aq.)	
	SYMPATENS-BS/230	PEG-23 stearate	'n' no. 23; sapon. no. 45	HLB 15.6; white-yellow solid; viscosity 65 mPas (60°C); pH 5-7 (1% aq.)	
	SYMPATENS-BS/300	PEG-30 stearate	'n' no. 30; sapon. no. 35	HLB 16.5; white-yellow solid; viscosity 80 mPas (60°C); pH 5-7 (1% aq.)	o/w emulsifier; refatting agent; viscosity modifier *personal care products; cosmetics*
	SYMPATENS-BS/400	PEG-40 stearate	'n' no. 40; sapon. no. 30	HLB 17.3; white-yellow solid; viscosity 105 mPas (60°C); pH 5-7 (1% aq.)	
	SYMPATENS-BS/500	PEG-50 stearate	'n' no. 50; sapon. no. 25	HLB 17.7; white-yellow solid; viscosity 150 mPas (60°C); pH 5-7 (1% aq.)	
	SYMPATENS-BS/1000	PEG-100 stearate	'n' no. 100; sapon. no. 10	HLB 18.8; white-yellow solid; viscosity 410 mPas (60°C); pH 5-7 (1% aq.)	o/w emulsifier; viscosity modifier *cosmetics*
	SYMPATENS-EDO/120	PEG-12 distearate	acid value < 20	white-yellow liquid; viscosity 30 mPas (60°C); pH ca. 5 (1% aq.)	
	SYMPATENS-EDO/1500	PEG-150 distearate	acid value < 10	white-yellow solid; viscosity 930 mPas (60°C); pH ca. 5 (1% aq.)	viscosity modifier; refatting agent *shampoos; bath preparations*
	SYMPATENS-EDS/080	PEG-8 distearate	acid value < 20	white-yellow solid; viscosity 25 mPas (60°C); pH ca. 5 (1% aq.)	
Protex	PROX-ONIC CC-05	POE (5) coco fatty acid	'n' no. 5; sapon. no. 135	liquid	emulsifier; additive; viscosity modifier
	PROX-ONIC CC-09	POE (9) coco fatty acid	'n' no. 9; sapon. no. 90	liquid	*metal working; textile industry; cosmetics; defoamers*
	PROX-ONIC CC-014	POE (14) coco fatty acid	'n' no. 14; sapon. no. 68	liquid	
	PROX-ONIC OL-1/05	POE (5) oleic fatty acid	'n' no. 5; sapon. no. 118	liquid	emulsifier
	PROX-ONIC OL-1/09	POE (9) oleic fatty acid	'n' no. 9; sapon. no. 85	liquid	*metal cleaning; textile industry; leather industry*

Supplier	Trade name	Chemical description	Composition	General properties	Functionality Application
Protex	PROX-ONIC OL-1/014	POE (14) oleic fatty acid	'n' no. 14; sapon. no. 65	liquid	
	PROX-ONIC OL-2/05	POE (5) oleic fatty acid	'n' no. 5; sapon. no. 118	liquid	emulsifier *metal cleaning; textile industry; leather industry*
	PROX-ONIC OL-2/09	POE (9) oleic fatty acid	'n' no. 9; sapon. no. 85	liquid	
	PROX-ONIC OL-2/014	POE (14) oleic fatty acid	'n' no. 14; sapon. no. 65	liquid	
	PROX-ONIC ST-05	POE (5) stearic fatty acid	'n' no. 5; sapon. no. 112	paste	
	PROX-ONIC ST-09	POE (9) stearic fatty acid	'n' no. 9; sapon. no. 87	paste	
	PROX-ONIC ST-014	POE (14) stearic fatty acid	'n' no. 14; sapon. no. 62	paste	
	PROX-ONIC ST-023	POE (23) stearic fatty acid	'n' no. 23; sapon. no. 43	solid	
	PROX-ONIC TA-1/08	POE (8) tall oil acid	'n' no. 8	HLB 10.5; liquid; cloud point <25°C (1% aq.)	
	PROX-ONIC TA-1/010	POE (10) tall oil acid	'n' no. 10	HLB 11.5; liquid; cloud point 25°C (1% aq.)	emulsifier; detergent *degreasing formulations; detergents*
	PROX-ONIC TA-1/016	POE (16) tall oil acid	'n' no. 16	HLB 13.4; liquid; cloud point 70°C (1% aq.)	
	PROX-ONIC TA-2/08	POE (8) tall oil acid	'n' no. 8	HLB 10.5; liquid; cloud point <25°C (1% aq.)	
	PROX-ONIC TA-2/010	POE (10) tall oil acid	'n' no. 10	HLB 11.5; liquid; cloud point 25°C (1% aq.)	
Rhone-Poulenc	ALKAMULS A	oleic acid ethoxylate	100% solids; 'n' no. 5	HLB 9.7; liquid	emulsifier; wetting agent
	ALKAMULS PE/220	PEG 220 monolaurate		liquid	viscosity modifier *emulsion polymerisation*
	ALKAMULS PE/400-l	PEG 400 monolaurate	100% solids; 'n' no. 9	liquid	
Seppic	SIMULSOL 2599	oleic acid 10EO	100% active; 'n' no. 10	HLB 12.5; liquid	emulsifier
	SIMULSOL A	oleic acid 5-6EO	100% active; 'n' no. 5-6	HLB 8; liquid	emulsifier
	SIMULSOL M 45	PEG-8 stearate	100% active; 'n' no. 8	HLB 11.1; wax	
	SIMULSOL M 49	PEG-20 stearate	100% active; 'n' no. 20	HLB 15; wax	
	SIMULSOL M 51	stearic acid 30EO	100% active; 'n' no. 30	HLB 16.5; wax	emulsifier; detergent *personal care products*
	SIMULSOL M 52	PEG-40 stearate	100% active; 'n' no. 40	HLB 16.9; flakes	
	SIMULSOL M 53	PEG-50 stearate	100% active; 'n' no. 50	HLB 18.4; wax	
	SIMULSOL M 59	PEG-100 stearate	100% active; 'n' no. 100	HLB 18.8; wax	
	SIMULSOL PS 20	stearic acid 25EO	100% active; 'n' no. 25	HLB 16; wax	emulsifier; detergent
Servo Delden	SERDOX NOG 200	oleic acid 4-5EO		liquid; cloud point 50-54°C (10% in 25% BDG)	
	SERDOX NOG 330	oleic acid 8EO		liquid; cloud point 62-65°C (10% in 25% BDG)	
	SERDOX NOG 440	oelic acid 10EO		liquid; cloud point 67-71°C (10% in 25% BDG)	
	SERDOX NOG 600	oleic acid 15EO		liquid	
	SERDOX NOG 880	oleic acid 20EO		paste; cloud point 55-59°C (1% in 10% NaCl soln.)	

Supplier	Trade name	Chemical description	Composition	General properties	Functionality *Application*
Servo Delden	SERDOX NSG 145	stearic acid 3EO		solid; cloud point 65°C (2% in 60% BDG)	
	SERDOX NSG 200	stearic acid 4·5EO		solid; cloud point 58-62°C (10% in 25% BDG)	
	SERDOX NSG 275	stearic acid 6EO		solid; cloud point 86-88°C (10% in 40% BDG)	
	SERDOX NSG 400	stearic acid 9EO		solid; cloud point 69-73°C (10% in 25% BDG)	
Stepan Europe	SECOSTER A	fatty acid ethoxylate; made-to-order		yellow to brown liquid	dispersant; emulsifier; low foam *household, institutional and industrial cleaners; oil industry*
Thor Chemicals	SOVATEX L4	coconut fatty acid ethoxylate	100% active	liquid	detergent; emulsifier; lubricant; antistatic agent; softener
	SOVATEX L10	coconut fatty acid ethoxylate	100% active	solid	
	SOVATEX O4	oleic acid ethoxylate	100% active	liquid	
	SOVATEX O10	oleic acid ethoxylate	100% active	solid	
	SOVATEX S4	stearic acid ethoxylate	100% active	solid	
	SOVATEX S10	stearic acid ethoxylate	100% active	solid	
Witco	EMULPON A6	oleic acid 6EO	100% active; 'n' no. 6	liquid	emulsifier *cutting oils*
	EMULPON A9	oleic acid 9EO	100% active; 'n' no. 9	liquid	
	ETHOTAL CH 5	ethoxylated tall oil	100% active	liquid	wetting agent *detergents; hard-surface cleaners; agrochemicals; paper industry*
	REWOLAN EO 70	oleic acid polyglycol ester	100% active	liquid	emulsifier *mineral oils; textile industry; metal working*
	REWOPAL M 365	ricinoleic acid polyglycol ester	100% active	liquid	additive *metal working*
Zschimmer & Schwarz	MULSIFAN RT 1	fatty acid ethoxylate	100% active	HLB 10; liquid	emulsifier *textile industry; metal working; mineral oils*
	MULSIFAN RT 2	fatty acid ethoxylate	100% active	HLB 10; liquid	
	MULSIFAN RT 113	fatty acid ethoxylate	100% active	HLB 9; liquid	

Glyceride ethoxylates

Supplier	Trade name	Chemical description	Composition	General properties	Functionality *Application*
Akcros Chemicals	ETHYLAN C12AH	castor oil ethoxylate	100% active; ethylene oxide content 35%	HLB 7.0; clear yellow liquid; pour point 3°C; viscosity 500 cSt; pH 5-7 (1% aq.)	emulsifier; mould release agent; lubricant *oils; metal working; plastics industry*
	ETHYLAN C30	castor oil ethoxylate; made-to-order	100% active; ethylene oxide content 58%	HLB 11.6; pale amber liquid; pour point 5°C; viscosity 550 cSt; pH 5-7 (1% aq.)	emulsifier; lubricant *chlorinated solvents; vegetable oils; textile industry; pesticides*
	ETHYLAN C35	castor oil ethoxylate	100% active; ethylene oxide content 62%	HLB 12.4; pale amber liquid; pour point 7°C; viscosity 560 cSt; pH 5-7 (1% aq.)	emulsifier; lubricant *chlorinated solvents; vegetable oils; textile industry; pesticides*
	ETHYLAN C404	castor oil ethoxylate	100% active; ethylene oxide content 65%	HLB 13.5; yellow liquid; pour point 6.5°C; viscosity 650 cSt; pH 5-7 (1% aq.)	emulsifier; lubricant *chlorinated solvents; vegetable oils; textile industry*
Akzo Nobel	ELFAPUR R 150	castor oil ethoxylate (15)	100% active	liquid	
Albright & Wilson	DEHSCOFIX CO SERIES	castor oil ethoxylates	100% active	liquid/paste	emulsifiers *agrochemicals*

Supplier	Trade name	Chemical description	Composition	General properties	Functionality / Application
Auschem Cesalpinia	CEREX EL 150	PEG-15 castor oil		HLB 9.0; liquid	
	CEREX EL 250	PEG-25 castor oil		HLB 11.0; liquid	
	CEREX EL 300	PEG-30 castor oil		HLB 11.8; liquid	
	CEREX EL 360	PEG-36 castor oil		HLB 13.2; liquid	
	CEREX EL 400	PEG-40 castor oil		HLB 14.0; liquid	o/w emulsifier; solubiliser
	CEREX EL 429	PEG-9 ricinoleate		HLB 11.4; liquid	shampoos; bath preparations; perfumes; vitamins
	CEREX ELS 250	PEG-25 hydrogenated castor oil		HLB 10.9; liquid	
	CEREX ELS 400	PEG-40 hydrogenated castor oil		HLB 14.0; liquid	
	CEREX ELS 450	PEG-45 hydrogenated castor oil		HLB 14.4; liquid	
	CEREX UL 60	PEG-6 olive oil		HLB 10.0; liquid	
	CHIMIPAL EL 15	PEG-15 castor oil	100% active; 'n' no. 15	HLB 9.0; liquid	solubiliser; o/w emulsifier; wetting agent; lubricant; dispersant / vitamins; zootechnical industry; perfumes; household and industrial detergents
	CHIMIPAL EL 25	PEG-25 castor oil	100% active; 'n' no. 15	HLB 11.0; liquid	
	CHIMIPAL EL 30	PEG-30 castor oil		HLB 11.8; liquid	solubiliser; o/w emulsifier / vitamins; zootechnical industry
	CHIMIPAL EL 36	PEG-36 castor oil		HLB 13.2; liquid	
	CHIMIPAL EL 40	PEG-40 castor oil	100% active; 'n' no. 40	HLB 14.0; liquid	solubiliser; o/w emulsifier; wetting agent; lubricant; dispersant / vitamins; zootechnical industry; perfumes; household and industrial detergents
	CHIMIPAL EL 200	PEG-200 castor oil	100% active; 'n' no. 200	HLB 18.7; waxy	
	CHIMIPAL EL 429	PEG-9 rinoleate		HLB 11.4; liquid	solubiliser; o/w emulsifier / vitamins; zootechnical industry
	CHIMIPAL ELH 6	POE (6) hydrogenated castor oil	100% active; 'n' no. 6	liquid	solubiliser; wetting agent; lubricant; dispersant / perfumes; household and industrial detergents
	CHIMIPAL ELH 25	POE (25) hydrogenated castor oil	100% active; 'n' no. 25	liquid	
	CHIMIPAL ELH 40	POE (40) hydrogenated castor oil	100% active; 'n' no. 40	paste	
	CHIMIPAL ELS 25	PEG-25 hydrogenated castor oil		HLB 10.9; liquid	solubiliser; o/w emulsifier / vitamins; zootechnical industry
	CHIMIPAL ELS 40	PEG-40 hydrogenated castor oil		HLB 14.0; thick liquid	
	CHIMIPAL U 6	POE (6) vegetable fatty glyceride	100% active; 'n' no. 6	liquid	solubiliser; wetting agent; lubricant; dispersant / perfumes; household and industrial detergents
	EMULSON EL	POE castor oil		HLB 14.4; waxy	
	EMULSON EL 25	POE castor oil		HLB 11.0; liquid	
	EMULSON EL 40	POE castor oil		HLB 13.1; liquid	emulsifier / pesticides
	EMULSON EL 50 C	POE castor oil		HLB 14.0; paste	
	EMULSON EL 81 C	POE castor oil		HLB 15.8; waxy	
	EMULSON ELH 25	POE hydrogenated castor oil		HLB 10.9; liquid	
	STEROL ELS 50	PEG-5 hydrogenated castor oil		HLB 4.9; liquid	co-emulsifier; emollient; lubricant; thickener / cosmetics; toiletries; hair care products; skin care products
BASF	EMULAN EL	castor oil ethoxylate	97% active	cloud point 71°C (in NaCl soln.)	emulsifier

Supplier	Trade name	Chemical description	Composition	General properties	Functionality Application
Chemax	CHEMAX CO-5	POE (5) castor oil	'n' no. 5; sapon. no. 145	HLB 3.8; liquid	additive; emulsifier; dispersant; re-wetting agent; softener *plastics industry; metal working; textile industry; paint industry; paper industry; leather industry*
	CHEMAX CO-16	POE (16) castor oil	'n' no. 16; sapon. no. 100	HLB 8.6; liquid	
	CHEMAX CO-25	POE (25) castor oil	'n' no. 25; sapon. no. 80	HLB 10.8; liquid	
	CHEMAX CO-30	POE (30) castor oil	'n' no. 30; sapon. no. 73	HLB 11.7; liquid	
	CHEMAX CO-36	POE (36) castor oil	'n' no. 36; sapon. no. 68	HLB 12.6; liquid	
	CHEMAX CO-40	POE (40) castor oil	'n' no. 40; sapon. no. 61	HLB 12.9; liquid	
	CHEMAX CO-80	POE (80) castor oil	'n' no. 80; sapon. no. 34	HLB 15.8; solid	
	CHEMAX CO-200/50	POE (200) castor oil 50%	'n' no. 200; sapon. no. 16	HLB 18.1; liquid	
	CHEMAX HCO-5	POE (5) hydrogenated castor oil	'n' no. 5; sapon. no. 142	HLB 3.8; liquid	
	CHEMAX HCO-16	POE (16) hydrogenated castor oil	'n' no. 16; sapon. no. 100	HLB 8.6; liquid	
	CHEMAX HCO-25	POE (25) hydrogenated castor oil	'n' no. 25; sapon. no. 80	HLB 10.8; liquid	
	CHEMAX HCO-200/50	POE (200) hydrogenated castor oil 50%	'n' no. 200; sapon. no. 17	HLB 18.1; liquid	
Croda	CRODURET 7 SPECIAL	PEG-7 hydrogenated castor oil		HLB 5.0; white to pale yellow viscous liquid	w/o emulsifier *cosmetics*
	CRODURET 10	PEG-10 hydrogenated castor oil		HLB 6.3; colourless liquid to white semi-solid	w/o emulsifier *cosmetics*
	CRODURET 25	PEG-25 hydrogenated castor oil		HLB 10.8; off-white hazy liquid	o/w emulsifier; wetting agent *cosmetics*
	CRODURET 40	PEG-40 hydrogenated castor oil		HLB 13.0; white semi-solid to liquid	
	CRODURET 50	PEG-50 hydrogenated castor oil		HLB 14.1; white paste	
	CRODURET 60	PEG-60 hydrogenated castor oil		HLB 14.7; white waxy solid	o/w emulsifier; solubiliser *cosmetics*
	CROVOL A40	PEG-20 almond glycerides	'n' no. 20	HLB 10.0; clear pale yellow liquid	emulsifier; wetting agent; dispersant; superfatting agent; emollient; plasticiser *skin care products; hair care products; bath preparations*
	CROVOL A70	PEG-60 almond glycerides	'n' no. 60	HLB 15.0; yellow paste/liquid	emulsifier; wetting agent; dispersant; superfatting agent; emollient; plasticiser; solubiliser *skin care products; hair care products; perfumes*
	CROVOL BA70G	PEG-42 babassu glycerides	'n' no. 42	HLB 15.0; clear to slightly hazy yellow liquid	
	CROVOL EP40	PEG-20 evening primrose glycerides	'n' no. 20	HLB 10.0; clear yellow oil	emulsifier; wetting agent; dispersant; superfatting agent; emollient; plasticiser *skin care products; hair care products; bath preparations*
	CROVOL EP70	PEG-60 evening primrose glycerides	'n' no. 60	HLB 15.0; pale yellow paste/liquid	emulsifier; wetting agent; dispersant; superfatting agent; emollient; plasticiser; solubiliser *skin care products; hair care products; perfumes*
	CROVOL M40	PEG-20 corn glycerides	'n' no. 20	HLB 10.0; clear to slightly hazy yellow liquid	emulsifier; wetting agent; dispersant; superfatting agent; emollient; plasticiser *skin care products; hair care products; bath preparations*
	CROVOL M70	PEG-40 corn glycerides	'n' no. 40	HLB 15.0; clear yellow liquid/soft paste	emulsifier; wetting agent; dispersant; superfatting agent; emollient; plasticiser; solubiliser *skin care products; hair care products; perfumes*

Supplier	Trade name	Chemical description	Composition	General properties	Functionality / Application
Croda	CROVOL PK40	PEG-12 palm kernel glycerides	'n' no. 12	HLB 9.0; colourless to pale yellow liquid	emulsifier; wetting agent; dispersant; superfatting agent; emollient; plasticiser; *skin care products; hair care products; bath preparations*
	CROVOL PK70	PEG-45 palm kernel glycerides	'n' no. 45	HLB 15.0; white to yellow soft paste/liquid	emulsifier; wetting agent; dispersant; superfatting agent; emollient; plasticiser; solubiliser; *skin care products; hair care products; perfumes*
	CROVOL S70G	PEG-45 safflower oil glycerides	'n' no. 45	HLB 15.0; yellow liquid	
	ETOCAS 5	PEG-5 castor oil	'n' no. 5	HLB 3.0; clear pale yellow liquid	
	ETOCAS 15	PEG-15 castor oil	'n' no. 15	clear to hazy yellow liquid	emulsifier; dispersant; solubiliser; *cosmetics; household products*
	ETOCAS 29	PEG-29 castor oil	'n' no. 29	HLB 11.7; clear yellow liquid	
	ETOCAS 40	PEG-40 castor oil	'n' no. 40	HLB 13.0; pale yellow liquid	
	ETOCAS 60	PEG-60 castor oil	'n' no. 60	HLB 14.7; white to pale yellow soft solid	
	GLYCEROX 767	PEG-6 caprylic/capric glycerides		HLB 13.2; clear pale yellow liquid	o/w emulsifier; solubiliser; emollient; *shampoos; foam baths; fragrances; skin care products; antiperspirants*
Dac International Surfactants	EMULGANTE EL 200	castor oil ethoxylate			antistatic agent; lubricant; *textile industry: polyester fibres*
	EMULGANTE EL 65	castor oil ethoxylate			dispersant; *textile industry: dyeing of fabrics*
Elf Atochem	REMCOPAL 40	castor oil 31EO	95% active; 'n' no. 31	HLB 11.9; liquid; cloud point 70°C (aq. soln. with 25% BDG)	
	REMCOPAL 40 S 3	castor oil 40EO	97% active; 'n' no. 40	HLB 13; liquid; cloud point 74°C (aq. soln. with 25% BDG)	
	REMCOPAL 40 S3 LE	castor oil 40EO	90% active; 'n' no. 40	HLB 13; liquid; cloud point 74°C (aq. soln. with 25% BDG)	
	REMCOPAL 4000	castor oil 31EO	100% active; 'n' no. 31	HLB 12.9; liquid; cloud point 70°C (aq. soln. with 25% BDG)	
	REMCOPAL 4018	castor oil 23EO	100% active; 'n' no. 23	HLB 10.4; liquid; cloud point 70°C (aq. soln. with 25% BDG)	
	REMCOPAL HC 7	hydrogenated castor oil 7EO; in course of development	100% active; 'n' no. 7	HLB 1.8; viscous liquid; cloud point 1.3°C	
	REMCOPAL HC 20	hydrogenated castor oil 20EO; in course of development	100% active; 'n' no. 20	HLB 9.5; paste; cloud point 69°C (aq. soln. with 25% BDG)	
	REMCOPAL HC 33	hydrogenated castor oil 33EO; in course of development	100% active; 'n' no. 33	HLB 12; cloud point 75°C (aq. soln. with 25% BDG)	
	REMCOPAL HC 40	hydrogenated castor oil 40EO; in course of development	100% active; 'n' no. 40	HLB 12.9; cloud point 77°C (aq. soln. with 25% BDG)	
	REMCOPAL HC 60	hydrogenated castor oil 60EO; in course of development	100% active; 'n' no. 60	HLB 14.6; cloud point 80°C (aq. soln. with 25% BDG)	
	SELLIG HR 18 100	castor oil 21EO; in course of development	100% active; 'n' no. 21	HLB 10.4; liquid; cloud point 70°C (aq. soln. with 25% BDG)	
	SELLIG R 20 100	castor oil 20EO	100% active; 'n' no. 20	HLB 9.5; liquid; cloud point 66°C (aq. soln. with 25% BDG)	

Supplier	Trade name	Chemical description	Composition	General properties	Functionality / Application
Elf Atochem	SELLIG R 3395 SP	castor oil 30EO	95% active; 'n' no. 30	HLB 11.6; liquid; cloud point 72°C (aq. soln. with 25% BDG)	
	SELLIG R 3395-C435	castor oil 32EO	95% active; 'n' no. 32	HLB 11.2; liquid; cloud point 72°C (aq. soln. with 25% BDG)	
	SELLIG R 3395	castor oil 33EO	95% active; 'n' no. 33	HLB 12; liquid; cloud point 69°C (aq. soln. with 25% BDG)	
	SELLIG R 4095	castor oil 40EO; in course of development	95% active; 'n' no. 40	HLB 12.9; liquid; cloud point 74°C (aq. soln. with 25% BDG)	
	SELLIG R 4495	castor oil 44EO; in course of development	95% active; 'n' no. 44	HLB 13.3; liquid; cloud point 70°C (aq. soln. with 50 g/l NaCl)	
Gattefosse	RICINION	ethoxylated castor oil	acid value 30-40; sapon. no. 60-80	HLB 12-13; liquid	solvent; emulsifier *pharmaceuticals; cosmetics*
Th. Goldschmidt	TAGAT R 40	PEG-40 hydrogenated castor oil	OH value 55-75; acid value 2 max.; iodine value 2 max.; 'n' no. 40; sapon. no. 45-65	HLB 13.0; ivory solid	solubiliser *baby care products; baby shampoos; hair care products; shampoos; hair conditioners; skin cleansing preparations; bath preparations; shower gels; bubble baths; liquid soaps; skin care products; shaving products; sunscreen products; deodorants; antiperspirants; cosmetics; perfumes*
	TAGAT R 60	PEG-60 hydrogenated castor oil	OH value 38-58; acid value 2 max.; iodine value 2 max.; 'n' no. 60; sapon. no. 38-58	HLB 15.0; ivory solid	
	TAGAT R 63	PEG-60 hydrogenated castor oil and propylene glycol	acid value 2 max.; 'n' no. 60	HLB 15.0; pale yellow liquid	
Henkel	EUMULGIN HRE 40	PEG-40 hydrogenated castor oil	'n' no. 40	lard-like solid	o/w emulsifier; solubiliser *emulsions; perfumes; essential oils*
	EUMULGIN HRE 60	PEG-60 hydrogenated castor oil	'n' no. 60	lard-like solid	
	EUMULGIN RO 35	PEG-35 castor oil	99-100% active; 'n' no. 35	liquid	solubiliser; o/w emulsifier *perfume*
	EUMULGIN RO 40	PEG-40 castor oil	99-100% active; 'n' no. 40	liquid	
	EUMULGIN RT 40	castor oil 40EO	99.8-100% active; 'n' no. 40	liquid	emulsifier *fats; oils; solvents*
Hoechst	EMULSOGEN EL 400	castor oil 40EO	100% active; 'n' no. 40		emulsifier; solubiliser *cosmetics; agrochemicals*
	EMULSOGEN EL	castor oil 36EO	100% active; 'n' no. 36		
Hüls	MARLOSOL R 70	castor oil 70EO	active detergent 100%; 'n' no. 70	solid	consistency improver *solid toilet cleaners*
	MARLOWET R 11	ethoxylated castor oil	active detergent 100%	liquid; cloud point 52°C (10% in 25% BDG soln.)	w/o emulsifier *animal and vegetable oils; leather industry*
	MARLOWET R 20	ethoxylated castor oil	active detergent 100%	liquid	emulsifier *solvents; cosmetics; textile industry; pesticides*
	MARLOWET R 22	ethoxylated castor oil	active detergent 100%	liquid	
	MARLOWET R 25	ethoxylated castor oil	active detergent 100%	liquid	
	MARLOWET R 32	ethoxylated castor oil	active detergent 100%	liquid	
	MARLOWET R 36	ethoxylated castor oil	active detergent 100%	liquid; cloud point 50°C (2% in 10% NaCl soln.)	
	MARLOWET R 40	ethoxylated castor oil	active detergent 100%	liquid; cloud point 50°C (2% in 10% NaCl soln.)	
	MARLOWET R 54	ethoxylated castor oil	active detergent 100%	liquid; cloud point 50°C (2% in 10% NaCl soln.)	
	SOFTIGEN 767	C₈/₁₀-fatty acid partial glyceride, ethoxylated	active detergent 100%	liquid	refatting agent *shampoos; liquid soaps; washing-up liquids*
ICI	ARLAMOL GM	ethoxylated glyceride		HLB 15.7; yellow liquid	emollient *personal care products*

Supplier	Trade name	Chemical description	Composition	General properties	Functionality Application
ICI	ARLATONE 285	POE castor oil		HLB 14.4; pale cream semi-solid	
	ARLATONE 289	POE hydrogenated castor oil		HLB 14.0; pale cream solid	
	ARLATONE 650	POE castor oil		HLB 12.5; pale yellow liquid	o/w emulsifier; solubiliser *personal care products*
	ARLATONE 827	POE castor oil		HLB 11.9; yellow liquid	
	ARLATONE 975	POE-(45)-hydrogenated castor oil		HLB 14.0; yellow liquid	
	ARLATONE 980	POE-(35)-hydrogenated castor oil		HLB 12.8; yellow liquid	
	ARLATONE G	POE-(25)-hydrogenated castor oil		HLB 10.8; yellow liquid	
	ATLAS G-1281	POE triglyceride		HLB 9.7; pale yellow liquid	
	ATLAS G-1284	POE triglyceride		HLB 13.1; pale yellow suspension	
	ATLAS G-1285	POE triglyceride		HLB 14.4; pale cream paste	*household and industrial applications; agrochemicals; textile industry*
	ATLAS G-1288	POE triglyceride		HLB 16.0; pale cream paste	
	ATLAS G-1292	POE triglyceride		HLB 10.8; pale yellow liquid	
	ATLAS G-1300	POE triglyceride		HLB 18.1; pale cream solid	
	ATLAS G-1304	POE triglyceride		HLB 18.7; pale cream solid	
	ATLAS G-4847	POE castor oil		HLB 11.9; yellow liquid	
	ATLAS G-5650	POE castor oil		HLB 12.5; light yellow liquid	
	ATLOX 1285	POE triglyceride		HLB 14.4; pale cream semi-solid	emulsifier; dispersant *agrochemicals*
	CIRRASOL ALN-WY	POE triglyceride		HLB 11.0; yellow liquid	*textile industry*
Kao Corporation	FINDET AR/30	polyethoxylated castor oil	'n' no. 18; sapon. no. 90–95	liquid	
	FINDET AR/45	polyethoxylated castor oil	'n' no. 33; sapon. no. 66–73	liquid	
	FINDET AR/52	polyethoxylated castor oil	'n' no. 40; sapon. no. 58–62	liquid/paste; cloud point 68–72°C	
	FINDET ARH/17	polyethoxylated hydrogenated castor oil	'n' no. 5	solid/paste	
	FINDET ARH/52	polyethoxylated hydrogenated castor oil	'n' no. 40	solid/paste	
Karlshamns	LIPEX 102 E-75 (50% ACTIVE)	ethoxylated mono, diglyceride of Shea butter	50% active; acid value 1 max.; moisture 48–52% max.; sapon. no. 18 max.	HLB 16–18; viscous liquid; colour 6 max. (Gardner); pH 6–8.5 (5% in deionised H_2O)	
	LIPEX 102 E-75 (100% ACTIVE)	ethoxylated mono, diglyceride of Shea butter	100% active; acid value 2 max.; moisture 1.5% max.; sapon. no. 30 max.	HLB 16–18; waxy solid; colour 6 max. (Gardner); pH 6–8.5 (5% in deionised H_2O)	
	LIPEX 106 E-75 (50% ACTIVE)	ethoxylated mono, diglyceride of Illipe butter	50% active; acid value 1 max.; moisture 48–52% max.; sapon. no. 18 max.	HLB 16–18; viscous liquid; colour 6 max. (Gardner); pH 6–8.5 (5% in deionised H_2O)	emulsifier *cosmetics; shampoos; bath preparations; hair care products; shaving products*
	LIPEX 106 E-75 (100% ACTIVE)	ethoxylated mono, diglyceride of Illipe butter	100% active; acid value 2 max.; moisture 2.0% max.; sapon. no. 30 max.	HLB 16–18; waxy solid; colour 6 max. (Gardner); pH 6–8.5 (5% in deionised H_2O)	
	LIPEX 203 E-70 (50% ACTIVE)	ethoxylated mono, diglyceride of Mango kernel oil	50% active; acid value 1 max.; moisture 48–52% max.; sapon. no. 12 max.	HLB 16–18; viscous liquid; colour 6 max. (Gardner); pH 6–8.5 (5% in deionised H_2O)	
	LIPEX 203 E-70 (100% ACTIVE)	ethoxylated mono, diglyceride of Mango kernel oil	100% active; acid value 2 max.; moisture 1.5% max.; sapon. no. 20 max.	HLB 16–18; waxy solid; colour 6 max. (Gardner); pH 6–8.5 (5% in deionised H_2O)	
Dr. W. Kolb	HEDIPIN-P/51	ethoxylated modified castor oil; $C_{16/18/18'}$ carbon chain		paste; cloud point 72°C (1% in deionised water)	*personal care products; household detergents; industrial and institutional cleaning; pharmaceuticals*
	HEDIPIN-PAR	ethoxylated modified rapeseed oil; $C_{14/16/18/18'}$ carbon chain		liquid	

Supplier	Trade name	Chemical description	Composition	General properties	Functionality Application
Dr. W. Kolb	HEDIPIN-PR	ethoxylated modified rapeseed oil; $C_{14/16/18}$' carbon chain		liquid	
	HEDIPIN-R/050	ethoxylated castor oil; $C_{16/18}$' carbon chain	'n' no. ca. 5.0	liquid; cloud point 30°C (10 g in 50 ml BDG 25%)	
	HEDIPIN-R/55	ethoxylated modified castor oil; $C_{16/18/18}$' carbon chain		liquid; cloud point 55°C (2% in 10% NaCl)	
	HEDIPIN-R/060	ethoxylated castor oil; $C_{16/18}$' carbon chain	'n' no. ca. 6.0	liquid; cloud point 18°C (5 g in 25 g BDG 25%)	
	HEDIPIN-R/085	ethoxylated castor oil; $C_{16/18}$' carbon chain	'n' no. ca. 8.5	liquid; cloud point 42°C (5 g in 25 g BDG 25%)	
	HEDIPIN-R/100	ethoxylated castor oil; $C_{16/18/18}$' carbon chain	'n' no. ca. 10.0	liquid; cloud point 44°C (5 g in 25 g BDG 25%)	
	HEDIPIN-R/150	ethoxylated castor oil; $C_{16/18/18}$' carbon chain	'n' no. ca. 15.0	liquid; cloud point 50°C (5 g in 25 g BDG 25%)	*personal care products; household detergents; industrial and institutional cleaning; pharmaceuticals*
	HEDIPIN-R/150 H	ethoxylated modified castor oil; $C_{16/18/18}$' carbon chain	'n' no. ca. 15.0	liquid	
	HEDIPIN-R/200	ethoxylated castor oil; $C_{16/18/18}$' carbon chain	'n' no. ca. 20.0	liquid; cloud point 58°C (5 g in 25 g BDG 25%)	
	HEDIPIN-R/250	ethoxylated castor oil; $C_{16/18/18}$' carbon chain	'n' no. ca. 25.0	liquid; cloud point 67°C (10% in BDG 25%)	
	HEDIPIN-R/250 H	ethoxylated modified castor oil; $C_{16/18/18}$' carbon chain	'n' no. ca. 25.0	liquid	
	HEDIPIN-R/300	ethoxylated castor oil; $C_{16/18/18}$' carbon chain	'n' no. ca. 30.0	liquid; cloud point 70°C (10% in BDG 25%)	
	HEDIPIN-R/450	ethoxylated castor oil; $C_{16/18/18}$' carbon chain	'n' no. ca. 45.0	liquid	
	HEDIPIN-R/1600	ethoxylated castor oil; $C_{16/18/18}$' carbon chain	'n' no. ca. 160.0	solid	
	HEDIPIN-R/2000	ethoxylated castor oil; $C_{16/18/18}$' carbon chain	'n' no. ca. 200.0	solid	
	HEDIPIN-RH/020	ethoxylated hydrogenated castor oil; $C_{16/18}$ carbon chain	'n' no. ca. 2.0	paste/solid	
	HEDIPIN-RH/070	ethoxylated hydrogenated castor oil; $C_{16/18}$ carbon chain	'n' no. ca. 3.0	liquid	
	HEDIPIN-RH/250	ethoxylated hydrogenated castor oil; $C_{16/18}$ carbon chain	'n' no. ca. 25.0	liquid	
	SYMPATENS-TAL/350	PEG-35 almond glycerides	OH value 60; 'n' no. 35	HLB 12.0; yellow paste; viscosity 190 mPas; pH ca. 7 (1% aq.)	
	SYMPATENS-TOL/400	PEG-40 olive glycerides	OH value 65; 'n' no. 40	HLB 13.0; yellow liquid; viscosity 230 mPas; pH ca. 7 (1% aq.)	w/o co-emulsifier; refatting agent *cosmetics*
	SYMPATENS-TSO/360	PEG-36 soya glycerides	OH value 60; 'n' no. 36	HLB 12.0; yellow paste; viscosity 100 mPas (40°C); pH ca. 7 (1% aq.)	
	SYMPATENS-TR/050	PEG-5 castor oil	OH value 145; 'n' no. 5	HLB 3.9; white-yellow liquid; viscosity 670 mPas; pH ca. 7 (1% aq.)	
	SYMPATENS-TR/100	PEG-10 castor oil	OH value 125; 'n' no. 10	HLB 6.4; white-yellow liquid; viscosity 610 mPas; pH ca. 7 (1% aq.)	
	SYMPATENS-TR/150	PEG-15 castor oil	OH value 105; 'n' no. 15	HLB 8.3; white-yellow liquid; viscosity 600 mPas; pH ca. 7 (1% aq.)	
	SYMPATENS-TR/200	PEG-20 castor oil	OH value 95; 'n' no. 20	HLB 9.7; white-yellow liquid; viscosity 600 mPas; pH ca. 7 (1% aq.)	o/w emulsifier; refatting agent; solubiliser *cosmetics; essential oils*
	SYMPATENS-TR/300	PEG-30 castor oil	OH value 75; 'n' no. 30	HLB 11.7; white-yellow liquid; viscosity 600 mPas; pH ca. 7 (1% aq.)	

Supplier	Trade name	Chemical description	Composition	General properties	Functionality *Application*
Dr. W. Kolb	SYMPATENS-TR/400	PEG-40 castor oil	OH value 65; 'n' no. 40	HLB 13.0; white-yellow liquid; viscosity 600 mPas; pH ca. 7 (1% aq.)	o/w emulsifier; refatting agent; solubiliser *cosmetics; essential oils*
	SYMPATENS-TRH/020	PEG-2 hydrogenated castor oil	OH value 165; 'n' no. 2	HLB 1.7; white-yellow paste/solid; viscosity 630 mPas (40°C); pH ca. 7 (1% aq.)	refatting agent *cosmetics*
	SYMPATENS-TRH/070	PEG-7 hydrogenated castor oil	OH value 135; 'n' no. 7	HLB 5.0; white-yellow liquid; viscosity 1200 mPas; pH ca. 7 (1% aq.)	w/o emulsifier *cosmetics*
	SYMPATENS-TRH/250	PEG-25 hydrogenated castor oil	OH value 85; 'n' no. 25	HLB 10.8; white-yellow liquid; viscosity 1200 mPas; pH ca. 7 (1% aq.)	o/w emulsifier; solubiliser; refatting agent *cosmetics*
	SYMPATENS-TRH/400	PEG-40 hydrogenated castor oil	OH value 65; 'n' no. 40	HLB 13.0; white-yellow liquid/paste; viscosity 1200 mPas; pH ca. 7 (1% aq.)	o/w emulsifier; solubiliser; refatting agent *cosmetics*
Nikko Chemicals	NIKKOL CO-3	PEG-3 castor oil		pale yellow liquid	w/o emulsifier *cosmetics*
	NIKKOL CO-10	PEG-10 castor oil		pale yellow liquid	emulsifier *cosmetics*
	NIKKOL CO-20TX	PEG-20 castor oil		pale yellow liquid	
	NIKKOL CO-40TX	PEG-40 castor oil		pale yellow liquid	solubiliser *cosmetics*
	NIKKOL CO-50TX	PEG-50 castor oil		pale yellow liquid	
	NIKKOL CO-60TX	PEG-60 castor oil		pale yellow liquid	
	NIKKOL HCO-5	PEG-5 hydrogenated castor oil		pale yellow liquid	w/o emulsifier *cosmetics*
	NIKKOL HCO-10	POE (10) hydrogenated castor oil		pale yellow liquid	
	NIKKOL HCO-20	PEG-20 hydrogenated castor oil		pale yellow liquid	emulsifier *cosmetics*
	NIKKOL HCO-30	PEG-30 hydrogenated castor oil		pale yellow liquid	
	NIKKOL HCO-40	PEG-40 hydrogenated castor oil		pale yellow liquid	solubiliser *cosmetics*
	NIKKOL HCO-40 PHARM	PEG-40 hydrogenated castor oil		pale yellow liquid	*cosmetics; pharmaceuticals*
	NIKKOL HCO-50 PHARM	PEG-50 hydrogenated castor oil		white paste	*cosmetics; pharmaceuticals*
	NIKKOL HCO-50	PEG-50 hydrogenated castor oil		white paste	solubiliser *cosmetics*
	NIKKOL HCO-60 PHARM	PEG-60 hydrogenated castor oil		white paste	*cosmetics; pharmaceuticals*
	NIKKOL HCO-60	PEG-60 hydrogenated castor oil		white paste	solubiliser *cosmetics*
	NIKKOL HCO-80	PEG-80 hydrogenated castor oil		pale yellow liquid	emulsifier *cosmetics*
	NIKKOL HCO-100	PEG-100 hydrogenated castor oil		white solid	
Protex	PROX-ONIC HR-05	POE (5) castor oil	'n' no. 5; sapon. no. 145	HLB 3.8; liquid	emulsifier; dispersant; re-wetting agent; softener *metal working; textile, paint, paper, plastics and leather industries*

Supplier	Trade name	Chemical description	Composition	General properties	Functionality Application
Protex	PROX-ONIC HR-016	POE (16) castor oil	'n' no. 16; sapon. no. 100	HLB 8.6; liquid	emulsifier; dispersant; re-wetting agent; softener *metal working; textile, paint, paper, plastics and leather industries*
	PROX-ONIC HR-025	POE (25) castor oil	'n' no. 25; sapon. no. 80	HLB 10.8; liquid	
	PROX-ONIC HR-030	POE (30) castor oil	'n' no. 30; sapon. no. 73	HLB 11.7; liquid	
	PROX-ONIC HR-036	POE (36) castor oil	'n' no. 36; sapon. no. 68	HLB 12.6; liquid	
	PROX-ONIC HR-040	POE (40) castor oil	'n' no. 40; sapon. no. 61	HLB 12.9; liquid	
	PROX-ONIC HR-080	POE (80) castor oil	'n' no. 80; sapon. no. 34	HLB 15.8; solid	
	PROX-ONIC HR-0200	POE (200) castor oil	'n' no. 200; sapon. no. 16	HLB 18.1; liquid	
	PROX-ONIC HR-0200/50	POE (200) castor oil 50%	'n' no. 200; sapon. no. 16	HLB 18.1; liquid	
	PROX-ONIC HRH-05	POE (5) hydrogenated castor oil	'n' no. 5; sapon. no. 142	HLB 3.8; liquid	
	PROX-ONIC HRH-016	POE (16) hydrogenated castor oil	'n' no. 16; sapon. no. 100	HLB 8.6; liquid	
	PROX-ONIC HRH-025	POE (25) hydrogenated castor oil	'n' no. 25; sapon. no. 80	HLB 10.8; liquid	
	PROX-ONIC HRH-0200	POE (200) hydrogenated castor oil	'n' no. 200; sapon. no. 17	HLB 18.1; liquid	
	PROX-ONIC HRH-0200/50	POE (200) hydrogenated castor oil 50%	'n' no. 200; sapon. no. 17	HLB 18.1; liquid	
	SURFAROX HR 25	25 mole castor oil	'n' no. 25		
	SURFAROX HR 40	40 mole castor oil	'n' no. 40		
	SURFAROX HRH 40C	40 mole EO hydrogenated castor oil	'n' no. 40		solubiliser *perfumes*
	SURFAROX HRH 60C	60 mole EO hydrogenated castor oil	'n' no. 60		solubiliser *perfumes*
Rhone-Poulenc	ALKAMULS EL SERIE	castor oil ethoxylates		HLB 12-14; liquid; cloud point 42-80°C (1% aq.)	foaming agent (moderate); emulsifier; wetting agent
Seppic	SIMULSOL 989	PEG-7 hydrogenated castor oil	100% active; 'n' no. 7	HLB 6.0; paste	solubiliser; w/o emulsifier *personal care products*
	SIMULSOL 1285	castor oil 60EO	100% active; 'n' no. 60	HLB 15.2; wax	solubiliser; emulsifier
	SIMULSOL 1292	PEG-25 hydrogenated castor oil	100% active; 'n' no. 25	HLB 11.3; liquid	
	SIMULSOL 1293	PEG-40 hydrogenated castor oil	100% active; 'n' no. 40	HLB 13.5; paste	solubiliser; emulsifier *personal care products*
	SIMULSOL 1294	PEG-60 hydrogenated castor oil	100% active; 'n' no. 60	HLB 15.2; paste	
	SIMULSOL 5817	PEG-35 castor oil	100% active; 'n' no. 30	HLB 12.5; liquid	
	SIMULSOL ELL	castor oil 40EO	100% active; 'n' no. 40	HLB 14.9; liquid	solubiliser; emulsifier
	SIMULSOL HR 33	castor oil 33EO	100% active; 'n' no. 33	HLB 12.9; liquid	solubiliser; emulsifier
	SIMULSOL OL 50	PEG-40 castor oil	100% active; 'n' no. 38	HLB 13.3; liquid	solubiliser; emulsifier *personal care products*
Servo Delden	SERVIROX OEG 45	castor oil 45% EO		liquid; cloud point 59-63°C (10% in 25% BDG)	
	SERVIROX OEG 49	castor oil 49% EO		liquid; cloud point 62-66°C (10% in 25% BDG)	
	SERVIROX OEG 55	castor oil 55% EO		liquid; cloud point 65-69°C (10% in 25% BDG)	

Supplier	Trade name	Chemical description	Composition	General properties	Functionality Application
Servo Delden	SERVIROX OEG 59	castor oil 59% EO		liquid; cloud point 61-63°C (2% in 25% BDG)	
	SERVIROX OEG 68.5	castor oil 68.5% EO		solid; cloud point 80-84°C (1% in water)	
	SERVIROX OEG 90	castor oil 90% EO		solid; cloud point 64-67°C (1% in 10% NaCl soln.)	
	SERVIROX OEG 90/50	castor oil 90% EO	dry content 50%	liquid	
Witco	EMULPON EL 20	castor oil 20EO	100% active; 'n' no. 20	liquid	emulsifier *textile softeners*
	EMULPON EL 33	castor oil 33EO	100% active; 'n' no. 33	liquid	emulsifier; levelling agent; solubiliser *agrochemicals; flavours; fragrances; pharmaceuticals*
	EMULPON EL 40	castor oil 40EO	100% active; 'n' no. 40		emulsifier; levelling agent *agrochemicals*
	REWODERM ES 90	coconut fatty acid monoglyceride polyglycol ether	100% active	liquid	emulsifier; super-fatting agent; solubiliser *cosmetics*
	REWODERM LI 48	tallow fatty acid monoglyceride polyglycol ether	100% active	wax	
	REWODERM LI 48-50	tallow fatty acid monoglyceride polyglycol ether	50% active	gel	
	REWODERM LI 63	coconut fatty acid monoglyceride polyglycol ether	100% active	paste	
	REWODERM LI 67	coconut fatty acid monoglyceride polyglycol ether	100% active	wax	
	REWODERM LI 67-75	coconut fatty acid monoglyceride polyglycol ether	75% active	liquid	thickener *shampoos; foam baths; baby shampoos*
	REWODERM LI 420-70	tallow fatty acid monoglyceride polyglycol ether	70% active	gel	
	REWODERM LI 420	tallow fatty acid monoglyceride polyglycol ether	100% active	wax	
	REWODERM LI S 75	modified fatty acid monoglyceride polyglycol ether; nitrogen free	75% active	gel	
	REWODERM LI S 80	modified palm oil monoglyceride polyglycol ether; nitrogen free	70% active	liquid	
	REWOPAL RO 40	castor oil ethoxylate	100% active	liquid	emulsifier *metal working; textile industry; agrochemicals*
	WITCONOL 5906	POE (30) castor oil		HLB 11.8; liquid; clear point 55°C; colour 2 (Gardner)	dispersant; o/w emulsifier; lubricant *personal care products; household and industrial applications*
	WITCONOL 5909	POE (40) castor oil		HLB 13.0; liquid; clear point 80°C; colour 2 (Gardner)	dispersant; o/w emulsifier; lubricant *personal care products; household and industrial applications*
Zschimmer & Schwarz	MULSIFAN RT 163	triglyceride ethoxylate	100% active	HLB 6; liquid	emulsifier *textile industry*
	MULSIFAN RT 7	triglyceride ethoxylate	100% active	HLB 10; liquid	
	OXYPON 288	olive oil ethoxylate	100% active	pasty	solubiliser; emollient; emulsifier *cosmetics; perfumes*
	OXYPON 328	jojoba oil ethoxylate	100% active	wax	solubiliser; emollient *cosmetics*
	OXYPON 365	avocado oil ethoxylate	100% active	liquid	
	OXYPON 2145	isostearic acid monoglyceride ethoxylate	100% active	liquid	emollient *cosmetics; bath preparations; cleansing preparations*

Miscellaneous ethoxylates

Supplier	Trade name	Chemical description	Composition	General properties	Functionality *Application*
Akros Chemicals	ETHYLAN FO30	fish oil ethoxylate	100% active; ethylene oxide content 30%	HLB 6.0; dark brown liquid; pour point −5°C; viscosity 113 cSt; pH 5-7 (1% aq.)	emulsifier; dispersant *anti-foam products*
	ETHYLAN FO60	fish oil ethoxylate	100% active; ethylene oxide content 54%	HLB 10.7; dark brown liquid; viscosity 155 cSt; pH 5-7 (1% aq.)	emulsifier; dispersant *oil slick dispersants*
	ETHYLAN HB1	aromatic ethoxylate	ethylene oxide content 36%	clear pale liquid; pour point <0%; viscosity 27 cSt; pH 5.5-7.5 (1% aq.)	solvent; plasticiser; film former; preservative *paint industry; copolymers*
	ETHYLAN HB4	aromatic ethoxylate	ethylene oxide content 69%	clear water-white liquid; cloud point 67°C (10% aq.); pour point <0°C; viscosity 64 cSt; pH 5.5-7.5 (1% aq.)	hydrotrope; solvent; penetrant *detergents; rinse aids; iodophors; glass cleaners*
Allied Colloids	ALCOPOL 650	ethoxylated detergent and wetting agent		liquid	detergent; wetting agent; scouring agent *textile industry*
	ALCOSIST PN FLAKE	ethoxylated alkyl aryl compound		flake	levelling agent; anti-precipitant *textile industry; dyeing of polyamides and wools*
	ALCOSPERSE RJL	polyethoxylated disperse dye levelling agent		liquid	levelling agent; dispersant; low foam *textile industry; dyeing of polyesters*
	ALCOSPERSE TL	polyethoxylated disperse dye levelling agent		liquid	levelling agent; solubiliser; dispersant *textile industry; dyeing of polyesters*
	ALCOSTAT S10	ethoxylated antistatic agent		flake	antistatic agent; lubricant; sizing agent *textile industry; carpet industry*
	ALCOSTAT S10 40	ethoxylated antistatic agent; a 40% version of Alcostat S10 as an aq. soln.		liquid	antistatic agent; lubricant *textile industry; carpet industry*
Auschem Cesalpinia	EMULSON 8 A	POE (8) tristirylphenol	100% active; 'n' no. 8	liquid	
	EMULSON 13 A	POE (13) tristirylphenol	100% active; 'n' no. 13	liquid	
	EMULSON 15 A	POE (15) tristirylphenol	100% active; 'n' no. 15	liquid	emulsifier; dispersant *pesticides*
	EMULSON 17 A	POE (17) tristirylphenol	100% active; 'n' no. 17	paste	
	EMULSON 24 A	POE (24) tristirylphenol	100% active; 'n' no. 24	waxy	
	MADEOL P 13 A	POE (13) tristirylphenol	50% active; 'n' no. 13	powder	
	TEWAX BW	POE (8) beeswax		HLB 5.0; solid	wetting agent; antistatic agent *pesticides*
BASF	EMULAN OSN	fatty alcohol ethoxylate and alkylphenol ethoxylate	100% active	cloud point 64°C (in NaCl soln.)	emulsifier *cosmetics*
Bayer	AVOLAN IWN	alkylarylamino polyglycol ether			emulsifier
	AVOLAN S	alkylarylamino polyglycol ether			levelling agent; low foam; dispersant *textile industry; dyeing of wools and acrylics*
	AVOLAN SC	alkylarylamino polyglycol ether			
	AVOLAN SCN 150	alkylarylamino polyglycol ether			levelling agent; low foam *textile industry; dyeing of wools*
	AVOLAN UL 75	alkylarylamino polyglycol ether sulfonate			

Supplier	Trade name	Chemical description	Composition	General properties	Functionality / Application
Bayer	EMULSIFIER PHN	alkylaryl polyglycol ether			dispersant; emulsifier / *textile industry; dyeing*
	EMULSIFIER WS	aryl polyglycol ether			emulsifier
	LEVALIN VKU	alkylaryl polyglycol ether		dark brown viscous liquid	padding auxiliary / *textile industry; dyeing*
Dr. Th. Boehme	EMULGATOR 6059	alkyl-alkylaryl polyglycol ether			
	EMULGATOR 7364	alkyl polyglycol ether/fatty acid polyglycol ester			spin finish / *textile industry; fibre production*
	EMULGATOR E 17	alkyl polyglycol ether/fatty acid polyglycol ester			
	EMULGATOR F3	alkyl-alkylaryl polyglycol ether			
Ciba	ALBEGAL B	ethoxylated fatty acid amine derivative		low viscosity yellowish brown opalescent liquid	levelling agent / *textile industry; dyeing of wool*
	ALBEGAL NF	based on ethoxylated fatty amine fatty alcohol derivatives		yellowish, clear to slightly turbid liquid	
	ALBEGAL SET	based on partially modified nitrogenous fatty and alcohol ethoxylates		yellowish brown, turbid, low viscosity liquid	levelling agent / *textile industry; dyeing*
Elf Atochem	REMCOPAL 625	ethoxylated bisphenol A	100% active	HLB 16.5; liquid; cloud point 30°C (aq. soln. with 25% BDG)	
	REMCOPAL 31310	bisphenol A 10EO	100% active; 'n' no. 10	HLB 12.9; liquid; cloud point 67°C	
	REMCOPAL N 6	beta naphthol 6EO; in course of development	100% active; 'n' no. 6	HLB 12.9; liquid; cloud point 47°C	
	REMCOPAL N 11	beta naphthol 11EO; in course of development	100% active; 'n' no. 11	HLB 15.4; liquid; cloud point 69°C	
Henkel	DEHYPON OCP 502	oleyl cetyl alcohol propylene glycol ether	99.7-100% active	liquid	anti-foam / *industrial cleaning*
	DISPONIL AAP 43	alkyl aryl polyglycol ether	ca. 70% active	HLB 17.9; liquid	
	DISPONIL AAP 307	alkyl aryl polyglycol ether	ca. 70% active	HLB 17.3; liquid	emulsifier / *plastics industry; coatings*
	DISPONIL AAP 436	alkyl aryl polyglycol ether	ca. 60% active	HLB 17.9; liquid	
	EPICOL G 2 G 10	ethoxylate of $C_{12/14}$ epoxide	99.5-100% active	liquid	emulsifier / *rapeseed oil*
	EUMULGIN 2142	blend of fatty acid polyglycol ester and fatty alcohol polyglycol ether	99.7-100% active	liquid	
	EW-POL 8021	aryl polyglycol ether	ca. 100% active	liquid	plasticiser; thickener / *plastics industry; coatings; adhesives*
	GENEROL 122 E 5	PEG-5 soya sterol	'n' no. 5	wax-like flakes	o/w co-emulsifier / *emulsions*
	GENEROL 122 E 10	PEG-10 soya sterol	'n' no. 10	wax-like flakes	o/w emulsifier / *emulsions*
	GENEROL 122 E 16	PEG-16 soya sterol	'n' no. 16	hard ivory-coloured wax	
	GERONOL 122 E 25	PEG-25 soya sterol	'n' no. 25	hard ivory-coloured wax	o/w emulsifier; solubiliser / *cosmetics*
	OSIMOL AL	fatty acid amide ethoxylate			crease inhibitor / *textile industry; dyeing*
	TEXADRIL LT 1285	polyglycol ether	ca. 85% active	liquid	solubiliser; emulsifier / *plastics industry; coatings*
Hoechst	GENAPOL L-3	coco ethoxylate	100% active		thickener / *cosmetics*

Supplier	Trade name	Chemical description	Composition	General properties	Functionality Application
Hüls	DIONIL OC	oleic acid monoethanolamide 3EO	active detergent 100%; 'n' no. 3	liquid; cloud point 67°C (10% in 25% BDG soln.)	refatting agent
	DIONIL SH 100	oleic acid monoethanolamide 6EO	active detergent 100%; 'n' no. 6	liquid/paste	*fine detergents; car washes*
	DIONIL W 100	oleic acid monoethanolamide 14EO	active detergent 100%; 'n' no. 14	solid	*fine detergents; car washes*
	MARLAZIN 7102	blend of fatty amine ethoxylate and fatty alcohol ethoxylate	active detergent 100%	liquid	*cleaners*
ICI	ATLAS G-389	POE-bisphenol A		HLB 11.0; amber viscous liquid	
	ATLAS G-1726	POE-(20)-sorbitol beeswax derivative		HLB 5.0; yellow brown solid	
	ATLAS G-2005	POE-(50)-sorbitol		yellow liquid	*household and industrial applications; agrochemicals; textile industry*
	ATLAS G-2240	POE-(6)-sorbitol		yellow liquid	
	ATLAS G-2250	POE-(10)-sorbitol		yellow liquid	
	ATLAS G-2330	POE-(30)-sorbitol		colourless liquid	
	ATLAS G-4849	POE-nonyl phenol methyl ether		white to pale yellow solid	
	ATLAS G-4978	mixture of POE fatty acids and POE alcohols		HLB 12.2; amber liquid	
	ATLOX 4848	POE alkyl ($C_{13/15}$) methyl ether		white to pale yellow liquid	*emulsifier; dispersant agrochemicals*
	ATLOX 4849	POE-nonylphenol methyl ether		white to pale yellow solid	
	ATLOX 4911	POE alkylaryl phenol		HLB 14.0; yellow-brown liquid	
	ATPET 545	polymeric ethoxylate		dark yellow liquid	*oil industry*
	ATSURF D309	ethoxylated phenolic resin in an aromatic solvent		amber viscous liquid	emulsifier *paint industry*
Kao Corporation	FINDET G/14	polyethoxylated glycerine	OH value 840-935; 'n' no. 2	liquid	
	FINDET G/19	polyethoxylated glycerine	OH value 390-400; 'n' no. 7	liquid	
Nikko Chemicals	NIKKOL BPS-5	POE (5) phytosterol		yellow paste	emollient *cosmetics*
	NIKKOL BPS-10	POE (10) phytosterol		yellow paste	emulsifier *cosmetics*
	NIKKOL BPS-20	POE (20) phytosterol		pale yellow solid	
	NIKKOL BPS-30	POE (30) phytosterol		pale yellow solid	emulsifier; dispersant; solubiliser *cosmetics*
	NIKKOL BPSH-25	POE (25) phytostanol		pale yellow solid	emulsifier; dispersant *cosmetics*
	NIKKOL GBW-25	POE (6) sorbit beeswax		yellow solid	emulsifier; stabiliser *cosmetics*
	NIKKOL GBW-125	POE (20) sorbit beeswax		yellow solid	
Protex	PROX-ONIC MG-010	methylgluceth 10	'n' no. 10	liquid	emollient
	PROX-ONIC MG-020	methylgluceth 20	'n' no. 20	liquid	emollient
	PROX-ONIC TM-010	POE (10) alkylmercaptan	'n' no. 10	HLB 13.9; liquid; cloud point 52°C (1% aq.)	wetting agent; emulsifier; detergent *metal cleaning; household and industrial cleaners; degreasing formulations; herbicides; insecticides*
	PROX-ONIC TM-06	POE (6) alkylmercaptan	'n' no. 6	HLB 11.0; liquid	
	PROX-ONIC TM-08	POE (8) alkylmercaptan	'n' no. 8	HLB 12.7; liquid; cloud point 28°C (1% aq.)	

Supplier	Trade name	Chemical description	Composition	General properties	Functionality / Application
Rhone-Poulenc	ALCODET HSC 1000	ethoxylated mercaptan; tertiary thioether		HLB 12.0; liquid; cloud point 19°C (1% aq.)	low foam; emulsifier; wetting agent
	SOPROPHOR 37	polystyrylphenol 16EO	100% solids; 'n' no. 16	liquid; surface tension 40 dynes/cm	emulsion polymerisation
	SOPROPHOR S/40-P	polystyrylphenol 40EO	100% solids; 'n' no. 40	powder; surface tension 41 dynes/cm	emulsion polymerisation
Rudolf Chemicals	RUCOGEN VIX	ethoxylation product		clear, oily liquid	low foam; scouring agent; detergent; wetting agent / feather washing
Sandoz	EKALINE F(70)	aliphatic polyglycol ether in aq. soln.		liquid	dispersant; levelling agent; scouring agent
	LANASAN LT	alkylaryl polyglycol ether in aq. soln.		liquid	textile industry; dyeing of wools
	LYOGEN V	alkylaryl polyglycol ether		liquid	emulsifier / solvents
	SANDOCLEAN MW	substituted polyglycol ether in aq. soln.		liquid	cleaning agent / industrial and institutional cleaning
	SANDOCLEAN RIG MOD	alkylaryl polyglycol ethers in aq. soln.		liquid	cleaning agent; degreaser; oil remover / industrial machinery cleaning
Servo Delden	SERDOX NKO 6·5	bisphenol A 6·5EO		liquid	
	SERDOX NLO 20	glycerine 20EO		liquid	
	SERDOX NSP 14	styrylphenol 14EO		liquid; cloud point 61°C (1% in water)	
	SERDOX NSP 50	styrylphenol 50EO		liquid; cloud point 72°C (1% in 10% NaCl soln.)	
Surfachem	SURFAC PEM	phenol ethoxylate			
	SURFAC RC	beta naphthol ethoxylate	100% active		
Witco	REWOPAL BN 13	naphthole 13EO	100% active	liquid	wetting agent / electroplating
	REWOPAL MPG 10	phenol polyglycol ether	100% active	liquid	solvent; solubiliser; preservative
	REWOPAL MPG 12	phenol polyglycol ether	100% active	liquid	solvent
	REWOPAL MPG 40	phenol polyglycol ether	100% active	liquid	solvent
Zschimmer & Schwarz	MULSIFAN RT 37	alkylaryl polyglycol ether	100% active	HLB 11; liquid	emulsifier / textile industry; metal working; mineral oils
	MULSIFAN RT 245	alkyl aryl polyglycol ether	100% active	liquid	solubiliser
	MULSIFAN RT 275	abietinic acid ethoxylate	100% active	liquid	additive / hair care products
	MULSIFAN RT 324	blend of alkyl and alkylaryl polyglycol ethers	100% active	HLB 8; liquid	emulsifier / white oils
	TISSOCYL CS	alkylaryl polyglycol ether		clear viscous liquid	emulsifier; washing agent; detergent / textile industry
	TISSOCYL CSB	combination of polyglycol ethers		viscous liquid	low foam; emulsifier; washing agent; detergent / textile industry

Other alkoxylates

Supplier	Trade name	Chemical description	Composition	General properties	Functionality / Application
Akcros Chemicals	ETHYLAN CD964	alcohol alkoxylate	100% active	clear colourless liquid; cloud point 30°C (1% aq.); pour point −10°C; viscosity 60 cSt; pH 5-7 (1% aq.)	wetting agent; detergent; low foam; scouring agent *dishwashing agents; rinse aids; metal cleaning; spray cleaning; textile industry*
	ETHYLAN CPG816	fatty alcohol alkoxylate	100% active	clear colourless to pale yellow liquid; cloud point 36°C (1% aq.); pour point −10°C; viscosity 48 cSt; pH 5-7 (1% aq.)	wetting agent; scouring agent; low foam; penetrant *rinse aids; metal cleaning; spray cleaning; textile industry; dyeing*
	ETHYLAN CPG833	fatty alcohol alkoxylate	100% active	clear colourless to pale yellow liquid; cloud point 35°C (1% aq.); pour point −8°C; viscosity 65 cSt; pH 5-7 (1% aq.)	wetting agent; scouring agent; low foam; penetrant *rinse aids; metal cleaning; spray cleaning; textile industry; dyeing*
	MONOLAN 1206/2	propoxylated diol	100% active	clear colourless liquid; pour point <0°C; viscosity 440 cSt; pH 7.0 (1% aq.)	defoaming agent *lubricants*
	MONOLAN OM48	high molecular weight propoxylate	100% active	clear pale yellow liquid; pour point <0°C; viscosity 171 cSt	defoaming agent *lubricants; hydraulic fluids; paint industry; polymer industry; paper industry*
	MONOLAN OM59	high molecular weight propoxylate	100% active	clear pale yellow liquid; pour point <0°C; viscosity 130 cSt	
	MONOLAN OM81	high molecular weight propoxylate	100% active	clear pale yellow liquid; pour point <0°C; viscosity 235 cSt	
Albright & Wilson	EMPILAN KCMP0703/F	fatty alcohol ethoxylate/propoxylate	100% active	liquid	dispersant; emulsifier; detergent; wetting agent; antistatic agent; compounding aid *agrochemicals; textile industry; emulsion polymerisation; polymer industry*
	EMPILAN KCMP0705/F	fatty alcohol ethoxylate/propoxylate	100% active	liquid	
	EMPILAN P7061	ethylene oxide/propylene oxide condensate	100% active	liquid	wetting agent; defoaming agent *agrochemicals; polymer industry*
	EMPILAN P7062	ethylene oxide/propylene oxide condensate	100% active	liquid	
	EMPILAN P7087	ethylene oxide/propylene oxide condensate	100% active	liquid	
Auschem Cesalpinia	CHIMIPAL PE 302	modified alkyl polyether	100% active	liquid	low foam; detergent; emulsifier; dispersant *industrial formulations*
	CHIMIPAL PE 402	modified alkyl polyether	100% active	liquid	
	CHIMIPAL PE 403	modified alkyl polyether	100% active	liquid	
	CHIMIPAL PE 405	modified alkyl polyether	100% active	liquid	
	CHIMIPAL PE 520	modified alkyl polyether	100% active	liquid	
	EMULSON 11	ethylene-propylene oxide adduct based on nonylphenol		HLB 9.1; paste	emulsifier; dispersant *pesticides*
	EMULSON 12	ethylene-propylene oxide adduct based on nonylphenol		HLB 12.5; waxy	
	EMULSON 13	ethylene-propylene oxide adduct based on nonylphenol		HLB 6.9; paste	
	EMULSON 14	ethylene-propylene oxide adduct based on nonylphenol		HLB 7.6; paste	
	EMULSON 2040	ethylene-propylene oxide adduct based on butanol		HLB 16.0; waxy	
	EMULSON PE	ethylene-propylene oxide adduct based on butanol		HLB 9.4; waxy	

Supplier	Trade name	Chemical description	Composition	General properties	Functionality / Application
Auschem Cesalpinia	ROLFOR EP 10	POE-POP linear chain alcohol	100% active	liquid	low foam
	ROLFOR EP 20	POE-POP linear chain alcohol	100% active	liquid	low foam
	ROLFOR EP 237	POE-POP short chain oxo alcohol	100% active	liquid	low foam; detergent; emulsifier; dispersant; *industrial formulations*
	ROLFOR EP 238	POE-POP short chain oxo alcohol	100% active	liquid	
	ROLFOR EP 239	POE-POP short chain oxo alcohol	100% active	liquid	
	ROLFOR EP 526	POE-POP short chain oxo alcohol	100% active	liquid	
BASF	DEGRESSAL SD 20	polypropoxylate	100% active		foam suppressor
	DEGRESSAL SD 21	fatty alcohol alkoxylate	100% active		
	DEGRESSAL SD 22	fatty alcohol alkoxylate	100% active		
	DEGRESSAL SD 23	alcohol alkoxylate	100% active		
	PLURAFAC LF 031	fatty alcohol alkoxylate, alkali-stable	95% active	cloud point 40°C (in H_2O)	
	PLURAFAC LF 120	fatty alcohol alkoxylate	100% active	cloud point 28°C (in H_2O)	
	PLURAFAC LF 131	fatty alcohol alkoxylate, alkali-stable	100% active	cloud point 35°C (in butyl digol)	
	PLURAFAC LF 132	fatty alcohol alkoxylate, alkali-stable	100% active	cloud point 30°C (in butyl digol)	
	PLURAFAC LF 220	fatty alcohol alkoxylate	95% active	cloud point 42°C (in H_2O)	
	PLURAFAC LF 221	fatty alcohol alkoxylate	95% active	cloud point 33°C (in H_2O)	
	PLURAFAC LF 223	fatty alcohol alkoxylate	100% active	cloud point 33°C (in butyl digol)	
	PLURAFAC LF 224	fatty alcohol alkoxylate	100% active	cloud point 27°C (in butyl digol)	
	PLURAFAC LF 231	fatty alcohol alkoxylate, alkali-stable	100% active	cloud point 28°C (in butyl digol)	
	PLURAFAC LF 400	fatty alcohol alkoxylate	100% active	cloud point 32°C (in H_2O)	
	PLURAFAC LF 401	fatty alcohol alkoxylate	100% active	cloud point 74°C (in H_2O)	low foam
	PLURAFAC LF 403	fatty alcohol alkoxylate	100% active	cloud point 41°C (in butyl digol)	
	PLURAFAC LF 404	fatty alcohol alkoxylate	100% active	cloud point 45°C (in butyl digol)	
	PLURAFAC LF 405	fatty alcohol alkoxylate	95% active	cloud point 55°C (in butyl digol)	
	PLURAFAC LF 431	fatty alcohol alkoxylate, alkali-stable	100% active	cloud point 38°C (in butyl digol)	
	PLURAFAC LF 500	fatty alcohol alkoxylate	100% active	cloud point 32°C (in butyl digol)	
	PLURAFAC LF 600	fatty alcohol alkoxylate	100% active	cloud point 55°C (in H_2O)	
	PLURAFAC LF 700	fatty alcohol alkoxylate	100% active	cloud point 29°C (in butyl digol)	
	PLURAFAC LF 711	fatty alcohol alkoxylate	100% active	cloud point 36°C (in H_2O)	
	PLURAFAC LF 1300	fatty alcohol alkoxylate	100% active	cloud point 21°C (in butyl digol)	
	PLURAFAC LF 1430	special alkoxylate	100% active	cloud point 35°C (in H_2O)	
Chemax	CHEMAL DA-4P2	alkoxylated branched alcohol		liquid; cloud point 19°C (1% aq.)	low foam; wetting agent; detergent; defoaming agent
	CHEMAL DA-5P8	alkoxylated branched alcohol		liquid; cloud point 18°C (1% aq.)	*dishwashing agents; metal cleaning; detergents*
	CHEMAL LF-25B	alkoxylated linear alcohol		liquid; cloud point 25°C (1% aq.)	
	CHEMAL LF-40B	alkoxylated linear alcohol		liquid; cloud point 40°C (1% aq.)	
	CHEMAL LFL-17	alkoxylated linear alcohol		liquid; cloud point 17°C (1% aq.)	
	CHEMAL LFL-19	alkoxylated linear alcohol		liquid; cloud point 19°C (1% aq.)	
	CHEMAL LFL-28	alkoxylated linear alcohol		liquid; cloud point 28°C (1% aq.)	
	CHEMAL LFL-47	alkoxylated linear alcohol		liquid; cloud point 47°C (1% aq.)	

Supplier	Trade name	Chemical description	Composition	General properties	Functionality / Application
Chemax	CHEMAL PS-1020E	alkoxylated linear alcohol		liquid; cloud point 45°C (1% aq.)	low foam; wetting agent; detergent; defoaming agent
	CHEMAL PS-1040I	alkoxylated linear alcohol		liquid; cloud point 25°C (1% aq.)	*dishwashing agents; metal cleaning; detergents*
	CHEMAL PS-1043I	alkoxylated linear alcohol		liquid; cloud point 24°C (1% aq.)	
Croda	PROCAS H3	PPG-3 hydrogenated castor oil		clear, colourless to pale yellow viscous liquid	emollient; lubricant; pigment dispersant
					cosmetics; bath preparations; skin care products
	PROCETYL 10	PPG-10 cetyl ether		clear, essentially colourless liquid	emollient; lubricant; co-solvent; plasticiser; superfatting agent
	PROCETYL 20	PPG-20 cetyl ether		clear, essentially colourless liquid	
	PROCETYL 30	PPG-30 cetyl ether		clear, essentially colourless liquid	*cosmetics; skin care products; hair care products; bath*
	PROCETYL 50	PPG-50 cetyl ether		clear, essentially colourless liquid	*preparations; antiperspirants; deodorants*
	PROCETYL AWS	PPG-5 ceteth-20		colourless, clear or slightly hazy liquid	emollient; emulsifier; plasticiser; solubiliser
					bath preparations; antiperspirants
	PROMYRISTYL PM3	PPG-3 myristyl ether		clear, essentially colourless liquid	emollient; lubricant; co-solvent; plasticiser; superfatting agent
	PROSTEARYL 15	PPG-15 stearyl ether		clear, essentially colourless liquid	*cosmetics; skin care products; hair care products; bath preparations; antiperspirants; deodorants*
Dac International Surfactants	BIODAC 11009	C$_{10}$/C$_{9/11}$ fatty alcohol ethoxylate/propoxylate			detergent; wetting agent; low foam
	BIODAC 11027	C$_{10}$/C$_{9/11}$ fatty alcohol ethoxylate/propoxylate			*textile industry; fermentation*
Ellis & Everard	CAFLON TLE	complex polyoxyalkalene derivative			
Henkel	AETHOXAL B	PPG-5 laureth-5	100% active	pale yellow oily liquid	super-fatting agent; solubiliser; spreading agent
					bath preparations; shampoos
	DEHYPON KE 2574	fatty alcohol C$_{10/12}$ 6EO 8PO	99.7-100% active	liquid	defoaming agent
					metal degreasers
	DEHYPON KE 2623 DEO	fatty alcohol propoxylate	99.7-100% active	liquid	defoaming agent
					liquid cleaners with APG
	DEHYPON LS 24	fatty alcohol C$_{12/14}$ 2EO 4PO	99.5-100% active	liquid	
	DEHYPON LS 36	fatty alcohol C$_{12/14}$ 3EO 6PO	99.5-100% active	liquid	
	DEHYPON LS 36 G	fatty alcohol C$_{12/14}$ 3EO 6PO, bleached	99.5-100% active	liquid	low foam
	DEHYPON LS 45	fatty alcohol C$_{12/14}$ 4EO 5PO	99.5-100% active	liquid	*industrial cleaning; dishwashing agent; laundry products*
	DEHYPON LS 45 G	fatty alcohol C$_{12/14}$ 4EO 5PO, bleached	99.5-100% active	liquid	
	DEHYPON LS 54	fatty alcohol C$_{12/14}$ 5EO 4PO	99.5-100% active	liquid	
	EUMULGIN L	PPG-2-ceteareth-9		liquid	solubiliser
					skin care products; hair care products; cosmetics; perfumes
Hoechst	DISSOLVER GX-5	speciality alkoxylate	100% active		emulsifier; solubiliser
					specialities
	EMULSOGEN P	speciality alkoxylate	100% active		emulsifier
					paraffin
	GENAPOL 2822	alcohol EO-PO	100% active		
	GENAPOL 2908	alcohol EO-PO	100% active		low foam
	GENAPOL 2909	alcohol EO-PO	100% active		*detergents*
	GENAPOL B	ethylene diamine alkoxylate	100% active		

Supplier	Trade name	Chemical description	Composition	General properties	Functionality Application
Hüls	MARLOWET 5001	C_{18}-alcohol polyalkylene glycol ether	active detergent 100%	liquid; cloud point 61°C (10% in 25% BDG soln.)	emulsifier; *paraffinic mineral oils; textile industry*
	MARLOWET 5165	alkyldiamine-alkylene oxide addition product	active detergent 100%	liquid; cloud point 39°C (10% in 25% BDG soln.)	foam regulator; *metal working*
	MARLOWET WOE	C_{18}-alcohol polyalkylene glycol ether	active detergent 100%	liquid; cloud point 69°C (10% in 25% BDG soln.)	emulsifier; *paraffinic mineral oils; textile industry*
	MARLOX 3000	butylglycol-ethylene oxide-propylene oxide addition product	active detergent 100%	liquid; cloud point 64°C (2% in 10% NaCl soln.)	*textile industry*
	MARLOX 3000E	butylglycol-ethylene oxide-propylene oxide addition product; low-electrolyte form	active detergent 100%	liquid; cloud point 64°C (2% in 10% NaCl soln.)	*textile industry*
	MARLOX FK 14	$C_{10/12}$-fatty alcohol-alkylene oxide addition product	active detergent 100%	liquid; cloud point 45°C (2% in H_2O)	
	MARLOX FK 14 E	$C_{10/12}$-fatty alcohol-alkylene oxide addition product; low-electrolyte form	active detergent 100%	liquid; cloud point 45°C (2% in H_2O)	
	MARLOX FK 64	$C_{10/12}$-fatty alcohol-alkylene oxide addition product	active detergent 100%	liquid; cloud point 55°C (10% in 25% BDG soln.)	
	MARLOX FK 64 E	$C_{10/12}$-fatty alcohol-alkylene oxide addition product; low-electrolyte form	active detergent 100%	liquid; cloud point 55°C (10% in 25% BDG soln.)	*low-foaming detergents and cleaners; dishwashing agents; industrial cleaners; textile industry*
	MARLOX FK 69	$C_{10/12}$-fatty alcohol-alkylene oxide addition product	active detergent 100%	liquid; cloud point 43°C (2% in H_2O)	
	MARLOX FK 69 E	$C_{10/12}$-fatty alcohol-alkylene oxide addition product; low-electrolyte form	active detergent 100%	liquid; cloud point 43°C (2% in H_2O)	
	MARLOX FK 86	$C_{10/12}$-fatty alcohol-alkylene oxide addition product	active detergent 100%	liquid; cloud point 22°C (2% in H_2O)	
	MARLOX FK 86 E	$C_{10/12}$-fatty alcohol-alkylene oxide addition product; low-electrolyte form	active detergent 100%	liquid; cloud point 22°C (2% in H_2O)	
	MARLOX FK 1614	$C_{10/12}$-fatty alcohol-alkylene oxide addition product	active detergent 100%	liquid; cloud point 26°C (5 g plus 25 g of 25% BDG soln.)	
	MARLOX L6	C_{12}-fatty alcohol-alkylene oxide addition product	active detergent 100%	liquid; cloud point 27°C (10% in 25% BDG soln.)	*textile industry*
	MARLOX MO 124	$C_{12/14}$-fatty alcohol-alkylene oxide addition product	active detergent 100%	liquid; cloud point 39°C (10% in 25% BDG soln.)	
	MARLOX MO 145	$C_{12/14}$-fatty alcohol-alkylene oxide addition product	active detergent 100%	liquid; cloud point 43°C (10% in 25% BDG soln.)	
	MARLOX MO 154 E	$C_{12/14}$-fatty alcohol-alkylene oxide addition product; low-electrolyte form	active detergent 100%	liquid; cloud point 42°C (5 g plus 25 g 25% BDG soln.)	*low-foaming detergents and cleaners; dishwashing agents; industrial cleaners; textile industry*
	MARLOX MO 154	$C_{12/14}$-fatty alcohol-alkylene oxide addition product	active detergent 100%	liquid; cloud point 42°C (5 g plus 25 g 25% BDG soln.)	
	MARLOX MO 174	$C_{12/14}$-fatty alcohol-alkylene oxide addition product	active detergent 100%	liquid; cloud point 41°C (2% in H_2O)	
	MARLOX MO 244	$C_{12/14}$-fatty alcohol-alkylene oxide addition product	active detergent 100%	liquid; cloud point 55°C (10% in 25% BDG soln.)	
	MARLOX MS 48	$C_{12/14}$-fatty alcohol-alkylene oxide addition product	active detergent 90%	liquid; cloud point 72°C (10% in 25% BDG soln.)	*textile industry*

Supplier	Trade name	Chemical description	Composition	General properties	Functionality / Application
Hüls	MARLOX NP 109	nonylphenol-alkylene oxide addition product	active detergent 100%	liquid; cloud point 29°C (2% in H_2O)	foam regulator
	MARLOX NP 109 E	nonylphenol-alkylene oxide addition product; low-electrolyte form	active detergent 100%	liquid; cloud point 29°C (2% in H_2O)	*low-foaming cleaners*
	MARLOX OD 105	C_8-fatty alcohol-alkylene oxide addition product	active detergent 100%	liquid; cloud point 20°C (2% in H_2O)	foam regulator
	MARLOX Q 286	$C_{16/18}$-fatty alcohol-alkylene oxide addition product	active detergent 100%	liquid; cloud point 62°C (10% in 25% BDG soln.)	*textile industry*
	MARLOX S 58	$C_{13/15}$-fatty alcohol-alkylene oxide addition product	active detergent 100%	liquid; cloud point 39°C (10% in 25% BDG soln.)	foam regulator
	MARLOX T 50/5	$C_{16/18}$-fatty alcohol-alkylene oxide addition product	active detergent 90%	liquid; cloud point 54°C (10% in 25% BDG soln.)	*textile industry*
ICI	ARLAMOL E	POP-(15)-stearyl alcohol		colourless liquid	emollient / *personal care products*
	ATLAS G-1350	POE-POP sorbitan linoleic phthalic ester		HLB 14.4; dark amber solid	
	ATLAS G-1652	POP bisphenol A		white solid	
	ATLAS G-1653	POP bisphenol A		white liquid	
	ATLAS G-1666	POP bisphenol A		white liquid	
	ATLAS G-2162	POE-(25)-oxypropylene monostearate		HLB 16.0; pale cream solid	*household and industrial applications; agrochemicals; textile industry*
	ATLAS G-3721	POP-POE alkanol		HLB 9.0; pale yellow liquid to paste	
	ATLAS G-3887	alkoxylated alcohol		colourless liquid	
	ATLAS G-4809	alkoxylated alkylphenol		HLB 17.0; colourless to pale yellow liquid	
	ATLAS G-4848	POE-alkyl ($C_{13/15}$) methyl ether		white to pale yellow liquid	
	ATLOX 4898	alkoxylated alkyl phenol		HLB 15.0; amber suspension	emulsifier; dispersant / *agrochemicals*
	ATLOX 5002	alkoxylated nonylphenol		pale yellow solid	
	SYNPERONIC 87K	alkoxylated synthetic primary $C_{13/15}$ alcohol		HLB 12.0; colourless liquid	
	SYNPERONIC LF/ RA30	alkoxylated primary alcohol		colourless liquid; cloud point 34°C (1% aq. soln.)	wetting agent; defoaming agent; emulsifier / *household, industrial and institutional cleaning*
	SYNPERONIC LF/ RA40	alkoxylated primary alcohol		colourless liquid; cloud point 24.5°C (1% aq. soln.)	
	SYNPERONIC LF/ RA43	alkoxylated primary alcohol		colourless liquid	
	SYNPERONIC LF/ RA50	alkoxylated primary alcohol		colourless liquid; cloud point 40°C (1% aq. soln.)	
	SYNPERONIC LF/ RA260	alkoxylated primary alcohol		colourless liquid; cloud point < 10°C (1% aq. soln.)	
	SYNPERONIC LF/ RA270	alkoxylated primary alcohol		colourless liquid; cloud point 14°C (1% aq. soln.)	
	SYNPERONIC LF/ RA280	alkoxylated primary alcohol		colourless liquid; cloud point 21°C (1% aq. soln.)	
	SYNPERONIC LF/ RA290	alkoxylated primary alcohol		colourless liquid; cloud point 28°C (1% aq. soln.)	
	SYNPERONIC LF/ RA310	alkoxylated primary alcohol		colourless liquid; cloud point 58°C (1% aq. soln.)	

Supplier	Trade name	Chemical description	Composition	General properties	Functionality / Application
ICI	SYNPERONIC LF/ RA320	alkoxylated primary alcohol		colourless liquid; cloud point 74.5°C (1% aq. soln.)	wetting agent; defoaming agent; emulsifier *household, industrial and institutional cleaning*
	SYNPERONIC LF/ RA343	alkoxylated primary alcohol		colourless liquid	
	SYNPERONIC NPE1800	alkoxylated nonylphenol		pale yellow solid	wetting agent *textile industry; agrochemicals; industrial applications*
	TEEFROTH A	short-chain alcohol alkoxylate		pale yellow liquid	
	TEEFROTH AN	short-chain alcohol alkoxylate		pale yellow liquid	foaming agent *mining industry*
	TEEFROTH AUN	short-chain alcohol alkoxylate		pale yellow liquid	
	TEEFROTH D	short-chain alcohol alkoxylate		pale yellow liquid	
	TEEFROTH G	short-chain alcohol alkoxylate		pale yellow liquid	
	TEEFROTH GUN	short-chain alcohol alkoxylate		pale yellow liquid	
	UKANIL 190	modified fatty alcohol alkoxylate		yellow/amber liquid; cloud point <10°C (1% aq. soln.)	
	UKANIL 2078	polyol alkoxylate		colourless liquid; cloud point <10°C (1% aq. soln.)	
	UKANIL 2090	polyamine alkoxylate		amber liquid; cloud point 23.3°C (1% aq. soln.)	
	UKANIL 2094	modified polyol alkoxylate		amber liquid; cloud point <10°C (1% aq. soln.)	defoaming agent; wetting agent *agrochemicals; textile industry; industrial and institutional cleaners*
	UKANIL 2111	modified fatty alcohol alkoxylate		colourless liquid; cloud point <10°C (1% aq. soln.)	
	UKANIL 2163	polyamine alkoxylate		colourless liquid; cloud point 21.8°C (1% aq. soln.)	
	UKANIL 2220N	polyol alkoxylate		colourless liquid; cloud point 41.0°C (1% aq. soln.)	
	UKANIL 2265	alcohol alkoxylate		colourless liquid; cloud point <10°C (1% aq. soln.)	
	UKANIL 2340	polyol alkoxylate		colourless liquid; cloud point 40.0°C (1% aq. soln.)	
	UKANIL 3000	alcohol alkoxylate		yellow/amber liquid; cloud point <10°C (1% aq. soln.)	
Dr. W. Kolb	IMBENTIN-AG/100/ 040 OFA	alkoxylated C_{10} alcohol	'n' no. ca. 4.0	liquid; cloud point 69°C (5 g in 25 g BDG 25%)	
	IMBENTIN-AG/100/ 35 OFA	alkoxylated C_{10} alcohol	'n' no. ca. 5.0	liquid; cloud point 56°C (5 g in 25 g BDG 25%)	
	IMBENTIN-AG/168/ 250/PEG	alkoxylated PEG-modified base; $C_{16/18}$ carbon chain	'n' no. ca. 25.0	solid/flakes	
	IMBENTIN-AGS/20 A	alkoxylated $C_{11/13}$ alcohol	'n' no. ca. 3.0	liquid; cloud point 11°C (3 g in 10 ml alcohol (30°C))	personal care products; household detergents; industrial and institutional cleaning; pharmaceuticals
	IMBENTIN-AGS/35	alkoxylated $C_{11/13}$ alcohol	'n' no. ca. 7.0	liquid; cloud point 36°C (1% in deionised water)	
	IMBENTIN-AGS/55	alkoxylated $C_{11/13}$ alcohol	'n' no. ca. 9.0	liquid; cloud point 59°C (1% in deionised water)	
	IMBENTIN-AGS/90	alkoxylated $C_{11/13}$ alcohol	'n' no. ca. 10.5	liquid; cloud point 90°C (1% in deionised water)	
	IMBENTIN-BD/200	alkoxylated butandiol; C_4 carbon chain	'n' no. ca. 20.0	paste/solid	
	IMBENTIN-BN/150	alkoxylated betanaphthol	'n' no. ca. 15.0	liquid; cloud point 61°C (1% in 10% NaCl)	

Supplier	Trade name	Chemical description	Composition	General properties	Functionality Application
Dr. W. Kolb	IMBENTIN-C/91/35 OFA	alkoxylated $C_{9/11}$ alcohol	'n' no. ca. 5.0	liquid; cloud point 53°C (1% in deionised water)	
	IMBENTIN-C/91/080 OFA	alkoxylated $C_{9/11}$ alcohol	'n' no. ca. 8.0	liquid; cloud point 84°C (1% in deionised water)	
	IMBENTIN-DNP/100	alkoxylated dinonylphenol	'n' no. ca. 10.0	liquid; cloud point 73°C (5 g in 25 g BDG 25%)	
	IMBENTIN-E/12/050 OFA	alkoxylated C_{12} alcohol	'n' no. ca. 7.0	liquid; cloud point 76°C (5 g in 25 g BDG 25%)	*personal care products; household detergents; industrial and institutional cleaning; pharmaceuticals*
	IMBENTIN-E/12/065 OFA	alkoxylated C_{12} alcohol	'n' no. ca. 7.0	liquid; cloud point 58°C (1% in deionised water)	
	IMBENTIN-KIB	alkoxylated n-butanol; C_4 carbon chain	'n' no. ca. 2.0	liquid	
	IMBENTIN-L/111/35 OFA	alkoxylated C_{11} alcohol	'n' no. ca. 5.5	liquid; cloud point 53°C (10% in deionised water)	
	IMBENTIN-NAP/27	alkoxylated nonylphenol		liquid; cloud point 27°C (0.1% in deionised water)	low foam
	IMBENTIN-NAP/40	alkoxylated nonylphenol		liquid; cloud point 40°C (0.1% in deionised water)	*personal care products; household detergents; industrial and institutional cleaning; pharmaceuticals*
	IMBENTIN-POA/050 OFA	alkoxylated $C_{16/18}$ alcohol	'n' no. ca. 7.0	liquid; cloud point 65°C (5 g in 25 g BDG 25%)	
	IMBENTIN-PPF	alkoxylated alcohol		liquid; cloud point 38°C (1% in deionised water)	
	IMBENTIN-SG/015/AG	alkoxylated alcohol		liquid; cloud point 21°C (5% in BDG 25%)	
	IMBENTIN-SG/29/AG	alkoxylated alcohol		liquid; cloud point 68°C (1% in deionised water)	
	IMBENTIN-SG/40/AG	alkoxylated alcohol		liquid; cloud point 45°C (3 g in 10 ml alcohol (30°C))	
	IMBENTIN-SG/43/AG	alkoxylated alcohol		liquid; cloud point 11°C (3 g in 10 ml alcohol (30°C))	
	IMBENTIN-SG/43/C	alkoxylated alcohol		liquid; cloud point 11°C (3 g in 10 ml alcohol (30°C))	
	IMBENTIN-SG/45/AG	alkoxylated alcohol		liquid; cloud point 27°C (1% in deionised water)	
	IMBENTIN-SG/66/AG	alkoxylated alcohol		liquid; cloud point 29°C (1% in deionised water)	*personal care products; household detergents; industrial and institutional cleaning; pharmaceuticals*
	IMBENTIN-SG/113/EB	endblocked alkoxylated alcohol		liquid; cloud point 24°C (1% in deionised water)	
	IMBENTIN-SG/128/EB	endblocked alkoxylated alcohol		liquid; cloud point 58°C (5% in BDG 25%)	
	IMBENTIN-SG/142/EB	endblocked alkoxylated alcohol		liquid; cloud point 37°C (10% in deionised water)	
	IMBENTIN-SG/525/AG	alkoxylated alcohol		solid; cloud point 80°C (1% in deionised water)	
	IMBENTIN-SG/541/EB	endblocked alkoxylated alcohol		liquid; cloud point 56°C (5 g in 25 g BDG 25%)	
	IMBENTIN-SG/918/AG	alkoxylated alcohol		solid; cloud point 53°C (1% in 10% NaCl)	low foam

Supplier	Trade name	Chemical description	Composition	General properties	Functionality Application
Dr. W. Kolb	IMBENTIN-SG/2940/AG	alkoxylated alcohol		liquid; cloud point 74°C (3 g in 10 ml alcohol (30°C))	
	IMBENTIN-SG/8917/AG	alkoxylated alcohol		liquid; cloud point 41°C (1% in deionised water)	low foam; personal care products; household detergents; industrial and institutional cleaning; pharmaceuticals
	IMBENTIN-SG/8922/AG	alkoxylated alcohol		liquid; cloud point 32°C (1% in deionised water)	
	IMBENTIN-SG/AGS/EB	endblocked alkoxylated alcohol		liquid; cloud point 55°C (5 g in 25 g BDG 25%)	
	IMBENTIN-SG/T/EB	endblocked alkoxylated alcohol		liquid; cloud point 55°C (5 g in 25 g BDG 25%)	personal care products; household detergents; industrial and institutional cleaning; pharmaceuticals
	IMBENTIN-SOR/010	alkoxylated sorbitol	'n' no. ca. 1.0	solid/paste	
	IMBENTIN-SOR/020	alkoxylated sorbitol	'n' no. ca. 2.0	solid/paste	
	IMBENTIN-SOR/060	alkoxylated sorbitol	'n' no. ca. 6.0	paste	
	IMBENTIN-SOR/200	alkoxylated sorbitol	'n' no. ca. 20.0	liquid	
	IMBENTIN-SOR/300	alkoxylated sorbitol	'n' no. ca. 30.0	liquid	
	IMBENTIN-TFA/015	alkoxylated tetrahydrofurfurylalcohol	'n' no. ca. 1.5	liquid; cloud point 66°C (5 g in 25 g BDG 25%)	coupling agent; emollient; refatting agent; cosmetics
	SYMPATENS-ASP/1 50	PPG-15 stearyl ether	OH value 50	colourless liquid; viscosity 85 mPas; pH 5-7 (1% aq.)	
Lonza	CARSONON 144-P	PPG-3 myristyl ether	100% active; OH value 130	liquid; colour 35 (APHA)	lubricant; emollient; solubiliser; cosmetics
	CARSONON 169-P	PPG-10 cetyl ether	100% active; OH value 75	liquid; colour 100 (APHA)	
Nikko Chemicals	NIKKOL PBC-31	PPG-4-ceteth-1		colourless liquid	
	NIKKOL PBC-33	PPG-4-ceteth-10		pale yellow paste	
	NIKKOL PBC-34	POE (20) POP (4) cetyl ether		white solid	emulsifier; cosmetics
	NIKKOL PBC-41	PPG-8-ceteth-1		colourless liquid	
	NIKKOL PBC-44	PPG-8-ceteth-20		pale yellow solid	
	NIKKOL PEN-4612	PPG-6-decyltetradeceth-12		pale yellow solid	
	NIKKOL PEN-4620	PPG-6-decyltetradeceth-20		pale yellow solid	solubiliser; cosmetics
	NIKKOL PEN-4630	PPG-6-decyltetradeceth-30		pale yellow solid	
Nippon Shokubai	SOFTANOL EP 3025	$C_{12/14}$ secondary alcohols reacted with ethylene oxide and propylene oxide	100% active	pale yellow liquid; pour point −48°C; viscosity 34 cSt; pH 7.5 (5% aq. soln.)	
	SOFTANOL EP 5035	$C_{12/14}$ secondary alcohols reacted with ethylene oxide and propylene oxide	100% active	pale yellow liquid; cloud point 21°C; pour point −46°C; viscosity 54 cSt; surface tension 30 dynes/cm (0.1%); Ross-Miles foam (0.1%): initial 13 mm, 5 min 5 mm; pH 7.5 (5% aq. soln.)	
	SOFTANOL EP 6035	$C_{12/14}$ secondary alcohols reacted with ethylene oxide and propylene oxide	100% active	pale yellow liquid; cloud point 27°C; pour point −30°C; viscosity 61 cSt; pH 7.5 (5% aq. soln.)	low foam; wetting agent; dishwashing agents; rinse aids; metal cleaners; textile industry; paper industry; leather industry; industrial cleaners
	SOFTANOL EP 7025	$C_{12/14}$ secondary alcohols reacted with ethylene oxide and propylene oxide	100% active	pale yellow liquid; cloud point 36°C; pour point −25°C; viscosity 64 cSt; surface tension 29 dynes/cm (0.1%); Ross-Miles foam (0.1%): initial 95 mm, 5 min 15 mm; pH 7.5 (5% aq. soln.)	
	SOFTANOL EP 7045	$C_{12/14}$ secondary alcohols reacted with ethylene oxide and propylene oxide	100% active	pale yellow liquid; cloud point 30°C; pour point −35°C; viscosity 70 cSt; surface tension 30 dynes/cm (0.1%); Ross-Miles foam (0.1%): initial 35 mm, 5 min 5 mm; pH 7.5 (5% aq. soln.)	

Supplier	Trade name	Chemical description	Composition	General properties	Functionality Application
Nippon Shokubai	SOFTANOL EP 7085	C$_{12/14}$ secondary alcohols reacted with ethylene oxide and propylene oxide	100% active	pale yellow liquid; cloud point 21°C; pour point −40°C; viscosity 84 cSt; surface tension 31 dynes/cm (0.1%); Ross-Miles foam (0.1%): initial 10 mm, 5 min 1 mm; pH 7.5 (5% aq. soln.)	low foam; wetting agent / *dishwashing agents; rinse aids; metal cleaners; textile industry; paper industry; leather industry; industrial cleaners*
	SOFTANOL EP 9050	C$_{12/14}$ secondary alcohols reacted with ethylene oxide and propylene oxide	100% active	pale yellow liquid; cloud point 37°C; pour point −26°C; viscosity 83 cSt; surface tension 31 dynes/cm (0.1%); Ross-Miles foam (0.1%): initial 90 mm, 5 min 5 mm; pH 7.5 (5% aq. soln.)	
Olin Chemicals	POLY-TERGENT S-305LF	alkoxylated linear alcohols		cloud point 19°C (1% aq. soln.); surface tension 31 dynes/cm (0.1%); Draves wetting 4 sec (0.25%); Ross-Miles foam (0.25%): initial 15 mm, 5 min 5 mm; freezing point 0°C	
	POLY-TERGENT S-405LF	alkoxylated linear alcohols		cloud point 28°C (1% aq. soln.); surface tension 32 dynes/cm (0.1%); Draves wetting 4 sec (0.25%); Ross-Miles foam (0.25%): initial 20 mm, 5 min 0 mm; freezing point 5°C	low foam
	POLY-TERGENT S-505LF	alkoxylated linear alcohols		cloud point 47°C (1% aq. soln.); surface tension 33 dynes/cm (0.1%); Draves wetting 7 sec (0.25%); Ross-Miles foam (0.25%): initial 120 mm, 5 min 10 mm; freezing point 8°C	
	POLY-TERGENT SL-42	alkoxylated linear alcohols	100% active	HLB 13.0; clear viscous liquid; cloud point 42°C (1% aq. soln.); pour point −5°C; surface tension 28 dynes/cm (0.1%); Ross-Miles foam (0.25%): initial 150 mm, 5 min 35 mm; freezing point −5°C; pH 7	wetting agent; penetrant; emulsifier; foaming agent / *laundry products; household cleaners; soaps; hand cleansing preparations; metal cleaners; floor cleaners; disinfectants*
	POLY-TERGENT SL-62	alkoxylated linear alcohols	100% active	HLB 14.0; clear viscous liquid; cloud point 62°C (1% aq. soln.); pour point 1°C; surface tension 29 dynes/cm (0.1%); Draves wetting 3 sec (0.25%); Ross-Miles foam (0.25%): initial 165 mm, 5 min 35 mm; freezing point 1°C; pH 7	
	POLY-TERGENT SL-92	alkoxylated linear alcohols	100% active	HLB 15.0; slush/solid; cloud point 92°C (1% aq. soln.); pour point 20°C; surface tension 31 dynes/cm (0.1%); Draves wetting 7 sec (0.25%); Ross-Miles foam (0.25%): initial 165 mm, 5 min 40 mm; freezing point 20°C; pH 7	
	POLY-TERGENT SLF-18	alkoxylated linear alcohols	100% active	cloud point 18°C (1% aq. soln.); pour point 3°C; colour 100 max. (APHA); surface tension 33 dynes/cm (0.1%); Draves wetting 5 sec (0.25%); Ross-Miles foam (0.25%): initial 20 mm, 5 min 0 mm; freezing point 3°C; pH 5.0-7.0 (1% aq. soln.)	low foam
PPG	MACOL 57	PPG-10 butanediol			emollient / *personal care products; toiletries; antiperspirants; shaving products*
	MACOL CA 30P	PPG-30 cetyl ether			emollient / *hair care products; skin care products*
	MAZEEN 173	alkoxylated diamine		liquid	chelating agent: cross-linking agent / *electroplating; metal working; plastics industry*
	MAZEEN 174	alkoxylated diamine		liquid	
	MAZEEN 241-3	alkoxylated diamine		liquid	chelating agent: dispersant / *electroplating; textile industry*

Supplier	Trade name	Chemical description	Composition	General properties	Functionality Application
Protex	PROX-ONIC MG-020 P	methylgluceth 20		liquid	co-solvent *perfume fixative*
	PROX-ONIC SA1-015/P	POP (15) stearyl alcohol		liquid	
Rhone-Poulenc	ANTAROX BL 214	linear alcohol base; EO/PO alcohol adduct		liquid; cloud point 14°C (1% aq.)	wetting agent
	ANTAROX BL 225	linear alcohol base; EO/PO alcohol adduct		liquid; cloud point 27°C (1% aq.)	
	ANTAROX BL 330	linear alcohol base; EO/PO alcohol adduct		liquid; cloud point 30°C (10% aq.)	low foam; wetting agent
	ANTAROX FM 33	linear alcohol base; EO/PO alcohol adduct		liquid; cloud point 34°C (1% aq.)	
	MIRAVON B12-DF	linear alcohol base; EO/PO alcohol adduct		liquid; cloud point 20°C (1% aq.)	wetting agent
	MIRAVON B79-R	linear alcohol base; EO/PO alcohol adduct		liquid; cloud point 33°C (1% aq.)	low foam; wetting agent
Sandoz	VELSAN D8P-3	aliphatic alcohol ethoxylate, propoxylate and carboxylate		liquid	emollient *cosmetics; toiletries*
	VELSAN D8P-16	aliphatic alcohol ethoxylate, propoxylate and carboxylate		liquid	emollient *cosmetics; toiletries*
Servo Delden	SERDOX NBOP 39/6E	tripropyleneglycol 39PO/6EO		liquid; cloud point 45°C (5% in 25% BDG)	
	SERDOX NBOT 17/9	tripropyleneglycol 17PO/9EO		liquid; cloud point 40-45°C (2% in water)	
	SERDOX NBSQ 3/2	$C_{9/11}$ oxoalcohol 3EO/2PO		liquid; cloud point 45-50°C (10% in 25% BDG)	
	SERDOX NBSQ 5/5	$C_{9/11}$ oxoalcohol 5EO/5PO		liquid; cloud point 23-26°C (1% in water)	
	SERDOX NEOT 2/5	butylglycol 2PO/5EO		liquid; cloud point 69°C (10% in water)	
	SERDOX NFSP 10/5	2-ethylhexanediol-1,3 10PO/5EO		liquid; cloud point 18-23°C (2% in water)	
	SERDOX NLAP 3	triethanolamine 3PO		liquid	
	SERDOX NRSQ 5/8	n-decylalcohol 5EO/8PO		liquid; cloud point 13°C (1% in water)	
Stepan Europe	BIO-SOFT EA 8/EA 10	fatty alcohol alkoxylate; made-to-order	100% active	water white liquid	detergent; wetting agent; emulsifier *household, institutional and industrial cleaners*
	BIO-SOFT PG4	polyalkoxylated compound; made-to-order	100% active	water white liquid	wetting agent; hydrotrope; low foam *dishwashing agents; detergents*
	MAKON NF 12	fatty alcohol alkoxylate	100% active	water white to turbid liquid	detergent; wetting agent; low foam *household, institutional and industrial cleaners; textile industry; oil industry*
	MAKON NI 10	alkylphenol alkoxylate	100% active	water white to yellow solid	emulsifier; dispersant *agrochemicals*
	MAKON NI 20	alkylphenol alkoxylate	100% active	water white to yellow solid	
	MAKON NI 30	alkylphenol alkoxylate	100% active	water white to yellow solid	
Union Carbide	TERGITOL MIN-FOAM 1X	mixture of $C_{11/15}$ secondary alcohols reacted with ethylene oxide and propylene oxide	100% active; OH no. 87	clear liquid; cloud point 40°C (1% aq. soln.); M.W. 645; pour point −38°C; viscosity 57 cP; colour 40 (Pt-Co; APHA); surface tension 29 dynes/cm (0.1% aq. soln.); pH 6.7 (1% aq. soln.)	wetting agent; cleansing agent *household and industrial cleaners; rinse aids; leather industry; paper industry; textile industry; dry cleaning*

Supplier	Trade name	Chemical description	Composition	General properties	Functionality / Application
Union Carbide	TERGITOL MIN-FOAM 2X	mixture of $C_{11/15}$ secondary alcohols reacted with ethylene oxide and propylene oxide	100% active; OH no. 89	clear liquid; cloud point 20°C (1% aq. soln.); M.W. 630; pour point −42°C; viscosity 49 cP; colour 50 (Pt-Co; APHA); surface tension 29 dynes/cm (0.1% aq. soln.); pH 6.6 (1% aq. soln.)	anti-foaming agent / *rinse aids; textile industry; metal cleaning; paper industry; water treatment; industrial cleaners*
	TRITON CF-10	low-foam surfactant	100% active	HLB 12.6; liquid; cloud point 28°C (1% aq.); pour point 16°C	low foam; detergent; rinsing aid / *dishwashing agents; hard-surface cleaners*
	TRITON CF-21	low-foam surfactant	100% active	HLB 12.9; liquid; cloud point 40°C (1% aq.); pour point −34°C	low foam; wetting agent / *textile industry; rinsing aids*
	TRITON CF-32	low-foam surfactant	95% active	HLB 11.0; liquid; cloud point 25°C (1% aq.); pour point −9°C	low foam; detergent / *dishwashing agents; food industry; industrial cleaners; bottle washing*
	TRITON CF-54	low-foam surfactant	100% active	HLB 13.6; liquid; cloud point 38°C (1% aq.); pour point 2°C	low foam / *food industry; metal cleaning; textile industry; industrial cleaners*
	TRITON CF-76	low-foam surfactant	100% active	HLB 12.6; liquid; cloud point 31°C (1% aq.); pour point 5°C	low foam / *metal cleaning; food industry; industrial cleaners*
	TRITON CF-87	low-foam surfactant	90% active	HLB 12.7; liquid; cloud point 32°C (1% aq.); pour point −1°C	low foam / *rinsing aids*
	TRITON CF-98/375	low-foam surfactant	75% active	liquid; cloud point 30°C (1% aq.)	low foam / *dishwashing agents; bottle washing*
	TRITON DF-12	low-foam surfactant	100% active	HLB 10.6; liquid; cloud point 17°C (1% aq.); pour point 18°C	low foam
	TRITON DF-16	low-foam surfactant	100% active	HLB 11.6; liquid; cloud point 36°C (1% aq.); pour point −6°C	low foam / *rinsing aids; metal cleaning*
	TRITON DF-18	low-foam surfactant	90% active	HLB 11.3; liquid; cloud point <0°C (1% aq.); pour point 18°C	low foam / *metal cleaning; food industry; industrial cleaners*
Witco	LAUROPAL 0227	fatty alcohol PO/EO	100% active	liquid	scouring agent / *metal cleaning; detergents*
	WITBREAK DRA-21	oxyalkylated phenolic resin		liquid; pour point −12.2°C; pH 11	
	WITBREAK DRA-22	oxyalkylated phenolic resin		liquid; pour point 10°C; pH 11	
	WITBREAK DRB-11	oxyalkylated phenolic resin		liquid; pour point −6.7°C; pH 7	
	WITBREAK DRB-127	oxyalkylated phenolic resin		liquid; pour point −6.7°C; pH 10	
	WITBREAK DRB-401	oxyalkylated phenolic resin		liquid; pour point −15°C; pH 11	
	WITBREAK DRC-163	oxyalkylated phenolic resin		liquid; pour point 1.7°C; pH 10	demulsifier / *petroleum industry*
	WITBREAK DRC-164	oxyalkylated phenolic resin		liquid; pour point 7.2°C; pH 10	
	WITBREAK DRC-165	oxyalkylated phenolic resin		liquid; pour point 4.4°C; pH 10	
	WITBREAK DRC-168	oxyalkylated phenolic resin		liquid; pour point −1.1°C; pH 10	
	WITBREAK DRC-232	oxyalkylated phenolic resin		liquid; pour point −6.7°C; pH 10	
	WITBREAK DTG-62	polyoxyalkylene glycol		paste; pour point 26.7°C; pH 10	
	WITBREAK GBG-3172	polyoxyalkylated modified resin		liquid; pour point −3.9°C; pH 7	
	WITCONOL 1206	alkyl polyoxyethylene glycol ether		liquid; colour 1 (Gardner); pH 7	detergent; o/w emulsifier; wetting agent / *personal care products; household and industrial applications*

Supplier	Trade name	Chemical description	Composition	General properties	Functionality / Application
Witco	WITCONOL APM	PPG-3 myristyl ether		liquid; colour 1 (Gardner); pH 7	coupling agent; emollient; lubricant; spreading agent *personal care products; household and industrial applications*
	WITCONOL APS	PPG-11 stearyl ether		liquid; colour 1 (Gardner); pH 7	
	WITCONOL NS 500K	polyalkoxylated butyl ether	moisture 0.2% max.	HLB 14.1; white waxy solid; softening point 40°C; pH 6.0-7.0 (10% in 10:6 isopropanol/H$_2$O)	emulsifier; wetting agent; dispersant *pesticides*
	WITCOSPERSE 500K	polyalkoxylated butyl ether adsorbed on silica	40% active; moisture 5.7%	white powder; pH 6.1 (3% in H$_2$O)	wetting agent; dispersant *wettable powders;*
Zschimmer & Schwarz	PROPETAL 99	fatty alcohol EO-PO adduct	100% active	liquid; cloud point 18-22°C (1% in distilled H$_2$O)	low foam; wetting agent *cleansing preparations; high-pressure cleaners*
	PROPETAL 241	fatty alcohol EO-PO adduct	100% active	liquid; cloud point 37-41°C (1% in distilled H$_2$O)	
	PROPETAL 281	fatty alcohol EO-PO adduct	100% active	liquid; cloud point 30-33°C (1% in distilled H$_2$O)	
	PROPETAL 340	fatty alcohol EO-PO adduct	100% active	liquid; cloud point 48-52°C (1% in distilled H$_2$O)	
	PROPETAL 341	fatty alcohol EO-PO adduct	100% active	liquid; cloud point 58-62°C (1% in distilled H$_2$O)	

Alkylolamides

Supplier	Trade name	Chemical description	Composition	General properties	Functionality / Application
Akcros Chemicals	ETHYLAN KELD	modified fatty alkylolamide		clear amber liquid; pour point 5°C; viscosity 967 cSt; pH 9.0 (1% aq.)	foam stabiliser; thickener; emulsifier; antistatic agent; detergent; dispersant *liquid detergents; hard-surface cleaners; general-purpose cleaners; shampoos; hand cleaning gels; plastics industry; industrial degreasers*
	ETHYLAN LD	coconut diethanolamide	free base 4%	clear amber liquid; pour point 15°C; viscosity 1408 cSt; pH 9.5 (1% aq.)	
	ETHYLAN LDA37	coconut diethanolamide	free base 25%	clear amber liquid; pour point −14°C; viscosity 1469 cSt; pH 9.5 (1% aq.)	
	ETHYLAN LDA48	coconut diethanolamide	free base 4%	clear amber liquid; pour point 5°C; viscosity 1342 cSt; pH 9.5 (1% aq.)	
	ETHYLAN LDG	coconut diethanolamide	free base 5%	clear amber liquid; pour point 5°C; viscosity 1303 cSt; pH 9.5 (1% aq.)	
	ETHYLAN LM	coconut monoethanolamide; made-to-order	free base <1.5%	pale waxy flake; pour point 63°C; viscosity 40 cSt (80°C); pH 9.5 (1% aq.)	
	ETHYLAN MLD	lauric diethanolamide		white waxy flake; pour point 40°C; viscosity 107 cSt (60°C); pH 8.5 (1% aq.)	
Akzo Nobel	LAURIDIT KD	coco diethanolamide	100% active	paste	
	LAURIDIT KDG	coco diethanolamide	100% active	liquid	
	LAURIDIT KM	coco monoethanolamide	100% active	flakes	
	LAURIDIT OD	oleyl diethanolamide	100% active	liquid	
	LAURIDIT SM	stearyl monoethanolamide	100% active	flakes	
Albright & Wilson	EMPILAN 2502	coconut diethanolamide	80% active	liquid	
	EMPILAN 2502/W	coconut diethanolamide	80% active	liquid/paste	
	EMPILAN CDE	coconut diethanolamide	90% active	liquid/paste	antistatic agent; softener *polymer industry; leather industry*
	EMPILAN CDX	coconut diethanolamide	65% active	liquid/paste	
	EMPILAN CIS	coconut monoisopropanolamide	92% active	waxy flakes	

Supplier	Trade name	Chemical description	Composition	General properties	Functionality / *Application*
Albright & Wilson	EMPILAN CME	coconut monoethanolamide	94% active	waxy flake	
	EMPILAN CME/T	coconut monoethanolamide	90% active	waxy flake	
	EMPILAN CSDE	coconut diethanolamide	92% active	waxy solid	
	EMPILAN LDE	lauric diethanolamide	90% active		
	EMPILAN LIS	lauric monoisopropanolamide	92% active		
	EMPILAN LIS/B	lauric isopropanolamide	92% active	waxy flakes	
	EMPILAN LME	lauric monoethanolamide	92% active		
Auschem Cesalpinia	ROLAMID CD	coconut oil diethanolamide	100% active	liquid	foaming agent; viscosity modifier *household and industrial detergents*
	ROLAMID CM	cocamide MEA	100% active	flakes	foam booster; foam stabiliser; viscosity modifier *cosmetics*
	ROLAMID MC	coconut oil monoethanolamide	100% active	flakes	foaming agent; viscosity modifier *household and industrial detergents*
	ROLAMID OLD	oleic diethanolamide	100% active	liquid	
Croda	CRILLON ODE	oleic diethanolamide	acid value 12-16; free ethanolamine 20-27%; H$_2$O 2% max.	amber liquid; pH 9.5-10.5	anti-corrosion; additive; emulsifier *cutting fluids*
	INCROMECTANT AMEA 70	acetamide MEA supplied as a 70% soln. in H$_2$O	70% active; H$_2$O content 30% max.; free monoethanolamide 1.5% max.	pale yellow liquid; viscosity 25 cPs; pH 6.0-8.5	humectant *skin care products; hair care products*
	INCROMECTANT LAMEA	lactamide MEA and acetamide MEA blend	95% active	viscous liquid; viscosity 600 cPs; colour 5 max. (Gardner); pH 5-8 (10% aq. soln.)	
	INCROMECTANT LMEA 100	lactamide MEA	95% active	viscous liquid; viscosity 2000 cPs; colour 8 max. (Gardner); pH 4-6 (10% aq. soln.)	
Dac International Surfactants	DACAMMID DC	coco diethanolamide			foam stabiliser
Ellis & Everard	CAFLON CD	coconut diethanolamide	90% active		
	CAFLON CD218	coconut diethanolamide	87% active		
Henkel	COMPERLAN 100	cocamide MEA	92-98% active	beads	thickener; foam stabiliser; viscosity modifier; pearlescent agent *shampoos; bubble baths; deodorants*
	COMPERLAN COD	cocamide DEA and glycerol	78% active	liquid	viscosity modifier *shampoos*
	COMPERLAN F	linoleamide DEA	86-94% active	liquid	viscosity modifier; thickener *cosmetics*
	COMPERLAN HS	stearamide MEA	90-97% active	flakes	consistency agent *cosmetics; lipsticks*
	COMPERLAN KD	cocamide DEA	90% active	liquid/solid	viscosity modifier; foam stabiliser *shampoos; bath preparations*
	COMPERLAN KM	cocamide MEA	28% active	paste	thickener; pearlescent agent
	COMPERLAN LD	lauramide DEA	90-97% active	solid	thickener; solubiliser; viscosity modifier *shampoos; bubble baths*
	COMPERLAN LM	lauramide MEA	92-98% active	beads	thickener; foam stabiliser; pearlescent agent; super-fatting agent *bubble baths; toiletries; soaps*

Supplier	Trade name	Chemical description	Composition	General properties	Functionality / Application
Henkel	COMPERLAN LMD	lauramide DEA	90-97% active	liquid/solid	viscosity modifier; foaming agent *cosmetics*
	COMPERLAN LP	lauramide MIPA	90% active	flakes	thickener; foam stabiliser; pearlescent agent *cosmetics; hair care products*
	COMPERLAN LS	cocamide DEA containing laureth-2	68% active	liquid	viscosity modifier; solubiliser *cosmetics; essential oils*
	COMPERLAN OD	oleamide DEA	87-94% active	liquid	viscosity modifier *shampoos; bath preparations*
	COMPERLAN PD	fatty acid polydiethanolamide	65-72% active	liquid	co-emulsifier
	COMPERLAN PVD	polydiethanolamide based on vegetable oils		liquid	*dishwashing agents; cleaners*
	COMPERLAN VOD	soyamide DEA	84% active	liquid	viscosity modifier; stabiliser; super-fatting agent *shampoos; bath preparations*
	EMID 6500	cocamide MEA		tan flaked solid	foam stabiliser; thickener *shampoos; bath preparations; hair care products*
	EMID 6515	cocamide DEA		yellow liquid	foaming agent; stabiliser; thickener *hair care products; shampoos; bubble baths; liquid soaps*
	EMID 6519	lauramide DEA		yellow liquid	foaming agent; stabiliser; thickener *hair care products; shampoos; liquid soaps; shaving products*
	STANDAMID CD	capramide DEA		amber viscous liquid	foaming agent; detergent *personal care products*
	STANDAMID KD	cocamide DEA		amber viscous liquid	foaming agent; thickener *shampoos; bubble baths*
	STANDAMID KDO	cocamide DEA		amber viscous liquid	viscosity modifier; foaming agent *shampoos; bath preparations; liquid soaps*
	STANDAMID KDS	lauramide DEA		light amber liquid	skin cleansing preparations; shampoos; bath preparations; shower preparations
	STANDAMID LD	lauramide DEA		light amber soft solid to liquid	foaming agent; stabiliser; emulsifier; detergent *shampoos; bubble baths; personal care products*
	STANDAMID LD 80/20	lauramide DEA and propylene glycol		clear liquid	foaming agent; stabiliser; emulsifier; detergent; viscosity builder; conditioner *shampoos; bubble baths; personal care products*
	STANDAMID LDO	lauramide DEA		amber liquid	foaming agent; viscosity modifier *shampoos; bath preparations*
	STANDAMID LDS	lauramide DEA		light amber liquid	foaming agent; stabiliser; emulsifier; detergent; conditioner *shampoos; bubble baths; personal care products*
	STANDAMID PD	cocamide DEA		amber liquid	foaming agent
	STANDAMID SD	cocamide DEA		light amber liquid	*personal care products*
	STANDAMID SDO	cocamide DEA		amber liquid	foaming agent; viscosity modifier; humectant *hair conditioner*
	STANDAMID SM	cocamide MEA		solid	*personal care products*
	TEXAMIN PD 1	fatty acid diethanolamide		liquid	corrosion protection agent *metal cleaners*
Hickson Manro	MANROMID 150-ADY	soya diethanolamide (1:1)	78% active	liquid	hair care products; bath preparations; household detergents; industrial cleaners; metal working

Supplier	Trade name	Chemical description	Composition	General properties	Functionality / Application
Hickson Manro	MANROMID 853	coconut diethanolamide (2:1)	70% active	liquid	*household detergents; industrial cleaners; metal working*
	MANROMID 1224	mixed acid diethanolamide (1:1)	82% active	liquid	*hair care products; bath preparations; household detergents; industrial cleaners*
	MANROMID CD	coconut diethanolamine (1:1)	92% active	liquid	*hair care products; bath preparations; household detergents; industrial cleaners; metal working*
	MANROMID CDG	coconut diethanolamine (1:1)	78% active	liquid	*hair care products; bath preparations; household detergents; industrial cleaners*
	MANROMID CDS	coconut diethanolamide (1:1)	85% active	liquid	*industrial cleaners*
	MANROMID CDX	coconut diethanolamide (2:1)	70% active	liquid	*household detergents; industrial cleaners; metal working*
	MANROMID CMEA	coconut monoethanolamide (1:1)	95% active	flake	*hair care products; bath preparations; household detergents; industrial cleaners*
	MANROMID LMA	lauric monoethanolamide (1:1)	95% active	flake	
Hüls	MARLAMID A 18	stearic acid alkanolamide	active detergent 100%	flakes	*fabric softeners*
	MARLAMID A 18 E	stearic acid alkanolamide	active detergent 100%	flakes	*fabric softeners*
	MARLAMID D 1218	coconut fatty acid diethanolamide	active detergent 100%	liquid	
	MARLAMID D 1885	oleic acid diethanolamide	active detergent 100%	liquid	
	MARLAMID DF 1218	coconut fatty acid diethanolamide	active detergent 100%	liquid	*thickener; refatting agent; foam stabiliser / shampoos; washing-up liquids*
	MARLAMID DF 1818	soya fatty acid diethanolamide	active detergent 100%	liquid	
	MARLAMID KL	coconut fatty acid alkanolamide	active detergent 100%	flakes	*silk lustre agent / shampoos; liquid soaps*
	MARLAMID KLP	blend of fatty alcohol ether sulfate with coconut fatty acid alkanolamide	active detergent 30%	pumpable liquid	
	MARLAMID M 1218	coconut fatty acid monoethanolamide	active detergent 100%	flakes	*thickener; refatting agent; soln. retarder / shampoos; solid cleaners*
	MARLAMID M 1618	tallow fatty acid monoethanolamide	active detergent 100%	flakes	
	MARLAMID PG 20	blend of fatty acid glycol ester with fatty acid alkanolamide	active detergent 21%	pumpable liquid	*pearlescent agent / shampoos; liquid soaps*
	MARLOWET 5459	fatty acid diethanolamide	active detergent 100%	liquid	*anti-corrosion / metal working*
	MARLOWET OCM	fatty acid alkylolamide polyethylene glycol ether	active detergent 100%	liquid	*anti-corrosion / lubricants*
	MARLOWET SDT	fatty acid diethanolamide	active detergent 100%	liquid	*anti-corrosion / metal working*
ISP	FOAMOLE A	linoleamide DEA	acid value 5.0 max.; alkali value 21-35	amber liquid	*conditioner; thickener; foam stabiliser / hair care products; shampoos*
	FOAMOLE B	minkamidopropyl dimethylamine	acid value 5.0 max.; alkali no. 125-140	amber liquid to flowing gel; pH 9.5-10.5 (1% aq. soln.)	*super-fatting agent / shampoos; hair care products*
	FOAMOLE M	cocamide MEA	acid no. 2.0 max.; alkali no. 12.0 max.	cream coloured flakes; m.p. 70-74°C	*foam builder; stabiliser; thickener; emulsifier / shampoos; bubble baths; detergents; hair care products*
Kao Corporation	AMIDET A/111	coco monoethanolamide	amine 2.5% max.; ester 4% max.; fatty acid 2% max.; amide ca. 85%; carbon chain composition C_8 5-9%, C_{10} 5-8%, C_{12} 46-52%, C_{14} 14-20%, C_{16} 8-10%, C_{18} 2-4%, C_{18}' 6-9%	solid	

Supplier	Trade name	Chemical description	Composition	General properties	Functionality / Application
Kao Corporation	AMIDET B/112	coco diethanolamide	amine 5% max.; ester 6% max.; fatty acid 1% max.; amide 80-85%; carbon chain composition C_8 5-9%, C_{10} 5-8%, C_{12} 46-52% C_{14} 14-20%, C_{16} 8-10%, C_{18} 2-4%, $C_{18'}$ 6-9%	liquid	
	AMIDET B/113	coco diethanolamide	amine 8-11%; ester 2% max.; fatty acid 1% max.; amide 75-80%; carbon chain composition C_8 5-9%, C_{10} 5-8%, C_{12} 46-52%, C_{14} 14-20%, C_{16} 8-10%, C_{18} 2-4%, $C_{18'}$ 6-9%	liquid	
	AMIDET B/125	undecilenic diethanolamide	amine ca. 24% fatty acid 7.7-9.5%; amide ca. 68%	liquid	
	AMIDET SB/13	coco superdiethanolamide	amine 5% max.; ester 3.5% max.; fatty acid 1% max.; amide ca. 90%; carbon chain composition C_8 5-9% C_{10} 5-8%; C_{12} 46-52%, C_{14} 14-20%, C_{16} 8-10%, C_{18} 2-4%, $C_{18'}$ 6-9%	liquid/solid	
	AMIDET SB/16	lauric superdiethanolamide	amine 3% max.; ester 1.5% max.; fatty acid 0.2% max.; amide ca. 95%; carbon chain composition C_{10} 2-5%, C_{12} 92-94%, C_{14} 2-7%	solid	
Lonza	CARSAMIDE AMEA	acetamide MEA	70% active	liquid	conditioner; antistatic agent *personal care products; hair conditioners*
	CARSAMIDE CA	cocamide DEA	100% active	liquid	foam booster *personal care products*
	CARSAMIDE CMEA	cocamide MEA	100% active	flakes	thickener *personal care products*
	CARSAMIDE SAC	cocamide DEA	100% active	liquid	viscosity modifier *personal care products*
	CARSAMIDE SAL-7	lauramide DEA	100% active	solid	
	CARSAMIDE SAL-9	lauramide DEA	100% active	solid	
	UNAMIDE C-5	PEG 6 cocamide	100% active	liquid	*industrial and household applications*
	UNAMIDE C-72-3	2:1 coco diethanolamide	100% active	liquid	
	UNAMIDE D-10	cocamide DEA	100% active	liquid	*industrial applications*
	UNAMIDE LDL	cocamide DEA	100% active	liquid	
McIntyre Group	MACKAMIDE AME-75	acetamide MEA	75% active	liquid; pH 7	thickener; foam booster; stabiliser; solubiliser; detergent; anti-corrosion; lubricant *industrial cleaners*
	MACKAMIDE AME-100	acetamide MEA	100% active	liquid; pH 7	
	MACKAMIDE C	cocamide DEA (1:1)	100% active	liquid; pH 10	
	MACKAMIDE CD	cocamide DEA (2:1)	100% active	liquid; pH 10	
	MACKAMIDE CD-8	cocamide DEA and mixed soaps	100% active	liquid; pH 9	
	MACKAMIDE CD-10	capramide DEA	100% active	liquid; pH 10	
	MACKAMIDE CD-25	cocamide DEA and tall oil soap	100% active	liquid; pH 9	
	MACKAMIDE CDM	cocamide DEA and DEA oleate	100% active	liquid; pH 9	
	MACKAMIDE CDS-80	cocamide DEA and DEA dodecylbenzene sulfonate	100% active	liquid; pH 9	

Supplier	Trade name	Chemical description	Composition	General properties	Functionality / *Application*
McIntyre Group	MACKAMIDE CDT	cocamide DEA and tall oil soap	100% active	liquid; pH 9	thickener; foam booster; stabiliser; solubiliser; detergent; anti-corrosion; lubricant *industrial cleaners*
	MACKAMIDE CMA	cocamide MEA	100% active	flake; pH 10	
	MACKAMIDE CS	cocamide DEA (1:1)	100% active	liquid; pH 10	
	MACKAMIDE ISA	isostearamide DEA	100% active	liquid; pH 10	
	MACKAMIDE L10	lauramide DEA	100% active	liquid; pH 10	
	MACKAMIDE L95	lauramide DEA (95% lauric)	100% active	solid; pH 10	
	MACKAMIDE LLM	lauramide DEA	100% active	liquid; pH 10	
	MACKAMIDE LMD	lauramide DEA (70% lauric)	100% active	solid; pH 10	
	MACKAMIDE LME	lactamide MEA	100% active	liquid; pH 5	
	MACKAMIDE LMM	lauramide MEA	100% active	flake; pH 10	
	MACKAMIDE LOL	linoleamide DEA	100% active	liquid; pH 10	
	MACKAMIDE MC	cocamide DEA (1:1)	100% active	liquid; pH 10	
	MACKAMIDE MO	oleamide DEA (1:1)	100% active	liquid; pH 10	
	MACKAMIDE NOA	oleamide DEA (1:1)	100% active	liquid; pH 10	
	MACKAMIDE O	oleamide DEA (2:1)	100% active	liquid; pH 10	
	MACKAMIDE ODM	oleamide DEA and DEA oleate	100% active	gel; pH 9	
	MACKAMIDE PK	palmkernelamide DEA	100% active	liquid; pH 10	
	MACKAMIDE PKM	palmkernelamide MEA	100% active	flake; pH 10	
	MACKAMIDE R	ricinoleamide DEA	100% active	liquid; pH 10	
	MACKAMIDE S	soyamide DEA (1:1)	100% active	liquid; pH 10	
	MACKAMIDE SD	soyamide DEA (2:1)	100% active	liquid; pH 10	
	MACKAMIDE SMA	stearamide MEA	100% active	flake; pH 10	
Millchem	ONYXOL 345	lauramide DEA	alkali no. 130-165; acid no. 8-13	liquid	foam stabiliser; wetting agent; dispersant; thickener *cosmetics*
	ONYXOL SD	cocoamide DEA	alkali no. 32 max.; acid no. 2 max.	liquid	foam stabiliser; thickener *shampoos; bubble baths; dishwashing agents*
	SUPER AMIDE GR	cocoamide DEA	alkali no. 32 max.; acid no. 10 max.	liquid	foam stabiliser; emulsifier; thickener *household and industrial formulations; cosmetics*
	SUPER AMIDE L9	lauramide DEA	alkali no. 45 max.; acid no. 9 max.	paste	thickener; foam stabiliser
	SUPER AMIDE L9C	lauramide DEA	alkali no. 16 max.; acid no. 2 max.	paste	*shampoos; bubble baths; dishwashing agents; industrial cleaners*
	SUPER AMIDE LL	lauramide DEA	alkali no. 37 max.; acid no. 2 max.	liquid	foam booster; stabiliser; viscosity modifier
	SUPER AMIDE LM	high activity lauric/myristic acid diethanolamine condensate		liquid	
	SURCO 128-T	amide condensate		liquid	foam stabiliser; thickener *cosmetics; industrial and household formulations*
	SURCO CMEA	cochin monoethanolamide		solid	
	SURCO WC CONCENTRATE	coconut fatty acid diethanolamide		liquid	

Supplier	Trade name	Chemical description	Composition	General properties	Functionality Application
Mona Industries	MONAMID 7-100	1:1 fatty acid diethanolamide	coconut fatty acid base; 100% active; acid value 0-2; alkali value 5-20	liquid; pH 8-9 (10% soln.)	foam booster; foam stabiliser; emulsifier; detergent; wetting agent; anti-corrosion; viscosity modifier; lubricant; dispersant *cosmetics; bubble baths; shampoos; dry cleaning; metal cleaning; dishwashing agents; soaps; waterless hand cleaners; leather industry; laundry products; hair conditioners; metal working*
	MONAMID 7-153CS	1:1 fatty acid diethanolamide	modified coconut fatty acid base, 100% active; acid value 0-2; alkali value 4-10	liquid; pH 8.5-9.5 (10% soln.)	
	MONAMID 15-70W	1:1 fatty acid diethanolamide	linoleic acid base; 100% active; acid value 0-1; alkali value 25-40	liquid; pH 10-11 (10% soln.)	
	MONAMID 150-AD	1:1 fatty acid diethanolamide	coconut fatty acid base; 100% active; acid value 0-3; alkali value 55-70	liquid; pH 9.8-10.8 (10% soln.)	
	MONAMID 150-ADD	1:2 fatty acid diethanolamide	coconut fatty acid base; 100% active; acid value 0-3; alkali value 58-68	liquid; pH 10-11 (10% soln.)	
	MONAMID 150-ADY	1:2 fatty acid diethanolamide	mixed fatty acid base; 100% active; acid value 0-1; alkali value 30-45	liquid; pH 10-11 (10% soln.)	
	MONAMID 150-CW	1:1 fatty acid diethanolamide	capric acid base; 100% active; acid value 0-2; alkali value 40-55	liquid (crystallises on aging); pH 10.3-11.3 (10% soln.)	
	MONAMID 150-DR	1:1 fatty acid diethanolamide	coconut fatty acid base; 100% active; acid value 0-5; alkali value 10-25	liquid; pH 9.0-10 (10% soln.)	
	MONAMID 150-GLT	1:1 fatty acid diethanolamide	lauric acid base; 100% active; acid value 0-1; alkali value 30-45	liquid (crystallises on aging); pH 10.3-11.3 (10% soln.)	
	MONAMID 150-LWW-C	1:1 fatty acid diethanolamide	70/30 lauric/myristic acid base; 100% active; acid value 0-1; alkali value 30-45	solid (supercools easily); pH 10.2-11.2 (10% soln.)	
	MONAMID 150-LW	1:1 fatty acid diethanolamide	lauric acid base; 100% active; acid value 0-1; alkali value 30-45	solid (supercools easily); pH 10-11 (10% soln.)	
	MONAMID 150-LWA	1:1 fatty acid diethanolamide	lauric acid base; 100% active; acid value 0-1; alkali value 10-25	solid (supercools easily); pH 9.5-10.5 (10% soln.)	
	MONAMID 150-MW	1:1 fatty acid diethanolamide	myristic acid base; 100% active; acid value 0-3; alkali value 35-50	solid; pH 9.5-10.5 (10% soln.)	
	MONAMID 716	1:1 fatty acid diethanolamide	modified lauric acid base; 100% active; acid value 0-3; alkali value 45-60	liquid; pH 10.0-11.0 (10% soln.)	
	MONAMID 718	1:1 fatty acid diethanolamide	stearic acid base; 100% active; acid value 21 ± 3; alkali value 45-65	solid; pH 9.3-10.3 (10% soln.)	
	MONAMID 770	1:1 fatty acid diethanolamide	modified coconut acid base; 85% active; acid value 0-1; alkali value 35-45	liquid; pH 9.2-10.2 (10% soln.)	
	MONAMID CMA	1:1 fatty acid monoethanolamide	coconut fatty acid base; 100% active; acid value 0-1; alkali value 6-12	granular; solidification point 63 ± 2°C; pH 9.4-10.8 (10% soln.)	
	MONAMID LIPA	1:1 fatty acid monoisopropanolamide	lauric acid base; 100% active; acid value 0-1; alkali value 12-22	granular; solidification point 55 ± 3°C; pH 10.3-11.3 (10% soln.)	
	MONAMID LMA	1:1 fatty acid monoethanolamide	lauric acid base; 100% active; acid value 0-1; alkali value 5-12	granular; solidification point 80 ± 2°C; pH 10-11 (10% soln.)	
	MONAMID LMMA	1:1 fatty acid monoethanolamide	lauric/myristic acid base; 100% active; acid value 0-1; alkali value 5-12	granular; solidification point 80 ± 2°C; pH 9.7-10.7 (10% soln.)	
	MONAMID R31-42	1:1 fatty acid diethanolamide	70/30 lauric/myristic acid base; 80% active; acid value 0-1; alkali value 25-35	liquid; pH 10-11 (10% soln.)	
	MONAMINE 150-IS	1:1 fatty acid diethanolamide	isostearic acid base; 100% active; acid value 5-10; alkali value 30-60	liquid; pH 8.8-9.8 (10% soln.)	

Supplier	Trade name	Chemical description	Composition	General properties	Functionality / Application
Mona Industries	MONAMINE AA-100	1:2 fatty acid diethanolamide	distilled coconut fatty acid base; 100% active; acid value 28-32; alkali value 165-185	liquid; pH 9.5-10.5 (10% soln.)	
	MONAMINE AC-100	1:2 fatty acid diethanolamide	mixed fatty acid base; 100% active; acid value 22-32; alkali value 170-190	liquid; pH 9.5-10.5 (10% soln.)	
	MONAMINE ACO-100	1:2 fatty acid diethanolamide	lauric acid base; 100% active; acid value 10-14; alkali value 180-200	paste; pH 9.5-10.5 (10% soln.)	
	MONAMINE AD-100	1:2 fatty acid diethanolamide	coconut fatty acid base; 100% active; acid value 2-8; alkali value 105-125	liquid; pH 9.5-10.5 (10% soln.)	
	MONAMINE ADD-100	1:2 fatty acid diethanolamide	coconut fatty acid base; 100% active; acid value 2-6; alkali value 105-125	liquid; pH 9.5-10.5 (10% soln.)	foam booster; foam stabiliser; emulsifier; detergent; wetting agent; anti-corrosion; viscosity modifier; lubricant; dispersant *cosmetics; bubble baths; shampoos; dry cleaning; metal cleaning; dishwashing agents; soaps; waterless hand cleaners; leather industry; laundry products; hair conditioners; metal working*
	MONAMINE ADS-100	1:2 fatty acid diethanolamide	mixed fatty acid base; 100% active; acid value 48-52; alkali value 110-125	liquid; pH 9.0-10.0 (10% soln.)	
	MONAMINE ADY-100	1:2 fatty acid diethanolamide	mixed fatty acid base; 100% active; acid value 0-2; alkali value 110-130	liquid; pH 10.5-11.5 (10% soln.)	
	MONAMINE ALX-80 SS	1:2 fatty acid diethanolamide	modified coconut fatty acid base; 80% active; acid value 52-60; alkali value 65-75	liquid; pH 8.5-9.5 (10% soln.)	
	MONAMINE ALX-100 S	1:2 fatty acid diethanolamide	modified coconut fatty acid base; 100% active; acid value 62-70; alkali value 70-90	liquid; pH 8.5-9.5 (10% soln.)	
	MONAMINE CF-100 M	1:2 fatty acid diethanolamide	mixed fatty acid base; 100% active; acid value 56-64; alkali value 110-120	liquid; pH 8.5-9.5 (10% soln.)	
	MONAMINE I-76	1:2 fatty acid diethanolamide	coconut fatty acid base; 100% active; acid value 45-55; alkali value 80-100	liquid; pH 8.5-9.5 (10% soln.)	
	MONAMINE LM-100	1:2 fatty acid diethanolamide	70/30 lauric/myristic acid base; 100% active; acid value 18-23; alkali value 160-175	liquid; pH 9.5-10.5 (10% soln.)	
	MONAMINE R8-26	1:2 fatty acid diethanolamide	mixed fatty acid base; 100% active; acid value 75-85; alkali value 177-187	liquid; pH 9.0-10.0 (10% soln.)	
	MONAMINE S	1:1 fatty acid monoethanolamide	stearic acid base; 100% active; acid value 0-1; alkali value 5-18	granular; solidification point 87 ± 2°C; pH 9.5-11.0 (10% soln.)	
	MONAMINE T-100	1:2 fatty acid diethanolamide	tall oil fatty acid base; 100% active; acid value 10-16; alkali value 100-120	liquid; pH 10.0-11.0 (10% soln.)	
Pentagon	PENTAMID C12	fatty acid polydiethanolamide; contains no sodium nitrite or other inorganic inhibitors	acid value 10.0-16.5; amine value 100-135	clear amber viscous liquid	anti-corrosion; low foam *metal working; lubricants*
	PENTAMID LD	linoleic diethanolamide	free diethanolamine 5.0% max.	clear amber liquid; colour 6 max. (Gardner); pH 9.5-11.0 (1% aq. soln.)	thickener; foam booster and stabiliser; emulsion stabiliser; super-fatting agent *shampoos; foam baths; liquid soaps*
PPG	LAROSTAT 902	alkanolamide			antistatic agent *polyolefins; polyethylene; polypropylene*
	LAROSTAT FPE	alkanolamide			antistatic agent *polyolefins*
	MAZAMIDE 62	lauramide DEA	free fatty acid 1.5% max.	liquid	*personal care products*
	MAZAMIDE 65	2:1 modified cocamide DEA	free fatty acid 25%; free amine 21%		emulsifier; thickener; foam booster; rust inhibitor; lubricant *metal working*
	MAZAMIDE 70	2:1 modified cocamide DEA	free fatty acid 3%; free amine 30%		
	MAZAMIDE 80	cocamide DEA	free fatty acid 0.5% max.; free amine 6% max.		thickener; foam stabiliser; solubiliser *shampoos; bubble baths; liquid soaps; toiletries*

Supplier	Trade name	Chemical description	Composition	General properties	Functionality / Application
PPG	MAZAMIDE CMEA	cocamide MEA	free fatty acid 0.5% max.; free amine 2% max.		viscosity modifier; foam stabiliser / *personal care products*
	MAZAMIDE O-20	2:1 oleamide DEA	free fatty acid 7%; free amine 24%		emulsifier; thickener; foam booster; rust inhibitor; lubricant / *metal working*
	MAZAMIDE SMEA	stearamide MEA	free fatty acid 0.5% max.; free amine 2% max.	flakes	stabiliser / *personal care products*
	MAZAMIDE SS-10	linoleamide DEA	free fatty acid 1.5-2%; free amine 6.5-7%		emulsifier; thickener; foam booster; rust inhibitor; lubricant / *metal working; shampoos; liquid soaps; shower gels*
	MAZAMIDE T-20	2:1 tall oil alkanolamide	free fatty acid 5%; free amine 21%		emulsifier; thickener; foam booster; rust inhibitor; lubricant / *metal working*
Rhone-Poulenc	ALKAMIDE KD	cocamide DEA 1:1		liquid	foaming agent (moderate); foam booster
Scher Chemicals	SCHERCOMID 1214	lauramide DEA and diethanolamine	amide 60% min.; acid value 12-16; alkali value 150-170	clear light amber liquid	thickener; foaming agent / *cosmetics*
	SCHERCOMID AME	acetamide MEA	amide 95% min.; acid value 10 max.; alkali value 15 max.	clear straw-coloured liquid	humectant; conditioner; coupling agent; wetting agent; dispersant; solubiliser / *skin care products; hair care products*
	SCHERCOMID CCD	cocamide DEA and diethanolamine	amide 60% min.; acid value 15-20; alkali value 140-160	clear amber liquid	emulsifier; thickener; detergent / *industrial cleaners; agrochemicals*
	SCHERCOMID CDA	cocamide DEA and diethanolamine	amide 60% min.; acid value 40-50; alkali value 150-170	clear light amber liquid	foam stabiliser; suspending agent; dispersant / *industrial and household cleaners*
	SCHERCOMID CDO-EXTRA	cocamide DEA and diethanolamine	amide 65% min.; acid value 5; alkali value 110-140	clear light amber liquid	wetting agent; detergent / *dishwashing agents; floor cleaners*
	SCHERCOMID EAC	modified coco amide	100% active	liquid	fulling agent; scouring agent / *textile industry*
	SCHERCOMID EAC-S	solubilised coco amide	30% active	liquid	
	SCHERCOMID HT-60	PEG-50 tallow amide	acid value 2 max.; alkali value 10 max.	hard tan wax	detergent; emulsifier; dispersant; thickener; foaming agent
	SCHERCOMID LME	lactamide MEA	amide 90% min.; acid value 20 max.; alkali value 20 max.	clear yellow liquid	humectant; conditioner; emollient / *skin care products; hair care products*
	SCHERCOMID ODA	oleamide DEA and diethanolamine	amide 60% min.; acid value 12-16; alkali value 120-140	clear light amber liquid	w/o emulsifier; thickener; conditioner; anti-corrosion / *household cleaners*
	SCHERCOMID OME	oleamide MEA	amide 85% min.; acid value 10 max.; alkali value 20 max.	tan wax	w/o emulsifier; conditioner; thickener
	SCHERCOMID OMI	oleamide MIPA	amide 85% min.; acid value 5-15 max.; alkali value 7-17 max.	clear amber liquid to soft solid	detergent; emulsifier; conditioner
	SCHERCOMID SAP	apricotamide DEA	amide 80% min.; acid value 3 max.; alkali value 20-40	clear amber liquid	thickener; foam stabiliser / *shampoos*
	SCHERCOMID SCE	cocamide DEA	amide 87% min.; acid value 1 max.; alkali value 20-40	clear light amber liquid	detergent; thickener; foam stabiliser / *cosmetics; household cleaners*
	SCHERCOMID SCO-EXTRA	cocamide DEA	amide 80% min.; acid value 3 max.; alkali value 20-40	clear light amber liquid	emulsifier; foam stabiliser; wetting agent / *household and industrial detergents*
	SCHERCOMID SL-EXTRA	lauramide DEA	amide 87% min.; acid value 1 max.; alkali value 20-40	white crystalline solid	thickener; foam stabiliser; wetting agent; detergent / *shampoos; skin care products; dishwashing agents*

Supplier	Trade name	Chemical description	Composition	General properties	Functionality Application
Scher Chemicals	SCHERCOMID SL-ML	lauramide DEA	amide 87% min.; acid value 1 max.; alkali value 20-40	clear light amber liquid	detergent; conditioner; foam stabiliser; foaming agent *hair care products; skin care products*
	SCHERCOMID SLE	linoleamide DEA	amide 87% min.; acid value 1 max.; alkali value 20-40	clear amber liquid	w/o emulsifier; stabiliser; thickener
	SCHERCOMID SLM-LC	lauramide DEA	amide 85% min.; acid value 1 max.; alkali value 30-50	clear amber liquid	wetting agent; thickener; foam stabiliser
	SCHERCOMID SLM-S	lauramide DEA	amide 87% min.; acid value 1 max.; alkali value 20-40	white crystalline solid	thickener; foam stabiliser; wetting agent; detergent *cosmetics; toiletries*
	SCHERCOMID SLS	soyamide DEA	amide 82% min.; acid value 2 max.; alkali value 20-40	clear amber liquid	w/o emulsifier; stabiliser; thickener
	SCHERCOMID SO-A	oleamide DEA	amide 85% min.; acid value 5 max.; alkali value 40-60	clear amber liquid	w/o emulsifier; lubricant; conditioner
	SCHERCOMID SO-T	tallamide DEA	amide 85% min.; acid value 15 max.; alkali value 40-50	clear amber liquid	w/o emulsifier
	SCHERCOMID SWG	wheatgermamide DEA	amide 80% min.; acid value 3 max.; alkali value 20-40	clear amber liquid	thickener; foam stabiliser *shampoos*
	SCHERCOMID TO-2	tallamide DEA and diethanolamine	amide 65% min.; acid value 16-19; alkali value 130-150	clear amber liquid	w/o emulsifier; thickener; lubricant; anti-corrosion
Seppic	ORAMIDE DL 200	coco diethanolamide $C_{8/18}$	100% active	soft paste	detergent; dispersant
	ORAMIDE DL 200 AF	coco diethanolamide $C_{12/18}$ (amine free)	100% active	soft paste	
	ORAMIDE DL 215	coco diethanolamide $C_{8/18}$	100% active	liquid	
	ORAMIDE DL 215 AF	cocamide DEA	95% active	liquid	*hair care products*
	ORAMIDE ML 115	coco monoethanolamide $C_{8/18}$	100% active	flakes	foam stabiliser; detergent
	ORAMIDE ML 200	coco monoethanolamide $C_{12/18}$	100% active	flakes	foam stabiliser; detergent
	ORAPOL DL 210	coco diethanolamide	100% active	liquid	detergent; dispersant
Servo Delden	SERDOLAMIDE PCE 83	cocos fatty acid monoethanolamide	acid no. 6 max.	solid	
	SERDOLAMIDE PCF 51	cocos fatty acid diethanolamide	acid no. 10 max.	liquid	
	SERDOLAMIDE POE 88	oleic acid monoethanolamide	acid no. 1·5 max.	solid	
	SERDOLAMIDE POF 56	oleic acid diethanolamide	acid no. 20-25	liquid	
	SERDOLAMIDE POF 61	oleic acid diethanolamide	acid no. 7 max.	liquid	
	SERDOLAMIDE PPF 67	cocofatty acid diethanolamide	acid no. 2-8 max.	liquid	
	SERDOLAMIDE PQF 54	soya oil diethanolamide + DEA/water	acid no. 5 max.	liquid	
	SERDOLMIDE PQF 74	soya oil diethanolamide	acid no. 5 max.	liquid	
	SERDOLAMIDE PQJ 48	soya oil diethanolamide + TEA/water	acid no. 13-15	liquid	

Supplier	Trade name	Chemical description	Composition	General properties	Functionality *Application*
Servo Delden	SERDOLAMIDE PSE 89	stearic acid monoethanolamide	acid no. 1·5 max.	solid	
	SERDOLAMIDE PVE 87	tallow fatty acid monoethanolamide	acid no. 5 max.	solid	
	SERDOLAMIDE PWO 84	stearic acid amidoethyletanolamine	acid no. 5-10	solid	
Stepan Europe	NINOL 1281	modified alkylolamide; made-to-order	100% active	amber viscous liquid	detergent; lubricant *household, institutional and industrial cleaners*
	NINOL 4821 F	coco diethanolamide	99% active	pale yellow liquid	emollient; thickener; foam booster; foam stabiliser *shampoos; bubble baths; liquid soaps; shower gels; household, institutional and industrial cleaners*
	NINOL CNR	coco monoethanolamide; made-to-order	98% active	yellow to ivory coloured wax	emollient; thickener; foam booster; foam stabiliser *shampoos; bubble baths; liquid soaps; shower gels; soaps*
	NINOL LDL 2	modified alkylolamide	96% active	orange/yellow viscous liquid	thickener; foam booster; foam stabiliser; detergent enhancer *liquid soaps; hard surface cleaners*
	NINOL LMP	coco monoethanolamide; made-to-order	93% active	white to beige powder	emollient; thickener; detergent; foam booster; foam stabiliser *shampoos; bubble baths; liquid soaps; shower gels; soaps; dishwashing agents*
	NINOL M10	mono isopropanolamide	96% active	white to beige flakes	thickener; emollient; anti-corrosion; foam booster; foam stabiliser *shampoos; bubble baths; liquid soaps; shower gels; soaps*
Surfachem	SURFAC CDE/G	coconut diethanolamide 10% glycerol			
	SURFAC CDE	coconut diethanolamide			
	SURFAC CDX	coconut diethanolamide			
	SURFAC CME	coconut monoethanolamide			
	SURFAC LDE	lauric diethanolamide			
	SURFAC LIS	lauric isopropanolamide			
	SURFAC LMD	lauric/myristic diethanolamide			
	SURFAC SFD	soya diethanolamide			
	SURFAC V	coconut diethanolamide 10% glycerol (conforms with European Cosmetic Directives)			
Thor Chemicals	SOVATEX LED	fatty acid ethanolamide	100% active	semi-liquid	
	SOVATEX OED	fatty acid ethanolamide	100% active	semi-liquid	
Unger Fabrikker	UFANON K-80	coconut fatty acid diethanolamide	amide content 48-56%; total free amine content (as diethanolamine) 25%	brown liquid; pH 9-10 (1% soln.)	foam stabiliser; viscosity modifier *liquid detergents; shampoos; bath preparations; leather industry*
	UFANON KD-S	coconut fatty acid diethanolamide	amide content 83%; total free amine content (as diethanolamine) 4% max.	pale yellow liquid; pH 9-10 (1% soln.)	foam stabiliser; viscosity modifier *liquid detergents; shampoos; bath preparations; leather industry*
Witco	REWOCID U 185	undecylenic acid monoethanolamide	100% active	flakes	fungicide; anti-mycotic agent
	REWOCOROS AC 28	special fatty acid alkylolamide	100% active	liquid	anti-corrosion *metal working*
	REWOLUB KSM 80	modified dicarboxylic acid diamide	83% active	liquid	lubricant *metal working*

Supplier	Trade name	Chemical description	Composition	General properties	Functionality Application
Witco	REWOMID C 212	coconut fatty acid monoethanolamide	100% active	flakes	additive; foam stabiliser; thickener *detergents*
	REWOMID DC 212	coconut fatty acid polydiethanolamide	100% active	liquid	lubricant; thickener; super-fatting agent *metal working; household products*
	REWOMID DC 212 S	coconut fatty acid diethanolamide	100% active	liquid	foam booster; thickener *shampoos; foam baths; all-purpose cleaners*
	REWOMID DL 240	coconut fatty acid polydiethanolamide	100% active	liquid	thickener; super-fatting agent *household products*
	REWOMID DO 280	oleic acid polydiethanolamide	100% active	liquid	
	REWOMID DO 280 SE	oleic acid diethanolamide	100% active	liquid	thickener; foam stabiliser; super-fatting agent
	REWOMID F	linoleic acid diethanolamide	100% active	liquid	super-fatting agent *personal care products*
	REWOMID IPE 280	oleic acid isopropanolamide	100% active	paste	foam stabiliser; super-fatting agent; thickener
	REWOMID IPL 203	lauric acid isopropanolamide	100% active	flakes	foam stabiliser; thickener *shaving products*
	REWOMID IPP 240	coconut fatty acid isopropanolamide	100% active	flakes	foam stabiliser; thickener
	REWOMID L 203	lauric acid monoethanolamide	100% active	flakes	thickener; foam stabiliser
	REWOMID S 280	stearic acid monoethanolamide	100% active	flakes	thickener; foam stabiliser; anti-inflammatory agent *synthetic soap bars*
	WITCAMIDE 61	oleamide MIPA	free amine 0.7%	paste	conditioner; lubricant; substantivity agent; viscosity modifier *personal care products; household and industrial applications*
	WITCAMIDE 70	stearamide MEA	free amine 1.3%	flake	conditioner; lubricant; opacifier; substantivity agent; viscosity modifier *personal care products; household and industrial applications*
	WITCAMIDE 128T	cocamide DEA	free amine 6%	liquid	conditioner; detergent; foam stabiliser; foaming agent; viscosity modifier; substantivity agent *personal care products; household and industrial applications*
	WITCAMIDE 272	fatty alkanolamide	100% active	liquid/paste	coupling agent; detergent; o/w emulsifier *personal care products; household and industrial applications; cleaners*
	WITCAMIDE 511	fatty alkanolamide	free amine 6%	liquid; pH 8.6	coupling agent; detergent; dispersant; o/w emulsifier *personal care products; household and industrial applications;*
	WITCAMIDE 512	fatty alkanolamide		liquid; pour point $-3.9°C$; pH 8.5	*petroleum industry*
	WITCAMIDE 1017	modified alkanolamide		liquid; pour point 0°C; pH 9.5	w/o emulsifier *petroleum industry*
	WITCAMIDE 5138	fatty alkanolamide	free amine 9.5%	liquid	dispersant; o/w emulsifier *personal care products; household and industrial applications*
	WITCAMIDE 6310	lauramide DEA	free amine 6%	paste	conditioner; detergent; foaming agent; substantivity agent; viscosity modifier *personal care products; household and industrial applications*
	WITCAMIDE 6445	modified cocamide DEA	free amine 18%	liquid	coupling agent; detergent; o/w emulsifier; lubricant *personal care products; household and industrial applications*

Supplier	Trade name	Chemical description	Composition	General properties	Functionality / Application
Witco	WITCAMIDE 6511	lauramide DEA	free amine 6%	paste	conditioner; detergent; foaming agent; substantivity agent; viscosity modifier / *personal care products; household and industrial applications*
	WITCAMIDE 6531	cocamide DEA	free amine 25%	liquid	detergent; foam stabiliser; foaming agent / *personal care products; household and industrial applications*
	WITCAMIDE 6546	oleamide DEA	free amine 7%	liquid	conditioner; coupling agent; detergent; o/w emulsifier; foam stabiliser / *personal care products; household and industrial applications*
	WITCAMIDE 6553	cocamide DEA	free amine 18%	liquid	detergent; dispersant; o/w emulsifier / *personal care products; household and industrial applications*
	WITCAMIDE 6625	modified cocamide DEA	free amine 10%	liquid	conditioner; detergent; foam stabiliser; foaming agent; viscosity modifier; substantivity agent / *personal care products; household and industrial applications*
	WITCAMIDE CD	cocamide DEA	free amine 27%	liquid	detergent; foam stabiliser; foaming agent / *personal care products; household and industrial applications*
	WITCAMIDE CDA	modified cocamide DEA	free amine 20%	liquid	detergent; dispersant; o/w emulsifier; lubricant / *personal care products; household and industrial applications*
	WITCAMIDE CDS	oleic DEA (1:2)	100% active	liquid	emulsifier; anti-corrosion / *cutting oils*
	WITCAMIDE CPA	coco IPA	100% active	flakes	foam stabiliser; refatting agent; viscosity modifier
	WITCAMIDE LDEA	lauramide DEA	free amine 6.5%	liquid	conditioner; detergent; foam stabiliser; foaming agent; substantivity agent; viscosity modifier / *personal care products; household and industrial applications*
	WITCAMIDE LDT/S	coco DEA	100% active	liquid	foam stabiliser; refatting agent; viscosity modifier / *cosmetics; shampoos*
	WITCAMIDE MAS	stearamide MEA-stearate	free amine 1%	flake	conditioner; lubricant; opacifier; substantivity agent; viscosity modifier / *personal care products; household and industrial applications*
	WITCAMIDE S771	modified cocamide DEA	free amine 14.5%	liquid	coupling agent; o/w emulsifier; lubricant / *personal care products; household and industrial applications*
	WITCAMIDE S780	modified cocamide DEA	free amine 13%	liquid	
	WITCAMIDE SSA	soyamide DEA	free amine 8%	liquid	conditioner; detergent; foam stabiliser; foaming agent; substantivity agent; viscosity modifier / *personal care products; household and industrial applications*
Zschimmer & Schwarz	PURTON CFD	coco fatty acid diethanolamide	100% active	liquid	emollient; thickener / *shampoos; bath preparations; dishwashing agents; cleansing preparations; all-purpose cleaners; liquid soaps*
	PURTON SFD	fatty acid diethanolamide	100% active	liquid	

Amine oxides

Supplier	Trade name	Chemical description	Composition	General properties	Functionality / Application
Akzo Nobel	AROMOX C/12-W	cocobis(2-hydroxyethyl)amine oxide	30-32% active	liquid	
	AROMOX DM14D-W 970	tetradecyldimethylamine oxide	24-26% active	liquid	
	AROMOX DMB-W	alkyldimethylamine oxide	29-31% active	liquid	

Supplier	Trade name	Chemical description	Composition	General properties	Functionality / Application
Akzo Nobel	AROMOX DMMCD-W	cocodimethylamine oxide (fractionated coco-alkyl)	30-32% active	liquid	
	AROMOX O/12	oleylbis(2-hydroxyethyl)amine oxide	57-59% active	liquid	
	AROMOX T/12	tallowbis(2-hydroxyethyl)amine oxide	49-51% active	paste	
	BEROL 305	coco amine ethoxylate N-oxide	dry content 28%	surface tension 31 dynes/cm (0.1% soln.)	foaming agent (high); antistatic agent; softener; foam booster *shampoos; hard-surface cleaners*
	LILAMINOX M4	tetradecyldimethylamine N-oxide	dry content 24-26%	surface tension 31 dynes/cm (0.1% soln.)	foaming agent (high); thickener *shampoos; cleaners; household bleaches*
	LILAMINOX M24	$C_{12/14}$ alkyldimethylamine N-oxide	dry content 30-32%	surface tension 32 dynes/cm (0.1% soln.)	foaming agent (high); thickener; foam booster *shampoos; hard-surface cleaners*
Albright & Wilson	EMPIGEN OB	alkyl dimethyl amine oxide	concentration 30%	liquid	compounding aid *polymer industry*
	EMPIGEN OC/B	alkyl dimethyl amine oxide	concentration 30%	liquid	
	EMPIGEN OH25	alkyl dimethyl amine oxide	concentration 25%	liquid	
Auschem Cesalpinia	CHIMIN CMO	cocoamidopropylamine oxide	30% active	liquid	foaming agent; detergent *liquid formulations*
	CHIMIN LMO	laurylamidopropylamine oxide	30% active	liquid	
Croda	INCROMINE OXIDE C	cocamidopropyl dimethylamine oxide supplied as a 30% aq. soln.	amine oxide 29.5-31.5%; free amine 1% max.	pale clear liquid; colour 1 max. (Gardner); pH 6.5-7.5 (5% aq. soln.)	emulsifier; viscosity modifier; foam stabiliser; foam booster; conditioner; softener *cosmetics; shampoos; hair conditioners; bath preparations*
Ellis & Everard	CAFLON 30AO	alkyl dimethyl amine oxide	30% active		
Fina Chemicals	RADIAMOX 6800	cocamine oxide	29-31% active; carbon chain composition $C_{8/10}$ 3%, C_{12} 48-58%, C_{14} 18-24%, C_{16} 8-12%, C_{18} 10-14%	liquid; colour 1 max. (Gardner); pH 7-8 (10% in water)	foam booster; emollient; detergent; antistatic agent; conditioner; viscosity modifier; levelling agent; thixotropic agent; emulsifier; wetting agent; anti-corrosion *cosmetics; toiletries; shampoos; personal care products; hair care products; bubble baths; shaving products; liquid detergents; paint industry; oil industry; metal working; textile industry; lubricants*
	RADIAMOX 6804	lauramine oxide	29-31% active; carbon chain composition $C_{8/10}$ 2%, C_{12} 68-73%, C_{14} 25-30%, C_{16} 2%	liquid; colour 1 max. (Gardner); pH 7-8 (10% in water)	
Th. Goldschmidt	AMINOXID WS 35	cocamidopropylamine oxide	ca. 35% active	amber liquid; pH 5-7	viscosity modifier; foam stabiliser *bath preparations; shower preparations; shampoos; hair conditioners; skin cleansing preparations; liquid soaps; shower gels; bubble baths*
Henkel	LAVIRON 118 S	fatty amine oxide			foaming agent *textile industry; finishing*
	STANDAMOX 01	oleamine oxide	50% active	clear liquid	thixotropic agent; softener; lubricant *personal care products*
	STANDAMOX CAW	cocamidopropylamine oxide	30% active	clear liquid	foaming agent; stabiliser; thickener; emollient *baby shampoos; bubble baths; skin care products*
Hickson Manro	MANRO AO25M	myristyl dimethylamine oxide	25% active	liquid	hair care products; bath preparations; household detergents; *industrial cleaners*
	MANRO AO30C	alkyl dimethylamine oxide	30% active	liquid	
Hoechst	DP1214	C_{10} dimethylamine oxide	30% active		*detergents*
	DP1332	soya dimethylamine oxide	50% active		*detergents*

Supplier	Trade name	Chemical description	Composition	General properties	Functionality Application
Hoechst	DP1368-1	coco (2-hydroxyethyl) amine oxide	30% active		*detergents; toiletries*
	DP1411	stearyl dimethylamine oxide	25% active		*toiletries*
	DP1425	coco amidopropylamine oxide	30% active		*toiletries; detergents*
	GENAMINOX CE	cetyl dimethylamine oxide	30% active		*detergents; toiletries*
	GENAMINOX CS	coco dimethylamine oxide	30% active		*thickener* *bleach; heavy-duty detergents; toiletries*
	GENAMINOX KC	coco dimethylamine oxide	30% active	clear yellow liquid	*toiletries*
	GENAMINOX LA	lauryl dimethylamine oxide	30% active		*toiletries; detergents; heavy-duty cleaners*
	GENAMINOX MY	myristyl dimethylamine oxide	25% active		*thickener* *bleach*
Kao Corporation	OXIDET DM4	amine oxide	29–31% active; carbon chain composition C_{12} 3%, C_{14} 95%, C_{16} 2%	liquid; M.W. 257; colour 50 max. (APHA); pH 6–8 (5%)	
	OXIDET DM20	amine oxide	29–31% active; carbon chain composition C_{10} 1%, C_{12} 97%, C_{14} 2%	liquid; M.W. 250; colour 100 max. (APHA); pH 6–8 (5%)	
	OXIDET DM48	amine oxide	28.8–31.2% active; carbon chain composition C_{10} 2%, C_{12} 48%, C_{14} 20%, C_{16} 12%, C_{18} 18%	liquid; M.W. 250; colour 50 max. (APHA); pH 6–8 (5%)	
	OXIDET DMC-LD	amine oxide	29–31% active; carbon chain composition C_8 4%, C_{10} 5%, C_{12} 46%, C_{14} 24%, C_{16} 11%, C_{18} 3%, $C_{18'}$ 7%	liquid; M.W. 240; colour 100 max. (APHA); pH 6–8 (5%)	
	OXIDET L-75	alkylamido propyl dimethyl amine oxide	29–31% active; carbon chain composition C_{10} 2%, C_{12} 56%, C_{14} 22%, C_{16} 10%, C_{18} 10%	liquid; M.W. 316; colour 70 max. (APHA); pH 6–8 (5%)	
Lonza	BARLOX 12	cocamine oxide	30% active	liquid; pour point 4°C; viscosity 45 cP's; surface tension 32.3 dynes/cm (0.1% active); Draves wetting 4 sec; pH 7.0 (1%)	viscosity modifier; emulsifier; conditioner *personal care products; industrial applications*
	BARLOX 14	myristamine oxide	solids 30%	pour point 2°C; viscosity 60 cP's; surface tension 31.0 dynes/cm (0.1% active); Draves wetting 5 sec; pH 7.0 (1%)	
	BARLOX 16	cetamine oxide	solids 30%	viscosity 27,000 cP's; surface tension 32.4 dynes/cm (0.1% active); Draves wetting 9 sec; pH 7.0 (1%)	
	BARLOX 16S	cetamine oxide	30% active	paste	
	BARLOX 18S	stearamine oxide	25% active	paste	viscosity modifier; emulsifier; conditioner *personal care products; industrial applications*
	BARLOX C	cocamidopropylamine oxide	30% active	liquid; pour point 4°C; viscosity 36 cP's; surface tension 35.4 dynes/cm (0.1% active); Draves wetting 27 sec; pH 7.0 (1%)	
McIntyre Group	MACKAMINE CAO	cocamidopropylamine oxide	30% active	liquid; pH 7	
	MACKAMINE CO	cocamine oxide	30% active	liquid; pH 7	
	MACKAMINE LAO	lauramidopropylamine oxide	30% active	liquid; pH 7	conditioner; viscosity modifier; foam booster; detergent; wetting agent *personal care products; industrial cleaners*
	MACKAMINE LO	lauramine oxide	30% active	liquid; pH 7	
	MACKAMINE O2	oleamine oxide	35% active	liquid; pH 7.5	
	MACKAMINE OAO	oleamidopropylamine oxide	50% active	gel; pH 7	
	MACKAMINE SO	stearamine oxide	25% active	paste; pH 7	
	MACKAMINE WGO	wheat germamidopropylamine oxide	30% active	gel; pH 7	

Supplier	Trade name	Chemical description	Composition	General properties	Functionality / Application
Millchem	AMMONYX CDO	coco amido propyl dimethyl amine oxide	amine oxide 29.5-31.5%	liquid	wetting agent; foaming agent; foam stabiliser; conditioner / *cosmetics; household and janitorial products; hair conditioners*
	AMMONYX CO	cetyl dimethyl amine oxide	amine oxide 29-31%	liquid	conditioner / *cosmetics; household and janitorial products*
	AMMONYX DMCD-40	lauryl dimethyl amine oxide	amine oxide 40-42%	liquid	wetting agent; foaming agent; foam stabiliser / *cosmetics; household and janitorial products*
	AMMONYX LO	lauryl dimethyl amine oxide	amine oxide 29-31%	liquid	wetting agent; foaming agent; foam stabiliser; grease emulsifier / *cosmetics; household and janitorial products*
	AMMONYX MCO	myristyl/cetyl dimethyl amine oxide	amine oxide 29-31%	liquid	wetting agent; foaming agent; foam stabiliser / *cosmetics; household and janitorial products*
	AMMONYX MO	myristyl dimethyl amine oxide	amine oxide 29-31%	liquid	
	AMMONYX SO	stearyl dimethyl amine oxide	amine oxide 24.5-26.5%	paste	conditioner; emulsifier / *cosmetics; household and janitorial products*
Pentagon	PENTANOX 4X	myristyl dimethyl amine oxide	24-26% active	clear viscous liquid; colour 1 max. (Gardner); pH 7.0-8.0 (5% aq. soln.)	foam booster; thickener / *toiletries; bleaches*
	PENTANOX 24X	alkyl ($C_{12/14}$) dimethyl amine oxide	29-31% active	clear liquid; pH 7.0-8.0 (10% aq. soln.)	foam booster / *toiletries; textile industry*
	PENTANOX CAP 30X	coco amido propyl dimethyl amine oxide	29.5-31.5% active	clear liquid; colour 200 max. (Hazen); pH 6.0-8.0 (10% aq. soln.)	
PPG	MAZOX CAPA	cocamidopropylamine oxide	amine oxide 30%; free amine 1% max.		foam booster; viscosity modifier / *personal care products*
	MAZOX CDA	palmitamine oxide	amine oxide 30%; free amine 1% max.		conditioner / *hair conditioners*
	MAZOX LDA	lauramine oxide	amine oxide 30%; free amine 1% max.		foam booster; viscosity modifier / *personal care products*
	MAZOX ODA-30	oleamine oxide	amine oxide 30%; free amine 1% max.		conditioner / *hair conditioners*
	MAZOX SDA	stearamine oxide	amine oxide 25%; free amine 1% max.		conditioner; emulsifier; low foam / *hair conditioners*
Rhone-Poulenc	RHODAMOX LO	lauramine oxide	30% solids	liquid	foaming agent (high); viscosity modifier; foam booster; emulsifier; wetting agent; thickener / *hard-surface cleaners*
Scher Chemicals	SCHERCAMOX CAA	cocamidopropylamine oxide	amine oxide 35% min.; free amine 0.5% max.; free peroxide 0.5% max.	clear to hazy liquid	wetting agent; detergent; conditioner; foaming agent; viscosity modifier / *shampoos; hair care products; bath preparations*
	SCHERCAMOX CMA	dihydroxyethyl cocamine oxide	amine oxide 38% min.; free amine 1.5% max.; free peroxide 0.5% max.	clear liquid	conditioner; foam stabiliser
	SCHERCAMOX DMA	myristamine oxide	amine oxide 29% min.; free amine 0.5% max.; free peroxide 0.3% max.	viscous liquid	wetting agent; foaming agent / *shampoos; bubble baths; dishwashing agents*
	SCHERCAMOX DMC	cocamine oxide	amine oxide 29% min.; free amine 1% max.; free peroxide 0.5% max.	clear yellow liquid	wetting agent; foam stabiliser; viscosity modifier
	SCHERCAMOX DML	lauramine oxide	amine oxide 29% min.; free amine 0.5% max.; free peroxide 0.3% max.	clear liquid	wetting agent; foaming agent; foam stabiliser / *shampoos; bath preparations; shaving products*
	SCHERCAMOX DMM	myristamine oxide	amine oxide 29% min.; free amine 0.5% max.; free peroxide 0.3% max.	clear liquid	wetting agent; foaming agent / *shampoos; bubble baths; dishwashing agents*

Supplier	Trade name	Chemical description	Composition	General properties	Functionality / Application
Scher Chemicals	SCHERCAMOX DMS	stearamine oxide	amine oxide 25% min.; free amine 1.5% max.; free peroxide 0.5% max.	white paste	conditioner; softener; foam stabiliser *shampoos; hair care products*
Stepan Europe	NINOX FCA	cocamidopropylamine oxide	33% active	water white to pale yellow liquid	foaming agent; thickener; foam booster; stabiliser; detergency enhancer; antistatic agent *liquid soaps; bubble baths; shower gels; household, institutional and industrial cleaners; scale-removing cleaners*
	NINOX L	laurylamine oxide	31% active	water white to pale yellow liquid	
	NINOX M	myristylamine oxide; made-to-order	30% active	water white to pale yellow liquid	
	NINOX SO	stearylamine oxide	24% active	white paste	thickener; conditioner; foam stabiliser; detergent; emollient *shampoos; liquid soaps; shower gels; household, institutional and industrial cleaners*
Surfachem	SURFAC AO30	alkyl dimethyl amine oxide			
	SURFAC AO100	alkyl ethoxy dimethyl amine oxide			
	SURFAC MAO	alkyl dimethyl amine oxide			
Tomah Products	TOMAH A0-14-2	ether amine oxide	50% active min.	clear liquid	degreaser; emulsifier; soil suspension characteristics *dishwashing agents; car washes; household cleaners*
	TOMAH A0728S	ether amine oxide	50% active min.	clear liquid	
Witco	EMCOL LO	lauramine oxide	amine oxide 30%; free amine 1%; free peroxide 0.1%	liquid	foam stabiliser; foaming agent; wetting agent *personal care products; household and industrial applications*
	REWOMINOX B 204	alkyl amidopropyl dimethylamine oxide	35% active	liquid	foam booster; thickener; antistatic agent *shampoos; foam baths; hair conditioner*
	REWOMINOX L 408	lauryl dimethylamine oxide	30% active	liquid	
	SOCHAMINE OX 30	cocodimethylamine oxide	30% active	liquid	foam booster *cosmetics; detergents; shampoos*

Esters

Aarhus Olie	CREMAO CE-34	C$_{16/18}$ triglycerides	iodine value 35; sapon. no. 195	slip m.p. 34°C	consistency regulator; emollient *cosmetics; toiletries*
	CREMAO CS-33	C$_{12/18}$ triglycerides	iodine value 7; sapon. no. 250	slip m.p. 33°C	
	CREMAO CS-34	C$_{12/18}$ triglycerides	iodine value 0; sapon. no. 250	slip m.p. 34°C	oxidative resistance; consistency regulator; emollient *cosmetics; toiletries*
	CREMEOL CW-31	C$_{12/18}$ triglycerides	iodine value 6; sapon. no. 248	slip m.p. 31°C	oxidative resistance; emollient *cosmetics; toiletries*
	CREMEOL FR-36	C$_{16/18}$ mono-, di-, triglycerides	iodine value 60; sapon. no. 178	slip m.p. 37°C	consistency regulator; co-emulsifier; stabiliser; emollient *cosmetics; toiletries*
	CREMEOL FR-57	C$_{16/18}$ mono-, di-, triglycerides	iodine value 2; sapon. no. 182	slip m.p. 57°C	oxidative resistance; consistency regulator; co-emulsifier; stabiliser; emollient *cosmetics; toiletries*
	CREMEOL HF-52	C$_{16/18}$ triglycerides	iodine value 26; sapon. no. 191	slip m.p. 52°C	oxidative resistance; consistency regulator; oil-adsorbing; stabiliser; emollient *cosmetics; toiletries*
	CREMEOL HF-62	C$_{16/24}$ triglycerides	iodine value 2; sapon. no. 175	slip m.p. 62°C	
	CREMEOL PFO	C$_{16/18}$ triglycerides	iodine value 138; sapon. no. 192	slip m.p. < 8°C	emollient *cosmetics; toiletries*
	CREMEOL PS-12	C$_{16/18}$ triglycerides	iodine value 87; sapon. no. 191	slip m.p. 12°C	oxidative resistance; emollient *cosmetics; toiletries*
	CREMEOL PS-17	C$_{16/18}$ triglycerides	iodine value 84; sapon. no. 190	slip m.p. 17°C	

Supplier	Trade name	Chemical description	Composition	General properties	Functionality / Application
Aarhus Olie	CREMEOL PW-41	$C_{12/18}$ triglycerides	iodine value 3; sapon. no. 242	slip m.p. 41°C	oxidative resistance; emollient *cosmetics; toiletries*
	CREMEOL SH	$C_{16/18}$ triglycerides	iodine value 34; sapon. no. 191	slip m.p. 30°C	oxidative resistance; consistency regulator; emollient *cosmetics; toiletries*
	CREMEOL SW-41	$C_{16/18}$ triglycerides	iodine value 66; sapon. no. 192	slip m.p. 41°C	oxidative resistance; emollient *cosmetics; toiletries*
	CREMEOL SW-42	$C_{16/18}$ triglycerides	iodine value 78; sapon. no. 199	slip m.p. 42°C	consistency regulator; emollient *cosmetics; toiletries*
Akcros Chemicals	ETHYLAN A2	PEG 200 oleate	100% active; ethylene oxide content 42%	HLB 8.4; light amber liquid; viscosity 75 cSt; pH 5-7 (1% aq.)	emulsifier; anti-foaming agent *kerosene; mineral oils*
	ETHYLAN A3	PEG 300 oleate	100% active; ethylene oxide content 52%	HLB 10.4; light amber liquid; viscosity 84 cSt; pH 5-7 (1% aq.)	emulsifier; anti-foaming agent; dispersant *kerosene; mineral oils; vegetable oils*
	ETHYLAN A4	PEG 400 oleate	100% active; ethylene oxide content 59%	HLB 11.8; light amber liquid; viscosity 120 cSt; pH 5-7 (1% aq.)	dispersant; antistatic agent; emulsifier *plastics industry; oils*
	ETHYLAN A6	PEG 600 oleate	100% active; ethylene oxide content 68%	HLB 13.6; light amber liquid; cloud point 42°C (1% aq.); viscosity 340 cSt; pH 5-7 (1% aq.)	
	ETHYLAN A10	PEG 1000 mono-oleate	100% active; ethylene oxide content 78%	HLB 15.6; cream waxy solid; pour point 36°C; viscosity 98 cSt (40°C); pH 5-7 (1% aq.)	dispersant; emulsifier *waxes*
	ETHYLAN C40AH	PEG unsaturated fatty acid ester	100% active; ethylene oxide content 65%	HLB 13.5; hazy yellow liquid; pour point 11°C (freeze-thaw recovery 16°C); viscosity 357 cSt (40°C); pH 5-7 (1% aq.)	emulsifier; levelling agent; solubiliser; dispersant; grinding aid *chlorinated solvents; vegetable oils; pesticides; dyes; perfumes; pigments*
	ETHYLAN C75AH	PEG unsaturated fatty acid ester	100% active; ethylene oxide content 78%	HLB 15.6; cream waxy solid; pour point 28°C; viscosity 259 cSt (60°C); pH 5-7 (1% aq.)	emulsifier; lubricant *oils; textile industry*
	ETHYLAN C160	PEG unsaturated fatty acid ester	100% active; ethylene oxide content 88%	HLB 17.6; cream waxy solid; pour point 39°C; viscosity 300 cSt (60°C); pH 5-7 (1% aq.)	emulsifier; lubricant *textile industry; cosmetics; pharmaceuticals*
	ETHYLAN CF71	coconut fatty acid ester	100% active; ethylene oxide content 71%	HLB 14.0; pale straw liquid; pour point 14°C; viscosity 55 cSt (40°C); pH 5-7 (1% aq.)	emulsifier; lubricant *textile industry; cosmetics; pharmaceuticals*
	ETHYLAN GL20	sorbitan monolaurate	100% active	HLB 8.0; amber viscous liquid; pour point 15°C; viscosity 5250 cSt	emulsifier; stabiliser; antistatic agent; lubricant; antifogging agent; softener *emulsion and suspension polymerisation; textile industry; cosmetics; pharmaceuticals; agrochemicals; plastic films*
	ETHYLAN GO80	sorbitan mono-oleate	100% active	HLB 4.3; amber viscous liquid; pour point −20°C; viscosity 1100 cSt	
	ETHYLAN GS60	sorbitan monostearate	100% active	HLB 4.7; tan waxy solid; pour point 50°C	
	ETHYLAN GT85	sorbitan trioleate	100% active	HLB 1.5; amber viscous liquid; pour point −10°C; viscosity 230 cSt	
	ETHYLAN VPK	PEG unsaturated fatty acid ester	98% active; ethylene oxide content 58%	HLB 12.4; clear pale straw liquid; pour point 0°C; viscosity 150 cSt; pH 5-7 (1% aq.)	emulsifier; lubricant *oils; textile industry*
Akzo Nobel	ARMOTAN ML	sorbitan laurate		HLB 8.5; liquid	
	ARMOTAN MP	sorbitan palmitate		HLB 7; flakes	
	ARMOTAN MS	sorbitan stearate		HLB 5; flakes	
	BEROL 251	fatty acid ester	100% active		emulsifier *degreasing products*
	ELFACOS C26	hydroxyoctacosanyl hydroxystearate		pellets	
	ELFAN L 310	glycol distearate		flakes	
	KESSCO BE	behenyl erucate		solid	

Supplier	Trade name	Chemical description	Composition	General properties	Functionality *Application*
Akzo Nobel	KESSCO BS	butyl stearate		liquid	
	KESSCO CP	cetyl palmitate		flakes	
	KESSCO DEGMS	diethylene glycol monostearate		flakes	
	KESSCO DO	decyl oleate		liquid	
	KESSCO EGDS	glycol distearate		flakes	
	KESSCO EGMS	glycol monostearate		flakes	
	KESSCO EHC	ethylhexyl cocoate		liquid	
	KESSCO EHP	ethylhexyl palmitate		liquid	
	KESSCO EO	ethyl oleate		liquid	
	KESSCO GMS	glyceryl stearate		flakes	
	KESSCO GMSSE	glyceryl stearate SE		flakes	
	KESSCO GT3C	caprylic/capric triglyceride		liquid	
	KESSCO ICS	isocetyl stearate		liquid	
	KESSCO IPM 95	isopropyl myristate		liquid	
	KESSCO IPP	isopropyl palmitate		liquid	
	KESSCO IPS	isopropyl stearate		liquid	
	KESSCO OE	oleyl erucate		liquid	
	KESSCO OHS	octyl hydroxystearate		liquid	
	KESSCO PEG 400 DS	PEG-8 distearate		HLB 8; solid	
	KESSCO PEG 400 ML	PEG-8 laurate		HLB 13; liquid	
	KESSCO PEG 400 MO	PEG-8 oleate		HLB 11.5; liquid	
	KESSCO PEG 400 MS	PEG-8 stearate		HLB 12; paste	
	KESSCO PEG 600 DS	PEG-12 distearate		HLB 10.5; paste	
	KESSCO PEG 600 MS	PEG-12 stearate		HLB 13.5; solid	
	KESSCO PEG 1000 MS	PEG-20 stearate		HLB 15.5; solid	
	KESSCO PEG 6000 DS	PEG-150 distearate		HLB 18.5; flakes	
	KESSCO PGMM	propylene glycol myristate		flakes	
	KESSCO PTTIS	pentaerythrityl tetraisostearate		liquid	
	KESSCO PTS	pentaerythrityl tetrastearate		flakes	
	KESSCO SYNSPER	synth. spermaceti NF		flakes	
Albright & Wilson	EMPILAN EGDS	ethylene glycol distearate	concentration 100%	waxy flake	
	EMPILAN EGMS	ethylene glycol monostearate	concentration 100%	waxy flake	

Supplier	Trade name	Chemical description	Composition	General properties	Functionality / Application
Amerchol	LANAMINE	mixed isopropanolamines myristate			*personal care products*
Auschem	ALGON CO 90	PEG-9 cocoate		HLB 11.5; liquid	*cosmetics*
Cesalpinia	ALGON DS 6000	PEG 600 distearate		flakes	thickener shampoos
	ALGON LA 40	PEG-4 laurate		HLB 9.2; liquid	
	ALGON LA 80	PEG-8 laurate		HLB 12.8; liquid	
	ALGON OL 60	PEG-6 oleate		HLB 9.7; liquid	
	ALGON OL 70	PEG-7 oleate		HLB 10.4; liquid	
	ALGON OL 90	PEG-9 oleate		HLB 11.2; liquid	
	ALGON ST 50	PEG-5 stearate		HLB 9.0; solid	*cosmetics*
	ALGON ST 80	PEG-8 stearate		HLB 11.0; solid	
	ALGON ST 100	PEG-10 stearate		HLB 11.5; solid	
	ALGON ST 200	PEG-20 stearate		HLB 15.5; flakes	
	ALGON ST 400	PEG-40 stearate		HLB 16.9; flakes	
	ALGON ST 500	PEG-50 stearate		HLB 17.9; flakes	
	ALGON ST 1000	PEG-100 stearate		HLB 18.8; flakes	
	EMULFIN LV	polyethyleneglycol esters of fatty acids	100% active	liquid	emulsifier / *artificial milk; zootechnical industry*
	EMULFIN LV 2	polyethyleneglycol esters of fatty acids	100% active	liquid	
	EMULFIN MSR	ricinoleate of glycerol polyethyleneglycol	100% active	liquid	*fatty feeds; zootechnical industry*
	EMULFIN MSR 2	ricinoleate of glycerol polyethyleneglycol	100% active	liquid	*fatty feeds; zootechnical industry*
	ROLFAT CO 9	PEG-9 coco fatty acid		HLB 11.5; liquid	
	ROLFAT OL 6	PEG-6 oleate		HLB 9.7; liquid	
	ROLFAT OL 7	PEG-7 oleate		HLB 10.4; liquid	
	ROLFAT OL 9	PEG-9 oleate		HLB 11.2; liquid	o/w emulsifier / *zootechnical industry*
	ROLFAT SO 6	PEG-6 soya fatty acid		HLB 11.0; liquid	
	ROLFAT ST 8	PEG-8 stearate		HLB 11.0; paste	
	ROLFAT ST 40	PEG-40 stearate		HLB 16.9; flakes	
	ROLFAT ST 100	PEG-100 stearate		HLB 18.8; flakes	
	SORBIROL ISM	sorbitan isostearate		HLB 4.7; liquid	
	SORBIROL L	sorbitan laurate		HLB 8.6; liquid	
	SORBIROL O	sorbitan oleate		HLB 4.3; liquid	
	SORBIROL P	sorbitan palmitate		HLB 6.7; solid or flakes	emulsifier / *cosmetics; pesticides; zootechnical industry*
	SORBIROL S	sorbitan stearate		HLB 4.7; flakes	
	SORBIROL SQ	sorbitan sesquiolate		HLB 3.7; liquid	
	SORBIROL TO	sorbitan trioleate		HLB 1.8; liquid	
	SORBIROL TS	sorbitan tristearate		HLB 2.1; flakes	
	STEROL CC 595	PEG-6 capryl-capric glyc.		HLB 10.7; liquid	emollient; solvent; solubiliser; co-emulsifier / *toiletries*
	STEROL GL 2	glycereth-2		liquid	

Supplier	Trade name	Chemical description	Composition	General properties	Functionality / Application
Auschem Cesalpinia	STEROL GMS	glyceryl stearate		HLB 3.5; flakes	emulsifier; stabiliser / *cosmetics; zootechnical industry*
	STEROL LG 491	PEG-7 glyceryl cocoate		HLB 11.0; liquid	emollient; solvent; solubiliser; co-emulsifier / *toiletries*
	STEROL ST 1	glycol stearate	100% active	flakes	pearlescent agent / *shampoos; bath preparations*
	STEROL ST 2	diethylene glycol stearate	100% active	flakes	
	STEROL TE 200	PEG-2 glycerides		HLB 3.8; liquid	co-emulsifier; emollient; lubricant; thickener / *cosmetics; toiletries; hair care products; skin care products*
	TEWAX TC 65	GMS and PEG-100 stearate		HLB 11.0; flakes	emulsifier
	TEWAX TC 80	glyceryl sorbitan oleostearate		HLB 4.5; wax	*cosmetics*
Chemax	CHEMAX PEG-200DT	PEG 200 ditallate	sapon. no. 142	HLB 5.4; liquid	emulsifier / *agriculture; cosmetics; household applications; leather industry; metal working; textile industry*
	CHEMAX PEG-400DO	PEG 400 dioleate	sapon. no. 116	HLB 8.8; liquid	
	CHEMAX PEG-400DT	PEG 400 ditallate	sapon. no. 118	HLB 8.5; liquid	
	CHEMAX PEG-600DO	PEG 600 dioleate	sapon. no. 102	HLB 10.3; soft paste	
	CHEMAX PEG-600DT	PEG 600 ditallate	sapon. no. 97	HLB 10.3; liquid	
	SORBAX SML	sorbitan monolaurate	sapon. no. 162	HLB 8.6; liquid	emulsifier / *agriculture; cosmetics; leather industry; metal working; textile industry*
	SORBAX SMO	sorbitan monooleate	sapon. no. 153	HLB 4.3; liquid	
	SORBAX SMP	sorbitan monopalmitate	sapon. no. 155	HLB 6.7; solid	
	SORBAX SMS	sorbitan monostearate	sapon. no. 152	HLB 4.7; solid	
	SORBAX STO	sorbitan trioleate	sapon. no. 180	HLB 1.8; liquid	
	SORBAX STS	sorbitan tristearate	sapon. no. 182	HLB 2.1; solid	
Condea	ISOFOL	guerbet alcohol esters			
	LINPLAST	linear alcohol esters			
Courtaulds Chemicals		polyethylene glycol 400 monooleate (PEG 400 MO)		HLB 11.3	emulsifier / *kerosine and mineral oil; oil slick dispersants; anti-foams; water treatment; cutting oils; textile lubricants*
		polyethylene glycol 600 monooleate (PEG 600 MO)		HLB 13.3	
		polyethylene glycol 1000 monooleate (PEG 1000 MO)		HLB 15.6	
		propylene glycol monooleate (PGMO)		HLB 2.9	viscosity modifier / *alkyd-based paints*
		polyethylene glycol 400 dioleate (PEG 400 DO)		HLB 7.7	
		polyethylene glycol 600 dioleate (PEG 600 DO)		HLB 10.2	emulsifier; thickener / *cosmetics; toiletries*
		polyethylene glycol 6000 distearate (PEG 6000 DS)		HLB 18.1	wetting agent; defoaming agent / *pigment grinding*
		polyethylene glycol 400 diricinoleate (PEG 400 DR)			

Supplier	Trade name	Chemical description	Composition	General properties	Functionality Application
Croda	ACETYL TRIBUTYL CITRATE	acetyl tributyl citrate	acid value 0.16 max.; sapon. no. 530 min.	water-white product; M.W. 402.5; pour point − 59°C; viscosity 40 cPs	plasticiser *food packaging; dairy products; resins; plastics industry*
	ANFOMUL O1T OX 1051E	2-heptadecenyl-4,4-bis hydroxymethyl oxazoline	acid value 5 max.	yellow hard wax; drop point ca. 60°C	
	ANFOMUL PLR PE 3861E	pentaerythritol ester of wax acid supplied as a 70% soln. in an oil solvent	acid value 5 max.; OH value 118-145; sapon. no. 95-100	clear, dark brown, viscous liquid	
	ANFOMUL PO PE 3064E	pentaerythritol ester of olefinic fatty acid	acid value 5 max.; sapon. no. 165-175	clear, yellow/amber liquid; colour 12 max. (Gardner)	
	ANFOMUL S4 SD 1400E	sorbitan monooleate	acid value 8 max.; OH value 193-209; sapon. no. 149-160	clear, amber viscous liquid; colour 10 max. (Gardner)	w/o emulsifier *explosives industry; hard rock blasting*
	ANFOMUL S6 SD 3616E	sorbitan monoisostearate	acid value 8 max.; OH value 220-250; sapon. no. 143-153	clear, yellow viscous liquid; colour 6 max. (Gardner)	
	ANFOMUL S43 SD 3293E	sorbitan sesquioleate	acid value 13 max.; OH value 188-210; sapon. no. 149-160	clear, amber viscous liquid; colour 8 max. (Gardner)	
	ANFOMUL S50 SD 3486E	sorbitan monooleate	acid value 7 max.; OH value 190-215; sapon. no. 145-165	clear, amber viscous liquid; colour 10 max. (Gardner)	
	CITHROL 2ML	monoester produced from polyethylene glycols of varying molecular weight and lauric acid	acid value 5.0 max.; iodine value 8.0 max.; sapon. no. 158-168	colour 15Y:2.5R (1 in. lovibond cell)	o/w emulsifier; detergent; wetting agent; thickener; solubiliser; super-fatting agent; foaming agent; dispersant; softener; lubricant *cosmetics; pharmaceuticals; textile industry; pesticides; leather industry*
	CITHROL 2MO	monoester produced from polyethylene glycols of varying molecular weight and oleic acid	acid value 5.0 max.; iodine value ca. 60; sapon. no. 110-120	amber liquid; colour 20Y:5R (1 in. lovibond cell)	
	CITHROL 2MS	monoester produced from polyethylene glycols of varying molecular weight and stearic acid	acid value 5.0 max.; iodine value 5.0 max.; sapon. no. 112-122	slip point 40°C; colour 3Y:1.5R (lovibond block)	
	CITHROL 3MO	PEG-6 oleate		HLB 9.3; amber liquid	o/w emulsifier *cosmetics*
	CITHROL 4DO	PEG-8 dioleate		HLB 8.3; amber liquid	
	CITHROL 4DS	PEG-8 distearate		HLB 9.3; off-white solid	
	CITHROL 4ML	monoester produced from polyethylene glycols of varying molecular weight and lauric acid	acid value 5.0 max.; iodine value 5.0 max.; sapon. no. 86-96	colourless to pale yellow liquid/solid; colour 3 (Gardner)	o/w emulsifier; detergent; wetting agent; thickener; solubiliser; super-fatting agent; foaming agent; dispersant; softener; lubricant *cosmetics; pharmaceuticals; textile industry; pesticides; leather industry*
	CITHROL 4MO	monoester produced from polyethylene glycols of varying molecular weight and oleic acid	acid value 3.0 max.; iodine value 50 max.; sapon. no. 85-93	amber liquid; colour 30Y:6R (1 in. lovibond cell)	o/w emulsifier; detergent; wetting agent; thickener; solubiliser; super-fatting agent; foaming agent; dispersant; softener; lubricant; antistatic agent *cosmetics; pharmaceuticals; textile industry; pesticides; leather industry*
	CITHROL 4MS	monoester produced from polyethylene glycols of varying molecular weight and stearic acid	acid value 5.0 max.; iodine value 5.0 max.; sapon. no. 95-105	off-white semi-solid; slip point 34°C; colour 1Y:0.5R (lovibond block)	o/w emulsifier; detergent; wetting agent; thickener; solubiliser; super-fatting agent; foaming agent; dispersant; softener; lubricant *industrial applications; cosmetics; pharmaceuticals; textile industry; pesticides; leather industry*
	CITHROL 6ML	monoester produced from polyethylene glycols of varying molecular weight and lauric acid	acid value 3.0 max.; iodine value 5.0 max.; sapon. no. 64-74	slip point 23°C; colour 4Y:1R (1 in. lovibond cell)	o/w emulsifier; detergent; wetting agent; thickener; solubiliser; super-fatting agent; foaming agent; dispersant; softener; lubricant *cosmetics; pharmaceuticals; textile industry; pesticides; leather industry*

Supplier	Trade name	Chemical description	Composition	General properties	Functionality Application
Croda	CITHROL 6MO	monoester produced from polyethylene glycols of varying molecular weight and oleic acid	acid value 4.0 max.; iodine value ca. 40 max.; sapon. no. 65-75	colour 10Y:2R (1 in. lovibond cell)	
	CITHROL 6MS	monoester produced from polyethylene glycols of varying molecular weight and stearic acid	acid value 4.0 max.; iodine value 5.0 max.; sapon. no. 68-76	off-white/cream paste; pour point slip point 30°C; colour 1Y:0.5R (lovibond block)	o/w emulsifier; detergent; wetting agent; thickener; solubiliser; super-fatting agent; foaming agent; dispersant; softener; lubricant *cosmetics; pharmaceuticals; textile industry; pesticides; leather industry*
	CITHROL 10ML	monoester produced from polyethylene glycols of varying molecular weight and lauric acid	acid value 5.0 max.; iodine value 5.0 max.; sapon. no. 51-55	slip point 38°C; colour 3Y:1.5R (lovibond block)	
	CITHROL 10MO	monoester produced from polyethylene glycols of varying molecular weight and oleic acid	acid value 5.0 max.; iodine value 30 max.; sapon. no. 35-45	colour 25Y:5R (1 in. lovibond cell)	
	CITHROL 10MS	monoester produced from polyethylene glycols of varying molecular weight and stearic acid	acid value 5.0 max.; iodine value 3.0 max.; sapon. no. 36-50	off-white solid; colour 0.5Y:0.2R (lovibond block)	
	CITHROL A	polyoxyethylene glycol oleate	iodine value 70 max.; acid value 3 max.; soap 4-6 %; moisture 2% max.; sapon. no. 85-93	amber-coloured liquid; colour 30Y:6R max. (1 in. cell)	antistatic agent; o/w emulsifier; dispersant *textile industry; metal cleaning; leather industry; paper industry; lacquers; rubber industry*
	CITHROL BUTYL LAURATE LB3119	n-butyl laurate	acid value 2.0 max.; iodine value 20 max.; sapon. no. 200-210	clear pale yellow liquid; colour 10Y:1.5R max. (1 in. lovibond cell)	
	CITHROL BUTYL MYRISTATE LB3568	n-butyl myristate	acid value 1.0 max.; iodine value 1 max.; sapon. no. 192-204	clear colourless liquid; colour 90 max. (HU)	*textile industry; lubricants; plastics industry; cosmetics; oil industry*
	CITHROL BUTYL OLEATE LB3191	n-butyl oleate	acid value 2.0 max.; iodine value 60 min.; sapon. no. 165-175	clear amber liquid; colour 30Y:10R max. (1 in. lovibond cell)	
	CITHROL BUTYL OLEATE LB3453	n-butyl oleate	acid value 0.2 max.; iodine value 60 min.; sapon. no. 164 min.	clear pale yellow liquid; colour 13Y:2.5R max. (5¼ in. lovibond cell)	
	CITHROL BUTYL OLEATE LB3823	n-butyl oleate	acid value 2.0 max.; sapon. no. 160-170	clear pale yellow liquid; colour 30Y:10R max. (1 in. lovibond cell)	
	CITHROL BUTYL STEARATE BS3094	n-butyl stearate	acid value 1.0 max.; sapon. no. 170-177	clear pale yellow liquid; colour 10Y:1.0R max. (5¼ in. lovibond cell)	
	CITHROL BUTYL STEARATE BS3200	n-butyl stearate	acid value 0.2 max.; sapon. no. 170-177	clear colourless liquid; colour 50 max. (HU)	
	CITHROL DEGD SN/E	PEG-2 distearate		HLB 3.0; cream to pale yellow solid	o/w emulsifier; pearlescent agent *shampoos; detergents*
	CITHROL DEGML N/E	PEG-2 laurate		HLB 6.1; pale yellow semi-solid	o/w emulsifier; pearlescent agent *shampoos; detergents*
	CITHROL DEGML S/E	PEG-2 laurate (S/E)		HLB 6.7; deep yellow semi-solid	emulsifier; stabiliser; bodying agent *skin care products; hair care products; cosmetics*
	CITHROL DGMS N/E	PEG-2 stearate		HLB 4.4; white to off-white solid	
	CITHROL EGDS N/E	glycol distearate		HLB 1.5; white to cream pastilles	o/w emulsifier; pearlescent agent *shampoos; detergents*
	CITHROL EGMS N/E	glycol stearate		HLB 2.0; white to cream pastilles	
	CITHROL GE 0041 N/E	glyceryl monooleate, non-emulsifying	monoglyceride 35%; free glycerine 5% soap content 1.0% max.; acid value 3.0 max.; sapon. no. 160-170	viscous pale yellow liquid	emulsifier; wetting agent; suspending agent; dispersant; stabiliser; antifogging agent; lubricant; plasticiser; softener *cosmetics; pharmaceuticals; metal lubricants; plastics industry; food packaging; rubber industry; toiletries; inks; paint industry; polishes*

Supplier	Trade name	Chemical description	Composition	General properties	Functionality Application
Croda	CITHROL GE 0802 S/E	glyceryl monostearate, self-emulsifying	monoglyceride 30%; free glycerine 7%; soap content 4.8-5.2%; acid value 6.0 max.; iodine value 3.0 max.; sapon. no. 152-162	wax-like off-white solid; m.p. ca. 57-61°C	emulsifier; emollient; dispersant; lubricant; stabiliser; plasticiser; antistatic agent *cosmetics; pharmaceuticals; paint industry; polishes; inks; plastics industry; textile industry; paper industry; rubber industry; metal lubricants; adhesives; toiletries*
	CITHROL GE 0803 N/E	glyceryl monostearate, non-emulsifying	monoglyceride 30%; free glycerine 5%; acid value 3.0 max.; iodine value 2.0 max.; sapon. no. 165-175	wax-like off-white solid; m.p. ca. 57-60°C	
	CITHROL GE 1148 N/E	glyceryl monolaurate, non-emulsifying	monoglyceride 35%; acid value 5.0 max.; iodine value 18 max.; sapon. no. 195 min.	colour 5Y:1R max. (1 in. lovibond)	emulsifier; spreading agent *cosmetics; pharmaceuticals; polishes; photography; toiletries; metal lubricants; plastics industry; rubber industry; textile industry*
	CITHROL GE 3047 S/E	glyceryl monoricinoleate, self-emulsifying	acid value 6.0 max.; iodine value 70-80 max.; sapon. no. 142-152	golden-coloured non-edible liquid; colour 2OY:4R max. (1 in. lovibond)	dispersant; antifogging agent; lubricant *cosmetics; rubber industry; pesticides; toiletries; paint industry; plastics industry*
	CITHROL GE 3202 S/E	glyceryl monostearate, self-emulsifying	monoglyceride 37%; free glycerine 6%; soap content 1.8-2.2%; acid value 3.0 max.; iodine value 2.0 max.; sapon. no. 160-170	wax-like off-white solid; m.p. ca. 57.5-60°C	emulsifier; emollient; dispersant; lubricant; stabiliser; plasticiser; antistatic agent *cosmetics; pharmaceuticals; paint industry; polishes; inks; plastics industry; textile industry; paper industry; rubber industry; metal lubricants; adhesive tapes; toiletries*
	CITHROL GE 3286 N/E	glyceryl monostearate, non-emulsifying	monoglyceride 40-50% free glycerine 6%; acid value 3.0 max.; iodine value 3.0 max.; sapon. no. 158-177	wax-like off-white solid; m.p. ca. 54-64°C	
	CITHROL GE 3342 N/E	glyceryl monostearate, non-emulsifying	monoglyceride 40% free glycerine 5%; acid value 6.0; iodine value 101-111; sapon. no. 155-175	viscous pale yellow liquid	emulsifier; wetting agent; suspending agent; dispersant; stabiliser; antifogging agent; lubricant; plasticiser; softener *cosmetics; pharmaceuticals; metal lubricants; plastics industry; food packaging; rubber industry; toiletries; inks; paint industry; polishes*
	CITHROL GE 3439 S/E	glyceryl monooleate, self-emulsifying	monoglyceride 35% free glycerine 6%; acid value 5.0 max.; iodine value 70-85; sapon. no. 160-175	viscous pale yellow liquid	
	CITHROL GE 3456 N/E	glyceryl monostearate, non-emulsifying	monoglyceride 90%; free glycerine 1%; soap content 0.1 max.; acid value 3.0 max.; iodine value 3.0 max.	wax-like off-white solid	emulsifier; emollient; dispersant; lubricant; stabiliser; plasticiser; antistatic agent *cosmetics; pharmaceuticals; paint industry; polishes; inks; plastics industry; textile industry; paper industry; rubber industry; metal lubricants; adhesives; toiletries*
	CITHROL GML N/E	glyceryl laurate		HLB 4.9; off-white semi-solid	w/o emulsifier; co-emulsifier *creams and lotions*
	CITHROL GMM	glyceryl myristate		white to off-white waxy semi-solid	
	CITHROL GMO N/E	glyceryl oleate		HLB 3.3; amber liquid/semi-solid	
	CITHROL GMR N/E	glyceryl ricinoleate		HLB 2.7; amber liquid	w/o emulsifier *antiperspirants; hair conditioners; astringents; creams and lotions*
	CITHROL GMR S/E	glyceryl ricinoleate (S/E)		HLB 3.6; amber liquid	emulsifier; stabiliser; bodying agent *skin care products; hair care products: cosmetics*
	CITHROL GMS A/S	glyceryl stearate and PEG-100 stearate		HLB 10.9; white waxy solid	o/w emulsifier *antiperspirants; hair conditioners; astringents; creams and lotions*
	CITHROL GMS N/E	glyceryl stearate		HLB 3.4; off-white solid	w/o emulsifier; co-emulsifier *creams and lotions*
	CITHROL GMS S/E	glyceryl stearate (S/E)		HLB 4.4; off-white to cream solid	emulsifier; stabiliser; bodying agent *skin care products; hair care products: cosmetics*
	CITHROL PGML N/E	propylene glycol laurate		HLB 2.7; yellow liquid	o/w emulsifier; pearlescent agent *shampoos; detergents*
	CITHROL PGMR N/E	propylene glycol ricinoleate		yellow to amber liquid	

Supplier	Trade name	Chemical description	Composition	General properties	Functionality / Application
Croda	CITHROL PGMR S/E	propylene glycol ricinoleate (S/E)		HLB 3.6; yellow to amber liquid	emulsifier; stabiliser; bodying agent / *skin care products; hair care products; cosmetics*
	CITHROL PGMS N/E	propylene glycol stearate		HLB 2.4; off-white solid	o/w emulsifier; pearlescent agent / *shampoos; detergents*
	CITHROL PGMS S/E	propylene glycol stearate (S/E)		HLB 3.2; yellow solid	emulsifier; stabiliser; bodying agent / *skin care products; hair care products; cosmetics*
	CITHROL PR	complex polyglycerol polyester of naturally occuring fatty acids	OH value 80-100; acid value 6 max.; iodine value 72-103; soap content 1.0-1.2%; sapon. no. 170-190	amber viscous liquid	emulsifier; dispersant / *cosmetics; pharmaceuticals*
	CRESTER KZ	mixed esters of polyglycerol and edible fatty acids	acid value 3 max.; iodine value 3 max.; moisture 2% max.; soap content 2% max.; sapon. no. 120-150	colour 20.0Y:3.0R max. (1 in. cell)	
	CRESTER L	mixed esters of polyglycerol and edible fatty acids	acid value 5 max.; moisture 1% max.; soap content 1%; sapon. no. 130-140	colour 30.0Y:10.0R (1 in. cell)	*food industry*
	CRESTER MG	mixed esters of monoglycerides and polyglycerol edible fatty acids	acid value 2 max.; OH value 230-276; sapon. no. ca. 160	buff-coloured block or pastilles	
	CRESTER PR	complex mixture of polyglycerol with linearly esterified fatty acids derived from castor oil	acid value 6 max.; iodine value 72-103; moisture 1.0% max.; OH value 80-100; soap content 1.0-1.2%; sapon. no. 170-190	amber liquid	
	CRILL 1	sorbitan laurate	acid value 7 max.; OH value 330-358; H_2O content 1.0% max.; sapon. no. 158-170	HLB 8.6; amber liquid; colour 9 max. (Gardner)	emulsifier; wetting agent; dispersant; antifoam; lubricant; antistatic agent / *cosmetics; pesticides; pharmaceuticals; plastics industry; polishes; cleaners; textile industry; paper industry; food industry*
	CRILL 2	sorbitan palmitate	acid value 4-7.5 max.; OH value 260-305; H_2O content 0.5% max.; sapon. no. 140-155	HLB 6.7; cream solid; colour 10 max. (Gardner)	emulsifier; wetting agent; dispersant / *cosmetics; polishes; cleaners; food industry*
	CRILL 3	sorbitan stearate	acid value 10 max.; OH value 235-260; H_2O content 1.5% max.; sapon. no. 147-157	HLB 4.7; cream/yellow solid; colour 8 max. (Gardner)	emulsifier; wetting agent; dispersant / *cosmetics; pesticides; metal working; pharmaceuticals; polishes; cleaners; textile industry; food industry*
	CRILL 4	sorbitan oleate	acid value 7 max.; OH value 193-209; H_2O content 0.5% max.; sapon. no. 149-160	HLB 4.3; amber liquid; colour 10 max. (Gardner)	emulsifier; wetting agent; dispersant; coupling agent / *cosmetics; pesticides; metal working; oil industry; paint industry; inks; pharmaceuticals; polishes; cleaners; textile industry; food industry*
	CRILL 6	sorbitan isostearate	acid value 8 max.; OH value 220-250; H_2O content 1.0% max.; sapon. no. 143-153	HLB 4.7; yellow liquid; colour 6 max. (Gardner)	emulsifier; wetting agent; dispersant / *cosmetics; pharmaceuticals; pigments*
	CRILL 35	sorbitan tristearate		HLB 2.1; pale tan waxy solid	w/o emulsifier; co-emulsifier / *cosmetics*
	CRILL 41	sorbitan tristearate	acid value 7 max.; OH value 60-80; H_2O content 1.0% max.; sapon. no. 172-185	HLB 2.1; cream/yellow solid; colour 8 max. (Gardner)	emulsifier; wetting agent; dispersant / *cosmetics; pesticides; metal working; pharmaceuticals; polishes; cleaners; textile industry; food industry*
	CRILL 43	sorbitan sesquioleate	acid value 8.5-13; OH value 188-210; H_2O content 1.0% max.; sapon. no. 149-160	HLB 3.7; amber liquid; colour 8 max. (Gardner)	emulsifier; wetting agent; dispersant; coupling agent / *cosmetics; pesticides; metal working; paint industry; inks; pharmaceuticals; polishes; cleaners*
	CRILL 45	sorbitan trioleate	acid value 10-14; OH value 56-68; H_2O content 1.0% max.; sapon. no. 172-186	HLB 1.8; amber viscous liquid; colour 8 max. (Gardner)	emulsifier; wetting agent; dispersant; coupling agent / *cosmetics; metal working; paint industry; inks; pharmaceuticals; food industry*
	CRILL 50	sorbitan monooleate, technical grade	acid value 2 max.		emulsifier; wetting agent; dispersant

Supplier	Trade name	Chemical description	Composition	General properties	Functionality / Application
Croda	CRILL 65	sorbitan tri-isostearate		HLB 1.8; yellow to pale yellow liquid	w/o emulsifier; co-emulsifier *cosmetics*
	CRODA GLS	glyceryl lactostearate	monoglyceride 20-26%		*food industry*
	CRODAMOL BS	butyl stearate		colourless liquid; viscosity 8.9 cS	emollient; solvent; plasticiser; diluent; super-fatting agent *cosmetics; cleansing preparations*
	CRODAMOL CAP	selected blend of branched chain esters		colourless liquid; viscosity 11.9 cS	emollient; lubricant; wetting agent; spreading agent *skin care products; hair care products; cosmetics; sunscreen products*
	CRODAMOL CL	cetyl lactate		pale straw soft solid; m.p. 24°C	wetting agent *skin care products; cosmetics*
	CRODAMOL CSS	cetostearyl stearate		white to off-white solid; m.p. 55°C	emollient *skin care products; hair care products; cosmetics*
	CRODAMOL DA	di-isopropyl adipate		colourless liquid; viscosity 3.9 cS	emollient; wetting agent; plasticiser; solvent *cosmetics; shaving products; bath preparations*
	CRODAMOL DO	decyl oleate		pale straw liquid; viscosity 15.7 cS	emollient; lubricant; wetting agent *cosmetics; skin care products*
	CRODAMOL DOA	di-2-ethylhexyl adipate		colourless to pale straw liquid; viscosity 12.5 cS	emollient *cosmetics; skin care products; antiperspirants*
	CRODAMOL GTCC	glyceryl tricaprylate/caprate	acid value 0.1 max.; iodine value 1 max.; sapon. no. 325-345	almost colourless low-viscosity liquid; cloud point − 5°C; colour 60 max. (Hazen)	emollient; super-fatting agent; diluent *cosmetics; food industry; pharmaceuticals; sunscreen products; hair care products; bath preparations; skin care products; skin cleansing preparations; baby care products*
	CRODAMOL ICS	isocetyl stearate		colourless liquid; viscosity 26.5 cS	wetting agent; spreading agent; emollient; lubricant *bath preparations; skin care products; cosmetics*
	CRODAMOL IPL	isopropyl laurate		colourless liquid; viscosity 3.7 cS	emollient; lubricant; penetrant; spreading agent *skin care products; cosmetics; bath preparations*
	CRODAMOL IPM	isopropyl myristate		colourless liquid; viscosity 5.3 cS	emollient; lubricant; plasticiser; penetrant; spreading agent *skin care products; cosmetics*
	CRODAMOL IPP	isopropyl palmitate		pale straw liquid; viscosity 7.7 cS	emollient; lubricant; penetrant; spreading agent *skin care products; cosmetics*
	CRODAMOL ISNP	isostearyl neopentanoate		pale straw liquid; viscosity 17.5 cS	emollient *skin care products*
	CRODAMOL JJ	oleyl erucate		pale yellow liquid; viscosity 34 cS	emollient; jojoba oil substitute
	CRODAMOL LL	lauryl lactate		colourless to pale straw liquid; viscosity 13.9 cS	wetting agent; emollient *hair care products; skin care products; antiperspirants*
	CRODAMOL ML	myristyl lactate		white to pale yellow liquid; viscosity 21.4 cS	wetting agent; emollient; lubricant *cosmetics; skin care products*
	CRODAMOL MM	myristyl myristate		white waxy solid; m.p. 37-39°C	emollient; lubricant *skin care products; hair care products; cosmetics*
	CRODAMOL OC	octyl cocoate		pale straw liquid	emollient *skin care products; cosmetics; toiletries*
	CRODAMOL OHS	octyl hydroxy stearate		pale straw liquid; viscosity 77 cS	emollient; super-fatting agent; additive *soaps; detergents; bath preparations*
	CRODAMOL OP	octyl palmitate		colourless liquid; viscosity 12.7 cS	emollient *cosmetics; skin care products; sunscreen products*

Supplier	Trade name	Chemical description	Composition	General properties	Functionality Application
Croda	CRODAMOL OSU	diester of succinic acid and 2-ethyl hexyl alcohol	acid value 1 max.; sapon. no. 320-333	water-white clear liquid; colour 30 max. (Hazen)	wetting agent; emollient; glossing agent; spreading agent; super-fatting agent; *cosmetics; skin care products; hair care products; deodorants*
	CRODAMOL PC	propylene glycol di-ester of coconut fatty acids		colourless to pale straw liquid; viscosity 9.3 cS	lubricant; emollient; spreading agent *cosmetics; hair care products; skin care products; bath preparations; antiperspirants; sunscreen preparations*
	CRODAMOL PMP	promyristyl propionate		colourless liquid; viscosity 8.9 cS	emollient; wetting agent; spreading agent *cosmetics; skin care products*
	CRODAMOL PTC	pentaerythritol tetra caprate/caprylate		pale straw liquid; viscosity 50 cS	lubricant; emollient; wetting agent; spreading agent *cosmetics; skin care products; sunscreen preparations*
	CRODAMOL PTIS	pentaerythritol tetraisostearate		amber liquid; viscosity 249 cS	emollient; lubricant; water repellent *skin care products; sunscreen preparations*
	CRODAMOL SS	synthetic spermaceti		white to off-white solid; m.p. 46°C	emollient; lubricant *hair care products; cosmetics*
	CRODAMOL W	stearyl heptanoate/octanoate		white to off-white solid; m.p. 26.5°C	wetting agent; spreading agent *cosmetics; skin care products*
	CRODAPEARL AM3417	modified glycol ester	OH value ca. 70; acid value 7 max; iodine value 2 max.; sapon. no. ca. 185	colour 1.5Y:0.8R max. (1 in. block); drop point 59-60°C; pH ca. 4.6 (3% soln.)	pearlescent agent; opacifier; super-fatting agent *shower gels; shampoos; foam baths; skin cleansing preparations; liquid soaps; cosmetics; pharmaceuticals; household cleaners; dishwashing agents*
	CRODAPEARL AM3843	modified glycol stearate		white to off-white solid	pearlescent agent *foam baths; shampoos; cosmetics*
	CRODAPEARL NI	modified glycol ester	H_2O content 80% max.	pearly white to off-white paste; pH 5.5-7.0 (5% soln.)	pearlescent agent; opacifier; super-fatting agent *shower gels; shampoos; foam baths; skin cleansing preparations; liquid soaps; cosmetics; pharmaceuticals; household cleaners; dishwashing agents*
	CRODESTA F10	sucrose distearate		HLB 3.0; off-white powder	o/w and w/o emulsifier; wetting agent *cosmetics*
	CRODESTA F20	sucrose distearate		HLB 3.0; off-white powder	
	CRODESTA F50	sucrose distearate		HLB 6.5; off-white powder	
	CRODESTA F70	sucrose distearate		HLB 3.0; off-white powder	
	CRODESTA F110	sucrose stearate and sucrose distearate		HLB 12.0; off-white powder	
	CRODESTA F140	sucrose stearate and sucrose distearate		HLB 13.0; off-white powder	
	CRODESTA F160	sucrose stearate		HLB 14.5; off-white powder	
	CRODESTA SL40	sucrose cocoate and H_2O		HLB 15.0; pale yellow liquid	o/w emulsifier; detergent; mildness additive *skin cleansing preparations; baby shampoos*
	CRODET 9OS40	PEG-40 stearate		white waxy solid	o/w emulsifier; wetting agent; solubiliser *skin care products; hair care products; antiperspirants; perfumes*
	CRODET S8	PEG-8 stearate		HLB 10.8; white soft solid	
	CRODET S40	PEG-40 stearate		HLB 16.7; white waxy solid	
	CRODET S50	PEG-50 stearate		white waxy solid	
	CRODET S100	PEG-100 stearate		HLB 18.8; white hard wax	
	PGMS 70	propylene glycol ester of stearic acid with other related acids present	monoester content 68% min.; free glycol 1.0% max.; acid value 2 max.; iodine value 2.0	pastilles; colour 20Y:3R (5 ¼ in. lovibond cell); m.p. 37-42°C; pH 5.3-6.8 (10% soln. at 45°C)	*food industry*

Supplier	Trade name	Chemical description	Composition	General properties	Functionality / Application
Croda	SUPERMOL L	pentaerythrityl isostearate/caprate/caprylate/ adipate		pale yellow liquid	emollient; moisturiser *skin care products; cosmetics*
	SUPERMOL S	pentaerythrityl stearate/caprate/caprylate/ adipate		pale yellow paste	emollient; moisturiser *skin care products; cosmetics*
	TRIBUTYL CITRATE	tributyl citrate		water-white product; M.W. 360.4; pour point −62°C; viscosity 32 cPs	plasticiser; defoaming agent *plastics industry; protein soln.s; resins*
	VYKAMOL 83G	blend of Crill 3 and Crillet 3	acid value 0.4 max.; sapon. no. 450 min.	HLB 6.2; creamy powder	*food industry*
Dac International Surfactants	DETAMMIDE 120	diethylenetriamine stearate	acid value 3.0–7.0; OH value 212–236; H₂O content 2.0% max.; sapon. no. 130–135		softener *textile industry*
	POLIDAC 4000 MS	polyethylene glycol monostearate			viscosity modifier
	STERAMIDE D	diethanolamide stearate			softener *textile industry*
Danisco Ingredients	ACIDAN BC	citric acid ester of monoglycerides; vegetable fat based	acid value 20–35; sapon. no. 225–255	flakes; dropping point ca. 63°C	stabiliser; emulsifier; anti-spattering agent *food industry; frying margarine; meat products*
	ACIDAN LC	citric acid ester of monoglycerides; edible fat based	acid value 25–40; sapon. no. 220–250	pellets; dropping point ca. 59°C	
	ACIDAN N 12	citric acid ester of monoglycerides; edible fat based	acid value 10–25; sapon. no. 220–250	fine powder; dropping point ca. 64°C	
	AMIDAN 30	cold-dispersible monoglyceride; vegetable fat based	monoester 85% min.; iodine value 2 max.; sodium palmitate and stearate content 6% max.	powder; dropping point ca. 68°C	emulsifier *food industry; bread; cakes; breakfast cereals; pasta products*
	AMIDAN B 727	cold-dispersible monoglyceride; edible fat based	monoester 85% min.; iodine value ca. 10	fine powder; dropping point ca. 65°C	
	AMIDAN SDM-T	cold-dispersible monoglyceride; vegetable fat based	monoester 90% min.; iodine value 35–45	fine powder; dropping point ca. 66°C	
	ARTODAN CF 40	stearoyl lactylate; vegetable fat based	lactic acid 20–24%; calcium 1.9–2.3%; ester value 125–150	flakes	emulsifier *food industry; bread; bakery goods; shortening; sour cream*
	ARTODAN CP 80	stearoyl lactylate; edible fat based	lactic acid 26–30%; calcium 3.0–3.6%; ester value 135–165	powder	
	ARTODAN SP 55	stearoyl lactylate; animal and vegetable fat based	lactic acid 31–34%; sodium 3.5–5.0%; ester value 150–190	beads	
	CETODAN 50-00	acetic acid ester of monoglycerides; animal and vegetable fat based	degree of acetylation 0.5; iodine value 2 max.	block; dropping point ca. 43°C	emulsifier; lubricant; release agent *food industry; topping powders; chewing gum base; coatings; cakes*
	CETODAN 70-00	acetic acid ester of monoglycerides; animal and vegetable fat based	degree of acetylation 0.7; iodine value 2 max.	block; dropping point ca. 40°C	
	CETODAN 90-40	acetic acid ester of monoglycerides; animal fat based	degree of acetylation 0.9; iodine value ca. 40	liquid; cloud point 12°C	
	DIMODAN BP	distilled monoglyceride; palm oil based	monoester 90% min.; iodine value ca. 45	block; dropping point ca. 60°C	emulsifier; aerating agent; stabiliser *food industry*
	DIMODAN CP	distilled monoglyceride; vegetable fat based	monoester 90% min.; iodine value ca. 80	plastic; dropping point ca. 52°C	
	DIMODAN LS	glyceryl linoleate; sunflower oil based	monoester 90% min.; iodine value ca. 105	HLB 4.3; soft plastic	emulsifier; lubricant; antistatic agent; antifogging agent *food industry; plastics industry*
	DIMODAN OT	distilled monoglyceride; soya bean oil based	monoester 90% min.; iodine value ca. 60	pellets; dropping point ca. 57°C	emulsifier; aerating agent; stabiliser *food industry*
	DIMODAN P	distilled monoglyceride; lard based	monoester 90% min.; iodine value 2 max.	beads; dropping point ca. 70°C	emulsifier *food industry*

Supplier	Trade name	Chemical description	Composition	General properties	Functionality / Application
Danisco Ingredients	DIMODAN PM	glyceryl stearate; animal fat based	monoester 90% min.; iodine value 2 max.	HLB 4.3; beads; dropping point ca. 70°C	emulsifier; lubricant; antistatic agent; antifogging agent; *food industry; plastics industry*
	DIMODAN PM 300	glyceryl stearate; animal fat based	monoester 90% min.; iodine value 2 max.	HLB 4.3; fine powder; dropping point ca. 70°C	emulsifier; *food industry; plastics industry*
	DIMODAN PS	distilled monoglyceride; soya bean oil based	monoester 90% min.; iodine value 2 max.	beads; dropping point ca. 72°C	emulsifier; aerating agent; stabiliser; *food industry*
	DIMODAN PV	distilled monoglyceride; vegetable fat based	monoester 90% min.; iodine value 2 max.	beads; dropping point ca. 72°C	emulsifier; *food industry*
	DIMODAN PVP	distilled monoglyceride; palm oil based	monoester 90% min.; iodine value 2 max.	beads; dropping point ca. 69°C	emulsifier; aerating agent; stabiliser; *food industry*
	DIMODAN S	lard monoglyceride	monoester 90% min.; iodine value ca. 50	HLB 4.3; block; dropping point ca. 56°C	emulsifier; lubricant; antistatic agent; antifogging agent; *food industry; plastics industry*
	DIMODAN TH	distilled monoglyceride; tallow based	monoester 90% min.; iodine value 2 max.	beads; dropping point ca. 70°C	emulsifier; *food industry*
	EMULDAN FP 40	glyceryl stearate		HLB 2.8; off-white fine powder	emulsifier; lubricant; antistatic agent; antifogging agent; *plastics industry*
	EMULDAN HA 32/S3	glyceryl stearate SE; edible fat based	monoester 30% min.; iodine value ca. 3; sodium palmitate and stearate content 3% max.	HLB 3.4; beads; dropping point ca. 64°C	emulsifier; lubricant; antistatic agent; antifogging agent; *food industry; plastics industry*
	EMULDAN HA 32/36	glyceryl stearate SE		HLB 3.7; off-white powder	emulsifier; lubricant; antistatic agent; anti-fogging agent; *plastics industry*
	EMULDAN HA 40	mono-diglyceride; edible fat based	monoester 40% min.; iodine value 2 max.	beads; dropping point ca. 63°C	emulsifier; stabiliser; *margarine; shortening; non-dairy creams; caramel; toffees*
	EMULDAN HA 60	mono-diglyceride; edible fat based	monoester 60% min.; iodine value 2 max.	beads; dropping point ca. 65°C	
	EMULDAN HLT 40	glyceryl stearate		HLB 2.8; off-white powder	emulsifier; lubricant; antistatic agent; antifogging agent; *plastics industry*
	EMULDAN HS 40	mono-diglyceride; vegetable fat based	monoester 40% min.; iodine value 2 max.	beads; dropping point ca. 65°C	emulsifier; stabiliser; *margarine; shortening; non-dairy creams; caramel; toffees*
	EMULDAN HV 60	mono-diglyceride; vegetable fat based	monoester 60% min.; iodine value 2 max.	beads; dropping point ca. 66°C	
	EMULDAN KS 60	mono-diglyceride; edible fat based	monoester 60% min.; iodine value ca. 70	soft plastic; dropping point ca. 47°C	
	EMULDAN PK 60	palm kernel glyceride		HLB 4.1; off-white block	emulsifier; lubricant; antistatic agent; antifogging agent; *plastics industry*
	EMULDAN TR 40	mono-diglyceride; edible fat base	monoester 55% min.; iodine value ca. 65	block; dropping point ca. 45°C	emulsifier; stabiliser; *food industry; margarine; shortening; non-dairy creams; caramel; toffees*
	FAMODAN MS	sorbitan ester of fatty acid; edible fat base	OH value 235-260; acid value 5-10	small beads; dropping point ca. 57°C	emulsifier; *food industry; cocoa butter substitutes and replacers; margarine; low-fat spreads; cooking oil; dry yeast*
	FAMODAN SMO	sorbitan monooleate		HLB 4.6; brown liquid	emulsifier
	FAMODAN TS 30	sorbitan ester of fatty acid; edible fat base	OH value 66-80; acid value 7 max.	small beads; dropping point ca. 59°C	emulsifier; *food industry; cocoa butter substitutes and replacers; margarine; low-fat spreads; cooking oil; dry yeast*
	FAMODAN TS	sorbitan tristearate		HLB 2.3; tan powder	emulsifier
	FAMODAN TSQ	sorbitan ester of fatty acid; vegetable fat base	OH value 66-80; acid value 2 max.	small beads; dropping point ca. 57°C	emulsifier; *food industry; cocoa butter substitutes and replacers; margarine; low-fat spreads; cooking oil; dry yeast*
	GATODAN 504	special blend; vegetable fat base	acid value 3 max.; sapon. no. 160-175	pellets; dropping point ca. 65°C	emulsifier
	GATODAN 505	special blend; vegetable fat base	acid value 3 max.; sapon. no. 145-160	pellets; dropping point ca. 65°C	emulsifier; *food industry; bakery goods*

Supplier	Trade name	Chemical description	Composition	General properties	Functionality Application
	GRINDTEK AML 60	acetylated palm kernel glyceride		HLB 2.1; light tan liquid	lubricant; plasticiser; co-solvent
	GRINDTEK DAT-S	DATA ester of glycerol monostearate		HLB 8; cream flakes	o/w emulsifier
	GRINDTEK ML 90	glyceryl laurate		HLB 5.3; off-white powder	
	GRINDTEK MM 90	glyceryl myristate		HLB 5.0; off-white powder	emulsifier; lubricant; antistatic agent; antifogging agent *plastics industry*
	GRINDTEK MOR 40	glyceryl oleate		HLB 32; yellow/brown liquid	
	GRINDTEK MOR 90	glyceryl oleate		HLB 4.3; off-white semi-solid	
	GRINDTEK MSP 52	glyceryl stearate		HLB 3.8; off-white powder	
	GRINDTEK P 40	glyceryl stearate		HLB 3.6; off-white powder	
	GRINDTEK PGE 55-6	polyglyceryl-3-stearate SE		HLB 7.4; tan block	o/w emulsifier; antifogging agent *plastics industry*
	GRINDTEK SML	sorbitan monolaurate		HLB 7.7; brown viscous liquid	emulsifier
	HOMODAN MO	emulsifier speciality: edible fat base	monoester ca. 40%	flakes; dropping point ca. 62°C	emulsifier *food industry; frying and puff pastry margarine*
	HOMODAN PT	polyglyceryl-3-ester of dimerised soya bean oil		HLB 2.8; yellow/brown viscous liquid	emulsifier; antifogging agent *plastics industry*
	HOMODAN RD	emulsifier speciality: vegetable fat base	monoester 30% min.	flakes; dropping point ca. 63°C	emulsifier *food industry; frying and puff pastry margarine*
	LACTODAN B 30	lactic acid ester of monoglycerides; edible fat base	total lactic acid 25-30%; sapon. no. 290-320	pellets; dropping point ca. 46°C	emulsifier
	LACTODAN F 15	lactic acid ester of monoglycerides; vegetable fat base	total lactic acid 13-16%; sapon. no. 245-265	pellets; dropping point ca. 50°C	emulsifier *food industry; topping powders; creams; cake margarine; shortening*
	LACTODAN P 22	lactic acid ester of monoglycerides; vegetable fat base	total lactic acid 20-25%; sapon. no. 270-300	small beads; dropping point ca. 45°C	
	LACTODAN Z 30	glyceryl stearate lactate		HLB 8.2; off-white block	emulsifier
	LECIDAN SB	special blend of emulsifier and special soya lecithin; vegetable fat base	monoester 40% min.	pellets; dropping point ca. 60°C	emulsifier
	LECIDAN SR	special blend of emulsifier and special soya lecithin; edible fat base	monoester 40% min.	pellets; dropping point ca. 60°C	emulsifier; anti-spattering agent *food industry; margarine; spreads*
	LECIDAN SV	special blend of emulsifier and normal soya lecithin; vegetable fat base	monoester 40% min.	pellets; dropping point ca. 62°C	
	LIPODAN HP 100	emulsifier speciality: vegetable fat base	iodine value 2 max.	beads; dropping point ca. 60°C	emulsifier
	LIPODAN PBS	emulsifier speciality: vegetable fat base	monoester 20% min.; iodine value 2 max.	beads; dropping point ca. 62°C	emulsifier *food industry; margarine; peanut butter; creams*
	PANODAN 80	diacetyl tartaric acid ester of monoglycerides; edible fat base	acid value (ester part) 85-110; calcium carbonate 20%; sapon. no. (ester part) 480-510	fine powder	emulsifier; stabiliser *food industry*
	PANODAN 90	diacetyl tartaric acid ester of monoglycerides; edible fat base	acid value (ester part) 80-100; tricalcium orthophosphate 10%; sapon. no. (ester part) 475-505	fine powder	
	PANODAN 120	diacetyl tartaric acid ester of monoglycerides; edible fat base	acid value (ester part) 80-100; tricalcium orthophosphate 5%; calcium carbonate 10%; enzyme-enriched soya flour 5%; sapon. no. (ester part) 475-505	fine powder	emulsifier *food industry*
	PANODAN 165	diacetyl tartaric acid ester of monoglycerides; vegetable fat base	acid value (ester part) 47-57; sapon. no. (ester part) 325-355	powder	emulsifier; stabiliser *food industry*

Supplier	Trade name	Chemical description	Composition	General properties	Functionality / Application
Danisco Ingredients	PANODAN 482	diacetyl tartaric acid ester of monoglycerides; edible fat base	acid value (ester part) 85-105; tricalcium orthophosphate 7.5%; calcium carbonate 7.5%; sapon. no. (ester part) 500-535	powder	emulsifier / food industry
	PANODAN 507	diacetyl tartaric acid ester of monoglycerides; vegetable fat base	acid value (ester part) 65-85; sapon. no. (ester part) 420-450	powder	emulsifier; stabiliser / food industry
	PANODAN AB 90	DATA ester of glycerol monooleate		HLB 7; yellow/brown viscous liquid	o/w emulsifier
	PANODAN AB 100	diacetyl tartaric acid ester of monoglycerides; edible fat base	acid value (ester part) 75-95; sapon. no. (ester part) 490-520	semi-liquid	emulsifier; stabiliser / food industry
	PANODAN AM	diacetyl tartaric acid ester of monoglycerides; edible fat base	acid value (ester part) 85-110; sapon. no. (ester part) 480-510	flakes; dropping point ca. 47°C	emulsifier; stabiliser / food industry
	PROMODAN 70	propylene glycol ester of fatty acids; edible fat base	monoester 67% min.; free propylene glycol 1% max.	small beads; dropping point ca. 40°C	emulsifier; aerating agent; foam stabiliser / food industry; topping powders; bakery goods; shortening
	PROMODAN SP	propylene glycol stearate		HLB 3.5; off-white block	w/o emulsifier
	PROMODAN SPV	propylene glycol ester of fatty acids; vegetable fat base	monoester 95% min.; free propylene glycol 1% max.	pellets; dropping point ca. 40°C	emulsifier; aerating agent; foam stabiliser / food industry; topping powders; bakery goods; shortening
	RELIFIER B	emulsifier speciality; vegetable fat base		viscous brown liquid	emulsifier / food industry; fin grease emulsion
	RELIFIER B 160	emulsifier speciality; vegetable fat base		viscous brown liquid	
	TRIODAN 20	polyglyceryl-3-oleate		HLB 5.5; yellow/brown liquid	o/w emulsifier; antifogging agent / plastics industry
	TRIODAN 55	polyglyceryl-3-stearate; vegetable fat base	iodine value 2 max.; sapon. no. 130-145	HLB 6.8; base; small beads; dropping point ca. 57°C	emulsifier; antifogging agent / food industry; plastics industry
	TRIODAN R 90	polyglycerol ester of fatty acids; ricinoleic acid base	iodine value 72-103; OH value 80-100	liquid	emulsifier / food industry; chocolate; low fat spreads; bakery goods; cake margarine
Eastman	MYVACET 5-07	distilled acetylated monoglycerides derived from hydrogenated vegetable oil	acetylation 48.5-51.5%; OH value 133-152; acid value 3 max.; iodine value 5 max.; sapon. no. 279-292	waxy solid; m.p. ca. 41-46°C	
	MYVACET 7-07	distilled acetylated monoglycerides derived from hydrogenated vegetable oil	acetylation 66.5-69.5%; OH value 80.5-95; acid value 3 max.; iodine value 5 max.; sapon. no. 316-331	waxy solid; m.p. ca. 37-40°C	
	MYVACET 9-08	distilled acetylated monoglycerides derived from hydrogenated coconut oil	acetylation 95% min.; OH value 20 max.; acid value 3 max.; iodine value 2 max.; sapon. no. 440-455	liquid; m.p. ca. -12 to -14°C	film former; stabiliser; plasticiser; lubricant; emulsifier / food industry
	MYVACET 9-45	distilled acetylated monoglycerides derived from partially hydrogenated soybean oil	acetylation 96% min.; OH value 0-15; acid value 3 max.; iodine value 43-53; sapon. no. 370-382	liquid; m.p. ca. 4-12°C	
	MYVAPLEX 600	concentrated glyceryl monostearate derived from hydrogenated soybean oil	monoester content 90% min.; glycerol content 1.2% max.; acid value 3 max.; iodine value 5 max.	small beads; m.p. ca. 69°C	lubricant; emulsifier; starch complexing agent / food industry
	MYVAPLEX 600P	concentrated glyceryl monostearate derived from hydrogenated soybean oil	monoester content 90% min.; glycerol content 1.2% max.; acid value 3 max.; iodine value 5 max.	powder; m.p. ca. 69°C	lubricant; emulsifier; starch complexing agent / food industry
	MYVATEX 3-50	distilled monoglycerides and distilled propylene glycol monoesters based on soybean oil	monoester content 90% min.; acid value < 3; iodine value 5 max.	beads; m.p. ca. 58°C	emulsifier / food industry
	MYVATEX 7-85	distilled monoglycerides with 30% cottonseed oil	monoester content 63% min.; acid value < 3; iodine value 91-101	plastic; m.p. ca. 49°C	

Supplier	Trade name	Chemical description	Composition	General properties	Functionality Application
Eastman	MYVATEX 8-06	distilled monoglycerides with 20% hydrogenated soybean oil or sunflower oil	monoester content 72% min.; acid value < 3; iodine value 24-30	beads; m.p. ca. 67°C	
	MYVATEX 8-16	distilled monoglycerides with 20% hydrogenated palm oil	monoester content 72% min.; acid value < 3; iodine value 28 max.	beads; m.p. ca. 61°C	emulsifier *food industry*
	MYVATEX 40-06S	distilled propylene glycol monoesters, distilled monoglycerides, lactylic esters of fatty acids (stearic), H₂O, and potassium sorbate; fat source soybean oil	solids 25%; acid value < 13; iodine value < 2	soft plastic	
	MYVATEX DO CONTROL	distilled succinylated monoglycerides and distilled monoglycerides derived from palm oil	monoester content 41% min.; acid value > 74; iodine value 4-7	powder; m.p. ca. 53°C	emulsifier; dough strengthener *food industry; bakery products*
	MYVATEX MIGHTY SOFT	distilled monoglycerides derived from soybean oil	monoester content 90% min.; acid value < 3; iodine value 19-36	powder; m.p. ca. 67°C	emulsifier; softener *food industry*
	MYVATEX MONOSET	distilled monoglycerides derived from rapeseed, cottonseed and palm oils	monoester content 18% min.; acid value < 4; iodine value 5 max.	beads; m.p. ca. 63°C	
	MYVATEX MSPS	distilled monoglycerides with 25% polysorbate 80 derived from soybean oil	monoester content 67.5% min.; acid value < 3; iodine value 6 max.	beads; m.p. ca. 69°C	emulsifier *food industry*
	MYVATEX SSH	distilled monoglycerides, lecithin, H₂O, and propionic acid; fat source soybean oil	solids 45%; acid value < 7.3; iodine value 6 min.	soft plastic	
	MYVATEX TEXTURE LITE	distilled monoglycerides, distilled propylene glycol monoester, sodium stearoyl lactylate with silicon dioxide; fat source soybean oil	monoester content 80% min.; acid value < 13; iodine value 5 max.	powder; m.p. ca. 55°C	
	MYVEROL 18-04	distilled monoglycerides derived from hydrogenated palm oil or palm stearine	monoester content 90% min.; acid value 3 max.; iodine value 5 max.	small beads; m.p. ca. 66°C	
	MYVEROL 18-06	distilled monoglycerides derived from hydrogenated soybean oil	monoester content 90% min.; acid value 3 max.; iodine value 5 max.	small beads; m.p. ca. 69°C	
	MYVEROL 18-07	distilled monoglycerides derived from hydrogenated vegetable oil or cottonseed oil	monoester content 90% min.; acid value 3 max.; iodine value 5 max.	small beads; m.p. ca. 68°C	
	MYVEROL 18-35	distilled monoglycerides derived from refined palm oil	monoester content 90% min.; acid value 3 max.; iodine value 36-45	plastic; m.p. ca. 60°C	emulsifier; defoaming agent; lubricant *food industry*
	MYVEROL 18-50	distilled monoglycerides derived from partially hydrogenated soybean oil	monoester content 90% min.; acid value 3 max.; iodine value 50-60	plastic; m.p. ca. 54°C	
	MYVEROL 18-85	distilled monoglycerides derived from cottonseed oil	monoester content 90% min.; acid value 3 max.; iodine value 85-95	plastic; m.p. ca. 46°C	
	MYVEROL 18-92	distilled monoglycerides derived from sunflower oil	monoester content 90% min.; acid value 3 max.; iodine value 105-115	semi-plastic; m.p. ca. 41°C	
	MYVEROL 18-99	distilled monoglycerides derived from canola oil (low erucic acid rapeseed oil)	monoester content 90% min.; acid value 3 max.; iodine value 90-95	semi-plastic; m.p. ca. 35°C	
	MYVEROL P-06	distilled propylene glycol monoester derived from hydrogenated soybean oil	monoester content 90% min.; acid value 3 max.; iodine value 5 max.	beads; m.p. ca. 45°C	aerating emulsifier *food industry*
	MYVEROL SMG	succinylated monoglyceride derived from hydrogenated palm oil or palm stearine	monoester content 12-20%; succinylated monoglyceride content 55% min.; acid value 70-120; iodine value 3 max.	beads; m.p. ca. 58°C	emulsifier; dough strengthener; anti-staling agent *food industry; bakery products*
Ellis & Everard	CAFLON GMS NSE40	glycerol monstearate (non-self emulsifying)	100% active		
	CAFLON GMS SE40	glycerol monostearate (self emulsifying)	100% active		

Supplier	Trade name	Chemical description	Composition	General properties	Functionality Application
Fina Chemicals	RADIAMULS ACETEM 2021	acetylated mono-diglyceride	70% acetylation; acid value 10 max; iodine value 2 max.	paste; colour 15Y/3R (1 in. lovibond cell); m.p. ca. 40°C	emulsifier; film former; *food industry*
	RADIAMULS ACETEM 2050	acetylated mono-diglyceride	50% acetylation; acid value 10 max.; iodine value 2 max.	paste; colour 15Y/3R (1 in. lovibond cell); m.p. ca. 43°C	emulsifier; film former; whipping agent; *food industry*
	RADIAMULS ACETEM 2134	acetylated mono-diglyceride	100% acetylation; acid value 2 max.; iodine value 10 max.	liquid; cloud point ca. −15°C; colour 20Y/3R (1 in. lovibond cell)	emulsifier; film former; lubricant; *food industry*
	RADIAMULS CITREM 2931	citrylated mono-diglyceride (citroglyceride)	acid value 20 max.; iodine value < 1	powder or flakes; colour 5Y/1R (1 in. lovibond cell); m.p. ca. 61°C	anti-spattering agent; emulsifier; *food industry; margarines; meat products*
	RADIAMULS CSL 2980	stearoyl lactylate salt	acid value 60-80; iodine value 3 max.; calcium 2–5%	powder or flakes; colour 30Y/6R (1 in. lovibond cell); m.p. ca. 50°C	
	RADIAMULS DATEM 2001	diacetyl-tartryl mono-diglycerides	acid value 80-100; iodine value 2 max.	flakes; colour 30Y/5R (1 in. lovibond cell); m.p. ca. 55°C	stabiliser; *bakery goods*
	RADIAMULS DATEM 2001	diacetyl-tartryl mono-diglycerides	acid value 80-100; iodine value 2 max.	flakes; colour 30Y/5R (1 in. lovibond cell); m.p. ca. 55°C	
	RADIAMULS DATEM 2008	diacetyl-tartryl mono-diglycerides (15% TCP)	acid value 68-85; iodine value 2 max.	powder	
	RADIAMULS DATEM 2013	diacetyl-tartryl mono-diglycerides	acid value 64-80; iodine value 2 max.	powder	emulsifier; *food industry*
	RADIAMULS LACTEM 2950	lactylated mono-diglyceride	acid value 7 max.; iodine value 2 max.	flakes; colour 5Y/2R (1 in. lovibond cell); m.p. ca. 44°C	
	RADIAMULS MCT 2106	middle chain triglycerides (C$_{8/10}$ esters)	acid value 0.5 max.; iodine value 1 max.	liquid; viscosity 25 cPs; colour 60 (APHA)	film former; mould release aid; lubricant; viscosity modifier; *food industry*
	RADIAMULS MCT 2108	middle chain triglycerides (C$_{8/10}$ esters)	acid value 1 max.; iodine value 1 max.	liquid; viscosity 25 cPs; colour 5Y/1R (1 in. lovibond cell)	
	RADIAMULS MCT 2375	glycerol triheptanoate	C$_7$ fatty acid 95% min.; acid value 0.3 max.; iodine value 1 max.	cloud point − 25°C; colour 40Y/4R (1 in. lovibond cell)	tracer; *dairy products*
	RADIAMULS MCT 2376	glycerol triheptanoate	C$_7$ fatty acid 95% min.; acid value 0.3 max.; iodine value 1 max.	cloud point − 25°C; colour 40Y/4R (1 in. lovibond cell)	tracer; *dairy products*
	RADIAMULS MG 2141	saturated mono-diglycerides of fatty acids	acid value 3 max.; iodine value < 2; alpha-monoglycerides 35% min.	powder or flakes; colour 5Y/1R (1 in. lovibond cell); m.p. ca. 58°C	
	RADIAMULS MG 2142	saturated mono-diglycerides of fatty acids	acid value 3 max.; iodine value < 2; alpha-monoglycerides 40% min.	HLB 3; powder or flakes; colour 5Y/1R (1 in. lovibond cell); m.p. ca. 58°C	
	RADIAMULS MG 2143	saturated mono-diglycerides of fatty acids; guaranteed of vegetable origin	acid value 3 max.; iodine value < 2; alpha-monoglycerides 40% min.	HLB 3; powder or flakes; colour 5Y/1R (1 in. lovibond cell); m.p. ca. 58°C	
	RADIAMULS MG 2152	unsaturated mono- and diglycerides of fatty acids	acid value 3 max.; iodine value 75-85; alpha-monoglycerides 40% min.	HLB 3; liquid; colour 15Y/3R (1 in. lovibond cell); m.p. ca. 15°C	
	RADIAMULS MG 2600	saturated mono-diglycerides of fatty acids	acid value 3 max.; iodine value < 2; alpha-monoglycerides 52% min.	HLB 3; powder or flakes; colour 5Y/1R (1 in. lovibond cell); m.p. ca. 59°C	*w/o* emulsifier; *food industry*
	RADIAMULS MG 2606	saturated mono- and diglycerides of fatty acids	acid value 3 max.; iodine value 50-60; alpha-monoglycerides 52% min.	HLB 3; paste; colour 20Y/3.5R (1 in. lovibond cell); m.p. ca. 37°C	
	RADIAMULS MG 2610	unsaturated mono-diglycerides of fatty acids; guaranteed of vegetable origin	acid value 3 max.; iodine value 78-92; alpha-monoglycerides 40% min.	HLB 3; liquid/paste; colour 10Y/2.5R (1 in. lovibond cell); m.p. ca. 25°C	
	RADIAMULS MG 2643	saturated mono-diglycerides of fatty acids; guaranteed of vegetable origin	acid value 3 max.; iodine value < 2; alpha-monoglycerides 52% min.	HLB 3; powder or flakes; colour 5Y/1R (1 in. lovibond cell); m.p. ca. 59°C	
	RADIAMULS MG 2900	saturated mono-diglycerides of fatty acids	acid value 3 max.; alpha-monoglycerides 90% min.	HLB 3; powder; colour 5Y/1R (1 in. lovibond cell); m.p. ca. 67°C	

Supplier	Trade name	Chemical description	Composition	General properties	Functionality Application
Fina Chemicals	RADIAMULS MG 2901	saturated mono-diglyceride of fatty acids; guaranteed of vegetable origin	acid value 3 max.; iodine value <2; alpha-monoglycerides 90% min.	HLB 3; powder; colour 5Y/1R (1 in. lovibond cell); m.p. ca. 70°C	w/o emulsifier *food industry*
	RADIAMULS POLY 2248	polyglycerol monostearate	acid value 3 max.; iodine value 2 max.	flakes; colour 20Y/4R (1 in. lovibond cell); m.p. ca. 58°C	emulsifier *food industry; yeast industry*
	RADIAMULS POLY 2253	polyglycerol polyricinoleate	acid value 2 max.; iodine value 75-95	liquid; colour 8 max. (Gardner)	w/o emulsifier; fluidity improver; stabiliser; lubricant *food industry; chocolate*
	RADIAMULS SORB 2145	sorbitan monostearate	acid value 10 max.; iodine value 1 max.	HLB 5; powder or flakes; colour 6 max. (Gardner)	
	RADIAMULS SORB 2157	polysorbate 80	acid value 2 max.; iodine value 30 max.	HLB 15; liquid; colour 10 max. (Gardner)	
	RADIAMULS SORB 2161	sorbitan monostearate; of vegetable origin	acid value 10 max.; iodine value 3 max.	powder or flakes; colour 6 max. (Gardner)	emulsifier; dispersant; anti-staling agent; solubiliser; whipping agent; dryness improver *food industry*
	RADIAMULS SORB 2166	sorbitan monostearate; of vegetable origin	acid value 10 max.; iodine value 1 max.	HLB 5; powder or flakes; colour 6 max. (Gardner)	
	RADIAMULS SORB 2344	sorbitan tristearate	acid value 5 max.; iodine value 1 max.	HLB 2.5; powder or flakes; colour 6 max. (Gardner)	
	RADIAMULS SORB 2345	sorbitan tristearate	acid value 15 max.; iodine value 1 max.	HLB 2.5; powder or flakes; colour 6 max. (Gardner)	
	RADIAMULS SSL 2990	stearoyl lactylate salt	acid value 60-80; iodine value 3 max.; sodium 2-5%	powder or flakes; colour 30Y/6R (1 in. lovibond cell); m.p. ca. 50°C	stabiliser *bakery goods*
	RADIASTAR CA 1060	calcium stearate	total ash 9.5-11%; free fatty acids 1% max.	fine, white fluffy powder; m.p. 150-160°C	tableting aid; anti-caking agent; anti-tacking agent; defoaming agent; w/o emulsifier *food industry*
	RADIASTAR MG 1100	magnesium stearate	total ash 7.2-7.8%; free fatty acids 4% max.	fine, white fluffy powder; m.p. 130-150°C	
Gattefosse	ANTISTATIQUE WL 879	sorbitan caprylate		oily liquid	antistatic agent; conditioner *shampoos; cosmetics*
	BEHENATE DE GLYCEROL WL 251	glyceryl behenate		waxy solid	thickener *cosmetics*
	COMPRITOL 888 ATO	tribehenin		fine powder	thickener *cosmetics; sunscreen preparations*
	COMPRITOL 888 STO	glyceryl behenate	acid value <4; iodine value <3; sapon. no. 145-165	HLB 2; drop point ca. 70°C	tabletting agent; lipophilic matrix *pharmaceuticals*
	COMPRITOL HD5 ATO	amphiphilic glyceryl behenate	acid value <4; iodine value <3; sapon. no. 105-125	HLB 2; drop point ca. 62°C	tabletting agent; lipophilic matrix *pharmaceuticals*
	D.P.P.G.	propylene glycol dipelargonate	acid value <4; iodine value <3; sapon. no. 305-325	HLB 2; liquid	superfatting agent *pharmaceuticals; cosmetics*
	GELEOL	glyceryl palmitostearate	acid value <3; iodine value <3; sapon. no. 160-180	HLB 3; drop point 54.5-58.5°C	consistency enhancer *pharmaceuticals; cosmetics*
	GELOT 64	glyceryl palmitostearate SE	acid value <6; iodine value <3; sapon. no. 105-125	HLB 10; drop point 55.5-62.5°C	self-emulsifying base *pharmaceuticals; cosmetics*
	GELUCIRE 33/01	hemi-synthetic glycerides	acid value <2; iodine value ≤3; sapon. no. 240-260	HLB 1; drop point 33-38°C	excipient *pharmaceuticals; hard gelatin capsules*
	GELUCIRE 35/10	saturated polyglycolised glycerides	acid value <2; iodine value <2; sapon. no. 120-135	HLB 10; drop point 29-34°C	

Supplier	Trade name	Chemical description	Composition	General properties	Functionality / Application
Gattefosse	GELUCIRE 37/02	saturated polyglycolised glycerides	acid value <2; iodine value <2; sapon. no. 200-215	HLB 2; drop point 34.5-39.5°C	
	GELUCIRE 42/12	saturated polyglycolised glycerides	acid value <2; iodine value <2; sapon. no. 95-115	HLB 12; drop point 41.5-46.5°C	
	GELUCIRE 44/14	saturated polyglycolised glycerides	acid value <2; iodine value <2; sapon. no. 75-95	HLB 14; drop point 42.5-47.5°C	
	GELUCIRE 46/07	saturated polyglycolised glycerides	acid value <2; iodine value <2; sapon. no. 125-140	HLB 7; drop point 47-52°C	excipient *pharmaceuticals; hard gelatin capsules*
	GELUCIRE 48/09	saturated polyglycolised glycerides	acid value <2; iodine value <2; sapon. no. 105-125	HLB 9; drop point 46-51°C	
	GELUCIRE 50/02	saturated polyglycolised glycerides	acid value <2; iodine value <2; sapon. no. 180-195	HLB 2; drop point 46.5-51.5°C	
	GELUCIRE 50/13	saturated polyglycolised glycerides	acid value <2; iodine value <2; sapon. no. 65-80	HLB 13; drop point 46-51°C	
	GELUCIRE 53/10	saturated polyglycolised glycerides	acid value <2; iodine value <2; sapon. no. 95-115	HLB 10; drop point 47.5-52.5°C	
	GELUCIRE 62/05	polyglycolised natural wax	acid value <5; iodine value <10; sapon. no. 70-90	HLB 5; drop point 59-70°C	
	HYDRINE	diethylene glycol palmitostearate	acid value <6; iodine value <3; sapon. no. 150-160	HLB 4-5; drop point 45.5-48.5°C	consistency enhancer *pharmaceuticals; cosmetics*
	HYDROLACTOL 70	propylene glycol isostearostearate S.E.	acid value <3; iodine value <5; sapon. no. 125-145	HLB 9-10; drop point ca. 36-45°C	self-emulsifying base *pharmaceuticals; cosmetics*
	HYDROPHILOL ISO	propylene glycol isostearate	acid value <6; iodine value <15; sapon. no. 150-170	HLB 3-4; liquid	superfatting agent *pharmaceuticals; cosmetics*
	HYDROVEGETOL C	glyceryl stearate, propylene glycol stearate, glyceryl isostearate and propylene glycol isostearate		doughy solid	thickener *cosmetics*
	ISO ISOSTEARYLE WL 3196	isostearyl isostearate	acid value <6; iodine value <8; sapon. no. 90-110	HLB 1-2; liquid	superfatting agent *pharmaceuticals; cosmetics*
	LABRAFAC HYDRO	saturated polyglycolised glycerides	acid value <2; iodine value <5; sapon. no. 265-285	HLB 4-5; liquid	amphiphilic agent *pharmaceuticals*
	LABRAFAC LIPO	caprylo-capric glycerides		liquid	superfatting agent *pharmaceuticals; cosmetics*
	LABRAFIL ISO	polyglycolised isostearic glycerides	acid value <2; iodine value <15; sapon. no. 145-165	HLB 3-4; liquid	amphiphilic agent *pharmaceuticals; drug delivery*
	LABRAFIL M 1944CS	unsaturated polyglycolised glycerides	acid value ≤2; iodine value 70-95; sapon. no. 150-165	HLB 3-4; liquid (40°C); colour <5 (Gardner); pH 4.5-6.0 (10% in H$_2$O)	
	LABRAFIL M 1969CS	unsaturated polyglycolised glycerides	acid value ≤2; iodine value 70-90; sapon. no. 150-165	HLB 3-4; liquid	
	LABRAFIL M 1980CS	unsaturated polyglycolised glycerides	acid value ≤2; iodine value 60-80; sapon. no. 50-170	HLB 3-4; liquid	amphililic agent *pharmaceuticals*
	LABRAFIL M 2125CS	unsaturated polyglycolised glycerides	acid value ≤2; iodine value 90-115; sapon. no. 150-170	HLB 3-4; liquid	
	LABRAFIL M 2130CS	saturated polyglycolised glycerides	acid value <2; iodine value <2; sapon. no. 185-205	HLB 4-5; drop point 33-38°C	

Supplier	Trade name	Chemical description	Composition	General properties	Functionality / Application
Gattefosse	LABRAFIL M 2130BS	saturated polyglycolised glycerides	acid value <2; iodine value <2; sapon. no. 162-176	HLB 5-7; drop point 30.5-35.5°C	amphiphilic agent; *pharmaceuticals*
	LABRAFIL M 2735CS	unsaturated polyglycolised glycerides	acid value ≤2; iodine value 70-85; sapon. no. 150-170	HLB 3-4; liquid	
	LABRASOL	saturated polyglycolised glycerides	acid value <1; iodine value <2; sapon. no. 85-105	HLB 14; liquid	amphiphilic agent; solvent; emulsifier; *pharmaceuticals; cosmetics*
	LAFIL WL 3254	polyglyceryl isostearostearate	acid value <1; iodine value ≤6; sapon. no. 155-175	HLB 3; drop point 35.5-40.5°C	superfatting agent; *pharmaceuticals; cosmetics*
	LAS	saturated polyglycolised glycerides	acid value <1; iodine value <2; sapon. no. 85-105	HLB 14; liquid	solvent; emulsifier; *pharmaceuticals; cosmetics*
	LIPOCIRE A	glycerides $C_{12/18}$	acid value <0.5; iodine value <2; sapon. no. 225-245	drop point 35-36.5°C	base; *lipsticks*
	LIPOCIRE CM	glycerides $C_{12/18}$	acid value <0.5; iodine value <2; sapon. no. 225-245	drop point 38-40°C	
	LIPOCIRE DM	glycerides $C_{12/18}$	acid value <0.5; iodine value <2; sapon. no. 215-235	drop point 43-45°C	
	M.O.D.	2-octyl dodecyl myristate	acid value ≤7; iodine value ≤7; sapon. no. 90-110	HLB 1; liquid	superfatting agent; *pharmaceuticals; cosmetics*
	MONOSTEOL	propylene glycol palmitostearate	acid value <6; iodine value <3; sapon. no. 165-175	HLB 4; drop point 33-36°C	consistency enhancer; *pharmaceuticals; cosmetics*
	MONTHYLE	ethylene glycol palmitostearate	acid value <6; iodine value <3; sapon. no. 170-180	HLB 2-3; drop point 55-58°C	
	OLEPAL ISO	polyglycol (300) isostearate	acid value <6; iodine value <15; sapon. no. 95-125	HLB 12-13; liquid	solvent; emulsifier; *pharmaceuticals; cosmetics*
	OVUCIRE WL 2558	hemi-synthetic glycerides	acid value <0.5; iodine value ≤3; sapon. no. 240-260	drop point 31-36°C	excipient; *pharmaceuticals; suppositories; pessaries*
	OVUCIRE WL 2944	hemi-synthetic glycerides	acid value <0.5; iodine value <3; sapon. no. 215-235	drop point 32.5-35.5°C	
	PALMITATE DE CETYLE	cetyl palmitate		waxy solid	thickener; *cosmetics*
	PECEOL ISO	glyceryl isostearate	acid value <4; iodine value <15; sapon. no. 150-170	HLB 3; liquid	superfatting agent; *pharmaceuticals; cosmetics*
	PLUROL DIISOSTEARIQUE	polyglyceryl-3 disostearate		viscous liquid	emulsifier; *skin care products; baby care products; sunscreen preparations*
	PLUROL ISO	polyglycol isostearate	acid value <6; iodine value <10; sapon. no. 115-135	HLB 8-10; liquid	solvent; emulsifier; *pharmaceuticals; cosmetics*
	PLUROL OLEIQUE	polyglyceryl oleate	acid value <6; iodine value 50-70; sapon. no. 110-140	HLB 10; liquid	
	PLUROL STEARIQUE	polyglyceryl palmitostearate	acid value <6; iodine value <3; sapon. no. 120-140	HLB 8-10; drop point 48-53°C	
	POLYSTATE C	polyglycol (300) palmitostearate SE	acid value <6; iodine value <3; sapon. no. 90-110	HLB 9-10; drop point 26.5-30.5°C	self-emulsifying base; *pharmaceuticals; cosmetics*
	PRECIFAC ATO	cetyl palmitate	acid value <6; iodine value <3; sapon. no. 95-120	HLB 2; drop point ca. 48-52°C	tabletting agent; lipophilic matrix; *pharmaceuticals*

Supplier	Trade name	Chemical description	Composition	General properties	Functionality Application
Gattefosse	PRECIROL ATO 5	glyceryl palmitostearate	acid value < 6; iodine value < 3; sapon. no. 175-195	HLB 2; drop point 53-57°C	tabletting agent; lipophilic matrix *pharmaceuticals*
	PRECIROL WL 2155 ATO	glyceryl stearate	acid value < 6; iodine value < 3; sapon. no. 180-190	HLB 2; drop point 63.5-67.5°C	tabletting agent; lipophilic matrix *pharmaceuticals; cosmetics*
	SEDEFOS 75	glyceryl stearate SE	acid value < 6; iodine value < 3; sapon. no. 100-115	drop point 45-50°C	self-emulsifying base *pharmaceuticals; cosmetics*
	STEARATE 400 WL 817	polyglycol (400) palmitostearate	acid value < 6; iodine value < 3; sapon. no. 70-95	HLB 11-12; drop point 29-34°C	
	STEARATE 1500	polyglycol (1500) palmitostearate	acid value < 6; iodine value < 3; sapon. no. 30-45	HLB 16; drop point 46-50°C	solvent; emulsifier *pharmaceuticals; cosmetics*
	STEARATE 6000 WL 1644	polyglycol (6000) palmitostearate	acid value < 2; iodine value < 3; sapon. no. 15-25	HLB 18; drop point 55-60°C	
	SUCRO-ESTER 7	saccharose distearate	acid value < 5	HLB 7; solid	tabletting agent; lipophilic matrix *pharmaceuticals; cosmetics*
	SUCRO-ESTER 11	saccharose mono-distearate	acid value < 5	HLB 11; solid	
	SUCRO-ESTER 15	saccharose monopalmitate	acid value < 5	HLB 15; solid	tabletting agent; lipophilic matrix; solvent; emulsifier *pharmaceuticals; cosmetics*
	SUPERPOLYSTATE	polyglycol (300) stearate SE	acid value < 6; iodine value < 3; sapon. no. 90-110	HLB 9-10; drop point 33-37°C	self-emulsifying base *pharmaceuticals; cosmetics*
	SUPPOCIRE AI	hemi-synthetic glycerides	acid value < 0.5; iodine value < 2; sapon. no. 225-245	drop point 33-35°C	
	SUPPOCIRE AIM	hemi-synthetic glycerides	acid value < 0.2; iodine value < 2; sapon. no. 230-250	drop point 33-35°C	
	SUPPOCIRE AIML	hemi-synthetic glycerides	acid value < 0.5; iodine value < 3; sapon. no. 230-250	drop point 33-35°C	
	SUPPOCIRE AIP	saturated polyglycolised glycerides	acid value < 1; iodine value < 1; sapon. no. 205-225	drop point 30-33°C	
	SUPPOCIRE AIX	hemi-synthetic glycerides	acid value < 0.5; iodine value < 2; sapon. no. 220-240	drop point 33-35°C	
	SUPPOCIRE AM	hemi-synthetic glycerides	acid value < 0.2; iodine value < 2; sapon. no. 225-245	drop point 35-36.5°C	
	SUPPOCIRE AML	hemi-synthetic glycerides	acid value < 0.5; iodine value < 3; sapon. no. 225-245	drop point 35-36.5°C	excipient *pharmaceuticals; suppositories; pessaries*
	SUPPOCIRE AP	saturated polyglycolised glycerides	acid value < 1; iodine value < 1; sapon. no. 200-220	drop point 33-35°C	
	SUPPOCIRE AS2	hemi-synthetic glycerides	acid value < 0.5; iodine value < 2; sapon. no. 225-245	drop point 35-36.5°C	
	SUPPOCIRE AS2X	hemi-synthetic glycerides	acid value < 0.5; iodine value < 2; sapon. no. 225-245	drop point 35-36.5°C	
	SUPPOCIRE B	hemi-synthetic glycerides	acid value < 0.5; iodine value < 2; sapon. no. 225-245	drop point 36-37.5°C	
	SUPPOCIRE BM	hemi-synthetic glycerides	acid value < 0.2; iodine value < 2; sapon. no. 225-245	drop point 36.5-37.5°C	
	SUPPOCIRE BML	hemi-synthetic glycerides	acid value < 0.5; iodine value < 3; sapon. no. 225-245	drop point 36-37.5°C	

Supplier	Trade name	Chemical description	Composition	General properties	Functionality / *Application*
Gattefosse	SUPPOCIRE BP	saturated polyglycolised glycerides	sapon. no. 200-220	drop point 35-37°C	
	SUPPOCIRE BS2	hemi-synthetic glycerides	acid value <0.5; iodine value <2; sapon. no. 225-245	drop point 36-37.5°C	
	SUPPOCIRE BS2X	hemi-synthetic glycerides	acid value <0.5; iodine value <2; sapon. no. 220-240	drop point 36-37.5°C	
	SUPPOCIRE C	hemi-synthetic glycerides	acid value <0.5; iodine value <2; sapon. no. 220-240	drop point 38-40°C	
	SUPPOCIRE CM	hemi-synthetic glycerides	acid value <0.2; iodine value <2; sapon. no. 225-245	drop point 38-40°C	
	SUPPOCIRE CP	saturated polyglycolised glycerides	acid value <1; iodine value <1; sapon. no. 195-215	drop point 37-39°C	excipient / *pharmaceuticals; suppositories; pessaries*
	SUPPOCIRE CS2	hemi-synthetic glycerides	acid value <0.5; iodine value <2; sapon. no. 225-245	drop point 38-40°C	
	SUPPOCIRE CS2X	hemi-synthetic glycerides	acid value <0.5; iodine value <2; sapon. no. 220-240	drop point 38-40°C	
	SUPPOCIRE DM	hemi-synthetic glycerides	acid value <0.2; iodine value <2; sapon. no. 220-240	drop point 42-45°C	
	SUPPOCIRE NA	hemi-synthetic glycerides	acid value <0.5; iodine value <2; sapon. no. 225-245	drop point 34.5-36.5°C	
	SUPPOCIRE NAI	hemi-synthetic glycerides	acid value <0.5; iodine value <2; sapon. no. 225-245	drop point 33.5-35.5°C	
	SUPPOCIRE NB	hemi-synthetic glycerides	acid value <0.5; iodine value <2; sapon. no. 225-245	drop point 36.5-38.5°C	
	SUPPOCIRE NC	hemi-synthetic glycerides	acid value <0.5; iodine value <2; sapon. no. 220-240	drop point 38.5-40.5°C	
	TEFOSE 63	polyglycol palmitostearate SE	acid value <6; iodine value <3; sapon. no. 100-120	HLB 9-10; drop point 46-53°C	self-emulsifying base / *pharmaceuticals; cosmetics*
	TEFOSE 1500	polyglycol palmitostearate SE	acid value <6; iodine value <3; sapon. no. 75-95	HLB 10-12; drop point 42-46°C	
	TEFOSE 2000	polyglycol palmitostearate SE	acid value <6; iodine value <3; sapon. no. 65-85	HLB 10-12; drop point 33-38°C	
	TEFOSE 2561	polyglycol palmitostearate SE	acid value <5; iodine value <3; sapon. no. 65-105	HLB 10-12; drop point 37-43°C	
	TESAL	propylene glycol palmitostearate SE	acid value 90-110; iodine value <5; sapon. no. 160-180	drop point 37-43°C	
Th. Goldschmidt	ANTIL 141 LIQUID	propylene glycol and PEG-55 propylene glycol oleate	acid value 5 max.; sapon. no. 10-22	HLB 3; pale yellow liquid	thickener / *shampoos; baby cleansing preparations; hair conditioners; shower gels; bubble baths; liquid soaps; intimate hygiene products*
	TEGIN	glyceryl stearate SE	total monoester 32-40%; acid value 32-36; iodine value 4 max.; sapon. no. 145-160	HLB 12; white/ivory powder; rise m.p. 57-62°C	o/w emulsifier / *cosmetics; pharmaceuticals*
	TEGIN 90	glyceryl stearate	total monoester >90%; acid value 3 max.; iodine value 2 max.; sapon. no. 155-170	HLB 4.5; white/ivory powder; rise m.p. 67-72°C	stabiliser / *cosmetics; pharmaceuticals*
	TEGIN 4011	glyceryl stearate	total monoester ca. 40%; acid value 3 max.; iodine value 3 max.; sapon. no. 162-173	HLB 3.8; white/ivory powder; rise m.p. 54-60°C	w/o emulsifier / *pharmaceuticals*

Supplier	Trade name	Chemical description	Composition	General properties	Functionality / Application
Th. Goldschmidt	TEGIN 4100	glyceryl stearate	total monoester ca. 45%; acid value 2 max.; iodine value 3 max.; sapon. no. 164-180	HLB 3.8; white/ivory powder; rise m.p. 58-63°C	stabiliser / cosmetics; pharmaceuticals
	TEGIN 4433	glyceryl stearate SE	total monoester 30-40%; acid value 18 max.; iodine value 8 max.; sapon. no. 148-156	HLB 12; white/ivory powder; rise m.p. 54-57°C	w/o emulsifier / pharmaceuticals
	TEGIN C 62	hydrogenated tallow glyceride citrate	iodine value 3 max.; sapon. no. 215-265	HLB 10; ivory powder; rise m.p. 58-64°C	o/w emulsifier / cosmetics
	TEGIN D 1102	PEG-3 stearate	total diester ca. 90%; acid value 5-7; iodine value 3 max.; sapon. no. 168-178	HLB 3.8; white/ivory powder; rise m.p. 47-52°C	pearlescent agent / shampoos; shower preparations; bath preparations
	TEGIN D 6100	PEG-2 stearate	total monoester 50-60%; acid value 5 max.; iodine value 3 max.; sapon. no. 145-160	HLB 2.8; white/ivory wax; rise m.p. 40-46°C	o/w emulsifier / cosmetics
	TEGIN G	glycol stearate SE	total monoester ca. 50%; iodine value 36-38; iodine value 3 max.; sapon. no. 150-165	HLB 12; white/ivory wax; rise m.p. 48-53°C	o/w emulsifier / cosmetics; pharmaceuticals
	TEGIN G 1100	glycol distearate	total diester ca. 90%; acid value 6 max.; iodine value 3 max.; sapon. no. 192-208	white/ivory powder; rise m.p. 59-63°C	pearlescent agent / shampoos; shower preparations; bath preparations
	TEGIN G 6100	glycol stearate	total monoester 50-60%; acid value 3 max.; iodine value 3 max.; sapon. no. 180-195	HLB 3.2; white/ivory powder; rise m.p. 53-58°C	o/w co-emulsifier / shampoos
	TEGIN ISO	glyceryl stearate	total monoester ca. 50%; acid value 2 max.; iodine value 15 max.; sapon. no. 150-165	HLB 3.4; ivory paste	w/o emulsifier / cosmetics
	TEGIN M	glyceryl stearate	total monoester ca. 60%; acid value 2 max.; iodine value 4 max.; sapon. no. 165-180	HLB 3.8; white/ivory powder; rise m.p. 58-63°C	stabiliser / cosmetics; pharmaceuticals
	TEGIN O	glyceryl oleate	total monoester ca. 60%; acid value 2 max.; iodine value 70-76; sapon. no. 158-175	HLB 3.3; pale yellow paste	w/o emulsifier / cosmetics; pharmaceuticals
	TEGIN P	propylene glycol stearate SE	total monoester ca. 50%; acid value 16-18; iodine value 3 max.; sapon. no. 150-170	HLB 12; white/ivory wax; rise m.p. 39-46°C	o/w emulsifier / cosmetics; pharmaceuticals
	TEGIN SPEZIAL	glyceryl stearate SE	total monoester 40-48%; acid value 16-20; iodine value 3 max.; sapon. no. 148-158	HLB 12; white/ivory powder; rise m.p. 54-60°C	
Henkel	CEGESOFT C 17	myristyl lactate		solid; melting range 29-34°C	emollient; fatting agent; solubiliser / cosmetics; stick preparations; antiperspirants; bath preparations
	CEGESOFT C 24	octyl palmitate		liquid	emollient / cosmetics
	CETIOL	oleyl oleate	iodine value 97	clear, pale yellow oil	emollient / cosmetics
	CETIOL 165	isocetyl stearate	iodine value 5	clear low viscosity oil	emollient / pharmaceuticals
	CETIOL 868	octyl stearate	iodine value 2	pale yellow oil	emollient / skin care products; bath preparations
	CETIOL A	hexyl laurate	iodine value 1	clear liquid	emollient / cosmetics; pharmaceuticals
	CETIOL B	dibutyl adipate	iodine value 1	colourless, oily liquid	emollient / bath preparations; skin care products
	CETIOL J 600	oleyl erucate	iodine value 92	clear, pale yellow oil	emollient / skin care products; baby care products; bath preparations
	CETIOL LC	coco-caprylate/caprate	iodine value 1	clear liquid	emollient; jojoba oil substitute / skin care products; cosmetics; hair care products
					emollient / skin care products; sunscreen preparations

Supplier	Trade name	Chemical description	Composition	General properties	Functionality / Application
Henkel	CETIOL MM	myristyl myristate	iodine value 1	wax-like mass; melting range 34-40°C	*cosmetics*
	CETIOL SN	cetearyl isononanoate	iodine value 1	colourless, oily liquid	emollient; *skin care products; sunscreen preparations*
	CETIOL V	decyl oleate	iodine value 60	yellow, low viscosity oil	emollient; fatting agent; *creams; lotions*
	CUTINA AGS	glycol distearate	sapon. no. 194-200	flakes	pearlescent agent; opacifier; *shampoos; shower and bath preparations*
	CUTINA CBS	mixture of glyceryl stearate, cetearyl alcohol, cetyl palmitate and coco-glycerides	100% active	white wax-like substance in flake form; m.p. 52-58°C	cream base; viscosity modifier; *creams; emulsions*
	CUTINA CP	cetyl palmitate		waxy coarse flakes; m.p. 50°C	emollient; viscosity modifier; spermaceti substitute; *cosmetics; stick preparations*
	CUTINA CP-A	cetyl palmitate		flakes	emollient; spermaceti substitute; *cosmetics; stick preparations*
	CUTINA EGMS	glycol stearate		flakes	pearlescent agent; opacifier; *shampoos; shower and bath preparations*
	CUTINA GMRO	glyceryl ricinoleate		viscous to pasty	mucous membrane protection agent; *suppositories; ointments; w/o emulsions*
	CUTINA GMS	glyceryl stearate		beads; m.p. 54-60°C	consistency factor; *cosmetics; pharmaceuticals*
	CUTINA HR	hydrogenated castor oil		white-to slightly yellow fine free-flowing powder; m.p. 80-88°C	additive; consistency factor; lubricant; thickener; *tablets; stick preparations; oils; pharmaceuticals*
	CUTINA KD 16	glyceryl stearate SE		granules	cream base; *creams; ointments*
	CUTINA MD-A	glyceryl stearate		white waxy compound in granular form; m.p. 52-58°C	emollient; consistency factor; *cosmetics; pharmaceuticals*
	CUTINA MD	glyceryl stearate		wax-like mass	consistency factor; *o/w emulsions; cosmetics; pharmaceuticals*
	CUTINA TS	PEG-3 stearate		powder	pearlescent agent; opacifier; *shampoos; shower and bath preparations*
	DEHYMULS SML	sorbitan laurate		liquid	
	DEHYMULS SMO	sorbitan oleate		liquid	w/o emulsifier; *creams; emulsions*
	DEHYMULS SMS	sorbitan stearate		liquid	
	DEHYMULS SSO	sorbitan sesquioleate		liquid	
	DISPONIL GMS 90	fatty acid glyceride	ca. 100% active	HLB 5.1; beads	dispersant; *plastics industry; coatings: PVC*
	DISPONIL SML 100 SPEC.	sorbitan monolaurate	100% active	HLB 8.0; liquid	
	DISPONIL SMO 100 SPEC.	sorbitan monooleate	100% active	HLB 4.5; liquid	*plastics industry; coatings*
	DISPONIL SMP 100 SPEC.	sorbitan monopalmitate	100% active	HLB 6.7; flakes	
	DISPONIL SMS 100 SPEC.	sorbitan monostearate	100% active	HLB 4.7; flakes	

Supplier	Trade name	Chemical description	Composition	General properties	Functionality / Application
Henkel	DISPONIL STS 100 SPEC.	sorbitan tristearate	100% active	HLB 2.1; flakes	plastics industry; coatings
	EMEREST 2310	isopropyl isostearate	iodine value 1	light yellow clear liquid	emollient *personal care products*
	EMEREST 2314	isopropyl myristate	iodine value 1	water white clear liquid	emollient *personal care products*
	EMEREST 2316	isopropyl palmitate	iodine value 1	pale yellow clear liquid	emollient *personal care products*
	EMEREST 2325	butyl stearate	iodine value 1	water white clear liquid	emollient; solubiliser *hair care products; personal care products*
	EMEREST 2350	glycol stearate		white waxy solid	emulsifier; thickener; stabiliser; opacifier; pearlescent agent *personal care products*
	EMEREST 2355	glycol distearate		white waxy solid	emulsifier; thickener; stabiliser; opacifier; pearlescent agent *personal care products*
	EMEREST 2380	propylene glycol stearate		white waxy solid	emulsifier; opacifier; pearlescent agent *personal care products*
	EMEREST 2384	propylene glycol isostearate	iodine value 3	light yellow liquid	emollient; stabiliser; lubricant *personal care products; bath preparations; skin care products*
	EMEREST 2388	propylene glycol dipelargonate	iodine value 1	water white liquid	emollient *personal care products*
	EMEREST 2407	glyceryl stearate SE		white waxy beads	emulsifier *personal care products*
	EMEREST 2410	glyceryl isostearate	iodine value 3	white to yellow opaque viscous liquid	emollient; lubricant *personal care products*
	EMEREST 2452	polyglyceryl-3 diisostearate		amber viscous liquid	emulsifier; emollient *personal care products*
	EMEREST 2486	pentaerythrityl tetrapelargonate		clear light yellow liquid	emollient *personal care products*
	EMEREST 2704	PEG-4 dilaurate		clear yellow liquid	emulsifier; lubricant; dispersant *bath preparations*
	EMEREST 2711	PEG-8 stearate		white waxy solid	emulsifier; thickener *personal care products*
	EMEREST 2712	PEG-8 distearate		white waxy solid	emulsifier; thickener *personal care products; hair conditioners*
	EMEREST 2715	PEG-40 stearate		white waxy solid	emulsifier; stabiliser; antigelling agent; lubricant *shampoos; cosmetics; personal care products; deodorants*
	GLUCOPON 225 CS UP	fatty alcohol $C_{8/10}$ polyglycoside	66-69% active	liquid	solubiliser; hydrotrope; foaming agent *cleaners; dishwashing agents; laundry products*
	GLUCOPON 225 DK	fatty alcohol $C_{8/10}$ polyglycoside	68-72% active	liquid	solubiliser; hydrotrope; foaming agent *cleaners; dishwashing agents; laundry products*
	GLUCOPON 600 CS UP	fatty alcohol $C_{12/14}$ polyglycoside	50-53% active	liquid/solid	foaming agent; foam stabiliser *dishwashing agents; cleaners; laundry products; powder detergents; hand cleansing preparations*
	GLUCOPON 600 EC	fatty alcohol $C_{8/14}$ polyglycoside	50-53% active	liquid/solid	foaming agent; foam stabiliser *dishwashing agents; cleaners; laundry products; powder detergents; hand cleansing preparations*
	GLUCOPON 650 EC	fatty alcohol $C_{8/14}$ polyglycoside	50-53% active	liquid	foaming agent; foam stabiliser *dishwashing agents; cleaners; laundry products; powder detergents; hand cleansing preparations*
	ISOPROPYL MYRISTAT	isopropyl myristate		liquid	emollient *skin care products; cosmetics*
	ISOPROPYL PALMITAT	isopropyl palmitate		liquid	emollient *skin care products; cosmetics; bath preparations; aerosol products*

Supplier	Trade name	Chemical description	Composition	General properties	Functionality Application
Henkel	ISOPROPYL STEARAT	isopropyl stearate		liquid	emollient *skin care products; cosmetics; bath preparations; aerosol products*
	LAMECREME DGE 18	polyglyceryl-2 PEG-4 stearate		wax-like mass	o/w emulsifier *o/w emulsions; o/w ointments*
	LAMEFORM TGI	polyglyceryl-3 diisostearate		viscous liquid	w/o emulsifier; emollient *w/o creams; w/o ointments*
	MONOMULS 90-L 12	glyceryl laurate		powder	refatting agent; consistency factor; co-emulsifier; viscosity modifier *shampoos; bath preparations; creams; lotions*
	MONOMULS 90-O 18	glyceryl oleate		white paste	w/o emulsifier *w/o creams; w/o ointments*
	MYRITOL 318	caprylic/capric triglyceride	iodine value 0.5	clear low viscosity liquid	emollient *pharmaceuticals; cosmetics*
	NOVATA 299	cocoglycerides		wax-like substance; rise m.p. 33.5–35.5°C	
	NOVATA A	cocoglycerides		wax-like substance; rise m.p. 33.5–35.5°C	
	NOVATA AB	cocoglycerides		wax-like substance; rise m.p. 29–31°C	
	NOVATA B	cocoglycerides		wax-like substance; rise m.p. 33–35.5°C	
	NOVATA BBC	cocoglycerides		wax-like substance; rise m.p. 34–36°C	
	NOVATA BC	cocoglycerides		wax-like substance; rise m.p. 33–35.5°C	
	NOVATA BCF	cocoglycerides		wax-like substance; rise m.p. 35–37°C	base *suppositories*
	NOVATA BD	cocoglycerides		wax-like substance; rise m.p. 33–35.5°C	
	NOVATA C	cocoglycerides		wax-like substance; rise m.p. 36–38°C	
	NOVATA D	cocoglycerides		wax-like substance; rise m.p. 40–42°C	
	NOVATA E	cocoglycerides		wax-like substance; rise m.p. 34–36°C	
	NOVATA PK	cocoglycerides		wax-like substance; rise m.p. 34–36°C	
	NOVATA PKS 37	cocoglycerides		wax-like substance; rise m.p. 36–37.5°C	
	NOVATA PKS 39	cocoglycerides		wax-like substance; rise m.p. 38–40°C	
	RILANIT BS	n-butyl palmitate/stearate		liquid; pour point <22°C	super-fatting agent *hand cleaning preparations*
	RILANIT EGS	ethylene glycol dipalmitate/stearate		solid; pour point 60–65°C	pearlescent agent *dishwashing agent; cleaners*
	RILANIT EHK	2-ethylhexyl C$_{8/14}$ fatty acid ester		liquid	*cleaners; hand cleaning preparations*
	RILANIT EHO	2-ethylhexyl oleate		liquid	super-fatting agent *hand cleaning preparations*
	RILANIT EHS	2-ethylhexyl palmitate/stearate		liquid; pour point <10°C	
	RILANIT GL 401	polyglycol monolaurate		liquid; pour point <9°C	co-emulsifier *technical emulsions*
	RILANIT GMO	glycerol monooleate		liquid/solid; pour point <4°C	super-fatting agent *hand cleaning preparations; soaps; lanolin substitute*
	RILANIT GMRO	glycerol monoricinoleate		liquid/solid	
	RILANIT GMS	glycerol mono C$_{16/18}$ fatty acid ester		solid; pour point 50–60°C	super-fatting agent *hand cleaning preparations*
	RILANIT GTO	glycerol trioleate		liquid	

Supplier	Trade name	Chemical description	Composition	General properties	Functionality / Application
Henkel	RILANIT HE	polyol partial ester		liquid/solid; pour point <0°C	
	RILANIT HL	n-hexyl laurate		liquid	super-fatting agent / hand cleaning preparations
	RILANIT IBO	i-butyl oleate		liquid	
	RILANIT IBS	i-butyl palmitate/stearate		liquid; pour point <19°C	
	RILANIT OLO	oleyloleate		liquid	
	SELBANA 2001 N	fatty acid esters	100% active		lubricant / textile industry; spinning
Hoechst	DP2331	trimethylolpropane trioleate	100% active		lubricant / metal working
	GENAPOL BPMS	butandiol PEG-monostearate	90% active		conditioner / toiletries
	GENAPOL PMS	glycol distearate	100% active		pearlescent agent
	GENAPOL TS POWDER	PEG-3 distearate	100% active		silk lustre agent
	HOSTACERIN DGI	polyglyceryl-2-sesquiisostearate	100% active		
	HOSTACERIN DGL	polyglyceryl-2-PEG-10 laurate	100% active		emulsifier / cosmetics
	HOSTACERIN DGS	polyglyceryl-2-PEG-4 stearate	100% active		
	HOSTACERIN DGSB	polyglyceryl-2-PEG-4 stearate	100% active		
	HOSTACERIN LSE	sucrose laurate	100% active		
	HOSTACERIN T-3	ceteareth-3	100% active		
	HUK 039	PEG-400 monoleate	100% active		emulsifier / textile industry
Hüls	MARLOSOL BS	PEG-600-distearate	active detergent 100%	solid	refatting agent; thickener / shampoos; textile industry; fabric softeners
	MARLOSOL F08	blend of PEG-600-dioleate with olein polyglycol ester	active detergent 100%	liquid	
	MARLOSOL FS	PEG-600 dioleate	active detergent 100%	liquid	
	MARLOWET 4702	C$_{18}$-fatty acid polyethylene glycol ester	active detergent 100%	liquid; cloud point 26°C (10% in 25% BDG soln.)	emulsifier / mineral oils; spindle oils; metal working; textile industry
	MARLOWET 4703	C$_{18}$-fatty acid polyethylene glycol ester	active detergent 100%	liquid; cloud point 39°C (10% in 25% BDG soln.)	emulsifier / mineral oils; spindle oils; metal working; textile industry
	MARLOWET LVS	C$_{18}$-fatty acid ester of ethoxylated castor oil	active detergent 100%	liquid	emulsifier / vegetable oils; metal working; leather industry; release agents
	MARLOWET LVX	blend of C$_{18}$-fatty acid esters	active detergent 100%	liquid	emulsifier / vegetable oils; leather industry
	MARLOWET OTS	C$_{18}$-fatty acid polyethylene glycol ester	active detergent 100%	liquid; cloud point 78°C (10% in 25% BDG soln.)	emulsifier / mineral oils; spindle oils; metal working; textile industry
	MARLOWET SLS	C$_{18}$-fatty acid ester of ethoxylated castor oil	active detergent 98%	liquid; cloud point 65°C (10% in 25% BDG soln.)	emulsifier / mineral oils; white spirit; bitumen; textile industry; cleaners
ICI	ARLACEL 20	sorbitan monolaurate		HLB 8.6; yellow amber liquid	
	ARLACEL 40	sorbitan monopalmitate		HLB 6.7; cream solid or sprayed	
	ARLACEL 60	sorbitan monostearate		HLB 4.7; white cream solid or sprayed	emulsifier / personal care products
	ARLACEL 80	sorbitan monooleate		HLB 4.3; yellow amber liquid	
	ARLACEL 83	sorbitan sesquioleate		HLB 3.7; clear yellow liquid	

Supplier	Trade name	Chemical description	Composition	General properties	Functionality / Application
ICI	ARLACEL 85	sorbitan trioleate		HLB 1.8; yellow amber liquid	
	ARLACEL 129	glycerol monostearate		HLB 3.2; white powder	
	ARLACEL 165	glyceryl mono and distearate, and POE-stearate		HLB 11.0; white solid, flaked or sprayed	
	ARLACEL 186	glycerol mono and dioleate, and propylene glycol		HLB 2.8; pale yellow liquid	
	ARLACEL 481	glycerol sorbitan oleostearate		HLB 4.5; yellow amber wax	
	ARLACEL 581	POE glycerol sorbitan oleostearate		HLB 5.0; amber waxy solid	
	ARLACEL 582	POE glycerol sorbitan isostearate		HLB 5.0; amber waxy solid	emulsifier
	ARLACEL 780	POP-POE glycerol sorbitan hydroxyisostearate		HLB 4.7; amber suspension	*personal care products*
	ARLACEL 986	glycerol sorbitan isostearate		HLB 4.5; yellow amber wax	
	ARLACEL 987	sorbitan monoisostearate		HLB 4.3; yellow amber liquid	
	ARLACEL 988	POE glycerol sorbitan isostearate		HLB 4.7; yellow wax	
	ARLACEL 989	POE fatty acid ester, saturated		HLB 4.9; yellow liquid, gel	
	ARLACEL 1689	sorbitan ester and polyglycerol ester		HLB 3.5; amber liquid	
	ARLACEL A	mannide monooleate		HLB 4.3; amber liquid	
	ARLACEL C	sorbitan sesquioleate		HLB 3.7; amber liquid	
	ARLAMOL DOA	diester of adipic acid		colourless liquid	
	ARLATONE 2121	sorbitan ester and sucrose ester		HLB 6.0; tan flakes	o/w emulsifier; solubiliser *personal care products*
	ATLAS G-100	dimethylisosorbide		colourless liquid	
	ATLAS G-695	glycerol mono and dioleate		HLB 3.0; yellow-amber viscous suspension	
	ATLAS G-950	sorbide dioleate		HLB 2.7; orange-red liquid	
	ATLAS G-1325	propylene glycol monostearate		white flakes	
	ATLAS G-1327	propylene glycol monostearate		white flakes	
	ATLAS G-1556	mixture of nonionic esters		HLB 11.2; yellow solid	*household and industrial applications; agrochemicals; textile*
	ATLAS G-1564	mixture of nonionic esters		yellow solid or flake	*industry*
	ATLAS G-1989	sorbitan monostearate		HLB 4.7; cream waxy beads	
	ATLAS G-4821	sorbitan tristearate		HLB 3.4; tan waxy beads	
	ATLAS G-4884	technical sorbitan monoleate		HLB 4.3; amber liquid	
	ATLAS G-4885	technical sorbitan trioleate		HLB 1.8; amber liquid	
	ATLAS G-4909	glycerol sorbitan dioleohydroxystearate		amber waxy solid	
	ATLAS G-4968	mixture of nonionic esters		HLB 6.8; cream waxy solid	
	ATLOX 4885	sorbitan trioleate		HLB 1.8; amber liquid	emulsifier; dispersant *agrochemicals*
	ATMOS 150	glycerol mono and distearate		HLB 3.2; white powder	lubricant; emulsifier *food industry; metal working; polymer industry*
	ATPET 80	sorbitan monoleate		HLB 4.3; amber liquid	
	ATPET 100	sorbitan monoleate		HLB 4.3; amber liquid	*oil industry*
	ATPET 200	sorbitan monoleate		HLB 4.3; brown-red liquid	

Supplier	Trade name	Chemical description	Composition	General properties	Functionality / Application
ICI	ATSURF 594	glycerol monooleate		HLB 2.8; pale yellow suspension	emulsifier; paint industry
	CIRRASOL BASE N20	mixture of nonionic esters		cream solid	textile industry
	CIRRASOL LN-GS	synthetic alcohol fatty acid ester		colourless liquid	textile industry
	SPAN 20	sorbitan monolaurate		HLB 8.6; red-brown liquid	
	SPAN 40	sorbitan monopalmitate		HLB 6.7; tan solid or sprayed	
	SPAN 60	sorbitan monostearate		HLB 4.7; pale cream solid or sprayed	emulsifier
	SPAN 65	sorbitan tristearate		HLB 2.1; cream solid or sprayed	personal care products; textile industry; industrial applications
	SPAN 80	sorbitan monooleate		HLB 4.3; amber liquid	
	SPAN 85	sorbitan trioleate		HLB 1.8; amber liquid	
Inolex	LEXEMUL 515	glyceryl stearate			
	LEXEMUL 561	glyceryl stearate and PEG-100 stearate			
	LEXEMUL AR	glyceryl stearate and stearamidoethyl diethylamine			
	LEXEMUL AS	glyceryl stearate and sodium lauryl sulfate			
	LEXEMUL GDL	glyceryl dilaurate			
	LEXEMUL P	propylene glycol stearate SE			
	LEXEMUL T	glyceryl stearate SE			
ISP	CERAPHYL 28	cetyl lactate	acid value 2 max.; sapon. no. 174-189	HLB 13-15; white solid	
	CERAPHYL 31	lauryl lactate	acid value 2 max.; sapon. no. 210-225	HLB 10; light yellow liquid	
	CERAPHYL 41	C$_{12/15}$ alkyl lactate	acid value 2 max.; sapon. no. 195-210	white to straw-coloured liquid	
	CERAPHYL 45	dioctyl malate	acid value 5 max.; sapon. no. 310 min.	HLB 12; clear, colourless to pale yellow liquid	
	CERAPHYL 50	myristyl lactate	acid value 2 max.; sapon. no. 166-181	HLB 12; water-white to pale yellow liquid or soft solid (depending on temperature)	
	CERAPHYL 55	tridecyl neopentanoate	acid value 2 max.; sapon. no. 190 min.	HLB 8; clear, light yellow liquid; M.W. 284	
	CERAPHYL 140	decyl oleate	acid no. 5 max.; iodine no. 57-60; sapon. no. 132-142	white to straw liquid	
	CERAPHYL 140-A	isodecyl oleate	acid no. 5 max.; iodine no. 50-65; sapon. no. 130-145	white to straw coloured liquid	
	CERAPHYL 230	diisopropyl adipate	acid no. 2 max.; sapon. no. 465-500	clear, colourless liquid	emollient; skin care products; hair care products; cosmetics; sunscreen preparations
	CERAPHYL 368	octyl palmitate	acid no. 3 max.; sapon. no. 146-156	water white liquid	
	CERAPHYL 375	isostearyl neopentanoate	acid value 2 max.; sapon. no. 144-165	clear, pale yellow liquid; colour 2 max. (Gardner)	
	CERAPHYL 424	myristyl myristate	acid value 3 max.; sapon. no. 120-130	HLB 8-10; white to slightly yellow waxy solid; m.p. 36-39°C	
	CERAPHYL 494	isocetyl stearate	acid value 5 max.; sapon. no. 95-110	HLB 8; white to light yellow liquid	
	CERAPHYL 791	isocetyl stearoyl stearate	acid value 10 max.; OH no. 15 max.; sapon. no. 312-148	light to straw coloured liquid	
	CERAPHYL 847	octyldodecyl stearoyl stearate	acid no. 10 max.; sapon. no. 115-135	HLB 6; light to straw coloured liquid	
	CERAPHYL GA-D	maleated soybean oil	acid no. 43-53; iodine value 107 max.; sapon. no. 220-250	amber-yellow, viscous, oily liquid	

Supplier	Trade name	Chemical description	Composition	General properties	Functionality Application
ISP	CERAPHYL ICA	isocetyl alcohol	acid no. 5 max.; OH no. 195-230; iodine value 10 max.; sapon. no. 10.0 max.	HLB 12-14; clear, colourless liquid; M.W. 242	emollient *skin care products; hair care products; cosmetics; sunscreen preparations*
	CERASYNT 303	diethylaminoethyl stearate	acid no. 30-40; alkaline no. 127-137; sapon. no. 150-160	straw to amber-coloured liquid to semi-solid; pH 9.5-10.5 (3% soln.)	
	CERASYNT 840	PEG-20 stearate	acid no. 5 max.; iodine no. 0.25 max.; sapon. no. 40-50	m.p. 39.5-42.5°C	
	CERASYNT 945	glyceryl stearate and laureth-23	acid no. 5 max.; iodine no. 0.5 max.; sapon. no. 142-152	white to cream flakes; m.p. 53-55°C	
	CERASYNT D	stearamide MEA stearate	acid no. 10-20; iodine no. 0.5 max.; sapon. no. 97-107	cream flakes; m.p. 76-82°C	
	CERASYNT GMS	glyceryl stearate	acid value 3 max.; iodine no. 2 max.; sapon. no. 162-175	white to cream flakes; m.p. 56-59°C	
	CERASYNT IP	glycol stearate and other ingredients	acid no. 5 max.; sapon. no. 174-184	white to cream flakes; m.p. 56.5-58.5°C	emulsifier *skin care products; hair care products*
	CERASYNT LP	glycol stearate, sodium laureth sulfate and hexylene glycol		white opaque pourable liquid; viscosity < 5000 cPs; pH 6.5	
	CERASYNT M	glycol stearate	acid value 5 max.; iodine value 0.5 max.; sapon. no. 185-195	HLB 3; white to cream waxy flakes; m.p. 56-60°C	
	CERASYNT MN	glycol stearate SE	acid no. 5 max.; iodine no. 0.5 max.; sapon. no. 181-191	white to cream flakes; m.p. 57-60°C	
	CERASYNT PA	propylene glycol stearate	acid value 5 max.; iodine no. 0.5 max.; sapon. no. 181-191	HLB 3; white to cream-coloured flakes; m.p. 35-38°C	
	CERASYNT Q	glyceryl stearate SE	acid no. 10 max.; iodine no. 1.0 max.; sapon. no. 150-160	white to cream flakes; m.p. 57-59°C	
	CERASYNT SD	glyceryl stearate	acid no. 2 max.; iodine no. 0.5 max.; sapon. no. 165-177	white to cream flakes; m.p. 55-57.5°C	
	CERASYNT WM	glyceryl stearate, stearyl alcohol and sodium lauryl sulfate	acid no. 5 max.; iodine no. 0.5 max.; sapon. no. 140-150	white to cream flakes; m.p. 55-57°C	
	EMULSYNT 1055	polyglyceryl-4-oleate and PEG-8 propylene glycol cocoate	acid value 5 max.; iodine no. 58-68; sapon. no. 142-152	light amber liquid	
	EMULSYNT GDL	glyceryl dilaurate	acid value 5 max.; sapon. no. 219-229	HLB 10-12; white to off-white solid	emollient *skin care products; hair care products; cosmetics; sunscreen preparations*
Kao Corporation	EMANON HE	polyol fatty acid ester	OH value 170-190; sapon. no. 90-100	liquid	
	EXCEPARL HO	n-hexadecyl 2-ethyl hexanoate	OH value 3 max.; sapon. no. 135-160	liquid	
	KAOPAN SP-L-10	sorbitan monolaurate	OH value 330-360; sapon. no. 157-171	liquid	
	KAOPAN SP-O-10	sorbitan monooleate	OH value 193-209; sapon. no. 149-160	liquid	
	KAOPAN SP-O-30	sorbitan trioleate	OH value 56-68; sapon. no. 179-190	liquid	
	KAOPAN SP-S-10	sorbitan monostearate	OH value 235-260; sapon. no. 146-158	solid	
Dr. W. Kolb	HEDIPIN-DITA	based on tridecanol and adipic acid		liquid	
	HEDIPIN-EDO/6	based on PEG 600 and oleic acid		liquid	personal care products; household detergents; industrial and institutional cleaning; pharmaceuticals
	HEDIPIN-EDS/4	based on PEG 400 and stearic acid		paste	
	HEDIPIN-EDT/4	based on PEG 400 and tallow acid		liquid	
	HEDIPIN-GML	glycerin monolaurate		paste	

Supplier	Trade name	Chemical description	Composition	General properties	Functionality Application
Dr. W. Kolb	HEDIPIN-GMO	glycerin monooleate		paste	personal care products; household detergents; industrial and institutional cleaning; pharmaceuticals
	HEDIPIN-GMS	glycerin monostearate		solid/flakes	
	HEDIPIN-IOC	based on 2-ethylhexanol and caprylic acid		liquid	
	HEDIPIN-ITS	based on tridecanol and stearic acid		liquid	
	HEDIPIN-PEDS	based on pentaerytrit and stearic acid		solid	
	HEDIPIN-S6/T2	based on sorbitol EO and tall oil acid		liquid	
	KOSTERAN-I/1	sorbitan isostearate	acid value < 8; OH value 235; sapon. no. 145	HLB 4.7; yellow-brown liquid; viscosity 2000 mPas (50°C)	
	KOSTERAN-L/1	sorbitan laurate	acid value < 8; OH value 345; sapon. no. 165	HLB ca. 8.6; yellow-brown liquid; viscosity 400 mPas (50°C)	
	KOSTERAN-O/1	sorbitan oleate	acid value < 8; OH value 200; sapon. no. 150	HLB 4.3; yellow liquid; viscosity 1200 mPas (50°C)	
	KOSTERAN-O/3	sorbitan trioleate	acid value <15; OH value 60; sapon. no. 180	HLB 1.8; yellow-brown liquid; viscosity 60 mPas (50°C)	w/o emulsifier; o/w co-emulsifier cosmetics; personal care products; household detergents; industrial and institutional cleaning; pharmaceuticals
	KOSTERAN-P/1	sorbitan palmitate	acid value < 8; OH value 290; sapon. no. 145	HLB ca. 6.7; white-yellow solid; viscosity 200 mPas (50°C)	
	KOSTERAN-S/1	sorbitan stearate	acid value <10; OH value 245; sapon. no. 150	HLB 4.7; white-yellow solid; viscosity 400 mPas (50°C)	
	KOSTERAN-S/3	sorbitan tristearate	acid value <15; OH value 75; sapon. no. 182	HLB 2.1; white solid; viscosity 100 mPas (50°C)	
	KOSTERAN-SQ/O	sorbitan sesquioleate	acid value <10; OH value 200; sapon. no. 160	HLB 3.7; yellow-brown liquid; viscosity 200 mPas (50°C)	
	KOSTERAN-T/1	sorbitan monotallate		HLB ca. 4.5; liquid	personal care products; household detergents; industrial and institutional cleaning; pharmaceuticals
	SYMPATENS-GMIS	glyceryl isostearate	acid value <3; sapon. no. 160	HLB ca. 3.5; white-yellow liquid; viscosity 375 mPas	
	SYMPATENS-GML	glyceryl laurate	acid value <3; sapon. no. 200	HLB ca. 5; white-yellow paste; viscosity 115 mPas (40°C)	w/o emulsifier; refatting agent; viscosity modifier cosmetics
	SYMPATENS-GMO	glyceryl oleate	acid value <3; sapon. no. 175	HLB ca. 3.5; yellow liquid; viscosity 55 mPas	
	SYMPATENS-GMS	glyceryl stearate	acid value <3; sapon. no. 170	HLB ca. 3.5; white-yellow solid; viscosity 45 mPas (60°C)	
Lonza	ALDO HMS KFG	glyceryl monostearate; Kosher food grade	acid value 2; sapon. no. 170	HLB 3; beads; colour 1 (Gardner); m.p. 64°C	emulsifier; softener food industry
	ALDO MLD	glyceryl monolaurate	acid value 5; sapon. no. 182	HLB 7; soft solid; colour 3 (Gardner); m.p. 23°C	defoaming agent; emulsifier food industry; cosmetics; personal care products; textiles industry; plastics industry
	ALDO MO	glyceryl monooleate	acid value 2; sapon. no. 171	HLB 3; liquid to soft solid; colour 2 (Gardner); m.p. 21°C	
	ALDO MR	glyceryl monoricinoleate	acid value 4; sapon. no. 127	HLB 6; liquid; colour 6 (Gardner)	emulsifier
	ALDO M5 FG	glyceryl monostearate; food grade	acid value 3; sapon. no. 163	HLB 4; beads; colour 3 (Gardner); m.p. 59°C	emulsifier food industry
	ALDO MSA	glyceryl monostearate	acid value 1; sapon. no. 95	HLB 11; beads; colour 2 (Gardner); m.p. 57°C	emulsifier cosmetics
	ALDO MSC	glyceryl monostearate	acid value 6; sapon. no. 172	HLB 3; beads; colour 4 (Gardner); m.p. 58°C	
	ALDO MSD FG	glyceryl monostearate; food grade	acid value 4; sapon. no. 140	HLB 6; beads; colour 2 (Gardner); pH 58°C	emulsifier
	ALDO MSLG FG	glyceryl monostearate; food grade	acid value 2; sapon. no. 162	HLB 3; beads; colour 1 (Gardner); m.p. 60°C	emulsifier food industry

Supplier	Trade name	Chemical description	Composition	General properties	Functionality / Application
Lonza	ALDO PGHMS KFG	propylene glycol monostearate; Kosher food grade	acid value 2	HLB 3; beads; colour 2 (Gardner); m.p. 39°C	emulsifier; whipping agent / *food industry*
	GLYCOMUL L	sorbitan monolaurate	acid value 5	HLB 9; liquid; colour 7 (Gardner)	w/o emulsifier / *food industry; cosmetics; household and industrial applications*
	GLYCOMUL O	sorbitan monooleate	acid value 7	HLB 4; liquid; colour 6 (Gardner)	
	GLYCOMUL S FG	sorbitan monostearate; food grade	acid value 5	HLB 5; beads; colour 5 (Gardner); m.p. 53°C	
	GLYCOMUL S KFG	sorbitan monostearate; Kosher food grade	acid value 5	HLB 5; beads; colour 5 (Gardner); m.p. 53°C	
	GLYCOMUL SOC	sorbitan sesquioleate	acid value 10	HLB 4; liquid; colour 7 (Gardner)	
	GLYCOMUL TO	sorbitan trioleate	acid value 12	HLB 2; liquid; colour 9 (Gardner)	
	GLYCOMUL TS KFG	sorbitan tristearate; Kosher food grade	acid value 14	HLB 2; beads; colour 2 (Gardner); m.p. 55°C	
	PEGOSPERSE 50 DS	ethylene glycol distearate	acid value 3; sapon. no. 195	HLB 1; beads; colour 1 (Gardner); m.p. 59°C	emulsifier; dispersant; opacifier; defoaming agent; viscosity modifier / *cosmetics; household applications; textile industry; plastics industry; water treatment*
	PEGOSPERSE 50 MS	ethylene glycol monostearate	acid value 5; sapon. no. 184	HLB 2; beads; colour 2 (Gardner); m.p. 58°C	
	PEGOSPERSE PMS CG	propylene glycol monostearate	acid value 3; sapon. no. 180	HLB 3; flakes; colour 2 (Gardner); m.p. 51°C	
	POLYALDO 256 KFG	hexaglyceryl distearate; Kosher food grade	acid value 2; sapon. no. 136	HLB 6.3; beads; colour <1 (Gardner); m.p. 57°C	aerating agent; whipping agent / *food industry*
	POLYALDO 2010 KFG	decaglyceryl dioleate; Kosher food grade	acid value 2; sapon. no. 85	HLB 11.5; liquid; colour <1 (Gardner)	emulsifier / *food industry*
	POLYALDO 2P10 KFG	decaglyceryl dipalmitate; Kosher food grade	acid value 2; sapon. no. 89	HLB 12.0; waxy solid; colour 1 (Gardner); m.p. 48°C	emulsifier / *food industry*
	POLYALDO DGDO KFG	decaglyceryl decaoleate; Kosher food grade	acid value 5; sapon. no. 175	HLB 3.0; liquid; colour 4 (Gardner)	dispersant / *food industry*
	POLYALDO HGDS KFG	hexaglyceryl distearate; Kosher food grade	acid value 8; sapon. no. 130	HLB 70; beads; colour 6 (Gardner); m.p. 56°C	emulsifier / *food industry*
	POLYALDO TGMS KFG	triglyceryl monostearate; Kosher food grade	acid value 5; sapon. no. 128	HLB 7.0; beads; colour 5 (Gardner); m.p. 55°C	aerating agent; whipping agent / *food industry*
McIntyre Group	MACKESTER EGDS	glycol stearate	100% active	flake	emollient; pearlescent agent; emulsifier; lubricant; defoaming agent; antistatic agent / *cosmetics; metal working; textile industry; plastics industry; paper industry*
	MACKESTER EGMS	glycol stearate	100% active	flake	
	MACKESTER IP	glycol stearate with other ingredients	100% active	flake	
	MACKESTER SP	glycol stearate and stearamide MEA	100% active	flake	
Nikko Chemicals	NIKKOL BATYL ALCOHOL EX	batyl alcohol		white powder	emollient; emulsifier; thickener / *cosmetics*
	NIKKOL BATYL ALCOHOL 100	batyl alcohol		white powder	emollient; emulsifier; thickener / *cosmetics*
	NIKKOL BM	butyl myristate		colourless liquid	*cosmetics*
	NIKKOL BS	butyl stearate		colourless liquid	*cosmetics*
	NIKKOL CHIMYL ALCOHOL 100	cetyl glyceryl ether		white powder	emollient; emulsifier; thickener / *cosmetics*
	NIKKOL CIO	cetyl octanoate		colourless liquid	*cosmetics*
	NIKKOL CS	cholesteryl stearate		white crystalline powder	stabiliser; emollient / *cosmetics*

Supplier	Trade name	Chemical description	Composition	General properties	Functionality / *Application*
Nikko Chemicals	NIKKOL DECAGLYN 1-IS	decaglyceryl monoisostearate		pale yellow viscous liquid	
	NIKKOL DECAGLYN 1-L	polyglyceryl-10 laurate		pale yellow viscous liquid	
	NIKKOL DECAGLYN 1-LN	decaglyceryl monolinoleate		pale yellow viscous liquid	
	NIKKOL DECAGLYN 1-M	polyglyceryl-10 myristate		pale yellow viscous liquid	
	NIKKOL DECAGLYN 1-O	polyglyceryl-10 oleate		pale yellow viscous liquid	emulsifier; dispersant *cosmetics*
	NIKKOL DECAGLYN 1-S	polyglyceryl-10 stearate		pale yellow plate	
	NIKKOL DECAGLYN 2-IS	decaglyceryl diisostearate		pale yellow viscous liquid	
	NIKKOL DECAGLYN 2-O	decaglyceryl dioleate		pale yellow viscous liquid	
	NIKKOL DECAGLYN 2-S	decaglyceryl distearate		pale yellow plate	
	NIKKOL DECAGLYN 3-IS	decaglyceryl triisostearate		pale yellow viscous liquid	emulsifier *cosmetics*
	NIKKOL DECAGLYN 3-O	decaglyceryl trioleate		pale yellow viscous liquid	
	NIKKOL DECAGLYN 3-S	decaglyceryl tristearate		white plate	
	NIKKOL DECAGLYN 5-IS	decaglyceryl pentaisostearate		pale yellow viscous liquid	
	NIKKOL DECAGLYN 5-O	decaglyceryl pentaoleate		pale yellow viscous liquid	
	NIKKOL DECAGLYN 5-S	decaglyceryl pentastearate		white flakes	w/o emulsifier *cosmetics*
	NIKKOL DECAGLYN 7-IS	decaglyceryl heptaisostearate		pale yellow viscous liquid	
	NIKKOL DECAGLYN 7-O	decaglyceryl heptaoleate		pale yellow viscous liquid	
	NIKKOL DECAGLYN 7-S	decaglyceryl heptastearate		white flakes	anti-crystallising agent *cosmetics*
	NIKKOL DECAGLYN 10-IS	decaglyceryl decaisostearate		pale yellow viscous liquid	super-fatting agent *cosmetics*
	NIKKOL DECAGLYN 10-O	polyglyceryl-10 decaoleate		pale yellow viscous liquid	super-fatting agent *cosmetics*
	NIKKOL DECAGLYN 10-S	polyglyceryl-10 decastearate		white flakes	anti-crystallising agent *cosmetics*
	NIKKOL DES-SP	diethyl sebacate		colourless liquid	*cosmetics*
	NIKKOL DGDO	diglyceryl dioleate		yellow liquid	*cosmetics*

Supplier	Trade name	Chemical description	Composition	General properties	Functionality / *Application*
Nikko Chemicals	NIKKOL DGMIS	diglyceryl monoisostearate		pale yellow liquid	w/o emulsifier / *cosmetics*
	NIKKOL DGMO-90	polyglyceryl-2 oleate		pale yellow liquid	*cosmetics*
	NIKKOL DGMO-C	polyglyceryl-2 oleate		yellow liquid	w/o emulsifier / *cosmetics*
	NIKKOL DGMS	diglyceryl monostearate		pale yellow solid	
	NIKKOL DGO-80	glyceryl dioleate		yellow liquid	
	NIKKOL DGS-80	glyceryl distearate		white powder	*cosmetics*
	NIKKOL DID	diisopropyl adipate		colourless liquid	
	NIKKOL DIS	diisopropyl sebacate		colourless liquid	
	NIKKOL EGDS	glycol distearate		pale yellow solid	pearlescent agent / *cosmetics*
	NIKKOL EGMS-70	glycol stearate		pale yellow flakes	pearlescent agent / *shampoos*
	NIKKOL EOO	ethyl olive oleate		pale yellow liquid	*cosmetics*
	NIKKOL ESTEPEARL 10	glycol distearate		white flakes	pearlescent agent / *shampoos*
	NIKKOL ESTEPEARL 15	glycol distearate		white flakes	
	NIKKOL ESTEPEARL 30	PEG-3 distearate		white flakes	
	NIKKOL GM-18IS	batyl isostearate		white paste	stabiliser / *cosmetics*
	NIKKOL GM-18S	batyl stearate		white solid	
	NIKKOL HEXAGLYN 1-L	hexaglyceryl monolaurate		pale yellow viscous liquid	
	NIKKOL HEXAGLYN 1-M	hexaglyceryl monomyristate		pale yellow viscous liquid	
	NIKKOL HEXAGLYN 1-S	hexaglyceryl monostearate		white plate	emulsifier / *cosmetics*
	NIKKOL HEXAGLYN 1-O	hexaglyceryl monooleate		pale yellow viscous liquid	
	NIKKOL HEXAGLYN 3-S	hexaglyceryl tristearate		white flakes	
	NIKKOL HEXAGLYN 5-O	hexaglyceryl pentaoleate		pale yellow liquid	w/o emulsifier / *cosmetics*
	NIKKOL HEXAGLYN 5-S	hexaglyceryl pentastearate		white flakes	anti-crystallising agent / *cosmetics*
	NIKKOL HEXAGLYN PR-15	hexaglyceryl polyricinoleate		yellow viscous liquid	w/o emulsifier / *cosmetics*
	NIKKOL ICIS	hexyldecyl isostearate		colourless liquid	
	NIKKOL ICM-R	isocetyl myristate		colourless liquid	*cosmetics*
	NIKKOL ICS-R	isocetyl stearate		colourless liquid	

Supplier	Trade name	Chemical description	Composition	General properties	Functionality / Application
Nikko Chemicals	NIKKOL IPIS	isopropyl isostearate		pale yellow liquid	
	NIKKOL IPM-100	isopropyl myristate		colourless liquid	
	NIKKOL IPM-EX	isopropyl myristate		colourless liquid	
	NIKKOL IPP	isopropyl palmitate		colourless liquid	*cosmetics*
	NIKKOL IPP-EX	isopropyl palmitate		colourless liquid	
	NIKKOL ISP	isostearyl palmitate		colourless liquid	
	NIKKOL MGIS	glyceryl isostearate		pale yellow liquid	
	NIKKOL MGM	glyceryl myristate		white flakes	
	NIKKOL MGO	glyceryl oleate		pale yellow paste	
	NIKKOL MGS-150	glyceryl stearate SE		white solid	
	NIKKOL MGS-A	glyceryl stearate		white flakes	emulsifier *cosmetics*
	NIKKOL MGS-ASE	glyceryl stearate SE		white flakes	
	NIKKOL MGS-B	glyceryl stearate		white flakes	
	NIKKOL MGS-BSE-C	glyceryl stearate SE		white flakes	
	NIKKOL MGS-C	glyceryl stearate		white flakes	
	NIKKOL MGS-DEX	glyceryl stearate SE		white flakes	
	NIKKOL MGS-F20	glyceryl stearate		white flakes	
	NIKKOL MGS-F40	glyceryl stearate		white flakes	
	NIKKOL MGS-F50	glyceryl stearate		white flakes	additive *cosmetics; food industry*
	NIKKOL MGS-F50SE	glyceryl stearate SE		white flakes	
	NIKKOL MGS-F75	glyceryl stearate		white flakes	
	NIKKOL MGS-TG	glyceryl stearate		white flakes	
	NIKKOL MGS-TGL	glyceryl stearate		white flakes	
	NIKKOL MM	myristyl myristate		white crystalline solid	
	NIKKOL N-SP	cetyl palmitate		white flakes	
	NIKKOL NEO-DECANOATE 20	octyldodecyl neodecanoate		colourless liquid	*cosmetics*
	NIKKOL ODM-100	octyldodecyl myristate		colourless liquid	
	NIKKOL PDD	propylene glycol didecanoate		colourless liquid	
	NIKKOL PEARL 1218	glyceryl stearate		white flakes	pearlescent agent *shampoos*
	NIKKOL PEARL 1222	glycol distearate		white flakes	pearlescent agent *skin cleaning preparations*
	NIKKOL PENTARATE-408	pentaerythritol tetraoctanoate		colourless liquid	*cosmetics*
	NIKKOL PMEA	palmitamide MEA		pale yellow powder	pearlescent agent *hair care products*
	NIKKOL PMS-1C	propylene glycol stearate		white solid	
	NIKKOL PMS-1CSE	propylene glycol stearate SE		white solid	emulsifier; stabiliser *cosmetics*
	NIKKOL PMS-FR	propylene glycol stearate		white solid	

Supplier	Trade name	Chemical description	Composition	General properties	Functionality / Application
Nikko Chemicals	NIKKOL SEFSOL 218	propylene glycol monocaprylate		colourless liquid	solubiliser / *cosmetics*
	NIKKOL SEFSOL 228	propylene glycol dicaprylate		colourless liquid	solubiliser / *cosmetics*
	NIKKOL SI-10R	sorbitan isostearate		yellow liquid	*cosmetics*
	NIKKOL SI-15R	sorbitan sesquiisostearate		pale yellow viscous liquid	*cosmetics*
	NIKKOL SL-10	sorbitan laurate		pale yellow liquid	emulsifier / *cosmetics*
	NIKKOL SO-10	sorbitan oleate		yellow liquid	
	NIKKOL SO-10R	sorbitan oleate		pale yellow liquid	
	NIKKOL SO-15	sorbitan sesquioleate		yellow liquid	w/o emulsifier / *cosmetics*
	NIKKOL SO-15EX	sorbitan sesquioleate		pale yellow liquid	
	NIKKOL SO-15R	sorbitan sesquioleate		pale yellow liquid	
	NIKKOL SO-30	sorbitan trioleate		yellow liquid	
	NIKKOL SO-30R	sorbitan trioleate		pale yellow liquid	
	NIKKOL SP-10	sorbitan palmitate		pale yellow flakes	
	NIKKOL SS-10	sorbitan stearate		white to pale yellow flakes	
	NIKKOL SS-15	sorbitan sesquistearate		white to pale yellow flakes	
	NIKKOL SS-30	sorbitan tristearate		white to pale yellow flakes	
	NIKKOL TETRAGLYN 1-O	polyglyceryl-4 oleate		pale yellow viscous liquid	emulsifier / *cosmetics*
	NIKKOL TETRAGLYN 1-S	tetraglyceryl monostearate		white flakes	
	NIKKOL TETRAGLYN 3-S	tetraglyceryl tristearate		white flakes	
	NIKKOL TETRAGLYN 5-O	tetraglyceryl pentaoleate		pale yellow liquid	*cosmetics*
	NIKKOL TETRAGLYN 5-S	tetraglyceryl pentastearate		white flakes	anti-crystallising agent / *cosmetics*
	NIKKOL TMGO-5	POE (5) glyceryl monooleate		yellow liquid	
	NIKKOL TMGO-15	PEG-15 glyceryl oleate		yellow liquid	emulsifier; dispersant / *cosmetics*
	NIKKOL TMGS-5	PEG-5 glyceryl stearate		pale yellow semi-solid	
	NIKKOL TMGS-15	POE (15) glyceryl monostearate		white solid	
	NIKKOL TRIALAN-308	trimethylolpropane trioctanoate		colourless liquid	*cosmetics*
	NIKKOL TRIALAN-318	trimethylolpropane triisostearate		pale yellow liquid	
	NIKKOL TRIFAT S-308	glyceryl trioctanoate		pale yellow liquid	stabiliser / *cosmetics*
	NIKKOL VF-E	ethyl linoleate		pale yellow liquid	
	NIKKOL VF-IP	isopropyl linoleate		colourless liquid	
PPG	LAROSTAT GMO	glycerol monooleate			antistatic agent

Supplier	Trade name	Chemical description	Composition	General properties	Functionality Application
PPG	LAROSTAT GMS	glycerol monostearate			antistatic agent
	MAPEG 200 DL	PEG 4 dilaurate		HLB 7.6; liquid	spreading agent; emulsifier *bath preparations; hair conditioners; skin care products*
	MAPEG 200 DO	PEG 4 dioleate		HLB 6.0; liquid; viscosity 80 cPs	lubricant; w/o emulsifier; dispersant; emollient *metal working; bath preparations; skin care products*
	MAPEG 200 DOT	PEG 200 ditallate		HLB 6.0; liquid; viscosity 80 cPs	lubricant; emulsifier; dispersant *metal working*
	MAPEG 200 ML	PEG 4 laurate		HLB 9.3; liquid	emollient; emulsifier; dispersant *cosmetics; personal care products*
	MAPEG 200 MO	PEG 200 monooleate		HLB 8.3; liquid; viscosity 80 cPs	lubricant; emulsifier; dispersant *metal working*
	MAPEG 200 MOT	PEG 200 monotallate		HLB 8.3; liquid; viscosity 85 cPs	
	MAPEG 400 DL	PEG 8 dilaurate		HLB 10.8; liquid	emulsifier *hair care products; hair conditioners; skin cleansing preparations; bath preparations*
	MAPEG 400 DO	PEG 8 dioleate		HLB 8.8; liquid; viscosity 110 cPs	lubricant; emulsifier; dispersant; emollient; plasticiser *metal working; bubble baths; bath preparations; skin care products; hair spray preparations*
	MAPEG 400 DOT	PEG 400 ditallate		HLB 8.8; liquid; viscosity 110 cPs	lubricant; emulsifier; dispersant *metal working*
	MAPEG 400 DS	PEG 8 distearate		HLB 8.1; solid	lubricant; emulsifier; dispersant *metal working; hair conditioners; hair care products; skin care products; skin cleansing preparations*
	MAPEG 400 ML	PEG 8 laurate		HLB 13.2; liquid	emulsifier *hair care products; hair conditioners; skin cleansing preparations; bath preparations*
	MAPEG 400 MO	PEG 8 oleate		HLB 11.8; liquid; viscosity 115 cPs	lubricant; emulsifier; dispersant; spreading agent *metal working; bath preparations; skin cleansing preparations; skin care products*
	MAPEG 400 MS	PEG 8 stearate		HLB 11.5; solid	lubricant; emulsifier; dispersant *metal working; hair conditioners; hair care products; skin care products; skin cleansing preparations*
	MAPEG 600 DO	PEG 600 dioleate		HLB 10.3; liquid; viscosity 160 cPs	lubricant; emulsifier; dispersant *metal working*
	MAPEG 600 DOT	PEG 600 ditallate		HLB 10.3; liquid; viscosity 165 cPs	
	MAPEG 600 DS	PEG 12 distearate		HLB 10.6; solid	lubricant; emulsifier; dispersant *metal working; hair conditioners; skin cleansing preparations; cosmetics; shaving products*
	MAPEG 600 MO	PEG 600 monooleate		HLB 13.6; liquid; viscosity 170 cPs	lubricant; emulsifier; dispersant *metal working*
	MAPEG 600 MOT	PEG 600 monotallate		HLB 13.6; liquid; viscosity 175 cPs	
	MAPEG 600 MS	PEG 12 stearate		HLB 13.6; solid	lubricant; emulsifier; dispersant *metal working; hair conditioners; skin cleansing preparations; cosmetics; shaving products*
	MAPEG 6000 DS	PEG 150 distearate		HLB 18.4; flakes	thickener *baby shampoos; bubble baths; shower gels*
	MAPEG EGDS	glycol distearate		HLB 1.4; cream flakes	emulsifier; thickener; opacifier; pearlescent agent *skin care products; shampoos; bubble baths; hair conditioners*
	MAPEG EGMS	glycol stearate		HLB 2.9; cream flakes	

Supplier	Trade name	Chemical description	Composition	General properties	Functionality / *Application*
PPG	MAPEG S-40K	PEG 40 stearate		HLB 17.2; flakes	emulsifier / *skin care products; hair care products*
	MAZOL 80 MGK	ethoxylated monodiglycerides; Kosher food grade	iodine value 1.0; acid value 2.0 max.; sapon. no. 65-75	HLB 13.5; amber paste; m.p. 25-27°C	emulsifier / *food industry*
	MAZOL 165C	glyceryl stearate and PEG-100 stearate		HLB 11.2; tan flakes	emulsifier; thickener; opacifier / *personal care products*
	MAZOL 300 K	glycerol monooleate; Kosher food grade	iodine value 80; acid value 2.0 max.; alpha mono 40%; sapon. no. 145-155	HLB 3.8; amber liquid; m.p. 20°C	co-emulsifier; defoaming agent / *food industry*
	MAZOL GMO	glycerol monooleate	acid value 2 max.; monoglyceride content 40% min.; sapon. no. 150-170		lubricant; w/o emulsifier; coupling agent / *metal working*
	MAZOL GMO 1	glycerol monooleate	acid value 3 max.; sapon. no. 145-175		
	MAZOL GMO IND	glycerol monooleate; industrial version of Mazol GMO 1	acid value 5 max.		
	MAZOL GMOK	glycerol monooleate; available as Kosher food grade	iodine value 90; acid value 2 max.; alpha mono 40%; free glycerine 2%; sapon. no. 160-170	HLB 3.8; tan paste; m.p. 25°C	co-emulsifier; defoaming agent; emollient / *food industry; personal care products; cosmetics*
	MAZOL GMSD	glycerol monostearate	acid value 3.5 max.; sapon. no. 145-160		lubricant; w/o emulsifier; coupling agent / *metal working*
	MAZOL GMSDK	glyceryl stearate SE		HLB 6.0; tan flakes	viscosity modifier; opacifier / *personal care products*
	MAZOL GMSK	glycerol monostearate; available as Kosher food grade	iodine value 5.0; acid value 3.0 max.; alpha mono 40%; free glycerine 2%; sapon. no. 165-176	HLB 3.9; tan flake; m.p. 61°C	emulsifier; viscosity modifier; opacifier / *food industry; personal care products*
	MAZOL PETO MOD 1	pentaerythritol tetraoleate			additive / *metal working*
	MAZOL PGMSK	propylene glycol monostearate; available as Kosher food grade	iodine value 2.0; acid value 3.0 max.; alpha mono 70%; sapon. no. 170-190	HLB 3.4; tan solid; m.p. 39-46°C	emulsifier; defoaming agent; opacifier; pearlescent agent / *food industry; personal care products*
	MAZOL PGO 31K	triglycerol monooleate; available as Kosher food grade	iodine value 78; acid value 3.0 max.; sapon. no. 140-150	HLB 6.2; dark liquid	solubiliser / *food industry; essential oils; flavours*
	MAZOL PGO 104	decaglycerol tetraoleate; Kosher food grade	iodine value 61; acid value 8.0 max.; sapon. no. 125-145	HLB 6.2; dark liquid	solubiliser / *food industry; essential oils; flavours*
	MAZON EE-1	benzyl laurate		HLB 8.0; amber liquid	emollient / *sunscreen preparations; skin care products*
	S-MAZ 20	sorbitan laurate			emollient; dispersant; emulsifier / *cosmetics; skin care products*
	S-MAZ 60 K	sorbitan monostearate; available as Kosher food grade	OH value 235-260; acid value 10 max.; sapon. no. 147-157	HLB 4.7; tan flake; colour 3 (Gardner); m.p. 50-55°C	emulsifier; defoaming agent / *food industry*
	S-MAZ 60 KHS	sorbitan monostearate; available as Kosher food grade	OH value 235-260; acid value 10 max.; sapon. no. 147-157	HLB 4.7; flake; colour 3 (Gardner); m.p. 50-60°C	emulsifier / *food industry*
	S-MAZ 80	sorbitan oleate		HLB 4.6; amber liquid; viscosity 1000 cPs	lubricant; rust inhibitor; penetrant; binder; emulsifier / *metal working; cosmetics; skin care products; skin cleansing preparations*
	S-MAZ 85	sorbitan trioleate		HLB 2.1; amber liquid; viscosity 200 cPs	
	S-MAZ 90	sorbitan monotallate		HLB 4.4; liquid; viscosity 1000 cPs	lubricant; rust inhibitor; penetrant / *metal working*
Protex	PROTE-SORB SML	sorbitan monolaurate	sapon. no. 162	HLB 8.6; liquid	emulsifier
	PROTE-SORB SMO	sorbitan monooleate	sapon. no. 153	HLB 4.3; liquid	*metal working; food industry; agriculture; cosmetics; household and industrial applications; textile industry; leather industry*
	PROTE-SORB SMP	sorbitan monopalmitate	sapon. no. 155	HLB 6.7; liquid	

Supplier	Trade name	Chemical description	Composition	General properties	Functionality Application
Protex	PROTE-SORB SMS	sorbitan monostearate	sapon. no. 152	HLB 4.7; liquid	emulsifier / metal working; food industry; agriculture; cosmetics; household and industrial applications; textile industry; leather industry
	PROTE-SORB STO	sorbitan trioleate	sapon. no. 180	HLB 1.8; liquid	
	PROTE-SORB STS	sorbitan tristearate	sapon. no. 182	HLB 2.1; solid	
	SURFASORB SERIES	sorbitan esters and ethoxylates			emulsifier / waxes; mineral oils; solvents
Quest International	ADMUL 2232	distilled monoglyceride	total monoglyceride 90%; free glycerol 1% max.; moisture 2% max.; sodium stearate 0.1% max.; acid value 3 max.; iodine value ± 10 max.	fine powder	
	ADMUL 2233	distilled monoglyceride ex vegetable oil	total monoglyceride 90%; free glycerol 1% max.; moisture 2% max.; sodium stearate 0.1% max.; acid value 3 max.; iodine value 3 max.	fine powder	emulsifier; anti-staling agent / food industry: margarine; bread
	ADMUL AC 2722 K	acetic acid ester of mono-diglyceride	50% acetylation; moisture 2% max.; acid value 3 max.; iodine value 5 max.; sapon. no. 280-295	block: m.p. 41-46°C	
	ADMUL AC 2723 K	acetic acid ester of mono-diglyceride	96% acetylation; moisture 2% max.; acid value 3 max.; iodine value ± 48 max.; sapon. no. 370-385	liquid; m.p. 4-12°C	emulsifier / food industry: edible coatings; whippable fat powders
	ADMUL AC 2724 K	acetic acid ester of mono-diglyceride	70% acetylation; moisture 2% max.; acid value 3 max.; iodine value 5 max.; sapon. no. 310-330	block: m.p. 37-40°C	
	ADMUL CSL 2010	calcium stearoyl lactylate	lactic acid 23-27%; calcium 3.0-5.0%; ester value 115-155; acid value 50-70	fine powder	emulsifier; volume improver; stabiliser / food industry: bread
	ADMUL DATEM 1075	diacetyl tartaric acid ester of mono-diglycerides	acid value 90-115; iodine value 2 max.; sapon. no. 490-530	flake	
	ADMUL DATEM 1117	diacetyl tartaric acid ester of mono-diglycerides	acid value 65-85; iodine value 2 max.; sapon. no. 405-445	microbead	
	ADMUL DATEM 1901	diacetyl tartaric acid ester of mono-diglycerides	acid value 70-95; iodine value ± 40 max.; sapon. no. 400-440	paste	
	ADMUL DATEM 1913	diacetyl tartaric acid ester of mono-diglycerides	acid value 60-85; iodine value ± 55 max.; sapon. no. 410-450	paste	
	ADMUL DATEM 1935	diacetyl tartaric acid ester of mono-diglycerides	acid value 70-95; iodine value 2 max.; sapon. no. 420-460	microbead	emulsifier; volume improver / food industry: bread
	ADMUL DATEM 1938	diacetyl tartaric acid ester of mono-diglycerides	tricalcium phosphate E341 20%; acid value 65-90; iodine value 2 max.; sapon. no. 415-455	powder	
	ADMUL DATEM 1951	diacetyl tartaric acid ester of mono-diglycerides	tricalcium phosphate E341 20%; acid value 80-105; iodine value 2 max.; sapon. no. 450-490	powder	
	ADMUL DATEM 1953	diacetyl tartaric acid ester of mono-diglycerides	tricalcium phosphate E341 20%; acid value 90-115; iodine value 2 max.; sapon. no. 490-530	powder	
	ADMUL EMULSPONGE 2213		emulsifier 35%; milk powder 65%; moisture 3% max.	powder	emulsifier / food industry: cake mixes
	ADMUL EMULSPONGE 5306		emulsifier 35%; starch 65%; moisture 3% max.	powder	
	ADMUL GLP 2033	lactic acid ester of mono-diglyceride	lactic acid 12-16.5%; acid value 4 max.; sapon. no. 245-275	flake; m.p. 47°C	emulsifier / food industry: whippable fat powders; synthetic cream
	ADMUL LEC 2251	lecithin product	lecithin E322 45%; mono-diglyceride E471 5%; modified milk solids 50%; moisture 3% max.	powder	emulsifier
	ADMUL LEC 2879	lecithin product	lecithin E322 100%; phospholipids 95% min.	powder	emulsifier / food industry: bakery products; milk replacers

Supplier	Trade name	Chemical description	Composition	General properties	Functionality / Application
Quest International	ADMUL MG 1524 K	unsaturated mono-diglyceride	total monoglyceride 55%; free glycerol 3% max.; moisture 1% max.; sodium stearate 0.3% max.; acid value 3 max.; iodine value ± 71 max.	paste	
	ADMUL MG 4123	self-emulsifying mono-diglyceride	total monoglyceride 36%; free glycerol 2.5% max.; moisture 2% max.; sodium stearate ± 2% max.; acid value 3 max.; iodine value 3 max.	bead	
	ADMUL MG 4143	self-emulsifying mono-diglyceride	total monoglyceride 36%; free glycerol 7% max.; moisture 2% max.; sodium stearate ± 4% max.; acid value 3 max.; iodine value 3 max.	bead	
	ADMUL MG 4163	self-emulsifying mono-diglyceride	total monoglyceride 36%; free glycerol 7% max.; moisture 2% max.; sodium stearate ± 6% max.; acid value 3 max.; iodine value 3 max.	bead	
	ADMUL MG 4203	saturated mono-diglyceride ex palm oil	total monoglyceride 40%; free glycerol 1.5% max.; moisture 2% max.; sodium stearate 0.3% max.; acid value 3 max.; iodine value 3 max.	bead	
	ADMUL MG 4223	self-emusifying mono-diglyceride ex palm oil	total monoglyceride 36%; free glycerol 2.5% max.; moisture 2% max.; sodium stearate ± 2% max.; acid value 3 max.; iodine value 3 max.	bead	emulsifier *food industry; margarine; ice cream*
	ADMUL MG 4304	unsaturated mono-diglyceride ex marine oil	total monoglyceride 40%; free glycerol 1.5% max.; moisture 2% max.; sodium stearate 0.3% max.; acid value 3 max.; iodine value ± 70 max.	paste	
	ADMUL MG 4904	unsaturated mono-diglyceride ex soyabean oil	total monoglyceride 40%; free glycerol 1.5% max.; moisture 2% max.; sodium stearate 0.3% max.; acid value 3 max.; iodine value ± 85 max.	paste	
	ADMUL MG 6203	saturated mono-diglyceride ex palm oil	total monoglyceride 60%; free glycerol 1.5% max.; moisture 2% max.; sodium stearate 0.3% max.; acid value 3 max.; iodine value 3 max.	bead	
	ADMUL MG 6404	unsaturated mono-diglyceride ex soyabean oil	total monoglyceride 60%; free glycerol 1.5% max.; moisture 2% max.; sodium stearate 0.3% max.; acid value 3 max.; iodine value ± 80 max.	paste	
	ADMUL MG 6504	unsaturated mono-diglyceride ex lard	total monoglyceride 60%; free glycerol 1.5% max.; moisture 2% max.; sodium stearate 0.3% max.; acid value 3 max.; iodine value ± 55 max.	paste	
	ADMUL PGE 1405	polyglycerol ester (vegetable)	moisture 2% max.; sodium stearate 0.5% max.; acid value 3 max.; iodine value 3 max.; sapon. no. 135-175	bead	emulsifier *food industry; margarine; synthetic cream*
	ADMUL PGE 1410	polyglycerol ester (vegetable)	moisture 2% max.; sodium stearate 1% max.; acid value 3 max.; iodine value ± 60 max.; sapon. no. 105-145	paste	emulsifier *food industry; margarine; synthetic cream*
	ADMUL PGMS 2103	propylene glycol ester	monoester 90% min.; free propylene glycol 1% max.; acid value 2 max.; iodine value 2 max.	block; m.p. 36-40°C	emulsifier *food industry*
	ADMUL SSL 1078	sodium stearoyl lactylate	lactic acid 31-35%; sodium 3.5-5.0%; ester value 150-190; acid value 60-80	powder	emulsifier; volume improver; stabiliser *food industry; bread*
	ADMUL SSL 2012	sodium stearoyl lactylate ex vegetable fatty acid	lactic acid 31-35%; sodium 3.5-5.0%; ester value 150-190; acid value 60-80	powder	emulsifier; volume improver; stabiliser *food industry; bread*
	ADMUL WFP 2413	whippable fat powder	lipid content 60% min.; carbohydrate 30% min.; protein 8.5%; moisture 4% max.	powder	emulsifier *food industry; desserts; ice-cream*

Supplier	Trade name	Chemical description	Composition	General properties	Functionality / Application
Quest International	ADMUL WFP 2415	whippable fat powder	lipid content 55% min.; carbohydrate 30% min.; protein 8.5%; moisture 4% max.	powder	emulsifier *food industry; desserts; ice-cream*
	ADMUL WFP 5100	whippable fat powder	lipid content 50% min.; carbohydrate 40% min.; protein 8.5%; moisture 4% max.	powder	
	ADMUL WOL 1403	polyglycerol polyricinoleate	moisture 2% max.; OH value 80-100; acid value 6 max.; iodine value 72-103; sapon. no. 170-210	liquid	emulsifier; viscosity modifier *food industry; chocolate*
	HYMONO 1103	saturated distilled monoglyceride ex lard or tallow	total monoglyceride 90%; free glycerol 1% max.; moisture 2% max.; sodium stearate 0.1% max.; acid value 3 max.; iodine value 3 max.	bead	
	HYMONO 1163	self-emulsifying distilled monoglyceride ex lard or tallow	total monoglyceride 70%; free glycerol 1% max.; moisture 2% max.; sodium stearate ± 6% max.; acid value 3 max.; iodine value 3 max.	bead	
	HYMONO 3203	saturated distilled monoglyceride ex lard	total monoglyceride 90%; free glycerol 1% max.; moisture 2% max.; sodium stearate 0.1% max.; acid value 3 max.; iodine value 3 max.	bead	
	HYMONO 4404	unsaturated distilled monoglyceride ex lard	total monoglyceride 90%; free glycerol 1% max.; moisture 2% max.; sodium stearate 0.1% max.; acid value 3 max.; iodine value ± 40 max.	paste	emulsifier; anti-staling agent *food industry; margarine; bread*
	HYMONO 7804	unsaturated distilled monoglyceride ex vegetable oil	total monoglyceride 90%; free glycerol 1% max.; moisture 2% max.; sodium stearate 0.1% max.; acid value 3 max.; iodine value ± 80 max.	paste	
	HYMONO 8803	saturated distilled monoglyceride ex soyabean oil	total monoglyceride 90%; free glycerol 1% max.; moisture 2% max.; sodium stearate 0.1% max.; acid value 3 max.; iodine value 3 max.	bead	
	HYMONO 8903	saturated distilled monoglyceride ex palm oil	total monoglyceride 90%; free glycerol 1% max.; moisture 2% max.; sodium stearate 0.1% max.; acid value 3 max.; iodine value 3 max.	bead	
Rhone-Poulenc	ALKAMULS S/20	sorbitan laurate	100% solids	liquid	suspending agent *emulsion polymerisation*
Sandoz	CERANINE NC LIQUID	fatty acid ester derivative in aq. dispersion		liquid	softener *cellulosic fibres*
	CERANINE NC W	fatty acid ester derivative in aq. dispersion		liquid	
	VELSAN P8-3	carboxylic acid ester derivative		liquid	emollient *cosmetics; toiletries*
Scher Chemicals	DIPSAL	dipropylene glycol salicylate	acid value 3 max.; sapon. no. 225-240	clear yellow liquid	
	ETHYLENE GLYCOL SALICYLATE	glycol salicylate	acid value 3 max.; sapon. no. 300-330	light amber crystalline solid	emollient *skin care products*
	SCHERCEMOL 65	isohexyl neopentanoate	acid value 2 max.; sapon. no. 290-305	clear liquid	emollient
	SCHERCEMOL 145	myristyl neopentanoate	acid value 2 max.; sapon. no. 180-200	cler liquid; m.p. 4°C	emollient *skin care products; hair care products*
	SCHERCEMOL 185	isostearyl neopentanoate	acid value 2 max.; iodine value 12 max.; sapon. no. 135-155	clear straw-coloured liquid	emollient; binder *bath preparations; skin care products; cosmetics*
	SCHERCEMOL 318	isopropyl isostearate	acid value 1 max.; iodine value 3.0 max.; sapon. no. 160-180	clear yellow liquid; m.p. − 28°C	emollient
	SCHERCEMOL 1688	cetearyl octanoate	acid value 1 max.; sapon. no. 135-150	clear colourless liquid; m.p. 3°C	emollient *bath preparations; skin care products*

Supplier	Trade name	Chemical description	Composition	General properties	Functionality / Application
Scher Chemicals	SCHERCEMOL 1818	isostearyl isostearate	acid value 2 max.; iodine value 13 max.; sapon. no. 95-110	clear yellow liquid; m.p. −5°C	emollient; solubiliser; solvent; *skin care products; perfumes*
	SCHERCEMOL BE	behenyl erucate	acid value 2 max.; iodine value 55 max.; sapon. no. 80-95	cream-coloured soft solid; m.p. 45°C	emollient
	SCHERCEMOL CO	cetyl octanoate	acid value 3 max.; sapon. no. 140-155	clear liquid; m.p. 10°C	solvent; *skin cleansing preparations*
	SCHERCEMOL DIA	diisopropyl adipate	acid value 2 max.; sapon. no. 480-500	clear colourless liquid; m.p. −1°C	emollient; solvent; *skin care products*
	SCHERCEMOL DID	diisopropyl dilinoleate	acid value 3 max.; iodine value 15 max.; sapon. no. 165-185	yellow liquid; m.p. −9°C	emollient; dispersant; *skin care products; cosmetics*
	SCHERCEMOL DIS	diisopropyl sebacate	acid value 1 max.; sapon. no. 380-400	clear colourless liquid; m.p. 0°C	emollient; solubiliser; coupling agent; *skin care products; bath preparations*
	SCHERCEMOL DISD	diisostearyl dilinoleate	acid value 5 max.; iodine value 20 max.; sapon. no. 90-110	yellow liquid; m.p. −3°C	emollient; binder; moisturiser; *skin care products; cosmetics*
	SCHERCEMOL DISF	diisostearyl fumarate	acid value 2 max.; sapon. no. 160-175	colourless liquid; m.p. −5°C	lubricant; conditioner
	SCHERCEMOL DO	decyl oleate	acid value 3 max.; iodine value 65 max.; sapon. no. 130-140	clear yellow liquid; m.p. −10°C	emollient; lubricant; penetrant; dispersant; *cosmetics; skin cleansing preparations*
	SCHERCEMOL EGMS	glycol stearate	acid value 5 max.; iodine value 1 max.; sapon. no. 170-190	white to cream-coloured flakes; m.p. 58°C	emulsifier; opacifier; pearlescent agent; *skin care products; hair care products*
	SCHERCEMOL GMIS	glyceryl isostearate	acid value 5 max.; iodine value 10 max.; sapon. no. 160-180	clear straw-coloured liquid to soft solid; m.p. 5°C	emulsifier; emollient; *skin care products*
	SCHERCEMOL ICS	isocetyl stearate	acid value 2 max.; sapon. no. 105-120	clear straw-coloured liquid; m.p. −5°C	emollient; *cosmetics; bath preparations; skin care products*
	SCHERCEMOL IDO	isodecyl oleate	acid value 5 max.; iodine value 65 max.; sapon. no. 130-140	clear liquid; m.p. 10°C	emollient; lubricant; penetrant; dispersant; *cosmetics; skin cleansing preparations*
	SCHERCEMOL MEL-3	myreth-3 laurate	acid value 3 max.; sapon. no. 100-120	clear straw-coloured liquid; m.p. 15°C	emollient; solubiliser; coupling agent; *bath preparations; skin care products*
	SCHERCEMOL MEM-3	myreth-3 myristate	acid value 3 max.; sapon. no. 95-115	clear straw-coloured liquid; m.p. 23°C	
	SCHERCEMOL MEP-3	myreth-3 palmitate	acid value 3 max.; sapon. no. 85-100	soft cream-coloured wax; m.p. 29°C	
	SCHERCEMOL MM	myristyl myristate	acid value 2 max.; sapon. no. 120-135	white waxy solid; m.p. 40°C	emollient; viscosity modifier; *skin care products*
	SCHERCEMOL MP	myristyl propionate	acid value 2 max.; sapon. no. 190-210	clear straw-coloured liquid	emollient; *antiperspirants; skin care products*
	SCHERCEMOL NGDC	neopentyl glycol dicaprate	acid value 3 max.; sapon. no. 255-270	clear liquid; m.p. 2°C	solvent; *skin cleansing preparations*
	SCHERCEMOL NGDL	neopentyl glycol dilaurate	acid value 3 max.; sapon. no. 230-250	clear yellow liquid; m.p. 6°C	emollient; conditioner; *skin care products*
	SCHERCEMOL NGDO	neopentyl glycol dioctanoate	acid value 3 max.; sapon. no. 290-310	clear liquid	emollient; solvent; *skin cleansing preparations*
	SCHERCEMOL OHS	octyl hydroxystearate	acid value 1.0 max.; iodine value 3.0 max.; sapon. no. 140-160	clear liquid; m.p. 20°C	emollient; *skin care products*
	SCHERCEMOL OLO	oleyl oleate	acid value 2 max.; iodine value 95 max.; sapon. no. 95-110	clear amber liquid; m.p. −5°C	emollient; solubiliser; solvent; *skin care products; perfumes*

Supplier	Trade name	Chemical description	Composition	General properties	Functionality / Application
Scher Chemicals	SCHERCEMOL OP	octyl palmitate	acid value 2 max.; sapon. no. 145-160	clear straw-coloured liquid; m.p. 0°C	emollient; *skin care products; anti-perspirants*
	SCHERCEMOL OPG	octyl pelargonate	acid value 2 max.; sapon. no. 200-215	clear straw-coloured liquid; m.p. −10°C	emollient; binder; *skin care products; cosmetics*
	SCHERCEMOL PGDP	propylene glycol dipelargonate	acid value 5 max.; sapon. no. 300-320	clear straw-coloured liquid; m.p. −25°C	emollient; solvent; *bath preparations; skin care products; perfumes*
	SCHERCEMOL PGML	propylene glycol laurate	acid value 5 max.; iodine value 1 max.; sapon. no. 225-240	clear yellow liquid; m.p. 10°C	emollient; solvent; *cosmetics; skin care products*
	SCHERCEMOL PGMS	propylene glycol stearate	acid value 4 max.; iodine value 1 max.; sapon. no. 175-190	white to cream-coloured solid; m.p. 35°C	emulsifier; *skin care products*
	SCHERCEMOL TISC	triisostearyl citrate	acid value 3 max.; iodine value 3 max.; sapon. no. 150-165	colourless liquid; m.p. −5°C	*cosmetics*
	SCHERCEMOL TIST	triisostearyl trilinoleate	acid value 5 max.; iodine value 30 max.; sapon. no. 90-110	dark amber liquid; m.p. −10°C	moisturiser; binder; emollient
Seppic	LANOL 14 M	myreth-3 myristate	100% active	paste	emollient; *cosmetics; personal care products*
	LANOL 1688	cetearyl octanoate	100% active	liquid	emollient; *personal care products*
	LANOL P	glycol palmitate	100% active	wax	emollient; *cosmetics; personal care products*
	LANOL SDG	diethylenegycol monostearate	100% active	wax	emollient
	LANOL SEG	ethyleneglycol monostearate	100% active	wax	
	LANOL SG	glycerol monostearate	100% active	flakes	*cosmetics*
	LANOL SPG	propylene glycol monostearate	100% active	wax	
	MONTANE 20	sorbitan laurate	100% active	HLB 8.6; liquid	
	MONTANE 40	sorbitan palmitate	100% active	HLB 6.7; flakes	
	MONTANE 60	sorbitan stearate	100% active	HLB 4.7; flakes	emulsifier; wetting agent; dispersant; *personal care products*
	MONTANE 65	sorbitan tristearate	100% active	HLB 2.1; flakes	
	MONTANE 70	sorbitan isostearate	100% active	HLB 4.3; liquid	
	MONTANE 80	sorbitan oleate	100% active	HLB 4.3; liquid	
	MONTANE 83	sorbitan sesquioleate	100% active	HLB 3.7; liquid	emulsifier; wetting agent; dispersant
	MONTANE 85	sorbitan trioleate	100% active	HLB 1.8; liquid	*personal care products*
	SIMULSOL 165	PEG 100-stearate and glyceryl stearate	100% active	flakes	self emulsifying base; *personal care products*
Sisterna	SISTERNA L70-C	sucrose esters of edible fatty acids	monoester content 70%	HLB 15; free-flowing, cream-like powder	emulsifier; detergent; emollient; *cosmetics; liquid soaps; baby shampoos; skin care products; baby care products*
	SISTERNA SP10	sucrose esters of edible fatty acids	monoester content 10%	HLB 2; cream-coloured, free-flowing powder	emulsifier; *food industry*
	SISTERNA SP30	sucrose esters of edible fatty acids	monoester content 30%	HLB 6; cream-coloured, free-flowing powder	
	SISTERNA SP01-C	sucrose esters of edible fatty acids	monoester content ca. 0%	HLB 1; free-flowing, cream-like powder	emulsifier; detergent; emollient; *cosmetics; liquid soaps; baby shampoos; skin care products; baby care products*
	SISTERNA SP10-C	sucrose esters of edible fatty acids	monoester content 10%	HLB 2; free-flowing, cream-like powder	
	SISTERNA SP20-C	sucrose esters of edible fatty acids	monoester content 20%	HLB 4; free-flowing, cream-like powder	

Supplier	Trade name	Chemical description	Composition	General properties	Functionality / *Application*
Sisterna	SISTERNA SP30-C	sucrose esters of edible fatty acids	monoester content 30%	HLB 6; free-flowing, cream-like powder	emulsifier; detergent; emollient; *cosmetics; liquid soaps; baby shampoos; skin care products; baby care products*
	SISTERNA SP40-C	sucrose esters of edible fatty acids	monoester content 40%	HLB 8; free-flowing, cream-like powder	
	SISTERNA SP50	sucrose esters of edible fatty acids	monoester content 50%	HLB 11; cream-coloured, free-flowing powder	emulsifier; *food industry*
	SISTERNA SP50-C	sucrose esters of edible fatty acids	monoester content 50%	HLB 11; free-flowing, cream-like powder	emulsifier; detergent; emollient; *cosmetics; liquid soaps; baby shampoos; skin care products; baby care products*
	SISTERNA SP60-C	sucrose esters of edible fatty acids	monoester content 60%	HLB 13; free-flowing, cream-like powder	
	SISTERNA SP70	sucrose esters of edible fatty acids	monoester content 70%	HLB 15; cream-coloured, free-flowing powder	emulsifier; *food industry*
	SISTERNA SP70-C	sucrose esters of edible fatty acids	monoester content 70%	HLB 15; free-flowing, cream-like powder	emulsifier; detergent; emollient; *cosmetics; liquid soaps; baby shampoos; skin care products; baby care products*
Stepan Europe	DREWMULSE 200K	glyceryl monostearate; made-to-order	100% active	white to beige solid	emulsifier; emollient; antistatic agent; stabiliser; viscosity modifier; opacifier; *hair conditioners; personal care products*
	DREWPOL 10.4.0	decaglyceryl tetraoleate; made-to-order	100% active	amber liquid	emulsifier; emollient; antistatic agent; stabiliser; viscosity modifier; opacifier; solubiliser; *hair conditioners; personal care products; pharmaceuticals*
	DREWPOL 10.10.0	decaglyceryl decaoleate; made-to-order	100% active	amber liquid	
	NEOBEE M5	caprylic/capric triglyceride	100% active	water white to pale yellow liquid	emollient; lubricant; *personal care products*
	NEOBEE M20	propylene glycol dicaprylate caprate	100% active	water white to pale yellow liquid	
	SECOSTER 874	pentaerythritol ester; made-to-order	100% active	water white liquid	lubricant; *metal working; marine and jet engine lubricants*
	SECOSTER 887	trimethylolpropane ester; made-to-order	100% active	water white liquid	
	SECOSTER DMS	glycol distearate; made-to-order	100% active	white to beige solid	pearlescent agent; opacifier; co-emulsifier; emollient; *shampoos; personal care products*
	SECOSTER DO 600	PEG 600 diolate	100% active	pale yellow to amber liquid	emulsifier; dispersant; *household, institutional and industrial cleaners; textile industry; oil industry*
	SECOSTER EMS	glycol monostearate; made-to-order	100% active	white acid	pearlescent agent; opacifier; co-emulsifier; emollient; *shampoos; personal care products*
	SECOSTER MA 300	PEG 300 monoabietate	100% active	brown liquid	emulsifier; sticking agent; *agrochemicals*
	SECOSTER ML 400	PEG 400 monolaurate; made-to-order	100% active	water white to pale yellow liquid	emulsifier; *skin cleansing preparations; personal care products*
	SECOSTER MO 400	PEG 400 monooleate	100% active	pale yellow to amber liquid	lubricant; antistatic agent; emulsifier; dispersant; *household, institutional and industrial cleaners; oil industry*
	SECOSTER SDG	glyceryl monostearate; made-to-order	100% active	white solid	antistatic agent; lubricant; *emulsion polymerisation*
Surfachem	SATINEX	magnesium myristate			
	SURFAC 2389	polyethylene glycol mono oleate			
	SURFAC EGMS	ethylene glycol monostearate			
	SURFAC GMS NSE40	glycerol monostearate (non-self emulsifying)			
	SURFAC GMS SE40	glycerol monostearate (self emulsifying)			
	SURFAC IPM	isopropyl myristate			

Supplier	Trade name	Chemical description	Composition	General properties	Functionality / Application
Thor Chemicals	ATOLEX AST/2	ester		liquid	antistatic agent; lubricant / *textile industry*
Unichema Chemicals	ESTOL 1407	glycerol monooleate	sapon. no. 165-175		
	ESTOL 1447	PEG 400 dioleate	sapon. no. 120-130	cloud point 10°C max.	emulsifier; defoaming agent; lubricant
	ESTOL 1461	glycerol monoostearate SE	sapon. no. 156-170		
	ESTOL 1474	glycerol monostearate	sapon. no. 168-184		
Witco	REWOMUL MG SE	glycerine monostearate (SE)	100% active	flakes	emulsifier
	REWOPAL PEG 6000 DS	polyethylene glycol distearate	100% active	flakes	thickener / *personal care products*
	REWOPAL PG 280	modified ethylene glycol distearate	100% active		pearlescent agent
	SOG 10	glycerol monostearate; food grade	100% active	powder	emulsifier / *food industry*
	SORBAN AL	sorbitan monolaurate	100% active	paste	defoaming agent
	SORBAN AO	sorbitan monooleate	100% active	liquid	emulsifier; anti-corrosion / *dry cleaning; metal working*
	SORBAN AST	sorbitan monostearate	100% active	flakes	emulsifier / *cosmetics; food industry; metal working*
	SORBAN CO	sorbitan trioleate	100% active	liquid	anti-corrosion; defoaming agent; emulsifier / *metal working*
	WITBREAK DGE-128A	glycol ester		liquid; pour point −12.2°C; pH 4	
	WITBREAK DGE-169	glycol ester		liquid; pour point −6.7°C; pH 4	
	WITBREAK DPG-15	polyoxyethylene glycol		liquid; pour point −6.7°C; pH 7	demulsifier / *petroleum industry*
	WITBREAK DPG-482	polyoxyethylene glycol		liquid; pour point −1.1°C; pH 7	
	WITBREAK DRE-8164	resin ester		liquid; pH 6	
	WITCONOL 14	polyglyceryl-4 oleate		HLB 9.4; liquid; colour 9 (Gardner)	emollient; o/w emulsifier; w/o emulsifier; lubricant; spreading agent / *personal care products; household and industrial applications*
	WITCONOL 2301	methyl oleate		liquid; pour point −16°C	defoaming agent; lubricant; moisture barrier / *personal care products; household and industrial applications*
	WITCONOL 2380	propylene glycol stearate		HLB 1.8; beads; colour 2 (Gardner); m.p. 36°C	o/w emulsifier; lubricant; opacifier / *personal care products; household and industrial applications*
	WITCONOL 2400	glyceryl stearate		HLB 3.9; beads; colour 1 (Gardner); m.p. 58°C	dispersant; o/w emulsifier; lubricant / *personal care products; household and industrial applications*
	WITCONOL 2407	glyceril stearate SE		HLB 5.1; beads; colour 3 (Gardner); m.p. 58°C	o/w emulsifier; lubricant / *personal care products; household and industrial applications*
	WITCONOL 2421	glyceryl oleate		HLB 3.4; liquid; colour 5 (Gardner); m.p. 19°C	defoaming agent; o/w emulsifier; lubricant; moisture barrier / *personal care products; household and industrial applications*

Supplier	Trade name	Chemical description	Composition	General properties	Functionality *Application*
Witco	WITCONOL 2500	sorbitan oleate		HLB 4.6; liquid; pour point <0°C; colour 8 (Gardner)	coupling agent; w/o emulsifier; lubricant *personal care products; household and industrial applications*
	WITCONOL 2503	sorbitan trioleate		HLB 2.1; liquid; pour point <0°C; colour 7 (Gardner)	coupling agent; w/o emulsifier; lubricant *personal care products; household and industrial applications*
	WITCONOL 2620	PEG-4 laurate		HLB 9.3; liquid; pour point 9°C; colour 1 (Gardner)	o/w emulsifier; lubricant *personal care products; household and industrial applications*
	WITCONOL 2648	PEG-8 dioleate		HLB 8.8; liquid; pour point 6°C; colour 4 (Gardner)	defoaming agent; o/w emulsifier; lubricant *personal care products; household and industrial applications*
	WITCONOL 2711	PEG-8 stearate		HLB 12.0; solid; colour 1 (Gardner); m.p. 32°C	o/w emulsifier; lubricant *personal care products; household and industrial applications*
	WITCONOL 2712	PEG-8 distearate		HLB 7.5; solid; colour 2 (Gardner); m.p. 36°C	
	WITCONOL EGMS	glycol stearate		HLB 2.2; beads; colour 2 (Gardner); m.p. 50°C	conditioner; opacifier *personal care products; household and industrial applications*
	WITCONOL H31A	PEG-8 oleate		HLB 12.5; liquid	defoaming agent; o/w emulsifier; lubricant *personal care products; household and industrial applications*
	WITCONOL MST	glyceryl stearate		HLB 3.9; flake; colour 3 (Gardner); m.p. 58°C	dispersant; o/w emulsifier; lubricant *personal care products; household and industrial applications*
	WITCONOL RHT	glycerl stearate SE		HLB 5.1; flake; colour 2 (Gardner); m.p. 58°C	o/w emulsifier; lubricant *personal care products; household and industrial applications*
Zschimmer & Schwarz	ACIDIT PA	ester of organic acids		clear, light yellowish liquid	acidifying agent *textile industry*
	CEFASOFT MHZ	fatty acid ester		liquid dispersion	softener *textile industry*
	COLLASOFT WE WAX	fatty acid ester		white flakes	sizing agent; afterwaxing agent *textile industry*

Miscellaneous nonionics

Air Products	SURFYNOL 61	acetylenic alcohol		clear colourless liquid	wetting agent; dispersant *coatings; inks; polishes; pigments; adhesives; agrochemicals; hard surface cleaners; glass cleaners; silicon wafer cleaners; textile industry; electroplating*
	SURFYNOL 82	acetylenic diol		flakes	wetting agent; nonfoaming; defoaming agent *hard-surface cleaners; glass cleaners; electroplating; agrochemicals*
	SURFYNOL 82S	acetylenic diol		free-flowing powder	wetting agent; low foam; defoaming agent *agrochemicals; coatings*
	SURFYNOL 104	nonionic surfactant	100% active	white waxy solid	wetting agent; defoaming agent *paint industry; coatings; adhesives; inks; cements; metal working; agrochemicals; dyes; oil industry; hard-surface cleaners; textile industry; electroplating; ceramics*
	SURFYNOL 104A	Surfynol 104 in 2-ethylhexanol	50% active	light yellow liquid	wetting agent; defoaming agent *agrochemicals; paper industry; inks; dyes; adhesives; hard-surface cleaners; emulsion polymerisation; oil industry; textile industry; ceramics; coatings; metal working*

Supplier	Trade name	Chemical description	Composition	General properties	Functionality / Application
Air Products	SURFYNOL 104BC	Surfynol 104 in 2-butoxyethanol	50% active	light yellow liquid	wetting agent; defoaming agent paint industry; coatings; adhesives; inks; dyes; pigments; cements; metal working; agrochemicals
	SURFYNOL 104E	Surfynol 104 in ethylene glycol	50% active	light yellow liquid	wetting agent; defoaming agent agrochemicals; coatings; inks; dyes; pigments; adhesives; emulsion polymerisation; oil industry; hard-surface cleaners; metal working; textile industry; cements; electroplating; ceramics
	SURFYNOL 104PA	Surfynol 104 in isopropyl alcohol	50% active	light yellow liquid	wetting agent; defoaming agent coatings; adhesives; inks; pigments; emulsion polymerisation; ceramics
	SURFYNOL 104PG-50	Surfynol 104 in propylene glycol	50% active		wetting agent; defoaming agent coatings; adhesives; inks; dyes; pigments; emulsion polymerisation; agrochemicals; oil industry; hard-surface cleaners; textile industry; electroplating; metal working
	SURFYNOL 104S	Surfynol 104 on amorphous silica	46% active	free-flowing powder	wetting agent; defoaming agent agrochemicals; coatings; oil industry; hard-surface cleaners; cements; ceramics
	SURFYNOL DF-37	non-silicone defoamer			defoaming agent; wetting agent; dispersant coatings; latex; inks; adhesives; paint industry; pigments; oil industry; textile industry
	SURFYNOL GA	blend of nonionic surfactants		clear liquid	wetting agent; low foam; grinding agent; dispersant coatings; inks; pigments; dyes; cements
	SURFYNOL SE-F	self-emulsifiable surfactant			non-foaming; wetting agent paint industry; coatings; inks; adhesives; agrochemicals; ceramics; metal working
	SURFYNOL TG	proprietary acetylenic blend		clear light yellow liquid	wetting agent; low foam; dispersant coatings; inks; adhesives; pigments; oil industry; hard-surface cleaners; glass cleaners; cements; ceramics
	SURFYNOL TG-E	EPA compliant version of Surfynol TG		clear liquid	wetting agent; low foam agrochemicals; pesticides
Akcros Chemicals	AGRILAN AEC123	speciality nonionic	100% active	white soft paste; cloud point 74°C (1% aq.); pour point 28°C; viscosity 402 cSt (40°C)	co-emulsifier agrochemicals
	AGRILAN AEC145	speciality nonionic	100% active	off-white paste/liquid; cloud point 55°C (1% aq.); pour point 14°C; viscosity 141 cSt (60°C)	co-emulsifier agrochemicals
	AGRILAN AEC167	speciality nonionic	100% active	clear colourless liquid; cloud point 44°C (1% aq.); pour point 8°C; viscosity 800 cSt	wetting agent; spreading agent; emulsifier agrochemicals
	AGRILAN AEC178	speciality nonionic	100% active	clear colourless liquid; cloud point 34°C (1% aq.); pour point < 0°C; viscosity 650 cSt	co-emulsifier agrochemicals
	AGRILAN D54	ether-linked nonionic	80% active	slightly hazy brown liquid; cloud point 60°C (5% in 20% BDG/H$_2$O); pour point − 17°C; viscosity 950 cSt	emulsifier; wetting agent; dispersant agrochemicals
	AGRILAN F502		100% active	soft white solid; pour point 35°C; viscosity 380 cSt (40°C); pH 7.0 (1% aq.)	wetting agent; dispersant; additive; low foam agrochemicals
	ETHYLAN BCD42		100% active	yellow liquid; cloud point 55°C (1% aq.); pour point 14°C; viscosity 141 cSt (60°C); pH 5.5-7.5 (1% aq.)	wetting agent; dispersant; grinding aid; rheology modifier aq. pigment systems; optical brightening agents

Supplier	Trade name	Chemical description	Composition	General properties	Functionality / *Application*
Akcros Chemicals	ETHYLAN BD10	nonionic surfactant		clear yellow liquid; cloud point 40°C (1% aq.); pour point 4°C; viscosity 125 cSt; pH 5-7 (1% aq.)	
	ETHYLAN BD21	nonionic surfactant		clear colourless liquid; cloud point 26°C (1% aq.); pour point 3°C; viscosity 144 cSt; pH 5-7 (1% aq.)	low foam; wetting agent / *dishwashing agents; rinse aids; metal cleaning; spray cleaning*
	ETHYLAN BD32	nonionic surfactant		clear colourless liquid; cloud point 24°C (1% aq.); pour point < 0°C; viscosity 121 cSt; pH 5-7 (1% aq.)	
	ETHYLAN BKL130		85% active; H_2O content ca. 15%	yellow liquid; cloud point 71°C (1% aq.); pour point < 0°C; viscosity 960 cSt; pH 5.5-7.5 (1% aq.)	wetting agent; dispersant; grinding aid; rheology modifier / *aq. pigment systems; optical brightening agents*
	ETHYLAN BV		100% active	soft paste; cloud point 89°C (1% aq.); pour point 20°C; viscosity 130 cSt (40°C); pH 5.5-7.5 (1% aq.)	
	ETHYLAN C40AH		100% active	hazy yellow liquid; cloud point 95°C (1% aq.); pour point 11°C; viscosity 357 cSt (40°C); pH 5.5-7.5 (1% aq.)	co-emulsifier / *agrochemicals*
	ETHYLAN KEO		100% active	clear pale yellow liquid; cloud point 54°C (1% aq.); pour point 2°C; viscosity 320 cSt; pH 5.5-7.5 (1% aq.)	
	ETHYLAN TB345	amine polyglycol condensate	100% active	amber liquid; cloud point 40°C (10% in 25% butyl dioxitol/H_2O); pour point − 5°C; viscosity 551 cSt; surface tension 37.2 dynes/cm (1%); pH 8.0 (1% aq.)	defoaming agent; wetting agent; low foam / *metal cleaning; rinse aids*
Akzo Nobel	BEROL 28	nonionic surfactant	100% active		emulsifier / *mineral and vegetable oils; hydrocarbon solvents; degreasing; industrial and vehicle cleaning*
	BEROL 79	nonionic surfactant	100% active		emulsifier / *cleaners; degreasers; industrial and vehicle cleaners*
	BEROL 303	nonionic surfactant	100% active		emulsifier / *mineral and vegetable oils; hydrocarbon solvents; degreasing; industrial and vehicle cleaning*
	BEROL 791	nonionic surfactant	100% active		emulsifier / *solvent based degreasers*
	BEROL 797	nonionic surfactant	100% active		emulsifier / *degreasers*
Allied Colloids	ELFACOS GT 282 L	talloweth-60 myristyl glycol		HLB 15; soln.	
	ELFACOS GT 282 S	talloweth-60 myristyl glycol		HLB 17; pellets	
	ALCOLUBE NV	fatty acid amide and silicone blend		liquid	softener / *textile industry; polyamide/viscose blends*
	ALCOPOL CPB			liquid	detergent; wetting agent; low foam / *textile industry; bleaching*
	ALCOPRINT PSM			emulsion	softener / *textile industry; pigment printing*

Supplier	Trade name	Chemical description	Composition	General properties	Functionality / *Application*
Allied Colloids	SOFTENING AGENT LNS	substituted amine based softening agent		liquid	softener / *textile industry*
	SOFTENING AGENT LNS 100 FLAKES	substituted amine based softening agent; concentrated version of Softening Agent LNS		cream flakes	softener / *textile industry*
Auschem Cesalpinia	EMULTEX 802	blend of nonionic surfactants	100% active	waxy	*natural and synthetic waxes*
	EMULTEX 1302	blend of nonionic surfactants	100% active	waxy	
	EMULTEX 1502	blend of nonionic surfactants	100% active	flakes	
	EMULTEX 1515	blend of nonionic surfactants	100% active	waxy	
	EMULTEX 1602	blend of nonionic surfactants	100% active	flakes	
	EMULTEX 1620	blend of nonionic surfactants	100% active	waxy	
	MADEOL AG 1210	blend of nonionic surfactants	100% active	powder	anti-caking agent; anti-dusting agent / *fertilisers*
	POLIROL 797	nonionic surfactants	100% active	liquid	stabiliser; homogeniser / *latices; polymerisation*
Bayer	DIADAVIN EWN	H_2O-free mixture of various polyglycol ethers with aliphatic components	50% active	yellow viscous liquid	emulsifier; scouring agent / *textile industry*
	DIADAVIN EWN 200%	H_2O-free mixture of various polyglycol ethers with aliphatic components	100% active	yellow viscous liquid	emulsifier; scouring agent / *textile industry*
CHEM-Y	AMINOL TEC 111	mixture of an ethoxylated fatty acid monoethanolamide with fatty acid, monoethanolamine and triethanolamine	amine value 160-180; H_2O 4.5-6.5%	clear pale yellow liquid; viscosity 250-500 mPas; colour 2 max. (Gardner); pH 10.5-12.0 (4% in water)	emulsifier; lubricant; anti-corrosion / *metal working*
Croda	COSMOWAX D	cetearyl alcohol and ceteareth-20		white waxy solid	
	COSMOWAX EM5483	cetearyl alcohol and ceteareth-20		white waxy solid	o/w emulsifier / *cosmetics; toiletries*
	COSMOWAX J	cetearyl alcohol and ceteareth-20		white waxy solid	
	CRODEX N	cetearyl alcohol and ceteareth-20		white waxy solid	o/w emulsifier / *skin care products; pharmaceuticals*
	CROMUL EMO685	ceteth-5 and ceteareth-7		HLB 10.2; white translucent plastic waxy solid	o/w emulsifier / *creams; lotions*
	EMULSIFIER EMO650	blend of nonionic emulsifiers		white waxy solid	o/w emulsifier / *wax emulsions*
	POLAWAX A31	emulsifying wax developed for use in aerosol quick-breaking foams	acid value 1.0 max.; sapon. no. 7.5-12.5	off-white to yellow pastilles; colour 4Y:1R max. (1 in. cell); drop point 43-50°C	emulsifier / *cosmetics; pharmaceuticals; hair care products; baby care products; sunscreen products; skin care products; deodorants*
	POLAWAX GP200	general purpose emulsifying wax	acid value 1-8 max.; sapon. no. 7.8-15.0	white powder or pastilles; colour 1Y:0.5R max. (1 in. cell); drop point 47-51°C	
	POLAWAX NF	general purpose emulsifying wax	OH value 178-192; iodine value 3.5 max.; sapon. no. 14 max.	yellow pastilles; colour 4Y:1R max. (1 in. cell); drop point 48-52°C; pH 5.5-7.0 (3%)	
	SPREADING AGENT ET0672	proprietary product		HLB 7.5; clear, colourless to pale yellow liquid	o/w emulsifier; spreading agent; dispersant / *bath preparations*
Elf Atochem	DETERFLO A 210	nonionic surfactant	100% active	liquid; cloud point 27-31°C (5% in H_2O); surface tension 31 dynes/cm (0.25 g/l); pH 2-3 (10% in H_2O)	non-foaming / *industrial detergents; degreasing preparations; metal working*

Supplier	Trade name	Chemical description	Composition	General properties	Functionality / *Application*
Ellis & Everard	CAFLON EMD-1	nonionic degreaser concentrate			degreaser
	CAFLON EMD-2	aq. dilution of Caflon EMD-1			
	CAFLON EMD-3	kerosene dilution of Caflon EMD-1			
Gattefosse	BASE WL 2586	defined nonionic blend of stearic esters and natural waxes		waxy solid	thickener; co-emulsifier / *cosmetics*
	HYDROLACTOL	glyceryl stearate, propylene glycol stearate, glyceryl isostearate, propylene glycol isostearate, oleth-25 and ceteth-25		waxy solid	emulsifier / *cosmetics; sunscreen preparations; shaving products; personal care products*
	SOLUBILISANT GAMMA 2420	mixture of nonionic surfactants		liquid	
	SOLUBILISANT GAMMA 2428	mixture of nonionic surfactants		liquid	
	TEFOSE VEGETAL 2561	PEG-6 stearate, ceteth-20, glyceryl stearate and steareth-20		waxy solid	emulsifier / *sunscreen preparations; skin care products; toiletries*
Hays Colours	ALTAF CA	nonionic surfactant			anti-foam
	DYLEV DA	polyglycol ether			*textile industry; dyeing*
	HYSOFT CRL	nonionic surfactant			softener / *textile industry; cellulosics and their blends*
	METAPEX DD	nonionic surfactant			detergent / *textile industry*
	METAPEX HC27	alkylene oxide nonionic surfactant			detergent; wetting agent / *textile industry*
	METAPEX HC100	alkylene oxide nonionic surfactant			
	METAPEX HCB20	nonionic surfactant			levelling agent; dispersant; scouring agent; emulsifier / *textile industry; dyeing*
	METAPEX HCB90	nonionic surfactant			
	METAPEX LF	nonionic surfactant, solvent free			low foam; scouring agent; dispersant / *textile industry; dyeing*
Henkel	ADALIN AN	fatty acid derivate with silicone			softener / *textile industry; finishing*
	ADALIN SO	polyalkylene glycol, mineral oil-free			compressive shrinking / *textile industry; finishing*
	BELSOFT 200	fatty acid amide with nonionic emulsifier			softener; additive; stabiliser; crease inhibitor / *textile industry*
	BELSOFT N	combination of fatty acid amide with nonionic surfactants			softener; additive / *textile industry*
	BELSOFT NEC	combination of fatty acid ester with nonionic surfactants			softener / *textile industry*
	BELSOFT PS	blend of fatty acid amide and a silicone compound			softener; additive / *textile industry*
	BREVIOL DM	blend of heterocyclic bases			dissolving agent; dispersant; penetration aid / *textile industry; dyeing*

Supplier	Trade name	Chemical description	Composition	General properties	Functionality / Application
Henkel	BREVIOL PAM	polyamide derivate			crease inhibitor / *textile industry; dyeing*
	BREVIOL SPEZIAL	combination of polyamide and fatty acid derivate			
	CLEROL 9006-B		100% active	liquid	defoaming agent / *sugar industry; rotary [Trelemont] diffusers*
	CLEROL DTX 389		100% active	liquid	defoaming agent / *bet sugar processing*
	CLEROL IV		100% active	liquid	
	CLEROL S-21		100% active	liquid	
	DEHYDRAN 1620	mixture of specially modified alcohols and a polysiloxane adduct	100% active	yellow, slightly cloudy liquid; viscosity 60-80 mPas	defoaming agent / *aq. coatings; printing inks; adhesives; UV-curing systems*
	DEHYPON LST 254	blend of modified fatty alcohols	99-100% active	liquid	defoaming agent / *bottle cleaners*
	DYMSOL B		45% active	dispersion	anti-blocking agent / *rubber; latex*
	DYMSOL L		100% active	liquid	stabiliser / *rubber; latex*
	DYMSOL N		97.5% active	liquid	
	EMERWAX 1266	cetearyl alcohol and ceteareth-20		white waxy solid	o/w emulsifier / *hair care products; personal care products; antiperspirants*
	EMULGADE 1000 NI	cetearyl alcohol and ceteareth-20		flakes	o/w base / *o/w emulsions*
	EMULGADE A	cetearyl alcohol and laureth-10		granules	o/w base / *o/w emulsions; o/w creams*
	EMULGADE CBN	mixture of cetearyl isononanoate, glyceryl stearate, PEG-20 glyceryl stearate, cetearyl alcohol, ceteareth-20 and cetyl palmitate		lard-like mass	o/w base / *o/w creams; o/w emulsions*
	EMULGADE F SPECIAL	cetearyl alcohol and PEG-40 castor oil		granules	o/w base / *creams; o/w lotions*
	EMULGADE TE	tallow alcohol, glyceryl stearate and PEG-32 stearate		granules	base / *o/w creams; o/w lotions; skin care products*
	FOAMASTER 267-A		100% active	liquid	defoaming agent / *paper industry; coatings*
	FOAMASTER 1407-50			fluidised paste	defoaming agent / *paper industry*
	FOAMASTER 8034		100% active	liquid	defoaming agent / *paint industry; paper industry; adhesives; rubber; latex*
	FOAMASTER B		100% active	liquid	defoaming agent / *paint industry; synthetic rubber; adhesives*
	FOAMASTER DD-72		100% active	liquid	defoaming agent / *paint industry; wallpaper coatings; printing inks*
	FOAMASTER DF-124-L		100% active	liquid	defoaming agent / *emulsion polymerisation*
	FOAMASTER DF-177-F		100% active	liquid	foam suppressor / *paper industry; coatings; adhesives*

Supplier	Trade name	Chemical description	Composition	General properties	Functionality Application
Henkel	FOAMASTER DF-178		100% active	liquid	foam suppressor *paper industry; coatings; adhesives*
	FOAMASTER DF-198-L		100% active	liquid	defoaming agent *paint industry*
	FOAMASTER DNH-1		100% active	liquid	defoaming agent *paint industry; polymer industry*
	FOAMASTER E75-C			cloudy yellow liquid; viscosity 400-1400 mPas	defoaming agent *paint industry; adhesives*
	FOAMASTER H2	mixture of aliphatic hydrocarbons, hydrophobic silica, emulsifiers	100% active	cloudy, yellowish liquid; viscosity 800-1500 mPas	defoaming agent *aq. systems; printing inks; adhesives*
	FOAMASTER JMY		100% active	liquid	foam control *adhesives*
	FOAMASTER KF SERIES			paste or fluidised paste	defoaming agent *paper industry*
	FOAMASTER LD SERIES		100% active	liquid	defoaming agent; emulsifier *paper industry*
	FOAMASTER NDW		100% active	liquid	defoaming agent *paint industry; adhesives; cement screeds; paper industry; rubber; latex*
	FOAMASTER NS-20		100% active	liquid	defoaming agent *paper industry; coatings*
	FOAMASTER NXZ		100% active	liquid	defoaming agent *paint industry; adhesives; paper industry; emulsion polymerisation*
	FOAMASTER P		100% active	liquid	defoaming agent *paper industry; adhesives; systems containing PV alcohol*
	FOAMASTER PD-1		66% active	powder	defoaming agent *concrete*
	FOAMASTER PD-14		66% active	powder	defoaming agent *concrete; paint industry*
	FOAMASTER TDB-1		100% active	liquid	defoaming agent *paint industry; water-soluble resins; adhesives*
	FOAMASTER VL		100% active	liquid	defoaming agent *paint industry; monomer stripping; latex; emulsion polymerisation*
	FORYL CP	combination of polyalkylene ether and alkylenepolyethylene glycolether			wetting agent; scouring agent; non-foaming *textile industry; pretreatment*
	FORYL FKN	alkylenepolyethylene glycolether with solvent			scouring agent *textile industry; pretreatment*
	FORYL FW	alkylenepolyethylene glycolether with odour inhibitor			
	FORYL SF	polyalkylene ether			wetting agent; scouring agent; non-foaming *textile industry; pretreatment*
	HYDROPALAT 3037	modified natural oil; silicone free	98-100% active	yellowish, slightly cloudy liquid	flow additive; wetting agent; emulsifier *aq. systems based on polymer emulsions or secondary emulsions*

Supplier	Trade name	Chemical description	Composition	General properties	Functionality / Application
Henkel	HYONIC PE-90		100% active	liquid	paint industry
	LAMECREME AOM	hydrogenated palm oil glycerides, PEG-20 hydrogenated palm oil glycerides and cetyl alcohol		solid	cream base / o/w creams; o/w ointments
	LANETTE AO	cetearyl alcohol and ceteareth-30		flakes	cream base / creams; o/w ointments
	NOPCO 1097-A +		55% active	dispersion	water repellant / concrete
	NOPCO 1225-L		100% active	liquid	viscosity modifier / plastics industry; PVC plastisols; paint industry
	NOPCO 9114		100% active	liquid	tackifier / adhesives
	NOPCO E3765		100% active	liquid	defoaming agent / beet sugar processing
	NOPCO-COLOUR-SPERSE 188-A		100% active	liquid	paint industry
	NOPCOFLOC 320-L			emulsion	machine drying and retention aid; sludge conditioner / paper industry
	SELBANA 722	fatty acid derivative			lubricant / textile industry; spinning
	SELBANA AC	combination of fatty acid esters, antistatics and emulsifiers			lubricant; antistatic agent / textile industry; spinning
	SELBANA CW	combination of fatty acid esters, antistatics and emulsifiers			lubricant / textile industry; spinning
	SELBANA STB	synthetic oils			
	SPREITAN 38/4	blend of mineral oil, emulsifiers and antistatics			coning oil / textile industry; spinning
	SPREITAN 418	combination of fatty acid esters and emulsifiers			
	SPREITAN VLS	combination of fatty acid esters and emulsifiers			
Hoechst	GENAPOL PGL	nonionic blend			pearlescent agent / toiletries; detergents
Hüls	DIONIL RS	blend	active detergent 80%	liquid	
	MARLAMID AS 18	blend of stearic acid alkanolamide and fatty alcohol polyglycol ether	active detergent 100%	flakes	
	MARLOSOFT A 18 M	blend of fatty acid alkanolamide acetate and fatty acid polyethylene glycol ester	active detergent 97%	flakes	fabric softeners
	MARLOSOFT B 18 M	blend of fatty acid alkanolamide acetate and fatty acid polyethylene glycol ester	active detergent 100%	flakes	fabric softeners
	MARLOWET MA	blend of carboxylic acid polyglycol esters and alkylolamides	active detergent 100%	liquid; cloud point 90°C (5% in 25% BDG soln.)	emulsifier / mineral oils; spindle oils; metal working; textile industry
	MARLOWET RNP	blend of carboxylic acid polyglycol ester, alkylaryl polyglycol ether and glycol derivatives	active detergent 80%	liquid; cloud point 68°C (2% in H_2O)	emulsifier / perfumes
	MARLOWET SAF	blend of alkylpolyglycol ethers and glycol derivative	active detergent 90%	liquid; cloud point 90°C (10% in 25% BDG soln.)	emulsifier / mineral oils; white oils; metal working; textile industry

Supplier	Trade name	Chemical description	Composition	General properties	Functionality / Application
Hüls	MARLOWET SW	blend of carboxylic acid polyglycol ester and alkyl polyglycol ether	active detergent 100%	liquid; cloud point 90°C (10% in 25% BDG soln.)	emulsifier; *mineral oils; metal working; textile industry*
	MARLOWET SWN	blend of carboxylic acid polyglycol ester and alkyl polyglycol ether	active detergent 100%	liquid; cloud point 70°C (10% in 25% BDG soln.)	emulsifier; *spindle and white oils; textile industry*
	MARLOX M 606/1	TEA-soap containing blends	active detergent 95%	liquid	*low-foaming detergents and cleaners*
	MARLOX M 606/2	TEA-soap containing blends	active detergent 95%	liquid	*low-foaming detergents and cleaners*
ICI	ARLAMOL M 812	short-chain glycerides		colourless liquid	emollient; *personal care products*
	ATLAS G-3969	specialty blend		HLB 17.0; white waxy solid	*household and industrial applications: agrochemicals; textile industry*
	ATLOX 2081B	isopropylamine alkylaryl sulfonate nonionic blend		HLB 13.6; red-brown liquid	emulsifier; dispersant; *agrochemicals*
	ATLOX 4870B	isopropylamine alkylaryl sulfonate nonionic blend		HLB 13.0; red-brown liquid	
	ATLOX 4902	calcium alkylaryl sulfonate nonionic blend		HLB 12.0; yellow-brown liquid	
	ATLOX 5325	mixture of nonionic surfactants		HLB 13.7; amber paste	
	ATLOX 5340	mixture of nonionic surfactants		yellow solid	
	CIRRASOL 185A	acid solubilised fatty acid amide		amber semi-solid	
	CIRRASOL ALN-TV	blend of nonionic lubricants and emulsifiers		pale yellow liquid	
	CIRRASOL BASE HK	fatty acid derivative of substituted amine		cream solid	*textile industry*
	CIRRASOL BASE N22	mixture of non-nitrogenous lubricants		amber liquid	
	CIRRASOL LC-FG	fatty acid derivative of polyamine		cream/yellow paste	
ISP	SURFADONE LP-100	caprylyl pyrrolidone	> 97% active	water-white, low viscosity liquid	wetting agent; low foam; thickener; foam stabiliser; *shampoos; hair care products*
	SURFADONE LP-300	lauryl pyrrolidone	> 97% active	water-white, low viscosity liquid	
Lakeland Laboratories	LAKEWAX 29	oxidised homopolymer wax emulsion	solids 31%; wax 27%; hardness of wax < 0.5 mm (ASTM D5)	off white emulsion; softening point 140°C; pH 8.5	*floor polishes; printing inks; temporary protective coatings; wood finishes; paint industry*
	LAKEWAX 37	oxidised homopolymer wax emulsion	solids 36%; wax 26%; hardness of wax 0.5 mm (ASTM D5)	off white emulsion; softening point 134°C; pH 9.0	*floor polishes; printing inks; wood finishes; paint industry; food industry*
	LAKEWAX 52	polyethylene copolymer wax emulsion	solids 25%; wax 20%; hardness of wax 2 mm (ASTM D5)	clear translucent emulsion; softening point 108°C; pH 10	*floor polishes; wood finishes; paint industry; food industry*
	LAKEWAX 60	oxidised homopolymer wax emulsion	solids 25%; wax 20%; hardness of wax 5.5 mm (ASTM D5)	off white translucent emulsion; softening point 104°C; pH 9.0	*floor polishes; food industry*
	LAKEWAX 65	oxidised homopolymer wax emulsion	solids 30%; wax 20%; hardness of wax 90 dmm (ASTM D5)	off white emulsion; softening point 85°C; pH 8.5	concrete curing; *temporary protective coatings*
Nicca Chemical	LIPOTOL TC-350	special nonionic active agent		brown liquid; pH ca. 7.5 (5% soln.)	soaping agent; dispersant; *textile industry; cotton fibres; cotton/polyester blends*
	SUNMORL WX-12	special nonionic surfactant mixture		slightly yellowish clear liquid; pH ca. 8	emulsifier; dispersant; scouring agent; low foam; *textile industry; synthetic fibres*

Supplier	Trade name	Chemical description	Composition	General properties	Functionality / Application
Nikko Chemicals	NIKKOL LECINOL LL-20	hydrogenated lysolecithin		pale yellow powder	emulsifier; stabiliser / *cosmetics*
	NIKKOL LECINOL S-10	hydrogenated lecithin		pale yellow powder	
	NIKKOL LECINOL S-10E	hydrogenated lecithin		pale yellow powder	
	NIKKOL LECINOL S-10EX	hydrogenated lecithin		pale yellow powder	anti-irritant; moisturiser; solubiliser; liposoming agent / *cosmetics*
	NIKKOL LECINOL S-10EM	hydrogenated lecithin		pale yellow powder	
	NIKKOL LECINOL S-30	hydrogenated lecithin		pale yellow powder	
	NIKKOL LECINOL SH	hydroxyl lecithin		pale yellow powder	emulsifier; stabiliser; moisturiser / *cosmetics*
Pentagon	PENTASIZE 8	alkenyl succinic anhydride	sapon. no. 300-325	clear amber liquid; colour 15 max. (Gardner)	anti-corrosion; viscosity modifier; gelling agent; chemical intermediate; sizing agent / *lubricants; paper industry*
	PENTASIZE 68	alkenyl succinic anhydride	100% active: sapon. no. 320-350	clear amber liquid; colour 15 max. (Gardner)	sizing agent / *paper and board industry*
PPG	ALUBRASOFT 77N		100% active	liquid	softener / *textile industry*
	ALUBRASOL NPL-5		40% active		
	ALUBRASPIN 40		100% active	liquid	lubricant / *textile industry; fibre production*
	ALUBRASPIN M-150-P		98% active	liquid	
	LAROSOL NFSD	modified nonionic		liquid	low foam; wetting agent; scouring agent / *dyeing; textile industry*
	LAROSOL PNC		100% active	liquid	non-foaming; levelling agent; lubricant / *dyeing; textile industry*
	MACOL 21			liquid; cloud point 24°C (1% aq.); viscosity 70 cPs	lubricant; wetting agent; solubiliser; detergent; low foam / *metal working*
	MACOL 30			liquid; cloud point 30°C (1% aq.); viscosity 65 cPs	lubricant; wetting agent; solubiliser; detergent; low foam / *metal working*
	MACOL 124	cetearyl alcohol and ceteareth 20	100% active	flakes; m.p. 52°C	emulsifier / *skin care products*
	MACOL 125	stearyl alcohol and ceteareth 20	100% active	flakes; m.p. 60°C	
	MACOL CPS	cetearyl alcohol, polysorbate 60, PEG 150 stearate and steareth 20	100% active	flakes; m.p. 51°C	emulsifier / *skin care products*
	MACOL LF 110			liquid; cloud point 10°C (1% aq.); viscosity 140 cPs	lubricant; wetting agent; solubiliser; detergent; low foam / *metal working*
	MACOL LF 111			liquid; cloud point 11°C (1% aq.); viscosity 100 cPs	lubricant; wetting agent; solubiliser; detergent; low foam / *metal working*
	MACOL LF 115			liquid; cloud point 15°C (1% aq.); viscosity 100 cPs	lubricant; wetting agent; solubiliser; detergent; low foam / *metal working*
	MACOL LF 120			liquid; cloud point 20°C (1% aq.); viscosity 140 cPs	lubricant; wetting agent; solubiliser; detergent; low foam / *metal working*

Supplier	Trade name	Chemical description	Composition	General properties	Functionality / *Application*
PPG	MAZAWAX 163SS	cetearyl alcohol and polysorbate 60	100% active	flakes; m.p. 51°C	emulsifier / *skin care products*
	MAZAWET 77		100% active	cloud point 45°C (1% aq.); surface tension 28.1 dynes/cm (0.1% aq.); Draves wetting 4 sec	wetting agent; defoaming agent; solubiliser / *metal working; paint industry; inks; detergents; industrial cleaners*
	MAZAWET DF		100% active	cloud point 60°C (1% aq.); surface tension 27.3 dynes/cm (0.1% aq.); Draves wetting 6 sec.	
	MAZCLEAN W				detergent; foaming agent (moderate) / *metal working*
	MAZCLEAN W-10				
	MAZCLEAN WRI				detergent; rust inhibitor / *metal working*
	SAM 185 N	surface active monomer	100% active	HLB 6.9; beige waxy solid; viscosity 247 cPs (50°C); m.p. 39-42°C	emulsifier / *latex products*
	SAM 186 N	surface active monomer	100% active	HLB 9.9; beige waxy solid; viscosity 277 cPs (50°C); m.p. 39-42°C	
	SAM 187 N	surface active monomer	100% active	HLB 14.0; beige waxy solid; viscosity 170 cPs (65°C); m.p. 49°C	
Protex	SURFAROX 10AH10C / SURFAROX 10AH7C / SURFAROX 1620 AH 11C	compound of ethoxylates and decomposable oxypropylates			low foam
Rudolf Chemicals	RUCOGEN DFL	ethylene oxide condensation product		yellowish, oily liquid	detergent; dispersant; emulsifier; foaming agent / *textile industry*
	RUCOGEN DWA	ethylene oxide condensation product		clear, colourless liquid	detergent; low foam; dispersant; fat-dissolving agent; emulsifier / *textile industry*
	RUCOGEN SAS			clear liquid	scouring agent; dispersant; low foam / *textile industry*
Sandoz	ANTIMUSSOL 93/083	aliphatic hydrocarbons (mineral oil)		liquid	anti-foaming agent
	ANTIMUSSOL SF	emulsion of aliphatic hydrocarbons (mineral oil); silicone-free		liquid	
	ANTIMUSSOL UPZP	fatty acid derivative in mineral oil		liquid	
	ANTIMUSSOL WLN	aliphatic ether and emulsifier		liquid	
	CERANINE NRM	fatty acid ester/polyglycol ether solvent containing		liquid	softener / *cellulosic fibres*
	DERMAFINISH LB	mineral oil containing emulsifier		liquid	liquoring agent / *textile industry*
	FLAMCIDE NW	inorganic salts and nitrogen compound in aq. soln.		liquid	flame-proofing agent
	LYOGEN WL	fatty acid peptide and alkyl polyglycol ether		liquid	wetting agent; retarding agent / *dyeing; textile industry*
	SANDOPUR DSC	oxidising agent and polyglycol ether in aq. soln.		liquid	oxidising agent / *sulfur dyeing*
	SIRRIX LUNA	fatty acid and phosphate ester with enzyme		paste	additive / *stone washing of denim*

Supplier	Trade name	Chemical description	Composition	General properties	Functionality / Application
Scher Chemicals	LUBRISOL	natural oil	100% active	liquid	lubricant / *textile industry*
Surfachem	SURFAC LF SERIES	nonionic surfactants (low foaming)			
Thor Chemicals	SOVATEX NI2 (B)	fatty acid ethanolamine condensates		liquid	scouring agent; wetting agent; foaming agent (high)
Troy Chemicals	TROYSOL CD1			liquid	wetting agent; dispersant / *paint industry*
	TROYSOL MS2			liquid	slip improver / *coatings*
	TROYSOL UGA			liquid	wetting agent; dispersant; grinding agent / *pigments; paint industry*
Union Carbide	TRITON CA	speciality	100% active	HLB 11.1; liquid; pour point −18°C	dispersant / *paint industry*
	TRITON X-155,90%	speciality	90% active	HLB 12.5; liquid; cloud point <10°C (1% aq.); pour point −29°C	stabiliser / *textile industry; emulsion polymerisation*
	TRITON X-363M	alkylarylpolyether ethanol	100% active	liquid; pour point −32°C	emulsifier / *agriculture*
Warwick International	WARCODET LE	fatty glyceride blend			detergent; scouring agent; levelling agent / *textile industry*
	WARCOSOFT NI 150				softener / *textile industry*
Witco	ADSEE 775	nonionic surfactant blend		liquid	spreader sticker; wetting agent / *agrochemicals*
	ADSEE 799	nonionic surfactant blend		liquid	penetrant / *agrochemicals*
	ADSEE 801	nonionic surfactant blend		liquid	wetting agent / *agrochemicals*
	ADSEE 1080	nonionic surfactant blend		liquid	
	ADSEE AK31-73	nonionic surfactant blend		liquid	emulsifier; foaming agent / *agrochemicals*
	BUBBLE BREAKER DMD-1	complex surfactant		liquid; pH 10.5	defoaming agent / *petroleum industry*
	BUBBLE BREAKER DMD-2	complex surfactant		liquid; pH 10.5	defoaming agent / *petroleum industry*
	REWOCOROS B 2045	alkenyl sulfosuccinic acid anhydride	100% active	liquid	anti-corrosion
	REWOLUB GSM	blend of nonionics	100% active	liquid	lubricant / *metal working*
	REWOPAL MT 65	fatty alcohol polyglycol methyl ether	100% active	liquid	low foam / *textile industry; cleaners*
	REWOPAL MT 2540	fatty alcohol polyalkylenoxide methyl ether	100% active	liquid	
	REWOPAL MT 5722	fatty alcohol polyalkylenoxide methyl ether	100% active	liquid	phase transfer catalyst / *textile industry*
	SPONTO 217	nonionic surfactant blend		liquid	*agrochemicals*

Supplier	Trade name	Chemical description	Composition	General properties	Functionality / *Application*
Witco	SPONTO AG-540	nonionic surfactant blend		liquid	emulsifier / *agrochemicals*
	SPONTO AG-1040	nonionic surfactant blend		liquid	*agrochemicals; orchard and crop oil*
	SPONTO AG-1265	nonionic surfactant blend		liquid	*agrochemicals; herbicides*
	SPONTO AG3-55T	nonionic surfactant blend		liquid	*agrochemicals; herbicides*
	SPONTO AK30-02BT	nonionic surfactant blend		liquid	*agrochemicals; insecticides*
	SPONTO AK31-53	nonionic surfactant blend		liquid	emulsifier / *agrochemicals*
	SPONTO AK31-56	nonionic surfactant blend		liquid	*agrochemicals; insecticides*
	SPONTO AK31-64	nonionic surfactant blend		liquid	foaming agent / *agrochemicals*
	SPONTO AK31-66	nonionic surfactant blend		liquid	*agrochemicals; insecticides*
	SPONTO AK31-69	nonionic surfactant blend		liquid	*agrochemicals; orchard and crop oil*
	SPONTO AL69-49	nonionic surfactant blend		liquid	*agrochemicals; insecticides*
	SPONTO CA-861	nonionic surfactant blend		liquid	compatibility agent / *agrochemicals*
	SPONTO H-44-C	nonionic surfactant blend		liquid	*agrochemicals; insecticides*
	SPONTO N-140 B	nonionic surfactant blend		liquid	*agrochemicals; insecticides*
	WITBREAK 772	surfactant blend		liquid; pour point −13.3°C; pH 6.5	demulsifier; slugging compound / *petroleum industry*
	WITBREAK 774	surfactant blend		liquid; pH 6.5	
	WITBREAK DRI-9020	polyol		liquid; pour point −15°C; pH 7	
	WITBREAK DRI-9037	polyol		liquid; pH 6	demulsifier / *petroleum industry*
	WITBREAK DRI-9038	polyol		liquid; pH 5	
	WITCOMUL 4016	surfactant blend		liquid; pour point −12.2°C; pH 7.3	oil slick dispersant / *petroleum industry*
	WITCONATE 93S	amine dodecylbenzenesulfonate	91% active; sulfate 1.5%	liquid; pour point −3.9°C; pH 4.5	coupling agent; detergent; dispersant; o/w emulsifier; wetting agent; demulsifier / *personal care products; household and industrial applications; petroleum industry*
	WITCONATE P-1020 BUST	nonionic surfactant blend		liquid	
	WITCONOL NS-108LQ	nonionic surfactant blend		liquid	
	WITCONOL NS-500K	nonionic surfactant blend		liquid	*agrochemicals*
	WITCONOL NS-500LQ	nonionic surfactant blend		liquid	
Zschimmer & Schwarz	CEFASOFT FK	fatty acid condensate		liquid dispersion	
	CEFASOFT NI	fatty acid condensate		liquid dispersion	softener / *textile industry*
	CEFASOFT SEZ	silicone elastomer and fatty acid condensate		liquid emulsion	
	CEFASOFT UK	fatty acid condensate		liquid dispersion	softener; antistatic agent / *textile industry*

Supplier	Trade name	Chemical description	Composition	General properties	Functionality / Application
Zschimmer & Schwarz	CELLANA 568	mineral oil with special emulsifiers		clear, light brown liquid	lubricant / textile industry
	CELLANA VAW	high-grade white oil with special emulsifiers		clear yellow liquid	
	COLLALUB GA	polyoxyalkylene		white flakes	sizing agent; afterwaxing agent / textile industry
	CONTRIPON OC	hydrocarbons and emulsifiers; silicone-free		clear colourless liquid	anti-foaming agent / textile industry
	CONTRIPON TE	hydrocarbons and emulsifiers; silicone-free		clear colourless liquid	
	ENSIMOL B KONZ.	ethylene oxide adduct		clear colourless liquid	
	ENSIMOL FW	ethylene oxide adduct with smoothing additives		clear yellow liquid	waxing agent / textile industry
	ENSIMOL HF	ethylene oxide adduct with special additives		clear yellow liquid	
	ENSIMOL KW KONZ.	ethylene oxide adduct		clear, straw-coloured liquid	waxing agent; antistatic agent / textile industry
	ENSIMOL NW	ethylene oxide adduct with special additives		light brown wax in bars or tablets	
	ENSIMOL OC	ethylene oxide adduct with special additives		clear colourless liquid	waxing agent / textile industry
	ENSIMOL PF	ethylene oxide adduct with special additives		yellow liquid	
	ENSIMOL WS	ethylene oxide adduct		white flakes	
	NEWALOL 1477	alkyl polyoxyethylene		colourless liquid	low foam; wetting agent / textile industry
	NEWALOL NF	polyalkylene oxide product		colourless liquid	foam free; wetting agent; scouring agent; dispersant / textile industry
	SETAVIN DA	ethylene oxide adduct		yellowish viscous liquid	levelling agent; dispersant; penetrant / textile industry
	SETAVIN EW	polyamide derivative		clear colourless liquid	anti-crease agent / textile industry
	SETAVIN KE	ethylene propylene oxide adduct		colourless clear liquid	levelling agent / textile industry
	SETAVIN KS	ethylene oxide adduct		clear yellowish liquid	levelling agent; dispersant / textile industry
	SETAVIN MSN	polyglycol ether		clear light-brown liquid	levelling agent / textile industry
	SETAVIN PE	ethylene oxide adducts		clear brown liquid	
	SETAVIN SU	combination of ethylene oxide adducts		clear yellowish liquid	
	SINCAL SW-T	special enzymes and surfactants		clear brown liquid	washing agent; detergent / textile industry
	TISSOCYL LF	polyalkyleneoxide		white pourable liquid	foam free; washing agent; detergent / textile industry
	TISSOCYL USB	combination of ethoxylates and emulsifiers		viscous liquid	low foam; washing agent; wetting agent; detergent / textile industry
	TORSINOL VAW	high-grade white oil with a special emulsifying system		clear, bright yellow liquid	textile industry
	TORSINOL ZSB	white oil with nonionic emulsifying agents		clear, bright yellow liquid	textile industry
	ZETESAL OL	ethylene oxide adduct		clear viscous liquid	binder; aftertreatment agent / textile industry
	ZETESAN DI	polyoxyethylate		viscous, clear yellowish liquid	dispersant; wetting agent; levelling agent / textile industry

5 Miscellaneous surfactants

Fluorochemicals

Supplier	Trade name	Chemical description	Composition	General properties	Functionality *Applications*
3M	FLUORAD FC-95	anionic fluorochemical surfactant based on potassium perfluoroalkyl sulfonate	100% active	free-flowing powder; decomposes at 390°C; pH 7.5 (0.1% aq. soln.)	wetting agent; foaming agent; foam stabiliser *metal working; plastics industry*
	FLUORAD FC-98	anionic fluorochemical surfactant based on potassium perfluoroalkyl sulfonate	100% active	free-flowing powder; decomposes at 390°C; pH 7.0 (0.1% aq. soln.)	
	FLUORAD FC-100	amphoteric fluorochemical surfactant	fluoroalkyl sulfonate, sodium salt 25%; diethylene glycol monobutyl ether 25%; H_2O 50%	dark amber viscous liquid; pour point 12°C	foaming agent *metallurgy; electrochemical industry*
DuPont	ZONYL FSA	anionic fluorosurfactant	solids 25%; diluent H_2O/isopropyl alcohol (37.5:37.5)	liquid; surface tension 22 dynes/cm (0.01% solids)	wetting agent; anti-corrosion *paint industry; waxes; polishes; adhesives; polymer industry; metal working*
	ZONYL FSD	cationic fluorosurfactant	solids 30%; diluent H_2O/hexylene glycol (10:57)	liquid; surface tension 19 dynes/cm (0.01% solids)	wetting agent; foaming agent; anti-corrosion *polymer industry; cleaning; hair care products*
	ZONYL FSE	anionic fluorosurfactant	solids 14%; diluent H_2O/ethylene glycol (62:24)	liquid; surface tension 27 dynes/cm (0.01% solids)	wetting agent *paint industry; waxes; polishes; polymer industry; hair care products*
	ZONYL FSJ	anionic fluorosurfactant	solids 40%; diluent H_2O/isopropyl alcohol (45:15)	liquid; surface tension 26 dynes/cm (0.01% solids)	wetting agent *paint industry; waxes; polishes; adhesives; graphic arts products; polymer industry; metal working; cleaning; hair care products*
	ZONYL FSN	nonionic fluorosurfactant	solids 40%; diluent H_2O/isopropyl alcohol (30:30)	liquid; surface tension 24 dynes/cm (0.01% solids)	wetting agent *paint industry; waxes; polishes; adhesives; graphics arts products; polymer industry; metal working; cleaners*
	ZONYL FSN-100	nonionic fluorosurfactant	solids 100%	thin paste; surface tension 24 dynes/cm (0.01% solids)	wetting agent
	ZONYL FSO	nonionic fluorosurfactant	solids 50%; diluent H_2O/ethylene glycol (25:25)	turbid liquid; surface tension 19 dynes/cm (0.01% solids)	wetting agent *paints industry; waxes; polishes; graphics arts products; polymer industry; cleaning*
	ZONYL FSO-100	nonionic fluorosurfactant	solids 100%	turbid liquid; surface tension 19 dynes/cm (0.01% solids)	wetting agent
	ZONYL FSP	anionic fluorosurfactant	solids 35%; diluent H_2O/isopropyl alcohol (45:20)	liquid; surface tension 24 dynes/cm (0.01% solids)	wetting agent; anti-foam; anti-corrosion *paint industry; adhesives; polymer industry; metal working; hair care products*
	ZONYL TBS	anionic fluorosurfactant	solids 33%; diluent H_2O/acetic acid (64:3)	slurry; surface tension 38 dynes/cm (0.01% solids)	wetting agent; foaming agent *graphic arts products; polymer industry; cleaning; hair care products*
	ZONYL UR	anionic fluorosurfactant	solids 100%	paste; surface tension 40 dynes/cm (for product neutralised to pH 7-8 NaOH); m.p. ca. 50°C	wetting agent; anti-foam; anti-corrosion *adhesives; polymer industry; metal working; cleaners*

Supplier	Trade name	Chemical description	Composition	General properties	Functionality Application
Elf Atochem	FORAFAC 1033 D	polyfluorosulfonic acid; anionic	dry matter 30%; solvent medium H_2O	liquid; surface tension 46 dynes/cm (0.01% aq.), 23.3 dynes/cm (0.1% aq.)	
	FORAFAC 1033	polyfluorosulfonic acid; anionic	dry matter 50%; solvent medium methanol	liquid; surface tension 34 dynes/cm (0.01% aq.), 18.7 dynes/cm (0.1% aq.)	wetting agent; emulsifier; antimisting agent; *chromium plating; metal working; polymerisation*
	FORAFAC 1072	polyfluorochloroalkyl phosphoric acid; anionic	dry matter 10%; solvent medium H_2O	liquid; surface tension 56 dynes/cm (0.01% aq.), 33.6 dynes/cm (0.1% aq.)	
	FORAFAC 1098	polyfluoroalkyl pyridinium salt; cationic	dry matter 50%; solvent medium methanol	liquid; surface tension 34.7 dynes/cm (0.01% aq.), 17 dynes/cm (0.1% aq.)	*fire-fighting*
	FORAFAC 1110 D	ethoxylated polyfluoroalcohol; nonionic	dry matter 40%; solvent medium H_2O/isopropanol	liquid; surface tension 23 dynes/cm (0.01% aq.), 17.5 dynes/cm (0.1% aq.)	wetting agent; spreading agent; antimisting agent *inks*
	FORAFAC 1157	polyfluoroalkyl betaine; amphoteric	dry matter 27%; solvent medium ethanol/H_2O	liquid; surface tension 20 dynes/cm (0.01% aq.), 15.8 dynes/cm (0.1% aq.)	*fire-fighting*
	FORAFAC 1157 N	polyfluoroalkyl betaine; amphoteric	dry matter 27%; solvent medium ethanol/H_2O	liquid; surface tension 17.2 dynes/cm (0.01% aq.), 16.5 dynes/cm (0.1% aq.)	*fire-fighting*
	FORAFAC 1159	amino polyfluorosulfonate; anionic	dry matter 12.5%; solvent medium FORANE 113	liquid	cleaning solvent additive *electronics industry*
	FORAFAC 1176	polyfluorosulfonic salt; anionic	dry matter 100%	powder; surface tension 45.5 dynes/cm (0.01% aq.), 23.2 dynes/cm (0.1% aq.)	wetting agent; emulsifier *chromium plating; metal working; polymerisation*
	FORAFAC 1179	ammonium polyfluoroalkyl salt; cationic	dry matter 40%; solvent medium ethanol/H_2O	liquid; surface tension 33.2 dynes/cm (0.01% aq.), 18 dynes/cm (0.1% aq.)	spreading agent *floor coatings*
	FORAFAC 1183	amine oxide with perfluorinated chain; amphoteric	dry matter 40%; solvent medium ethanol/H_2O	liquid; surface tension 17.8 dynes/cm (0.01% aq.), 15 dynes/cm (0.1% aq.)	*fire-fighting*
	FORAFAC 1185 D	polyfluorosulfonic acid salt; anionic	dry matter 40%; solvent medium ethanol/H_2O	liquid; surface tension 59 dynes/cm (0.01% aq.), 30 dynes/cm (0.1% aq.)	wetting agent *chromium plating*
	FORAFAC 1187	polyfluorosulfonic acid salt; anionic	dry matter 20%; solvent medium H_2O	liquid; surface tension 45 dynes/cm (0.01% aq.), 24.2 dynes/cm (0.1% aq.)	spreading agent *floor coatings*
	FORAFAC 1191	fluorinated acrylic polymer; nonionic	dry matter 28%; solvent medium methyl ethyl ketone	liquid	wetting agent; spreading agent *inks; paint industry*
	FORAFAC 1199 D	fluorinated telomer; nonionic	dry matter 20%; solvent medium H_2O	liquid; surface tension 27 dynes/cm (0.01% aq.), 17.5 dynes/cm (0.1% aq.)	*fire-fighting*
Hoechst	FLUOWET EA	perfluoro-alkyl ethanol	99% active		chemical intermediate
	FLUOWET EA93	perfluoro-alkyl ethanol	92-94% active		
	FLUOWET EA612N	perfluoro-alkyl ethanol	96-98% active		
	FLUOWET EA800	perfluoro-alkyl ethanol	99% active		
	FLUOWET EA812	perfluoro-alkyl ethanol	96-98% active		
	FLUOWET EI800	perfluoro-alkyl ethyl iodide	99% active		
	FLUOWET OL	anionic fluorinated surfactant	50% active		flow agent *polishes*
	FLUOWET OTN	nonionic fluorinated surfactant	100% active		
	FLUOWET PL80	perfluorinated phosphoric acid product	80% active		wetting agent *acid baths*
	FLUOWET PL80B	perfluorinated phosphoric acid product	80% active		
	FLUOWET SB	anionic fluorinated surfactant	30% active		flow agent; spreading agent *polishes*

Supplier	Trade name	Chemical description	Composition	General properties	Functionality / *Application*
Sandoz	ANTIMUSSOL CAN	perfluoro organic acids in aq. soln.		liquid	anti-foaming agent

Glucosides

Supplier	Trade name	Chemical description	Composition	General properties	Functionality / *Application*
Amerchol	GLUCAM E-10	methyl gluceth-10		pale yellow medium syrup	humectant; emollient; moisturiser *shampoos; shaving products*
	GLUCAM E-20	methyl gluceth-20		pale yellow thin syrup	emollient; humectant; foam booster; lubricant *personal care products; soaps*
	GLUCAM E-20 DISTEARATE	methyl gluceth-20 distearate			emollient; lubricant *soaps; hair care products*
	GLUCAM P-10	PPG-10 methyl glucose ether		pale yellow heavy syrup	emollient; humectant; foam booster *shampoos; hair care products; cosmetics*
	GLUCAM P-20	PPG-20 methyl glucose ether	OH value 160-180	pale yellow medium syrup	emollient; lubricant; humectant *perfumes; shampoos; shaving products; cosmetics; hair care products*
	GLUCAM P-20 DISTEARATE	PPG-20 methyl glucose ether distearate			emollient; moisturiser; dispersant; binder *hair care products; cosmetics*
	GLUCAMATE DOE-120	PEG-120 methyl glucose dioleate			thickener; solubiliser *perfumes; personal care products*
	GLUCATE DO	methyl glucose dioleate			lubricant; emulsifier; dispersant *antiperspirants; cosmetics*
	GLUCATE SS	methyl glucose sesquistearate			w/o emulsifier *cosmetics; pharmaceuticals*
	GLUCOMATE SSE-20	PEG-20 methyl glucose sesquistearate		pale yellow soft liquid	o/w emulsifier; solubiliser *cosmetics; pharmaceuticals; perfumes*
	GLUCQUAT 100	lauryl methyl gluceth-10 hydroxypropyl dimonium chloride			*personal care products*
Auschem Cesalpinia	AEG 102/61	alkyl ethoxy glucoside	60% active	liquid	solubiliser; hydrotrope [
	AEG 124/41	alkyl ethoxy glucoside	50% active	liquid	*household and industrial detergents* [
	EUCAROL APG EC	alkyl polyglucoside citrate salt	30% active	liquid	detergent [
	EUCAROL APG ET	alkyl polyglucoside tartrate salt	30% active	liquid	*dishwashing agents; laundry products* [
	EUCAROL APG SS	alkyl polyglucoside sulfosuccinate	50% active	liquid	
BASF	LUTENSOL GD 70	alkyl polyglucoside	70% active	cloud point > 100°C (in NaCl soln.)	
Henkel	APG 300 CS	decyl polyglucose	50% active	clear viscous liquid	foaming agent [
	APG 325 CS	decyl polyglucose	50% active	clear viscous liquid	*shampoos; cleansing preparations* [
	APG 600 CS	lauryl polyglucose	50% active	clear viscous liquid	viscosity modifier; thickener [
	APG 625 CS	lauryl polyglucose	50% active	clear viscous liquid	*shampoos; cleansing preparations* [
	PLANTAREN 1200	lauryl polyglycose	48-52% active	viscous	

Supplier	Trade name	Chemical description	Composition	General properties	Functionality / Application
Henkel	PLANTAREN 2000	decyl polyglucose	48-52% active	viscous	
	PLANTAREN PS 10	lauryl polyglucose and sodium laureth sulfate	ca. 60% active	paste	
ICI	ATLAS G-2541	alkyl glucoside		HLB 14.0; dark brown liquid	household and industrial applications; agrochemicals; textile industry
PPG	MAZON 40	alkyl glucoside			coupling agent / metal working; cleaners
	MAZON 40A	modified alkyl glucoside			wetting agent; coupling agent / metal working; cleaners
Seppic	MONTANOL 68	cetearylglucoside	100% active	flakes	emulsifier / personal care products
	ORAMIX CG 110-60	capryryl/capryl glucoside	60% active	liquid	hair care products
	ORAMIX NS 10	decyl glucoside	55% active	liquid	hair care products
Union Carbide	TRITON BG-10	alkylpolyglucoside	70% active	liquid; cloud point > 100°C (1% aq.); pour point −12°C	bottle washing; metal cleaning; steam cleaning
	TRITON CG-110	alkylpolyglucoside	60% active	liquid; cloud point > 100°C (1% aq.); pour point −20°C	personal care products

Lanolin and wool wax derivatives

Supplier	Trade name	Chemical description	Composition	General properties	Functionality / Application
Amerchol	ACETULAN	cetyl acetate and acetylated lanolin alcohol	sapon. no. 180-200	light straw-coloured water-thin liquid	emollient; spreading agent; penetrant; lubricant / personal care products; cosmetics; pharmaceuticals
	AMERCHOL 400	petrolatum, lanolin alcohol, cetyl alcohol, lanolin and stearone	sapon. no. 8	anhydrous light cream soft solid	emollient; moisturiser; emulsifier; stabiliser / cosmetics; personal care products
	AMERCHOL BL	lanolin, mineral oil and lanolin alcohol	sapon. no. 60-70	yellow-amber semi-solid	w/o emulsifier / personal care products
	AMERCHOL C	petroleum, lanolin and lanolin alcohol	sapon. no. 10-20	pale anhydrous oil-miscible soft solid	w/o emulsifier; emollient / personal care products
	AMERCHOL CAB	petrolatum and lanolin alcohol	sapon. no. 1 max.	very pale anhydrous oil-miscible soft solid	w/o emulsifier; stabiliser; moisturiser / personal care products
	AMERCHOL H-9	petrolatum, lanolin and lanolin alcohol	sapon. no. 15-27	anhydrous oil-miscible soft solid	w/o emulsifier; emollient / pharmaceuticals; personal care products
	AMERCHOL L-99	mineral oil and lanolin alcohol	sapon. no. 2 max.	light yellow low-viscosity fluid	w/o emulsifier; stabiliser / skin care products; hair care products
	AMERCHOL L-101	mineral oil and lanolin alcohol	sapon. no. 1 max.	pale straw-coloured, anhydrous low-viscosity fluid	w/o emulsifier; moisturiser; penetrant; emollient / cosmetics; pharmaceuticals
	AMERCHOL L-500	mineral oil, lanolin alcohol and octyldodecanol	sapon. no. 5 max.	hazy, medium-viscosity yellow liquid	moisturiser; conditioner; w/o emulsifier; stabiliser / personal care products; hair care products; baby care products; pharmaceuticals
	AMERCHOL RC	petrolatum, lanolin alcohol, stearyl alcohol and stearone	sapon. no. 10 max.	cream-coloured soft solid	moisturiser; conditioner / cosmetics; personal care products

Supplier	Trade name	Chemical description	Composition	General properties	Functionality / Application
Amerchol	AMERLATE LFA	lanolin acid	sapon. no. 170-190	tan plastic wax	wetting agent; dispersant; emulsifier *cosmetics; personal care products*
	AMERLATE P	isopropyl lanolate	sapon. no. 130-155	pale yellow soft buttery solid	wetting agent; spreading agent; penetrant; lubricant; moisturiser; emollient; dispersant *cosmetics*
	AMERLATE W	isopropyl lanolate		yellow buttery solid	wetting agent; dispersant; lubricant *cosmetics; personal care products*
	AMERLATE WFA	lanolin acid	sapon. no. 165	tan waxy solid	wetting agent; dispersant; emollient *cosmetics; hair care products; personal care products*
	CERALAN	lanolin alcohol			
	ISOPROPYLAN 33	isopropyl palmitate and lanolin oil			
	ISOPROPYLAN 50	isopropyl palmitate and lanolin oil			*personal care products*
	LANOCERIN	lanolin wax			
	LANOGEL 21	PEG-27 lanolin			
	LANOGEL 31	PEG-40 lanolin			
	LANOGEL 41	PEG-75 lanolin			
	LANOGEL 61	PEG-85 lanolin			
	MODULAN	acetylated lanolin	sapon. no. 95-120	anhydrous yellow-amber semi-solid	emollient; foam booster *cosmetics; pharmaceuticals; personal care products; baby care products; soaps*
	OHLAN	hydroxylated lanolin		yellow to light tan waxy solid	emulsifier; stabiliser; wetting agent; dispersant; moisturiser *pharmaceuticals; personal care products; soaps*
	POLYLAN	oleyl linoleate and lanolin linoleate	sapon. no. 90-105	liquid	emollient; conditioner *pharmaceuticals; cosmetics; bath preparations*
	SOLULAN 5	laneth-5, ceteth-5, oleth-5 and steareth-5	OH value 120-135; 'n' no. 5	yellow-amber semi-solid	emulsifier; wetting agent; dispersant; conditioner; solubiliser; plasticiser *cosmetics; toiletries; pharmaceuticals; hair care products*
	SOLULAN 16	laneth-16, ceteth-16, oleth-16 and steareth-16	'n' no. 16	light tan waxy solid; cloud point 64-70°C (1% in 5% aq. NaCl)	emulsifier; softener; conditioner; solubiliser; plasticiser *cosmetics; toiletries; pharmaceuticals; hair care products*
	SOLULAN 25	laneth-25, ceteth-25, oleth-25 and steareth-25	'n' no. 25	light tan waxy solid; cloud point 82-88°C (1% in 5% aq. NaCl)	emulsifier; wetting agent; dispersant; conditioner; solubiliser; plasticiser *cosmetics; toiletries; pharmaceuticals; hair care products*
	SOLULAN 75	PEG-75 lanolin	'n' no. 75	light yellow-amber soft waxy solid; cloud point 80-87°C (1% in 5% aq. NaCl)	emulsifier; wetting agent; dispersant; solubiliser; plasticiser *cosmetics; toiletries; pharmaceuticals; hair care products*
	SOLULAN 97	polysorbate 80 acetate, cetyl acetate and acetylated lanolin alcohol	'n' no. 9; sapon. no. 110-130	light amber viscous liquid	o/w emulsifier; dispersant; solubiliser; plasticiser *cosmetics; toiletries; pharmaceuticals; hair care products*
	SOLULAN 98	polysorbate 80, cetyl acetate and acetylated lanolin alcohol	'n' no. 10; sapon. no. 65-75	light amber viscous liquid	o/w emulsifier; dispersant; solubiliser; plasticiser *cosmetics; toiletries; pharmaceuticals; hair care products*
	SOLULAN C-24	choleth-24 and ceteth-24	'n' no. 24	light yellow firm, waxy solid; cloud point 91°C (1% in 5% aq. NaCl)	o/w emulsifier; wetting agent; solubiliser; plasticiser *pharmaceuticals; cosmetics; hair care products*
	SOLULAN L-575	PEG-75 lanolin	'n' no. 75	50% aq. clear yellow-amber liquid; cloud point 80-87°C (2% in 5% aq. NaCl)	o/w emulsifier; wetting agent; dispersant; conditioner; solubiliser; plasticiser *shampoos; bubble baths; hair care products; soaps*

Supplier	Trade name	Chemical description	Composition	General properties	Functionality Application
Amerchol	SOLULAN PB-2	PPG-2 lanolin alcohol ether	OH value 127.5	amber soft liquid	emollient; emulsifier; solubiliser; plasticiser; *hair care products; cosmetics*
	SOLULAN PB-5	PPG-5 lanolin alcohol ether	OH value 98.5	light amber liquid	wetting agent; emulsifier; plasticiser; solubiliser; conditioner; *personal care products; hair care products*
	SOLULAN PB-10	PPG-10 lanolin alcohol ether	OH value 76.5	clear straw-coloured fluid	spreading agent; wetting agent; emollient; emulsifier; solubiliser; plasticiser; *bath preparations; bubble baths; hair products*
	SOLULAN PB-20	PPG-20 lanolin alcohol ether	OH value 57.0	light straw-coloured clear liquid	spreading agent; conditioner; emulsifier; solubiliser; plasticiser; *hair care products; bath preparations; cosmetics*
Auschem Cesalpinia	LANOTEX 730	ethoxylated lanolin		liquid	emollient; super-fatting agent
	STEROL LN 50	laneth-50	60% active	HLB 14.0; liquid	co-emulsifier; emollient; lubricant; thickener; *cosmetics; toiletries; hair care products; skin care products*
Cosmetochem	LANOCOS W	lanolin polyethylene glycol ether with 75-80EO		yellow, highly viscous liquid	solubiliser; emulsifier; *shampoos; bath preparations; perfumes*
Croda	ACYLAN	acetylated lanolin; non-allergenic		yellow soft solid	emollient; conditioner; *cosmetics; baby care products*
	CHOLESTEROL USP	cholesterol		HLB 4.0; white crystalline powder	w/o emulsifier; emollient; conditioner; moisturiser; *cosmetics; hair conditioners; skin care products*
	CORONA PURE NEW LANOLIN	lanolin		HLB 4.0; yellow soft mass	emollient; w/o emulsifier; conditioner; moisturiser; superfatting agent; *cosmetics; skin care products; hair conditioners; soaps*
	CORONET	lanolin		HLB 4.0; yellow stiff mass	
	CRESTALAN A	isopropyl myristate and lanolin oil		clear yellow liquid	emollient; plasticiser; conditioner; pigment dispersant; *cosmetics; hair care products; skin care products*
	CRESTALAN B	isopropyl myristate and lanolin oil		clear yellow liquid	
	CRODALAN AWS	polysorbate 80, cetyl acetate and acetylated lanolin alcohol		clear yellow liquid	emollient; superfatting agent; emulsifier; *cosmetics; shampoos; bath preparations; antiperspirants*
	CRODALAN C24	proprietary product		off-white to pale yellow waxy solid	o/w emulsifier; emollient; conditioner; *skin care products; cosmetics*
	CRODALAN IPL	isopropyl lanolate		pale yellow, soft buttery solid	emollient; conditioner; *cosmetics*
	CRODALAN LA	cetyl acetate and acetylated lanolin alcohol		pale yellow liquid and/or crystalline solid	emollient; spreading agent; superfatting agent; *cosmetics; skin care products; hair care products*
	CROSTEROL SFA	petrolatum, mineral oil, lanolin alcohol and glyceryl laurate		pale yellow paste	emollient; superfatting agent; *soaps*
	FLUILAN	lanolin oil		clear golden viscous liquid	emollient; w/o emulsifier; conditioner; moisturiser; superfatting agent; pigment dispersant; *cosmetics; skin care products; hair care products; soaps*
	HARTOLAN	lanolin alcohol		HLB 4.0; golden brown hard wax	emollient; w/o emulsifier; conditioner; moisturiser; superfatting agent; *cosmetics; soaps*
	LANEXOL	PPG-12 PEG-50 lanolin		slightly turbid liquid	o/w emulsifier; emollient; plasticiser; *shampoos; detergents; hairspray preparations*
	LANOLIC ACID	lanolic acid		yellow waxy solid	emollient; w/o emulsifier; conditioner; moisturiser; superfatting agent; *skin care products; hair care products; cosmetics*

Supplier	Trade name	Chemical description	Composition	General properties	Functionality / *Application*
Croda	LANOSTEROL	lanosterol		white to pale yellow powder	emollient; gelling agent / *cosmetics; skin care products*
	LANPOLS	PEG-(n) lanolate		soft yellow solid	o/w emulsifier; solubiliser; wetting agent / *skin care products; cosmetics*
	PENTALAN	pentaerythritol ester of lanolin fatty acids		dark brown solid	anti-corrosion; lubricant / *leather industry*
	POLYCHOLS	laneth-(n)		golden yellow solid	o/w emulsifier; emollient; solubiliser; wetting agent / *cosmetics; skin care products*
	SATULAN	hydrogenated lanolin		white to off-white solid	emollient; w/o emulsifier; conditioner; moisturiser / *cosmetics*
	SOLAN E50	PEG-75 lanolin and H_2O		yellow to amber liquid	o/w emulsifier; wetting agent; solubiliser; emollient / *skin care products; hair care products*
	SUPER HARTOLAN	lanolin alcohol		HLB 4.0; yellow hard wax	emollient; w/o emulsifier; conditioner; moisturiser; superfatting agent / *cosmetics*
	SUPER STEROL ESTER	$C_{10/30}$ cholesterol/lanosterol esters		white or off-white soft solid	emollient / *moisturising preparations; cosmetics*
	WHITE SWAN	lanolin		HLB 4.0; yellow soft mass	emollient; w/o emulsifier; conditioner; moisturiser; superfatting agent; lubricant / *cosmetics; skin care products; hair conditioners; soaps; leather industry*
	YEOMAN	lanolin		HLB 4.0; yellow soft mass	emollient; w/o emulsifier; conditioner; moisturiser; superfatting agent / *cosmetics; skin care products; hair conditioners; soaps*
Elf Atochem	SELLIG LANO 30	lanolin 30EO; in course of development	100% active; 'n' no. 30	paste; cloud point 62°C (aq. soln. of 50 g/l NaCl)	
Henkel	ACETOL 1706	cetyl acetate and acetylated lanolin alcohol		light yellow liquid	emollient; penetrant; lubricant; solvent / *hair care products; cosmetics; baby care products; bath preparations; sunscreen preparations*
	EMERY 1650	anhydrous lanolin		yellow amber paste	emulsifier; emollient; lubricant; conditioner / *cosmetics; hair care products; sunscreen preparations*
	EMERY 1656	anhydrous lanolin		yellow amber paste	emulsifier; emollient; conditioner; lubricant / *ointments; veterinary products; industrial hand cleansers*
	EMERY 1660	ultra anhydrous lanolin		yellow amber paste	emulsifier; emollient; conditioner; lubricant / *cosmetics*
	EMERY 1730	liquid absorption base		yellow to straw-coloured liquid	emulsifier; emollient; penetrant; stabiliser / *personal care products; cosmetics; shampoos*
	EMERY 1732	liquid absorption base		yellow to straw-coloured liquid	emollient; stabiliser; moisturiser; emulsifier / *personal care products; hair care products; cosmetics*
	EMERY 1740	solid absorption base		yellow to straw-coloured soft solid	emollient; emulsifier; moisturiser / *personal care products; hair care products*
	EMERY 1747	solid absorption base		yellow to straw-coloured soft solid	emulsifier; emollient; moisturiser / *hair care products; skin care products*
	EMERY 1780	lanolin alcohol		straw yellow waxy solid	w/o emulsifier; emollient; viscosity modifier / *personal care products; hair care products; cosmetics*

Supplier	Trade name	Chemical description	Composition	General properties	Functionality / Application
Henkel	EMERY HP-2050	high-purity anhydrous lanolin		yellow amber paste	emulsifier; emollient; conditioner; lubricant / *personal care products; cosmetics; hair care products; skin care products; baby care products*
	EMERY HP-2060	high-purity ultra anhydrous lanolin (low odour)		pale yellow amber paste	moisturiser / *cosmetics; lipsticks*
	ETHOXYLAN 1685	PEG-75 lanolin		dark amber/brown waxy solid	emollient; emulsifier; dispersant; plasticiser; foam stabiliser / *antiperspirants; personal care products; cosmetics; hair care products; pharmaceuticals*
	ETHOXYLAN 1686	PEG-75 lanolin; 50% aq. soln. of Ethoxylan 1685		yellow/light amber liquid	
	ETHOXYOL 1707	emulsifying acetate ester		yellow to straw-coloured liquid	emollient; lubricant; penetrant / *personal care products; cosmetics; shampoos; liquid soaps*
	LANFRAX 1776	lanolin wax		light yellow waxy solid	w/o emulsifier; stabiliser; emollient; crystallisation inhibitor; film former / *personal care products; cosmetics; sunscreen preparations*
	LANFRAX 1779	lanolin wax (low odour)		light yellow waxy solid	
	LANOQUAT 1756	lanolin quaternary		clear amber viscous liquid	o/w emulsifier; emollient; conditioner / *liquid soaps; personal care products; shaving products; hair care products*
	LANOQUAT 1757	lanolin quaternary		clear amber viscous liquid	emollient; conditioner; emulsifier / *personal care products; cosmetics; shampoos; hair care products*
	LANTROL 1673	lanolin oil		amber viscous liquid	emollient; moisturiser; dispersant / *personal care products; hair care products; bath preparations; pharmaceuticals*
	LANTROL 1674	lanolin oil		amber viscous liquid	
	LANTROL AWS 1692	PPG-12-PEG-65 lanolin oil		light amber viscous liquid	emollient; plasticiser; solubiliser; conditioner / *hair care products; shaving products; personal care products; perfumes; antiperspirants*
	LANTROL HP-2073	high purity lanolin oil		amber viscous liquid	emollient; moisturiser; dispersant / *cosmetics; personal care products; hair care products; bath preparations; pharmaceuticals*
	LANTROL HP-2074	ultra high-purity lanolin oil (low odour)		pale amber viscous liquid	*skin care products; cosmetics*
	PROPOXYOL 1695	PPG-5 lanolin wax glyceride		amber waxy solid	*household and industrial applications; agrochemicals; textile industry*
ICI	ATLAS G-1441	POE-(40)-sorbitol lanolin derivative	'n' no. 20	HLB 14.0; amber paste	
	ATLAS G-1790	POE(20)-lanolin		HLB 11.0; brown waxy solid	
Dr. W. Kolb	SYMPATENS-LAN/300	PEG-30 lanolin	OH value 35; 'n' no. 30	HLB 13.0; yellow paste; viscosity 440 mPas (50°C); pH ca. 7 (1% aq.)	
	SYMPATENS-LAN/600	PEG-60 lanolin	OH value 20; 'n' no. 60	HLB 15.9; yellow solid; viscosity 465 mPas (50°C); pH ca. 7 (1% aq.)	w/o co-emulsifier; refatting agent / *cosmetics*
	SYMPATENS-LAN/750	PEG-75 lanolin	OH value 15; 'n' no. 75	HLB 16.7; yellow solid; viscosity 830 mPas (50°C); pH ca. 7 (1% aq.)	
Nikko Chemicals	NIKKOL BWA-5	laneth-5		yellowish brown paste	*cosmetics*
	NIKKOL BWA-10	laneth-10		yellowish brown solid	*cosmetics*
	NIKKOL BWA-20	laneth-20		yellowish brown solid	emulsifier; solubiliser / *cosmetics*
	NIKKOL BWA-40	laneth-40		yellowish brown solid	
	NIKKOL TW-10	POE lanolin		yellowish brown paste	emulsifier; super-fatting agent / *shampoos; hair care products*
	NIKKOL TW-20	POE lanolin		yellowish brown paste	
	NIKKOL TW-30	POE lanolin		yellowish brown paste	
Seppic	LANOXAL 75	lanolin 75EO	100% active; 'n' no. 75	HLB 18; wax	emollient

Supplier	Trade name	Chemical description	Composition	General properties	Functionality Application
Westbrook	ACETADEPS	acetylated lanolin	acid value 3.0 max.; OH value 8 max.; sapon. no. 107-127	colour 8-14/1.0-1.8 (yellow/red units in ¼ in. lovibond cell); m.p. 33-39°C	emollient; moisturiser; binder *antiperspirants; deodorants; baby care products; skin cleansing preparations; skin care products; cosmetics; shampoos; hair conditioners; sunscreen preparations*
	ALBALAN	lanolin wax	paraffins 1.0% max.; acid value 1.0 max.; iodine value 18-32; sapon. no. 96-114	colour 6.9/0.6-1.6 (yellow/red units in ¼ in. lovibond cell); m.p. 38-48°C	emulsifier; emollient; moisturiser; binder *skin care products; cosmetics; hair conditioners; sunscreen preparations; soaps*
	ANHYDROUS LANOLIN GRADE 1	anhydrous lanolin (pharmaceutical quality which can be produced to any pharmacopoeia requirement in the world)	paraffins 1.0% max.; acid value 1.0 max.; sapon. no. 90-105	colour 9-13/1.0-1.8 (yellow/red units in ¼ in. lovibond cell); m.p. 38-44°C	emulsifier; emollient; moisturiser; binder *skin cleansing preparations; skin care products; shampoos; shaving products; sunscreen preparations*
	ANHYDROUS LANOLIN GRADE 2	anhydrous lanolin (pharmaceutical quality which can be produced to any pharmacopoeia requirement in the world)	paraffins 1.0% max.; acid value 1.0 max.; sapon. no. 90-105	colour 14-20/1.5-3.0 (yellow/red units in ¼ in. lovibond cell); m.p. 38-44°C	emulsifier; emollient; moisturiser; binder *shampoos; cosmetics; sunscreen preparations; soaps*
	ANHYDROUS LANOLIN P.80	ultra-pure lanolin	paraffins 1.0% max.; acid value 1.0 max.; total pesticides 3.0 ppm max.; sapon. no. 90-105	colour 8-14/0.8-2.0 (yellow/red units in ¼ in. lovibond cell); m.p. 38-44°C	emulsifier; emollient; moisturiser *baby care products; skin cleansing preparations; skin care products; cosmetics; sunscreen preparations*
	ANHYDROUS LANOLIN P.95	ultra-pure lanolin	paraffins 1.0% max.; acid value 1.0 max.; total pesticides 1.0 ppm; sapon. no. 90-105	colour 8-14/0.8-2.0 (yellow/red units in ¼ in. lovibond cell); m.p. 38-44°C	
	ANHYDROUS LANOLIN P.95RA	ultra-pure lanolin; hypoallergenic grade	paraffins 1.0% max.; acid value 1.0 max.; free alcohol 3.0% max.; total pesticides 1.0 ppm max.; sapon. no. 90-105	colour 8-14/0.8-2.0 (yellow/red units in ¼ in. lovibond cell); m.p. 38-44°C	
	ANHYDROUS LANOLIN SUPERFINE	anhydrous lanolin (pharmaceutical quality which can be produced to any pharmacopoeia requirement in the world)	paraffins 1.0% max.; acid value 1 max.; sapon. no. 90-105	colour 5-8/0.5-1.0 (yellow/red units in ¼ in. lovibond cell); m.p. 38-44°C	emulsifier; emollient; moisturiser; binder *baby care products; skin cleansing preparations; cosmetics; personal care products; shampoos; hair conditioners; shaving products; soaps*
	ANHYDROUS TECHNICAL LANOLIN	anhydrous technical lanolin	acid value 1.5 max.; sapon. no. 83-108	colour 25-50 (yellow units ¼ in. lovibond cell); m.p. 36-44°C	
	AQUABASE		acid value 4.0 max.; iodine value 6.0% max.; sapon. no. 6-17	wax; m.p. 46-53°C; pH 4.0-8.0 (in aq. soln.)	emulsifier *baby care products; skin cleansing preparations; skin care products; cosmetics; sunscreen preparations*
	AQUABASE NF		iodine value 3.5% max.; OH value 178-192; sapon. no. 14 max.	wax; m.p. 50-54°C; pH 5.5-7.0 (in aq. soln.)	
	AQUALOSE L30	ethoxylated lanolin	acid value 5.0 max.; 'n' no. 30; sapon. no. 24-40	cloud point 68-75°C (aq); colour 13-18/1.3-2.0 (yellow/red units in ¼ in. lovibond cell); drop point 40-48°C; pH 3.5-7.0 (5% aq. soln.)	emollient; moisturiser *skin cleansing preparations; shampoos; cosmetics; soaps*
	AQUALOSE L75	ethoxylated lanolin	acid value 5.0 max.; 'n' no. 75; sapon. no. 10-26	cloud point 75-83°C (aq); colour 8-12/1.0-1.8 (yellow/red units in ¼ in. lovibond cell); drop point 45-52°C; pH 3.5-7.0 (5% aq. soln.)	emollient; moisturiser; solubiliser *shaving products; skin cleansing preparations; bubble baths; shampoos; hair conditioners; soaps*
	AQUALOSE LL100		acid value 4.0 max.; sapon. no. 5-20	cloud point 32-40°C (aq.); colour 5-10/0.7-1.8 (yellow/red units in ¼ in. lovibond cell); pH 3.5-8.0 (5% aq. soln.)	emollient; moisturiser; solubiliser; emulsifier; plasticiser *shaving products; skin cleansing preparations; bubble baths; hairspray preparations*
	AQUALOSE SLT	contains unreacted lanolin oil in a water soluble base	acid value 2.0 max.; sapon. no. 16-34	colour 3-7/0.3-1.5 (yellow/red units in ¼ in. lovibond cell); drop point 35-45°C; pH 3.5-8.0 (5% aq. soln.)	emollient; moisturiser; solubiliser; carrier *shaving products; skin cleansing preparations; shampoos; cosmetics*
	AQUALOSE SLW	contains unreacted lanolin oil in a water soluble base	acid value 5.0 max.; sapon. no. 20-36	soft, golden yellow wax; colour 12-18/1.0-2.7 (yellow/red units in ¼ in. lovibond cell); drop point 42-50°C; pH 3.5-8.0 (5% aq. soln.)	emollient; moisturiser; solubiliser; carrier *shaving products; skin cleansing preparations; shampoos; cosmetics*

Supplier	Trade name	Chemical description	Composition	General properties	Functionality/Application
Westbrook	AQUALOSE W15	ethoxylated lanolin alcohol	acid value 6.0 max.; sapon. no. 12.0 max.	cloud point 68-76°C (aq.); colour 18-25/1.0-3.0 (yellow/red units in ¼ in. lovibond cell); drop point 40-50°C; pH 3.5-8.0 (5% aq. soln.)	emollient; moisturiser; solubiliser; carrier antiperspirants; deodorants; bubble baths; shampoos; cosmetics
	AQUALOSE W20	ethoxylated lanolin alcohol	acid value 6.0 max.; 'n' no. 20; sapon. no. 6-18	cloud point 74-88°C (aq.); colour 18-25/1.0-3.0 (yellow/red units in ¼ in. lovibond cell); drop point 41-49°C; pH 3.5-7.0 (5% aq. soln.)	emollient; moisturiser; solubiliser; carrier shaving products; antiperspirants; deodorants; skin cleansing preparations; bubble baths; shampoos; cosmetics
	AQUAPHIL K		acid value 1.0 max.; free cholesterol 6% min.	drop point 36-42°C; m.p. 36-42°C	emollient; emollient; moisturiser baby care products; skin care products; cosmetics
	ARGOBASE 125	lanolin-derived free cholesterol with other sterols and high molecular weight alcohols	acid value 1.0 max.; iodine value 12.0% max.; OH value 9-16; sapon. no. 2.0 max.	clear liquid; colour 2.8/0.2-1.2 (yellow/red units in a 1 in. lovibond cell)	emollient; moisturiser; lubricant baby care products; bubble baths; shampoos; skin care products; sunscreen preparations
	ARGOBASE 500		acid value 2.0 max.; OH value 80-95; sapon. no. 5.0 max.	colour pale yellow	emulsifier; emollient; moisturiser; lubricant baby care products; skin care products; cosmetics; sunscreen preparations
	ARGOBASE EST		acid value 0.5 max.; sapon. no. 1.5 max.	m.p. 42-55°C	emulsifier; emollient; moisturiser baby care products; skin care products; cosmetics
	ARGOBASE L1		acid value 2.0 max.; iodine value 12.0% max.; sapon. no. 14-32	golden yellow viscous liquid; colour 2.6/0.2-1.0 (yellow/red units in ¼ in. lovibond cell)	emollient; moisturiser baby care products; skin care products; cosmetics; soaps
	ARGOBASE L2		acid value 1.0 max.; sapon. no. 32-42	colour pale yellow (yellow/red units in ¼ in. lovibond cell); drop point 30-38°C; m.p. 30-38°C	emollient; emollient; moisturiser baby care products; skin care products; cosmetics; shaving products
	ARGOBASE MS5		acid value 2.0 max.; OH value 12-28; sapon. no. 20-30	colour pale yellow	emulsifier; emollient; moisturiser baby care products; skin care products; cosmetics
	ARGOBASE S1	blend of anhydrous lanolin, long chain fatty lanolin alcohols	acid value 2.0 max.; iodine value 12-28%; sapon. no. 55-75	colour 6-13/0.7-1.6 (yellow/red units in ¼ in. lovibond cell); drop point 32-40°C; m.p. 32-40°C	emulsifier; emollient; moisturiser baby care products; skin care products; cosmetics
	ARGONOL 40	lanolin	acid value 2.0 max.; OH value 16 max.; sapon. no. 112-132	clear golden yellow liquid; cloud point 17.0°C max.; pour point 15.0°C; colour 6-11/0.8-1.7 (yellow/red units in ¼ in. lovibond cell)	emollient; moisturiser shaving products; antiperspirants; deodorants; baby care products; skin cleansing preparations; bubble baths; cosmetics; sunscreen preparations
	ARGONOL 50	lanolin	acid value 2.0 max. sapon. no. 92-118	liquid; cloud point 21.0°C max.; pour point 18.0°C; colour 6-11/0.9-1.7 (yellow/red units in ¼ in. lovibond cell)	emollient; moisturiser; emulsifier; binder antiperspirants; deodorants; baby care products; skin cleansing preparations; skin care products; cosmetics; bubble baths; sunscreen preparations
	ARGONOL 60	lanolin	acid value 2.0 max.; sapon. no. 85-105	liquid; cloud point 32.0°C max.; pour point 18.0°C max.; colour 6-11/0.9-1.7 (yellow/red units in ¼ in. lovibond cell)	emulsifier; emollient; moisturiser; solubiliser baby care products; bubble baths; cosmetics; sunscreen preparations
	ARGONOL ACE 5	lanolin derivative	acid value 1.0 max.; OH value 8 max.; iodine value 8-12%; sapon. no. 180-200	pale yellow liquid; colour 2-6/0.4-1.4 (yellow/red units in 1 in. lovibond cell)	emollient; moisturiser; lubricant; binder baby care products; skin care products; cosmetics; bubble baths; hairspray preparations; sunscreen preparations
	ARGONOL ISO	lanolin derivative	acid value 1.0 max.; iodine value 11-26%; sapon. no. 140-170	clear pale yellow liquid; cloud point 13.0°C max.; colour 2.7/0.2-1.4 (yellow/red units in ¼ in. lovibond cell)	emollient; moisturiser; plasticiser shaving products; baby care products; bubble baths; hairspray preparations; cosmetics; sunscreen preparations
	ARGONOL RIC 2	lanolin derivative	acid value 1.0 max.; iodine value 23-40%; sapon. no. 140-170	liquid; colour 2-6/0.2-1.0 (yellow/red units in ¼ in. lovibond cell)	emollient; moisturiser; lubricant; binder skin cleansing preparations; hair conditioners; cosmetics
	ARGOWAX COSMETIC SUPER		acid value 3.0 max.; cholesterol 10.0% min.; sapon. no. 15.0 max.	colour 6-10/0.6-1.5 (yellow/red units in ¼ in. lovibond cell); m.p. 50.0°C	emulsifier; emollient; moisturiser; plasticiser baby care products; skin care products; cosmetics; hairspray preparations; shaving products; sunscreen preparations

Supplier	Trade name	Chemical description	Composition	General properties	Functionality Application
Westbrook	ARGOWAX DISTILLED	distilled wools alcohol B.P.	acid value 2.0 max.; OH value 120-180; cholesterol 30.0% min.; sapon. no. 12.0 max.	hard semi-crystalline wax; colour 10-15/1-2 (yellow/red units in ¼ in. lovibond cell); m.p. 58.0°C min.	emulsifier; emollient; moisturiser; plasticiser *baby care products; skin care products; cosmetics; hairspray preparations; sunscreen preparations*
	ARGOWAX L.F.A. DISTILLED		acid value 138-152; OH value 24-46; sapon. no. 170-190	m.p. 50-60°C	emollient; moisturiser *shaving products; soaps*
	ARGOWAX STANDARD	wools alcohol B.P.	acid value 2.0 max.; OH value 120-180; cholesterol 30.0% min.; sapon. no. 12.0 max.	hard yellow wax; m.p. 58.0°C min.	emulsifier; emollient; moisturiser; plasticiser *baby care products; skin care products; cosmetics; hairspray preparations; sunscreen preparations*
	LANESTA G	glyceryl lanolate ester; produced from natural raw materials	acid value 3.0 max.; OH value 165 min.; sapon. no. 140-160	colour pale yellow; m.p. 42-50°C	emulsifier; emollient; moisturiser *baby care products; skin care products; sunscreen preparations*
	LANESTA GR	modified glyceryl lanolate	acid value 5.0 max.; sapon. no. 140-180	colour dark yellow; m.p. 40-55°C	emulsifier; emollient; moisturiser *skin care products; cosmetics; sunscreen preparations*
	LANESTA L	isopropyl lanolate	acid value 7.0 max.; sapon. no. 160-190	pale yellow liquid; pour point 10.0°C; colour 4-12/0.4-1.5 (yellow/red units in ¼ in. lovibond cell)	emollient; moisturiser; lubricant *baby care products; skin cleansing preparations; bubble baths; sunscreen preparations*
	LANESTA P	isopropyl lanolate	acid value 4.0 max.; iodine value 6-20%; sapon. no. 125-155	golden yellow, soft buttery solid; colour 24-30/3.0-3.9 (yellow/red units in ¼ in. lovibond cell); m.p. 28-38°C	emollient; moisturiser; lubricant *baby care products; skin cleansing preparations; cosmetics; soaps*
	LANESTA S	refined isopropyl lanolate	acid value 5.0 max.; iodine value 6-20%; sapon. no. 115-145	very pale yellow solid; colour 10-18/1.0-2.5 (yellow/red units in 1/2 in. lovibond cell); m.p. 30-39°C	emollient; moisturiser; lubricant; binder *cosmetics; sunscreen preparations; soaps*
	LANESTA SA-30	isopropyl lanolate	acid value 18.0 max.; OH value 48-68; sapon. no. 140-160	colour 8-15/0.8-2.0 (yellow/red units in ¼ in. lovibond cell); m.p. 30-40°C	emollient; moisturiser; binder; lubricant *baby care products; skin care products; cosmetics; soaps*
	SEBASE	isopropyllanolate	acid value 1.5 max.; sapon. no. 6-24	pale yellow, pearly substance; m.p. 30-38°C	emulsifier; emollient; moisturiser; carrier; binder *skin cleansing preparations; skin care products; cosmetics; sunscreen preparations*
Witco	LANOXYL 30	ethoxylated lanolin	100% active	wax	softener *household detergents; textile industry*
	REWOLAN 5	lanolin sulfosuccinate	50% active	liquid	super-fatting agent *cosmetics; skin care products*
	REWOLAN AWS	ethoxylated lanolin	100% active	liquid	super-fatting agent *shower preparations; foam baths; shampoos; hair spray preparations; hair care products*
	REWOLAN E 50	lanolin polyglycol ether	50% active	liquid	super-fatting agent *shower preparations; foam baths; shampoos; liquid soaps*
	REWOLAN E 100	lanolin polyglycol ether	100% active	wax	super-fatting agent *shower preparations; foam baths; liquid soaps*
	REWOLAN LP	isopropyllanolate	100% active	liquid	lipid *cosmetics; skin care products*

Polymer-based products

Akzo Nobel	ELFACOS E200	methoxy PEG-22/dodecyl glycol copolymer		HLB 8; paste	
	ELFACOS OW 100	methoxy PEG-17/dodecyl glycol copolymer		HLB 15; paste	

Supplier	Trade name	Chemical description	Composition	General properties	Functionality / Application
Akzo Nobel	ELFACOS ST 9	PEG-45/dodecyl glycol copolymer		HLB 6.5; paste	
	ELFACOS ST 37	PEG-22/dodecyl glycol copolymer		HLB 7.5; liquid	
BASF	SOKALAN CP 2	maleic acid-methylvinyl ether copolymer, sodium salt (polycarboxylate)	35% active	M.W. 70,000	
	SOKALAN CP 2 POWDER	maleic acid-methylvinyl ether copolymer, sodium salt (polycarboxylate)	92% active	powder; M.W. 70,000	
	SOKALAN CP 5	maleic acid-acrylic acid copolymer, sodium salt (polycarboxylate)	40% active	M.W. 70,000	
	SOKALAN CP 5 GRANULES	maleic acid-acrylic acid copolymer, sodium salt (polycarboxylate)	92% active	granules; M.W. 70,000	
	SOKALAN CP 5 POWDER	maleic acid-acrylic acid copolymer, sodium salt (polycarboxylate)	92% active	powder; M.W. 70,000	
	SOKALAN CP 7	polycarboxylate	40% active	M.W. 50,000	
	SOKALAN CP 7 GRANULES	polycarboxylate	92% active	granules; M.W. 50,000	
	SOKALAN CP 7 POWDER	polycarboxylate	92% active	powder; M.W. 50,000	
	SOKALAN CP 9	maleic acid-olefin copolymer, sodium salt (polycarboxylate)	25% active	M.W. 12,000	
	SOKALAN CP 10	modified polyacrylic acid, sodium salt (polycarboxylate)	45% active	M.W. 4,000	
	SOKALAN CP 10 S	modified polyacrylic acid (polycarboxylate)	50% active	M.W. 4,000	dispersant
	SOKALAN CP 12 S	maleic acid-acrylic acid copolymer (polycarboxylate)	50% active	M.W. 3,000	
	SOKALAN CP 13 S	modified polyacrylic acid (polycarboxylate)	25% active	M.W. 20,000	
	SOKALAN CP 45	partially neutralised polycarboxylate	45% active	M.W. 70,000	
	SOKALAN CP 45 GRANULES	partially neutralised polycarboxylate	92% active	granules; M.W. 70,000	
	SOKALAN CP 45 POWDER	partially neutralised polycarboxylate	92% active	powder; M.W. 70,000	
	SOKALAN DCS	mixture of aliphatic dicarboxylic acids (polycarboxylate)			
	SOKALAN HP 22	polycarboxylate	20% active	M.W. 24,000	
	SOKALAN HP 25	modified polycarboxylate	45% active	M.W. 3,000	
	SOKALAN HP 50	polyvinylpyrrolidone (polycarboxylate)	95% active	M.W. 40,000	
	SOKALAN HP 53	polyvinylpyrrolidone (polycarboxylate)	30% active	M.W. 40,000	
	SOKALAN PA 13 PN	partially neutralised polyacrylic acid, sodium salt (polycarboxylate)	54% active	M.W. 1,000	
	SOKALAN PA 15	polycarboxylate	45% active	M.W. 1,200	
	SOKALAN PA 20	polycarboxylate	45% active	M.W. 2,500	
	SOKALAN PA 25 PN	partially neutralised polycarboxylate	54% active	M.W. 4,000	

Supplier	Trade name	Chemical description	Composition	General properties	Functionality / Application
BASF	SOKALAN PA 25 PN GRANULES	partially neutralised polycarboxylate	92% active	granules; M.W. 4,000	
	SOKALAN PA 30 CL	polycarboxylate	45% active	M.W. 8,000	
	SOKALAN PA 30 CL GRANULES	polycarboxylate	92% active	granules; M.W. 8,000	
	SOKALAN PA 40	polycarboxylate	35% active	M.W. 15,000	
	SOKALAN PA 40 POWDER	polycarboxylate	92% active	powder; M.W. 15,000	dispersant
	SOKALAN PA 50	polycarboxylate	40% active	M.W. 30,000	
	SOKALAN PA 70 PN	partially neutralised polycarboxylate	30% active	M.W. 70,000	
	SOKALAN PA 80 S	polyacrylic acid (polycarboxylate)	35% active	M.W. 100,000	
	SOKALAN PA 110 S	polycarboxylate	35% active	M.W. 250,000	
	SOKALAN PM 10	maleic acid copolymer, sodium salt (polycarboxylate)	45% active	M.W. 1,200	
Hüls	POLYCARBOXYLATE AMC 60	acrylic acid-maleic acid copolymer	active detergent 40%	liquid	co-builder / *laundry products*
ICI	ATLAS G-3965	low molecular weight carboxylated polymer		straw-coloured viscous liquid	*household and industrial applications; agrochemicals; textile industry*
	ATLAS G-3975	low molecular weight carboxylated polymer		straw-coloured viscous liquid	
	ATLAS G-7264	aq. dispersion of a hydrophilic copolymer		cream liquid	
	ATLOX 4912	nonionic polymeric surfactant		HLB 5.5; red-brown waxy solid	emulsifier; dispersant / *agrochemicals*
	ATLOX 4913	nonionic polymeric surfactant		HLB 11.5; red-brown liquid	
	HYPERMER 1083	nonionic polymeric surfactant		HLB 5.0; light brown liquid	emulsifier / *polymer industry*
	HYPERMER 1599A	nonionic polymeric surfactant		HLB 5.5; red-brown paste	
	HYPERMER 2296	nonionic polymeric surfactant		HLB 5.0; brown liquid	
	HYPERMER A60	modified polyester nonionic surfactant		HLB 6.0; dark brown liquid	
	HYPERMER A95	modified polyester nonionic surfactant		HLB 6.0; yellow-brown liquid	
	HYPERMER A109	modified polyester nonionic surfactant		HLB 14.0; light brown liquid	emulsifier / *polymer industry; metal working*
	HYPERMER A200	modified polyester nonionic surfactant		HLB 7.0; yellow-brown liquid	
	HYPERMER A394	modified polyester nonionic surfactant		HLB 8.0; yellow-brown liquid	
	HYPERMER A409	modified polyester nonionic surfactant		HLB 9.0; yellow-brown liquid	
	HYPERMER CG6	aq. alcohol soln. of acrylic copolymer		HLB 12.0; red-brown liquid	dispersant / *paint industry; inks*
	HYPERMER D477	hydrocarbon soln. of polymeric surfactant		HLB 4.0; yellow-brown liquid	
	HYPERMER E475	nonionic polymeric surfactant		HLB 6.0; dark brown liquid	dispersant / *metal working*
	HYPERMER E476	nonionic polymeric surfactant		HLB 7.0; dark brown liquid	
	HYPERMER FP1	polymeric dispersant		amber-brown waxy paste	dispersant
	HYPERMER FP2	polymeric dispersant		light brown liquid	
	HYPERMER FP3	polymeric dispersant		amber-coloured solid	
	HYPERMER FP4	polymeric dispersant		grey-coloured solid	

Supplier	Trade name	Chemical description	Composition	General properties	Functionality Application
ICI	HYPERMER KD1	polymeric dispersant		amber-coloured solid	
	HYPERMER KD2	polymeric dispersant		amber-coloured liquid	
	HYPERMER KD3	polymeric dispersant		amber-light brown waxy paste/viscous liquid	
	HYPERMER KD4	polymeric dispersant		amber-light brown waxy paste/viscous liquid	
	HYPERMER KD5	polymeric dispersant		amber-brown waxy paste	
	HYPERMER LP1	polymeric dispersant		amber-light brown waxy paste/viscous liquid with slight couldiness	
	HYPERMER LP2	polymeric dispersant		blue powder	
	HYPERMER LP3	polymeric dispersant		amber-light brown waxy paste	
	HYPERMER LP4	polymeric dispersant		amber-light brown waxy paste/viscous liquid	
	HYPERMER LP5	polymeric dispersant		amber-light brown waxy paste/viscous liquid with slight cloudiness	
	HYPERMER LP6	polymeric dispersant in high-boiling petroleum fraction	36% active	amber liquid	dispersant
	HYPERMER LP7	polymeric dispersant in dioctylphthalate	45% active	amber liquid	
	HYPERMER LP8	polymeric dispersant in toluene	37% active	amber liquid	
	HYPERMER MT1	polymeric dispersant		amber-coloured solid	
	HYPERMER MT2	polymeric dispersant		amber coloured liquid	
	HYPERMER OM1	polymeric dispersant		amber waxy paste or liquid	
	HYPERMER OM2	polymeric dispersant		amber waxy paste or liquid	
	HYPERMER OM3	polymeric dispersant		amber light brown waxy paste/viscous liquid	
	HYPERMER OM4	polymeric dispersant		amber waxy paste or liquid	
	HYPERMER PS1	polymeric dispersant		blue powder	
	HYPERMER PS2	polymeric dispersant		light brown liquid	
	HYPERMER PS3	polymeric dispersant		amber-coloured solid	
	HYPERMER PS4	polymeric dispersant		grey-coloured solid	
Kao Corporation	AQUAKREEN KC 550	alyl-maleic copolymer	48-52% active	pH 4-6 (10%)	desalting agent *water treatment*
	CELLESH 50	acrylic homopolymer	39-41% active	pH 8-9	chelating agent; dispersant *textile industry; detergents*
	CELLESH 100	acrylic homopolymer	39-41% active	pH 8-9	chelating agent; dispersant; antiscale agent *textile industry; detergents; water treatment*
	CELLESH 200 E	anhydride maleic derivative copolymer	39-41% active	pH 8-9	chelating agent; dispersant; protective colloid *textile industry; detergents*
	CELLESH 200	acrylic-maleic copolymer	39-41% active	pH 8-9	chelating agent; dispersant; antiscale agent *textile industry; detergents; water treatment*
	POIZ 50	acrylic homopolymer	39-41% active	pH 8-9	pigment dispersant
	POIZ 100	acrylic homopolymer	39-41% active	pH 8-9	pigment dispersant
	POIZ 200	acrylic-maleic copolymer	39-41% active	pH 8-9	*paint industry*

Supplier	Trade name	Chemical description	Composition	General properties	Functionality / *Application*
PPG	LAROSOL 214-A	polymeric polyester/polyether			non-foaming; lubricant; soaping off agent / *dyeing; textile industry*
	LAROSOL 405-HS	polymeric polyester/polyether			non-foaming; lubricant; antistatic agent; soaping off agent / *dyeing; textile industry*
Rhone-Poulenc	BEVALOID 111	maleic copolymer	solids 25%	M.W. 12,000; viscosity 100 cP	dispersant / *paint industry*
	BEVALOID 116	acrylic acid homopolymer	solids 92%	powder; M.W. 10,000	detergent builder / *liquid and powder detergents*
	BEVALOID 202/25	acrylic acid homopolymer	solids 25%	M.W. 60,000; viscosity 250 cP	
	BEVALOID 207	acrylic acid homopolymer	solids 45%	M.W. 4500; viscosity 600 cP	scale inhibitor; detergent builder / *oil industry; liquid and powder detergents*
	BEVALOID 207/20	acrylic acid homopolymer	solids 26%	M.W. 4500; viscosity 10 cP	scale inhibitor / *industrial water treatment*
	BEVALOID 211	acrylic acid homopolymer	solids 44%	M.W. 3300, viscosity 240 cP	dispersant / *bulk slurry preparations; paper industry; paint industry*
	BEVALOID 211C	acrylic acid homopolymer	solids 44%	M.W. 3000; viscosity 250 cP	dispersant / *paper industry; bulk slurry preparations*
	BEVALOID 223	acrylic acid homopolymer	solids 44%	M.W. 3500; viscosity 275 cP	scale inhibitor; detergent builder / *oil industry; industrial water treatment; liquid and powder detergents*
	BEVALOID 226/35	acrylic copolymer	solids 35%	M.W. 3500; viscosity 60 cP	dispersant / *paper industry; paint industry*
	BEVALOID 245G	acrylic acid homopolymer	solids 92%	granular; M.W. 4500	detergent builder / *liquid and powder detergents*
	BEVALOID 6703	acrylic copolymer	solids 40%	M.W. 2500; viscosity 200 cP	dispersant / *bulk slurry preparations*
	BEVALOID 6708	acrylic terpolymer	solids 30%	M.W. 16,000; viscosity 125 cP	dispersant / *paper industry*
	BEVALOID 6744	acrylic terpolymer	solids 30%	M.W. 16,000; viscosity 125 cP	dispersant / *oil industry*
	BEVALOID 6755	acrylic copolymer	solids 40%	M.W. 5000; viscosity 50 cP	dispersant / *paint industry*
	BEVALOID 6770	acrylic copolymer	solids 50%	M.W. 4000; viscosity 1000 cP	scale inhibitor; dispersant / *oil industry; industrial water treatment; paint industry*
	BEVALOID 6777	acrylic acid homopolymer	solids 50%	M.W. 4500; viscosity 500 cP	scale inhibitor / *industrial water treatment*
	BEVALOID 6778	acrylic acid homopolymer	solids 54%	M.W. 2500; viscosity 1200 cP	scale inhibitor / *oil industry*
	BEVALOID 6780	acrylic acid homopolymer	solids 48%	M.W. 4500; viscosity 700 cP	detergent builder / *liquid and powder detergents*
	BEVALOID 6783	acrylic copolymer	solids 44%	M.W. 3000; viscosity 150 cP	scale inhibitor / *oil industry*
	BEVALOID 6785	acrylic terpolymer	solids 44%	M.W. 3500; viscosity 300 cP	
Zeneca	SOLSPERSE 3000	polymeric hyperdispersant	100% active; acid value 30-35	waxy paste/viscous liquid; pour point 35°C; viscosity ca. 40 poise; colour 11-14 (Gardner)	hyperdispersant / *inks; paint industry; plastics industry*
	SOLSPERSE 13240	polymeric hyperdispersant	40% soln. in toluene; acid value 6-8	amber liquid; viscosity ca. 1.5 poise; colour 5-8 (Gardner)	

Supplier	Trade name	Chemical description	Composition	General properties	Functionality / Application
Zeneca	SOLSPERSE 13650	polymeric hyperdispersant	50% soln. in diundecylphthalate; acid value 6-8	amber liquid; viscosity ca. 120 poise; colour 10 (Gardner)	hyperdispersant / *inks; paint industry; plastics industry*
	SOLSPERSE 13940	polymeric hyperdispersant	40% soln. in 240/260 aliphatic distillate; acid value 6-8	amber liquid; viscosity ca. 10 poise; colour 9 (Gardner)	
	SOLSPERSE 17000	polymeric hyperdispersant	100% active; acid value 17-23	waxy paste/viscous liquid; pour point 35°C; viscosity ca. 20 poise; colour 14 (Gardner)	
	SOLSPERSE 17270	polymeric hyperdispersant	70% soln. in toluene; acid value 12-16	amber liquid; viscosity ca. 10 poise; colour 12 (Gardner)	
	SOLSPERSE 17940	polymeric hyperdispersant	40% soln. in 240/260 aliphatic distillate; acid value 7-9	amber liquid; viscosity ca. 5 poise; colour 12 (Gardner)	
	SOLSPERSE 20000	polymeric hyperdispersant	100% active	amber liquid; viscosity ca. 3 poise; colour 8 (Gardner)	
	SOLSPERSE 21000	polymeric hyperdispersant	100% active; acid value 70-75	waxy paste/viscous liquid; pour point 20°C; viscosity ca. 20 poise; colour 11-13 (Gardner)	
	SOLSPERSE 24000SC	polymeric hyperdispersant	100% active; acid value 21-28	amber granules; colour 3 (20% soln.; Gardner); m.p. ca. 47°C	
	SOLSPERSE 27000	polymeric hyperdispersant	100% active	viscous yellow liquid; viscosity ca. 5 poise; colour 8 (Gardner); pH 6-8	

Protein-based surfactants

Supplier	Trade name	Chemical description	Composition	General properties	Functionality / Application
Croda	AMINOFOAM C	TEA-lauroyl collagen amino acids and H_2O		clear yellow liquid; M.W. 550	detergent; foaming agent (high) / *shampoos; skin cleansing preparations*
	AMINOFOAM W	potassium lauroyl wheat amino acids and H_2O		pale yellow viscous soln.; M.W. 400	detergent / *skin care products; hair care products*
	CRODACEL QL	PG-hydroxyethylcellulose lauryldimonium chloride		yellow clear viscous soln.	film former / *hair care products; skin care products*
	CRODACEL QM	PG-hydroxyethylcellulose cocodimonium chloride		yellow clear viscous soln.	
	CRODACEL QS	PG-hydroxyethylcellulose stearyldimonium chloride		opaque viscous liquid	
	CROQUAT K	lauryldimonium hydroxypropyl hydrolysed keratin		clear, dark amber soln.; M.W. 3000	conditioner; bodying agent / *hair care products*
	CROQUAT L	lauryldimonium hydroxypropyl hydrolysed collagen		clear yellow viscous liquid; M.W. 2500	
	CROQUAT M	cocodimonium hydroxypropyl hydrolysed collagen		clear yellow viscous liquid; M.W. 2500	softener; bodying agent / *hair conditioners*
	CROQUAT S	stearyldimonium hydroxypropyl hydrolysed collagen		opaque yellow semi-solid; M.W. 2700	
	CROQUAT SOYA	lauryl dimonium hydroxypropyl hydrolysed soy protein		viscous amber liquid; M.W. 5000	conditioner; moisturiser; bodying agent / *hair care products; bath and shower preparations*

Supplier	Trade name	Chemical description	Composition	General properties	Functionality / *Application*
Croda	CROQUAT WKP	cocodimonium hydroxypropyl hydrolysed keratin		clear amber soln.; M.W. 1000	conditioner; bodying agent / *hair care products*
	CROSILKQUAT	cocodimonium hydroxypropyl silk amino acids		yellow to pale amber clear soln.; M.W. 450	conditioner; moisturiser; bodying agent / *skin care products*
	CROTEIN Q	hydroxypropyl trimonium hydrolysed collagen		off-white powder; M.W. 12,000	
	HYDROTRITICUM QL	lauryldimonium hydroxypropyl hydrolysed wheat protein		clear light amber liquid; M.W. 4000	
	HYDROTRITICUM QM	cocodimonium hydroxypropyl hydrolysed wheat protein		clear light amber liquid; M.W. 4000	conditioner / *hair care products*
	HYDROTRITICUM QS	stearyldimonium hydroxypropyl hydrolysed wheat protein		opaque viscous dispersion	
	HYDROTRITICUM WQ	hydroxypropyl trimonium hydrolysed wheat protein		clear amber liquid; M.W. 3500	conditioner / *skin care products*
Henkel	BREVIOL WSM	protein hydrolysate with additives			wool protecting agent / *textile industry; dyeing*
	EGALISAL CS	protein hydrolysate			
	LAMEPON 4SK	potassium coco-hydrolysed animal protein	36% active	light yellow clear liquid	foaming agent / *shampoos; bath preparations; skin cleansing preparations*
	LAMEPON A	fatty acid protein condensate			dispersant / *textile industry; dyeing*
	LAMEPON PA-K	potassium abietoyl hydrolysed collagen	30-32% active	liquid	shampoos for greasy hair
	LAMEPON PA-TR	TEA-abietoyl hydrolysed collagen	30-32% active	clear amber liquid	shampoos for greasy hair
	LAMEPON S-TR	TEA-cocoyl-hydrolysed collagen	31.5-32.5% active	liquid	shampoos; skin cleansing preparations
	LAMEPON S	potassium cocoyl-hydrolysed collagen	31.5-32.5% active	clear light yellow liquid	shampoos; skin cleansing preparations; baby shampoos; bath preparations; shower preparations
	LAMEPON ST 40	TEA-cocoyl-hydrolysed collagen	40-41% active	clear light yellow liquid	shampoos; skin cleansing preparations; bath preparations; shower preparations
	LAMEPON UD	potassium undecylenoyl hydrolysed collagen	31-32%	clear amber liquid	shampoos; antidandruff shampoos
	LAMEQUAT L	lauryldimonium hydroxypropyl hydrolysed collagen	34-36%	clear, slightly viscous liquid	antistatic agent; conditioner / *hair care products; skin cleansing preparations*
	LAMESOFT 156	hydrogenated tallow glycerides and potassium cocoyl hydrolysed collagen	24-25% active	liquid	turbidity agent / *shampoos; foam baths*
	LAMESOFT LMG	glyceryl laurate and TEA-cocoyl hydrolysed collagen	24-26% active	liquid	refatting agent; thickener / *shampoos; shower preparations; foam baths*
	TEPHAL FL	protein fatty acid condensate			aftersoaping agent / *textile industry; dyeing*
McIntyre Group	MACKPRO KLP	quaternium-79 hydrolysed keratin	35% active	liquid; pH 5	
	MACKPRO MLP	quaternium-79 hydrolysed milk protein	35% active	liquid; pH 5	
	MACKPRO NLP	quaternium-79 hydrolysed collagen	40% active	liquid; pH 5	
	MACKPRO NLW	quaternium-79 hydrolysed wheat protein	35% active	liquid; pH 5	conditioner / *hair care products; skin care products*
	MACKPRO NSP	quaternium-79 hydrolysed silk	33% active	liquid; pH 5	
	MACKPRO SLP	quaternium-79 hydrolysed soy protein	35% active	liquid; pH 5	
	MACKPRO WLW	wheatgermamidopropyl hydroxypropyl dimonium hydrolysed wheat protein	37% active	liquid; pH 5	

Supplier	Trade name	Chemical description	Composition	General properties	Functionality / Application
McIntyre Group	MACKPRO WWP	wheatgermamidopropyl dimethylamine hydrolysed wheat protein	35% active	liquid; pH 5	conditioner / *hair care products; skin care products*
Nikko Chemicals	NIKKOL CCK-40	potassium salt of collagen peptide coco fatty acid condensate; aq. soln.		yellow liquid	foaming agent; cleansing agent / *cosmetics*
	NIKKOL CCN-40	sodium salt of collagen peptide coco fatty acid condensate; aq. soln.		yellow liquid	foaming agent; cleansing agent / *cosmetics*
Seppic	MONTEINE LCK	lipoprotein, potassium salt	40% active	liquid	foaming agent
	MONTEINE LCK 32	potassium coco-hydrolysed animal protein	30% active	liquid	*hair care products*
	MONTEINE LCS 30	sodium coco-hydrolysed animal protein	30% active	liquid	*hair care products*
	MONTEINE LCQ	cocamidopropyldimethylaminohydroxypropyl hydrolysed animal protein	30% active	liquid	conditioner / *hair conditioners; hair care products*
	MONTEINE LCT	TEA-coco-hydrolysed animal protein	40% active	liquid	foaming agent / *hair care products*
	MONTEINE PCO	palmitoyl animal collagen amino acids	100% active	flakes	anti-inflammatory / *personal care products*
	MONTEINE V	TEA-coco-hydrolysed animal protein and sorbitol	70% active	liquid	*hair care products*
	ORAMIX L 30	sodium lauroyl sarcosinate	30% active	liquid	foaming agent / *hair care products*
	ORAMIX O	oleylsarcosine, acid salt	93–95% active	liquid	co-emulsifier
	SIMULSOL SA 20 NA	laurylsarcosinate, sodium salt	30% active	liquid	anti-corrosion
	SIMULSOL SA 80	oleylsarcosine, acid salt	93–95% active	liquid	anti-corrosion
	SOMEPON T25	sodium methyl cocoyl taurate	30% active	paste	foaming agent / *hair care products*
Stepan Europe	BIO-SOFT MT 40	TEA cocohydrolysed animal protein; made-to-order	42% active	amber liquid	conditioner; foaming agent / *shampoos; personal care products; baby care products*
Witco	REWOTEIN CPK	protein fatty acid condensate, potassium salt	32% active	liquid	*shampoos; foam baths; shower gels; skin cleansing preparations*
	REWOTEIN CPT	protein fatty acid condensate, triethanolamine salt	32% active	liquid	*shampoos; foam baths; shower gels; skin cleansing preparations*

Silicone-based products

Supplier	Trade name	Chemical description	Composition	General properties	Functionality / Application
Auschem Cesalpinia	ANTISCHIUMA AB	silicone emulsion	8% active	emulsion	antifoaming agent
	ANTISCHIUMA L7	silicone emulsion	15% active	emulsion	*general purpose; textile, paper and paint industries; detergents*
Dr. Th. Boehme	TEBELUB	silicone product			
Chemax	CHEMAX DF-10	silicone defoamer	10% silicone additives	emulsion	defoaming agent; antifoaming agent / *paper industry; textile industry; paint industry; effluent treatment; cleaners; adhesives; metal working*
	CHEMAX DF-10A	silicone defoamer	10% silicone additives	emulsion	
	CHEMAX DF-30	silicone defoamer	30% silicone additives	emulsion	
	CHEMAX DF-100	silicone compound	100% silicone additives	liquid	

Supplier	Trade name	Chemical description	Composition	General properties	Functionality / Application
Th. Goldschmidt	ABIL B 8843	dimethicone copolyol		pale yellow liquid	plasticiser
	ABIL B 8851	dimethicone copolyol		pale yellow liquid	*hair care products; hairspray preparations; skin care products; deodorants; antiperspirants*
	ABIL B 8852	dimethicone copolyol		pale yellow liquid	
	ABIL B 8863	dimethicone copolyol		amber liquid	
	ABIL B 88183	dimethicone copolyol		pale yellow liquid	anti-cracking agent
	ABIL B 88184	dimethicone copolyol		pale yellow liquid	*soaps*
	ABIL WE 09	cetyl dimethicone copolyol, polyglyceryl-4 isostearate and hexyl laurate		pale yellow liquid	w/o emulsifier; *skin care products*
	ABIL WS 08	cetyl dimethicone copolyol, cetyl dimethicone, polyglyceryl-3 oleate and hexyl laurate		pale yellow liquid	w/o emulsifier; *skin care products*
	POLYMEKON 1-85	silicone based antifoam	10% active		antifoam; *water treatment; polymer industry*
	POLYMEKON 2-57	antifoam based on organo-modified siloxanes	20% active		antifoam; *polymer industry*
	POLYMEKON 2-80	antifoam based on organo-modified siloxanes	20% active		antifoam; *textile industry; polymer industry*
	POLYMEKON 2-82	antifoam based on organo-modified siloxanes	20% active		
	POLYMEKON 50	silicone based antifoam	50% active		antifoam; *textile industry; latexes; metal working*
	POLYMEKON 67	silicon based antifoam	10% active		antifoam; *textile industry; low foam cleaners; industrial applications*
	POLYMEKON 730	silicone based antifoam	30% active		antifoam; *textile industry; water treatment; industrial applications*
	POLYMEKON 1435	silicone based antifoam	20% active		antifoam; *textile industry; powder detergents*
	POLYMEKON 1488	antifoam based on organo-modified siloxanes	20% active		antifoam; *polymer industry; textile industry; metal working*
	POLYMEKON WM 20	silicone based antifoam	20% active		antifoam; *liquid and powder detergents; cleaners*
	TEGO ANTIFOAM 2-86	antifoam based on organo-modified siloxanes	20% active		antifoam; *polymer industry*
	TEGO ANTIFOAM 2-87	antifoam based on organo-modified siloxanes	20% active		antifoam; *polymer industry; textile industry; metal working*
	TEGOPREN 3012	copolymer consisting of polymethyl siloxanes and ethylene oxide/propylene oxide segments		cloud point 63°C; surface tension 27 dynes/cm (0.1% in H_2O)	
	TEGOPREN 3020	copolymer consisting of polymethyl siloxanes and ethylene oxide/propylene oxide segments		cloud point 32°C; surface tension 27 dynes/cm (0.1% in H_2O)	
	TEGOPREN 3021	copolymer consisting of polymethyl siloxanes and ethylene oxide/propylene oxide segments		cloud point 38°C; surface tension 28 dynes/cm (0.1% in H_2O)	dispersant
	TEGOPREN 3022	copolymer consisting of polymethyl siloxanes and ethylene oxide/propylene oxide segments		cloud point 45°C; surface tension 30 dynes/cm (0.1% in H_2O)	
	TEGOPREN 3070	copolymer consisting of polymethyl siloxanes and ethylene oxide/propylene oxide segments		cloud point 38°C; surface tension 30 dynes/cm (0.1% in H_2O)	levelling agent

Supplier	Trade name	Chemical description	Composition	General properties	Functionality / Application
Th. Goldschmidt	TEGOPREN 3110	polymethyl siloxane modified by long-chain alkyl groups		solid; m.p. 20-30°C	
	TEGOPREN 3130	polymethyl siloxane modified by long-chain alkyl aryl groups		liquid; viscosity 75 mPas	wetting agent; adhesion enhancer; film former / *plastics industry; lubricants*
	TEGOPREN 5830	copolymer consisting of polymethyl siloxanes and ethylene oxide/propylene oxide segments		cloud point <25°C; surface tension 30 dynes/cm (0.1% in H_2O)	o/w emulsifier
	TEGOPREN 5840	copolymer consisting of polymethyl siloxanes and ethylene oxide/propylene oxide segments		cloud point 30°C; surface tension 22 dynes/cm (0.1% in H_2O)	wetting agent / *agrochemicals*
	TEGOPREN 5842	copolymer consisting of polymethyl siloxanes and ethylene oxide/propylene oxide segments		cloud point 80°C; surface tension 30 dynes/cm (0.1% in H_2O)	o/w emulsifier
	TEGOPREN 5843	copolymer consisting of polymethyl siloxanes and ethylene oxide/propylene oxide segments		cloud point >85°C; surface tension 29 dynes/cm (0.1% in H_2O)	o/w emulsifier
	TEGOPREN 5847	copolymer consisting of polymethyl siloxanes and ethylene oxide/propylene oxide segments		cloud point 59°C; surface tension 23 dynes/cm (0.1% in H_2O)	wetting agent
	TEGOPREN 5851	copolymer consisting of polymethyl siloxanes and ethylene oxide/propylene oxide segments		liquid; cloud point 64°C; surface tension 30 dynes/cm (0.1% in H_2O)	defoaming agent / *oil industry*
	TEGOPREN 5852	copolymer consisting of polymethyl siloxanes and ethylene oxide/propylene oxide segments		cloud point <25°C	defoaming agent; dispersant / *paper industry*
	TEGOPREN 5863	copolymer consisting of polymethyl siloxanes and ethylene oxide/propylene oxide segments		cloud point 42°C; surface tension 30 dynes/cm (0.1% in H_2O)	defoaming agent; o/w emulsifier; softener; flow improver; wetting agent / *paper industry; fabric softener*
	TEGOPREN 5873	copolymer consisting of polymethyl siloxanes and ethylene oxide/propylene oxide segments		cloud point 31°C; surface tension 27 dynes/cm (0.1% in H_2O)	
	TEGOPREN 5878	copolymer consisting of polymethyl siloxanes and ethylene oxide/propylene oxide segments		cloud point <25°C; surface tension 21 dynes/cm (0.1% in H_2O)	wetting agent / *agrochemicals*
	TEGOPREN 5884	copolymer consisting of polymethyl siloxanes and ethylene oxide/propylene oxide segments		cloud point 71°C	softener / *fabric softener*
	TEGOPREN 6800	polymethyl siloxane modified by long-chain alkyl groups		solid/liquid; m.p. 24-26°C	
	TEGOPREN 6801	polymethyl siloxane modified by long-chain alkyl groups		liquid; viscosity 20 mPas; m.p. 8-12°C	wetting agent; coating agent / *plastics industry; metal working*
	TEGOPREN 6814	polymethyl siloxane modified by long-chain alkyl groups		liquid; viscosity 300 mPas; m.p. <7°C	
	TEGOPREN 6846	polymethyl siloxane modified by long-chain alkyl groups		solid; m.p. ca.60°C	
	TEGOPREN 6920	copolymer consisting of polymethyl siloxanes and organic quaternary groups	50% active	liquid; viscosity 350 mPas	antistatic agent; softener; wetting agent / *fabric softener; textile industry*
	TEGOPREN 6922	copolymer consisting of polymethyl siloxane and organic quaternary groups	50% active	liquid; viscosity 750 mPas	antistatic agent; softener; wetting agent / *fabric softener; textile industry*
	TEGOPREN 6950	copolymer consisting of polymethyl siloxane and betainic nitrogen groups	30% active	liquid; viscosity 75 mPas; surface tension 26 dynes/cm (0.1% in H_2O)	
	TEGOPREN 7006	polymethyl siloxane modified by long chain alkyl and polyether groups		liquid; viscosity 900 mPas; m.p. 0°C	dispersant; w/o emulsifier; softener; film former / *fabric softener; metal working / fabric softeners; metal working; textile industry; lubricants*
	TEGOSIPON 793	antifoam based on organo-modified siloxanes	100% active		antifoam / *textile industry; metal working; fermentation processes*

Supplier	Trade name	Chemical description	Composition	General properties	Functionality / Application
Th. Goldschmidt	TEGOSIPON 3062	antifoam based on organo-modified siloxanes	100% active		antifoam *textile industry; polymer industry; metal working*
	TEGOSIPON D 90	antifoam based on organo-modified siloxanes	100% active		antifoam *polymer industry*
	TEGOSIPON ESMK 93	antifoam based on organo-modified siloxanes	100% active		antifoam *textile industry; metal working; fermentation processes*
	TEGOSIPON N	silicone based antifoam	100% active		antifoam *pharmaceuticals; metal working; fermentation processes*
	TEGOSIPON T 5	silicone based antifoam	100% active		antifoam *pharmaceuticals; industrial applications*
	TEGOSIPON WM 100	silicone based antifoam	100% active		antifoam *liquid detergents; cleaners*
Hays Colours	ALTAF NFA	silicone based nonionic			antifoam
	ALTAF NFA/4	silicone based nonionic			antifoam
Henkel	BREVIOL JET	self-emulsifying silicone oil			defoaming agent *textile industry; dyeing*
	DEHYDRAN 1293	soln. of a special modified polydimethyl siloxane		clear, light yellow liquid; viscosity 5-15 mPas	defoaming agent; degassing agent *aq. coatings; emulsion and suspension polymerisation; plastic industry*
	LAMESPUMIN HW	silicone derivate			defoaming agent *textile industry; dyeing*
PPG	MASIL 263 FLUID	alkyl methyl polysiloxane	active silicone 98%	colourless liquid; viscosity 1200 cSt	
	MASIL 264 FLUID	alkyl methyl polysiloxane	active silicone 100%	colourless liquid; viscosity 825 cSt	
	MASIL 265 FLUID	mixed alkyl methyl polysiloxanes	active silicone 100%	clear yellow liquid; viscosity 1500 cSt; surface tension 28.0 dynes/cm	lubricant *rubber industry; plastics industry; textile industry; personal care products*
	MASIL 265HV FLUID	aryl methyl polysiloxane	active silicone 100%	clear yellow liquid; viscosity 2000 cSt; surface tension 22.5 dynes/cm	
	MASIL 280		non-volatiles at 105°C 98.5%	clear, colourless to pale amber; cloud point 72°C (1% aq.); viscosity 600 cSt; colour 2 max. (Gardner); surface tension 27.0 dynes/cm (1% aq.); Draves wetting > 500 sec (1% soln.); Ross-Miles foam (1% aq.): initial 107 mm, 5 min 101 mm; pH 5-7 (3% aq.)	
	MASIL 280 LP		non-volatiles at 105°C 98.5%	clear, colourless to pale amber; cloud point 63°C (1% aq.); viscosity 400 cSt; colour 2 max. (Gardner); surface tension 27.3 dynes/cm (1% aq.); Draves wetting 112 sec (1% soln.); Ross-Miles foam (1% aq.): initial 132 mm, 5 min 103 mm; pH 5-7 (3% aq.)	wetting agent; levelling agent; flow control agent; antistatic agent; antifogging agent; lubricant; emollient; anticaking agent *cosmetics; personal care products; industrial coatings; textile industry; glass cleaners; inks*
	MASIL 1066C		non-volatiles at 105°C 98.5%	clear, colourless to pale amber; cloud point 43°C (1% aq.); viscosity 1800 cSt; colour 2 max. (Gardner); surface tension 25.9 dynes/cm (1% aq.); Draves wetting > 500 sec (1% soln.); Ross-Miles foam (1% aq.): initial 49 mm, 5 min 43 mm; pH 5-7 (3% aq.)	

Supplier	Trade name	Chemical description	Composition	General properties	Functionality / Application
PPG	MASIL 1066D		non-volatiles at 105°C 98.5%	clear, colourless to pale amber; cloud point 39°C (1% aq.); viscosity 1200 cSt; colour 2 max. (Gardner); surface tension 25.2 dynes/cm (1% aq.); Draves wetting 229 sec (1% soln.); Ross-Miles foam (1% aq.): initial 28 mm, 5 min 23 mm; pH 5-7 (3% aq.)	wetting agent; levelling agent; flow control agent; antistatic agent; antifogging agent; lubricant; emollient; anticaking agent / *cosmetics; personal care products; industrial coatings; textile industry; glass cleaners; inks*
	MASIL 2132		non-volatiles at 105°C 98.5%	clear, colourless to pale amber; cloud point 42°C (1% aq.); viscosity 1500 cSt; colour 2 max. (Gardner); surface tension 25.6 dynes/cm (1% aq.); Draves wetting > 500 sec (1% soln.); Ross-Miles foam (1% aq.): initial 57 mm, 5 min 45 mm; pH 5-7 (3% aq.)	
	MASIL EM 201 B	reactive silicone		liquid emulsion	softener / *textile industry*
	MASIL EM 266 EMULSION	based on Masil 265 Fluid	active silicone 50%	milky white emulsion; viscosity 100 cSt	lubricant / *rubber industry; plastics industry; textile industry; personal care products*
	MASIL EM 266(35) EMULSION	based on Masil 265 Fluid	active silicone 35%	milky white emulsion; viscosity 80 cSt	
	MASIL EM 266HV EMULSION	based on Masil 265HV Fluid	active silicone 50%	milky white emulsion; viscosity 120 cSt	
	MASIL EM 401 A	reactive silicone		slightly viscous liquid emulsion	softener / *textile industry*
	MASIL SF 19		non-volatiles at 105°C 98.5%	clear, colourless to pale amber; cloud point < 2°C (1% aq.); viscosity 45 cSt; colour 2 max. (Gardner); surface tension 20.4 dynes/cm (1% aq.); Draves wetting 2 sec (1% soln.); Ross-Miles foam (1% aq.): initial 23 mm, 5 min 18 mm; pH 5-7 (3% aq.)	wetting agent; levelling agent; flow control agent; antistatic agent; antifogging agent; lubricant; emollient; anticaking agent / *cosmetics; personal care products; industrial coatings; textile industry; glass cleaners; inks*
	MAZU DF 210 SX	10% silicone emulsion			defoaming agent / *metal working*
Sandoz	ANTIMUSSOL APXW	emulsion of polydimethyl siloxane and silica		liquid	
	ANTIMUSSOL DW	polysiloxane in aq. emulsion		liquid	antifoam
	ANTIMUSSOL ED5	aq. emulsion of polydimethyl siloxane and silica		liquid	antifoam
	ANTIMUSSOL SI	polysiloxane in aq. emulsion		liquid	
	ANTIMUSSOL SIF	polysiloxane in aq. emulsion		liquid	antifoam / *food industry*
	CERANINE HFC LIQUID	polydimethyl siloxane in aq. emulsion		liquid	fabric softeners
	SANDOPERM FE	polysiloxane in aq. soln.		liquid	finishing agent / *textile industry*
	SANDOPERM FV	methyl hydrogen polysiloxane in aq. soln.		liquid	cross-linking agent / *textile industry*
	SANDOPERM HSS	polyether modified polysiloxane		liquid	softener / *textile industry*
Surfachem	SURFOAMER 17	silicone emulsion			antifoam

Supplier	Trade name	Chemical description	Composition	General properties	Functionality / *Application*
Thor Chemicals	SOVATEX AFA	highly concentrated aq. nonionic emulsion of a silicone anti-foam compound		viscous white liquid, dispersible in H_2O; pH 6.6-6.8 (1% soln.)	antifoam / *textile industry; leather industry; inks; latex; paper industry; oil industry; gas industry*
	SOVATEX AFA/3	aq. nonionic emulsion of a silicone anti-foam compound	active silicone defoamer 10%	viscous white emulsion, dispersible in H_2O; viscosity 300-1000 cPs; pH 4.0-4.6 (1% soln.)	antifoam / *food industry; medical industry*
	SOVATEX AFA/4	aq. nonionic emulsion of a silicone anti-foam compound		viscous white liquid, dispersible in H_2O; pH 6.6-6.8 (1% soln.)	antifoam / *textile industry; leather industry; inks; latex; paper industry; oil industry; gas industry*
	SOVATEX AFA/5	aq. nonionic emulsion of a silicone anti-foam compound		viscous white liquid, dispersible in H_2O; pH 6.6-6.8 (1%)	
	SOVATEX AFA/B	aq. nonionic emulsion of a silicone anti-foam compound		viscous white liquid, dispersible in H_2O; pH 6.6-6.8 (1% soln.)	
Zschimmer & Schwarz	AVIVAGE LS 10	silicone compound; nonionic		yellowish viscous liquid	softener / *textile industry*
	CONTRASPUM W CONC.	thixotropic silicone emulsion	25% active	liquid	defoaming agent
	CONTRIPON M-P	silicone emulsion; nonionic		white viscous emulsion	antifoam / *textile industry*
	CONTRIPON NK-P	silicone emulsion; nonionic		white liquid	
	CONTRIPON W CONC.	thixotropic silicone emulsion	33% active	liquid	defoaming agent
	CONTRIPON W-P	silicone emulsion; nonionic		white viscous liquid	antifoam / *textile industry*
	PRODUKT ZS 2945	polydimethyl siloxane	100% active	liquid	release agent / *syndet bars*
	SKINOTAN S 10	oxalkylated polysiloxane	100% active	liquid	additive / *hair care products*

Other surfactants including blends

Supplier	Trade name	Chemical description	Composition	General properties	Functionality / *Application*
Air Products	SURFYNOL CT-136	nonionic/anionic pigment grind aid		liquid	grinding aid; dispersant / *inks; pigments*
	SURFYNOL DF-58	self-emulsifying organically modified silicone defoamer	100% active		defoaming agent; wetting agent / *inks; coatings; pigments; adhesives; cements; latex; oil industry; textile industry*
	SURFYNOL DF-75	silicone-free additive			defoaming agent / *latex; inks; adhesives; pigments; coatings; paper industry; oil industry; cements*
	SURFYNOL DF-110	proprietary non-silicone defoamer based on acetylenic chemistry		white waxy solid	defoaming agent; deairentrainment aid / *inks; metal working; ceramics; adhesives; cements; coatings; oil industry; hard-surface cleaners; textile industry; electroplating*
	SURFYNOL DF-110D	proprietary non-silicone defoamer based on acetylenic chemistry			defoaming agent; deairentrainment aid / *inks; metal working; ceramics; adhesives; cements; coatings; paper industry; dyes; emulsion polymerisation; oil industry; textile industry*

Supplier	Trade name	Chemical description	Composition	General properties	Functionality / Application
Air Products	SURFYNOL DF-110S	proprietary non-silicone defoamer based on acetylenic chemistry		free-flowing powder	defoaming agent; deairentrainment aid; wetting agent inks; metal working; ceramics; adhesives; cements; oil industry; coatings
	SURFYNOL DF-210	silicone-free defoamer			defoaming agent adhesives; inks; paint industry; coatings; metal working
Akcros Chemicals	AGRILAN A	anionic/nonionic blend	100% active	brown liquid; pour point <0°C; viscosity 2400 cSt; pH 4.5-6.5 (1% aq.)	
	AGRILAN AEC189	anionic/nonionic blend	83% active	amber viscous liquid; pour point 1°C; viscosity 978 cSt (40°C); pH 5.0-8.0 (1% aq.)	
	AGRILAN AEC200	anionic/nonionic blend	85% active	amber viscous liquid; pour point 12°C; viscosity 468 cSt; pH 5.0-8.0 (1% aq.)	emulsifier agrochemicals
	AGRILAN AEC211	anionic/nonionic blend	85% active	amber viscous liquid; pour point 6°C; viscosity 930 cSt (40°C); pH 5.0-8.0 (1% aq.)	
	AGRILAN AEC310	anionic/nonionic blend	88% active	amber liquid; pour point <0°C; viscosity 740 cSt; pH 5.0-8.0 (1% aq.)	
	AGRILAN EA14		80% active; co-solvent content 20%	hazy white liquid; pour point <0°C; viscosity 32 cSt; pH 7.0 (1% aq.)	
	AGRILAN EA25		100% active	pale amber liquid; pour point <0°C; viscosity 155 cSt; pH 7.0 (1% aq.)	
	AGRILAN EA36		100% active	clear colourless liquid; pour point <0°C; viscosity 58 cSt; pH 7.0 (1% aq.)	
	AGRILAN EA47		100% active	clear straw liquid; pour point <0°C; viscosity 195 cSt; pH 9.6 (1% aq.)	activity optimiser agrochemicals
	AGRILAN EA58		100% active	hazy straw liquid; pour point 5°C; viscosity 58 cSt; pH 9.6 (1% aq.)	
	AGRILAN EA69		100% active	brown liquid; pour point 0°C; viscosity 250 cSt; pH 9.5 (1% aq.)	
	AGRILAN EA80		100% active	pale yellow liquid; pour point <0°C; viscosity 366 cSt; pH 6.5 (1% aq.)	
	AGRILAN F460		60% active	clear pale yellow liquid; pour point −12°C; viscosity 230 cSt; pH 6.7 (1% aq.)	wetting agent; additive agrochemicals
	AGRILAN FS101	epoxidised vegetable oil	100% active	clear liquid; pour point −13°C; viscosity 350 cSt	stabiliser agrochemicals
	AGRILAN FS112	epoxidised vegetable oil	100% active	clear liquid; pour point −13°C; viscosity 800 cSt	
	AGRILAN WP101		85% active; inorganic salt content 11%	off-white powder/flake; pH 9.5 (10% aq.)	wetting agent agrochemicals
	AGRILAN WP112		60% active	white powder; pH 6.6 (10% aq.)	
	AGRILAN WP145		84% active; inorganic salt content 11%	tan powder; pH 9.5 (10% aq.)	dispersant agrochemicals
	AGRILAN WP156		87% active; inorganic salt content 6.5%	tan powder; pH 9.5 (10% aq.)	
	AGRILAN WP167		75% active; inorganic salt content 20%	off-white powder; pH 9.0 (10% aq.)	wetting agent agrochemicals
	AGRILAN WP178		96% active	off-white powder; pH 6.5 (10% aq.)	dispersant agrochemicals

Supplier	Trade name	Chemical description	Composition	General properties	Functionality / Application
Akcros Chemicals	AGRISOL PX401			clear liquid; pour point − 15°C; viscosity 27 cSt; pH 5.5-7.5 (1% aq.)	co-solvent; penetrant; flow promoter *agrochemicals*
	AGRISOL PX413			clear liquid; pour point − 6°C; viscosity 64 cSt; pH 5.5-7.5 (1% aq.)	co-solvent; penetrant; flow promoter *agrochemicals*
	VERSILAN MX134	electrolyte soluble surfactant	100% active	clear straw viscous liquid; pour point 3°C; viscosity 1170 cSt; surface tension 28.1 dynes/cm (0.1%); pH 2.2 (1% aq.)	wetting agent; foaming agent (moderate) *steam cleaning; vehicle cleaning; general-purpose cleaners; descalers; concrete cleaners; aluminium cleaners; de-rusting agents*
	VERSILAN MX332	electrolyte soluble surfactant	100% active	clear straw viscous liquid; pour point < 0°C; viscosity 9920 cSt; surface tension 32.0 dynes/cm (0.1%); pH 2.2 (1% aq.)	wetting agent; low foam *plate cleaning; metal cleaning; dishwashing agents; descalers; concrete cleaners; de-rusting agents*
Akzo Nobel	BEROL 223	surfactant blend	100% active		emulsifier
	BEROL 225	cationic/nonionic blend based on nonylphenol ethoxylates	100% active		*hard-surface cleaners; industrial cleaners; vehicle cleaners*
	BEROL 226	cationic/nonionic blend	100% active		*hard-surface cleaners; industrial cleaners; vehicle cleaners*
	BEROL 239	nonionic/cationic blend	100% active		emulsifier *degreasing products*
	BEROL 242	surfactant blend	100% active		emulsifier *microemulsions based on n-paraffins*
	BEROL 784	anionic/nonionic blend	100% active		*car washes; washing-up liquids; all-purposes cleaners*
Albright & Wilson	EMPICOL 0627/F	pearlising/opacifying concentrate	concentration 25%	paste	pearlescent agent; opacifier
	EMPICOL 9060X	pearlising/opacifying concentrate	concentration 40%	paste	pearlescent agent; opacifier
	EMPICOL XC35/M	pearlised shampoo concentrate			pearlescent agent *shampoos*
	EMPICOL XP40	pearlising/opacifying concentrate	concentration 40%	paste	pearlescent agent; opacifier
	EMPICOL XP40H	pearlising agent		liquid	pearlescent agent
	MARCHON DC1102	cuttings wash cleaner	concentration 81%	liquid	*oil industry*
	NANSA BMC		concentration 27%	liquid	air entraining agent *mortar; cement*
Allied Colloids	ALCOPOL BLF	anionic/nonionic blend		liquid	detergent; wetting agent; dispersant; emulsifier; scouring agent *textile industry*
	ALCOPRINT PEM			liquid	emulsifier *textile industry; pigment printing*
	ALCOSPERSE BOD	anionic/nonionic blend of buffers and dispersants		liquid	buffer; dispersant; sequestration agent *textile industry; dyeing of polyesters*
	ALCOSPERSE BT	anionic/nonionic blend of levelling and dispersing agents		liquid	levelling agent; dispersant; wetting agent *textile industry*
Angus Chemie	ALKATERGE E	amine type compound with surface active characteristics	oxazoline content ca. 70%	clean, free from suspended matter	wetting agent *paper industry; textile industry; metal cleaning; emulsion stabilisation; pigment dispersion; foam control*
	ALKATERGE T	weakly cationic amine-type surface active agent	oxazoline content ca. 60% min.	firm, buff to brown waxy solid	w/o emulsifier; anti-corrosion *foam control; pigment grinding and dispersion; prevention of sludge formation in oil tanks*

Supplier	Trade name	Chemical description	Composition	General properties	Functionality / *Application*
Angus Chemie	AMP 95	2-amino-2-methyl-1-propanol	ca. 95% solids	colourless, clear liquid	pigment dispersant; anti-corrosion / *latex paints; emulsification of wax and polyethylene*
Auschem Cesalpinia	ANTISCHIUMA 7011	polyol ester; non-silicone	100% active	liquid	antifoam / *general purpose*
	ANTISCHIUMA 7342	polyol ester; non-silicone	100% active	liquid	
	ANTISCHIUMA F1	phenoxyethanol; non-silicone	100% active	liquid	
	ANTISCHIUMA N	complex blend	100% active	liquid	
	BIOROL CM	complex blend, without sulfates/sulfonates	27% active	liquid	detergent / *hard-surface cleaners*
	BIOROL CP	complex blend, without sulfates/sulfonates	60% active	liquid	detergent / *dishwashing agents*
	BIOROL CS	complex blend, without sulfates/sulfonates	65% active	liquid	detergent / *textile industry; wool cleaners*
	BIOROL EB	complex blend	50% active	liquid/paste	detergent / *washing-up liquids*
	BIOROL NT 1	complex enzymatic blend	60% active	liquid	detergent / *laundry products*
	BIOROL SHA	complex blend	46% active	liquid	detergent / *self-drying detergent preparations*
	CERAL LA 75	complex blend	100% active	liquid	polishes
	CEROL AG 1520	compound of natural substances, polymer surfactants and fatty amines	100% active	waxy; solidification point ca. 60°C	anti-caking agent; anti-dusting agent / *fertilisers*
	CEROL AG 1520 A	compound of natural substances, polymer surfactants and fatty amines	100% active	waxy; solidification point ca. 40°C	
	CEROL AG 1641	compound of organic substances with 50% of inert components suspended	100% active	fluid paste; solidification point ca. 5°C	
	CEROL AG 2143 A	modified urea resin without free formaldehyde	60% active	liquid; solidification point ca. <0°C	
	CEROL AG 3405	sulfonated salts with natural substances	100% active	liquid; solidification point ca. 5°C	
	CEROL AG 4023	complex blend	75% active	liquid; solidification point < −5°C	
	CEROL AG 4555 A	compound of natural substances and fatty amines	100% active	waxy; solidification point ca. 40°C	antifoam / *fertilisers*
	CEROL AG 4921	natural compound	100% active	paste; solidification point ca. <30°C	anti-caking agent; anti-dusting agent / *fertilisers*
	CEROL AG 5133	fatty amines in oil	100% active	waxy; solidification point ca. 55°C	anti-dusting agent / *fertilisers*
	CEROL AG 5624	compound of surfactants and soluble polymers	40% active	liquid; solidification point ca. 5°C	anti-caking agent; anti-dusting agent / *fertilisers*
	CESMETIC 4 W	hydroxypropyl polygalactomannan	100% active	powder	thickener; stabiliser; gelling agent / *cosmetics; shampoos*
	CESMETIC 6 W	hydroxypropyl polygalactomannan	100% active	powder	
	CESMETIC DP 2/92	hydroxypropyl trimetilammonium chloride polygalactomannan	100% active	powder	conditioner; antistatic agent

Supplier	Trade name	Chemical description	Composition	General properties	Functionality / Application
Auschem Cesalpinia	CESMETIC DP 4/92	hydroxypropyl trimetilammonium chloride polygalactomannan	100% active	powder	conditioner; antistatic agent
	CESMETIC DP 5/92	hydroxypropyl trimetilammonium chloride polygalactomannan	100% active	powder	conditioner; antistatic agent
	CHIMIN P 20	alkyl polyphosphate	100% active	flakes	antifoaming agent / industrial detergents
	CHIMIN RI	alkylpolyphosphate	50% active	liquid	dishwashing agents
	CHIMIPAL WS	polyol ester	100% active	liquid	insecticides; deodorants
	CHIMIPON SK	soft soap	35% active	liquid	liquid detergents
	COSMOPON AT	fully formulated concentrate	47% active	liquid	baby care products; specialised mild preparations
	COSMOPON BS	fully formulated concentrate	50% active	liquid	foam baths; shower gels
	COSMOPON BV	fully formulated concentrate	55% active	liquid	foam baths; shower gels
	COSMOPON HC	fully formulated concentrate	47% active	liquid	liquid soaps
	COSMOPON ID	fully formulated concentrate	47% active	liquid	personal hygiene products
	COSMOPON MP	fully formulated concentrate	30% active	liquid	shampoos; bath preparations
	COSMOPON MT	fully formulated concentrate	33% active	liquid	shampoos; bath preparations
	COSMOPON SES	2-ethyl-hexylic sulfate sodium salt	30% active	liquid	wetting agent; penetrant
	COSMOPON XM	fully formulated concentrate	30% active	liquid	liquid soaps
	EMULSON 126 B	special blend	80% active	HLB 12.0; liquid	emulsifier / pesticides
	EMULSON 155 L	special blend	80% active	HLB 8.8; liquid	
	EMULSON 603 HB	special blend	86% active	HLB 11.3; liquid	
	EMULSON 2300	special blend	91% active	HLB 11.8; paste	
	EMULSON 2650	special blend	65% active	HLB 9.0; liquid	
	EMULSON 2950	special blend	88% active	HLB 12.8; liquid	
	EMULSON 3020	special blend	92% active	HLB 14.5; waxy	
	EMULSON 3050	special blend	80% active	HLB 13.0; paste	
	EMULSON 3080	special blend	68% active	HLB 11.5; liquid	
	EMULSON 3490	special blend	85% active	HLB 10.9; liquid	
	EMULSON 3720	epoxy soya-bean oil	100% active	liquid	
	EMULSON 4020	special blend	92% active	HLB 13.5; paste	
	EMULSON 4050	special blend	80% active	HLB 12.3; liquid	
	EMULSON 4080	special blend	68% active	HLB 11.2; liquid	
	EMULSON 4132	special blend	86% active	HLB 11.8; paste	
	EMULSON 4475	special blend	68% active	HLB 9.6; liquid	
	EMULSON 5190	special blend	76% active	HLB 13.3; liquid	
	EMULSON 5512	special blend	99% active	HLB 13.9; paste	
	EMULSON 5590	special blend	72% active	HLB 9.2; liquid	
	EMULSON 5790	special blend	75% active	HLB 12.3; liquid	

Supplier	Trade name	Chemical description	Composition	General properties	Functionality / Application
Auschem Cesalpinia	EMULSON 7545	special blend	100% active	liquid	activator / *pesticides*
	EMULSON AGC	special blend	100% active	paste	
	EMULSON DM 101 B	special blend	97% active	HLB 12.5; liquid	emulsifier / *pesticides*
	EMULSON EP	special blend	84% active	HLB 11.6; liquid	
	EMULSON FN	special blend	88% active	HLB 12.2; liquid	
	EMULSON GPE	special blend	100% active	liquid	activator / *pesticides*
	EMULSON MP	special blend	62% active	HLB 11.2; liquid	emulsifier / *pesticides*
	EMULSON TRF 24	special blend	81% active	HLB 11.2; liquid	
	MADEOL DO 65	special blend	60% active	powder	wetting agent; dispersant / *pesticides*
	PEARLANTE 503	anionic surfactants and pearling agents	45% active	fluid paste	pearlescent agent / *shampoos; bath preparations*
	PEARLANTE CR	nonionic surfactants and pearling agents	30% active	fluid paste	
	POLIROL 1 BS	alkyl polyoxy alkyleneglycol	100% active	paste	wetting agent / *polymerisation*
	POLIROL 215/20	alkyl polyether	20% active	liquid	stabiliser / *latices; polymerisation*
	POLIROL 3413	anionic and nonionic surfactants	48% active	liquid	emulsifier / *emulsion polymerisation*
	POLIROL F4	aryl polyethyleneglycol	100% active	liquid	low foam; wetting agent / *latex; polymerisation*
	POLIROL SA	soap and esters	100% active	liquid	antifoam / *polymerisation*
	PRODOTTO CSG	sulfonated cresol	32% active	liquid	electroplating
	PRODOTTO T 8455	complex blend	100% active	liquid	thickener / *hydrochloric acid*
	ROLPON AB	complex blend		liquid/paste	detergent / *abrasive pastes*
	ROLPON FC	complex blend	40% active	liquid	detergent / *floor cleaners*
	ROLPON HD	complex blend	42% active	liquid	detergent / *laundry products*
	ROLPON HDE	complex blend with enzyme	56% active	liquid	
	ROLPON HDM	complex blend	42% active	liquid	
	ROLPON HDZ	complex blend	50% active.	liquid	
	ROLPON HL	complex blend	85% active	liquid	
	ROLPON LDN	complex blend	53% active	liquid	detergent / *dishwashing agents*
	ROLPON LSL	complex blend	85% active	liquid	detergent / *laundry products*
	STEROL BST	butyl stearate	85% active	liquid	co-emulsifier; emollient; lubricant; thickener / *cosmetics; toiletries; hair care products; skin care products*

Supplier	Trade name	Chemical description	Composition	General properties	Functionality / Application
Auschem Cesalpinia	STEROL GTC	tricaprin		HLB 1.8; liquid	co-emulsifier; emollient; lubricant; thickener *cosmetics; toiletries; hair care products; skin care products*
	STEROL GTS	triisostearin		HLB 1.5; liquid	
	STEROL ITB	dioctyl sebacate		liquid	
	STEROL ITL	isotridecyl laurate		liquid	
	STEROL LLT	lauryl lactate		liquid	
	STEROL TG 8	caprylic-capric triglyceride		liquid	
	TEWAX TC 1	cet. alcohol and POE fatty ethers		flakes	emulsifier *hair care products*
	TEWAX TC 2	cet. alcohol and POE fatty ethers		flakes	
	TEWAX TC 60	cet. alcohol and PEG-100 stearate		HLB 11.0; flakes	emulsifier *cosmetics*
	TEWAX TC 81	glycer. sorb. oleost. and beeswax		HLB 4.5; wax	
	TEWAX TC 82	POE (2) glyceryl sorb. isostearate		HLB 5.0; waxy solid	
	TEWAX TC 840	POE fatty ethers/esters and lanolin		flakes	emulsifier *hair care products*
	TEWAX TL 1	blend of POE fatty ethers		liquid	
BASF	DEGRESSAL SD 30	carboxylic ester	100% active		foam suppressor
	DEGRESSAL SD 40	organophosphate	100% active		
	DEGRESSAL SNC	modified phosphic monoester	100% active		
	KORANTIN AST	triethanolamine salt of a sulfonamidocarboxylic acid	100% active		
	KORANTIN BH LIQUID	butyne diol	33% active	liquid	
	KORANTIN BH SOLID	butyne diol	100% active	solid	
	KORANTIN CD	fatty acid-diethanolamine condensation product	98% active		anti-corrosion
	KORANTIN LUB	acid polyether phosphate	100% active		
	KORANTIN MAT	alkanolamine salt of a nitrogenous organic acid	100% active		
	KORANTIN PA	nitrogenous organic acid	80% active		
	KORANTIN PAT	triethanolamine salt of a nitrogenous organic acid	80% active		
	KORANTIN SH	fatty acid condensation product	100% active		
	KORANTIN SMK	phosphoric monoester	100% active		
	KORANTIN TD	fatty acid-diethanolamine condensation product	100% active		
	LUTENSIT AN 10	anionic/nonionic surfactant combination based on an alkylphenol ethoxylate	100% active		
	LUTENSIT AN 30	anionic/nonionic surfactant combination based on a fatty alcohol ethoxylate	100% active		
	LUTENSIT AN 40	mixture of nonionic surfactants and alkylcarboxylic acids	70% active		
	SEDIPOL PN	mixture of aliphatic hydroxy compounds in an aq. emulsion	100% active		foam suppressor
	SEDIPUR AF 100		100% active		flocculant

Supplier	Trade name	Chemical description	Composition	General properties	Functionality Application
BASF	SEDIPUR AF 203		100% active		
	SEDIPUR AF 402		100% active		
	SEDIPUR AF 403		100% active		
	SEDIPUR AF 404		100% active		
	SEDIPUR AF 430		25% active		
	SEDIPUR AF 701		100% active		
	SEDIPUR AF 900		100% active		
	SEDIPUR CF 104		100% active		
	SEDIPUR CF 303		100% active		
	SEDIPUR CF 304		100% active		
	SEDIPUR CF 403		100% active		
	SEDIPUR CF 404		100% active		flocculant
	SEDIPUR CF 503		100% active		
	SEDIPUR CF 504		100% active		
	SEDIPUR CF 505		100% active		
	SEDIPUR CF 603		100% active		
	SEDIPUR CF 604		100% active		
	SEDIPUR CF 803		100% active		
	SEDIPUR CL 341		35% active	emulsion	
	SEDIPUR CL 540		42% active	emulsion	
	SEDIPUR CL 640		44% active	emulsion	
	SEDIPUR CL 750		47% active	emulsion	
	SEDIPUR CL 930		40% active	soln.	
	SEDIPUR CL 950		50% active	soln.	
	SEDIPUR NF 104		100% active	emulsion	
	SEDIPUR NL 130		30% active		
	SEDIPUR NL 190		100% active	emulsion	
	SEDIPUR TF 2-TR		100% active		
Bayer	AVOLAN DR LIQUID	condensation product of aromatic sulfonic acid		liquid	dispersant; *textile industry; dyeing of polyesters and triacetates*
	AVOLAN IS	condensation product of aromatic sulfonic acids			levelling agent; dispersant; *textile industry; dyeing of wools; dyeing of polyesters and triacetates*
	AVOLAN IS LIQUID	condensation product of aromatic sulfonic acids		liquid	
	DIADAVIN WTS	mixture of alkylaryl polyglycol ester with aliphatic sodium sulfonate and chlorinated hydrocarbons		yellow-brown viscous fluid	scouring agent; spotting agent; *textile industry*
	EDOLAN PAW LIQUID	aromatic sulfonates		liquid	reserving agent; *textile industry; dyeing of wool/polyamide blends*
	EMULSIFIER U	polyglycol ester of high molecular weight carboxylic acids		yellow-brown flakes	emulsifier

Supplier	Trade name	Chemical description	Composition	General properties	Functionality Application
Bayer	ERKANTOL RN	sodium salt of a substituted sulfo fatty acid ester		brown liquid	*dyestuffs*
	ERKANTOL WH	isoalkyl sulfate preparation			padding auxiliary / *textile industry; dyeing*
	LEVALIN ACE	nitril with an araliphatic alcohol and polyglycol ethers			accelerant / *textile industry; dyeing of acrylics*
	LEVALIN APS	ester and polyglycol ether			padding auxiliary / *textile industry; dyeing of acrylics*
	LEVALIN CB	condensation products of aromatic sulfonic acids			padding auxiliary / *textile industry; dyeing of cellulosic fibres*
	LEVALIN CDP	fatty alcohol derivative			padding auxiliary / *textile industry; dyeing of acrylics*
	LEVALIN DS	acid amide, aliphatic ether sulfate and a naphthylmethane sulfonate			padding auxiliary / *textile industry; dyeing*
	LEVALIN TM	polyacrylate			thickener / *textile industry; dyeing*
	LEVAPON A	mixture of alkyl sulfonate and alkylaryl polyglycol		yellowish liquid	wetting agent; milling agent; scouring agent / *textile industry; wools; synthetic fibres*
	LEVAPON AN	alkyl sulfonate and alkylaryl polyglycol ether			wetting agent; scouring agent; emulsifier; re-wetting agent
	LEVAPON AN LIQUID	alkyl sulfonate and alkylaryl polyglycol ether		liquid	*textile industry; all fibres*
	LEVAPON CA	mixture of an aliphatic sodium sulfonate and an amine sulfate		clear, yellow to brown viscous fluid	milling agent; scouring agent; wetting agent; penetrant; stain removing agent / *textile industry; wools; synthetic fibres*
	LEVAPON ME	alkyl sulfonate	76% active	white paste	scouring agent; wetting agent
	LEVAPON ME LIQUID	alkyl sulfonate	30% active	colourless clear liquid	*textile industry; all fibres*
	LEVAPON TH	sodium salt of aliphatic sulfo acid and alkyl polyglycol ether		brownish paste	wetting agent; emulsifier; dispersant
	LEVEGAL EPA	polyglycol ether derivative with sulfonates			levelling agent; non-foaming / *textile industry; dyeing of polyamides*
	LEVEGAL ER LIQUID	aromatic sulfonates		liquid	reserving agent; levelling agent / *textile industry; dyeing of wool/cellulosic blends*
	LEVEGAL FTS	polyglycol ether derivative with sulfonates			levelling agent / *textile industry; dyeing of polyamides*
	LEVEGAL HTC	halogen aromatics with alkyl polyglycol ether and alkylaryl sulfonates			levelling agent / *textile industry; dyeing of synthetic fibres*
	LEVEGAL HTN	alkyl phenol and fatty acid polyglycol ether			levelling agent; dispersant / *textile industry; dyeing of polyesters*
	LEVEGAL MKN	methylene-linked condensation product of alkylaryl sulfonic acids, sodium salt			levelling agent / *textile industry; dyeing of polyamides*
	LEVEGAL MPE	fatty acid polyglycol ether and alkylaryl sulfonates			levelling agent; dispersant / *textile industry; dyeing of polyesters*

Supplier	Trade name	Chemical description	Composition	General properties	Functionality / Application
Bayer	LEVEGAL MSF	fatty acid glycerine ester			levelling agent; non-foaming / *textile industry; dyeing of synthetic fibres*
	MESITOL HWS LIQUID	condensation product of high molecular aromatic sulfonic acids		liquid	reserving agent; after-treating agent / *textile industry; dyeing*
	MESITOL NBS	condensation product of high molecular aromatic sulfonic acids		liquid	
	MESITOL NBS NEW LIQUID 40%	condensation product of high molecular aromatic sulfonic acids		liquid	
	MESITOL P	formaldehyde condensation product of naphthalene sulfonic acids		finely dispersed yellow-brown powder	*dyestuffs*
Dr. Th. Boehme	SYNTHESIN CSN	sarcoside			spin finish / *textile industry; fibre production*
	SYNTHESIN L 63	alkyl ether adduct			
	SYNTHESIN OS	sarcoside			
	SYNTHESIN W	phosphoric acid ester and nonionic lubricant			
	TEBECID 7090				emulsifier; lubricant; anti-corrosion
	TEBECID O				emulsifier; lubricant; dispersant
	TEBECID RF				solubiliser; emulsifier
	TEBECID S				emulsifier; softener
	TEBECLEAN				antistatic agent; cleaning agent; wetting agent
	TEBECOR				emulsifier; detergent
	TEBEDYN AS				wetting agent; foam stabiliser; cleaning agent
	TEBEFOM				defoaming agent
	TEBEMINE C				antistatic agent; emulsifier; wetting agent; anti-corrosion / *industrial cleaners*
	TEBENAL C				emulsifier; wetting agent / *mineral oils*
	TEBENAL L				emulsifier
	TEBESOL 383				lubricant; smoothing agent
	TEBESOL BE				emulsifier / *fats and waxes*
	TEBESOL X 5				co-solvent
Chemax	CHEMAX DFO-133	organic defoamer		liquid	defoaming agent; antifoaming agent / *paper industry; textile industry; paint industry; effluent treatment; cleaners; adhesives; metal working*
	CHEMAX DFO-155	organic defoamer		liquid	
	CHEMSULF SBO/65	sulfated butyl oleate	65% active	liquid	softener; wetting agent; additive; emulsifier; solubiliser / *textile industry; metal working*
	CHEMSULF SCO/75	sulfated castor oil	75% active	liquid	
Ciba	ALBATEX FFC	based on alkanol, containing alkylarylethoxylates and derivatives of methyl polysiloxane		white viscous liquid	de-aerating agent; penetration accelerant; foam suppressant / *textile industry; dyeing*
	ALBATEX PON CONC.	polyglycol ether derivative of sulfonated alkyl benzimidazole		yellowish brown, turbid low viscosity liquid	levelling agent; penetrant; dispersant / *textile industry; dyeing*
	ALBEGAL A	alkylamine polyglycol ether sulfate		yellow/brown opalescent liquid of low viscosity	dispersant; levelling agent / *textile industry; dyeing*

Supplier	Trade name	Chemical description	Composition	General properties	Functionality / Application
Ciba	ALBEGAL FFA	alkylaryl polyglycol ether sulfate containing de-aerating and solubilising agents		white, high viscosity liquid	de-aerating agent; penetration accelerant; anti-foam *textile industry*
	ALBEGAL FFD	alkylaryl polyglycol ether sulfate containing de-aerating and solubilising agents		white, high viscosity liquid	de-aerating agent; penetration accelerant; anti-foam *textile industry; dyeing*
	CIBATEX LFN	benztriazole derivative containing a dispersant		white, low viscosity emulsion	light fastness improver *textile industry; dyeing of polyester*
	CIBATEX PA	aromatic sulfonic acids condensed with formaldehyde		clear, brown low viscosity liquid	wet fastness improver; reserving agent *textile industry; dyeing of polyamides*
	CIBATEX RN	aryl sulfonate condensed with formaldehyde		clear, dark brown low viscosity liquid	wet fastness improver; reserving agent; levelling agent *textile industry; dyeing of wool/polyamide blends*
	COPRANTEX BN	low-molecular amine formaldehyde condensation products containing a copper salt		fine, slightly hygroscopic, greyish yellow powder	aftertreatment agent *textile industry; dyeing of cellulosic fibres*
	ERIONAL RF	aromatic sulfonic acids formaldehyde condensation product		brown, clear, low viscosity liquid	levelling agent; reserving agent *textile industry; dyeing of wool/polyamide blends*
	ERIOPON NM	mixture of anionic and nonionic surfactants		turbid, low viscosity liquid	non-foaming; afterclearing agent; soaping agent *textile industry; dyeing*
	INVADINE MC NEW	based on sodium alkanol sulfonate		clear pale yellow liquid	wetting agent *causticising and mercerising liquors*
	INVADINE MR	based on sodium alkanol sulfonate plus alkyldiol derivative		clear pale yellow liquid	low foam; wetting agent *causticising and mercerising liquors*
	INVADINE NF	mixture of anionic and nonionic modified fatty alcohol ethoxylates		clear colourless liquid	non foaming; wetting agent *exhaust pretreatment of cotton in long liquors*
	INVALON HTC	based on alkylbenzenes, emulsifiers and solubilising agents		yellow, clear low viscosity liquid containing no water	levelling accelerant *textile industry; dyeing of triacetates*
	IRGAPADOL AS	alkylaryl polyglycol ether sulfate and solvent		gelatinous yellow liquid	detergent *textile industry; dyeing*
	IRGAPADOL FFU	alkylaryl ether sulfate and polyethylene glycol ether		pale yellow medium viscosity liquid	de-aerating agent; penetrating accelerant; wetting agent; foam-suppresant *textile industry; dyeing*
	IRGAPADOL PAC	fatty acid amine derivatives and antifoam		yellow-brown medium viscosity liquid	padding auxiliary *textile industry; acrylic fibres*
	IRGAPADOL PN	based on fatty acid amide and alkyl polyglycol sulfate		clear yellowish medium viscosity liquid	detergent; wetting agent; levelling agent; penetrant *textile industry; printing and continuous dyeing*
	IRGASOL AR	mixture of alkylphenol polyglycol ether and fatty amine		clear, faintly yellow liquid	emulsifier; dispersant *textile industry; cleaning dyeing machines*
	IRGASOL CO	based on anionic and nonionic surfactants		dark brown low viscosity liquid	dispersant; protective colloid *textile industry; dyeing cellulosic fibres and their blends*
	IRGASOL DAM	formaldehyde condensation product of naphthalene sulfonic acid		pale beige granules	dispersant; levelling agent *textile industry; dyeing*
	IRGASOL HTW	surfactant-containing preparation based on a heterocyclic N-methylol compound		clear colourless liquid	fibre protectant *textile industry; dyeing*
	IRGASOL OD	based on an aromatic polycarboxylic acid derivative and alkylaryl ether sulfate		white slightly viscous liquid	prevention of oligomer deposits and frosting *textile industry*
	PETOLON U	polyester condensate containing oxalkylene groups		yellowish medium viscosity liquid	lubricant *textile industry; cellulosic fibres and their blends*

Supplier	Trade name	Chemical description	Composition	General properties	Functionality / Application
Ciba	TINEGAL MGE	mixture of cyclo-aliphatic and araliphatic ammonium salts		clear, yellowish brown liquid	levelling agent / textile industry; dyeing
	TINEGAL MR	n-tetra-alkylammonium halide		clear, almost colourless liquid	retardant / textile industry; dyeing of acrylics and blends
	TINEGAL W	aq. soln. of a polyglycol ether and an aliphatic amine		pale yellow, turbid low viscosity liquid	levelling agent; wetting agent; dispersant / textile industry; acrylics and wool
	ULTRAVON GPN	combination of sodium alkylarylsulfonates and ethoxylated fatty alcohols		clear colourless liquid	detergent; cleaning agent / pretreatment of cotton
	ULTRAVON W H.C.	benzimidazole derivative		pale brown powder	dispersant; scouring agent / washing and scouring of wool
	UNIVADINE DP	ammonium salt of an acid modified polyhydroxy alkylene glycol ether		yellowish brown, clear low viscosity liquid	dispersant; levelling agent / textile industry; dyeing of polyester
	UNIVADINE PA	combination of surfactants		clear, pale brown low viscosity liquid	levelling agent / textile industry; dyeing of polyamides
Cosmetochem	EMULSIFIER W/O-83	sorbitan glycerol oleate, micro-crystalline waxes, magnesium stearate, BHT, ascorbyl palmitate and citric acid	acid value 30-40; OH value 105-130; sapon. no. 130-150	light brownish wax; drop point ca. 50-60°C	emulsifier / w/o emulsions; personal care products
	EMULSIFIER B-6	ceteareth-6 and stearyl alcohol	H_2O <1%; OH value 115-135	white waxy mass; m.p. ca. 42°C	emulsifier / liquid o/w emulsions; o/w creams
	SOFT-CREAM-EMULSIFIER COS-136/6-A	hydrogenated tallow glyceride, cetyl alcohol, TEA isostearyl hydrolysed animal protein, PEG 20 glyceryl laurate, propylparaben, methylparaben, phenoxyethanol and BHT	acid value 12-18; OH value 220-250; sapon. no. 110-130	slightly yellowish waxy material; drop point 50-55°C	emulsifier / o/w creams
Croda	CRESTAWHIP 100	based on GMS products with polyglycerol and polysorbate emulsifiers			emulsifier; stabiliser / food industry; confectionery; bakery goods
	CRESTAWHIP 200	based on GMS products with polyglycerol emulsifiers			
	CRESTAWHIP 300	based on polyglycerol and polysorbate emulsifiers			
	CRODACREME	combined emulsifier and stabiliser			emulsifier; stabiliser / food industry; ice cream
	CRODASCOOP	combined emulsifier, stabiliser and freeze point depressant			emulsifier; stabiliser; freeze point depressant / food industry; ice cream
	CROLACTIL SSL	acyl lactylate derived from stearic isostearate acid (sodium salt)	acid value 60-90; sapon. no. 210-280	HLB 6.5; off-white powder	emulsifier / food industry; colognes; deodorants
	GMS 400	mono and diglycerides of fatty acids	sapon. no. 160-170		
	GMS 400 V	mono and diglycerides of fatty acids	sapon. no. 160-170		food industry
	GMS 402	mono and diglycerides of fatty acids	sapon. no. 158-165		
	GMS 402 V	mono and diglycerides of fatty acids	sapon. no. 158-165		
	GMS 900	mono and diglycerides of fatty acids	sapon. no. 150-165		
	HBF	hardened bread fat	acid value 2.5 max.; monoglyceride content 2.50% max.; sapon. no. 193-198	powder	food industry; snack food coatings
	INCROMEGA 3E50	Omega 3 essential fatty acid derivatives	total Omega 3 50-55%; total combined eicosapentaenoic acid and docosahexaenoic acid 45-50%; acid value 1 max.	clear, pale yellow liquid; colour <7 (Gardner)	pharmaceuticals; dietary supplements; health foods

Supplier	Trade name	Chemical description	Composition	General properties	Functionality Application
Croda	INCROMEGA 3E60	Omega 3 essential fatty acid derivatives	total Omega 3 65-70%; total combined eicosapentaenoic acid and docosahexaenoic acid 60-65%; acid vlue 5 max.	clear, pale yellow liquid; colour < 7 (Gardner)	
	INCROMEGA 3E70	Omega 3 essential fatty acid derivatives	total Omega 3 > 70%; eicosapentaenoic acid > 50%; acid value 5 max.	clear, pale yellow liquid; colour < 7 (Gardner)	
	INCROMEGA 3F50	Omega 3 essential fatty acid derivatives	total Omega 3 50-55%; total combined eicosapentaenoic acid and docosahexaenoic acid 45-50%; acid value 175-185	clear, pale liquid; colour < 7 (Gardner)	
	INCROMEGA 3F60	Omega 3 essential fatty acid derivatives	total Omega 3 65-70%; total combined eicosapentaenoic acid and docosahexaenoic acid 60-65%; acid value 180-190	clear, pale liquid; colour < 7 (Gardner)	*pharmaceuticals; dietary supplements; health foods*
	INCROMEGA 3TG30	Omega 3 essential fatty acid derivatives	total Omega 3 30-35%; total combined eicosapentaenoic acid and docosahexaenoic acid 25-30%; acid value 1 max.	clear, pale liquid; colour < 3 (Gardner)	
	INCROMEGA 3TG40	Omega 3 essential fatty acid derivatives	total Omega 3 40-45%; total combined eicosapentaenoic acid and docosahexaenoic acid 35-40%; acid value 1 max.	clear, pale liquid; colour < 7 (Gardner)	
	INCROMEGA 3TG60	Omega 3 essential fatty acid derivatives	total Omega 3 60-65%; total combined eicosapentaenoic acid and docosahexaenoic acid 50-55%; acid value 1 max.	clear, pale liquid; colour < 7 (Gardner)	
	ROCSOL B	oxidised hydrocarbon wax		maleable yellow/amber wax	provides gloss; confers buffability; aids solvent retention *polishes*
	ROCSOL C	oxidised hydrocarbon wax		white microbeads or powder	
	SYNCROWAX AWIC	$C_{18/36}$ acid		yellow/fawn solid	structure modifier *cosmetics; toiletries; polishes*
	SYNCROWAX BB4	synthetic beeswax		pale yellow solid	beeswax alternative *cosmetics; toiletries; furniture, shoe, floor and leather care applications*
	SYNCROWAX BB5	synthetic beeswax		yellow solid	beeswax alternative *cosmetics; toiletries*
	SYNCROWAX ERLC	$C_{18/36}$ acid glycol ester		pale yellow solid	gloss improver; structure modifier *cosmetics; toiletries; furniture, shoe, floor and leather care applications*
	SYNCROWAX HGLC	$C_{18/36}$ acid triglyceride		light tan solid	confers strength and rigidity; gloss provider *cosmetics; toiletries; furniture, shoe, floor and leather care applications*
	SYNCROWAX HRC	tribehenin		off-white solid	gloss improver and builder *cosmetics; toiletries*
	SYNCROWAX HRSC	tribehenin and calcium behenate		beige solid	gellant; stabiliser; solvent binder *cosmetics; toiletries; furniture, shoe, floor and leather care applications*
Dac International Surfactants	APA 0/55-69-96	a blend of surfactants			*pesticides*
	APA 100	a blend of surfactants			*pesticides*
Daniel Products Company	DAPRO DF 1181	dispersion of non-reactive olefinic solids in a paraffin liquid carrier		creamy-white free flowing, free from gel specs and coarse particles	defoaming agent *aq. dip coatings*
	DAPRO U-99	as for Dapro W-77 but contains more butyl cellosolve; silicone-free	ca. 50% active (non-volatile)		interfacial tension modifier *water and solvents systems*

Supplier	Trade name	Chemical description	Composition	General properties	Functionality Application
Daniel Products Company	DAPRO W-77	anionic/nonionic blend containing butyl cellosolve and H$_2$O (20/30); silicone-free	ca. 50% active (non-volatile)		interfacial tension modifier / *water systems*
	DISPERSE-AYD 1	reaction product of high molecular weight surfactants with a long oil alkyd	ca. 75% solids		dispersant; wetting agent; stabiliser / *solvent-thinned coatings; tinting pastes*
	DISPERSE-AYD 8	composite dispersant/film former	ca. 65% solids		wetting agent; dispersant; stabiliser / *carbon blacks in solvent-thinned coatings*
	DISPERSE-AYD W-22	multi-functional synergistic blend of surface active agents	ca. 35% solids		wetting agent; de-flocculating agent; stabiliser / *gloss and semi-gloss emulsion systems; water thinnable alkyds, acrylics, polyesters and epoxies*
	DISPERSE-AYD W-28	multi-functional synergistic blend of surface active agents	ca. 55% solids		wetting agent; de-flocculating agent; stabiliser / *gloss and semi-gloss emulsion systems; water thinnable alkyds, acrylics, polyesters and epoxies*
Ellis & Everard	CAFLON BCC50	fully formulated cream cleanser			*cream cleansers*
	CAFLON BT25	bleach thickener surfactant blend			thickener / *bleaches*
	CAFLON CC55	cream cleanser surfactant blend			*cream cleansers*
	CAFLON CF CONC	liquid detergent concentrate			*liquid detergents*
	CAFLON CF30	liquid detergent concentrate			
	CAFLON CF56	liquid detergent concentrate			
	CAFLON CF95	liquid detergent concentrate			
	CAFLON DC60	disinfectant concentrate			*disinfectants*
	CAFLON DC60PG	disinfectant concentrate			*disinfectants*
	CAFLON DP440				deicing chemical
	CAFLON DP893	liquid laundry detergent			laundry products
	CAFLON F570/30	fabric softener concentrate			fabric softeners
	CAFLON HDL40B	liquid laundry detergent			laundry products
	CAFLON LL20	liquid laundry detergent			laundry products
	CAFLON OSD	oil slick dispersant			dispersant / *oil industry*
	CAFLON PS502	shampoo and bubble bath concentrate			*shampoos; bubble baths*
	CAFLON SM27	shampoo and bubble bath concentrate			*shampoos; bubble baths*
	EMSOL 30				cleaner / *removing heavy wax deposits*
	VANTOCIL IB	polymeric biguanide hydrochloride			biocide; preservative
Gattefosse	ANTIOXYGENE WL 774	defined blend of antioxidants in a vegetable oil soln.		oily liquid	antioxidant / *cosmetics*
	ANTIOXYGENE WL 3036	defined blend of antioxidants in a glyceryl tricaprylocaprate soln.		oily liquid	antioxidant / *cosmetics*
	APIFAC	beeswax SE	acid value < 6; iodine value < 10; sapon. no. 90-110	HLB 6; waxy solid; colour ≤ 8 (Gardner); drop point 57-65°C	self-emulsifying base / *pharmaceuticals; cosmetics; hair care products; skin care products*

Supplier	Trade name	Chemical description	Composition	General properties	Functionality Application
Gattefosse	APIFIL	beeswax SE	acid value <5; iodine value <10; sapon. no. 70-90	HLB 5; drop point 59-70°C	self-emulsifying base pharmaceuticals; cosmetics
	BASE CB 8047	defined blend of fatty acid esters and waxes		waxy solid	ready-to-use lipstick base cosmetics; sunscreen preparations; lipstick
	BASE PL 1916	defined blend of natural vegetable oils and fatty acid esters		waxy solid	
	BRILLANCE 515	apricot kernel oil PEG-6 esters and ethylcellulose		viscous liquid	brightening agent cosmetics
	FONDIX G BIS	defined blend of preservatives in a propylene glycol soln.		liquid	antiseptic; fungicide; preservative cosmetics
	TRANSCUTAL	purified ethyldiglycol		liquid	amphiphilic agent; solvent; emulsifier pharmaceuticals; cosmetics
Th. Goldschmidt	EMULGATOR E 2149	blend of steareth-7 and stearyl alcohol	acid value 2 max.; iodine value 2 max.; sapon. no. 3 max.	HLB 11; white ivory wax; rise m.p. 41-47°C	o/w emulsifier; stabiliser cosmetics; baby care products; hair conditioners; hair care products; skin care products; skin cleansing preparations; shaving products; sunscreen products; deodorants; antiperspirants; pharmaceuticals
	EMULGATOR E 2155	blend of stearyl alcohol, steareth-7 and steareth-10	acid value 2 max.; iodine value 2 max.; sapon. no. 3 max.	HLB 11; white/ivory wax; rise m.p. 49-55°C	
	EMULGATOR E 2568	ceteareth-25	acid value 2 max.; iodine value 1 max.; sapon. no. 3 max.	HLB 16; white/ivory wax; rise m.p. 43-49°C	
	PROTEGIN W	blend of petrolatum, ozokerite, hydrogenated castor oil, glyceryl isostearate and polyglyceryl-3 oleate	OH value 18-28; acid value 1 max.; sapon. no. 18-28	HLB 3; ivory wax; thaw point 75-82°C	
	PROTEGIN WX	blend of petrolatum, ozokerite, hydrogenated castor oil, glyceryl isostearate and polyglyceryl-3 oleate	OH value 32-42; acid value 1 max.; sapon. no. 27-37	HLB 3; ivory wax; thaw point 76-83°C	w/o emulsifier baby care products; skin care products; sunscreen products; deodorants; antiperspirants; cosmetics; pharmaceuticals
	PROTEGIN X	blend of mineral oil, petrolatum, ozokerite, glyceryl oleate and lanolin alcohol	OH value 25-38; acid value 1 max.; sapon. no. 10-16	HLB 3; ivory wax; thaw point 58-65°C	
	PROTEGIN	blend of mineral oil, petrolatum, ozokerite, glyceryl oleate and lanolin alcohol	OH value 18-30; acid value 1 max.; sapon. no. 8-12	HLB 3; ivory wax; thaw point 58-65°C	
	TEGINACID	glyceryl stearate and ceteareth-20	total monoester 50-60%; acid value 3 max.; iodine value 4 max.; sapon. no. 153-165	HLB 12; ivory powder; rise m.p. 58-63°C	
	TEGINACID H	glyceryl stearate and ceteth-20	total monoester ca. 20%; acid value 3 max.; iodine value 3 max.; sapon. no. 55-70	HLB 11; ivory powder; rise m.p. 45-51°C	o/w emulsifier cosmetics; baby creams; hair creams; hair grooming preparations; skin care products; skin cleansing preparations; shaving products; sunscreen products; deodorants; antiperspirants
	TEGINACID SPEZIAL	glyceryl stearate and sodium lauryl sulfate	total monoester 35-45%; acid value 16-18; iodine value 3 max.; sapon. no. 160-172	HLB 12; ivory powder; rise m.p. 53-59°C	
	TEGINACID X	glyceryl stearate and ceteareth-20	total monoester 45-55%; acid value 3 max.; iodine value 4 max.; sapon. no. 145-160	HLB 12; ivory powder; rise m.p. 55-61°C	
	TEGO ANTIFOAM KS 6	organic antifoam concentrate	100% active		antifoam paper industry; polymer industry; water treatment
	TEGO ANTIFOAM KS 10	organic antifoam concentrate	100% active		antifoam polymer industry; metal working
	TEGO ANTIFOAM KS 53	organic antifoam concentrate	100% active		antifoam polymer industry; adhesives; water treatment
	TEGO ANTIFOAM KS 88	organic antifoam concentrate	100% active		antifoam polymer industry; adhesives; water treatment

Supplier	Trade name	Chemical description	Composition	General properties	Functionality / *Application*
Th. Goldschmidt	TEGO ANTIFOAM KS 83	organic antifoam concentrate	100% active		antifoam / *fermentation processes*
	TEGO ANTIFOAM KS 95	organic antifoam concentrate	100% active		antifoam / *polymer industry; adhesives; water treatment*
	TEGO ANTIFOAM KS 911	organic antifoam concentrate	100% active	cloud point 16°C (1% in H_2O)	antifoam / *fermentation processes*
	TEGO-CARE 150	blend of glyceryl stearate, steareth-25, ceteth-20 and stearyl alcohol	acid value 3 max.; sapon. no. 90-106	HLB 12; ivory powder; rise m.p. 52-58°C	o/w emulsifier / *cosmetics; baby care products; hair care products; skin care products; shaving products; sunscreen products; deodorants; antiperspirants*
Grillo-Werke	GRILLOTEN LSE 65 K	sucrose cocoate combined with coconut fatty acid esters, manufactured from natural renewable raw materials	iodine value 10 max.; H_2O 7.5 ± 1.5%; sapon. no. 68 ± 5	light yellow solid; pH 7.5 ± 0.5 (5% in methanol/H_2O)	anti-irritant; viscosity modifier; re-fatting agent; moisturiser; foam stabiliser / *hair care products; skin care products*
	GRILLOTEN LSE 65 K SOFT	sucrose cocoate combined with coconut fatty acid esters, manufactured from natural renewable raw materials	iodine value 5 max.; H_2O 35 ± 2%; sapon. no. 52 ± 5	light yellow paste; pH 7.0 ± 0.5 (5% in methanol/H_2O)	
	GRILLOTEN LSE 87 K	sucrose cocoate combined with coconut fatty acid esters, manufactured from natural renewable raw materials	iodine value 10 max.; H_2O 6.0 ± 1.5%; sapon. no. 87 ± 7	light yellow solid; pH 7.3 ± 0.6 (5% in methanol/H_2O)	
	GRILLOTEN LSE 87 K SOFT	sucrose cocoate combined with coconut fatty acid esters, manufactured from natural renewable raw materials	iodine value 5 max.; H_2O 53 ± 2.0%; sapon. no. 45 ± 5	light yellow paste; pH 7.5 ± 0.5 (5% in methanol/H_2O)	
	GRILLOTEN PSE 141 G PELLETS	sucrose stearate with palmitic/stearic fatty acid esters, manufactured from natural renewable raw materials	iodine value 10 max.; H_2O 1.5-4.5%; sapon. no. 90 ± 6	light yellow pellets; pH 9.0 ± 0.5 (5% in methanol/H_2O)	anti-irritant; moisturiser; re-fatting agent / *skin care products*
Hays Colours	CARADIL LR	anionic/nonionic self emulsifiable carrier, based on ethers and hydrocarbons			carrier; levelling agent; stripping agent / *textile industry; dyeing of polyester and blends*
	DYLEV SDA	solvent based anionic/nonionic			low foam / *textile industry; dyeing*
Henkel	METAPEX 26	anionic/nonionic blend			detergent / *textile industry*
	METAPEX 38	anionic/nonionic blend			
	ARBYL 18/50	polyhydroxyalkyl			scouring agent; dyeing agent / *textile industry; dyeing*
	ARBYL DH	alkylsulfonate with fatty alcohol polyglycolether			wetting agent / *textile industry; dyeing*
	BREVIOL ACN	arylsulfonate with alkylamine polyglycolether			levelling agent / *textile industry; dyeing*
	BREVIOL AFM	polyamide derivate			crease inhibitor / *textile industry; dyeing*
	BREVIOL LC	blend of lecithins			dispersant; levelling agent; crease inhibitor / *textile industry; dyeing*
	CAPCURE 3-800	mercaptan	100% active	liquid	additive / *plastics industry; epoxy resins*
	CAPCURE 40 SEC HV		100% active	liquid	
	CAPCURE WR-6		100% active	liquid	/ *plastics industry; coatings; adhesives*

Supplier	Trade name	Chemical description	Composition	General properties	Functionality / Application
Henkel	CAPCURE WR-35		100% active	liquid	*plastics industry; coatings; adhesives*
	CUTINA BW	mixture of glyceryl hydroxystearate, cetyl palmitate, microcrystalline wax and trihydroxystearin		pale yellow wax-like substance; m.p. 61-66°C	consistency factor; beeswax substitute *cosmetics; stick preparations*
	CUTINA LM	castor oil, glyceryl ricinoleate, octyldodecanol, carnuba wax, candelilla wax, microcrystalline wax, cetyl alcohol, beeswax and mineral oil		wax-like mass	*cosmetics; lipsticks*
	DEHYDOL HD-FC 4	blend of fatty alcohol polyglycol ethers and fatty acids	90-92% active	liquid	low foam *powder detergents; heavy-duty liquid detergents*
	DEHYDOL HD-FC 6	blend of fatty alcohol polyglycol ethers and fatty acids	90-92% active	liquid	low foam *powder detergents; heavy-duty liquid detergents*
	DEHYDOL HD-TP 1	powder compound based on zeolite-containing surfactants	ca. 20% active	powder	*powder detergents*
	DEHYDRAN 520	silicone-free	ca. 100% active	liquid	defoaming agent; foam inhibitor *polymerisation; plastics industry; coatings*
	DEHYDRAN 1208		solids 22-25%	clear, yellow liquid	defoaming agent *industrial coatings; alkyd resins*
	DEHYDRAN 1227	contains traces of silicone	ca. 100% active	liquid	defoaming agent; foam inhibitor *polymers; plastics industry; coatings*
	DEHYDRAN 1513	mixture of special alcohols, emulsifiers and a small quantity of silicone	100% active	colourless, slightly cloudy liquid; viscosity 400-500 mPas	defoaming agent *coatings; printing inks*
	DEHYDRAN C	silicone-free	ca. 100% active	liquid	
	DEHYDRAN F	silicone-free	ca. 100% active	liquid	defoaming agent; foam inhibitor *plastics industry; coatings*
	DEHYDRAN G	contains traces of silicone	ca. 100% active	liquid	
	DEHYDRAN P 4	silicone-free	ca. 100% active	liquid	
	DEHYDRAN P 11	contains traces of silicone	ca. 100% active	liquid	
	DEHYDRAN P 12	silicone-free	ca. 100% active	liquid	
	DEHYDRAN P 7215		ca. 35% active	emulsion	defoaming agent; foam inhibitor *polymers; plastics industry; coatings*
	DEHYMULS E	mixture of higher molecular esters, principally of pentaerythritol fatty acid esters and citric acid alcohol ester		yellowish wax-like mass	w/o emulsifier *creams; ointments*
	DEHYMULS F	mixture of higher molecular fatty acid esters, fatty acid salts and oil binding additives		white to yellowish wax-like mass	w/o emulsifier *creams; ointments*
	DEHYMULS K	mixture of Dehymuls E, fatty alcohol, fatty acid esters and mineral fats		ointment-like mass	cream base *w/o creams; w/o ointments*
	DEHYPON KONZ.	combination of fatty alcohol polyglycol ethers with special additives	48-51% active	paste	low foam *cleaners*
	DEHYQUART KE 2415 POWDER	alkylamido ammonium lactate	≥ 97% active	powder	*fabric softeners*
	DISPONIL MGS 65	mixture of highly effective surface-active substances	ca. 44% active	liquid	emulsifier *plastics industry; coatings*
	DISPONIL MGS 156	mixture of highly effective surface-active substances	ca. 50% active	liquid	emulsifier *plastics industry; coatings*

Supplier	Trade name	Chemical description	Composition	General properties	Functionality / Application
Henkel	EMERWAX 1257	emulsifying wax		white waxy solid	*personal care products*
	EMULGADE C	mixture of high fatty alcohols with nonionic emulsifiers		flaked waxy solid	emulsifier / *personal care products*
	EUMULGIN 550	blend of polyglycol ethers	99.5-100% active	liquid	emulsifier / *mineral oils; natural oils*
	FOAMASTER 340	combination of hydrocarbons, special additives and emulsifiers			defoaming agent / *textile industry; pretreatment*
	FOAMASTER AP	hydrophobic silicic acid, low silicone	ca. 100% active	opaque tan liquid; viscosity 200-600 mPas; pH 6-8 (10%)	defoaming agent; foam inhibitor / *polymers; plastics industry; coatings; paint industry; adhesives*
	FOAMASTER DF-124-L	silicone free defoamer		light yellow opaque liquid; viscosity 200-800 mPas	antifoaming agent; defoaming agent / *inks; lacquers; paint industry*
	FOAMASTER ENA 227			hazy amber liquid; viscosity 600 cP's; pH 7.0	defoaming agent / *paint industry; adhesives*
	FOAMASTER TMC 1	hydrophobic silicic acid, low silicone	ca. 100% active	liquid	defoaming agent; foam inhibitor / *polymers; plastics industry; coatings*
	FOAMASTER VC	hydrophobic silicic acid, low silicone	ca. 100% active	liquid	defoaming agent; foam inhibitor / *polymers; plastics industry; coatings*
	FOAMASTER VP 3063	blend of hydrocarbons and nonionic surfactants; silicone free	100% active	yellowish, opaque liquid; pH 5-7.5 (10%)	defoaming agent / *paint industry; aq. coatings*
	FOAMASTER VT	hydrophobic silicic acid, low silicone	ca. 100% active	liquid	defoaming agent; foam inhibitor / *polymers; plastics industry; coatings*
	FORYL 333	fatty alcohol polyglycolether with anionic additives			scouring agent / *textile industry; pretreatment*
	FORYL KS	combination of alkylpolyglycolether with alcohol derivatives			scouring agent; low foam / *textile industry; pretreatment*
	FUMAN 623	alkylpolyglycolether with anionic additives			wetting agent; scouring agent; bleaching agent; boiling-off agent / *textile industry; pretreatment*
	GRUNAU PILLENWACHS 685/715	wax with emulsifiers			additive; afterwaxing agent / *textile industry; sizing*
	HYDROPALAT 3043	aq. soln. of sodium polyacrylate	solids 34-36%	clear, yellowish, low viscous liquid; viscosity 100-200 mPas; pH 6.5-7.5 (10% aq. soln.)	wetting agent; dispersant / *paint industry; pigment preparations*
	HYDROPALAT 3051	aq. soln. of ammonium polyacrylate	solids 39-41%	clear, yellowish, low viscous liquid; viscosity 100-200 mPas; pH 6-7 (10% aq. soln.)	wetting agent; dispersant / *paint industry; pigment preparations*
	HYDROPALAT VP 3204	partially neutralised chelating agent	solids 49-51%	clear, pale yellow liquid; pH 6-7 (10% aq. soln.)	dispersant; chelating agent / *paint industry; pigment preparations*
	LAMEFIN K	fatty acid amide polyglycolether, paraffin			yarn smoothing agent / *textile industry; dyeing*
	LAMEFIX 680	ethylene oxide adducts/polyglycolether			fixation accelerator / *textile industry; printing*
	LAMEPHAN 303	fatty acid amide polyglycolether			softener / *textile industry; dyeing*
	LAMEPHAN NFL	fatty acid alkanolamide polyglycolether			crease inhibitor; softener / *textile industry; dyeing*

Supplier	Trade name	Chemical description	Composition	General properties	Functionality / Application
Henkel	LAMEPON KN	high molecular sulfonic acid derivative with anionic detergent			dispersant; padding agent / *textile industry; dyeing*
	LAMEPON PO	phosphoric acid derivative			dispersant; sequestration agent / *textile industry; dyeing*
	LAMEPON RE	high molecular sulfonic acid derivate			dispersant; padding agent / *textile industry; dyeing*
	LAMESPUMIN E	hydrocarbons with emulsifiers; silicone free			defoaming agent / *textile industry; dyeing*
	NOPCO 8034 E		ca. 100% active	liquid	defoaming agent; foam inhibitor / *latex paints*
	NOPCO 8034 FA		ca. 100% active	liquid	
	NOPCO NDW		ca. 100% active	liquid	
	NOPCO NXZ	fatty acid base, low silicone	ca. 100% active	liquid	
	NOPCOCIDE 121-B			powder	preservative / *paper industry; coatings; size press treatments*
	NOPCOCIDE 130			liquid	slimicide / *paper industry*
	NOPCOCIDE 170			liquid	
	NOPCOCIDE N-54-D		54% active	dispersion	fungicide / *paint industry*
	NOPCOCIDE N-96		100% active	powder	
	NOPCOSPERSE SERIES	organic dispersant systems			dispersant / *paper industry*
	NOPCOTE C-104-HS	dispersion of calcium stearate		liquid	flow and levelling improver / *paper industry*
	NOPCOWAX SIZES 1055 SERIES	range of wax emulsions			*paper industry; sizing*
	OLINOR 308 AS	fatty acid glyceride			additive / *textile industry; sizing*
	OLINOR 712	combination of polyglycolether and phosphoric acid ester			waxing agent; afterwaxing agent / *textile industry; sizing*
	OLINOR 4007	combination of polyglycolether and antistatics			afterwaxing agent / *textile industry; sizing*
	OLINOR NW 81	wax with emulsifier			additive; afterwaxing agent / *textile industry; sizing*
	OLINOR PA	high molecular polyglycolether			additive / *textile industry; sizing*
	OSIMOL 110	alkylarylsulfonate with alkylaminepolyglycolether			levelling agent / *textile industry; dyeing*
	OSIMOL SF	high molecular sulfonic acid			dispersant / *textile industry; dyeing*
	POLYMUL SERIES	range of polyethylene emulsions			dispersant / *paper industry; coatings*
	PRODUCT VP 3122	modified polyester with surface-active properties in xylene/butyl acetate (1:1)	50% active	light brown viscous oil	wetting agent; dispersant / *pigments; coatings; resins*
	SELBANA 4554	combination of fatty acid esters, antistatics and emulsifiers			lubricant / *textile industry; spinning*

Supplier	Trade name	Chemical description	Composition	General properties	Functionality / Application
Henkel	SELBANA 4611	polyalkoxyether			lubricant / *textile industry; spinning*
	STABIFIX NY	condensate of high molecular sulfonic acid			retarding agent; aftertreatment agent / *textile industry; dyeing*
	STANDAPOL 7088	ammonium myreth sulfate and cocamide MEA	38% active	clear liquid	*bath preparations; cleansing preparations*
	STANDAPOL 7092	sodium laureth sulfate and glycol stearate	35% active	pearly white viscous liquid	*shampoos; bath preparations*
	STANDAPOL AP BLEND	sodium laureth sulfate, cocamide DEA and cocamidopropyl betaine	38% active	light amber liquid	foaming agent / *shampoos; bath preparations; cleansing preparations*
	STANDAPOL CAT	glycol stearate, lauramine oxide, propylene glycol and ceteareth 20	45% active	pearly liquid	*shampoos; bubble baths; hair conditioners*
	STANDAPOL CS PASTE	sodium lauryl sulfate, sodium cetyl sulfate and laureth 3	58% active	white creamy paste	*shampoos; bubble baths*
	STANDAPOL EA-K	ammonium myreth sulfate and cocamide DEA	60% active	light amber liquid	foaming agent / *shampoos; bath preparations; cleansing preparations*
	STANDAPOL LIS	mixture of special alkylether sulfates, amphoterics and alkanolamides	38% active	clear viscous liquid	*shampoos; bath preparations*
	STANDAPOL PEARL CONC. 7130	glycol distearate, sodium laureth sulfate, propylene glycol, cocamide MEA and laureth-9	38% active	pearly liquid	*shampoos; bath preparations; cleansing preparations*
	STANDAPOL S	sodium lauryl sulfate, sodium laureth sulfate, lauramide MIPA, cocamide MEA and glycol	38% active	pearlescent liquid	*skin cleansing preparations; shampoos*
	TEXAPHOR 963	electroneutral salt of a polycarboxylic acid with amine derivatives	ca. 50% active	clear, brownish liquid	dispersant; grinding aid / *coatings; paint industry; printing inks; resins; rubber industry*
	TEXAPON GL 20 UP	blend of fatty alcohol ether sulfate with alkyl polyglycoside	ca. 38% active	paste	*dishwashing agents; cleaners*
	TEXAPON P	mixture of highly effective surface-active substances	27-29% active	liquid	emulsifier / *plastics industry; coatings; dishwashing agents*
	TEXAPON P KONZ.	combination of anionic and nonionic substances	49-51% active	liquid	*dishwashing agents; cleaners; light-duty detergents; hand cleaning preparations*
	TEXAPON SP 60 N	blend of LAS, fatty alcohol ether sulfate, fatty alcohol ethoxylate, fatty acid alkanolamide	59-60% active	paste	
	TEXAPON SP 100	blend of LAS, nonionic surfactants, fatty acid alkanolamides	98-100% active	liquid	*dishwashing agents; cleaners; light-duty detergents; hand cleaning preparations*
	TEXAPON SP 100 A	blend of LAS, nonionic surfactants, fatty acid alkanolamides	98-100% active	liquid	
	TEXAPON SP KONZ.	combination of anionic and nonionic substances	49.5-50.4% active	paste	
	TEXAPON WW 99	MIPA-laureth sulfate, laureth-3 and cocamide DEA	90% active	liquid	base / *cosmetics; bath preparations*
Hickson Manro	MANRO DB30	ternary detergent blend	30% active	viscous liquid	
	MANRO DB56	ternary detergent blend	56% active	soft gel	
	MANRO DB98	ternary detergent concentrate	98% active	viscous liquid	*household detergents; industrial cleaners*
	MANRO MA53	ternary detergent concentrate	30% active	liquid	
	MANRO PSC	pearlised shampoo concentrate	35% active	white opaque gel	pearlescent agent / *shampoos; hair care products*

Supplier	Trade name	Chemical description	Composition	General properties	Functionality / *Application*
Hickson Manro	MANRO PSC40		40% active	white opaque gel	pearlescent agent / *hair care products; bath preparations*
	TENSIOREX BND57		40% active	viscous liquid	
Hoechst	EMULSOGEN BB	alkenyl succinic amide	80% active		emulsifier; anti-corrosion
	EMULSOGEN IT	complex nonionic/anionic blend			emulsifier / *agrochemicals*
	EMULSOGEN T	speciality blend with anionics	100% active		emulsifier / *specialities*
	EMULSOGEN V 4114	alkenyl succinic amide	90% active		emulsifier; anti-corrosion
	EMULSOGEN V 4165	alkenyl succinic ester	85% active		
	EMULSOGEN V 4169	alkenyl succinic ester	60% active		
	GENAMIN KSE	cationic/nonionic blend			base / *hair conditioners*
	GENAPOL PGC	nonionic/betaine blend			pearlescent agent / *toiletries; detergents*
	GENAPOL PGM CONC	anionic/nonionic blend			pearlescent agent / *toiletries*
	GENAPOL TSM	nonionic/anionic blend			silk lustre agent / *toiletries*
	GENAPUR ASA	alkenyl succinic anhydride	100% active		anti-corrosion; detergent; hydrotrope
	HOE S 3971	alkenyl succinic imide	100% active		emulsifier; anti-corrosion
	HOE S 3981	alkenyl succinic imide	80% active		emulsifier; anti-corrosion
	HOSTACERIN CG	anionic/nonionic blend; self-emulsifying			base / *cosmetics*
	HOSTACERIN WO	complex blend			emulsion concentrate / *cosmetics*
	HOSTACERIN WOL	complex blend			
	HOSTACOR L	alkenyl succinic anhydride	100% active		anti-corrosion
	HOSTACOR V 3928	alkenyl succinic amide	80% active		anti-corrosion
Hüls	AVISTIN PN	fatty acid amide amine derivative	active detergent 100%	flakes	cationic preparation agents; reviving agents
	MARLOWET BIK	blend of carboxylic acid amine salt and alkylaryl polyglycol ether	active detergent 100%	liquid	emulsifier / *mineral oils; hydrocarbons; bitumen; metal working; road making*
	MARLOWET IHF	blend of n-alkylbenzene sulfonate, alkyl polyglycol ether and isopropanol	active detergent 80%	liquid	emulsifier / *agrochemicals*
	MARLOWET OFA	blend of n-alkylbenzene sulfonate, alkyl polyglycol ether and carboxylic acid polyglycol ester	active detergent 100%	liquid	emulsifier / *solvents; textile industry; pesticides; metal cleaning*
	MARLOWET RA	blend of n-alkylbenzene sulfonate, carboxylic polyglycol ester	active detergent 70%	liquid	emulsifier / *perfumes; cosmetic*
ICI	ARLAMOL D4	octamethylcyclotetrasiloxane		colourless liquid	emollient / *personal care products*
	ARLAMOL S3	PPG-(15)-stearyl ether and cyclomethicone		colourless liquid	

Supplier	Trade name	Chemical description	Composition	General properties	Functionality Application
ICI	ARLAMOL S7	PPG-(15)-stearyl ether and cyclomethicone		colourless liquid	emollient *personal care products*
	ARLASOLVE DMI	dimethyl isosorbide		colourless liquid	solubiliser *personal care products*
	ARLATONE UVB	octyl dimethyl PABA		HLB 10.0; pale yellow liquid	o/w emulsifier; solubiliser *personal care products*
	ATLAS G-1530	mixture of cationic and nonionic surfactants		amber liquid	*household and industrial applications; agrochemicals; textile industry*
	ATLAS G-1970	mixture of nonionic and cationic surfactants		HLB 11.5; red-brown liquid	
	ATLAS G-2684	blend of SMO and POE esters of mixed fatty acids and rosin esters		amber liquid	
	ATLAS G-2960	specialty blend		yellow suspension	
	ATLAS G-4702	aq. alcohol-based oil spill dispersant		yellowish liquid	dispersant *oil industry*
	ATLAS G-4704	hydrocarbon-based oil spill dispersant		yellowish liquid	
	ATLAS G-5000	polyalkylene glycol ether		HLB 16.9; cream solid	*household and industrial applications; agrochemicals; textile industry*
	ATLAS G-72148	vinyl acetate aq. dispersion		milky white liquid	
	ATLOX 3335B	mixture of anionic and nonionic surfactants		HLB 12.6; red-amber liquid	
	ATLOX 3386B	mixture of anionic and nonionic surfactants		HLB 9.4; amber liquid	
	ATLOX 3387BM	mixture of anionic and nonionic surfactants		HLB 11.0; amber suspension	
	ATLOX 3400B	mixture of anionic and nonionic surfactants		HLB 10.9; red-amber liquid	
	ATLOX 3403FB	mixture of anionic and nonionic surfactants		HLB 12.5; amber liquid	
	ATLOX 3404FB	mixture of anionic and nonionic surfactants		HLB 10.0; amber liquid	
	ATLOX 3409FB	mixture of anionic and nonionic surfactants		HLB 11.9; amber suspension	
	ATLOX 4851B	mixture of anionic and nonionic surfactants		HLB 13.2; yellow suspension	
	ATLOX 4853B	mixture of anionic and nonionic surfactants		HLB 11.9; yellow suspension	
	ATLOX 4855B	mixture of anionic and nonionic surfactants		HLB 9.2; red-amber liquid	
	ATLOX 4856B	mixture of anionic and nonionic surfactants		HLB 8.7; yellow liquid	
	ATLOX 4857B	mixture of anionic and nonionic surfactants		HLB 12.3; red-amber suspension	emulsifier; dispersant *agrochemicals*
	ATLOX 4858B	mixture of anionic and nonionic surfactants		HLB 11.6; red-amber liquid	
	ATLOX 4868B	mixture of anionic and nonionic surfactants		HLB 11.9; yellow-brown liquid	
	ATLOX 4878B	mixture of anionic and nonionic surfactants		HLB 10.5; amber liquid	
	ATLOX 4880B	mixture of anionic and nonionic surfactants		HLB 12.8; yellow liquid	
	ATLOX 4890B	mixture of anionic and nonionic surfactants		HLB 12.7; yellow-brown liquid	
	ATLOX 4899B	mixture of anionic and nonionic surfactants		HLB 12.4; pale yellow liquid	
	ATLOX 4990B	mixture of anionic and nonionic surfactants		HLB 10.8; amber liquid	
	ATLOX 5330	mixture of anionic and nonionic surfactants		amber liquid	
	ATLOX 5380B	mixture of anionic and nonionic surfactants		HLB 11.7; yellow liquid	
	ATLOX 5390B	mixture of anionic and nonionic surfactants		HLB 10.7; yellow liquid	
	ATLOX 5966B	mixture of anionic and nonionic surfactants		HLB 13.0; yellow liquid	
	ATLOX 5967B	mixture of anionic and nonionic surfactants		HLB 12.0; yellow liquid	

Supplier	Trade name	Chemical description	Composition	General properties	Functionality Application
ICI	ATLOX 6404B	mixture of anionic and nonionic surfactants		HLB 10.4; yellow-brown liquid	emulsifier; dispersant *agrochemicals*
	ATLOX 6415B	mixture of anionic and nonionic surfactants		HLB 14.0; pale yellow liquid	
	ATMOS 300	glycerol mono and dioleate and propylene glycol		HLB 2.8; pale yellow liquid	lubricant; emulsifier *food industry; metal working; polymer industry*
	ATPET 787	oil spill dispersant concentrate		yellow liquid	dispersant *oil industry*
	ATPET 790	hydrocarbon-based oil spill dispersant		yellowish liquid	
	ATPET 900	oil spill dispersant concentrate		yellow-brown liquid	
	ATRUST 9013	ashless organo-nitrogen compound		light amber liquid	additive *oils; greases*
	ATRUST 9052	ashless organo-nitrogen compound		orange-brown semi-solid	
	ATRUST 9100	polyether amine		amber to light brown waxy paste/liquid	
Kao Corporation	DANOX C-008	surfactant blend	44-47% dry matter	paste; pH 9-10 (5%)	pearlescent agent
	DANOX P-15	surfactant blend	39-41% dry matter	paste; pH 8-10 (5%)	
	DANOX PAC-T	surfactant blend	34-36% dry matter	paste; pH 8.5-10 (5%)	
	TETRANYL DTHL/65	chemical specialities	67-71% active; carbon chain composition C_{14} 5%, C_{16} 32%, C_{18} 61%, $C_{18'}$ 2%	flakes; M.W. 650; colour 4 max. (Gardner)	
	TETRARYL 2S/90	chemical specialities	7.3-7.7% active; carbon chain composition C_{14} 5%, C_{16} 32%, C_{18} 61%, $C_{18'}$ 2%	liquid; M.W. 570	
	TETRARYL 511 DMS	chemical specialities	74-75% active; carbon chain composition C_{12} 1%, C_{14} 3%, C_{16} 30%, C_{18} 20%, $C_{18'}$ 40%, $C_{18''}$ 6%	liquid; M.W. 875; colour 7 max. (Gardner) H 6-8 (5%)	
Dr. W. Kolb	SYMPATENS-O/120	modified glycerol-PEG-stearate	acid value < 2; OH value 120; sapon. no. 95	HLB 11.0; white-yellow solid; viscosity 85 mPas (75°C)	o/w emulsifier *cosmetics*
	SYMPATENS-O/200	modified glycerol-PEG-stearate	acid value < 2; OH value 140; sapon. no. 117	HLB 8.5; white-yellow solid; viscosity 30 mPas (50°C)	o/w emulsifier *cosmetics*
	SYMPATENS-S/160	modified sorbitan-castor-laurate	acid value < 2; OH value 150; sapon. no. 55	HLB 14.0; yellow-brown liquid; viscosity 55 mPas	solubiliser *essential oils; perfumes*
	SYMPATENS-S/210	modified castor-hydroxystearate	acid value < 2; OH value 65; sapon. no. 50	HLB 14.0; yellow liquid; viscosity 630 mPas (50°C)	solubiliser *essential oils; perfumes*
	SYMPATENS-W/015	modified sorbitan-castor-isostearate	acid value < 5; OH value 135; sapon. no. 120	HLB 5.0; yellow-brown liquid; viscosity 220 mPas (50°C)	w/o emulsifier *cosmetics*
	SYMPATENS-W/130	modified sorbitan-castor-oleostearate	acid value < 22; OH value 155; sapon. no. 150	HLB 4.5; yellow-brown waxy solid; viscosity 25 mPas (50°C)	w/o emulsifier; viscosity modifier *cosmetics*
	SYMPATENS-W/140	modified sorbitan-castor-oleate	acid value < 8; OH value 130; sapon. no. 120	HLB 5.0; yellow-brown waxy solid; viscosity 50 mPas (75°C)	w/o emulsifier; viscosity modifier *cosmetics*
	SYMPATENS-W/150	modified sorbitan-castor-isostearate	acid value < 8; OH value 120; sapon. no. 120	HLB 5.0; yellow-brown waxy solid	w/o emulsifier *cosmetics*
	SYMPATENS-W/230	modified sorbitan-castor-stearate	acid value < 22; OH value 140; sapon. no. 145	HLB 4.5; yellow-brown waxy solid	
Lakeland Laboratories	LAKEWAX 20	montan ester wax emulsion; nonionic/anionic blend	solids 20%; wax 17%	milky white emulsion; softening point 82-88°C; pH 4.5	*packaging film slip finish wash*
Lonza	CARSOFOAM BS-1	blend of water, PEG-80 sorbitan laurate, sodium trideceth sulfate, PEG-150 distearate, disodium lauroamphodiacetate, cocoamidopropylhydroxysultaine and sodium laureth-13 carboxylate	40% solids		*baby shampoos*

Supplier	Trade name	Chemical description	Composition	General properties	Functionality Application
Lonza	CARSOFOAM MSP	blend of TEA lauryl sulfate, cocamide DEA, cocamidopropyl betaine and methyl paraben	50% solids		*shampoos*
McIntyre Group	MACKADET 40K	formulated product in concentrated form containing potassium coconate	38% active	liquid; pH 9	*skin cleansing preparations*
	MACKADET BBC	formulated product in concentrated form	35% active	liquid; pH 6.5	*bubble baths*
	MACKADET BGC	formulated product in concentrated form	57% active	flowable gel; pH 6.5	*bath preparations*
	MACKADET BSC	formulated product in concentrated form	45% active	liquid; pH 7	*baby shampoos*
	MACKADET CA	formulated product in concentrated form	42% active	liquid; pH 7	*shampoos*
	MACKADET CBC	formulated product in concentrated form	100% active	flakes; pH 4	*conditioner*
	MACKADET INC	formulated product in concentrated form	16.5% active	liquid; pH 4.5	
	MACKADET LCB	formulated product in concentrated form	30% active	liquid; pH 3	
	MACKADET OHIA	formulated product in concentrated form	100% active	liquid; pH 6	*fabric softeners*
	MACKADET RS	formulated product in concentrated form	35% active	liquid; pH 7	*carpet cleaners*
	MACKADET SBC-8	formulated product in concentrated form	46% active	liquid; pH 6.5	*shampoos; soaps; bubble baths*
	MACKADET WGS	formulated product in concentrated form containing mixed wheat germ oil and coconut oil soap	40% active	liquid; pH 9	*shampoos; skin cleansing preparations*
	MACKADET WHC	formulated product in concentrated form	100% active	liquid; pH 8	*skin cleansing preparations*
Millchem	MAPROFIX AEG	anionic/amphoteric/alkanolamide blend	solids 40-44%	liquid; pH 4.5-6.0	concentrate *shampoos; soaps*
	MAPROLYTE LX	anionic/amphoteric blend	solids 33-37%	liquid; pH 7.5-8.5	concentrate; foaming agent *shampoos; soaps; bubble baths*
	ONYXIDE 200	hexahydro-1,3,5-tris(2-hydroxyethyl)-s-triazine	solids 78.5% min.	liquid	preservative; bactericide *metal working; oil industry*
Nicca Chemical	PITCHRUN L-50	combination of plant extract and special nonionic and anionic surfactants		clear yellowish liquid; viscosity 60 cPs (20°C); 170 cPs (5°C); pH ca. 10	solubiliser; scouring agent; emulsifier; dispersant *textile industry: synthetic fibres*
	PITCHRUN L-100	plant extract in a unique blend of nonionic and anionic surfactants		light yellow transparent liquid; pH ca. 8 (2% soln.)	scouring agent; emulsifier; dispersant; solubiliser; low foam *textile industry: natural fibres*
	SUNMORL BH-6	special nonionic/anionic surfactant compound		light yellow transparent liquid; pH ca. 12 (5% soln.)	scouring agent; emulsifier; dispersant; chelating agent *textile industry: cotton knit fabrics*
	SUNMORL BL-600	mixture of anionic and nonionic surfactants		straw-coloured liquid; pH ca. 9 (1% aq. soln.)	detergent; emulsifier; dispersant; scouring agent *textile industry: synthetic fibres*
	SUNMORL SX-10	compounds of special type nonionic and anionic active agents		viscous liquid, slightly yellowish and transparent; pH 7.5	scouring agent *textile industry: synthetic fibres*
Pentagon	PENTAMID BD		ca. 80% active	clear, viscous liquid; freezing point ca. 10°C; pH 9.0-11.0 (1% aq. soln.)	anti-corrosion; low foam *mineral oil-free coolant lubricants*
	PENTAMID KH	methyl n-octadecyl terephthalamate	sapon. no. 110-140	cream coloured flakes; m.p. ca. 120°C	gelling agent (after saponification) *organic liquids*
	PENTASEPT M	mixture of 15% cetrimide and 1.5% chlorhexidine digluconate		clear yellow liquid; pH 5.0-7.0	disinfectant
	PENTATERIC CADP		solids 29-31%	clear yellow liquid; colour 250 max. (20% soln.; Hazen); pH 6.0-7.0 (20% aq. soln.)	*industrial cleaners*

Supplier	Trade name	Chemical description	Composition	General properties	Functionality / Application
Pentagon	PENTONATE DB	C$_{12/15}$ alcohol benzoate	acid value 0.5 max.; sapon. no. 175-185	clear, almost colourless liquid; colour 125 max. (Hazen)	solubiliser; extender; fixative; suspending agent; emollient / *perfumes; antiperspirants; sunscreen preparations*
PPG	ALUBRASOL 50-PI	nonionic lubricant and phosphate ester antistat blend	46% active		lubricant; antistatic agent / *textile industry; fibre production*
	ALUBRASOL APL-5	nonionic lubricant and cationic antistat blend	57% active		
	ALUBRASPIN 100-P	blend of nonionic lubricant with phosphate ester antistat	85% active	liquid	
	LAROSTAT 92				antistatic agent / *textile industry*
	LAROSTAT 96				antistatic agent / *PVC; textile industry*
	LAROSTAT 265-199				antistatic agent
	LAROSTAT 300I				antistatic agent / *textile industry; carpet industry*
	LAROSTAT 451				antistatic agent / *fibreglass*
	LAROSTAT 477				antistatic agent / *floor finishes*
	LAROSTAT 519				antistatic agent / *rubber industry; polyolefins*
	LAROSTAT 903				antistatic agent / *PVC*
	LAROSTAT 904				antistatic agent / *fibreglass; plastics industry*
	LAROSTAT 905				antistatic agent / *textile industry; PVC; plastics industry*
	LAROSTAT 906				antistatic agent / *fibreglass; PVC; plastics industry*
	LAROSTAT 1084				antistatic agent / *fibreglass; textile industry*
	LAROSTAT PVC				antistatic agent / *PVC*
	LAROSTAT U-101				antistatic agent / *polyurethane*
	LAROSTAT U-102				
	LAROSTAT U-103				
	LAROSTAT U-104				
	LAROSTAT U-105				
	LAROSTAT XXXS				antistatic agent
	MAZCLEAN EP				emulsifier / *metal working; cleaners*
	MAZON 114			liquid; viscosity 9500 cPs	
	MAZON RI 6			liquid; viscosity 2500 cPs	rust inhibitor
	MAZON RI 37			liquid; viscosity 100 cPs	*metal working*

Supplier	Trade name	Chemical description	Composition	General properties	Functionality / Application
PPG	MAZTREAT 246				defoaming agent / metal working
	MAZU DF 197			opaque liquid	
	MAZU DF 204			clear liquid	
	MAZU DF 255				defoaming agent / metal working; institutional cleaners
Protex	ACRYLRON A10	acrylic based			dispersant
	NOVEPROX TAN 117				emulsifier
	OLEOPROX	phenolic based			demulsifier
	PROTEPON KP				degreaser
	PROTEPON MTF				non-foaming; wetting agent / textile industry
	PROTEPON SCFN				non foaming; dispersant; wetting agent; detergent
	PROTEPON SCI				stain removing agent
	PROTEPON SLV1				detergent / solvent medium
	PROTEPON SLX				degreaser / cement
	PROTESOL RLN2	alkali/peroxide based			low foam; wetting agent
Rhone-Poulenc	ABEX 18-S		35% solids	liquid; surface tension 48 dynes/cm	emulsifier
	ABEX 26-S		33% solids	liquid; surface tension 55 dynes/cm	emulsion polymerisation
	ABEX JKB		30% solids	liquid; surface tension 40 dynes/cm	
	ABEX VA-50		46% solids	liquid; surface tension 44 dynes/cm	
	ANTAROX VRO SERIE			liquid	iodophor; foaming agent (moderate)
	BEVALOID 681F				foam control agent / adhesives; carpet industry; paint industry
	BEVALOID 691				foam control agent / adhesives; building; carpet industry; cleaners; detergents; leather industry; paint industry; printing inks; textile industry
	BEVALOID 770DD				foam control agent / agriculture; building
	BEVALOID 981				foam control agent / SBr stripping
	BEVALOID 2553				foam control agent / cutting fluids
	BEVALOID 2561				foam control agent / effluents; sugar production
	BEVALOID 581B				foam control agent / adhesives; agriculture; building; cleaners; cutting fluids; paint industry; paper industry; polymerisation; textile industry

Supplier	Trade name	Chemical description	Composition	General properties	Functionality / Application
Rhone-Poulenc	BEVALOID 6002				foam control agent / *paint industry*
	BEVALOID 6368				foam control agent / *effluents*
	BEVALOID 6420				foam control agent / *distillation solvents; leather industry; paint industry*
	BEVALOID 6475				foam control agent / *paint industry; polymerisation*
	BEVALOID 6624				foam control agent / *cleaners; polymerisation; SBr stripping*
	BEVALOID 6681				foam control agent / *adhesives; agriculture; cleaners; cutting fluids; detergents; leather industry; paint industry; paper industry; printing inks; SBr stripping*
	BEVALOID 6685				foam control agent / *adhesives; paint industry*
	BEVALOID 6686(W)				foam control agent / *effluents; fermentation processes; paper industry*
	BEVALOID 6695				foam control agent / *fermentation processes; vegetable processing*
	BEVALOID 6698				foam control agent / *adhesives; paint industry; paper industry*
	BEVALOID GENSIL 400 SERIES				foam control agent / *cleaners; fermentation processes; sugar production; vegetable processing*
	BEVALOID GENSIL 400T SERIES				foam control agent / *cleaners; effluents; textile industry*
	MIRASHEEN			paste	pearlescent agent; foaming agent (moderate); viscosity modifier
	RHODASPEC SERIE				
	RHODATERGE SERIE				
Sandoz	DILATIN NAN	nonionic/anionic emulsion containing mixture of aromatic hydrocarbons		liquid	accelerator / *textile industry; dyes*
	EMULSIFIER LDR	anionic/nonionic mixture of alkylaryl sulfonate, alkylaryl polyglycol ether and polyalkyl benzenes		liquid	emulsifier / *oils*
	IMACOL C	ester in aq. emulsion; anionic/nonionic		liquid	lubricant; low foam / *textile industry; cellulosic fibres*
	IMEROL AFR	blend of aliphatic esters and aliphatic ethoxylates		liquid	
	IMEROL SM	anionic/nonionic blend of solvents and emulsifier		liquid	emulsifier; wetting agent; detergent
	SANDOLIX LFE	sulfited natural and synthetic oils		liquid	
	SANDOLIX PFE	phosphate ester and oils, containing solvent		liquid	
	SANDOLUBE SVN	polyalkylene adduct plus fat derivative		liquid	

Supplier	Trade name	Chemical description	Composition	General properties	Functionality / Application
Sandoz	SANDOPAN D-PA LIQUID	carboxymethylated alkylaryl ethoxylate, sodium salt; chlorine-stable		colourless to pale straw liquid	*hypochlorite soln.*
	SANDOPAN D-PB LIQUID	carboxymethylated alkylaryl ethoxylate, sodium salt; chlorine-stable		colourless to pale straw liquid	
	SANDOPAN D-PC LIQUID	carboxymethylated alkylaryl ethoxylate, sodium salt; chlorine-stable		colourless to pale straw liquid	
	SANDOPAN DTC LINEAR P ACID LIQUID	carboxymethylated alkoxylate, free acid		clear, colourless liquid	
	SANDOPAN DTC LINEAR P PASTE	nonionic/anionic carboxymethylated alkoxylate, sodium salt		opaque, colourless gel	detergent; wetting agent *hair care products*
	SANDOPAN JA-36 ACID LIQUID	nonionic/anionic carboxymethylated alkoxylate, free acid		slightly cloudy, colourless liquid	detergent; wetting agent *hair care products*
	SANDOPAN LS-24 PASTE	nonionic/anionic carboxymethylated alkoxylate, sodium salt		colourless to pale straw paste	detergent *skin care products; hair care products*
	SANDOTAN APS	condensation product of phenolic sulfonates with aluminium salts		powder	
	SANDOZIN NA	triisobutyl phosphate in organic/aq. soln.; anionic/nonionic		liquid	wetting agent; de-aerating agent; anti-foaming agent
	SIRRIX 2UD	organic acids in aq. soln.		liquid	buffering agent; sequestering agent
	STABILIZER BLF	polyglycol ether, fatty acid salt and aliphatic alcohol		liquid	stabiliser; low foam *peroxide bleaching*
	STABILIZER JET	polyglycol ether, fatty acid salt and aliphatic alcohol		liquid	low foam; detergent *peroxides*
	STABILIZER SIFA	polycarboxylic acid in aq. soln.		liquid	stabiliser *hydrogen peroxide*
	TERGOLIX CA	aliphatic polyglycol ether in aq. soln.; nonionic/cationic		liquid	degreaser
Seppic	BASE NACRANTE 1100 AD	sodium laureth sulfate, cocamide DEA and glycol stearate	35% active	fluid paste	
	BASE NACRANTE 2078	cocamidopropyl betaine, cocamide DEA and glycol stearate	45% active	fluid paste	pearlescent agent *hair care products*
	BASE NACRANTE 6030 CP	sodium coceth-2 sulfate, PEG-3 distearate and glycol stearate	36% active	fluid paste	
	BASE NACRANTE 9578	sodium laureth sulfate, cocamide DEA and glycol stearate	40% active	fluid paste	
	MONTANE 481	sorbitan oleate, beeswax and stearic acid	100% active	soft wax	w/o emulsifier *personal care products*
	SEPIGEL 305	polyacrylamide, $C_{13/14}$ isoparaffin and laureth-7	48% active	liquid	thickener *personal care products*
Servo Delden	SERDAS GNA	blend		liquid	
	SERFAX AK290	fatty acid amidopropyltrimethylammonium chloride	50% active	liquid	
	SERFAX AK330	fatty acid amidopropyldimethylamine	95% active	liquid	

Supplier	Trade name	Chemical description	Composition	General properties	Functionality / Application
Servo Delden	SERVON XGF	glycerolformal	99% active	liquid	
Stepan Europe	BIO-SOFT 9283	surfactant blend; made-to-order	90% active	orange to brown viscous liquid	detergent / *industrial and household detergents*
	BIO-SOFT D 233	surfactant blend	100% active	water white to pale yellow liquid	detergent; wetting agent; thickening agent / *scale-removing cleaners*
	BIO-SOFT ERM	surfactant blend	27% active	orange yellow viscous liquid	concentrated soap / *degreasing soaps*
	BIO-SOFT LF 77A	surfactant blend	77% active	yellow liquid	low foam; detergent / *household, institutional and industrial cleaners*
	BIO-STEP SERIES	anionic/nonionic surfactant blends		liquid	emulsifier / *agrochemicals*
	MICRO-STEP SERIES	surfactant blends		pale yellow to amber liquid	
	NEOBEE 18	hybrid safflower oil	100% active	pale yellow liquid	emollient; lubricant / *personal care products*
	SECOMIX E40	surfactant blend	23% active	yellow viscous liquid	*industrial and household detergents*
	SEVESTAT ML 300	surfactant blend; made-to-order	42% active	yellow amber liquid	lubricant; antistatic agent; emulsifier / *textile industry*
	SEVESTAT NDE	phosphoric ester and nonionic blend; made-to-order	100% active	viscous liquid	antistatic agent / *textile industry*
	STEP-FLOW SERIES	nonionic/anionic copolymer blends; made-to-order		liquid to paste	dispersant; stabiliser / *agrochemicals*
	STEPAN PEARL SERIES	surfactant blends; made-to-order		white paste	pearlescent agent; opacifier; conditioner / *shampoos; liquid soaps; bubble baths; shower gels*
	STEPAN TAB-2	*N,N*-dialkyl phthalic acid amide	99% active	beige to pale yellow solid	emulsifier; suspending agent / *two-in-one shampoos and conditioners*
	STEPAN-MILD LSB	surfactant blend; made-to-order		water white to pale yellow liquid	foaming agent; thickener / *shampoos; bubble baths; liquid soaps; shower gels; bath preparations*
	STEPANHOLD EXTRA	PVP/ethyl methacrylate/methacrylic acid terpolymer; made-to-order	40% active	pale yellow liquid	hairspray preparations
	STEPOSOL CA 207	surfactant blend; made-to-order	50% active	pale yellow liquid	foaming agent / *oil industry*
	STEPSPERSE SERIES	surfactant blends; made-to-order		liquid or powder	dispersant / *agrochemicals*
	STEPWET SERIES	anionic and/or nonionic surfactant blends; made-to-order		liquid or powder	wetting agent / *agrochemicals*
	TOXIMUL 60 SERIES	surfactant blends		white liquid to paste	emulsifier / *agrochemicals*
	TOXIMUL 360 A	surfactant blend; made-to-order		HLB 10.4; dark brown liquid	emulsifier / *agrochemicals; pesticides*
	TOXIMUL 360 B	surfactant blend; made-to-order		HLB 13.5; dark brown liquid	
	TOXIMUL 374	surfactant blend; made-to-order		HLB 8; pale yellow viscous liquid	emulsifier / *agrochemicals*
	TOXIMUL 500	surfactant blend; made-to-order		HLB 10.5; dark brown liquid	
	TOXIMUL 705/707	surfactant blend; made-to-order		HLB 8; water white to pale yellow liquid	

Supplier	Trade name	Chemical description	Composition	General properties	Functionality / Application
Stepan Europe	TOXIMUL D	surfactant blend		HLB 10.5; dark brown liquid	emulsifier; dispersant; stabiliser / *agrochemicals*
	TOXIMUL FF	surfactant blend		liquid to paste	emulsifier / *agrochemicals*
	TOXIMUL H	surfactant blend		dark brown liquid	
	TOXIMUL MP	surfactant blend		dark brown liquid	
	TOXIMUL MP 10	surfactant blend; made-to-order		dark brown liquid	
	TOXIMUL MP 26	surfactant blend; made-to-order		HLB 12.5; dark brown liquid	emulsifier; dispersant; stabiliser / *agrochemicals*
	TOXIMUL R	surfactant blend		HLB 10.5; dark brown liquid	
	TOXIMUL S	surfactant blend		HLB 13; dark brown liquid	
	TOXIMUL SF SERIES	surfactant blends		paste	emulsifier / *agrochemicals*
Surfachem	WECOBEE SERIES	triglyceride from vegetable oils	100% active		personal care products; pharmaceuticals
	SURFAC 272	non-silicone emulsion		white to beige paste	antifoam
	SURFAC PEARL 40	pearlising agent			pearlescent agent
	SURFAWAX SW/BP	self emulsifying wax			
Troy Chemicals	TROYKYD D44	particle-free		liquid	defoaming agent / *paint industry; pigments; adhesives*
	TROYKYD D55	blend		liquid	defoaming agent / *paint industry; pigments; adhesives; inks; metal working*
	TROYKYD D126	silicone-free		powder	defoaming agent
	TROYKYD D333	silicone-free		liquid	defoaming agent
	TROYKYD D666	silicone-free		liquid	defoaming agent / *paint industry; adhesives; personal care products; resins; textile industry; detergents; metal working*
	TROYKYD D777			liquid	defoaming agent / *paint industry*
	TROYKYD D999	silicone-free		liquid	defoaming agent
	TROYSOL 307			liquid	anti-floating agent; defoaming agent / *polymer industry*
	TROYSOL AFL	silicon-free		liquid	
	TROYSOL AFP			powder	anti-floating agent
	TROYSOL S366			liquid	anti-cratering agent / *industrial applications*
	TROYTHIX 21BA	chemically-treated processed polymerised oil		liquid	
	TROYTHIX 42BA	chemically-treated processed polymerised oil		liquid	
	TROYTHIX A	polymerised organic ester		powder	
	TROYTHIX ANTI-SETTLE	silicate compound		powder	rheology control agent
	TROYTHIX ANTI-SETTLE SPECIAL	processed and stearated clay		powder	
	TROYTHIX LLBA	polymeric		liquid	
	TROYTHIX XYZ	polymerised organic ester		powder	

Supplier	Trade name	Chemical description	Composition	General properties	Functionality / Application
Unger Fabrikker	UFABLEND HDL	anionic/nonionic concentrated blended product	100% active	yellow viscous liquid; pH 6.5-7.0 (1% soln.)	*laundry products*
	UFABLEND HS 70	anionic/nonionic concentrated blended product	70% active	pale yellow viscous liquid; pH 7-8 (1% soln.)	*shampoos; bath preparations*
	UFABLEND MPL	anionic/nonionic concentrated blended product	30% active	white pearlised viscous liquid; pH 6.5-7.0 (1% soln.)	*pearlescent agent* *shampoos; bath preparations; soaps*
	UFABLEND W	anionic/nonionic concentrated blended product	32% active	pale yellow liquid; pH 6.5-7.5 (1% soln.)	*dishwashing agents; hard-surface cleaners*
	UFABLEND W 100	anionic/nonionic concentrated blended product	100% active	yellow viscous liquid; pH 6.5-7.5 (1% soln.)	
	UFABLEND W CONC.	anionic/nonionic concentrated blended product	65% active	pale yellow viscous liquid; pH 6.5-7.5 (1% soln.)	
	UFABLEND W CONC. 60	anionic/nonionic concentrated blended product	60% active	pale yellow viscous liquid; pH 6.5-7.5 (1% soln.)	
Warwick International	MYKON OES	synergistic blend of nonionic and anionic surfactants			softener *textile industry*
	WARCODET ASW	special blend of anionic and nonionic surfactants			scouring agent; wetting agent; foaming agent *textile industry*
	WARCODET NAS	synergistic blend of anionic and nonionic surfactants			scouring agent *textile industry*
	WARCODET SCS CONC	special blend of anionic and nonionic surfactants; concentrated version of Warcodet ASW			scouring agent; wetting agent; foaming agent *textile industry*
	WARCOSOL 446	special blend of nonionic and anionic surfactants			detergent *textile industry*
Witco	EMCOL 14		100% active	liquid	o/w emulsifier; anti-corrosion aerosols
	EMULGATEUR SO		100% active	liquid	emulsifier *aromatic oils; petroleum white spirit; metal working*
	MORWET EFW POWDER		95.0% active min.; moisture 2.0% max.; inorganic (sodium sulfate) 2.5% max.	water dispersible granules; surface tension 33.0 dynes/cm (0.25% in H_2O); Draves wetting 5.7 sec (0.25% in H_2O); pH 7.5-10.0 (5% solids)	wetting agent *agrochemicals*
	NEOPON 33	AOS/LES	33% active	liquid	*liquid detergents; shampoos; bubble baths*
	REWOMID RE	blend of anionics and alkylolamides	100% active	liquid	emulsifier; anti-corrosion *metal working*
	REWOPOL FBR	blend of nonionic surfactants with alkylolamides	98% active	liquid	*floor cleaners; all-purpose cleaners; metal cleaners*
	REWOPON IM AN	imidazoline derivative	100% active	liquid/paste	water repellent *car washes; pickling baths*
	REWOPON IM OA	imidazoline derivative	100% active	liquid	anti-corrosion; emulsifier; antistatic agent
	SPONTO 101	anionic/nonionic surfactant blend		liquid	*agrochemicals*
	SPONTO 140T	anionic/nonionic surfactant blend		liquid	*agrochemicals; insecticides*
	SPONTO 150T	anionic/nonionic surfactant blend		liquid	*agrochemicals*
	SPONTO 203	anionic/nonionic surfactant blend		liquid	*agrochemicals*
	SPONTO 232T	anionic/nonionic surfactant blend		liquid	*agrochemicals; insecticides*
	SPONTO 234 T	anionic/nonionic surfactant blend		liquid	*agrochemicals; insecticides*
	SPONTO 300T	anionic/nonionic surfactant blend		liquid	*agrochemicals*
	SPONTO 500T	anionic/nonionic surfactant blend		liquid	*agrochemicals*

Supplier	Trade name	Chemical description	Composition	General properties	Functionality / Application
Witco	SPONTO 710T	anionic/nonionic surfactant blend		liquid	
	SPONTO 712T	anionic/nonionic surfactant blend		liquid	*agrochemicals; herbicides*
	SPONTO 714T	anionic/nonionic surfactant blend		liquid	
	SPONTO 723T	anionic/nonionic surfactant blend		liquid	
	SYNTARYL 35	LAS/LES	35% active	liquid	*liquid detergents*
	SYNTARYL 48	LAS/LES/amide	50% active	liquid	*liquid detergents*
	SYNTARYL AMO	priopietary blend	30% active	liquid	*hard-surface cleaners*
	WITCODET 100	detergent concentrate, anionic/nonionic blend	98% solids	liquid; viscosity 400 cP's; colour 5 (Gardner); pH 8	detergent / *personal care products; household and industrial applications; all-purpose cleaners; dishwashing agents; degreasing products; laundry products*
	WITCODET 804	detergent concentrate, anionic/nonionic blend	50% solids	liquid; viscosity 1000 cP's; colour 2 (Gardner); pH 7	detergent / *personal care products; household and industrial applications; bubble baths; soaps; shampoos*
	WITCODET AE	detergent concentrate, pearlised anionic/nonionic/amphoteric blend	43% solids	liquid; viscosity 4000 cP's; pH 5.5	detergent; pearlescent agent / *personal care products; household and industrial applications; bubble baths; soaps; shampoos*
	WITCODET AEG	detergent concentrate, anionic/nonionic/amphoteric blend	42% solids	liquid; viscosity 3000 cP's; colour 3 (Gardner); pH 5.5	detergent / *personal care products; household and industrial applications; bubble baths; soaps; shampoos*
	WITCODET B-1	detergent concentrate, anionic/nonionic blend	50% solids	liquid; viscosity 200 cP's; colour 4 (Gardner); pH 5.8	detergent / *personal care products; household and industrial applications; dishwashing agents*
	WITCODET CWC	detergent concentrate, anionic/nonionic blend	52% solids	liquid; viscosity 300 cP's; colour 4 (Gardner); pH 8.5	detergent / *personal care products; household and industrial applications; car washes*
	WITCODET DC-47	detergent concentrate, anionic/nonionic blend	47% solids	liquid; viscosity 500 cP's; colour 5 (Gardner); pH 8	detergent / *personal care products; household and industrial applications; dishwashing agents*
	WITCODET DLC-47	detergent concentrate, anionic/nonionic blend	50% solids	liquid; viscosity 500 cP's; colour 5 (Gardner); pH 8	
	WITCODET SC	detergent concentrate, anionic/nonionic blend	57% solids	liquid; viscosity 400 cP's; colour 3 (Gardner); pH 6.5	
	WITCOMUL 3107	surfactant blend		liquid; pH 7	o/w emulsifier / *petroleum industry*
	WITCOMUL 3154	surfactant blend		liquid; pour point − 17.8°C; pH 7	w/o emulsifier / *petroleum industry*
	WITCOMUL 3158	surfactant blend		liquid; pour point − 2.2°C; pH 10.5	
	WITCONATE 68KN	anionic/nonionic blend	67% active; sulfate 1%	liquid; pH 7	detergent; foaming agent; wetting agent / *personal care products; household and industrial applications*
	WITCOR CI-1	cationic/nonionic complex surfactant		liquid; pH 12	anti-corrosion / *petroleum industry*
Zeneca	SOLSPERSE 5000	synergist hyperdispersant		blue powder; m.p. > 250°C	hyperdispersant / *inks; paint industry; plastics industry*
	SOLSPERSE 12000	synergist hyperdispersant		blue powder	
	SOLSPERSE 22000	synergist hyperdispersant		yellow powder; m.p. > 250°C	

Supplier	Trade name	Chemical description	Composition	General properties	Functionality Application
Zschimmer & Schwarz	AUTOPUR WK 4121	nonionic and cationic surfactants	40% active	liquid	car washes
	CONTRASPUM 1010 WEN	emulsifier, white oil, fatty acid derivatives, silicone	100% active	liquid	defoaming agent paint industry
	CONTRASPUM 1010 WN	emulsifier, white oil, fatty acid derivatives, silicone	100% active	liquid	defoaming agent paint industry
	CONTRASPUM 210	emulsifier, fatty acid, paraffin oil	100% active	liquid	defoaming agent yeast production; alcohol production
	CONTRASPUM 300	alkyl polyalkylene glycol ether	100% active	liquid	defoaming agent detergents; cleansing preparations
	CONTRASPUM A 4050	alkyl polyalkylene glycol ether, fatty alcohol	100% active	liquid	defoaming agent
	CONTRASPUM M 4053	emulsifier, white oil, fatty acid derivatives	100% active	liquid	defoaming agent
	EXTRAKT 52	special blend of anionic, amphoteric and nonionic surfactants	50% active	liquid	bath preparations; shampoos; skin care products; baby care products
	INHIBITOR 212	fatty acid alkanol amide	100% active	liquid	anti-corrosion
	INHIBITOR 4000	modified boric acid diethanolamide	73% active	liquid	
	INHIBITOR 7114	boric acid diethanolamide	73% active	liquid	
	LUMOROL 4153	blend of anionic and nonionic surfactants with solvents	60% active	liquid	cleansing preparations; floor cleaners; all-purpose cleaners
	LUMOROL 4154	blend of anionic and nonionic surfactants	49% active	liquid	cleansing preparations; dishwashing agents; all-purpose cleaners; floor cleaners; detergents
	LUMOROL 4192	blend of anionic and nonionic surfactants with solvents	48% active	liquid	cleansing preparations; floor cleaners; all-purpose cleaners
	LUMOROL 4290	blend of anionic surfactants with refatting agent	30% active	liquid	cleansing agent; all-purpose cleaners; dishwashing agents
	LUMOROL GG 65	blend of anionic and nonionic surfactants, refatting agent and foam stabiliser	65% active	paste	dishwashing agents; cleansing agents
	LUMOROL K 5019	blend of sodium lauryl sulfosuccinate and lauryl sulfoacetate	40% active	liquid	shampoos; bath preparations; skin care products
	LUMOROL RK	blend of anionic and amphoteric surfactants with refatting agent	28% active	liquid	cleansing preparations; skin care products; all-purpose cleaners; dishwashing agents
	LUMOROL W 5157	blend of anionic and nonionic surfactants	83% active	liquid	liquid detergents
	MULSIFAN K 326 SPEZIAL	blend of anionic and nonionic surfactants	85% active	HLB 10; liquid	emulsifier solvents; floor cleaners
	MULSIFAN RT 231	blend of polyalkylene glycol ethers	100% active	HLB 12; wax	emulsifier waxes
	MULSIFAN RT 237	alkyl aryl polyalkylene glycol ether	100% active	liquid	spreading agent waxes
	PERLGLANZMITTEL GM 4006	blend of nonionic substances with fatty alcohol ether sulfate	30% active	fluid	pearlescent agent shampoos; bath preparations
	PERLGLANZMITTEL GM 4055	blend of fatty acid glycol ester with fatty alcohol ether sulfate	38% active	fluid	opacifier shampoos; bath preparations

Supplier	Trade name	Chemical description	Composition	General properties	Functionality Application
Zschimmer & Schwarz	PERLGLANZMITTEL GM 4175	blend of fatty alcohol ether sulfate with pearlerscent substances	42% active	fluid	pearlescent agent *shampoos; bath preparations*
	PRODUKT 2058	fatty alcohol fatty acid ester	100% active	liquid	degreaser *cleaners*
	PRODUKT GM 4210	monoisopropanolamine fatty alcohol ether sulfate in 1,2-propylene glycol	90% active	liquid	*car washes*
	TRUBUNGSMITTLE 1	fatty acid amidoalkyl ether	100% active	flakes	opacifier *cosmetics*
	WACHSEMULSION 1864	wax emulsion with nonionic emulsifier	20% active	liquid	cleansing agent; glossing agent *car washes; floor cleaners*
	ZETESAP 813 A	blend of surfactants, plasticisers and fillers	50% active	noodle-shaped chips	syndet bars
	ZETESAP 5165	blend of surfactants (isethionate and sulfosuccinate), plasticers and fillers	40% active	noodle-shaped chips	syndet bars
	ZETESOL 100	blend of fatty alcohol ether sulfate and nonionic surfactants	98% active	liquid	*bubble baths; liquid soaps*

Index of company addresses

This section lists the names, addresses, telephone numbers and fax numbers of the supplier companies whose products are listed in this directory. The list includes company headquarters, subsidiary addresses, sales offices and agents throughout Europe. The list was correct, as far as is known, at the time of going to press.

3M

3M Oesterreich Ges mbh
Brunnerfeldstrasse 63
A-2380 Perchtoldsdorff
PO Box 611
A-1011 Vienna
Austria

3M France
Boulevard de l'Oise
F-95006 Cergy Pointoise Cedex
France

3M Deutschland GmbH
Carl-Schurz-Strasse 1
D-41453 Neuss
Germany

3M Svenska AB
Bollstanaesvaegen 3
S-191 89 Sollentuna
Saffans Vag 4
Sweden

3M (Schweiz) AG
Eggstrasse 93
CH-8803 Rueschlikon
Zurich
Switzerland

3M United Kingdom plc
3M House
PO Box 1
Bracknell
Berkshire RG12 1JU
United Kingdom
TEL: +44 1344 858000
FAX: +44 1344 858278

3M United Kingdom plc
3M House
28 Great Jackson Street
Manchester M15 4PA
United Kingdom
TEL: +44 161 2368500
FAX: +44 161 2376136

3M United Kingdom plc
Hudson Road
Bedford MK41 0HR
United Kingdom
TEL: +44 1234 268868
FAX: +44 1234 229499

AARHUS OLIE

Dr Hans Werba OHG
Postfach 868
Lugeck 1-2
1011 Wien 1
Austria
TEL: +43 1 51 49 00
FAX: +43 1 51 49 020

Merkantile
Trg. Marsala Tita8
Svacicev TRG 8
41000 Zagreb
Croatia
TEL: +385 41 411439
      +385 41 444827
FAX: +385 41 444924

Selectchemie AG
Etzelstrasse 42
CH-8038 Zurich
Switzerland
TEL: +41 1 4829611
FAX: +41 1 4823842

Aarhus Oliefabrik A/S
27, M.P. Bruuns Gade
PO Box 50
DK-8100 Aarhus C
Denmark
TEL: +45 86 12 60 00
FAX: +45 86 18 38 39

Scantho Oy
Lampikuja 2
SF-34600 Ruovesi
Finland
TEL: +358 34 47 63 380
FAX: +358 34 47 63 390

Societe Nouvelle SACI
9, rue Richepanse
F-75008 Paris
France
TEL: +33 1 42 60 12 83
FAX: +33 1 42 61 42 79

Helmut Kiesow
Chemikallen und Rohstoffe
Bellershelmer Strasse 28
D-35410 Hungen
(OT Trals-Horlof)
Germany
TEL: +49 6402 3005
FAX: +49 6402 3042

I. Ignatiadas Trading Co.
A. Ignatiades - T. Kostavaras Ltd.
PO Box 19010
117 10 Athens
Greece
TEL: +30 92 19 3 70
FAX: +30 19 23 50 42

Aako B.V.
PO Box 205
3830 Leusden
Holland
TEL: +31 33 94 84 94
FAX: +31 33 94 80 44

Berk Ltd.
34 Lower Leeson St.
Dublin 2
Ireland
TEL: +353 1 767915
FAX: +353 1 6612172

Eltra Chemicals
P.O. Box 5271
Kiryat Ha'Asakim
Derech Hebron 1
Pinat Yosef Haburekal
Beer Sheva
Israel
TEL: +972 7 280 186
FAX: +972 7 277 019

Prodotti Gianni S.p.A.
Via M.F. Quintiliano, 30
I-20138 Milano
Italy
TEL: +39 2 50 97 1
FAX: +39 2 50 97 313

Store & Salveson
Osterndalen 8
PO Box 54
N-1345 Oslo
Norway
TEL: +47 67 149 270
FAX: +47 67 149 360

Consorima
Rua do Cadavao, Ap. 135
Vllar do Paraiso - V.N. Gaia
4.408 Valadares Codex
Portugal
TEL: +351 2 711 55 24/41/48
FAX: +351 2 711 55 70

Bionord
Granstigen 2
S-444 41 Stenungsund
Sweden
TEL: +46 303 818 60
FAX: +46 303 818 66

Bernadi, S.C.P.
Balmes Street, 341
08006 Barcelona
Spain
TEL: +34 3 212 16 67
FAX: +34 3 418 58 48

Zeus Quimica
Frelxa, 32
E-08021 Barcelona
Spain
TEL: +34 3201 94 00
FAX: +34 3200 59 02

Kima Kimyevl Maddeler
Pazarlama Ve Ticaret AS
P.O.B. Slsli 602
Istanbul
Turkey
TEL: + 90   1 24 85 09 0
FAX: + 90   1 24 75 18 9

Efkay Chemicals Ltd.
204 Banderway House
156-162 Kilburn High Road
London NW6 4JD
United Kingdom
TEL: + 44   171 625 4445
FAX: + 44   171 328 9101

AIR PRODUCTS

Keyser & Mackay
Avenue Plasky 144, Box 8/9
B-1040 Brussels
Belgium
TEL: + 32   2 7354072
FAX: + 32   2 7347600

Oy Esope AB
Postipuuntie 2, PO Box 49
SF-02601 Helsinki
Finland
TEL: + 358   0 519088
FAX: + 358   0 677576

Air Products
Le Forum Est
48, Bd de Coquibus
F-91000 Evry
France
TEL: + 33   1 69364402
FAX: + 33   1 69364390

W. Biesterfeld & Co. GmbH
Ferdinandstrasse 41
D-20095 Hamburg
Germany
TEL: + 49   40 320081
FAX: + 49   40 32008340

E + N Macrakis & Co. Ltd.
26, Mylopotamoustreet
11526 Athens
Greece
TEL: + 30   1 6480278
FAX: + 30   1 6482867

Gadot Chemical Terminals
PO Box 3196
22 Shalom Aleichem Street
Tel Aviv
Israel
TEL: + 972   3 286262
FAX: + 972   3 5282717

Eigenmann & Veronelli SpA
Via della Mosa, 6
I-20017 RHO (Milano)
Italy
TEL: + 39   2 935391
FAX: + 39   2 93539300

Kraft Trading A.S.
PO Box 1673 Vika
N-0120 Oslo 1
Norway

TEL: + 47   2 833330
FAX: + 47   2 834124

Salmon & Cia LDA
Rua Cova da Moura 2-6
P-1300 Lisbon
Portugal
TEL: + 351   1 608181
FAX: + 351   1 3967591

Quimigranel S.A.
Via Augusta, 143
E-08021 Barcelona
Spain
TEL: + 34   3 2000511
FAX: + 34   3 2000789

Air Products & Chemicals Inc.
c/o World Trade Centre
Maessans gata 10 - Box 5264
S-40225 Gothenburg
Sweden
TEL: + 46   31 357511
FAX: + 46   31 164200

Chemische Fabrik Schweizerhall
PO Box 401
CH-4013 Basel
Switzerland
TEL: + 41   61 3268111
FAX: + 41   61 3225022

Air Products Nederland B.V.
Kanaalweg 15
PO Box 3193
3502 GD Utrecht
The Netherlands
TEL: + 31   30 857100
        (2857100 after 10-10-95)
FAX: + 31   30 857111
        (2857111 after 10-10-95)

Stokvis Chemicalien BV
Postbus 287
NL-2990 AG Barendrecht
The Netherlands
TEL: + 31   1806 56222
FAX: + 31   1806 56333

H. Levyi Co.
PO Box 677
Karakoy/Istanbul
Turkey
TEL: + 90   212 2442281
FAX: + 90   212 525278

Air Products plc
Clayton Lane
Clayton
Manchester M11 4SR
United Kingdom
TEL: + 44   161 8778633
FAX: + 44   161 8778580

AKCROS CHEMICALS

Telko Oy
Hitsaajankatu 9
FIN-00811 Helsinki
Finland
TEL: + 358   0 615500
FAX: + 358   0 780064

Akcros Chemicals France S.A.
Zone Industrielle BP 40
41220 Saint-Laurent-Nouan
France
TEL: + 33   54 87 73 89
FAX: + 33   54 87 79 40

Akcros Chemicals GmbH & Co KG
Postfach 10 01 32
52301 Düren
Germany
TEL: + 49   2421 492 221
FAX: + 49   2421 492 470

S.B. Piperas E.E.
7 Leontiou Street
Athens 117 45
Greece
TEL: + 30   1 92 25 311
        + 30   1 92 20 947
FAX: + 30   1 92 25 311

Akcros Chemicals Italia Srl
Via Christina Belgioioso 13/15
20021 Baranzate Di Bollate (MI)
Italy
TEL: + 39   2 38200484
FAX: + 39   2 38200443

Akcros Chemicals v.o.f.
Haagen Site
Molenweg 10
PO Box 44
6040 AA Roermond
The Netherlands
TEL: + 31   4750 91777
FAX: + 31   4750 17489

Akcros Chemicals Nordic ApS
Vesterbrogade 14 A
DK-1620 Copenhagen V
Scandinavia
Denmark
TEL: + 45   3121 4200
FAX: + 45   3121 4227

Warwick Benbassat S.A.
Comte d'Urgell, 240
08036 Barcelona
Spain
TEL: + 34   322 06 53
        + 34   322 37 12
FAX: + 34   322 09 93

Akcros Chemicals
Lankro House
Silk Street
PO Box 1
Eccles
Manchester M30 0BH
United Kingdom
TEL: + 44   161 788 7300
FAX: + 44   161 788 7886

AKZO NOBEL SURFACE CHEMISTRY

Akzo Nobel Chemicals A/S
Literbuen 9
P.O. Box 40
DK-2740 Skovlunde
Denmark
TEL: + 45   44 538600
FAX: + 45   44 538601

Akzo Nobel Chemicals Oy
Mikonkatu 8
00100 Helsinki
Finland
TEL: + 358   0 624611
FAX: + 358   0 628503

Akzo Nobel Surface Chemistry S.A.
Z.I. de Venette
P.O. Box 217
60202 Compiegne, Cedex
France
TEL: + 33   44 903000
FAX: + 33   44 903044

Akzo Nobel Chemicals GmbH
27, Philippstrasse
P.O. Box 100132
52301 Düren
Germany
TEL: + 49   2421 49201
FAX: + 49   2421 492487

Akzo Nobel Chemicals GmbH
Wettensteinstrasse 4
82024 Taufkirchen
Germany
TEL: + 49   89 6120080
FAX: + 49   89 61200825

Akzo Nobel Chemicals SpA
80, via E. Vismara
20020 Arese (Milano)
Italy
TEL: + 39   2 935251
FAX: + 39   2 9380816

Akzo Nobel Chemicals BV
Stationsplein 4
P.O. Box 247
3800 AE Amersfoort
The Netherlands
TEL: + 31   33 676324
FAX: + 31   33 676178

Akzo Nobel Surface Chemistry A/S
Stabburveien 2
P.O. Box 85
N-1601 Fredrikstad
Norway
TEL: + 47   69 398099
FAX: + 47   69 398710

Akzo Nobel Chemicals SA
Autovia de Castelldefels. km 4.65
08820 El Prat de Llobregat (Barcelona)
Spain
TEL: + 34   3 4784411
FAX: + 34   3 4780734

Akzo Nobel Surface Chemistry AB
S-444 85 Stenungsund
Sweden
TEL: + 46   303 85225
FAX: + 46   303 770281

Akzo Nobel Surface Chemistry Ltd.
23 Grosvenor Road
St Albans
Hertfordshire AL1 3AW
United Kingdom

TEL: + 44   1727 841421
FAX: + 44   1727 841529

ALBRIGHT & WILSON

Albright & Wilson BV
Square Vergote 1 Box 1
1200 Brussels
Belgium
TEL: + 32   2 735 2072
FAX: + 32   2 736 8616

Albright & Wilson Ltd.
Hermanova 44
170 00 Praha 7 - Holesovice
Czech Republic
TEL: + 42   2 2423 2190
FAX: + 42   2 2423 2223

Albright & Wilson Denmark A/S
Siestavej 7;1
2600 Glostrup
Denmark
TEL: + 45   43 44 01 00
FAX: + 45   43 44 03 33

Albright & Wilson
Myyntikonttori
Lönnrotinkatu 25A
FI-00180 Helsinki
Finland
TEL: + 358   0 6931551
FAX: + 358   0 6931429

Albright & Wilson Saint Mihiel SA
Boite Postale 19
F-55300 St Mihiel
Han-sur-Meuse, Meuse
France
TEL: + 33   29 917300
FAX: + 33   29 917399

Albright & Wilson GmbH
City Center, Frankfurter Strasse 181
63263 Neu-Isenburg
Germany
TEL: + 49   6102 27051
FAX: + 49   6102 25286

Albright & Wilson Ltd.
Naiadon 45, 175-61
P. Faliro - Athens
Greece
TEL: + 30   1 9833984
FAX: + 30   1 9833985

Albright & Wilson Castiglione Srl
Via Cavour 50
Casella Postale No 142
I-46043 Castiglione delle Stiviere
(Mantova)
Italy
TEL: + 39   376 6371
FAX: + 39   376 637323

Albright & Wilson Patrica Srl
Strada Morolense KM 10, 100
I-03010 Patrica
(Frosinone)
Italy
TEL: + 39   775 2961
FAX: + 39   775 296237

Surphos Chemicals Benelux BV
Plein Eendragt 17
Postbus 238
3100 AE Schiedam
The Netherlands
TEL: + 31   10 426 0024
FAX: + 31   10 427 0322

Albright & Wilson A/S
PO Box 170
N-2040 Klofta
Norway
TEL: + 47   63 981410
FAX: + 47   63 981809

Albright & Wilson
ul Filtrowa 71A/5A 02-055
Warsaw
Poland
TEL: + 48   22 253335
FAX: + 48   22 252175

Albright & Wilson
Quimica Portuguesa Lda
PO Box 12
Santo Antonio 2780
Oeiras
Portugal
TEL: + 351   1 4863203
FAX: + 351   1 4863167

Albright & Wilson Iberica S.A.
Poligono Zona Franca
Sector F, Calle 43, No 10
08040 Barcelona
Spain
TEL: + 34   3 3350100
FAX: + 34   3 3363610

Albright & Wilson AB
Hälleflundregatan 12
S-426 58 Vastra Frolunda
Gothenburg
Sweden
TEL: + 46   31 698060
FAX: + 46   31 698222

Albright & Wilson UK Ltd.
Corporate Headquarters
PO Box 3
210-222 Hagley Road West
Oldbury, Warley
West Midlands B68 0NN
United Kingdom
TEL: + 44   121 429 4942
FAX: + 44   121 420 5151

ALLIED COLLOIDS

N.V. Allied Colloids Belgium S.A.
Zoning Sud
60 Chaussee De Namur
B-1400 Nivelles
Belgium
TEL: + 32   67 211701
FAX: + 32   67 219893

Oy Suomen Allied Colloids
Hyttipojankuja 2
PO Box 21
SF 02781 Espoo
Finland

TEL: + 358   0 812255
FAX: + 358   0 810603

Bez S.A.
43 Rue Du Landy
F-93210 La Plaine Saint Denis
France
TEL: + 33   1 49175300
FAX: + 33   1 49175309

Allied Colloids GmbH
PO Box 620107
D-2401 Hamburg
Germany
TEL: + 49   40 527208
FAX: + 49   40 5270915

Allied Colloids Ltd.
Mohai Ut 37
1116, Budapest
Hungary
TEL: + 36   1 1852664
FAX: + 36   1 2092877

Allied Colloids Poland
Ul. Krolowej Aldony 12 M 2
03-928 Warsaw
Poland
TEL: + 48   2 6173747
FAX: + 48   2 6174817

Allied Colloids
PO Box 38
Low Moor
Bradford
West Yorkshire BD12 0JZ
United Kingdom
TEL: + 44   1274 671267
FAX: + 44   1274 606499

AMERCHOL

Amerchol Europe
Havenstraat 84
B-1800 Vilvoorde
Belgium

Comptoir Francais des Produits
Aromatiques
379 Avenue du President Wilson
93210 La Plaine Saint-Denis
France

Nordmann, Rassmann & Co.
Kajen 2
2 Hamburg 11
Germany

Kyrlakos Yialantzis
P.O.B. 610.056
151.10 Marousl
Athens
Greece

Variati & Co, S.P.A.
Via A Pestalozza No. 16
20131 Milan
Italy

Lusifar Quimica Comercial, LDA.
Rua Conde De Almoster 22
1500 Lisboae 4
Portugal

Ricorvi, S.A.
Liussa 28, Plgois 20-22
Barcelona 28
Spain

Kemi-Intressen AB
Vintergatan 1
Box 6018
172 06 Sundbyberg 6
Sweden

Heinz Schlegel & Co.
Industriestrasse
4313 Mohlin
Switzerland

Ellis & Everard (UK) Ltd.
Anstead International
Radford Way
Billericay
Essex CM12 0DE
United Kingdom
TEL: + 44   1277 630063
FAX: + 44   1277 631356

ANGUS CHEMIE

Angus Chemie GmbH
4530 Ibbenburen
Germany
TEL: + 49   5459 560
FAX: + 49   5459 5641

Angus Chemie GmbH
Information Office
Unit 7
Rotunda Business Centre
Thorncliffe Park Estate
Chapeltown
Sheffield
South Yorkshire S30 4PH
United Kingdom
TEL: + 44   114 2571322
FAX: + 44   114 2571336

Samuel Banner & Co. Ltd.
59-61 Sandhills Lane
Liverpool L5 9XL
United Kingdom
TEL: + 44   151 922 7871
FAX: + 44   151 922 0407

AUSCHEM CESALPINIA

(see Cesalpinia Chemicals)

BASF

BASF Oesterreich Ges. m.b.H.
Hietzinger Haupstrasse 199
Postfach 1000
A-1131 Wien
Austria
TEL: + 43 1 8 789100
FAX: + 43 1 8 7890110

BASF Belgium S.A.
Avenue Hamoir 14
B-1180 Brussels
Belgium
TEL: + 32   2 373 1211
FAX: + 32   2 373 1220

BASF EOOD
Iskarsko Chaussee 5
1528 Sofia
Bulgaria
TEL: + 359   242 6018
FAX: + 359   242 7081

BASF Representation
PR Masherova 5
220004 Minsk
Byelorussia
TEL: + 7   172 23 90 42
FAX: + 7   172 23 90 13

BASF AG
Predstavnistvo u Hrvatskoj
Vlaska 40
41000 Zagreb
Croatia
TEL: + 381   41 431349
FAX: + 381   41 435464

BASF spol s.r.o.
Korunovacni 6/103
17000 Praha 7
Czech Republic
TEL: + 42   2 3808
FAX: + 42   2 37 8445

BASF Danmark A/S
Ved Stadsgraven 15
Postboks 1734
DK-2300 Kobenhavn S
Denmark
TEL: + 45   31 57 00 11
FAX: + 45   31 57 22 02

BASF Representation
Kr Valdemara 23 Rm 301
1010 Riga
Latvia
TEL: + 371   233 2775
FAX: + 358   4910 5537 (via Finland)

BASF Oy
Maistraatinportti 4
Postfach 5
SF-00241 Helsinki
Finland
TEL: + 358   0 615 981
FAX: + 358   0 615 98 250

BASF France SA
49 Avenue Georges Pompidou
F-92593 Levallois-Perret Cedex
France
TEL: + 33 1 49 645000
FAX: + 33 1 49 545050

BASF AG
ESA-J550
67056 Ludwigshafen
Germany
TEL: + 49   621 60 99603
FAX: + 49   621 60 41787

Dr D A Delis AG
Paleologou Benizelou 5
GR-10556 Athinai
Greece
TEL: + 30   1 3 29 72 22
FAX: + 30   1 3 23 05 50

BASF Hungaria Kft
Seregely u 1-5
H-1034 Budapest
Hungary
TEL: + 36   1 250 4111
FAX: + 36   1 250 4660

BASF Ireland
Enterprise House
Frascati Road
Blackrock
Co Dublin
Ireland
TEL: + 353   1 2 83 28 88
FAX: + 353   1 2 83 26 71

BASF Italia Spa
Via Marconato 8
1-20031 Cesano Maderno MI
Italy
TEL: + 39   13 62 51 21
FAX: + 39   13 62 55 27 80

BASF Representation
ul Tole bi 69
480091 Almaty
Kazakhstan
TEL: + 7   32 72 62 8447
FAX: + 7   32 72 62 0172

Hemomak Export-Import Co.
Dane Grujev 5
91000 Skopje
Macedonia
TEL: + 389   22 6222
FAX: + 389   22 6249

BASF Nederland BV
Locatie Kadestraat
Kadestraat 1
NL-6811 CA Arnhem
Netherlands
TEL: + 31   85 71 71 71
FAX: + 31   85 71 72 46

BASF Norge AS
Leanbukta 40
Postboks 233
N-1371 Asker
Norway
TEL: + 47 66 90 46 00
FAX: + 47 66 90 47 55

BASF Polska Sp z.o.o.
Aleje Jerozolimskie 154
02-326 Warszawa
Poland
TEL: + 48 2659 2142
FAX: + 48 2658 3534

BASF SRL
155 Calea Victoriei
Bl D1 Tronson 7-8 et 6
RO-71102 Bucuresti - sector 1
Rumania
TEL: + 40   1659 0535
FAX: + 40   1211 4195

BASF Representation
B Gnezdnikovskij per 7
103009 Moscow
Russian Federation

TEL: + 7   502 221 1213
FAX: + 7   502 221 1214

BASF Slovensko spol s.r.o.
Frana Krala 35
81105 Bratislava
Slovak Republic
TEL: + 42   749 8755
FAX: + 42   749 8839

BASF AG
Predstavnistvo BASF AG Sloveniji
Kersnikova 5
61000 Ljubljana
Slovenia
TEL: + 386   6131 6688
FAX: + 386   6131 5987

BASF Espanola S.A.
Paseo de Gracia 99
E-08008 Barcelona
Spain
TEL: + 34   3 215 13 54
FAX: + 34   3 488 20 20

BASF Svenska AB
Kungsportsavenyen 31-35
Box 53008
S-40014 Gothenburg
Sweden
TEL: + 46   31 63 98 00
FAX: + 46   31 18 76 81

BASF (Schweiz) AG
Appital
Postfach 99
CH-8820 Waedenswil/Au
Switzerland
TEL: + 41   1 7 81 91 11
FAX: + 41   1 7 81 93 88

BASF Representation
ul Ljuteranskaja 20
252601 Kiev
Ukraine
TEL: + 7   044 228 6374
FAX: + 7   044 228 6585

BASF plc
P.O. Box 4
Earl Road
Cheadle Hulme
Cheadle
Cheshire SK86QG
United Kingdom
TEL: + 44   161 485 6222
FAX: + 44   161 486 0891

BASF Representation
Uzbechistan
Beethoven Strasse 3
700079 Tashkent
Uzbechistan
TEL: + 7 3712 54 5789
FAX: + 7 3712 55 1381

BASF AG
Predstavnistvo u Jugoslaviji
Djure Djakovica 78
11000 Belgrade
Yugoslavia
TEL: + 381   11 772 999
FAX: + 381   11 751 743

BAYER

Bayer plc
Bayer House
Strawberry Hill
Newbury
Berkshire RG13 1JA
United Kingdom
TEL: + 44   1635 39000
FAX: + 44   1635 563393

DR TH. BOEHME

Dr Th. Boehme Chem. Fabrik Ges m.b.H.
A-2602 Blumau
Austria
TEL: + 43   2628 8205/6/7
FAX: + 43   2628 8816

Dr Th. Boehme Belgium N.V.
Keizer Karelstraat, 62
B-8930 Menen
Belgium
TEL: + 32   56 51 4092
FAX: + 32   56 51 6815

U STADION U 2187
CS-58001 Havl. Brod
Czech Republic
TEL: + 42   451 2 18 23
FAX: + 42   451 2 18 23

Bodotex International A/S
Ulvevaenget 10
DK-7100 Vejle
Denmark
TEL: + 45   75 82 57 66
FAX: + 45   75 72 42 26

Ari Ilmakunnas Oy
Hatamestarinkatu 11
SF-00520 Helsinki
Finland
TEL: + 358   0 1404 11
FAX: + 358   0 1404 56

Tanmat Oy
Ilvestie 6
37800 Toijala
Finland
TEL: + 358   37 2 16 31
FAX: + 358   37 2 25 50

Cevmi-Chimie
15, Avenue Paul-Doumer
F-75116 Paris
France
TEL: + 33   1 47 27 89 51
FAX: + 33   1 47 55 63 52

Boehme France s.a.r.l.
7, Place de la Gare
57200 Sarreguemines
France
TEL: + 33   87 98 48 32
FAX: + 33   63 95 99 01

Boehme France s.a.r.l.
Zone Artisanale
Rue Arago
81300 Graulhet
France
TEL: + 33   63 34 83 31
FAX: + 33   63 34 25 29

Dr Th. Boehme KG., Chemische
Fabrik GmbH & Co.
Isardamm 79-83
82538 Geretsried
Germany
TEL: + 49   8171 6280
FAX: + 49   8171 628388

Hellas-Chemie GmbH
Nikolaos Charissis
12, Panteli Makri Str.
11143 Athens
Greece
TEL: + 30   1 2 52 14 36
FAX: + 30   1 2 52 14 56

Vaxair Hellas
P. Vaxevanakis & Sons s.A.
21, Eponiton & Fokeas Str.
Piraeus 18547
Greece
TEL: + 30   1 4 81 24 92
FAX: + 30   1 4 81 92 50

b.v. v/h L. J. Volkers
Industrieweg 1
1231 KG Nieuw Loosdrecht
Holland
TEL: + 31   21 5824704
FAX: + 31   21 5824364

Hollandia Color B.V.
Postbus 28
Klokkenlaan 3 A
5175 NV Loon op Zand
Holland
TEL: + 31   41661902
FAX: + 31   41662716

F.T.R. Forniture Tessili Riunite S.p.A.
Via Galvani, 12
24061 Albano S/Alessandro (BG)
Italy
TEL: + 39   35 58 12 36
FAX: + 39   35 58 21 31

Andrea Parducci
Via C.A. Carlone, 1
20147 Milano
Italy
TEL: + 39   2 407 5244

Samia s.a.s.
3a Z. Ind.
10a Strada 15/17
36071 Arzignano (Vicenza)
Italy
TEL: + 39   444 45 16 51
FAX: + 39   444 45 04 91

De Bernardi Cesare
Via Labriola
Casella Postale 125
56029 S. Croce sull'Arno (Pisa)
Italy
TEL: + 39   5 71 3 10 15
FAX: + 39   5 71 36 60 98

Italchimica s.r.l.
Via Panoramica Turci
83029 Solofra (AV)
Italy

TEL: + 39   8 25 53 47 57
FAX: + 39   8 25 58 19 81

Kirsebom & Hurum A/S
Krokfaret 2
1340 Bekkestua
Norway
TEL: + 47   2 53 90 64
FAX: + 47   2 58 09 97

Americo da Oliveira Sengo
Rua da Mina, 438-Canelas
4405 Valadares
Portugal
TEL: + 351   2 762 64 06/56/31
FAX: + 351   2 762 63 81

Kemicna industrija IPLAS o.sol.o.
Krpanova 2
66000 Koper
Slovenia
TEL: + 386   66 6 3857
FAX: + 386   66 2 2379

P. Arguelles S.A.
Poligono Industrial
Torrente Estadella N° 23-25
08030 Barcelona (30)
Spain
TEL: + 34   3 346 15 51
FAX: + 34   3 311 02 10

Korsväga Kemi AB
Strandvägen 16
51300 Fristad
Sweden
TEL: + 46   33 26 68 30
FAX: + 46   33 26 91 85

Lea Trading
Backgatan 12
61400 Soderkoping
Sweden
TEL: + 46   121 116 61

Albert Isliker & Co. AG
Ringstrasse 72
CH-8050 Zurich
Switzerland
TEL: + 41   1 3 18 66 66
FAX: + 41   1 3 18 66 88

L + S AG
Departement Farben, Chemikalien
Mühlhauser Strasse 111
4056 Basel
Switzerland
TEL: + 41   3 22 43 00
FAX: + 41   3 22 43 76

Dr Th. Boehme (UK) Ltd.
Moorland House
Low Green
Rawdon
Leeds LS19 6HA
United Kingdom
TEL: + 44   113 2391402
FAX: + 44   113 2506868

BORREGAARD

Johan Foge Jensen A/S
Skelbaekgade 2
1717 Kobenhavn V
Denmark
TEL: + 45   31 212020
FAX: + 45   31 213120

LignoTech Finland Oy
PO Box 572
FIN-33101 Tampere
Finland
TEL: + 358   31 3472711
FAX: + 358   31 3464431

LignoTech Verkaufsgesellschaft mbH
Hansa Allee 159
Postfach 110732
D-40507 Dusseldorf
Germany
TEL: + 43   211 526020
FAX: + 43   211 5260222

Anorgachim Corporation
9 Mourozi Str.
Athens 138
Greece
TEL: + 30   1 7236676
        + 30   1 7236700
FAX: + 30   1 7236410

Scanac
Budapest II
BEM Ter 1, III EM.22
Hungary
TEL: + 36   1 1355464
FAX: + 36   1 1350874

Eltra Chemicals Ltd.
Boks 5271
84152 Beer Sheva
Israel
TEL: + 972   7280166
FAX: + 972   7277019

Benfer - Scheller SpA
Viale Monza 265
I-20126 Milano
Italy
TEL: + 39   2 257701
FAX: + 39   2 257701

LignoTech Norway
PO Box 162
N-1702 Sarpsborg
Norway
TEL: + 47   69118000
FAX: + 47   69118210

Salmon & Cia Lda.
Rua Cova da Moura 2-6, Andar
Apartado 2722
1118 Lisboa Codex
Portugal
TEL: + 351   1 608181
FAX: + 351   1 3967591

Curtex S.A.
Carretera del Medio 219
08907 l'Hospitalet
Spain

TEL: + 34   93 3352100
FAX: + 34   93 3352358

LignoTech Sweden AB
S-468 82 Vargon
Sweden
TEL: + 46   521 277500
FAX: + 46   521 223186

Kapudag Kollektif Sirketi
Husrev Gerede Caddesi
Baran Apt. 58/60 Diare 12
Tsvikiye-Istanbul
Turkey
TEL: + 90   1 614316
       + 90   1 473823
FAX: + 90   1 1587261

LignoTech (UK) Ltd.
Clayton Road
Birchwood
Warrington
Cheshire WA36QQ
United Kingdom
TEL: + 44   1925 824511
FAX: + 44   1925 812186

CESALPINIA CHEMICALS

(formerly Auschem Cesalpinia)

Cesalpinia Chemicals spa
Via Della Moia Angolo
Via Marconi
20020 Arese
Milano
Italy
TEL: + 39   2   939941
FAX: + 39   2   93581269

CHEMAX

Keyser and Mackay
144 Avenue Plasky
10040 Brussels
Belgium
TEL: + 32   2 7354072
FAX: + 32   2 7347600

Keyser and Mackay
94/102 rue de Buzenval
75020 Paris
France
TEL: + 33   1 44938500
FAX: + 33   1 40099166

Keyser and Mackay
Ringstrasse 46
50996 Köln
Germany
TEL: + 49   221 392044
FAX: + 49   221 395839

Finco, S.p.A.
Viale Bianca Maria, 13
20122 Milano
Italy
TEL: + 39   2 760 13701
FAX: + 39   2 783 701

Keyser and Mackay
Leidsegracht 19
1017 NA Amsterdam
The Netherlands

TEL: + 31   20 626 3323
FAX: + 31   20 625 4559

R. Monti
Badenerstrasse 701
CH-8048 Zurich
Switzerland
TEL: + 41   1 431 6161
FAX: + 41   1 432 8026

Alfa Chemicals Ltd.
Alfa House
15 Moor Park Avenue
Preston
Lancashire PR1 6AS
United Kingdom
TEL: + 44   1772 258969
FAX: + 44   1772 253207

Alfa Chemicals Ltd.
Arc House
Terrace Road South
Binfield
Berkshire RG12 5PZ
United Kingdom
TEL: + 44   1344 861800
FAX: + 44   1344 862010

CHEMVIRON SPECIALITY CHEMICALS

Schoeller Chemie Produkte GmbH
Markgraf Ruediger Strasse 6
A-1150 Vienna
Austria
TEL: + 43   1 985 5555
FAX: + 43   1 982 767116

Sepulcre S.A.
Avenue des Nenuphars 19/B.6
1160 Brussels
Belgium
TEL: + 32   2 672 2335
FAX: + 32   2 673 6782

Chemviron Speciality Chemicals
Brusselsesteenweg 359
3090 Overijse
Belgium
TEL: + 32   2 689 0911
FAX: + 32   2 687 3464

Mogens Nielsen ApS
Norregade 43
DK-165 Copenhagen K
Denmark
TEL: + 45   33 143 040
FAX: + 45   33 149 775

Groupe Arnaud
Dept. Schmitt-Jourdan
Av. du Gen. Michel Bizot 68
F-75012 Paris
France
TEL: + 33   1 4473 1000
FAX: + 33   1 4473 1078

Chemviron Speciality Chemicals
Garbersweg 74
D-21077 Hamburg 90
Germany
TEL: + 49   40 760 9595
FAX: + 49   40 760 8119

Dichem Chemicals Ltd.
364 Achamon & 10 Glaraki Str.
Eleptherios
Athens 1145 AG
Greece
TEL: + 30   1 202 8003
FAX: + 30   1 202 8382

Eigenmann & Veronelli SpA.
Via della Mosa 6
20017 Rho
(Milano)
Italy
TEL: + 39   2 935 391
FAX: + 39   2 935 39300/9

Heybroek B.V.
Leidsegracht 38-40
PO Box 555
NL-100 AN Amsterdam
The Netherlands
TEL: + 31   20 624 6973
FAX: + 31   20 625 5994

Masso y Carol S.A.
Casetes de Can Via s/n
E-08690 Sta Coloma de Cervello
Barcelona
Spain
TEL: + 34   3 634 0125
FAX: + 34   3 634 0116

AB CDM
Reningsverksgatan 5
Box 37
S-42121 Vastra Frolunda
Sweden
TEL: + 46   31 893 900
FAX: + 46   31 493 723

Dolder AG
Immengasse 9
CH-4004 Basel
Switzerland
TEL: + 41   61 326 6600
FAX: + 41   61 326 6204

Chemviron Speciality Chemicals Ltd.
No. 4 The Pavilion Business Park
Royds Hall Road
Leeds LS12 6AJ
United Kingdom
TEL: + 44   113 2319233
FAX: + 44   113 2319245

CHEM-Y

M Winzierl GmbH & Co. KG
Himbergerstr. 68-70
A-2320 Schwechat
Austria
TEL: + 43   1 7077628
FAX: + 43   1 7071343

Lassen Chemical Agency
Enghaven 9
DK-3450 Allerod
Denmark
TEL: + 45   48143500
FAX: + 45   48143483

East West Co. Ltd.
Tulppaanitie 4
SF-01390 Vantas
Finland
TEL: + 358   8252011
FAX: + 358   8252126

Lambert Riviere S.A.
17 Avenue Lou son Bobet
Val de Fontenay
F-94132 Fontenay-Sous-Bois Cedex
France
TEL: + 33   1 49748080
FAX: + 33   1 49748111

Dichem Chemicals Ltd.
364, Acharnon & 10 Glaraki St.
GR-11145 Athens
Greece
TEL: + 30   1 2028003
FAX: + 30   1 2028382

U.C.E. S.p.A.
Societa Industriale e Commerciale
Via Caldera, 21
I-20153 Milano
Italy
TEL: + 39   2 452771
FAX: + 39   2 4525810

Promol A.S.
PO Box 364
N-1701 Sarpsborg
Norway
TEL: + 47   69 121530
FAX: + 47   69 150770

Crimolara Productos Quimicos, Ida.
Campo Grande, 30,3.-H
P-1700 Lisboa
Portugal
TEL: + 351   7966168
FAX: + 351   7954762

Kao Corporation S.A.
Puig dels Tudons, 10
Apartado de Correos 74
E-08210 Barbera Del Valles
Barcelona
Spain
TEL: + 34   3 7399300
FAX: + 34   3 7399333

Dolder AG
Immengasse 9
CH-4004 Basel
Switzerland
TEL: + 41   61 3266600
FAX: + 41   61 3266204

Blagden Chemicals Ltd.
Osprey House
Black Eagle Square
Westerham
Kent TN16 1PA
United Kingdom
TEL: + 44   1959 562000
FAX: + 44   1959 565111

CIBA

Ciba Dyes and Chemicals
Hulley Road
Macclesfield
Cheshire
SK10 2NX
United Kingdom
TEL: + 44   1625 618585
FAX: + 44   1625 888006

CONDEA

Condea Chemie Benelux
Rubenslei 2
B-2018 Antwerp
Belgium
TEL: + 32   3 226 4248
FAX: + 32   3 227 0268

Condea Chimie S.A.R.L.
14 Rue Christine de Pisan
F-75017 Paris
France
TEL: + 33   1 44010520
FAX: + 33   1 47662425

Condea GMBH
Uberseering 40
22297 Hamburg
Germany
TEL: + 49   40 6375 0
FAX: + 49   40 6375 3595

DAC International Surfactants S.P.A.
V.le Forlani, 65
20138 Milano
Italy
TEL: + 39   2 70101711
FAX: + 39   2 719673

Condea Chemicals UK Ltd.
Millennium House
21 Eden Street
Kingston-upon-Thames
Surrey KT1 1BL
United Kingdom
TEL: + 44   181 5473006
FAX: + 44   181 5473608

Efkay Chemicals Ltd.
204 Banderway House
156-162 Kilburn High Road
London NW6 4JD
United Kingdom
TEL: + 44   171 6254445
FAX: + 44   171 3289101

COSMETOCHEM

Cosmetochem AG
Riedstrasse 7
CH-6330 Cham
Switzerland

Cosmetochem (UK) Ltd.
Cunningham House
Westfield Lane
Kenton
Harrow HA3 9ED
United Kingdom
TEL: + 44   181 9077779
FAX: + 44   181 9091053

COURTAULDS

Courtaulds Chemicals
Fine Chemicals
Macclesfield Road
Leek
Staffordshire ST13 8UZ
United Kingdom
TEL: + 44   1538 399100
FAX: + 44   1538 399025

CPB-Companhia Petroquimica do Barreiro

CPB-Companhia Petroquimica do Barreiro
Apt. 31 Lavradio - 2830
Barreiro
Portugal
TEL: + 351   207 9272
FAX: + 351   207 2923/0644

CRODA

Croda France SA
78193 Trappes Cedex
France
TEL: + 33   1 3062 6662
FAX: + 33   1 3064 7013

Croda GmbH
Postfach 2050
4054 Nettetal 2
Germany
TEL: + 49   2157 1016 7 8
FAX: + 49   2157 2350

Croda Italiana SpA
Via Grocco, N917 27036
27036 Mortara
Italy
TEL: + 39   28492701
FAX: + 39   39384 91973

Croda Chemicals Ltd.
Cowick Hall
Snaith
Goole
North Humberside DN14 9AA
United Kingdom
TEL: + 44   1405 860551
FAX: + 44   1405 860205

Croda Universal Ltd.
Cowick Hall
Snaith
Goole
North Humberside DN14 9AA
United Kingdom
TEL: + 44   1405 860551
FAX: + 44   1405 860205

CYTEC

Cytec Industries Inc.
Bowling Park Drive
Bradford
West Yorkshire BD4 7TT
United Kingdom
TEL: + 44   1274 733891
FAX: + 44   1274 734770

DAC INTERNATIONAL SURFACTANTS

Dac International Surfactants S.p.A
V.le Forlani 65
20134 Milano
Italy

TEL: + 39   2 70101711
FAX: + 39   2 719673

DANIEL PRODUCTS COMPANY

Firma Klaus Kail
A-1030 Wien
Neulinggasse 12
Austria
TEL: + 43   1 7132392
FAX: + 43   1 71323929

Socomer SA
Nieuwbrugstraat 73
B-1830 Machelen
Brussels
Belgium
TEL: + 32   2 2544789
FAX: + 32   2 2544619

Scancolor A/S
17, Tjoernebakken
DK-2800 Lyngby
Denmark
TEL: + 45   45873717
FAX: + 45   45874737

Stenberg Marketing Oy
PO Box 38
SF-07901 Loviisa
Finland
TEL: + 358   15 514408
FAX: + 358   15 514428

Ets. B. Rossow & Cie
6/8 Place Jean Zay
92300 Levallois-Perret
France
TEL: + 33   1 41066300
FAX: + 33   1 41066306

Krahn Chemie GmbH
Grimm 10
20457 Hamburg
Germany
TEL: + 49   40 32920
FAX: + 49   40 3292322

Ety - Keleris S.A.
119 Tritis Septemvriou Str.
GR-11251 Athens
Greece
TEL: + 30   1 883 0536/7/8
FAX: + 30   1 883 0019

DSM Italia s.r.l.
Via Rodi 5
24040 Filago (Bergamo)
Italy
TEL: + 39   35 997212
FAX: + 39   35 4942546

Profiltra BV
PO Box 1072
1300 BB Almere-Buiten
The Netherlands
TEL: + 31   36 5324228
FAX: + 31   36 5324268

Arcon A/S
PO Box 4296 Torshov
N-0401 Oslo 4
Norway

TEL: + 47   22 373130
FAX: + 47   22 375431

Improquime
Rua Dr. Afonso Cordeiro, 683-3°Tras.
4450 Matosinhos
Portugal
TEL: + 351   2 9372292
        + 351   2 9370548
FAX: + 351   2 9373813

Meplas HBC d.o.o.
Smartinska 152
61103 Ljubljana
Slovenia
TEL: + 38   61 440423
FAX: + 38   61 440423

DSM Resins Espana SA
Edificio Euro-3
Poligono Sant Joan Despi
c/Frederic Mompou 5
Planta 7, Puerta 3° y 4°
08960 Sant Just Desvern
Barcelona
Spain
TEL: + 34   3 4731100
FAX: + 34   3 4736350

A B Trebec
PO Box 2009
S-183 02 Taby
Sweden
TEL: + 46   8 7680090
FAX: + 46   8 7925020

Mineral-Chemie AG
Hardturmstrasse 175
PO Box 8037
Zurich
Switzerland
TEL: + 41   1 2727746
FAX: + 41   1 2731844

Tracomme AG
Webereistrasse 68
Postfach
8134 Adliswil
Switzerland
TEL: + 41   1 7090707
FAX: + 41   1 7090770

Cornelius Chemical Co. Ltd.
Cornelius House
Woodside
Dunmow Road
Bishops Stortford
Hertfordshire CM23 5RG
United Kingdom
TEL: + 44   127 9714300
FAX: + 44   127 9714320

DANISCO INGREDIENTS

Danisco Ingredients Austria Ges.m.b.H.
(Grindsted Division)
Hungerbergstrasse 1A
A-1190 Vienna
Austria
TEL: + 43   1 327799
FAX: + 43   1 327773

Danisco Ingredients Belgium B.V.B.A.
(Grindsted Division)
Keizerstraat 7
B-2000 Antwerp
Belgium
TEL: + 32   3 2310010
FAX: + 32   3 2341816

World Headquarters Danisco Ingredients
(Grindsted Division)
Edwin Rahrs Vej 38
DK-8220 Brabrand
Denmark
TEL: + 45   89 50 00
FAX: + 45   86 25 10 77

Danisco Ingredients France S.A.R.L.
(Grindsted Division)
Parc d'Activites de Pissaloup
Avenue Jean d'Alembert
F-78190 Trappes
France
TEL: + 33   1 30 66 08 08
FAX: + 33   1 30 66 75 08

Danisco Ingredients Deutschland G.m.b.H.
(Grindsted Division)
Robert Bosch Strasse 20-24
D-25451 Quickborn
Germany
TEL: + 49   4106 70960
FAX: + 49   4106 709666

Danisco Ingredients Deutschland G.m.b.H.
(Grindsted Division)
Buro Eberswalde
Carl-von-Ossietzky-Strass 14
D-16225 Eberswalde-Finow
Germany
TEL: + 49   33 34212235
FAX: + 49   33 34212733

Danisco Ingredients Holland B.V.
(Grindsted Division)
Zuideinde 80
NL-2991 LK Barendrecht
Holland
TEL: + 31   1806   20311
FAX: + 31   1806   12018

Danisco Ingredients Italia S.R.L.
(Grindsted Division)
Centro Direzionale Colleoni
Palazzo Andromeda, 3
Via Paracelso, 20
I-20041 Aerate Brianza (MI)
Italy
TEL: + 39   6899 647
FAX: + 39   6899 663

Danisco Ingredients Espana S.A.
(Grindsted Division)
Calle Balmes, 440, Entlo. G.
E-08022 Barcelona
Spain
TEL: + 34   3 211 97 00
FAX: + 34   3 418 53 33

Danisco Ingredients (Switzerland) AG
(Grindsted Division)
Loretohohe 16
CH-6301 Zug
Switzerland

TEL: + 41   222 658
FAX: + 41   210 922

Danisco Ingredients (UK) Ltd.
(Grindsted Division)
Northern Way
Bury St Edmunds
Suffolk IP32 6NP
United Kingdom
TEL: + 44   1284 769631
FAX: + 44   1284 760839

DOW CHEMICAL

Dow Deutschland Inc.
Am Kronberger Hang 4
65824 Schwalbach
Germany
TEL: + 49   6196 567 0
FAX: + 49   6196 567 406

Dow Europe S.A.
Bachtobelstrasse 4
8810 Horgen
Switzerland
TEL: + 41   17283181
FAX: + 41   17283060

Dow Chemical Company Ltd.
Regional Headquarters
Lakeside House
Stockley Park
Uxbridge
Middlesex UB11 1BE
United Kingdom
TEL: + 44   181 8488688
FAX: + 44   181 8485400

DUPONT

DuPont (UK) Ltd.
Head Office
Wedgwood Way
Stevenage
Hertfordshire SG14QN
United Kingdom
TEL: + 44   1438 734000
FAX: + 44   1438 734550

DuPont (UK) Ltd.
Hemel Hempstead & IDAC
Maylands Avenue
Hemel Hempstead
Hertforshire HP2 7DP
United Kingdom
TEL: + 44   1442 218500
FAX: + 44   1442 249463

Alfa Chemicals Ltd.
Arc House
Terrace Road South
Binfield
Berkshire RG12 5PZ
United Kingdom
TEL: + 44   1344 861800
FAX: + 44   1344 862010

Alfa Chemicals Ltd.
Alfa House
15 Moor Park Avenue
Preston
Lancashire PR1 6AS
United Kingdom
TEL: + 44   1772 258969
FAX: + 44   1772 253207

EASTMAN

Eastman Chemical (UK) Ltd.
Brindley House
Comer Hall
Lawn Lane
Hemel Hempstead
Hertfordshire HP3 9YT
United Kingdom
TEL: + 44   1442 241171
FAX: + 44   1442 241177

Eastman Chemical AG
PO Box 3263
CH-6300 Zug 3
Switzerland
TEL: + 41   42 232525
FAX: + 41   42 211252

ELF ATOCHEM

Elf Atochem
Centre d'Application de Levallois
95 Rue Danton
92300 Levallois-Perret
France
TEL: + 33   1 475912 34
FAX: + 33   1 475914 41

Elf Atochem UK Ltd.
Colthrop Way
Thatcham
Newbury
Berkshire RG13 4LW
United Kingdom
TEL: + 44   1635 870057
FAX: + 44   1635 870050

ELLIS & EVERARD

Ellis & Everard (UK) Ltd.
Anstead International
Radford House
Radford Way
Billericay
Essex CM12 0DE
United Kingdom
TEL: + 44   1277 630063
FAX: + 44   1277 631356

ENICHEM AUGUSTA

Enichem Austria GmbH
Diefenbachgasse 35
A-1150 Wien
Austria
TEL: + 43   1 891880
FAX: + 43   1 89188240

Enichem Benelux SA
8, Rue de L'Industrie
B-1400 Nivelles
Belgium
TEL: + 32   67 880611
FAX: + 32   67 210587

Enichem SpA
Bulgaria Liaison Office
TD Interpred
Bld. Balgarosavetska
Drujba 16
B-1057 Sofia
Bulgaria
TEL: + 359   2 703473
         + 359   2 704393
FAX: + 359   2 709207

EniChem SpA
Praha Liaison Office
Via Vinohradska 10
C-12147 Praha 2
Czech Republic
TEL: + 42   2 2354075/6/7/8/9
FAX: + 42   2 2354402

EniChem Norden A/S
Ny Ostergade 20, 2 fl.
DK-1101 Copenhagen K
Denmark
TEL: + 45   33 150 888
FAX: + 45   33 156 471

EniChem France SA
11, Rue de L'Abreuvoir
F-92411 Courbevoie Cedex
France
TEL: + 33   1 469 132 00
         + 33   1 469 130 50
FAX: + 33   1 478 803 20

EniChem Deutschland AG
Kolner Strasse 3ª
Postfach 5626
D-6236 Eschborn bei Frankfurt
Germany
TEL: + 49   6196 492 0
FAX: + 49   6196 492 218

EniChem Hellas SA
46 Thisseos Avenue & Frinis
PO Box 75123
G-17676 Kallithea
Athens
Greece
TEL: + 30   1 952 20 91/2/3/4/5/6
FAX: + 30   1 952 20 97

EniChem SpA
Hungary Liaison Office
Bajcsy Zsilinszky Ut. n. 12
VL fl.-601 - 602 - 603
H-1051 Budapest
Hungary
TEL: + 36   1 138 41 06
FAX: + 36   1 138 20 49

EniChem Augusta Industriale S.r.L.
Via Medici del Vascello, 26
20138 Milan
Italy
TEL: + 39   2 5201
FAX: + 39   2 52029558
         2 52029903

EniChem SpA
Poland Liaison Office
Ul. Stawki 2-24° p
Warszawa 00193
Poland
TEL: + 48   2 65564600
         + 48   2 65565754
FAX: + 48   2 6357602

EniChem Portugal
Industria Quimica SA
Avenida da Boavista 1681
2° Floor-Sala 3
P-4100 Porto
Portugal

TEL: + 351   2 606 41 47
       + 351   2 696 461
FAX: + 351   2 695 511
       + 351   2 691 772

EniChem Portugal
Industria Quimica SA
Rua Duque de Palmela, 30
4° C
P-1200 Lisboa
Portugal
TEL: + 351   1 352 84 30/39/49
FAX: + 351   521 030

EniChem Iberica SA
Avenida Diagonal 652 - 656
Edificio B 3°
E-08034 Barcelona
Spain
TEL: + 34   3401 23 00
FAX: + 34   3280 30 50

EniChem Iberica SA
Pl. de Espana 18
Piso 19 - Puerta 3
Torre de Madrid
E-28013 Madrid
Spain
TEL: + 34   1247 64 06
FAX: + 34   1247 64 86

EniChem Norden A/S
Svensk Filial
Box 11115
S 404-23 Goteborg
Sweden
TEL: + 46   31 806 490
FAX: + 46   31 150 260

EniChem Suisse SA
Seestrasse 42
CH-8802 Kilchberg
Switzerland
TEL: + 41   1 7161111
FAX: + 41   1 7150335

EniChem UK Ltd.
EniChem House
111 Upper Richmond Road
Putney
London SW15 2TJ
United Kingdom
TEL: + 44   181 7802000
FAX: + 44   181 7802852

FINA CHEMICALS

Fina Chemicals
Seidlgasse 22/5
A-1030 Wien
Austria
TEL: + 43   1 712 21 38
FAX: + 43   1 713 41 38

Fina Chemicals
Rue de l'Industrie, 52
Nijverheldsstraat 52
B-1040 Brussels
Belgium
TEL: + 32   2 2889970
FAX: + 32   2 2889550

Andreas Jennow A/S
Kongevejen 270
PO Box 330
DK-2830 Virum
Denmark
TEL: + 45   4583 1200
FAX: + 45   4583 1230

Berner OY
Etelaranta 4B
P.O. Box 15
SF-00131 Helsinki
Finland
TEL: + 358   0 134511
FAX: + 358   0 611684

Fina Chemicals
Rue Henri Sainte-Claire Deville, 8
F-92563 Rueil-Malmaison Cedex
France
TEL: + 33   1 47103157
FAX: + 33   1 47103139

Fina Chemicals
Postfach 10 08 62
D-60008 Frankfurt
Germany
TEL: + 49   69 2198440/441
FAX: + 49   69 2198443

M. Axiotis & Co, s.a.
PO Box 52880
GR-14601 N. Erythrea
Greece
TEL: + 30 1 807 8546/8712/8939
FAX: + 30 1 807 79 11

Corcoran Chemicals Ltd.
Kingsbridge House
17-22 Parkgate Street
IE-Dublin, 8
Ireland
TEL: + 353   1 778 163
FAX: + 353   1 679 3521

Fina Chemicals
Viale Premuda, 27
I-20129 Milano
Italy
TEL: + 39   2 7759489
FAX: + 39   2 76015030

Fina Chemicals
c/o Necarbo B.V.
PO Box 621
NL-1940 AP Beverwijk
Netherlands
TEL: + 31   0 2510 78300
FAX: + 31   0 2510 14611

Hydro Chemicals Norge a.s.
Postboks 6405, Etterstad
N-0604 Oslo
Norway
TEL: + 47   22432400
FAX: + 47   22432402
       + 47   22432403

Sepulchre Poland Ltd.
Ul. Gtowackiego 1/4
PL-47-220 Kedzierzyn-Kozle
Poland

TEL: + 48   794 382 87
       + 48   794 382 88
FAX: + 48   794 382 89

Iersa Lda
Av. Columbano Bordalo Pinheiro 94-2°-F.D.
P-1000 Lisboa
Portugal
TEL: + 351   1 72 67 676
FAX: + 351   1 72 67 905

PQI s.a.
Tuset 20-24,7°
E-08006 Barcelona
Spain
TEL: + 34   3 2181758
FAX: + 34   3 4154848

H. Duner Handels AB
Karleksgatan 2A
S-211 45 Malmo
Sweden
TEL: + 46   40 354850
FAX: + 46   40 118700

Tensochema A.G.
Rotelstrasse 28, CH-8006 Zurich
PO Box 429, CH-8042 Zurich
Switzerland
TEL: + 41   1 3626161
FAX: + 41   1 3626165

Fina Chemicals
Fina House
1 Ashley Avenue
Epsom
Surrey KT18 5AD
United Kingdom
TEL: + 44   1372 726 226
FAX: + 44   1372 744 737

GATTEFOSSE

Gattefosse France
58 rue du General de Gaulle
95880 Enghien les Bains
France
TEL: + 33   134 127050
       + 33   134 129900
FAX: + 33   139 896014

Gattesfosse (Deutschland) GmbH
Rheincenter
Haupstrasse 435
79576 Weil-Am-Rhein
Germany
TEL: + 49   76 21 720 07
FAX: + 49   76 21 792 293

Gattefosse Italia S.R.L.
Via Derganino, 20
20158 Milano
Italy
TEL: + 39   2 39 322 425/736/802
FAX: + 39   2 66 200 440

Gattefosse Espana S.A.
Diagonal 460
08006 Barcelona
Spain
TEL: + 34   3416 0520/1295
FAX: + 34   3415 3546

Gattefosse Espana S.A.
C/Padilla, 73, 1°B
28006 Madrid
Spain
TEL: + 34   1402 1311/1612
FAX: + 34   1402 00 82

Gattefosse AG
Haldenstrasse 11
CH-6006 Luzern
Switzerland
TEL: + 41   4151 49 66
FAX: + 41   4151 35 53   ·

Alfa Chemicals Ltd.
Alfa House
15 Moor Park Avenue
Preston
Lancashire PR1 6AS
United Kingdom
TEL: + 44   1772 258969
FAX: + 44   1772 253207

Alfa Chemicals Ltd.
Arc House
Terrace Road South
Binfield
Bracknell
Berkshire RG12 SPZ
United Kingdom
TEL: + 44   1344 861800
FAX: + 44   1344 862010

TH. GOLDSCHMIDT

Th. Goldschmidt Ges.m.b.H.
Erlaaer Strasse 61
A-1232 Vienna
Austria
TEL: + 43   2 22670531/0
FAX: + 43   2 2267053126

N.V. Th. Goldschmidt S.A.
Kapucijnenlaan 1
B-1030 Brussels
Belgium
TEL: + 32   2 2418750
FAX: + 32   2 2415452

Goldschmidt Scandinavia ApS
Tjaerepleten60, Strip
DK-5500 Middelfart
Denmark
TEL: + 45   64 401818
FAX: + 45   64 401018

Goldschmidt France S.A.
3 Avenue des Chaumes - Z.A.O.
F-78180 Montigny le Bretonneux
France
TEL: + 33   1 30 12 01 50
FAX: + 33   1 30 43 26 63

Th. Goldschmidt AG
Head Office
Postfach 101461
Goldschmidtstrasse 100
D-45127 Essen
Germany
TEL: + 49   2 01173 1
FAX: + 49   2 01224916

Goldchem E.P.E.
Leoforos Kifisias 7
GR-15123 Marousi
Greece
TEL: + 30   1 6825802
FAX: + 30   1 6834537

Th. Goldschmidt Ireland
Unit 81
Baldoyle Industrial Estate
Baldoyle
Dublin 13
Ireland
TEL: + 353   1 8320154
FAX: + 353   1 8320157

TEGO Italiana S.r.I.
Viale Campania, 33
I-20133 Milano
Italy
TEL: + 39   2 76111067
FAX: + 39   2 71564302

Th. Goldschmidt S.A.
Comte d'Urgell, 240, 3.°-C
08036 Barcelona
Spain
TEL: + 34   3 4190044
FAX: + 34   3 4107516

Th. Goldschmidt Ltd.
Tego House
Chippenham Drive
Kingston
Milton Keynes
Buckinghamshire MK10 0AE
United Kingdom
TEL: + 44   1908 582250
FAX: + 44   1908 582254

GRILLO-WERKE

See Th. Goldschmidt

HAMPSHIRE CHEMICAL

Hampshire Chemical GmbH
Waldhofer Strasse 17
D-6900 Heidelberg
Germany
TEL: + 49   6221 700111
FAX: + 49   6221 700115

Hampshire Chemical AB
Box 2066
S-25002 Helsingborg
Sweden
TEL: + 46   42 1473000
FAX: + 46   42 148385

Hampshire Chemical Ltd.
Northdale House
North Circular Road
London NW10 7UH
United Kingdom
TEL: + 44   181 961 9366
FAX: + 44   181 963 0928

Surfachem Ltd.
Wellington Park House
Thirsk Row
Leeds LS1 4DP
United Kingdom

TEL: + 44   113 2342636
FAX: + 44   113 2445910

HAYS COLOURS

Hays Colours
High Level Way
Queens Road
Halifax
West Yorkshire HX1 4PS
United Kingdom
TEL: + 44   1422 358431
FAX: + 44   1422 330867
         + 44   1422 381422

HENKEL

Henkel Austria GmbH
Erdbergstrasse 29
Postfach 180/181
1030 Wien
Austria
TEL: + 43   1 71104 0
FAX: + 43   1 71104620

Henkel Belgium S.A.
66, Avenue du Port
Havenlaan 65
1210 Bruxelles
Belgium
TEL: + 32   2 423 1711
FAX: + 32   2 424 3025

P. Broste A/S
Lundtofte Gaardsvej 95
DK-2800 Lyngby
Denmark
TEL: + 45   45 933333
FAX: + 45   45 931388

Henkel Nopco Oy
Lyhtytie 3
P.O. Box 20
00750 Helsinki 75
Finland
TEL: + 358   34 680
FAX: + 358   34 61 916

Sidobre Sinnova
185 Avenue de Fontainebleau
BP 4
77981 St-Fargeau-Ponthierry
France
TEL: + 33   1 60652100
FAX: + 33   1 60652101

Henkel KGaA
P.O. Box 10 11 00
D-40191 Dusseldorf
Germany
TEL: + 49   2117972289
FAX: + 49   2117987696

Dr. Demetrios A.Delis A.G.
Paleologou-Benizelou- 5
10556 Athens
Greece
TEL: + 30   1 3 29 70
FAX: + 30   1 32 30 550

Henkel (Ireland) Ltd.
Western Industrial Estate
Naas Road
Dublin 12
Ireland
TEL: + 353   1 4505622
FAX: + 353   1 4503649

Henkel Chimica S.p.A.
Via Scalabrini 24
22073 Fino Moranasco (Co)
Italy
TEL: + 39   31 884201
FAX: + 39   31 884360

Henkel Oleochemicals B.V.
Postbus 247
1170 AE Badhoevedorp
Niederlande
TEL: + 31   2968 97506
FAX: + 31   2065 98711

Henkel Nopco A/S
Postboks 2040
Stromse
3003 Drammen
Norway
TEL: + 47   3220 2200
FAX: + 47   3288 0701

Pulcra S.A.
Pasaje Mariner, no. 9
08025 Barcelona
Spain
TEL: + 34   3 2904760
FAX: + 34   3 2904879

Impag AG
Feldeggstrasse 26
8034 Zurich
Switzerland
TEL: + 41   1 38 36 843
FAX: + 41   1 38 36 289

Turk Henkel A.S.
Koresehitleri Cad. No. 39
80300 Zincirlikuyu-Instanbul
Turkey
TEL: + 90   212 2113480
FAX: + 90   212 2113467

Henkel Ltd.
Henkel House
292-308 Southbury Road
Enfield
Middlesex EN1 1TS
United Kingdom
TEL: + 44   181 8043343
FAX: + 44   181 4432777

HICKSON MANRO

Hickson Manro S.A.
Rue Renory 28
B4102 Ougree
Belgium
TEL: + 32   41301600
FAX: + 32   41301503

Hickson Manro Ltd.
Bridge House
Bridge Street
Stalybridge
Cheshire SK15 1PH
United Kingdom

TEL: + 44   161 3385511
FAX: + 44   161 3032991

HOECHST

Hoechst Austria AG
Postfach 14
A-1121 Wien
Austria

Hoechst Danmark A/S
Islevdalvej 110
DK-2610 Rodovre
Denmark

Oy Hoechst Fennica AB
Postboks 237
SF 00101 Helsinki
Finland

Hoechst AG
Postfach 80 03 20
D-65926 Frankfurt-am-Main
Germany

Svenska Hoechst AB
Box 5415
S-402 29 Goteborg
Sweden

Pluess-Staufer AG
CH-4665 Oftringen
Switzerland

Hoechst UK Ltd.
Stainland Works
Holywell Green
Halifax HX4 9DL
United Kingdom
TEL: + 44   1422 375522
FAX: + 44   1422 371689

HÜLS

Hüls-Austria Gesellschaft m.b.H.
Postfach 6
1011 Wien
Austria
TEL: + 43   1 5353561
FAX: + 43   1 53535616

N.V. Hüls Belgien S.A.
Meiboomstraat 26
1500 Halle
Belgium
TEL: + 32   2 3630811
FAX: + 32   2 3630849

Hüls Danmark A/S
Postbox 26 05
2100 Kopenhagen
Denmark
TEL: + 45   35 269111
FAX: + 45   35 269011

Oy Suomen Hüls Ab
Sinimaentie 10 A. PL 92
02631 Espoo
Finland
TEL: + 358   0 5022600
FAX: + 358   0 5022805

Hüls France S.A.
49-51, Quai de Dion Bouton
92815 Puteaux Cedex
France
TEL: + 33   1 49065500
FAX: + 33   1 47739765

Hüls Aktiengesellschaft
D-45764 Marl
Germany
TEL: + 49   2365 491
FAX: + 49   2365 492000

Hüls Ireland Ltd.
Stradbrook House
Stradbrook Road
Blackrock
Co. Dublin
Ireland
TEL: + 353   1 2800666
FAX: + 353   1 2800660

Hüls Italia S.p.A.
Viale Marelli 155
20099 Sesto San Giovanni (MI)
Italy
TEL: + 39   2 262361
FAX: + 39   2 26222653
        + 39   2 26222654

Hüls-Nederland B.V.
Postbus 40
3620 AA Breukelen
The Netherlands
TEL: + 31   3462 65811
FAX: + 31   3462 64854

Hüls Norge A/S
P.B. 396
1473 Skaarer
Norway
TEL: + 47   67 906210
FAX: + 47   67 970015

Hüls Espanola S.A.
Apartado de Correos 5375
08080 Barcelona
Spain
TEL: + 34   3 4871330
FAX: + 34   3 2157715

Hüls Sverige AB
Box 132
431 22 Molndal
Sweden
TEL: + 46   31 7765200
FAX: + 46   31 270809

Hüls Sverige AB - Abt. Polymere
Box 357
20 123 Malmo
Sweden
TEL: + 46   4075520
FAX: + 46   40302202

Hüls Sverige AB - Abt. Svenska Polystyren
Box 2022
231 02 Trelleborg
Sweden
TEL: + 46   410 43150
FAX: + 46   410 12650

Hüls (Schweiz) AG
Kohirainstrasse 1
8700 Küsnacht (ZH)
Switzerland
TEL: + 41   1 9104411
FAX: + 41   1 9104478

Hüls-Stockhausen Kimya Ticaret A.S.
Ankara Asfaiti Uzeri
Kanli Mandira Mevkii
Tuzla Kavsagi
81700 Tuzla-Istanbul
Turkey
TEL: + 90   216 3957207
FAX: + 90   216 3959965

Hüls (UK) Ltd.
Featherstone Road
Wolverton Mill South
Milton Keynes
Buckinghamshire MK12 5TB
United Kingdom
TEL: + 44   1908 226444
FAX: + 44   1908 224950

ICI SURFACTANTS

ICI Surfactants
Everslaan 45
B-3078 Everberg
Belgium
TEL: + 32   2 7589361
FAX: + 32   2 7589686

ICI Surfactants
196 Rue Houdan
F-92330 Sceaux
France
TEL: + 33   1 41133250
FAX: + 33   1 41133252

ICI Surfactants
Goldschmidtsrasse 100
D-4300 Essen 1
Germany
TEL: + 49   201 17304
FAX: + 49   201 227350

ICI Surfactants
Via Mazzini 58
I-21020 Ternate (VA)
Italy
TEL: + 39   332 941255
FAX: + 39   332 941264

ICI Surfactants
Ctra de Hostalric a Tossa
Km 1.8
08490 Fogars de Todera
Barcelona
Spain
TEL: + 34   3 7669858
FAX: + 34   3 7669806

ICI Surfactants
P.O. Box 90, Wilton Centre
Middlesbrough
Cleveland TS6 8JE
United Kingdom
TEL: + 44   1642 437476
FAX: + 44   1642 437374

Surfachem Ltd.
Wellington Park House
Thirsk Row
Leeds LS1 4DP
United Kingdom
TEL: + 44   113 2342636
FAX: + 44   113 2445910

INOLEX

S. Black (Import and Export) Ltd.
The Colonnade
High Street
Cheshunt
Hertfordshire EN8 0DJ
United Kingdom
TEL: + 44   1992 630751
FAX: + 44   1992 622838

ISP

ISP (Osterreich) GmbH
Belvederegasse 18/1
A-1040 Wien
Austria
TEL: + 43   222 50476210
FAX: + 43   222 5058944

ISP (Belgium) NV
Hoogkamerstraat 42
B-9100, Sint-Niklaas
Belgium
TEL: + 32   3 7781210
FAX: + 32   3 7780776

ISP (France) SA
ZAC Paris Nord IL
BP 50007, 95945 Roissy
CDG, Cedex
France
TEL: + 33   1 49905800
FAX: + 33   1 49905805

ISP Global Technologies Deutschland GmbH
Rudolf-Diesel-Strasse 25
Postfach 1380
50203 Frechen
Germany
TEL: + 49   2234 1050
FAX: + 49   2234 105211

ISP Hungaria
Kenese u.8
H-1113 Budapest, XI
Hungary
TEL: + 36   1 1662550
FAX: + 36   1 1662550

ISP (Italia) Srl
Via Ripamonti 66
20141 Milan
Italy
TEL: + 39   2 57403366
FAX: + 39   2 57403391

ISP Global Technologies (Nederland)
Olivier van Noorstraat 7
Postbus 4005
3102 GA Schiedam
The Netherlands
TEL: + 31   10 4705044
FAX: + 31   10 4712218

ISP Russia
Leninsky Prospekt
95A, Room 503
117313 Moscow
Russia
TEL: + 7   95   9362675
FAX: + 7   95   9362676

ISP International
Corp. Sucursal en Espana
C/La Selva, s/n
(Edificio Gemini) 1° A2
08820 El Prat de Llobregat
Barcelona
Spain
TEL: + 34   93 4787969
FAX: + 34   93 4785221

ISP (Norden) AB
Box 824
S-120 03 Arsta
Sweden
TEL: + 46   8 188830
FAX: + 46   8 189645

ISP (Switzerland) AG
Zugerstrasse 77
CH-6340 Baar Postfach 4354
CH-6304 Zug
Switzerland
TEL: + 41   42 301060
FAX: + 41   42 331586

ISP Europe
40 Alan Turing Road
Surrey Research Park
Guildford
Surrey GU2 5YF
United Kingdom
TEL: + 44   1483 301757
FAX: + 44   1483 302175

ISP (Great Britain) Company Ltd.
Tilson Road
Wythenshawe
Manchester M23 9PH
United Kingdom
TEL: + 44   161 9981122
FAX: + 44   161 9986218

KAO CORPORATION

Kao Corporation SA
C./Puig dels Tudons 10
08210 Barbera Del Valles
Barcelona
Spain
TEL: + 34   3 7399300
FAX: + 34   3 7399333

KARLSHAMNS

Karlshamns
PO Box 17
NL-1540 AA Koog aan de Zaan
The Netherlands
TEL: + 31   75 278400
FAX: + 31   75 285050

Karlshamns
S-374 82 Karlshamn
Sweden
TEL: + 46   454 82000
FAX: + 46   454 19692

Karlshamns
189-197 Wincolmlee
Hull
North Humberside HU10QA
United Kingdom
TEL: + 44   1482 586747
FAX: + 44   1482 587004

DR W KOLB AG

Dr W Kolb AG
Chemische Fabrik
CH-8908 Hedingen
Switzerland
TEL: + 41   1 7611646   762 4646
FAX: + 41   1 7611757

Libra Chemicals Ltd.
Northbank Industrial Park
Irlam
Manchester M44 5AX
United Kingdom
TEL: + 44   161 7551888
FAX: + 44   161 7779109

LAKELAND LABORATORIES

Krogh AS
17A Hellerup Vej
Copenhagen
DK 2900 Hellerup
Denmark
TEL: + 45   31 629808
FAX: + 45   31 625088

Unipex SA
30 Rue de Fort BP 150
92504 Ruel Malamaison
France
TEL: + 33   147 329293
FAX: + 33   147 490235

Espachem
Vivaldiveg 11
3752 HA Bunschoten
Holland
TEL: + 31   3499   81983
FAX: + 31   3499   88469

Paz Oil Co Ltd.
4 Hagefen Str
PO Box 434
Israel
TEL: + 972   4 352095
FAX: + 972   4 670006

Benfer-Scheller SPA
Viale Monza 265
20126 Milano
Italy
TEL: + 39   2 257701
FAX: + 39   2 2553273

Swedakjemi AS
Postboks 227
1601 Frederikstadt
Norway
TEL: + 47   9 323010
FAX: + 47   9 321070

Swedakjemie AS
Postboks 49
Bryn
N-0611 Oslo
Norway
TEL: + 47   881600
FAX: + 47   720052

Eurolabo SA
C/Industria
169-089 Badalona
Barcelona
Spain
TEL: + 34   93 600882
FAX: + 34   93 3830058

M B Sveda
Bruksgatan 3
S-21122 Malmo
Sweden
TEL: + 46   40 352800
FAX: + 46   40 520091

M B Sveda
Flemminggatan 30
Box 16
60102 Norrkoping
Sweden
TEL: + 46   11 106420
FAX: + 46   11 102621

M B Sveda
P O Box 1050
Kistagangen 19
S16421 Kista
Sweden
TEL: + 46   8 920410
FAX: + 46   8 7520091

M B Sveda
Marieholmsgatan 56
Box 48
40120 Goteborg
Sweden
TEL: + 46   31 838000
FAX: + 46   31 843980

Polygon Chemie AG
CH 4601 Olten
Switzerland
TEL: + 41   62 327989
FAX: + 41   62 327482

Lakeland Laboratories Ltd.
Peel Lane
Astley Green
Tyldesley
Manchester M29 7FE
United Kingdom
TEL: + 44   1942 873555
FAX: + 44   1942 884409

LIBRA CHEMICALS

Libra Chemicals Ltd.
Martens Road
North Bank Industrial Park
Irlam
Manchester M30 5AX
United Kingdom
TEL: + 44   161 7751888
FAX: + 44   161 7779109

LONZA

Lonza Biotec sro
Okruzni 134
CZ 281 61 Kourim
Czech Republic
TEL: + 42   321 83601
FAX: + 42   321 83610

Hoechst Danmark A/S
Islevdalvej 110
DK-2610 Rodovre
Denmark
TEL: + 45   44 888200
FAX: + 45   42 915052

Oy Hoechst Fennica
PL 237
SF 00101 Helsinki
Finland
TEL: + 35   80 870991
FAX: + 35   80 8709755

Lonza France Sarl
55 rue Aristide Briand
F-92300 Levallois Perret
France
TEL: + 331   40 899904
FAX: + 331   40 899911

Lonza GmbH
Morianstrasse 32
D-42103 Wuppertal
Germany
TEL: + 49   202 2453833
FAX: + 49   202 2453830

Ilforma Hellas
25, Antipatrou & Pyladou Street
GR-11853 Athens
Greece
TEL: + 30   1 346 0298
FAX: + 30   1 347 9764

Lonza Ltd. Representation Office
Labdarugo u.19
H-1047 Budapest
Hungary
TEL: + 36   1 266 46 88
FAX: + 36   1 266 45 62

John Aikman
Borgartun 23
IS-121 Reykjavik
Iceland
TEL: + 35   41 683555
FAX: + 35   41 683530

AAKON S.r.l.
Cso Magenta 43
I-20123 Milan
Italy
TEL: + 39   2 4695800
FAX: + 39   2 4986096

Lonza Benelux BV
Aluminiumstraat 1
NL-4823 Al Breda
The Netherlands
TEL: + 31   76 425100
FAX: + 31   76 424070

Norske Hoechst A/S
Okernveien 145
N-0509 Oslo
Norway
TEL: + 47   22 727842
FAX: + 47   22 631826

Alusuisse-Lonza Warszawa
ul. Smolenskiego 4 m 15
PL-01-698 Warsaw
Poland
TEL: + 48   22 338745
FAX: + 48   22 331239

Hoechst Portugesa S.A.
Av. Sidonio Pais, 379, Apartado 1041
P-4101 Porto
Portugal
TEL: + 351   12 6067051
FAX: + 351   12 6004881

Alusuisse-Lonza Products Ltd.
Ulitsa Usacheva
Rus-Moscow 119048
Russia
TEL: + 7   095 2455042
FAX: + 7   095 2442966

Commerce PO
Einspielerjeva 6
Sl-61109 Ljubljana
Slovenia
TEL: + 386   61 1322241
FAX: + 386   61 329378

Lonza Iberica SA
Milanesado, 25-27 4°1ª
E-08017 Barcelona
Spain
TEL: + 34   3 2805022
FAX: + 34   3 2056755

Svenska Hoechst AB
Box 414
S-402 29 Goteborg
Sweden
TEL: + 46   31 678500
FAX: + 46   31 877917

Lonza Ltd.
Münchensteinerstrasse 38
CH-4002 Basel
Switzerland
TEL: + 41   61 3168111
FAX: + 41   61 3168318

Lonza (UK) Ltd.
Imperial House
Lypiatt Road
Cheltenham
Gloucestershire GL502QL
United Kingdom
TEL: + 44   1242 513211
FAX: + 44   1242 222294

MCINTYRE GROUP

Jan Dekker France
B.P. 205
78104 Saint-Germain
En Laye Cedex
France
TEL: + 33   34 518511
FAX: + 33   39 735311

Jan Dekker
P.O. Box 10
1520 AA Wormerveer
The Netherlands
TEL: + 31 75 278278
FAX: + 31 75 213883

TensoChema
Rotelstrasse 28
CH-8006 Zurich
Switzerland
TEL: + 41   1 3626161
FAX: + 41   1 3626165

Jan Dekker U.K. Ltd.
Suite 3
Hillbrow House
Linden Drive
Liss
Hampshire GU33 7RJ
United Kingdom
TEL: + 44   1730 895511
FAX: + 44   1730 895111

MILLCHEM

Devan Chemicals SA
539 Chaussee de Ninovesteenweg
9600 Ronse-Renaix
Belgium
TEL: + 32   55 215165
FAX: + 32   55 219879

Chr Krogh A/S
Hellerupvaj 17A
DK-2900 Hellerup
Copenhagen
Denmark
TEL: + 45   31 629808
FAX: + 45   31 625088

J Haarla Trading & Co.
Pyhajarvenkatu 10A
33200 Tampere 20
Finland
TEL: + 358   31 2331822
FAX: + 358   31 2133870

A & L Rondot
9 Rue Jean Elysee Dupuy
B.P. 55
69410 Champagne Au Mont D'or
France
TEL: + 33   78 438786
FAX: + 33   78 352923

Intertex OE
22 Chalkidos Str.
Ano Patissia
111 42 Athens
Greece
TEL: + 30   1 2180600
FAX: + 30   1 2523633

Keyser & Mackay
PO Box 3899
1001 AR
Amsterdam
Netherlands
TEL: + 31   20 6263323
FAX: + 31   20 6254559

Santos Costa & Irmao LDA
Apartado 2123
Travessa Do Monte Louro 50-1
4202 Porto Codex
Portugal
TEL: + 351   2 590 272
FAX: + 351   2 550 4073

Cades S.A.
Avenida Bon Pastor 1-17
08930 Sant Adria De Besos
Barcelona
Spain
TEL: + 34   3 3812112
FAX: + 34   3 3819074

Vendico Chemical AB
PO Box 16039
20025 Malmo
Sweden
TEL: + 46   40 290070
FAX: + 46   40 290075

Sugro Ag
Sevogelstrasse 21
Postfach
CH 4002 Basle
Switzerland
TEL: + 41   61 3173214
FAX: + 41   61 3173200

Millchem U.K. Ltd.
Broseley House
81 Union Street
Oldham
Lancashire OL1 1PF
United Kingdom
TEL: + 44   161 6246415
FAX: + 44   161 6270329

MONA INDUSTRIES

Intercont Relations Group A.S. spol. s.r.o.
Na Cihadle 63
160 00 Praha 6
Czechoslovakia
TEL: + 42   2 3116458

Labochem Ltd.
Falirou 6
Athens 117 42
Greece
TEL: + 30   1 9231927
FAX: + 30   1 9219974

I.G.P.
J Van Stolbarglaan 114
1412 BK Naarden
Holland
TEL: + 31   215944098
FAX: + 31   215948641

Komest SRL
Via Soperga
4-20127 Milano
Italy
TEL: + 39   2 66981542
FAX: + 39   2 6690829

Ricorvi, S.A.
Calle Llussa
08028 Barcelona
Spain

TEL: + 34   3 3395300
FAX: + 34   3 3392162

Kemi-Intressen AB
Vintergatan 1
P.O. Box 2018
S-171 02 Solna
Sweden
TEL: + 46 8 289160
FAX: + 46 8 287221

Ellis & Everard (UK) Ltd.
Anstead International
Radford House
Radford Way
Billericay
Essex CM12 0DE
United Kingdom
TEL: + 44   1277 630063
FAX: + 44   1277 631356

NIACET CORPORATION

S.P.C.I.
Rue Cristino Garcia 43
93211 La Plaine St Denis
France
TEL: + 33   149333131
FAX: + 33   142438223

Brenntag Eurochem
Alfredstrasse 81
D-45130 Essen
Germany
TEL: + 49   201 87870
FAX: + 49   201 8781158

Vendico Chemical AB
PO Box 16039
200 25 Maimo
Sweden
TEL: + 46   40290070
FAX: + 46   40290075

Twinstar Chemical
Cunningham House
Westfield Lane
Kenton
Harrow HA3 9ED
United Kingdom
TEL: + 44   181 9072944
FAX: + 44   181 9091286

NICCA CHEMICAL

LJ Specialities Ltd.
Cale Lane
New Springs
Wigan WN2 1JR
United Kingdom
TEL: + 44   1942 821755
FAX: + 44   1942 821740

NIKKO CHEMICALS

Chesham Chemicals Ltd.
Cunningham House
Westfield Lane
Kenton
Harrow HA3 9ED
United Kingdom
TEL: + 44   181 9077779
FAX: + 44   181 9091053

NIPPON SHOKUBAI

Honeywill & Stein Ltd.
Times House
Throwley Way
Sutton
Surrey SM1 4AF
United Kingdom
TEL: + 44   181 7703442
FAX: + 44   181 7707295

OLIN CHEMICALS

Olin S.A.
ZAC Paris Nord II
209 Avenues des Nations
BP 60019
95970 Roissy cdg Cedex
France
TEL: + 33   1 4863 2166
FAX: + 33   1 4863 2225

Olin (UK) Ltd.
Suite 7
Kidderminster Road
Cutnall Green
Droitwich
Worchestershire WR9 0NS
United Kingdom
TEL: + 44   1299 851 561
FAX: + 44   1299 851 378

PENTAGON

Kemat
Rue de la Sablonniere, 7
B-1000 Brussels
Belgium
TEL: + 32   2219 4811
FAX: + 32   2219 4658

Antra ApS
Skelstien 2
Tune
DK-4000 Roskilde
Denmark
TEL: + 45   42 139466
FAX: + 45   42 139200

Telko Starkiohann Oy
PO Box 59
SF-00811 Helsinki
Finland
TEL: + 358 0615 500
FAX: + 358 0780 064

Unipex
30 Rue du Fort BP150
92504 Rucil-Malmaison Cedex
France
TEL: + 33   147329293
FAX: + 33   147490235

Trigon Chemie GmbH
Alte Hohenzeller Str 20
Postfach 1431
6490 Schluctern 2
Germany
TEL: + 49   66614088
FAX: + 49   66613886

I. Ignatiades Trading Corp.
7 Sofianou Str & Kallirrios 88
PO Box 2510 Makrygianni
Athens 404/1
Greece
TEL: + 30   19219370
FAX: + 30   19235042

Heterochem (Dist) Ltd.
Unit H
Baldoyle Industrial Estate
Baldoyle
Dublin 13
Ireland
TEL: + 353 18393201
FAX: + 353 18325746

Shiran Chemical Agencies Ltd.
PO Box 1420
47100 Ramat Hasharon
Israel
TEL: + 972 3540 3641
FAX: + 972 3540 3643

Eigenmann & Veronelli SpA
Via Della Mosa
6-20017 RHO (MI)
Italy
TEL: + 39   2935 39343
FAX: + 39   2935 39309

L J Volkers BV
Industrieweg 1
1231 KG Loosdrect
The Netherlands
TEL: + 31 2158 24704
FAX: + 31 2158 24064

Univete LDA
Rua Tristao VA2 59-3L
1400 Lisbon
Portugal
TEL: + 351 1301 4531
FAX: + 351 1301 4532

Julia Parrera
Tuset 8y10
08006 Barcelona
Spain
TEL: + 34 3218 4404
FAX: + 34 3218 3602

Croda Nordica AB
Tegnergatan 74
Box 30521
S-200 62 Malmo
Sweden
TEL: + 46 4016 3290
FAX: + 46 4015 0570

Kimetsan Co Ltd.
Sehit Adem Yavuz Sokak No 11/5
Kizilay
06440 Ankara
Turkey
TEL: + 90   312 4174977
FAX: + 90   312 4185617

Pentagon Chemicals Ltd.
Northside
Workington
Cumbria
CA14 1JJ

TEL: + 44   1900 604371
FAX: + 44   1900 66943

VRS Chemicals Ltd.
10 St Paul's House
Holme Valley Business Centre
Armitage Bridge
Huddersfield IID4 7NR
United Kingdom
TEL: + 44   1484 666447
FAX: + 44   1484 665253

PPG

PPG Ouvrie S.A.
64, rue Faidherbe
B.P. 127 59811
Lesquin Cedex
France
TEL: + 33   20 862080
FAX: + 33   20 875631

PPG Industries (U.K.) Ltd.
Speciality Chemicals, Chemicals Group
Carrington Business Park
Carrington
Urmston
Manchester M31 4DD
United Kingdom
TEL: + 44   161 7779203
FAX: + 44   161 7779064

PROTEX

Protex Warenhandels GmbH
Hainburger Strasse 19
1030 Wien
Austria
TEL: + 43   222 7393864
FAX: + 43   222 7137552

Protebel S.A.
Mechelsesienweg 163
Bus 1, 2018 Antwerpen
Belgium

Prox Kemi
Hovegaden 9
4622 Havdrup
Denmark
TEL: + 45   42 386204
FAX: + 45   42 386360

Protext-Extrosa
Postfach 1415
7850 Lorrach
Germany
TEL: + 49   7621 84772
FAX: + 49   7621 12429

Proxitalia
Via Fagnana 19
20090 Buccinasco
Italy
TEL: + 39   2 48841947
FAX: + 39   2 48842853

Protex Nederland
Broekhovenseweg 130 R
5021 LJ Tilburg
The Netherlands
TEL: + 31   13 367477
FAX: + 31   13 366592

Prox LDA
Rua do Outeiro 110 - Gemunde
4470 Mala
Portugal
TEL: + 35   941 50119
        + 35   941 50120
        + 35   941 50121
FAX: + 35   941 5022

Productos Protex S.A.
Comte Borrell
184 1 er 4a
Barcelona 15
Spain
TEL: + 34   3234913
        + 34   3234003

Protex Sverige AB
Box 320
S-26123 Landskrona
Sweden
TEL: + 46   418 58470
FAX: + 46   418 58788

Protex Chemie Basel AG
Schützenrainweg 11
4125 Riehen (CH)
Switzerland
TEL: + 41   61 673144
FAX: + 41   61 672705

Protex Chemicals Ltd.
Astley Lane Industrial Estate
Astley Way
Swillington
Leeds LS26 8XT
United Kingdom
TEL: + 44   113 2876002
FAX: + 44   113 2875003

QUEST INTERNATIONAL

Quest International France S.A.
42-50, Rue de la Reine Henriette
92704 Colombes Cedex
France
TEL: + 33   1 47847475
FAX: + 33   1 47811595

Quest International Deutschland GmbH
Postfach 1153
D-50362 Erftstadt-Lechenich
Germany
TEL: + 49   2235 79010
FAX: + 49   2235 7901160

Quest International Italia S.p.A.
Via Andrea Costa 112
40067 Bologna-Rastignano
Italy
TEL: + 39   51 744127
FAX: + 39   51 744373

Quest International N.V.
Fragrances, Flavours & Food Ingredients
Huizerstraatweg 28
1411 GP Naarden
The Netherlands
TEL: + 31   2159 99111
FAX: + 31   2159 46067

Quest International Nederland B.V.
PO Box 2
1400 CA Bussum
The Netherlands

TEL: + 31   2159 92802
FAX: + 31   2159 53437

Quest International Espana S.A.
Edificio Conata I
c/Fructuos Gelabert, 2-4
08970 Sant Joan Despi
Barcelona
Spain
TEL: + 34   3 4770188
FAX: + 34   3 4770322

Quest International Sverige AB
Box 66
S-240 21 Löddeköpinge
Sweden
TEL: + 46   46 709285
FAX: + 46   46 705029

Quest International UK Ltd.
Bromborough Port
Wirral
Merseyside L62 4SU
United Kingdom
TEL: + 44   151 645 2060
FAX: + 44   151 643 4300

RASCHIG

Raschig-France S.A.R.L.
49, Avenue de Versailles
F-75016 Paris
France
TEL: + 33   1 45240636
FAX: + 33   1 45208408

Raschig AG
Mundenheimer Strasse 100
D-6700 Ludwigshafen/Rhine
Germany
TEL: + 49   621 56180
FAX: + 49   621 532885

Raschig UK Ltd.
124 Doveleys Road
Salford M68 QW
United Kingdom
TEL: + 44   161 7458811
FAX: + 44   161 7458371

RHONE-POULENC

Rhone-Poulenc
Handelsgesellschaft
Grosse Neufasse 8
A-1040 Wien
Austria
TEL: + 43   222 563626

Rhone-Poulenc Belgique SA
Boulevard Sylvain Dupuis 243
B-1070 Bruxelles
Belgium
TEL: + 32   2 5294511
FAX: + 32   2 5294599

Rhone-Poulenc S.R.O.
Zeleny Pruh 99
140 50 Praha 4
Prague
Czech Republic
TEL: + 42   2 6121 4830
FAX: + 42   2 6121 4829

Rhone-Poulenc Danmark A/S
Gladsaxevej 378
DK-2860 Soborg
Denmark
TEL: + 45   1 56 32 00
FAX: + 45   1 67 36 26

Rhone-Poulenc Oy
BP 53 Viljatie 4
SF-00701 Helsinki
Finland
TEL: + 358   0 358411

Societe Francaise d'Organo-Synthese
17 Rue de l'Hotel-de-ville
9220 Neuilly-sur Seine
France
TEL: + 33   747 5890

European Headquarters
Rhone-Poulenc
Les Miroirs
Cedex 29
92097 Paris La Defense
France
TEL: + 33   1 47681234
FAX: + 33   1 47691911

Rhone-Poulenc GmbH
Staedelstrasse 10
Postfach 70 0862
D-6000 Frankfurt/Main 70
Germany
TEL: + 49   69 60930
FAX: + 49   69 6093333

Rhone-Poulenc Hellas
308 Messoghio et 2, rue Arkadiou
GR-15562 Holargos
Athens
Greece
TEL: + 30   1 6532492
FAX: + 30   1 6532497

Rhone-Poulenc SA
Palya U.9
H-1012 Budapest
Hungary
TEL: + 36   1 2015599
FAX: + 36   1 1382172

Rhone-Poulenc Ireland Ltd.
21 Cookstown Industrial Estate
Tallaght
Dublin 24
Ireland
TEL: + 353   1 4510033
FAX: + 353   1 4527874

RP Geronazzo Headquarters
Via Milano 78
I-20021 Ospiate di Bollate
Milano
Italy
TEL: + 39   2 3503212
FAX: + 39   2 3501770

Rhone-Poulenc Nederlands B.V.
Bovenkerkerweg 6-8
1185 XE Amstelveen
The Netherlands

TEL: + 31   20 5473911
FAX: + 31   20 6477445

Rhone-Poulenc Norge A/S
Kongens Gt. 14
Postboks 44
N-Oslo-1
Norway
TEL: + 47   2 415543

Rhone-Poulenc Polska Sarl
Grzybowska 80/82
00-844 Varsovie
Poland
TEL: + 48   2 6615501
FAX: + 48   2 6615570

Rhone-Poulenc Portugal
Rua Antonio Enes 25 - 2
P-1000 Lisboa
Portugal
TEL: + 351   1 3530974
FAX: + 351   1 525272

Rhone-Poulenc Quimica SA
Diagonal, 640 2
E-08017 Barcelona
Spain
TEL: + 34   3 4304900
FAX: + 34   3 4197341

Rhone-Poulenc Sverige AB
Box 41189
Folkugagatan 122
S 102-62 Stockholm 4
Sweden
TEL: + 46   8 237480
FAX: + 46   8 418863

Rhone-Poulenc Swisse S.A.
63 Rue de Lausanne
Case Postale 1211
Geneva 21
Switzerland
TEL: + 41   22 315950

Rhone-Poulenc Chem. Ltd., Perf. Prods. Group
Woodley
Stockport
Cheshire SK6 1PQ
United Kingdom
TEL: + 44   161 9101500
FAX: + 44   161 4304364

Rhone-Poulenc Chemicals Ltd.
Oak House
Reeds Crescent
Watford
Hertfordshire WD11QH
United Kingdom
TEL: + 44   1923 211700
FAX: + 44   1923 211580

RUDOLF CHEMICALS

Rudolf Chemie
Kortrijksesteenweg 867
9000 Gent
Belgium
TEL: + 32   9 2201842
FAX: + 32   9 2201977

Rudolf Sarl
183a rue de l'Illberg
68200 Mulhouse
France
TEL: + 33   89 600460
FAX: + 33   89 601051

Rudolf GmbH & Co KG
Altunterstrasse 58-64
D-82538 Gerestrieal
Germany
TEL: + 49   8171530
FAX: + 49   817153191

Rudolf Produtos Quimicos Ltd.
Rua do Castanhal 159
Apt. 2091 Castelo da Maia
4472 St. M. de Avioso Cedex
Portugal
TEL: + 35   2 9412257
FAX: + 35   2 9411888

Rudolf Chemicals Ltd.
Keys Road
Alfreton Trading Estate
Alfreton
Derbyshire DE5 7FQ
United Kingdom
TEL: + 44   1773 832703
FAX: + 44   1773 520092

SANDOZ

Sandoz GmbH
Brunerstr 59/Obj.5
A-1235 Wien
Austria
TEL: + 43 1 8632080
FAX: + 43 1 86320832

Sandoz Chemicals A/S
Kogevej 51
DK-4653 Karise
Denmark
TEL: + 45 53 688118
FAX: + 45 53 688265

Sandoz OY
PL 19 PB
0211 Helsinki 21
Finland
TEL: + 358   0 68 26 81
FAX: + 358   0 682 6950

Sandoz Chimie S.A.
Boite postale 313
F-92500 Rueil-Malmaison
France
TEL: + 33 1 47100211
FAX: + 33 1 47100234

Sandoz Quinn-Produckte GmbH
Postfach 2280
D-79540 Lorrach
Germany
TEL: + 49   7621 4170
FAX: + 49   7621 4170 150

Sandoz Chimica S.p.A
Casella postale 1876
I-20100 Milano
Italy
TEL: + 39 2 57951
      + 39 2 5795 484

Sandoz Chemicals Norge
Postboks 573
N-1301 Sandvika
Norway
TEL: + 47   22 541001
FAX: + 47   22 546150

Sandoz Quimica S.A.E.
Apartado 708
E-08080 Barcelona
Spain
TEL: + 34 3 2451700
FAX: + 34 3 2452127

Sandoz Chemicals AB
Flojelbergsgatan 7B
S-431 37 Moelndal
Sweden
TEL: + 46 31 878115
FAX: + 46 31 279876

Anthrachinon-Produkte BASF-Sandoz AG
Postfach
CH 4002 Basel
Switzerland
TEL: + 41   61 3241111
FAX: + 41   61 3248001

Sandoz Chemicals (UK) Ltd.
Calverley Lane
Horsforth
Leeds LS18 4RP
United Kingdom
TEL: + 44   113 2584646
FAX: + 44   113 2591632

SCHER CHEMICALS

Les Produits Alambic
60, Avenue de la Jonction
B-1060 Bruxelles (Forest)
Belgium
TEL: + 32   2 3449160
FAX: + 32   2 3442927

Seppic
75, Quai D'Orsay
75321 Paris Cedex 07
France
TEL: + 33   1 40625555
FAX: + 33   1 40625253

Nordmann, Rassmann GMBH & Co.
Kajen 2
20459 Hamburg
Germany
TEL: + 49   40 3687354
FAX: + 49   40 3687412

S.A.I. Societa Alcan Italia S.P.A.
via Antonello da Messina N.5
20146 Milano
Italy
TEL: + 39   2 40090202
FAX: + 39   2 40090544

Isisa
Rocafort, 241-243
08029 Barcelona
Spain
TEL: + 34   3 3220453
FAX: + 34   3 4106978

Dr Marc Steinfels
Rotelstrasse 28
CH-8006 Zurich
Switzerland
TEL: + 41   1 3626161
FAX: + 41   1 3626165

Chesham Chemicals Ltd.
Cunningham House
Westfield Lane
Kenton
Harrow HA3 9ED
United Kingdom
TEL: + 44   181 9077779
FAX: + 44   181 9091053

SEPPIC

Seppic
75 Quai d'Orsay
75321 Paris Cedex 07
France
TEL: + 33   140 625555
FAX: + 33   140 625253

Chesham Chemicals Ltd.
Cunningham House
Westfield Lane
Kenton
Harrow HA3 9ED
United Kingdom
TEL: + 44   181 9077779
FAX: + 44   181 9091053

SERVO DELDEN

Servo Delden B.V.
PO Box 1
7490 AA Delden
The Netherlands
TEL: + 31   5407 75000
          (7437 75000 after 10.10.95)
FAX: + 31   5407 75075
          (7437 75075 after 10.10.95)

SHELL CHEMICALS

Shell Austria AG
Postfach 174
A-1011 Vienna
Austria
TEL: + 43   1 79797
FAX: + 43   1 79797401

Belgian Shell SA
47 Cantersteen
B-1000 Brussels
Belgium
TEL: + 32   2 5089111
FAX: + 32   2 5110571

Shell Bulgaria OOD
PO Box 18
2 Louias Hailler Street
Sofia 1404
Bulgaria
TEL: + 359   2 576814/27
FAX: + 359   2 576712

Shell Uvoz-Izvoz Nafte I Kemikija Doo
TRG JF Kennedya 11/I
4100 Zagreb
Croatia
TEL: + 38   41 231444/7
FAX: + 38   41 236914

Shell Czech Republic
Na Strzi 63
140 62 Prague 4
Czech Republic
TEL: + 42   2 611 42891
FAX: + 42   2 611 5344

Shell Kemi A/S
Kampmannsgade 2
DK-1780 Copenhagen
Denmark
TEL: + 45   33 39 53 40
FAX: + 45   33 13 16 11

Shell Eesti: AS
81 Petersburg Street
EE0014 Tallinn
Estonia
TEL: + 372   232 1996
FAX: + 372   232 1996

Shell AB, Shell Chemicals
PO Box 16
SF-01301 Vantaa
Finland
TEL: + 358   0 85701
FAX: + 358   0 8570 2299

Shell Chimie
89 Boulevard Franklin Roosevelt
92564 Rueil Malmaison
France
TEL: + 33   1 47 14 71 00
FAX: + 33   1 47 14 82 99

Deutsche Shell Chemie GmbH
Postfach 5220
65727 Eschborn 1/Taunus
Germany
TEL: + 49   6196 4740
FAX: + 49   6196 474502

Shell Company (Hellas) Ltd.
PO Box 3499
GR102.10 Athens
Greece
TEL: + 30   1 929 5911
FAX: + 30   1 922 2804

Shell Hungaria Ker KFT
PO Box 37
H-1300 Budapest
Hungary
TEL: + 36   1 250 0954
FAX: + 36   1 250 2151

Shell Chemicals Ireland Ltd.
Shell House
Beech Hill
Clonskeagh
Dublin 4
Ireland
TEL: + 353   1 260 2177
FAX: + 353   1 260 2185

Shell Italia S.p.A.
Via Chiese 74
20126 Milan
Italy
TEL: + 39   2 661601
FAX: + 39   2 66160247

Shell Latvia SIA
2 Elizabetes Street
LV 1340 Riga
Latvia
TEL: + 371 833 0383
FAX: + 371 833 0383

Shell Lietuva UAB
Gedimino Pr. 3A-2
Vilnius 2600
Lithuania
TEL: + 370 2 611815
FAX: + 370 2 225785

Shell Nederland Chemie BV
PO Box 3030
3190 G H Hoogvliet
Rotterdam
The Netherlands
TEL: + 31 10 231 7000
FAX: + 31 10 231 7180

A/S Norske Shell, Shell Chemicals
PO Box 1154
N-0107 Oslo Sentrum
Norway
TEL: + 47 22 66 50 00
FAX: + 47 22 66 57 67

Shell Polksa Sp Zoo
ul Stawki 2
Intraco 1, 11 Pietro
00-193 Warsaw
Poland
TEL: + 48 2 635 5387/1047
FAX: + 48 2 635 1387

Shell Portuguesa SARL
Caixa Postal 2008
1101 Lisboa Codex
Portugal
TEL: + 351 1 3559155
FAX: + 351 1 578956

Shell Romania SRL
Diplomat Hotel
Apartment 110B
Sevastapol St 13-17
Bucharest 1, 78 118
Romania
TEL: + 40 1 650 5046
     + 40 1 659 6652
FAX: + 40 1 659 4242

Shell Russia A/O
30A Trubnikovsky Per
121069 Moscow
Russia
TEL: + 7 503 956 6000
FAX: + 7 503 956 6010

Shell Slovakia SRO
Zahradnicka 91
820 09 Bratislava
Slovakia
TEL: + 42 7 680 67
FAX: + 42 7 656 69

Shell Slovenia d.o.o.
10th Floor
World Trade Centre
156 Dunajska Cesta
61000 Ljubljana
Slovenia

TEL: + 386 61 1687070
FAX: + 386 61 1688033

Shell Espana SA
Apartado 652
28080 Madrid
Spain
TEL: + 34 1 5370100
FAX: + 34 1 5370106

AB Svenska Shell
Shell Chemicals Europe
S-171 79 Sol
Solna
Sweden
TEL: + 46 8 730 8000
FAX: + 46 8 735 47 07

Shell (Switzerland)/Chemie
Baarermatte
6340 Baar
Switzerland
TEL: + 46 42 34 44 44
FAX: + 46 42 35 54 20

The Shell Company of Turkey Ltd.
PK 24 Mecidiyekoy
Istanbul
Turkey
TEL: + 90 1 275 3410
FAX: + 90 1 267 2453

Shell International Petroleum Co Ltd.
Representative Office
Room 106
39/41 Shota Rustaveli
Kiev 252023
Ukraine
TEL: + 7 44 220 6041
FAX: + 7 44 220 5711

Shell Chemicals UK Ltd.
Heronbridge House
Chester Business Park
Chester CH49QA
United Kingdom
TEL: + 44 1244 685000
FAX: + 44 1244 685010

SISTERNA

Sisterna B.V.
P.O. Box 1308
4700 BH Roosendaal
The Netherlands
TEL: + 31 1650 82590/1
FAX: + 31 1650 58304

STEPAN EUROPE

Stepan Europe
Chemin Jonkind
BP127
38320 Voreppe
France
TEL: + 33 76 505100
FAX: + 33 76 567165

Stepan Europe
St James Court
The Executive Suite
Widerspool Causeway
Warrington WA4 6PS
United Kingdom

TEL: + 44 1925 231 511
FAX: + 44 1925 445 687

K & K Greeff Ltd. *→ Now Univar Fax 0208 256 1337*
Suffolk House
George Street
Croydon CR93QL
United Kingdom
TEL: + 44 181 6860544
FAX: + 44 181 7816874

SURFACHEM

Surfachem Group plc
Wellington Park House
Thirsk Row
Leeds LS1 4DP
United Kingdom
TEL: + 44 113 2342636
FAX: + 44 113 2420411

THOMAS SWAN & CO.

Thomas Swan & Co. Ltd.
Crookhall
Consett
County Durham DH8 7ND
United Kingdom
TEL: + 44 1207 505131
FAX: + 44 1207 590467

THOR CHEMICALS

Thor Chemicals (UK) Ltd.
Earl Road
Cheadle Hulme
Cheshire SK86QP
United Kingdom
TEL: + 44 161 4861051
FAX: + 44 161 4884125

TOMAH PRODUCTS

Caldic (Belgium) NV
Terlochtweg 1
B-2620 Hemiksem
Belgium
TEL: + 32 3870 4811
FAX: + 32 3870 4829

Brenntag-Eurochem GmbH
Alfredstrasse 81
45130 Essen
Germany
TEL: + 49 201 8781460
FAX: + 49 201 8781640

Eigenmanne Veronelli Spa
Via Della Mosa
6-20017 Rho
Milan
Italy
TEL: + 39 293 539265
FAX: + 39 293 539320

Allan Chemical AB
Drottninggatan
S-254 33 Helsingborg
Sweden
TEL: + 46 42 181430
FAX: + 46 42 181534

Tomah Products
Cadogan House
Cadogan Park
Woodstock
Oxfordshire OX20 1UW
United Kingdom
TEL: + 44   1993 812714
FAX: + 44   1993 812709

Chemlink Specialities Ltd.
Carrington Business Park
Carrington
Urmston
Manchester M31 4ZU
United Kingdom
TEL: + 44   161 7764303
FAX: + 44   161 7757163

TROY CHEMICALS

Unimex m.b.H.
Rotgasse 2
A-10-10 Wien
Austria
TEL: + 43   1 5354127
FAX: + 43   1 53541279

Socomer N.V.
Nieuwbrugstraat 73
B-1830 Machelen
Belgium
TEL: + 32   2 2544611
FAX: + 32   2 2544619

Gropa A/S
Bredgade 29
1260 Copenhagen K
Denmark
TEL: + 45   33 136711
FAX: + 45   33 138542

BV Chemie Pharmacie Holland
Friedman Building
Hogehilweg 4
1101 CC Amsterdam
TEL: + 31   0 206972818
FAX: + 31   0 206974419

Bang & Bonsomer
Italahdenkatu 18A
00210 Helsinki
Finland
TEL: + 358   0 681081
FAX: + 358   0 6924174

Pierre Langlois-Berthelot
9-31 Rue Boussingault
75013 Paris
France
TEL: + 33   1 45654711
FAX: + 33   1 45881399

Troy Chemie GmbH
Uerdingerstrasse 541
47800 Krefeld 1
Germany
TEL: + 49   2151 590338/39
FAX: + 49   2151 598145

Euro-Nexus Ltd.
Meredek u. 60
H-112 Budapest
Hungary

TEL: + 36   1 1858436
FAX: + 36   1 1859141

John Karl Andresson
P.O. Box 1249
Reykjavik
Iceland
TEL: + 354   1 27655
FAX: + 354   1 621918

The National Chem. Comp. Ltd.
NCC House
42 Lower Leeson Street
Dublin 2
Ireland
TEL: + 353 1 6764866
FAX: + 353 1 6616676

Troy Italia
Via Tolstoy, 58
20146 Milano
Italy
TEL: + 39   2 475054
FAX: + 39   2 471712

Messers Knud Knudsen & Sonn
Vestjordet 28
N-1345 Osteras
Norway
TEL: + 47   67149310
FAX: + 47   67140762

Focor
Travessa da Prelada, 613
4200 Porto
Portugal
TEL: + 35   1 2827762
FAX: + 35   1 2824227

Mrs M Horvat
Gornji Trg 5
61000 Ljubljana
Slovenia
TEL: + 386   61 1259088
FAX: + 386   61 1259104

DSM Resins Espana SA
Edificio Euro-3
Poligono Sant Joan Despi
c/Frederic Mompou 5
Planta 7, Puertas 3a y 4a
08960 Sant Just Desvern (Barcelona)
Spain
TEL: + 34   3 4731100
FAX: + 34   3 4736350

AB Lexum
Box 5012
S-165 10 Hässelby
Sweden
TEL: + 46   8 894790
FAX: + 46   8 890062

Chemische Fabrik Schweizerhall
Postfach
Elasserstrasse 231-245
CH-4013 Basel
Switzerland
TEL: + 41   61 3268111
FAX: + 41   61 3225022

Troy UK
Zenith House
Northolme Road
Louth
Lincolnshire LN11 0HQ
United Kingdom
TEL: + 44   1507 609606
       + 44   1507 607017
FAX: + 44   1507 607107

UNGER FABRIKKER

Unger Fabrikker A.S.
P.O. Box 254
N-1601 Fredrikstad
Norway
TEL: + 47   9 320020
FAX: + 47   9 323735

Jauk GmbH
Bauermarkt 6/5/10
A-1010 Vienna
Austria

Rode & Rode
Radhusplasden 4
DK-1550 Copenhagen V
Denmark

Oy Letra A.B.
Formannsgatan 4
SF-02600 Esbo
Finland

Unger Fabrikker A.S.
Zweigniderlassung Dusseldorf
Karl-Arnold Platz 2 (Handelcenter Norwegen)
D-4000 Dusseldorf 30
Germany

UNICHEMA CHEMICALS

Unichema France SA
148 Boulevard Haussemann
75008 Paris
France
TEL: 1 45630863
FAX: 1 42563188

Unichema Chemicals Ltd.
Bebington
Wirral
Merseyside L62 4UF
United Kingdom
TEL: + 44   151 6452020
FAX: + 44   151 6459197

Unichema Chemie GmbH
Postfach 1280
D-4240 Emmerich
Germany
TEL: + 49   0 2822720
FAX: + 49   0 282272276

UNION CARBIDE

Union Carbide Austria GMBH
Storechengasse 1
A-1150 Vienna
Austria
TEL: + 44   222 891500
FAX: + 44   222 8915030

Union Carbide
Benelux N.V.
Noorderlaan 147
B-2030 Antwerp
Belgium
TEL: + 32 3 5430211
FAX: + 32 3 5430224

N.V. Sepulchre S.A.
Avenue des Nenuphars, 19 B6
Waterlelienlaan
19 B6
B-1160 Brussels
Belgium
TEL: + 32 2 6722335
FAX: + 32 2 6736782

P Munch Nielsen A/S
Jernbane Alle 55
DK-3060 Espergaerde
Denmark
TEL: + 45 42 234111
FAX: + 45 42 234249

Aspo Oy
Suolakivenkatu 10
SF-00810 Helsinki
Finland
TEL: + 358 075951
FAX: + 358 0781071

Union Carbide France S.A.
4 Place Des Etats-Unis,
Silic 220
F-94518 Rungis-Cedex
France
TEL: + 33 1 45122071
FAX: + 33 1 46872924

Lambert Riviere S.A.
17, Av Louison-Bobet,
Val de Fontenay
F-94132 Fontenay Bois Cedex
France
TEL: + 33 1 49748080
FAX: + 33 1 49748111

Union Carbide Deutschland GmbH
Moersenbroicher WEG 200
D-40470 Dusseldorf
Germany
TEL: + 49 211 90850
FAX: + 49 211 9085267

Brenntag Eurochem GMBH
Alfredstrasse 81
D-45130 Essen
Germany
TEL: + 49 201 87810
FAX: + 49 201 87816 40

Union Carbide Italia S.r.l
via Durini 28
I-20122 Milano
Italy
TEL: + 39 2 76001465
FAX: + 39 2 76009006

Nymco S.P.A.
via Dei Giovi 6
I-1200332 Cormano
Milano
Italy

TEL: + 39 2 6150621
FAX: + 39 2 6152979

Thor's Kemiske Fabrikker A/S
P.O. Box 7
N-1473 Skarer
Norway
TEL: + 47 2 704100
FAX: + 47 2 709183

Quimigranel Productos
Quimicos LDA
Rua Vale Formoso 114A Porta 19
APT. 8046
P-1801 Lisboa, Codex
Portugal
TEL: + 351 1 8581564
FAX: + 351 1 8583710

Julia-Parrera S.A.
Tuset, 8 y 10
08006 Barcelona
Spain
TEL: + 34 3 2184404
FAX: + 34 3 2183602

M.B. Sveda
Adelgaten 5
P.O. Box 45
S-201 20 Malmo
Sweden
TEL: + 46 40 352800
FAX: + 46 40 125172

Christ AG (Chemie)
Neuhofweg 53
CH-4147 Aesch
Switzerland
TEL: + 41 61 708111
FAX: + 41 61 786511

Union Carbide (Europe) SA
15 Chemin Louis-Dunant
CH-1211 Geneva
Switzerland
TEL: + 41 22 9196111
FAX: + 41 22 9196545

Union Carbide Ltd.
Union Carbide House
95 High Street
Rickmansworth
Hertfordshire WD3 1RB
United Kingdom
TEL: + 44 1923 713000
FAX: + 44 1923 713040

Surfachem Ltd.
Wellington Park House
Thirsk Row
Leeds LS1 4DP
United Kingdom
TEL: + 44 113 2342636
FAX: + 44 113 2445910

WARWICK INTERNATIONAL

Warwick Chimilux S.A.
26 Rue Vignon
75009 Paris
France
TEL: + 33 1 44515000
FAX: + 33 1 44515019

Warwick Italia
Via San Rocco 4
15049 Vignale Monferrato
Italy
TEL: + 39 142 933004
FAX: + 39 142 933512

Warwick Portugal
Rua dos Bragas
208 2° S-13/14/15
4000 Porto
Portugal
TEL: + 351 0 220322418
FAX: + 351 0 2202081738

Warwick Benbassat
Comte D'Urgell, 240
08036 Barcelona
Spain
TEL: + 34 3 3220653
FAX: + 34 3 3220993

Warwick International Specialities Ltd.
Wortley Moor Road
Leeds LS12 4JE
United Kingdom
TEL: + 44 113 2637331
FAX: + 44 113 2311559

WESTBROOK

Jakob Pavel Handelsagentur
P.O. Box 45
Lammgasse 12
A-1082 Vienna
Austria
TEL: + 43 1 4061147
FAX: + 43 1 4062172

Westbrook Lanolin S.A. Belge
4-6 Rue Beribou
B 4800 Verviers
Belgium
TEL: + 32 87 336121
FAX: + 32 87 316552

Dagentas A/S
Blokken 21
DK 3460 Birkerod
Copenhagen
Denmark
TEL: + 45 45 828212
FAX: + 45 45 828202

Hanstrom Trading Company
Mannerheimintie 93L267
SF-00270 Helsinki
Finland
TEL: + 358 0 479683
FAX: + 358 0 479698

Lambert Riviere S.A.
17 Avenue Louison Bobet
Val de Fontenay
94132 Fontenay Sous Bois Cedex
France
TEL: + 33 1 49748080
FAX: + 33 1 49748111

Deutsche Lanolin Gesellschaft
Parmentier & Co.
Eichendorffstrasse 37
60320 Frankfurt am Main
Germany

TEL: +49 69 561034
FAX: +49 69 568518

H G & C Blau GmbH
Glockengiesserwall 20
2000 Hamburg 1
Germany
TEL: +49 40 321693
FAX: +49 40 338004

Avel Ltd.
3-7 Irinis Street
153 42 Agia Paraskevi
Athens
Greece
TEL: +30 1 6000860
FAX: +30 1 6003218

Charles Tennant & Co Ltd.,
Unit 71
Cookstown Industrial Estate
Belgard Road
Dublin 24
Ireland
TEL: +353 1 514099
FAX: +353 1 514702

J.F. Behr Ltd.
40 Sokolov Street
47100 Ramat Hasharon
Tel Aviv
Israel
TEL: +972 3 5403641
FAX: +972 3 5403643

Vailati E.C. SRL
Via Stephenson 43/A
20157 Milan
Italy
TEL: +39 2 33200900
FAX: +39 2 33200845

Danby Chemie B.V.
Leidsestraat 106-108
1017 PG Amsterdam
Netherlands
TEL: +31 20 6221616
FAX: +31 20 6253458

Paus & Paus A.S.
P.O. Box 281 Kongensgt 14
N-0153 Oslo 1
Norway
TEL: +47 2 415060
FAX: +47 2 417354

Sociedade Imbrasex Ltd.a.
Calcado do Carmo
1200 Lisbon
Portugal
TEL: +351 1 3420653
FAX: +351 1 3464850

Chemo Iberica S.A.
'Edificion Ofipinar'
C/Caleruega 102, 6 Izq.
Madrid 28033
Spain
TEL: +34 1 3021560
FAX: +34 1 7668963

M.B. Sveda
Marieholmsgaten 56
P.O. Box 48
S-401 20 Gothenburg
Sweden
TEL: +46 31 838000
FAX: +46 31 843980

Paul Brem AG
Ringstrasse 72
Postfach CH-8050
Zurich
Switzerland
TEL: +41 1 318 6633
FAX: +41 1 318 6622

Kima Kimyevi Maddeler
P.O.B. 602
80225 Sisli
Istanbul
Turkey
TEL: +90 1 2485090
FAX: +90 1 2475189

Westbrook Lanolin Company
Argonaut Works
Laisterdyke
Bradford BD4 8AU
United Kingdom
TEL: +44 1274 663331
FAX: +44 1274 667665

WITCO

Witco SA
20 rue de la Ville l'Eveque
75008 Paris
France
TEL: +33 1 44510505
FAX: +33 1 42656761

The Baxenden Chemical Co. Ltd.
Applied Chemicals Division
Union Lane
Droitwich
Worcestershire WR9 9BB
United Kingdom
TEL: +44 1905 794795
FAX: +44 1905 794002

ZENECA

Zeneca Osterreich GmbH
Postfach 153
Schwarzenbergplatz 7
A-1037 Vienna
Austria
TEL: +43 1 711310
FAX: +43 1 71131-221

Zeneca Specialities
Everslaan 45
B-3078 Everberg
Belgium
TEL: +32 2 7589211
FAX: +32 2 7589649

Zeneca Minsk
Ul. Kazintsa, 62
220108 Minsk
Belnii Pochvovedeniia i Agrohkimii
Byelorussia
Belarus

TEL: +7 172 773754
FAX: +7 172 783018

Zeneca Bulgaria
36 Dragan Tsankov Blvd
1040 Sofia
Bulgaria
TEL: +359 2 705227
+359 2 705266
+359 2 722106
FAX: +359 2 722053
+359 2 4920124

Zeneca Croatia Ltd.
Jukiceva 12
41000 Zagreb
Croatia
TEL: +38 541 445416
FAX: +38 541 446038

Zeneca (CZ) s.r.o.,
Cinska 4,
160 00 Praha 6
Czech Republic
TEL: +42 2 325 00/3117 200
FAX: +42 2 311 2149
+42 2 311 4117

Zeneca AS
Islands Brygge 41
DK2300 Copenhagen S
Denmark
TEL: +45 32 888100
FAX: +45 32 888275

Zeneca Oy
Mannerheiminitie 160A
00300 Helsinki
Finland
TEL: +358 0 6130011
FAX: +358 0 61300300

Zeneca S.A.
11 rue Felix Brun
69364 Lyon Cedex 07
France
TEL: +33 16 78696120
FAX: +33 16 78696139

Zeneca Ltd.
BP 5
38370 St Clair du Rhone
France
TEL: +33 16 74314000
FAX: +33 16 74563718

Zeneca GmbH,
Otto-Hahn-Strasse
68723 Plankstadt
Germany
TEL: +49 6202 9900
FAX: +49 6202 992000

Zeneca GmbH,
Emil-von-Behring-Str 2
60439 Frankfurt/M
Germany
TEL: +49 69 580102
FAX: +49 69 5801683

Zeneca Hellas S.A.
231 Syngrou Avenue
17121 Athens
Greece
TEL: + 30   1 9358302
FAX: + 30   1 9349964

Zeneca b.v.
Voorn 47
PO Box 4204
2980 GE Ridderkerk
Holland
TEL: + 31   180 450350
FAX: + 31   180 431660

Zeneca Hungary kft
Budapest 1539
PF. 672
Hungary
TEL: + 36   1 202 3082
       + 36   1 202 3134
FAX: + 36   1 202 3582

Zeneca Ireland Ltd.
College Park House
20 Nassau Street
Dublin 2
Ireland
TEL: + 353   1 6795799
FAX: + 353   1 6796650

Zeneca S.p.A.
Via Francesco Sfoza
Palazzo Volta
20080 Basiglio (MI)
Italy
TEL: + 39   2 904541
FAX: + 39   2 90755615

Zeneca Almaty
Ul. Abaya
8 kor. 3, kom. 324
480021 Almaty
Kazakhstan
TEL: + 7   3272 641 466
FAX: + 7   3272 650 792

Zeneca AS,
Drammensveien 126A
PB 173 Skoyen
N-0121 Oslo
Norway
TEL: + 47   22 926000
FAX: + 47   22 926095

Zeneca Polska Sp z.o.o.
ul. Nieklanska 23
skr poczt 73
03-924 Warsawa
Poland
TEL: + 48   26 170658
       + 48   26 176242
FAX: + 48   26 172205

Zeneca Romania
29 Tache Ionescu Street
Bucharest
Romania
TEL: + 40   1 312 3684
       + 40   1 312 6952
FAX: + 40   1 312 1859

AO Zeneca Russia
Office 201
Bolshoi Strochenovski Per. 22/25
113054 Moscow
Russia
TEL: + 7   95 2306111
FAX: + 7   95 2306119

Zeneca Slovakia s.r.o.
Stefanikova 31
811 05 Bratislava
Slovakia
TEL: + 42   333457
       + 42   333427
FAX: + 42   331024

Zeneca International Ltd.
Einspielerjeva 6
61113 Ljubljana
Slovenia
TEL: + 386   61 122074
       + 386   61 191886
FAX: + 386   61 121208

Zeneca Specialties SA,
Avda de la Granvia 179
08908 Hospitalet
Barcelona
Spain
TEL: + 34   3 4038000
FAX: + 34   3 2633115

Zeneca AB,
Drakegatan 10
PO Box 453
S-401 27 Gothenburg
Sweden
TEL: + 46   31 7737300
FAX: + 46   31 7737375

Zeneca AG,
Landenbergstrasse 34
Postfach CH 6002
Luzern
Switzerland
TEL: + 41   400000
FAX: + 41   400234

Zeneca Colours
PO Box 42
Hexagon House
Blackley
Manchester M9 8ZS
United Kingdom
TEL: + 44   161 7401460
FAX: + 44   161 7956005

ZSCHIMMER & SCHWARZ

Dr Hans Werba OHG
Lugeck 1-2
A-1011 Wien
Austria
TEL: + 43   222 514900
FAX: + 43   222 5149020

Bionord Aps
Rygaards Allee 131
DK-2900 Hellerup
Denmark
TEL: + 45   31183588
FAX: + 45   39292778

Zschimmer & Schwarz France S.A.
10, rue Saint-Marc
F-75002 Paris
France
TEL: + 33   1 42331033
FAX: + 33   1 40262381

Ashkenazi Ltd.
24 Hatehia Street
Tel Aviv
Israel
TEL: + 972   3 6818865
FAX: + 972   3 6818696

Zschimmer & Schwarz Italiana S.p.A.
Casella Postale N. 1
I-13038 Tricerro-Vercelli
Italy
TEL: + 39   161 821421
FAX: + 39   161 807335

Integrated Chemicals BV
Postbus 302
NL-2160 AH Lisse
The Netherlands
TEL: + 31   252119020
FAX: + 31   252115483

Lemmel S.A.
C/La Coruna, 9 - 11 bjs
E-08026 Barcelona
Spain
TEL: + 34   3 4473750
FAX: + 34   3 2654951

Aval Chemi AB
Billdals Häggväg 1
S-42700 Billdal
Sweden
TEL: + 46   31 910720
FAX: + 46   31 913273

Z&S Handel AG
Postfach Z 1222
CH-8302 Kloten
Switzerland
TEL: + 41   1 8030434
FAX: + 41   1 8030657

Unkim Chem. Ind. and Trad. Inc.
Fenerolu
Camtepe Sok. No. 14/1 D.4
TR-81040 Kadikoy
Istanbul
Turkey
TEL: + 90   1 3581495
FAX: + 90   1 3605498

Chemag AG
Hietzinger Hauptstr. 119
A-1130 Wien
Austria
TEL: + 43   222 822616
FAX: + 43   222 822610

Surfachem Ltd.
Wellington Park House
Thirsk Row
Leeds LS1 4DP
United Kingdom
TEL: + 44   113 2342636
FAX: + 44   113 2445910

Index of companies

This section contains an alphabetical list of companies whose products are listed in the directory, together with a reference to the page numbers on which the various entries occur. Details of associate companies or representatives who supply the products in the countries covered by this edition may be found in the Index of company addresses.

The company name as printed in each case is sufficient for identification purposes. The correct name in the various countries is given in the Index of company addresses.

Index of trade names

This section contains an alphabetical list of the trade names under which products are listed in the directory together with a reference to the page on which the product details are to be found.